# Contents

# Business Mathematics and Statistics

Andre Francis BSc MSc

Perinatal Institute
Birmingham

Andre Francis works as a medical statistician. He has previously taught Mathematics, Statistics and Information Processing to students on business and professional courses. His teaching experience has covered a wide area, including training students learning basic skills through to teaching undergraduates. He has also had previous industrial (costing) and commercial (export) experience and served for six years in statistical branches of Training Command in the Royal Air Force.

**Sixth Edition**

Australia • Canada • Mexico • Singapore • Spain • United Kingdom • United States

THOMSON

**Business Maths and Statistics, 6$^{th}$ edition**



For more information, contact Thomson Learning, High Holborn House, 50-51 Bedford Row, London WC1R 4LR or visit us on the World Wide Web at:
http://www.thomsonlearning.co.uk

---



*British Library Cataloguing-in-Publication Data*
A catalogue record for this book is available from the British Library

---

**ISBN-13: 978-1-84480-128-2**
**ISBN-10: 1-84480-128-4**

First edition 1986
Second edition 1988; Reprinted 1990, 1991
Third edition 1993; Reprinted 1993
Fourth edition 1995; Reprinted 1996, 1997
Fifth edition 1998
**Sixth edition 2004 published by Thomson Learning**
**Reprinted 2004, 2005 and 2006 by Thomson Learning**

Typeset in Nottingham, UK by Andre Francis
Printed in China by C&C Offset Printing Co., Ltd

**Acknowledgements**

The author would like to express thanks to the many students and teachers who have contributed to the text in various ways over the years.
In particular he would like to thank the following examining bodies for giving permission to reproduce selected past examination questions:

*Chartered Association of Certified Accountants (ACCA)*
*Chartered Institute of Management Accountants (CIMA)*
*Institute of Chartered Secretaries and Administrators (ICSA)*
*Chartered Institute of Insurance (CII)*
*Association of Accounting Technicians (AAT)*

Each question used is cross referenced to the appropriate Institute or Association.

# Preface

## 1. Aims of the book

The general aim of the book is to give a thorough grounding in basic Mathematical and Statistical techniques to students of Business and Professional studies. No prior knowledge of the subject area is assumed.

## 2. Courses covered

a) The book is intended to support the courses of the following professional bodies:
Chartered Association of Certified Accountants
Chartered Institute of Management Accountants
Institute of Chartered Secretaries and Administrators

b) The courses of the following bodies which will be supported by the book to a large extent:
Chartered Institute of Insurance
Business and Technical Education Council (National level)
Association of Accounting Technicians

c) The book is also meant to cater for the students of any other courses who require a practical foundation of Mathematical and Statistical techniques used in Business, Commerce and Industry.

## 3. Format of the book

The book has been written in a standardised format as follows:

a) There are TEN separate parts which contain standard examination testing areas.
b) Numbered chapters split up the parts into smaller, identifiable segments, each of which have their own Summaries and Points to Note.
c) Numbered sections split the chapters up into smaller logical elements involving descriptions, definitions, formulae or examples.

At the end of each chapter, there is a Student Self Review section which contains questions that are meant to test general concepts, and a Student Exercise section which concentrates on the more practical numerical aspects covered in the chapter.

At the end of each part, there is

a) a separate section containing examination examples with worked solutions and
b) examination questions from various bodies. Worked solutions to these questions are given at the end of the book.

## 4. How to use the book

Chapters in the book should be studied in the order that they occur.

After studying each section in a chapter, the Summaries and Points to Note should be checked through. The Student Self Review Questions, which are cross-referenced to appropriate sections, should first be attempted unaided, before checking the answers with the text. Finally the Student Exercises should be worked through and the answers obtained checked with those given at the end of the book.

After completing a particular part of the book, the relevant section of the examination questions (at the end of the book) should be attempted. These questions should be considered as an integral part of the book, all the subject matter included having been covered in previous chapters and parts. Always make some attempt at the questions before reading the solution.

## 5. The use of calculators

Examining bodies permit electronic calculators to be used in examinations. It is therefore essential that students equip themselves with a calculator from the beginning of the course.

Essential facilities that the calculator should include are:

a) a square root function, and
b) an accumulating memory.

Very desirable extra facilities are:

c) a power function (labelled '$x^y$'),
d) a logarithm function (labelled 'log x'), and
e) an exponential function (labelled '$e^x$').

Some examining bodies exclude the use (during examinations) of programmable calculators and/or calculators that provide specific statistical functions such as the mean or the standard deviation. Students are thus urged to check on this point before they purchase a calculator. Where relevant, this book includes sections which describe techniques for using calculators to their best effect.

Andre Francis, 2004

# 1 Introduction to business mathematics and statistics

## 1. Introduction

This chapter serves as an introduction to the whole book. It describes the main areas covered under the heading 'Business Mathematics and Statistics' and introduces the idea of a statistical investigation.

## 2. Differences in terminology

The title of this book is Business Mathematics and Statistics. However, many other terms are used in business and by Professional bodies to describe the same subject matter. For example, Quantitative Methods, Quantitative Techniques and Numerical Analysis.

## 3. Business mathematics and statistics

A particular problem for management is that most decisions need to be taken in the light of incomplete information. That is, not everything will be known about current business processes and very little (if anything) will be known about future situations. The techniques described in 'Business Mathematics and Statistics' enable structures to be built up which help management to alleviate this problem. The main areas included in the book are: (a) Statistical Method; (b) Management Mathematics; and (c) Probability.

These areas are described briefly in the following sections.

## 4. Statistical method

*Statistical method* can be described as:

a) the selection, collection and organisation of basic facts into meaningful data, and then

b) the summarizing, presentation and analysis of data into useful information.

The gap between facts as they are recorded (anywhere in the business environment) and information which is useful to management is usually a large one. (a) and (b) above describe the processes that enable this gap to be bridged. For example, management would find percentage defect rates of the fleets of lorries in each branch more useful than the daily tachometer readings of individual vehicles. That is, management generally require summarized values which represent large areas under their control, rather than detailed figures describing individual instances which may be untypical.

Note that the word 'Statistics' can be used in two senses. It is often used to describe the topic of Statistical Method and is also commonly used to describe values which summarize data, such as percentages or averages.

## 5. Management mathematics

The two areas covered in this book which can be described as Management Mathematics are described as follows:

a) *The understanding and evaluation of the finances involved in business investments.* This involves considering interest, depreciation, the worth of future cash flows (present value), various ways of repaying loans and comparing the value of competing investment projects;

b) *Describing and evaluating physical production processes in quantitative terms.* Techniques associated with this area enable the determination of the level of production and prices that will minimise costs or maximise the revenue and profits of production processes;

Involved in both of the above are the manipulation of algebraic expressions, graph drawing and equation solving.

## 6. Probability

*Probability* can be thought of as the ability to attach limits to areas of uncertainty. For example, company profit for next year is an area of uncertainty, since there will never be the type of information available that will enable management to forecast its value precisely. What can be done however, given the likely state of the market and a range of production capacity, is to calculate the limits within which profit is likely to lie. Thus calculations can be performed which enable statements such as 'there is a 95% chance that company profit next year will lie between £242,000 and £296,000' to be made.

## 7. Statistical investigations

Management decisions are based on numerous pieces of information obtained from many different sources. They may have used one, some or all of the techniques which have been described as Statistical Method, Management Mathematics or Probability. What the decisions will all have in common however is that they are the final product of a general structure (or set of processes) known as an *investigation* or *survey*. Some significant factors are listed as follows.

a) Investigations can be fairly trivial affairs, such as looking at today's orders to see which are to be charged to credit or cash. Others can be major undertakings, involving hundreds of staff and a great deal of expense over a number of years, such as the United Kingdom Population Census (carried out every ten years).

b) Investigations can be carried out in isolation or in conjunction with others. For example, the calculation of the official monthly Retail Price Index involves a major (ongoing) investigation which includes using the results of the Family Expenditure Survey (which is used also for other purposes). However, the information needed for first line management to control the settings of machines on a production line might depend only on sampling output at regular intervals.

c) Investigations can be regular (routine or ongoing) or 'one-off'. For example, the preparation of a company's trial balance as against a special investigation to examine the calculation of stock re-order levels.

d) Investigations are carried out on populations. A *population* is the entirety of people or items (technically known as *members*) being considered. Thus if a company wanted information on the time taken to complete jobs, the population would consist of all jobs started in the last calendar year say. Sometimes complete populations are investigated, but often only representative sections of the population, or *samples*, are surveyed due to time, manpower and resource restrictions.

## 8. Stages in an investigation

However small or large an investigation is, there are certain landmarks or identifiable stages, through which it should normally pass. These are listed as follows.

a) *Definition of target population and objectives of the survey*. Who? (e.g. does the term 'workers' include temporary part-timers?) Why? (Answering this correctly will ensure that unnecessary questions are not asked and essential questions are asked.)
b) *Choice of method of data collection*. Sometimes a survey will dictate which method is used and in other cases there will be a choice. A list of the most common methods of data collection is given in the following chapter.
c) *Design of questionnaire* or the specification of other criteria for data measurement.
d) *Implementation of a pilot (or trial) survey*. A pilot survey is a small 'pre-survey' carried out in order to check the method of data collection and ensure that questions to be asked are of the right kind. Pilot surveys are normally carried out in connection with larger investigations, where considerable expenditure is involved.
e) *Selection of population members to be investigated*. If the whole (target) population is not being investigated, then a method of sampling from it must be chosen. Various sampling methods are covered fully in the following chapter.
f) *Organisation of manpower and resources to collect the data*. Depending on the size of the investigation, there are many factors to be considered. These might include: training of interviewers, transport and accommodation arrangements, organisation of local reporting bases, procedures for non-responses and limited checking of replies.
g) *Copying, collation and other organisation* of the collected data.
h) *Analyses* of data (with which much of the book is concerned).
i) *Presentation* of analyses and preparation of reports.

## 9. Summary

a) The subject matter of this book, Business Mathematics and Statistics, is sometimes described as Quantitative Methods, Quantitative Techniques or Numerical Analysis.
b) The subject matter attempts to alleviate the problem of incomplete information for management under the three broad headings: Statistical Method, Management Mathematics and Probability.
c) The gap between facts as they are recorded and the provision of useful information for management is bridged by Statistical Method. This covers:
   i. the selection, collection and organisation of basic facts into meaningful data, and
   ii. the summarizing, presentation and analysis of data into useful information.

d) The extent of the Management Mathematics that is covered in this book is concerned with:
   i. the understanding and evaluation of the finances involved in business investments, and
   ii. describing and evaluating physical production processes in quantitative terms.

e) Probability can be thought of as the ability to attach limits to areas of uncertainty.

f) Statistical investigations can be considered as the logical structure through which information is provided for management. They can be: trivial or major; carried out in isolation or in conjunction with other investigations; regular or 'one-off'. Investigations are carried out on populations, which can be described as the entirety of people or items under consideration.

g) The stages in an investigation could be some or all of the following, depending on their size and scope.
   i. Definition of target population and survey objectives.
   ii. Choice of method of data collection.
   iii. Design of questionnaire or the specification of other criteria for data measurement.
   iv. Implementation of a pilot survey.
   v. Selection of population members to be investigated.
   vi. Organisation of manpower and resources to collect the data.
   vii. Copying, collation and other organisation of the collected data.
   viii. Analyses of data.
   ix. Presentation of analyses and preparation of reports.

## 10. Student self review questions

1. What does management use Business Mathematics and Statistics for? [3]
2. What is Statistical Method and what purpose does it serve? [4]
3. Describe the two main areas covered under the heading of Management Mathematics. [5]
4. What is Probability? [6]
5. What is the particular significance of a statistical investigation to management information? [7]
6. What is meant by the term 'population'? [7]
7. List the stages of a statistical investigation. [8]

# Part 1 Data and their presentation

This part of the book deals with the origins, organisation and presentation of statistical data.

Chapter 2 describes methods of selecting data items for investigation (using censuses and samples) and the various ways in which data can be collected.

Data are classified in chapter 3 and some aspects of their accuracy, including rounding, is discussed.

Chapter 4 covers various forms of frequency distributions, which are the main method of organising numerical data into a form which is convenient for either graphical presentation or analysis. Charts used to display frequency distributions include histograms and Lorenz curves.

Chapter 5 describes the many types of charts and graphs that are used to describe non-numeric data and data described over time. These include several types of bar charts, pie charts, and line diagrams.

# 2 Sampling and data collection

## 1. Introduction

This chapter is concerned with the various methods employed in choosing the subjects for an investigation and the different ways that exist for collecting data. Primary data sources (censuses and samples) are described in depth and include:

a) advantages and disadvantages in their use, and
b) data collection techniques.

Secondary data sources, mainly official publications, are covered later in the chapter.

## 2. Primary and secondary data

a) *Primary data* is the name given to data that are used for the specific purpose for which they were collected. They will contain no unknown quantities in respect of method of collection, accuracy of measurements or which members of the population were investigated. Sources of primary data are either censuses or samples and both of these are described in the following sections.
b) *Secondary data* is the name given to data that are being used for some purpose other than that for which they were originally collected. Summaries and analyses of such data are sometimes referred to as *secondary statistics*. The main sources of secondary data are described in later sections of the chapter.

Statistical investigations can use either primary data, secondary data or a combination of the two. An example of the latter follows. Suppose that a national company is planning to introduce a new range of products. It might refer to secondary data on rail and road transport, areas of relevant skilled labour and information on the production and distribution of similar goods from tables provided by the Government Statistical Service to site their new factory. The company might also have carried out a survey to produce their own primary data on prospective customer attitudes and the availability of distribution through wholesalers.

## 3. Censuses

A *census* is the name given to a survey which examines *every member* of a population.

a) A firm might take a census of all its employees to find out their opinions on the possible introduction of a new incentive scheme.
b) The Government Statistical Service carries out many official censuses. Some of them are described as follows.
    i. A *Population Census* is taken every ten years, obtaining information such as age, sex, relationship to head of household, occupation, hours of work, education, use of a car for travel to work, number of rooms in place of dwelling etc for the whole population of the United Kingdom.
    ii. A *Census of Distribution* is taken every five years, covering virtually all retail establishments and some wholesalers. It obtains information on numbers of employees, type of goods sold, turnover and classification etc.
    iii. A *Census of Production* is taken every five years, covering manufacturing industries, mines and quarries, building trades and public utility produc-

tion services. The information obtained and analysed includes distribution of labour, allocation of capital resources, stocks of raw materials and finished goods and expenditure on plant and machinery.

A census has the obvious advantages of completeness and being accepted as representative, but of course must be paid for in terms of manpower, time and resources. The three government censuses described above involve a great deal of organisation, with some staff needed permanently to answer queries on the census form, check and correct errors and omissions and extensively analyse and print the information collected. Forms can take up to a year to be returned with a further gap of up to two years before the complete results are published.

## 4. Samples

In practice, most of the information obtained by organisations about any population will come from examining a small, representative subset of the population. This is called a *sample*. For example:

i. a company might examine one in every twenty of their invoices for a month to determine the average amount of a customer order;

ii. a newspaper might commission a research company to ask 1000 potential voters their opinions on a forthcoming election.

The information gathered from a sample (i.e. measurements, facts and / or opinions) will normally give a good indication of the measurements, facts and / or opinions of the population from which it is drawn. The *advantages* of sampling are usually smaller costs, time and resources. A general *disadvantage* is a natural resistance by the layman in accepting the results as representative. Other disadvantages depend on the particular method of sampling used and are specified in later sections, when each sampling method is described in turn.

## 5. Bias

*Bias* can be defined as the tendency of a pattern of errors to influence data in an unrepresentative way. The errors involved in the results of investigations that have been subject to bias are known as systematic errors.

The main types of bias are now described.

a) *Selection bias*. This can occur if a sample is not truly representative of the population. Note that censuses cannot be subject to this type of bias. For example, sampling the output from a particular machine on a particular day may not adequately represent the nature and quality of the goods that customers receive. Factors that could be involved are: there may be other machines that perform better or worse; this machine might be manned by more or less experienced operators; this day's production may be under more or less pressure than another day's.

b) *Structure and wording bias*. This could be obtained from badly worded questions.

   For example, technical words might not be understood or some questions may be ambiguous.

c) *Interviewer bias*. If the subjects of an investigation are personally interviewed, the interviewer might project biased opinions or an attitude that might not gain the full cooperation of the subjects.

d) *Recording bias*. This could result from badly recorded answers or clerical errors made by an untrained workforce.

## 6. Sampling frames

Certain sampling methods require each member of the population under consideration to be known and identifiable. The structure which supports this identification is called a *sampling frame*. Some sampling methods require a sampling frame only as a listing of the population; other methods need certain characteristics of each member also to be known. Sampling frames can come in all shapes and sizes. For example:

i. A firm's customers can be identified from company records.
ii. Employees can be identified from personnel records.
iii. A sampling frame for the students at a college would be their enrolment forms.
iv. The relevant telephone book would form a sampling frame of people who have telephones in a certain area.
v. Stock items can be identified from an inventory file.

Note however that there are many populations that might need to be investigated for which no sampling frame exists. For example, a supermarket's customers, items coming off a production line or the potential users of a new product. Sampling techniques are often chosen on the basis of whether or not a sampling frame exists.

## 7. Sampling techniques

The sampling techniques most commonly used in business and commerce can be split into three categories.

a) *Random sampling*. This ensures that each and every member of the population under consideration has an equal chance of being selected as part of the sample. Two types of random sampling used are:
   i. Simple random sampling (see section 9), and
   ii. Stratified (random) sampling (see section 12).

b) *Quasi-random sampling*. (Quasi means 'almost' or 'nearly'.) This type of technique, while not satisfying the criterion given in a) above, is generally thought to be as representative as random sampling under certain conditions. It is used when random sampling is either not possible or too expensive to consider. Two types that are commonly used are:
   i. Systematic sampling (see section 13), and
   ii. Multi-stage sampling (see section 14).

c) *Non-random sampling*. This is used when neither of the above techniques are possible or practical. Two well-used types are:
   i. Cluster sampling (see section 15), and
   ii. Quota sampling (see section 16).

Before covering each of the above sampling methods in turn, it is necessary to describe some associated concepts and structures.

## 8. Random sampling numbers

The two types of random sampling, listed in section 7 above and described in sections 9 and 12 following, normally require the use of *random sampling numbers*. These consist of the ten digits from 0 to 9, generated in a random fashion (normally from a computer) and arranged in groups for reading convenience. The term 'generated in a random fashion' can be interpreted as 'the chance of any one digit occurring in any position in the table is no more or less than the chance of any other digit occurring'.

Appendix 2 shows a typical table of such numbers, blocked into groups of five digits. The table is used to ensure that any random sample taken from some sampling frame will be free from bias. The following section describes the circumstances under which the tables are used.

## 9. Simple random sampling

*Simple random sampling*, as described earlier, ensures that each member of the population has an equal chance of being chosen for the sample. It is necessary therefore to have a sampling frame which (at the least) lists all members of the target population. Examples of where this method might be used are:

a) by a large company, to sample 10% of their orders to determine their average value;
b) by an auditor, to sample 5% of a firm's invoices for completeness and compatibility with total yearly turnover;
c) by a professional association, to sample a proportion of its members to determine their views on a possible amalgamation with another association.

Each of these three would have obvious, ready-made sampling frames available.

It is generally accepted that the best method of drawing a simple random sample is by means of random sampling numbers. Example 1, which follows, demonstrates how the tables are used.

The *advantages* of this method of sampling include the selection of sample members being unbiased and the general acceptance by the layman that the method is fair.

*Disadvantages* of the method include:

i. the need for a population listing,
ii. the need for each chosen subject to be located and questioned (this can take time), and
iii. the chance that certain significant attributes of the population are under or over represented.

For example, if the fact that a worker is part-time is considered significant to a survey, a simple random sample might only include 25% part-time workers from a population having, say, a 30% part-time work force.

## 10. Example 1 (Use of *random sampling numbers*)

An auditor wishes to sample 29 invoices out of a total of 583 received in a financial year. The procedure that could be followed is listed below.

1. Each invoice would be numbered, from 001 through to 583.
2. Select a starting row or column from a table of random sampling numbers and begin reading groups of three digits sequentially. For example, using the random sampling numbers at Appendix 2, start at row 6 (beginning 34819 80011 17751 03275 ...etc). This gives the groups of three as: 348 198 001 117 751 032 ... etc.
3. Each group of three digits represents the choice of a numbered invoice for inclusion in the sample. Any number that is greater than 583 is ignored as is any repeat of a number. Using the illustration from 2. above, invoice numbers 348, 198, 001, 117, 032, etc would be accepted as part of the sample, while number 751 would be rejected as too large.
4. As many groups of three digits as necessary are considered until 29 invoices have been identified. This forms the required sample.

**Notes:**

a) Random sampling numbers can be generated by a computer or pre-printed tables can be obtained.

b) The number of digits to be read in groups will always depend upon how many members there are in the population. If there were 56,243 members, then digits would need to be read in fives; groups of four digits would be read if a population being sampled had 8771 members.

c) The choice of a starting row or column for reading groups of digits should be selected randomly.

## 11. Stratification of a population

*Stratification* of a population is a process which:

i. identifies certain attributes (or strata levels) that are considered significant to the investigation at hand;

ii. partitions the population accordingly into groups which each have a unique combination of these levels.

For example, if whether or not heavy goods vehicles had a particular safety feature was thought important to an investigation, the population would be partitioned into the two groups 'vehicles with the feature' and 'vehicles without the feature'. On the other hand, if whether an employee was employed full or part-time, together with their sex, was felt to be significant to their attitudes to possible changes in working routines, the population would be partitioned into the four groups: male / full-time; female / full-time; male / part-time and female / part-time.

Populations that are stratified in this way are sometimes referred to as *heterogeneous,* meaning that they are composed of diverse elements or attributes that are considered significant.

## 12. Stratified sampling

*Stratified random sampling* extends the idea of simple random sampling to ensure that a heterogeneous population has its defined strata levels taken account of in the sample. For example, if 10% of all heavy goods vehicles have a certain safety feature, and this is considered significant to the investigation in hand, then 10% of a sample of such vehicles must have the safety feature.

The general procedure for taking a stratified sample is:

a) Stratify the population, defining a number of separate partitions.
b) Calculate the proportion of the population lying in each partition.
c) Split the total sample size up into the above proportions.
d) Take a separate sample (normally simple random) from each partition, using the sample sizes as defined in (c).
e) Combine the results to obtain the required stratified sample.

Stratification of a population can be as simple or complicated as the situation demands. Some surveys might warrant that a population be split into many strata. A major investigation into car safety could identify the following significant factors having some bearing on safety: saloon and estate cars; radial and cross-ply tyres; two and four-door models; rear passenger safety belts (or not). The sampling frame in this case would have to be split into sixteen separate partitions in order to take account of all the combinations possible from (i) to (iv) above (for example, saloon/radial/2-door/belts and saloon/radial/2-door/no belts are just two of the partitions).

*Advantages* of this method of sampling include the fact that the sample itself (as well as the method of selection) is free from bias, since it takes into account significant strata levels (attributes) of a population considered important to the investigation.

*Disadvantages* of stratified sampling include:

i. an extensive sampling frame is necessary;
ii. strata levels of importance can only be selected subjectively;
iii. increased costs due to the extra time and manpower necessary for the organisation and implementation of the sample.

## 13. Systematic sampling

*Systematic sampling* is a method of sampling that can be used where the population is listed (such as invoice values or the fleet of company vehicles) or some of it is physically in evidence (such as a row of houses, items coming off a production line or customers leaving a supermarket). The technique is to choose a random starting place and then systematically sample every 40th (or 12th or 165th) item in the population, the number (40, say) having been chosen based on the size of sample required. For example, if a 2% sample was needed from a population, every 50th item would be selected, after having started at some random point.

This is because 2% = 2 in 100 = 1 in 50.

Systematic sampling is particularly useful for populations that (with respect to the investigation to hand) are of the same kind or are uniform. These are referred to as *homogeneous* populations. For example, the invoices of a company for one financial year would be considered as a homogeneous population by an auditor, if their value or relationship to type of goods ordered was of no consequence to the investigation. Thus, a systematic sample could be used.

Care must be taken however, when using this method of sampling, that no set of items in the population recur at set intervals. For example, if four machines are producing identical products at the same rate and these are being passed to a single conveyer, it could happen that the products form natural sets of four (one from each machine). A systematic sample, examining every n-th item (where n is a factor of 4), might well be selecting products from the same machine and therefore be biased.

*Advantages* of this method include:

i. ease of use;
ii. the fact that it can be used where no sampling frame exists (but items are physically in evidence).

The main *disadvantage* of systematic sampling is that bias can occur if recurring sets in the population are possible.

This method of sampling is not truly random, since (once a random starting point has been selected) all subjects are pre-determined. Hence the use of the term 'quasi-random' to describe the technique.

## 14. Multi-stage sampling

Where a population is spread over a relatively wide geographical area, random sampling will almost certainly entail travelling to all parts of the area and thus could be prohibitively expensive. *Multi-stage sampling*, which is intended to overcome this particular problem, involves the following.

a) Splitting the area up into a number of regions;
b) Randomly selecting a small number of the regions;
c) Confining sub-samples to these regions alone, with the size of each sub-sample proportional to the size of the area. For example, the United Kingdom could be split up into counties or a large city could be split up into postal districts;
d) The above procedure can be repeated for sub-regions within regions... and so on.

Once the final regions (or sub-regions etc) have been selected, the final sampling technique could be (simple or stratified) random or systematic, depending on the existence or otherwise of a sampling frame.

The main *advantage* of this method is that less time and manpower is needed and thus it is cheaper than random sampling.

*Disadvantages* of multi-stage sampling include:

i. possible bias if a very small number of regions is selected;
ii. the method is not truly random, since, once particular regions for sampling have been selected, no member of the population in any other region can be selected.

## 15. Cluster sampling

*Cluster sampling* is a non-random sampling method which can be employed where no sampling frame exists, and, often, for a population which is distributed over some geographical area. The technique involves selecting one or more geographical areas and sampling *all* the members of the target population that can be identified.

For example, suppose a survey was needed of companies in South Wales who use a computerized payroll. First, three or four small areas would be chosen (perhaps two of these based in city centres and one or two more in outlying areas). Each company, in each area, might then be phoned, to identify which of them have computerized systems. The survey itself could then be carried out.

The *advantages* of cluster sampling include:

i it is a good alternative to multi-stage sampling where no sampling frame exists;

ii. it is generally cheaper than other methods since little organisation or structure is needed in the selection of subjects.

The main *disadvantage* of the method is the fact that sampling is not random and thus selection bias could be significant. (Non-response is not normally considered to be a particular problem.)

## 16. Quota sampling

A sampling technique much favoured in market research is *quota sampling*. The method uses a team of interviewers, each with a set number (quota) of subjects to interview. Normally the population is stratified in some way and the interviewer's quota will reflect this. This method places a lot of responsibility onto interviewers since the selection of subjects (and there could be many strata involved) is left to them entirely. Ideally they should be well trained and have a responsible, professional attitude.

The *advantages* of quota sampling include:

i. stratification of the population is usual (although not essential);

ii. no non-response;

iii. low cost and convenience.

The main *disadvantages* of this method are:

i. sampling is non-random and thus selection bias could be significant;

ii. severe interviewer bias can be introduced into the survey by inexperienced or untrained interviewers, since all the data collection and recording rests with them.

## 17. Precision

Clearly the best way of obtaining information about a population is to take a census. This will ensure (barring any bias and clerical errors) that the information obtained about the population is accurate. However, sampling is a fact of life and the information about a population that is derived from a sample will inevitably be imprecise. The error involved is sometimes known as sampling error. One technique that is often used to compensate for this is to state limits of error for any sample statistics produced. Particular precision techniques are just outside the scope of this book.

## 18. Sample size

There is no universal formula for calculating the size of a sample. However, as a starting point, there are two facts that are well known from statistical theory and should be remembered.

1. The larger the size of sample, the more precise will be the information given about the population.
2. Above a certain size, little extra information is given by increasing the size.

All that can be deduced from the above two statements, together with some other points made in earlier sections of the chapter, is that a sample need only be large enough to be reasonably representative of the population. Some general factors involved in determining sample size are listed below.

a) *Money and time available.*

b) *Aims of the survey*. For example, for a quick market research exercise, a very small sample (perhaps just 50 or 100 subjects) might suffice. However if the opinions of the workforce were desired on a major change of working structures, a 20 or 30% sample might be in order.

c) *Degree of precision required.* The less precise the results need to be, the smaller the sample size.

   For example, to gauge an approximate market reaction to one of their new products, a firm would only need a very small sample. On the other hand, if motor vehicles were being sampled for exhaustive safety tests at a final production stage, the sample would need to be relatively large.

d) *Number of sub-samples required.* When a stratified sample needs to be taken and many sub-samples are defined, it might be necessary to take a relatively large total sample in order that some smaller groups contain significant numbers.

   For example, suppose that a small sub-group accounted for only 0.1% of the population. A total sample size as large as 10,000 would result in a sample size of only 10 (0.1%) for this sub-group, which would probably not be large enough to gain any meaningful information.

## 19. Methods of primary data collection

*Data collection* can be thought of as the means by which information is obtained from the selected subjects of an investigation. There are various data collection methods which can be employed. Sometimes a sampling technique will dictate which method is used and in other cases there will be a choice, depending on how much time and manpower (and inevitably money) is available. The following list gives the most common methods.

a) *Individual (personal) interview.*

   This method is probably the most expensive, but has the advantage of completeness and accuracy. Normally questionnaires will be used (described in more detail in the following section).

   Other factors involved are:

   i. interviewers need to be trained;
   ii. interviews need arranging;

iii. can be used to advantage for pilot surveys, since questions can be thoroughly tested;
iv. uniformity of approach if only one interviewer is used;
v. an interviewer can see or sense if a question has not been fully understood and it can be followed-up on the spot.

This form of data collection can be used in conjunction with random or quasi-random sampling.

b) *Postal questionnaire.*

This is a much cheaper method than the personal interview since manpower (one of the most expensive resources) is not used in the data collection. However, much more effort needs to be put into the design of the questionnaire, since there is often no way of telling whether or not a respondent has understood the questions or has answered them correctly (both of these are generally no problem in a personal interview).
Other factors involved are:

i. low response rates (although inducements, such as free gifts, often help);
ii. convenience and cheapness of the method when the population is scattered geographically;
iii. no prior arrangements necessary (unlike the personal interview);
iv. questionnaires sent to a company may not be filled in by the correct person.

This method can be used in conjunction with most forms of sampling.

c) *Street (informal) interview.*

This method of data collection is normally used in conjunction with quota sampling, where the interviewer is often just one of a team. Some factors involved are:

i. possible differences in interviewer approach to the respondents and the way replies are recorded;
ii. questions must be short and simple;
iii. non-response is not a problem normally, since refusals are ignored and another subject selected;
iv. convenient and cheap.

d) *Telephone interview.*

This method is sometimes used in conjunction with a systematic sample (from the telephone book). It would generally be used within a local area and is often connected with selling a product or a service (for example, insurance). It has an in-built bias if private homes are being telephoned (rather than businesses), since only those people with telephones can be contacted and interviewed. It can cause aggravation and the interviewer needs to be very skilled.

e) *Direct observation.*

This method can be used for examining items sampled from a production line, in traffic surveys or in work study. It is normally considered to be the most accurate form of data collection, but is very labour-intensive and cannot be used in many situations.

## 20. Questionnaire design

If a questionnaire is used in a statistical survey, its design requires careful consideration. A badly designed questionnaire can cause many administrative problems and may cause incorrect deductions to be made from statistical analyses of the results. One of the major reasons why pilot surveys are carried out is to check typical responses to questions. Some important factors in the design of questionnaires are given below.

a) The questionnaire should be as short as possible.

b) Questions should:

i. be simple and unambiguous.

ii. not be technical.

iii. not involve calculations or tests of memory.

iv. not be personal, offensive or leading.

c) As many questions as possible should have simple answer categories (so that the respondent has only to choose one). For example:

How many employees are there in your company?

| Under 10 | 10 to 24 | 25 to 49 | 50 or over |
|---|---|---|---|
| ☐ | ☐ | ☐ | ☐ |

Do you find the equipment supplied:

| | | | |
|---|---|---|---|
| Very easy to use? | ☐ | Fairly easy to use? | ☐ |
| Fairly difficult to use? | ☐ | Very difficult to use? | ☐ |

d) Questions should be asked in a logical order.

A useful check on the adequacy of the design of a questionnaire can be given by conducting a *pilot survey*.

## 21. The use of secondary data

Secondary data are generally used when:

a) the time, manpower and resources necessary for your own survey are not available (and, of course, the relevant secondary data exists in a usable form), or

b) it already exists and provides most, if not all, of the information required.

The *advantages* of using secondary data are savings in time, manpower and resources in sampling and data collection. In other words, somebody else has done the 'spade work' already.

The *disadvantages* of using secondary data can be formidable and careful examination of the source(s) of the data is essential. Problems include the following.

i. Data quality might be questionable. For example, the sample(s) used may have been too small, interviewers may not have been experienced or any questionnaires used may have been badly designed.
ii. The data collected might now be out-of-date.
iii. Geographical coverage of the survey may not coincide with what you require. For example, you might require information for Liverpool and the secondary data coverage is for the whole of Merseyside.
iv. The strata of the population covered may not be appropriate for your purposes. For example, the secondary data might be split up into male/female and full-time/part-time workers and you might consider that, for your purposes, whether part-time workers are permanent or temporary is significant.
v. Some terms used might have different meanings. Common examples of this are:
   - Wages (basic only or do they include overtime?)
   - Level of production (are rejects included?)
   - Workers (factory floor only or are office staff included?)

## 22. Sources of secondary data and their use

Secondary data sources fall broadly into two categories: those that are *internal* and those *external* to the organisation conducting the survey.

Some examples of *internal* secondary data sources and uses are:

a) a customer order file, originally intended for standard accounting purposes, could have its addresses and typical goods amounts used for route planning purposes;
b) using information on raw material type and price (originally collected by the purchase department to compare manufacturers) for stock control purposes;
c) information on job times and skills breakdown, originally compiled for job costing, used for organising new pay structures.

Some examples of *external* secondary data sources are:

d) the results of a survey undertaken by a credit card company, to analyse the salary and occupation of its customers, might be used by a mail order firm for advertising purposes;
e) a commercially produced car survey giving popularity ratings and buying intentions, might be used by a garage chain to estimate stock levels of various models.

Without doubt, the most important external secondary data sources are official statistics supplied by the Central Statistical Office and other government departments. These are listed and briefly described in the next section.

## 23. Official secondary data sources

The following list gives the major publications of the Central Statistical Office.

a) *Annual Abstract of Statistics*. This publication is regarded as the main general reference book for the United Kingdom and has been published for nearly 150 years. Its tables cover just about every aspect of economic, social and industrial life. For example: climate; population; social services; justice and crime;

education; defence; manufacturing and agricultural production; transport and communications; finance.

b) *Monthly Digest of Statistics*. A monthly abbreviated version of the Annual Abstract of Statistics. Gives the facts on such topics as: population; employment; prices; wages; social services; production and output; energy; engineering; construction; transport; retailing; finance and the weather. It has runs on monthly and quarterly figures for at least two years in most tables and annual figures for longer periods. An annual supplement gives definitions and explanatory notes for each section. An index of sources is included.

c) *Financial Statistics*. A monthly publication bringing together the key financial and monetary statistics of the United Kingdom. It is the major reference document for people and organisations concerned with government and company finance and financial markets generally. It usually contains at least 18 monthly, 12 quarterly or 5 annual figures on a wide variety of topics. These include: financial accounts for sectors of the economy; Government income and expenditure; public sector borrowing; banking statistics; money supply and domestic credit expansion; institutional investment; company finance and liquidity; exchange and interest rates. An annual explanatory handbook contains notes and definitions.

d) *Economic Trends*. Published monthly, this is a compilation of all the main economic indicators, illustrated with charts and diagrams. The first section (Latest Developments) presents the most up-to-date statistical information available during the month, together with a calendar of recent economic events. The central section shows the movements of the key economic indicators over the last five years or so. Finally there is a chart showing the movements of four composite indices over 20 years against a reference chronology of business cycles. In addition, quarterly articles on the national accounts appear in the January, April, July and October issues, and on the balance of payments in the March, June, September and December issues. Occasional articles comment on and analyse economic statistics and introduce new series, new analyses and new methodology. *Economic Trends* also publishes the release dates of forthcoming important statistics. An annual supplement gives a source for very long runs, up to 35 years in some cases, of key economic indicators. The longer runs are annual figures, but quarterly figures for up to 25 years or more are provided.

e) *Regional Trends*. An annual publication, with many tables, maps and charts, it presents a wide range of government statistics on the various regions of the United Kingdom. The data covers many social, demographic and economic topics. These include: population, housing, health, law enforcement, education and employment, to show how the regions of the United Kingdom are developing and changing.

f) *United Kingdom National Accounts (The Blue Book)*. Published annually, this is the essential data source for those concerned with macro-economic policies and studies. The principal publication for national accounts statistics, it provides detailed estimates of national product, income and expenditure. It covers industry, input and output, the personal sector, companies, public corporations, central and local government, capital formation and national accounts. Tables of statistical information, generally extending over eleven years, are supported by definitions and detailed notes. It is a valuable indicator of how the nation makes and spends its money.

g) *United Kingdom Balance of Payments (The Pink Book)*. This annual publication is the basic reference book of balance of payments statistics, presenting all the statistical information (both current and for the preceding ten years) needed by those who seek to assess United Kingdom trends in relation to those of the rest of the world.

h) *Social Trends*. One of the most popular and colourful annual publications, it has (for over ten years) provided an insight into the changing patterns of life in Britain. The chapters provide accurate analyses and breakdowns of statistical information on population, households and families, education and employment, income and wealth, resources and expenditure, health and social services and many other aspects of British life and work.

i) *Guide to Official Statistics*. A periodically produced reference book for all users of statistics. It indicates what statistics have been compiled for a wide range of commodities, services, occupations etc, and where they have been published. Some 1000 topics are covered and about 2500 sources identified with an index for easy use. It covers all official and significant non-official sources published during the last ten years.

The following publication is compiled by the Department of Employment.

j) *Employment Gazette*. Published monthly, it is a summary of statistics on: employment, unemployment, numbers of vacancies, overtime and short time, wage rates, retail prices, stoppages. Each publication includes one or more 'in-depth' article and details of arbitration awards, notices, orders and statutory instruments.

The following publication is compiled by the Department of Industry.

k) *British Business*. Published weekly, the main topics are production, prices and trade. It includes information on: the Census of Production, industrial materials, manufactured goods, distribution, retail and service establishments, external trade, prices, passenger movements, hire purchase, entertainment.

Other important business publications include: HSBC Holdings plc Annual Review, NatWest Bank Quarterly Review, Lloyds TSB Annual Report, Barclays Review (quarterly), International Review (Barclays, quarterly), Three Banks Review (quarterly), Journal of the Institute of Bankers (bi-monthly), Financial Times (daily), The Economist (weekly) and The Banker (monthly).

## 24. Summary

a) Data that are used for the specific purpose for which they are collected are called primary data. Secondary data is the name given to data that are being used for some purpose other than that for which they were originally collected.

b) A census is a survey which examines every member of the population. Three important official censuses are the Population Census, the Census of Distribution and the Census of Production.

c) A sample is a relatively small subset of a population with advantages over a census that costs, time and resources are much less. The main disadvantage is that of acceptability by the layman.
d) Bias is the tendency of a pattern of errors to influence data in an unrepresentative way. Bias can be due to selection procedures, structure and wording of questions, interviewers or recording.
e) A sampling frame is a structure which lists or identifies the members of a population.
f) Random sampling numbers are tables of randomly generated digits, used to ensure that the selection of the members of a sample is free from bias.
g) Simple random sampling is a technique which ensures that each and every member of a population has an equal chance of being chosen for the sample.
h) Stratified random sampling ensures that every significant group in the population is represented in proportion in the sample using a stratification process. An extensive sampling frame is needed with this method.
i) Systematic (quasi-random) sampling involves selecting a random starting point and then sampling every n-th member of the population. The value of n is chosen based on the size of sample required. It can be biased if certain recurring cycles exist in the population, but can often be used where no sampling frame exists.
j) Multi-stage (quasi-random) sampling is normally used with homogeneous populations spread over a wide area. It involves splitting the area up into regions, selecting a few regions randomly and confining sampling to these regions alone. It is cheaper than random sampling.
k) Cluster (non-random) sampling involves exhaustive sampling from a few well chosen areas. It is a cheap method, useful for populations spread over a wide geographical area where no sampling frame exists.
l) Quota (non-random) sampling normally involves teams of interviewers who obtain information from a set quota of people, the quota being based on some stratification of the population. It is commonly used in market research.
m) The precision of some statistic obtained from a sample can be measured by describing the limits of error with a given degree of confidence.
n) Some factors involved in determining the size of a sample are: money and time available, survey aims, degree of precision or number of sub-samples required. Generally, the larger the sample the better, but small samples can give relatively accurate information about a population.
o) Main methods of primary data collection are:
   i. Individual (personal) interview.
   ii. Postal questionnaire.
   iii. Street (informal) interview.
   iv. Telephone interview.
   v. Direct observation.
p) The main points in questionnaire design are: questionnaire to be as short as possible; questions to be simple, non-ambiguous, non-technical, not to be

personal or offensive and not to involve calculations or tests of memory; answer categories to be given where possible; questions asked in a logical order.

q) Secondary data can be used where the facilities for your own survey are not available or where the secondary data gives all the information you require. Disadvantages are: data might not be of an acceptable quality or out-of-date; geographical or strata coverage may not be appropriate; there may be differences in the meaning of terms.

r) Some of the main sources of external secondary data are contained in the following official publications:

Annual Abstract of Statistics; Monthly Digest of Statistics;
Financial Statistics; Economic Trends;
Regional Trends;
United Kingdom National Accounts (Blue Book);
United Kingdom Balance of Payments (Pink Book);
Social Trends; Employment Gazette;
British Business.

## 25. Student self review questions

1. Explain the difference between primary and secondary data. [2]
2. Give the meaning of a census and give some examples of official censuses. [3]
3. What are the major factors involved when deciding between a sample and a census? [3,4]
4. Describe what bias is and give some examples of how it can arise. [5]
5. Give at least four examples of a sampling frame. [6]
6. What is a random sample? [7]
7. What is quasi-random sampling and under what conditions might it be used? [7]
8. What are random sampling numbers and how are they used in simple random sampling? [8,10]
9. What does the term 'stratification of a population' mean and how is it connected with stratified sampling? [11,12]
10. What are the advantages and disadvantages of stratifed sampling when compared with simple random sampling? [12]
11. What is the difference between homogeneous and heterogeneous populations? [11,13]
12. Give an example of a situation where a systematic sample could be taken:
    a) where a sampling frame exists;
    b) where no sampling frame exists. [13]
13. What are the differences between multi-stage and cluster sampling methods? [14,15]
14. In what type of situation is quota sampling most commonly used and what are its main merits? [16]
15. How can the precision of a sample estimate be expressed? [17]
16. What are the factors involved in determining the size of a sample? [18]
17. List the main methods of collecting primary data [19]

18. What are the advantages and disadvantages of a postal questionnaire over a personal interview? [19]
19. Give some important considerations in the design of a questionnaire. [20]
20. Under what conditions might secondary data be used and what are its possible disadvantages compared with the use of primary data? [21]
21. Name some of the major official statistical publications. [22]

## 26. Student exercises

1. *MULTI-CHOICE*. Which one of the following is NOT a type of random sampling technique:
   a) Quota sampling b) Systematic sampling
   c) Stratified sampling d) Multi-stage sampling
2. *MULTI-CHOICE*. A 2% random sample of mail-order customers, each with a numeric serial number, is to be selected. A random number between 00 and 49 is chosen and turns out to be 14. Then, customers with serial numbers 14, 64, 114, 164, 214, ... etc are chosen as the sample. This type of sampling is:
   a) simple random b) stratified c) quota d) systematic.
3. A large company is considering a complete reshaping of its pay structures for production workers. What data might be collected and analysed, other than technical details, to help the management come to a decision? Consider both primary and secondary sources.
4. What factors would govern the use of a sample enquiry rather than a census if information was required about shopping facilities throughout a large city.
5. *MULTI-CHOICE*. A sample of 5% of the employees working for a large national company is required. Which one of the following methods would provide the best simple random sample?
   a) Wait in the car park in a randomly selected branch and select every tenth employee driving in to work .
   b) Use random number tables to select 1 in 20 of the branches and then select all the employees.
   c) Select a branch randomly and use personnel records to choose 1 in 20 randomly.
   d) Select 5% of all employees from personnel records at head office randomly.
6. Suggest an appropriate method of sampling that could be employed to obtain information on:
   a) passengers' views on the adequacy of a local bus service;
   b) the attitudes to authority of the workforce of a large company;
   c) the percentage of defects in finished items from a production line;
   d) the views of Welsh car drivers on the wearing seat belts;
   e) the views of schoolchildren on school meals.
7. A national survey has revealed that 40% of non-manual workers travel to work by public transport while one-half use their own transport. For all workers, 47.5% use public transport and one in every ten use methods other than their own or public transport. A statistical worker in a large factory (which is known to have about

three times as many manual workers as non-manual workers) has been asked to sample 200 employees for their views on factory-provided transport. He decides to take a quota sample at factory gate B at five o'clock one evening.

a) How many manual workers will there be in the sample?
b) How many workers who travel to work by public transport will be interviewed?
c) Calculate the quota to be interviewed in each of the six sub-groups defined.
d) Point out the limitations of the sampling technique involved and suggest a better way of collecting the data.

8. The makers of a brand of cat food 'Purrkins' wish to obtain information on the opinions of their customers and include a short questionnaire on the inside of the label as follows:
   1. Do you like Purrkins?
   2. Why do you buy Purrkins?
   3. Have you tried our dog food?
   4. What amount of Purrkins do you normally buy?
   5. When did you start using Purrkins?
   6. What type of house do you live in?

   Criticise the questions.
9. Design a short questionnaire to be posted to a sample of customers to obtain their views on your company's delivery service.
10. A proposal was received by the Local Authority Planning Office for a motel, public house and restaurant to be built on some private land in the city suburbs. Following an article by the builder in the local paper, the office received 300 letters of which only 28 supported the proposal. What conclusions can the Planning Officer draw from these statistics? Describe what action could be taken to gauge people's views further.

# 3 Data and their accuracy

## 1. Introduction

This chapter is concerned with the forms that data can take, how data are measured and the errors and approximations that are often made in their description.

## 2. Data classification

In order to present and analyse data in a logical and meaningful way, it is necessary to understand some of the natural forms that they can take. There are various ways of classifying data and these are now listed.

a) *By source*. Data can be described as either primary or secondary, depending on their source. This area has already been covered in the previous chapter.

b) *By measurement*. Data can be measured in either numeric (or quantitative) or non-numeric (qualitative) terms. This might sound very obvious, but the difference is important since the forms of both presentation and analysis differ markedly in these two cases. Presentation of numeric data is covered in chapter 4 and non-numeric data in chapter 5. For the business and accounting courses that this book covers, only numeric data is analysed and this is done from chapter 7 onwards.

c) *By preciseness*. Data can either be measured precisely (described as *discrete*) or only ever be approximated to (described as *continuous*). The differences between the two are described more fully in sections 3 and 4 in this chapter.

d) *By number of variables*. Data can consist of measurements of one or more variables for each subject or item. *Univariate* is the name given to a set of data consisting of measurements of just one variable, *bivariate* is used for two variables, and for two or more variables the data is described as *multivariate*. Some examples of these different types are given in example 1.

## 3. Discrete data

Discrete data can be described as data that can be measured precisely. One way of obtaining discrete data is by counting. For example:

i. the number of components produced from an assembly line over a number of consecutive shifts:

   45, 51, 44, 44, 43, 50, 46, 43, ... etc;

ii. the number of employees working in various offices of a company:

   12, 32, 8, 13, 8, 6, 11, 24, ... etc.

Discrete data can also be obtained from situations where counting is not involved. For example:

iii. shoe sizes of a sample of people:

   8, 10, 10, $6\frac{1}{2}$, 9, 9, $9\frac{1}{2}$, $8\frac{1}{2}$ ... etc;

iv. weekly wages (in £) for a set of workers:

   121.45, 162.85, 133.37, 108.32, ... etc.

A particular characteristic of discrete data is the fact that possible data values progress in definite steps, i.e. shoe sizes are measured as 6 or $6\frac{1}{2}$ or 7 or $7\frac{1}{2}$ ... etc or there are 1 or 2 or 3 ... etc people (and not 3.5 or 4.67).

## 4. Continuous data

The most significant characteristic of continuous data is the fact that they cannot be measured precisely; their values can only be approximated to. Examples of continuous data are dimensions (lengths, heights); weights; areas and volumes; temperatures; times.

How well continuous values are approximated to depends on the situation and the quality of the measuring instrument. It might be adequate to measure peoples' heights to the nearest inch, whereas spark plug end gaps would need to be measured to perhaps the nearest tenth of a millimetre. Time card punching machines only record times in hours and minutes while sophisticated process control computers, dealing with volatile chemicals, would need to measure both time and temperature very finely.

Although continuous values cannot be identified exactly, they are often recorded as if they were precise and this is normally acceptable. For example:

i. clock-in times of the workers on a particular shift:
   8:23, 8:28, 8:28, 8:32, 8:28, 8:26, ... etc;
ii. diameters (in mm) of a sample of screws from a production run:
   4.11, 4.10, 4.10, 4.10, 4.15, 4.09, 4.12, ... etc;
iii. weights (in gm) of the contents of a selection of cans of baked beans:
   446.8, 447.0, 446.8, 447.2, 447.0, 447.1, ... etc.

## 5. Example 1 (Demonstrations of various *classifications* of data)

a) Table 1 shows the ages and annual salaries of a sample of qualified certified accountants. Since each member of the sample is being measured in terms of two variables, age and salary, this is an example of bivariate data. Both variables are numeric with salary discrete (since each is an exact value) and age continuous (since age is really a particular type of time measurement) and approximated to years only.

*Table 1 Age and salary for a sample of certified accountants*

| Sample member | 1 | 2 | 3 | 4 | 5 | 6 | 7 | 8 | 9 |
|---|---|---|---|---|---|---|---|---|---|
| Age | 32 | 30 | 25 | 28 | 25 | 30 | 49 | 26 | 56 |
| Salary | 8800 | 11900 | 6000 | 8200 | 5800 | 12500 | 9650 | 7200 | 16450 |

b) Number of defects found in samples of 100 items taken by a quality control section from batches of finished products.

| Batch | 1 | 2 | 3 | 4 | 5 | 6 | 7 | 8 | 9 | 10 |
|---|---|---|---|---|---|---|---|---|---|---|
| Number of defectives | 2 | 0 | 2 | 5 | 1 | 3 | 0 | 0 | 1 | 0 |

The data are being described in terms of one variable (number of defectives in a sample of 100) and thus are univariate. They are also discrete, since the values have been obtained by counting, and numeric.

c) Policies handled by an agent for a particular insurance company:

| Policy | Type | Annual premium (£) |
|---|---|---|
| A | Motor (3rd party) | 86 |
| B | Motor (3rd party) | 124 |
| C | Life | 185 |
| D | Motor (comprehensive) | 120 |
| E | Disability | 24 |
| F | House contents | 49 |
| G | Motor (comprehensive) | 252 |

The data given for the various policies is described in terms of two variables, type of policy and annual premium, and thus is bivariate. Type of policy is non-numeric, annual premium is numeric and both variables are discrete.

## 6. Rounding and its conventions

a) Data are normally rounded for one of two reasons.

i. If they are continuous, rounding is the only way to give single values which will represent the magnitude of the data.

ii. If they are discrete, the values given may be too detailed to use as they stand. For example, the annual profits of a plc might have been calculated precisely as £14,286,453.88, but could be quoted on the stock exchange as £14 million.

b) As should already be familiar, *fair rounding* is the technique of cutting off particular digits from a given numeric value and, depending on whether the first digit discarded has value 5 or more (or not), adding 1 to the last of the remaining digits or not (known as rounding up or down).

There are two conventions used for displaying the results of fair rounding.

i. *By decimal place*. This is the most common form of rounding. For example, if the price of a car was given as £4684.45, it could be rounded as:

£4684.5 (to 1 decimal place or 1D)

£4684 (to the nearest whole number or n1)

£4680 (to the nearest 10 or n10) - note the final zero

£4700 (to the nearest 100 or n100)

£5000 (to the nearest 1000 or n1000)

ii. By *number of digits.* This convention is sometimes used as an alternative to decimal place rounding. For example, if a company's profit for the past financial year was £682,056.39, it could be rounded as:

£682,056.4 (to 7 significant digits or 7S)

£682,060 (to 5 significant digits or 5S)

£682,000 (3S) and, finally, £700,000 (1S)

## 7. Errors and their causes

Any data, whatever their source (international, national, company or personal), can be subject to errors. The causes of errors are numerous, but some of the more important of them can be classified into two main groups.

a) *Unpredictable errors*. These are errors that occur due to:

   i. Incomplete or incorrect records.

   ii. Ambiguous or over-complicated questions asked as part of questionnaires.

   iii. Data being obtained from samples.

   iv. Mistakes in copying data from one form to another.

   All that can be done to minimise this type of error is to ensure that: investigation procedures are carried out in a professional, logical and consistent way; questionnaires are well designed and tested; samples are as representative as possible; and so on.

b) *Planned (predictable) errors*. These are errors that were referred to in section 6 (a) and are due to:

   i. Measuring continuous data.

   ii. Rounding discrete data for the purposes of overall perspective.

   This is the type of error that can be taken account of and is discussed in the rest of the chapter.

## 8. Maximum errors

If a number is given and known to be subject to an error, then the error must be unknown (otherwise there would be no need to consider it). However, what can be determined often is the largest value that the error could possibly take. This is known as the *maximum error*. From now on, unless otherwise stated, any error referred to will be assumed to be a maximum error.

## 9. Methods of describing errors

A number that is subject to some unknown error is sometimes called an *approximate number*. Approximate numbers can be written in three different forms, described as follows.

a) *An interval*. This takes the form of a range of values within which the true number being represented lies. The range is normally given as a low and high value, separated by a comma and written within square brackets. For example: [19.5,20.5]. This would mean that the true value of the number being represented lies between 19.5 and 20.5.

b) An *estimate with a maximum absolute error*. In this form the real value of the number being represented is given as an estimate, together with the maximum error. It takes the form:

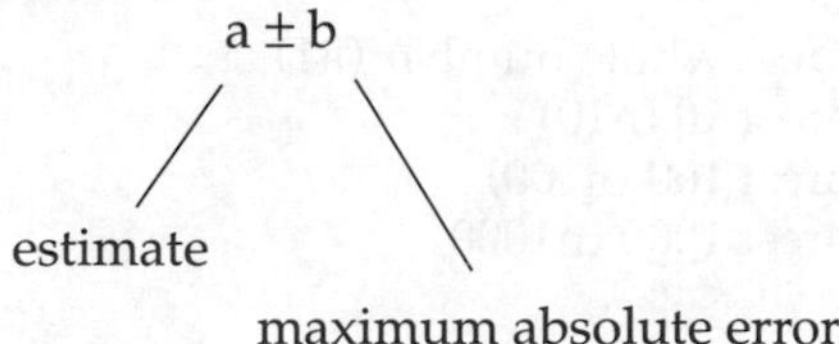

For example: 20 ± 0.5. This can be translated as 'the true value of the number being represented lies within 0.5 either side of 20'. Of course this is exactly equivalent to the example given in a) above.

c) *An estimate with a maximum relative error*. This is a slight adaptation of b) above, in that the maximum error is expressed in relative (i.e. proportional or percentage) terms. That is:

$$\text{maximum relative error} = \frac{\text{maximum absolute error}}{\text{estimate}} \times 100\%$$

For example: 20 ± 2.5%. This can be translated as 'the true value of the number being represented lies within 2.5% either side of 20'. Since 2.5% of 20 is just 0.5, it should now be realised that the examples of the three forms of expressing an approximate number shown above are equivalent.

## 10. Example 2 (Expressing a number *subject to error* in different forms)

If the annual output of a mine is given as 15 million tonnes, subject to a maximum error of 1.5 million tonnes, the real value of the output can be written as:

15 million tonnes ± 1.5 million tonnes

or: $$15 \pm \frac{5}{236} \times 100\% = 15 \text{ million tonnes} \pm 10\%$$

or: [15–1.5,15+1.5] million tonnes = [13.5,16.5] million tonnes.

## 11. Rounding errors

As soon as a numeric value is subjected to fair rounding, an error is introduced and an approximate number is thus defined. As an example, suppose the length of a bolt is measured as 14 mm to the nearest mm. This means that the true length of the component must lie between the values 13.5 and 14.5 mm (because any value that lies between these two would have been rounded to 14 mm). Hence, the maximum error can be expressed as ± 0.5 mm.

In fact, any value that is rounded to the nearest whole number will have a maximum error of ± 0.5. Similarly, any value that is rounded to 1D will have a maximum error of ± 0.05. Any value rounded to the nearest 10 will have a maximum error of ± 5. This pattern is tabulated in Table 2.

*Table 2 Maximum errors in fair rounded numbers*

| Degree of rounding | Maximum error |
|---|---|
| ... | ... |
| ... | ... |
| 3D | ± 0.0005 |
| 2D | ± 0.005 |
| 1D | ± 0.05 |
| nearest whole number (n1) | ± 0.5 |
| nearest 10 (n10) | ± 5 |
| nearest 100 (n100) | ± 50 |
| nearest 1000 (n1000) | ± 500 |
| ... | ... |
| ... | ... |

Thus, for example:

i. 64.23 (2D) = $64.23 \pm 0.005 = [64.225, 64.235]$;
ii. 12 (n1) = $12 \pm 0.5 = [11.5, 12.5]$;
iii. 78400 (n100) = $78400 \pm 50 = [78350, 78450]$;
iv. 24.6 (3S) = 24.6 (1D) = $24.6 \pm 0.05 = [24.55, 24.65]$;
v. 1530 (3S) = 1530 (n10) = $1530 \pm 5 = [1525, 1535]$.

## 12. Errors in expressions

An *arithmetic expression* is any combination of numbers and arithmetic operations (+, -, x and ÷).

For example: 12+6, (42–6)×3 and $\frac{72 - 12 \times 4}{4 + 8}$ are examples of arithmetic expressions.

If all the numbers involved in an expression are exact, then the expression will be able to be evaluated exactly. If, however, at least one of the numbers involved in an expression is subject to an error, then the value of the expression itself will be subject to error.

As an example of identifying the range of error involved in the value of an expression, suppose the first of the above expressions, 12+6, is to be evaluated, where both 12 and 6 have been *rounded to the nearest whole number*.

Now, 12 can be represented as [11.5, 12.5] (where 11.5 is the least value and 12.5 is the greatest value).

Similarly, 6 can be represented as [5.5, 6.5] (where 5.5 and 6.5 are the least and greatest values respectively).

So that: the *least* value of 12+6 = least value of 12 + least value of 6

= 11.5 + 5.5 = 17

Similarly, the *greatest* value of 12 + 6 is 12.5 + 6.5 = 19

Therefore the range of possible values for 12 + 6 is [17,19].

Finding the range of values for the multiplication of two numbers (each subject to an error) follows a similar line, but the rule for subtraction and division needs special consideration. The following section gives the rules for these four basic operations.

## 13. Error rules

The range of error for the addition, subtraction, multiplication and division of two numbers, each of which is subject to error, is now given.

**Ranges of error**

If the value of the number $X$ lies in the range $[a, b]$ and $Y$ lies in the range $[c, d]$, then:

$X + Y$ lies in the range $[a + c, b + d]$
$X \times Y$ lies in the range $[a \times c, b \times d]$
$X - Y$ lies in the range $[a - d, b - c]$
$X \div Y$ lies in the range $[a \div d, b \div c]$

As examples of the use of the above, suppose that $X$ has value 12 ($n$1) and $Y$ has value 5 (± 1). We can write $X = [11.5,12.5]$ and $Y = [4,6]$.

Therefore,

$$X + Y = [11.5,12.5] + [4,6] = [11.5+4,12.5+6] = [15.5,18.5]$$
$$X \times Y = [11.5,12.5] \times [4,6] = [11.5\times4,12.5\times6] = [46,75]$$
$$X - Y = [11.5,12.5] - [4,6] = [11.5\text{-}6,12.5\text{-}4] = [5.5,8.5]$$
$$X \div Y = [11.5,12.5] \div [4,6] = [11.5\div6,12.5\div4] = [1.92,3.13] \text{ (2D)}$$

The least and greatest values of any expression can now be calculated. The technique is to work through the expression, bit by bit, obeying the above rules. Remember, however, to obey also the usual rules of arithmetic when working through the expression (i.e. '×' and '÷' are considered before '+' and '-', and brackets to be evaluated first), sometimes known as the 'BODMAS' rule. Examples 3 and 4 demonstrate the technique.

## 14. Example 3 (The *range* in which the value of an arithmetic expression lies)

*Question*

Calculate the range of possible values for the expression: $\dfrac{4.12 - 8.3}{0.8}$, where each term has been rounded.

*Answer*

4.12 is measured to 2D = 4.12 ± 0.005 = [4.115,4.125];
8.3 is measured to 1D = 8.3 ± 0.05 = [8.25,8.35];
0.8 is measured to 1D = 0.8 ± 0.05 = [0.75,0.85].
Thus the expression can be written as:

$$\frac{[4.115,4.125] + [8.25,8.35]}{[0.75,0.85]} = \frac{[4.115+8.25,4.125+8.35]}{[0.75,0.85]}$$
$$= \frac{[12.365,12.475}{[0.75,0.85]}$$
$$= \left[\frac{12.365}{0.85},\frac{12.475}{0.75}\right] \text{ (division rule)}$$
$$= [14.55,16.63] \text{ (2D).}$$

## 15. Example 4 (The *value range* for an expression in a business situation)

*Question*

A machine can produce 2000 (± 25) items per day, each of which can weigh between 5 and 7 grammes. At the end of each day, the production from eight similar machines is loaded into at least 10 (but no more than 15) equally weighted shipping crates. Find the lower and upper limits of the weight of one loaded crate (to the nearest gramme).

*Answer*

Number of items produced per machine = 2000 ± 25 = [1975,2025].

Total weight of production per machine per day

$$= [1975,2025] \times [5,7]$$
$$= [1975{\times}5, 2025{\times}7] \text{ (using multiplication rule)}$$
$$= [9875,14175] \text{ grammes.}$$

Total weight of production per day from eight machines

$$= 8 \times [9875,14175]$$
$$= [79000,113400] \text{ grammes.}$$

Therefore, the weight range of one loaded crate

$$= \frac{[79000,113400]}{[10,15]}$$
$$= \left[\frac{79000}{15}, \frac{113400}{10}\right] \text{ (using division rule)}$$
$$= [5267,11340] \text{ grammes (n1).}$$

## 16. Fair and biased rounding

Only fair rounding has been considered so far. However, sometimes rounding is performed in one direction only. For instance:

i. When people's ages are quoted (in years), they are usually rounded down. Thus, if someone's age was given as 31 years, their actual age could be as low as 31 years and 0 days or as high as 31 years and 364 days.
ii. Suppose a job in the factory needed 83 bolts and the stores only issue bolts in sets of 10. Clearly the 83 would be rounded up to 90. In this, and similar situations, rounding would be performed upwards, since the tendency is to slightly overstock rather than to understock (and run the risk of not being able to satisfy the requirements of a job or a customer order).

This is called *biased* rounding. Maximum errors involved in biased rounding could be up to twice the size of the errors involved when using fair rounding. For example, an age quoted as 25 (which will have been rounded down) could be representing an actual age which is as high as 25 years and 364 days. In other words, the maximum error (to all intents and purposes) is 1 year. Compare this with a maximum error of only 0.5 years (either + or -) which would have resulted from fair rounding (i.e. to the nearest year).

A similar table to that shown for fair rounded numbers (in section 11) can be drawn up for biased rounded numbers and is shown in Table 3.

*Table 3 Maximum errors in biased rounded numbers*

**Maximum errors in biased rounded numbers**

| *Degree of rounding* | *Maximum error* |
|---|---|
| ... | ... |
| ... | ... |
| 3D | ± 0.001 |
| 2D | ± 0.01 |
| 1D | ± 0.1 |
| lowest or highest whole number | ± 1 |
| lowest or highest 10 | ± 10 |
| lowest or highest 100 | ± 100 |
| lowest or highest 1000 | ± 1000 |
| ... | ... |
| ... | ... |

## 17. Compensating and systematic errors

a) When numbers are rounded fairly, the errors involved are known as *compensating errors*. This name is used because, in the long run, we would expect half the errors to be on one side (i.e. negative) and half on the other (positive), compensating for each other. For example:

| | | | | | | Total |
|---|---|---|---|---|---|---|
| Real value | 15123 | 23375 | 32914 | 76089 | 23547 | 171048 |
| Rounded value (nearest 1000) | 15000 | 23000 | 33000 | 76000 | 24000 | 171000 |
| Error | +123 | +375 | -86 | +89 | -453 | +48 |

When numbers subject to compensating errors are added, the errors should (roughly speaking) cancel each other out, leaving the total relatively error-free. This can be seen from the above data, where the relative error in the rounded total

$$= \frac{48}{171000} \times 100\%$$

$$= 0.3\%.$$

b) When numbers are subject to biased rounding, the errors involved are known as *systematic* (or *biased* or *one-sided*). The example below shows the numbers used in a) rounded down.

| | | | | | | Total |
|---|---|---|---|---|---|---|
| Real value | 15123 | 23375 | 32914 | 76089 | 23547 | 171048 |
| Rounded value (lowest 1000) | 15000 | 23000 | 32000 | 76000 | 23000 | 169000 |
| Error | +123 | +375 | +914 | +89 | +547 | +2048 |

When numbers subject to systematic errors are added, the errors quickly accumulate in relative terms. The relative error in the total of rounded values above is $\frac{2048}{169000} \times 100\% = 12.1\%$ (much higher than the 0.3% obtained in a)).

> **Types of errors in numbers**
>
> *Compensating errors* are the errors involved when numbers are subject to fair rounding. Totals of these type of numbers will be relatively error-free.
>
> *Systematic errors* are errors involved when numbers are subject to biased rounding. Totals of these type of numbers will have a relatively high error.

## 18. Example 5 (Effects of adding numbers subject to *compensating / systematic* errors)

*Question*

The number of minor industrial accidents of a particular type reported per month over six successive months were 41, 62, 87, 96, 32, 39. Calculate the absolute and relative errors for the six-monthly totals of rounded figures, if rounding is performed: (a) to the nearest 10 (b) to the highest 10 (c) to the lowest 10.

*Answer*

The calculations are shown in Table 4.

**Note:** Relative error = $\frac{\text{Absolute error}}{\text{Total of rounded values}} \times 100\%$

The relative errors in the totals of the two columns subject to systematic errors can be seen to be approximately ten times the relative error in the total of the column subject to compensating errors.

*Table 4*

| | True values | Rounded to the nearest 10 (subject to compensating errors) | Rounded to the highest 10 | Rounded to the lowest 10 |
|---|---|---|---|---|
| | | | (subject to systematic errors) | |
| | 41 | 40 | 50 | 40 |
| | 62 | 60 | 70 | 60 |
| | 87 | 90 | 90 | 80 |
| | 96 | 100 | 100 | 90 |
| | 32 | 30 | 40 | 30 |
| | 39 | 40 | 40 | 30 |
| Totals | 357 | 360 | 390 | 330 |
| Absolute errors | | +3 | +33 | –27 |
| Relative errors | | +0.8% | +8.5% | –8.2% |

## 19. Avoiding errors when using percentages

Percentages are normally used with business statistics in order to eliminate actual units so that comparisons can be made. They are particularly useful for measuring increases (or decreases) in sets of values, but care must be taken when calculating and interpreting them. The following three notes highlight areas where errors in calculation or interpretation often occur.

a) *Do not confuse percentage and actual values.* As an example:

| | Number of new orders | | Absolute increase | Percentage increase |
|---|---|---|---|---|
| | Period 1 | Period 2 | | |
| Firm A | 10 | 50 | 40 | 400 |
| Firm B | 350 | 451 | 101 | 29 |

Notice that although firm A has a smaller actual increase in orders, its percentage increase is much larger than that of firm A since it is based on a very small period 1 value.

b) *When calculating percentage increases involving a set of values over time, the base time period for the increase must be clearly stated.*
As an example:

| Year | 1 | 2 | 3 | 4 |
|---|---|---|---|---|
| Number unemployed | 200,000 | 252,000 | 310,000 | 376,000 |

i. % age increase in year 2 (based on year 1) $= \frac{252{,}000-200{,}000}{200{,}000} \times 100\%$
$= 26\%$
% age increase in year 3 (based on year 2) = 23%
% age increase in year 4 (based on year 3) = 21%

ii. % age increase in year 2 (based on year 1) = 26%
% age increase in year 3 (based on year 1) = 55%
% age increase in year 4 (based on year 1) = 88%

c) *Percentages should not be added (or averaged) in the usual way*, since each is normally derived from a different base. An overall percentage must be calculated *using grand total figures*. Table 5 gives an example of this.

*Table 5 Calculating an average percentage*

| | Units of output | | Percentage increase |
|---|---|---|---|
| | Year 1 | Year 2 | |
| Factory A | 30,000 | 30,500 | 1.7 |
| Factory B | 15,000 | 16,000 | 6.7 |
| Factory C | 16,000 | 20,000 | 25.0 |
| Total | 61,000 | 66,500 | 9.0 |

Notice that the overall percentage increase in output has been calculated on 66,500 compared with the previous 61,000, and NOT a combination of 1.7, 6.7 and 25.0.

## 20. Summary

a) Data can be classified:
   i. by source - primary or secondary
   ii. by measurement - numeric or non-numeric
   iii. by preciseness - discrete or continuous
   iv. by number of variables - univariate (one variable) or bivariate (two variables) or multivariate (many variables).

b) Discrete data can be measured precisely and progresses from one value to another in definite steps. Continuous data cannot be measured precisely.

c) 'Fair' rounding is the process of rounding either up or down according to the value of the first digit of those that are ignored. Rounding can be performed by decimal place or number of significant digits.

d) Errors in data can be uncontrollable (and only minimized by organisational methods) or planned (by rounding).

e) Errors in numbers can be described using:
   i. intervals, giving the lowest and highest possible values that a number can take, in the form [a,b];
   ii. estimates, giving the maximum absolute error, in the form estimate ± error;
   iii. estimates, giving the maximum relative error, in the form estimate ± error%.

f) A number that has been rounded can be put into any one of the forms given in e) above. Any arithmetic expression involving at least one number subject to an error will itself be subject to error.

g) If X = [a,b] and Y = [c,d], then:

$$X+Y = [a+c, b+d]; \qquad X \times Y = [a \times c, b \times d];$$

$$X-Y = [a-d, b-c]; \qquad \frac{X}{Y} = \left[\frac{a}{d}, \frac{b}{c}\right].$$

h) Biased rounding is where the rounding is always one-sided (i.e. always up or always down). Maximum errors involved here are always twice those involved with fair rounding.

i) Compensating errors are made when numbers are rounded fairly and always have relatively small totals. Systematic errors are made if the rounding is biased and they have relatively large totals.

## 21. Student self review questions

1. What is meant by discrete data? Give some examples. [3]
2. What is meant by continuous data? Give some examples. [4]
3. Give an example each of univariate and bivariate data. [5]
4. What is fair rounding and what are the two main conventions used to display rounded values? [6]

5. Give some examples of 'unpredictable' errors in data and say how they might be minimized. [7]
6. Describe an 'approximate' number and give its three main forms. [9]
7. Explain what biased rounding means and give some examples. [16]
8. What is the difference between a compensating and systematic error and how do they affect the totals of approximate numbers? [17]
9. Why cannot percentages be added or averaged in the normal way? [19]

## 22. Student exercises

1. *MULTI-CHOICE*. Which one of the following would constitute a set of discrete data?
   a) Time taken to travel to work each day over one year.
   b) Weights of a consignment of tins of plum tomatoes.
   c) Number of cars passing a census point each minute over a 3-hour period.
   d) Age of applicants applying for catering jobs over a 3-month period in a large hotel chain.
2. *MULTI-CHOICE*. The value 8.2 has been rounded to 1 decimal place while 16 has been rounded to the nearest whole number. What is the largest value that the sum of the two numbers could possibly be?
   a) 24.75 b) 24.50 c) 24.30 d) 24.25
3. Classify the following sets of data in as many ways as possible.
   a) The times that a number of separate jobs have taken to complete.
   b) The job title and number of years experience of a sample of office workers.
   c) The location and number of employees of a set of textile firm's head offices.
   d) The departments to which a set of employees belong.
   e) The average weekly wages of manual and non-manual workers broken down by sex and industry over three separate years.
4. Round the following numbers to the level stated.
   a) £148,356.78 (nearest £10)
   b) 23,345 tons (to nearest 1000 tons)
   c) 3.245 mm (1D)
   d) £16.42 (to nearest £)
   e) 23 months (to highest 10 months)
   f) £18,625 (to lowest £100).
5. *MULTI-CHOICE*. Systematic errors are made when:
   a) numbers have been rounded fairly
   b) numbers have been rounded in a biased fashion
   c) numbers have been obtained by stratified sampling
   d) numbers have been obtained by systematic sampling
6. *MULTI-CHOICE*. Two groups of stock products, Kappa and Lambda, are valued. Kappa is valued at £100,000 ± 5% and Lambda at £200,000 ± 10%. The *maximum* percentage error in the combined stock valuation of £300,000 is closest to:
   a) 7% b) 8% c) 10% d) 15%

7. Give a range of error, in the form [a,b], for the value of the following expressions:
   a) 143 (± 4) + 56 (± 3).
   b) 12 (± 1) × 4 (± 1) + 27 (± 4).
   c) 31.4 (1D) + 12.23 (2D) × 11 (n1) to 2D.
   d) {31.4 (1D) + 12.23 (2D)} × 11 (n1) to 2D.
   e) $\dfrac{140\ (n10) - 130\ (n10)}{14\ (n1) + 13.2\ (1D)}$ to 2D.
8. A businessman estimates that he can buy 200 (± 20) tables at a cost of £15 (± £1) each and sell them at £18 (± £2) each.
   a) Calculate his estimated (expected) profit on the whole deal.
   b) Give a range of values within which his total profit will lie.
9. A firm works a nominal 38-hour week but with overtime and short time its actual working week varies by as much as 1 hour from the nominal figure. The firm produces 50 (± 2) articles per hour. The production cost and selling price are £2 and £3 per unit respectively, rounded off to the nearest 10 pence. Assuming that all production is sold, calculate the following.
   a) The range of production per week.
   b) The range of:
      i. weekly production costs
      ii. weekly revenue.
   c) The range of weekly profit.
   d) The expected weekly costs and hence the minimum and maximum percentage profit to 1D (based on expected costs).
10. In a single financial year, a building society lends £75 million (2S) to house buyers. The ratio of lending to income for the society for the year is 71.4% (1D). Calculate the range of income (to the nearest £1000) for the society over the year.
11. The stock movements of a particular stores item over a quarter is given below: Given that deliveries have been rounded to the lowest 10, issues have been rounded to the highest 10 and the balance at January 1st is exact, calculate the range of values for the balance of stock at April 1st. What is the expected balance at April 1st?

*Balance at January 1st:* 33

| | Deliveries | Issues |
|---|---|---|
| January | 80 | 90 |
| February | 110 | 90 |
| March | 140 | 130 |

# 4 Frequency distributions and charts

## 1. Introduction

This chapter is concerned with the organisation and presentation of 'numeric (univariate)' data. It describes how numeric data can be organised into frequency distributions of various types, their graphical presentation and their interpretation and use.

## 2. Raw statistical data

Before the data obtained from a statistical survey or investigation have been worked on they are known as *raw data*. Table 1 gives an example of a set of raw data.

*Table 1 Hours worked by employees*

**Hours worked in one week by employees in a company's production department**

| | | | | | | | | | |
|---|---|---|---|---|---|---|---|---|---|
| 46.3 | 45.1 | 45.6 | 45.6 | 46.1 | 45.0 | 43.5 | 39.2 | 39.2 | 39.1 |
| 39.2 | 42.3 | 39.6 | 39.5 | 38.9 | 44.4 | 43.4 | 43.2 | 43.8 | 39.1 |
| 44.2 | 43.5 | 42.0 | 43.1 | 42.4 | 42.4 | 42.8 | 42.9 | 43.1 | 39.8 |
| 41.3 | 40.0 | 39.6 | 39.7 | 42.1 | 39.8 | 44.3 | 46.2 | 41.3 | 40.8 |

## 3. Data arrays

Raw data normally yield little information as they stand. This is particularly true if there are many hundreds or even thousands of data items. One simple way of extracting some information from the data is by arranging them into size order. This is known as a *data array*. The raw data of Table 1 have been arrayed and are shown in Table 2.

*Table 2 Raw data of Table 1 put into an array*

| | | | | | | | | | |
|---|---|---|---|---|---|---|---|---|---|
| 38.9 | 39.1 | 39.1 | 39.2 | 39.2 | 39.2 | 39.5 | 39.6 | 39.6 | 39.7 |
| 39.8 | 39.8 | 40.0 | 40.8 | 41.3 | 41.3 | 42.0 | 42.1 | 42.3 | 42.4 |
| 42.4 | 42.8 | 42.9 | 43.1 | 43.1 | 43.2 | 43.4 | 43.5 | 43.5 | 43.8 |
| 44.2 | 44.3 | 44.4 | 45.0 | 45.1 | 45.6 | 45.6 | 46.1 | 46.2 | 46.3 |

Certain information can now be identified from the table. For example, the lowest number of hours worked is 38.9; the highest is 46.3. A typical number of hours worked could be quoted as 42.4, since this value lies half way along the array.

## 4. Simple frequency distributions

Some sets of raw data contain a limited number of data values, even though there may be many occurrences of each value. In this type of situation, the standard form into which the data is organised is known as a simple *frequency distribution*. A simple frequency distribution consists of a list of data values, each showing the number of items having that value (called the *frequency*). This type of structure is normally applicable to discrete raw data (i.e. where values have usually been obtained by counting), since data values are quite likely to be repeated many times. A simple frequency distribution is not normally suitable for continuous data (i.e. where values have been measured), since the likelihood of repeated values is small.

Example 1 shows a simple frequency distribution and its interpretation. Example 2 demonstrates how a simple frequency distribution is constructed with the aid of a *tally chart*, which is used to record the occurrence of repeated values systematically.

## 5. Example 1 (*Simple frequency distribution* and its interpretation)

The following simple frequency distribution resulted from examining the time cards of the employees of a firm for one complete month.

| Number of days late for work | 0 | 1 | 2 | 3 | 4 | 5 | 6 | 7 | 8 |
|---|---|---|---|---|---|---|---|---|---|
| Number of employees | 45 | 52 | 18 | 11 | 5 | 2 | 4 | 0 | 1 |

In this case, 'number of days late for work' is describing the data and 'number of employees' is the frequency of occurrence of each value. This distribution gives certain useful information about the nature of absences for the employees. For example:

i. the majority of employees were late for work at most on only one day during the month;
ii. very few employees were late more than four days in the month;
iii. the most common number of days late for work was 1;
iv. no employee was late for work more than 8 times.

## 6. Example 2 (Constructing a simple frequency distribution with a *tally chart*)

### *Question*

The following data record the number of children in the families of the 47 workers in a company.

113 202 012 213 524 002 411 220
300 213 602 103 222 100 113 14

Compile a simple frequency distribution table for these data, using a tally chart.

### *Answer*

The tally chart in Table 3 has been constructed by examining each value and recording its occurrence with a stroke and from this the frequency distribution table, shown as Table 4, has been constructed.

*Tables 3 and 4*

*Tally chart*

| Data value | Tally marks | Total |
|---|---|---|
| 0 | ||||| ||||| | | 11 |
| 1 | ||||| ||||| || | 12 |
| 2 | ||||| ||||| ||| | 13 |
| 3 | ||||| | | 6 |
| 4 | ||| | 3 |
| 5 | | | 1 |
| 6 | | | 1 |

*Frequency distribution table*

| Number of children in family | Number of workers |
|---|---|
| 0 | 11 |
| 1 | 12 |
| 2 | 13 |
| 3 | 6 |
| 4 | 3 |
| 5 | 1 |
| 6 | 1 |

## 7. Grouped frequency distributions

The main aim of a frequency distribution is to summarize numeric data in a logical manner that enables an overall perspective of the data to be obtained quickly and easily.

When the number of distinct data values in a set of raw data is large (20 or more, say), a simple frequency distribution is not appropriate, since there will be too much information, not easily assimilated. In this type of situation, a *grouped frequency distribution* is used. A grouped frequency distribution organizes data items into groups (or classes) of values, each showing how many items have values included within the group (known as the *class frequency*). Note however that once items have been grouped in this way, their individual values are lost. This is the cost of a better perspective of the data.

**Grouped Frequency Distribution**
A grouped frequency distribution summarises data into groups of values, each showing the number of items having values in the group.
**Note**: individual data values cannot be identified with this type of structure.

There are various ways of forming the groups. Some of these are described in the following examples.

## 8. Example 3 (Illustrations of *grouped frequency distributions*)

a) *The values of properties handled by a property dealer over a six-month period.* These data are shown in Table 5. Notice that the value 15 (£000) is included in the second group *but not the first*. Any value up to *but excluding* 15 (£000) is catered for, (with a practical upper limit of £14999.99). An equivalent way of describing the classes would be '10 but under 15'; '15 but under 20'; ... and so on.

b) *The distribution of the length of life of a sample of bad debts of a company.* These data are shown in Table 6. In this distribution there is a gap between the upper limit of the first group (5) and the lower limit of the second group (6) and the same is true of the other groups. Since data values between 5 and 6 are thus not catered for, it must be assumed that the data has been measured to whole numbers only.

*Tables 5 and 6*

| Value of property (£000) | Number of properties |
|---|---|
| 10 and less than 15 | 2 |
| 15 and less than 20 | 6 |
| 20 and less than 25 | 14 |
| 25 and less than 30 | 21 |
| 30 and less than 35 | 33 |
| 35 and less than 40 | 19 |
| 40 and less than 45 | 5 |

| Number of working days | Number of bad debts |
|---|---|
| 1-5 | 44 |
| 6-10 | 50 |
| 11-15 | 42 |
| 16-20 | 27 |
| 21-25 | 16 |

c) *The range of weekly income of a sample of 1000 accountants.* These are shown in Table 7. Notice that the first class has no defined lower limit and the last class has no defined upper limit. These are known as *open-ended* classes and are normally included where lower and upper limits are either not known or may change with time or circumstances.

*Table 7 The range of weekly income of a sample of 1000 accountants*

| Range of income (£) | Number of accountants |
|---|---|
| under 100 | 116 |
| 100 and under 120 | 89 |
| 120 and under 150 | 184 |
| 150 and under 170 | 142 |
| 170 and under 200 | 159 |
| 200 and under 250 | 186 |
| 250 and over | 124 |

## 9. Definitions associated with frequency distribution classes

a) *Class limits* are the lower and upper values of the classes as physically described in the distribution.

*Table 8 Class limits and boundaries*

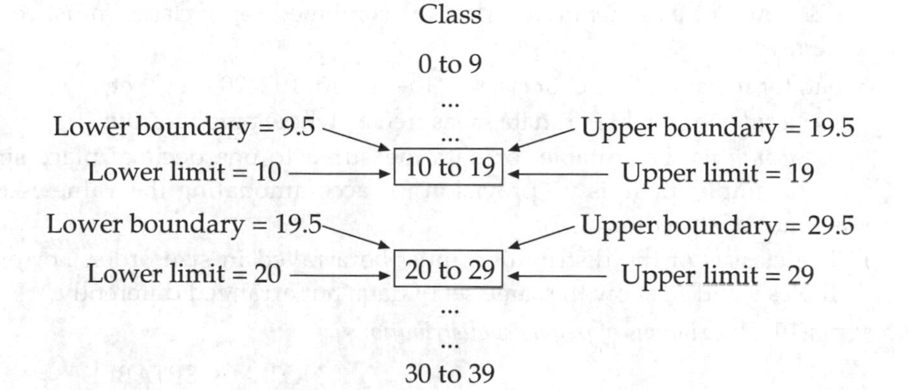


b) *Class boundaries* (sometimes called *mathematical limits*) are the lower and upper values of a class that mark common points between classes. The class structure and descriptions in Table 8 demonstrate the difference between limits and boundaries. Notice that the upper boundary of one class coincides with the lower boundary of the next class. Where there is a gap in the values between classes (i.e. between 9 and 10, or 19 and 20, etc, as in the structure demonstrated above), the boundaries must be fixed at exactly half way along the gap. Even if the data being described is discrete (counted), boundaries still exist (albeit artificially) and need to be identified, since they are used in later statistical analyses.

Sometimes limits and boundaries will coincide, as in the structure '12 and up to 13', '13 and up to 14', '14 and up to 15', ... etc.

c) *Class widths* (or *lengths*) are the numerical differences between lower and upper class boundaries (and not class limits).

d) *Class mid-points* are situated in the centre of the classes. They can be identified as being mid-way between the upper and lower boundaries (or limits).

For example: Class 10 to 19 has a mid-point of 14.5 (not 15); class width = 10 (19.5 minus 9.5)

Class 20 to 29 mid-point = 24.5; class width = 10

Class 30 to 39 mid-point = 34.5; class width = 10 ... etc.

However: Class 30 and up to 40 has a mid-point = 35; class width = 10

Class 40 and up to 50 mid-point = 45; class width = 10 ... etc.

A particular use of class mid-points is in estimating the totals of all the items lying in the class. This can be done by multiplying the class mid-point by the class frequency. Thus if a class is described as 10 to 20 (mid-point = 15) with a frequency of 6, an estimate of the total of all the items in the class is 15×6 = 90.

## 10. Rules and practices for compiling grouped frequency distributions

The following list gives the rules that frequency distributions *should obey* in order to present data logically and consistently.

a) All data values being represented must be contained within one (and only one) class. Thus overlapping classes (such as '10–15', '14–18', '17–21', ... etc in a single distribution) *must not* occur. Also, the combined set of classes must contain all items.

So that, for instance, the set of classes '10–14', '15–19', '20–24', ... etc:

i. *would* be suitable for data measured as whole numbers, but

ii. *would not* be suitable for data measured to one decimal place since, for example, there is no provision for accommodating the value 14.6 in the above structure.

b) The classes of the distribution must be arrayed in size order. For example, Tables 9 and 10 show the same set of data, but arranged differently.

*Tables 9 and 10 Descriptions of frequency distributions*

| ACCEPTABLE | | NOT ACCEPTABLE (since classes are not arranged in size order) | |
|---|---|---|---|
| Range of data | Frequency | Range of data | Frequency |
| 10 and up to 15 | 2 | 10 and up to 15 | 2 |
| 15 and up to 20 | 12 | 30 and up to 40 | 8 |
| 20 and up to 25 | 31 | 20 and up to 25 | 31 |
| 25 and up to 30 | 27 | 15 and up to 20 | 12 |
| 30 and up to 40 | 8 | 25 and up to 30 | 27 |
| 40 and over | 3 | 40 and over | 3 |

The following list gives the characteristics that *demonstrate good practice* when describing frequency distributions.

c) There should normally be between 8 and 10 classes in total, with not less than 5 or more than 15. Too few classes are thought not to give a good overall summary of the nature of the data. That is, there is too little information being shown. On the other hand, too many classes normally give too much information to comprehend quickly the overall nature of the data.

d) Class descriptions should be easy to assimilate with ranges that naturally describe the data being presented. Thus, for instance, annual salaries could be presented in classes of £500, £1000, £2000 or £5000.

*As an example of bad practice,* Table 11 shows a distribution, obtained from the quality control department of a manufacturing company, describing the number of rejects found in each of 180 batches of components.

*Tables 11 and 12*

| Number of rejects | Number of batches |
|---|---|
| 13 to 17 | 20 |
| 18 to 23 | 31 |
| 24 to 32 | 40 |
| 33 to 43 | 32 |
| 44 to 47 | 27 |
| 48 to 54 | 15 |
| 55 to 60 | 7 |
| 61 to 87 | 3 |

| | |
|---|---|
| 10 to 19 | 10 and up to 20 |
| 20 to 29 | 20 and up to 30 |
| 30 to 39 | 30 and up to 40 |
| | ... |
| ... | ... |
| ... | ... |
| *is* correct | *is not strictly* correct |

The information contained in the data is difficult to assimilate, since the lower and upper limits of classes follow no regular pattern and classes are of differing widths.

e) Discrete data should be represented within classes having limits which the data can attain. For example, when representing discrete (counted data), the right hand structure in Table 12 will generate errors when calculating statistical measures (covered in chapters 15 to 20).

f) Frequency distributions having equal class widths throughout are preferable, but where this is not possible, classes with smaller or larger widths can be used. Open-ended classes are acceptable only at the ends of a distribution.

g) When comparing two frequency distributions having the same class structure but different frequencies, it is often more illuminating if the actual frequencies are expressed as percentages (or proportions) of their respective totals. These adapted distributions are sometimes known as *relative frequency distributions*.

## 11. Formation of grouped frequency distributions

Given a set of raw statistical data, there is no single grouped frequency distribution which is uniquely 'correct' in representing them; many different structures of classes could be set up to describe the data. As long as the rules and practices given in the previous section are adhered to, the resulting frequency distribution will normally be acceptable. The way that classes of a grouped frequency distribution are arrived at is not important (and not particularly looked for by an examiner). However, the following procedure gives a step-by-step approach to forming the classes.

STEP 1 Calculate the range of values covered by the data, ignoring (if necessary and only temporarily) any extreme values at either end of the data set. The range is calculated as the highest value minus the lowest value (after extremes have been temporarily ignored). The identification of extreme values is a matter of judgement and / or experience.

STEP 2 Divide the range obtained by 10, and adjust this value upwards to obtain a standard class width for the distribution which is appropriate for the data concerned. Class widths of 1, 2 or 5 (or multiples, such as 10, 20 or 50 etc) are best, since these will span bands of values that form natural groups and are easily comprehended.

STEP 3 The distribution classes can now be constructed. The first class should contain the lowest value (ignoring extremes) and the last class should contain the highest value. If any extreme values are present, they can be taken account of by making the first and / or last class open-ended.

STEP 4 The frequency distribution table can now be formed, using a tally chart.

Example 4 demonstrates the use of this procedure.

## 12. Example 4 (Forming a *grouped frequency distribution*)

*Question*

The data in Table 13 on the opposite page describe the number of orders received by a company each week over a period of forty weeks. Compile a frequency distribution.

*Answer*

STEP 1 Lowest value = 9; highest value = 46. Range = 46–9 = 37.

STEP 2 Class width = $\frac{37}{10} = 3.7$, which is adjusted upwards to 5.

STEPS 3 and 4

The structure of classes is shown in Table 14, which has taken into account that the data is discrete and the first class must contain 9.

## 13. Histograms

A frequency distribution can be represented pictorially by means of a *histogram*. A histogram is a chart consisting of a set of vertical bars and is constructed as follows.

a) Each bar represents just one class; the bar width corresponds to the class width and the bar height generally corresponds to the class frequency (this point is described in more detail in section 28 b)).
b) The bars are joined together (reinforcing the fact that classes have common boundaries).
c) The vertical axis (representing frequency) and the horizontal axis (representing data values) must both be scaled and labelled clearly.
d) The chart as a whole must have a title.

*Tables 13 and 14 From original data to frequency distribution*

| | | | | | | | |
|---|---|---|---|---|---|---|---|
| 24 | 13 | 28 | 15 | 25 | 29 | 15 | 46 |
| 9 | 10 | 17 | 22 | 23 | 17 | 16 | 32 |
| 11 | 12 | 18 | 20 | 13 | 27 | 18 | 22 |
| 20 | 14 | 26 | 14 | 19 | 19 | 40 | 31 |
| 17 | 21 | 23 | 26 | 18 | 24 | 21 | 27 |

*Table 13*

| | |
|---|---|
| 5 - 9 | \| |
| 10 - 14 | \|\|\|\|\|\|\| |
| 15 - 19 | \|\|\|\|\|\|\|\|\|\|\| |
| 20 - 24 | \|\|\|\|\|\|\|\|\|\| |
| 25 - 29 | \|\|\|\|\|\|\| |
| 30 - 34 | \|\| |
| 35 - 39 | |
| 40 - 44 | \| |
| 45 - 49 | \| |

| Number of orders received | Number of weeks |
|---|---|
| 5 - 9 | 1 |
| 10 - 14 | 7 |
| 15 - 19 | 11 |
| 20 - 24 | 10 |
| 25 - 29 | 7 |
| 30 - 34 | 2 |
| 35 - 39 | 0 |
| 40 - 44 | 1 |
| 45 - 49 | 1 |
| Total | 40 |

*Table 14*

Figure 1 shows a histogram for the simple frequency distribution of days late for work from Example 1, while Figure 2 shows a histogram for the grouped frequency distribution of weekly orders, compiled in Example 4.

*Figure 1 Histogram for number of days late for work*

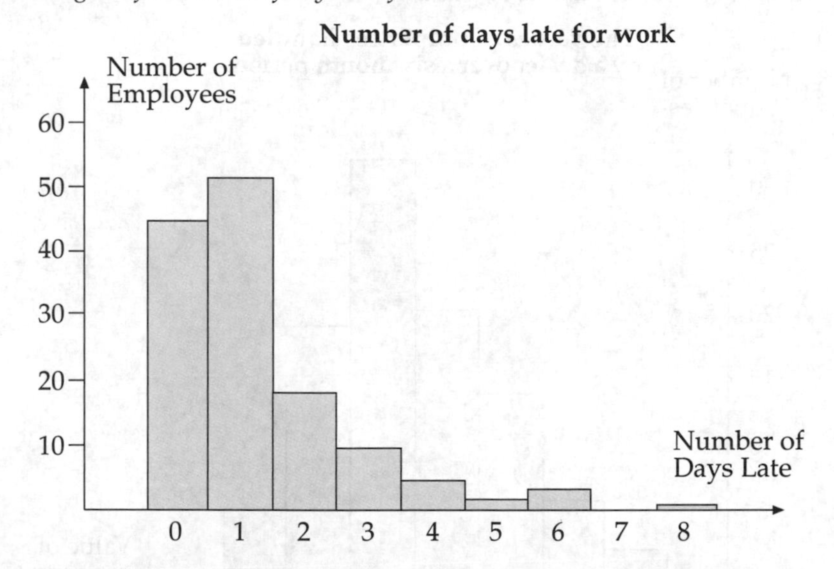


Two frequency distributions having the same class structure can be compared using histograms only if the two respective sets of bars are drawn and shaded *very carefully*. This technique is demonstrated in Example 6.

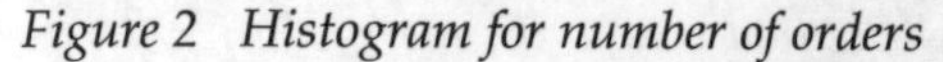

*Figure 2 Histogram for number of orders*

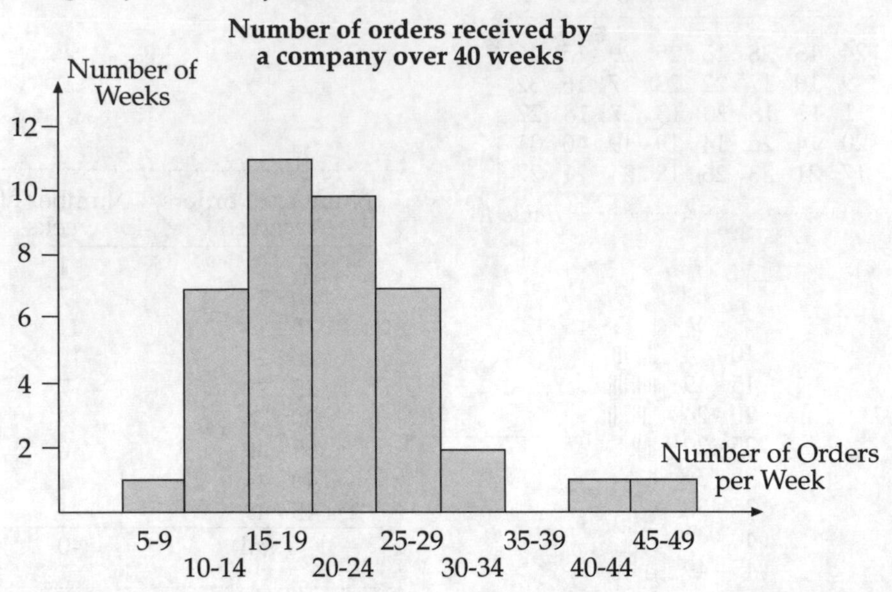


## 14. Example 5 (Displaying a frequency distribution by means of a *histogram*)

The values of properties handled by a property dealer over a six-month period (from Example 3) are displayed using a histogram and are shown in Figure 3.

*Figure 3 Values of properties handled by a dealer*

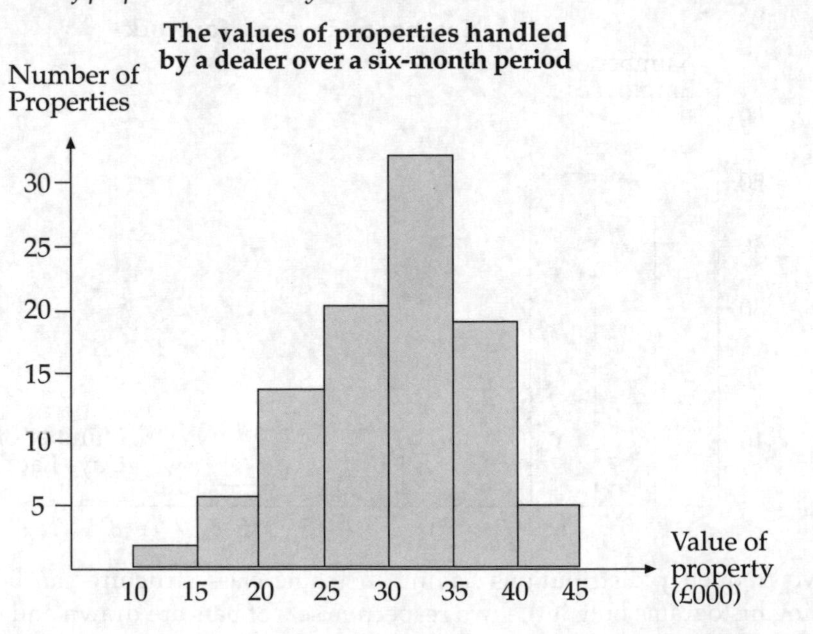

## 15. Example 6 (Comparison of frequency distributions using *comparative histograms*)

*Question*

The values of the orders received for two separate companies over one financial year are given in Table 15 below. Compare the two distributions diagrammatically and comment on the results.

*Table 15 and Figure 4 Values of orders received from two companies*

**The values of orders received in one year for two companies**

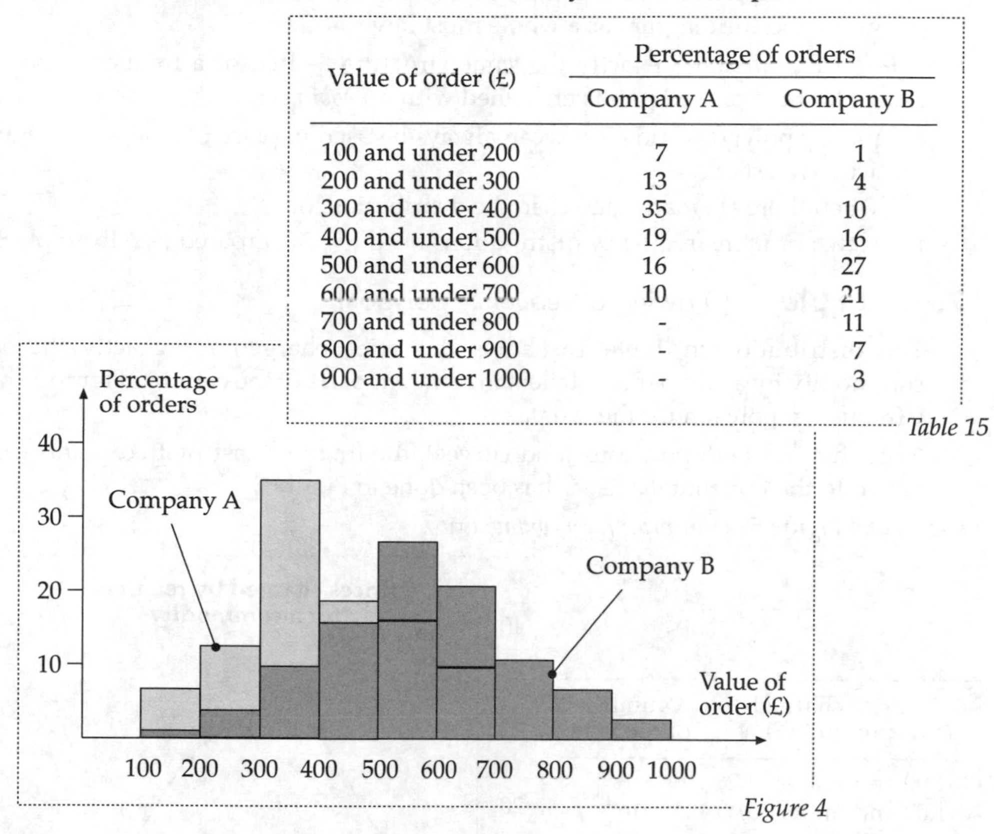

| Value of order (£) | Percentage of orders | |
|---|---|---|
| | Company A | Company B |
| 100 and under 200 | 7 | 1 |
| 200 and under 300 | 13 | 4 |
| 300 and under 400 | 35 | 10 |
| 400 and under 500 | 19 | 16 |
| 500 and under 600 | 16 | 27 |
| 600 and under 700 | 10 | 21 |
| 700 and under 800 | - | 11 |
| 800 and under 900 | - | 7 |
| 900 and under 1000 | - | 3 |

*Table 15*



*Figure 4*

*Answer*

The two distributions are compared, using comparative histograms, in Figure 4. It can be seen from the histograms that company B's orders are generally higher in value than those of company A. Also, company B's orders are spread over a greater range of values (£100 to £1000) than company A's (£100 to 700), and thus they are less consistent. Note that this method of comparison of two frequency distributions is acceptable only when the 'shapes' of the two distributions are different (as in the above case). Note however that if two distributions have frequencies of roughly the same value, this type of presentation can be confusing.

## 16. Frequency polygons and curves

A frequency distribution can also be represented pictorially using a *frequency polygon*. Instead of drawing vertical bars, a line chart is constructed. The description of a frequency polygon is given below.

a) Each class is represented by a single point. The height of the point represents the class frequency; the position of the point must be directly above the corresponding class mid-point.
b) The points are joined by *straight lines*.
c) The two axes are labelled and described in the same way as those for a histogram and the diagram as a whole must have a title.

A *frequency curve* has exactly the same structure as that of a frequency polygon except that the plotted points are joined with a *smooth curve*.

Frequency polygons and curves can always be used in place of histograms, but are particularly useful:

i. when there are many classes in the distribution, or
ii. if two or more frequency distributions need to be compared (see Example 8).

## 17. Example 7 (Drawing a *frequency polygon*)

The distribution in Table 16 shows the price charged for exactly the same commodity by a number of retailers in different parts of the country. Figure 5 shows a frequency polygon for these data.

Note that, for both polygons (and curves), the first and last plotted points can be joined to the horizontal axis, as has been done in Figure 5.

*Table 16 and Figure 5 Unit price for a commodity*

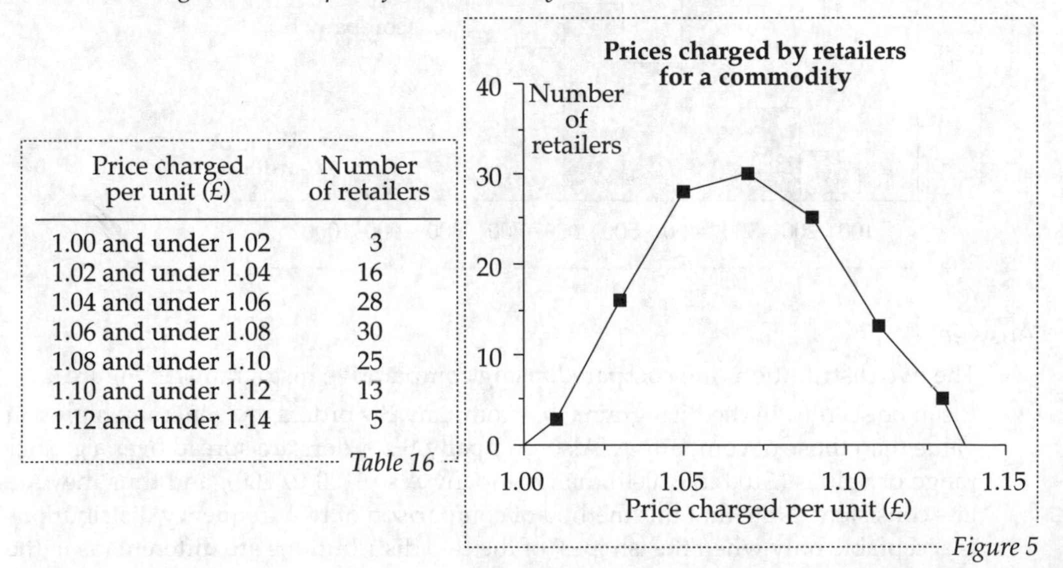

| Price charged per unit (£) | Number of retailers |
|---|---|
| 1.00 and under 1.02 | 3 |
| 1.02 and under 1.04 | 16 |
| 1.04 and under 1.06 | 28 |
| 1.06 and under 1.08 | 30 |
| 1.08 and under 1.10 | 25 |
| 1.10 and under 1.12 | 13 |
| 1.12 and under 1.14 | 5 |



*Table 16*

*Figure 5*

## 18. Example 8 (Comparison of frequency distributions using *frequency polygons*)

The two frequency distributions of Example 6 (Table 15) are compared diagrammatically, using frequency polygons. See Figure 6.

*Figure 6 Frequency polygons for numbers of orders*

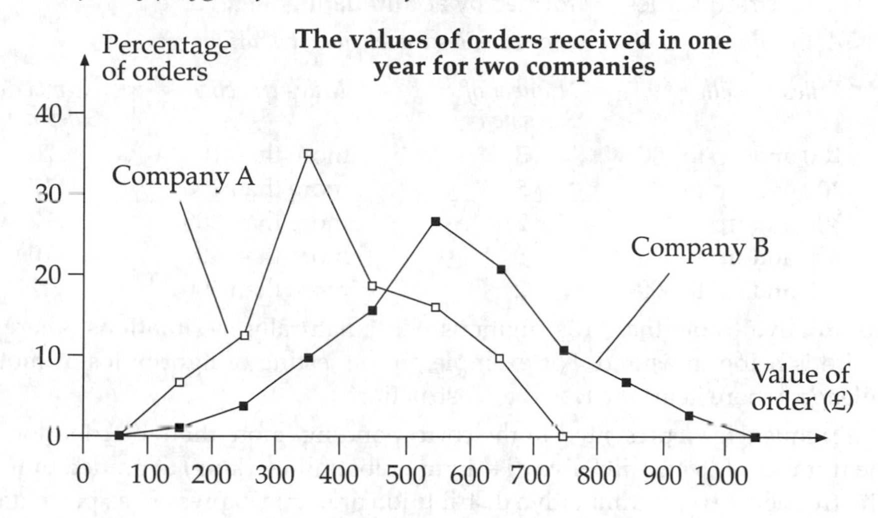


## 19. Cumulative frequency distributions

Any frequency distribution can be adapted to form what is known as a *cumulative frequency distribution*. Whereas ordinary frequency distributions describe a particular class of values according to how many items lie within it, cumulative frequency distributions describe the number of items that have values either above or below a particular level. Cumulative frequency distributions come in two different forms, described as follows.

a) *Less than distributions*. Here, a set of item values is listed (normally class 'upper boundaries'), with each one showing the number of items in the distribution having values less than this. Table 17 shows an ordinary frequency distribution transformed into a cumulative (less than) distribution. Note carefully, in the case of the transformed distribution, how:

   i. the descriptions of classes need to be changed
   ii. the frequencies are accumulated.

*Table 17 Actual and cumulative miles travelled distributions*

| *Miles travelled* | *Number of Salesmen* | *Miles travelled* | *Number of Salesmen* |
|---|---|---|---|
| 100 and up to 200 | 3 | less than 200 | 3 |
| 200 and up to 300 | 5 | less than 300 | 8 |
| 300 and up to 400 | 2 | less than 400 | 10 |
| 400 and up to 500 | 8 | less than 500 | 18 |
| 500 and up to 600 | 2 | less than 600 | 20 |

b) *More than distributions*. Here, a set of item values is listed (normally class 'lower boundaries'), with each one showing the number of items in the distribution having values greater than this. Table 18 shows the distribution of Table 17 transformed into a cumulative (more than) distribution. Note again how:
   i. the descriptions of classes need to be changed
   ii. the frequencies are formed by accumulating 'in reverse'.

*Table 18 Actual and (reverse) cumulative miles travelled distributions*

| *Miles travelled* | *Number of Salesmen* | *Miles travelled* | *Number of Salesmen* |
|---|---|---|---|
| 100 and up to 200 | 3 | more than 100 | 20 |
| 200 and up to 300 | 5 | more than 200 | 17 |
| 300 and up to 400 | 2 | more than 300 | 12 |
| 400 and up to 500 | 8 | more than 400 | 10 |
| 500 and up to 600 | 2 | more than 500 | 2 |

Cumulative (more than) distributions occur naturally in situations where length of life is being measured. For example, in the testing of light bulbs or motor tyre mileage, where items are tested to destruction.

The frequencies represented in the corresponding 'more than' distribution will be the number of items still 'alive' (i.e. light bulbs still working) at graded time points. The frequencies of a cumulative distribution are often expressed as percentages.

## 20. Cumulative frequency polygons and curves

Cumulative frequency distributions are graphed using either polygons or curves. The plotted points for the graphs are identified in the following way.

a) For *less than* distributions. Accumulated frequencies are plotted against class *upper* boundaries.
b) For *more than* distributions. Reverse accumulated frequencies are plotted against class *lower* boundaries.

To form a cumulative frequency polygon, the points are joined with straight lines. To form a cumulative frequency curve (traditionally called an ogive), the points are joined with as *smooth a curve as possible*. These diagrams can be used for estimation purposes, as demonstrated in Example 9 which follows.

## 21. Example 9 (*Cumulative percentage (less than) frequency polygon*)

The data of section 19 (Table 17) are shown in Table 19, with the cumulative frequencies expressed as percentages. Figure 7 shows the corresponding cumulative percentage (less than) frequency polygon.

*Table 19 and Figure 7 Cumulative distribution of miles travelled by salesmen*

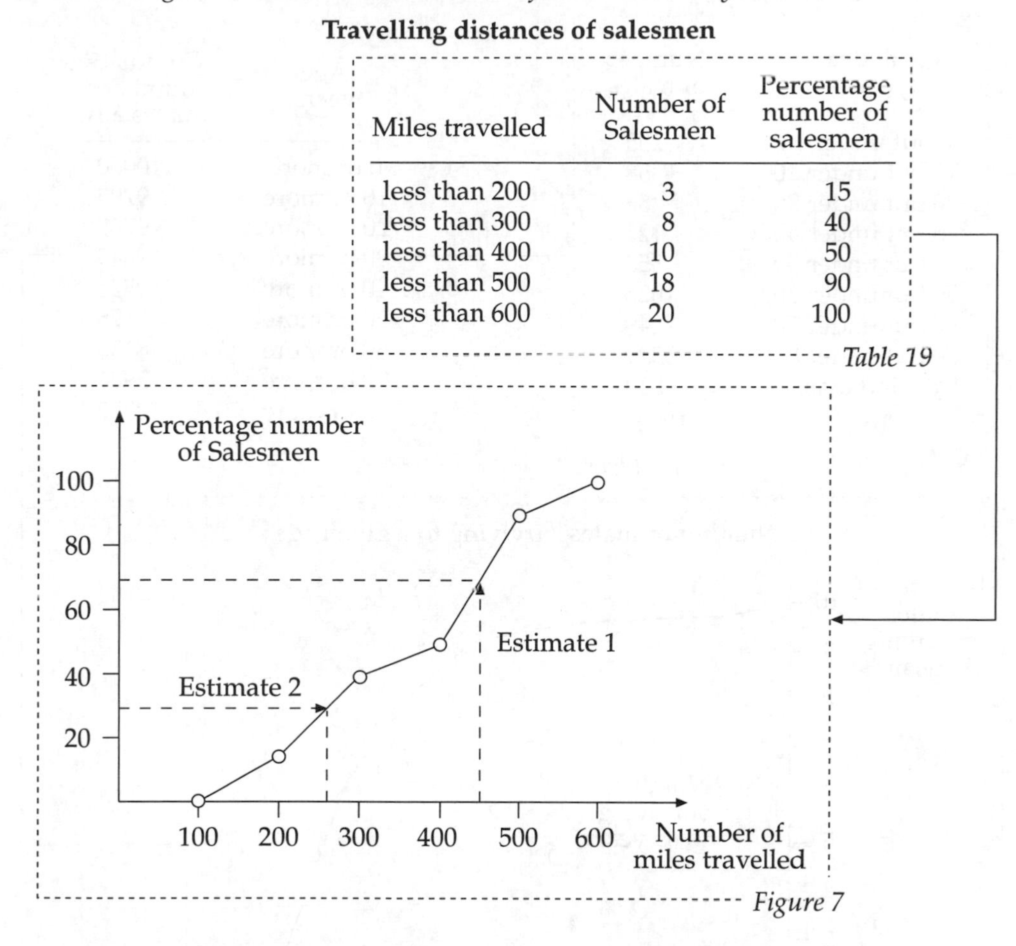

**Travelling distances of salesmen**

| Miles travelled | Number of Salesmen | Percentage number of salesmen |
|---|---|---|
| less than 200 | 3 | 15 |
| less than 300 | 8 | 40 |
| less than 400 | 10 | 50 |
| less than 500 | 18 | 90 |
| less than 600 | 20 | 100 |

*Table 19*



*Figure 7*

## 22. Example 10 (*Cumulative (more than) frequency curve*)

*Question*

Draw a cumulative (more than) frequency curve for the distribution given in Table 20 and use the curve to estimate the number of new born males who will survive to 65 years of age. Note that a diagram of this type is often referred to as a *survival chart*.

Hence estimate the proportion of males aged 65 who will survive to age 70.

*Answer*

The cumulative (more than) frequency distribution is shown as an intermediate step in the calculations between the original data of Table 20 and the corresponding cumulative frequency polygon in Figure 8.

*Table 20 and Figure 8 Construction of survival chart*

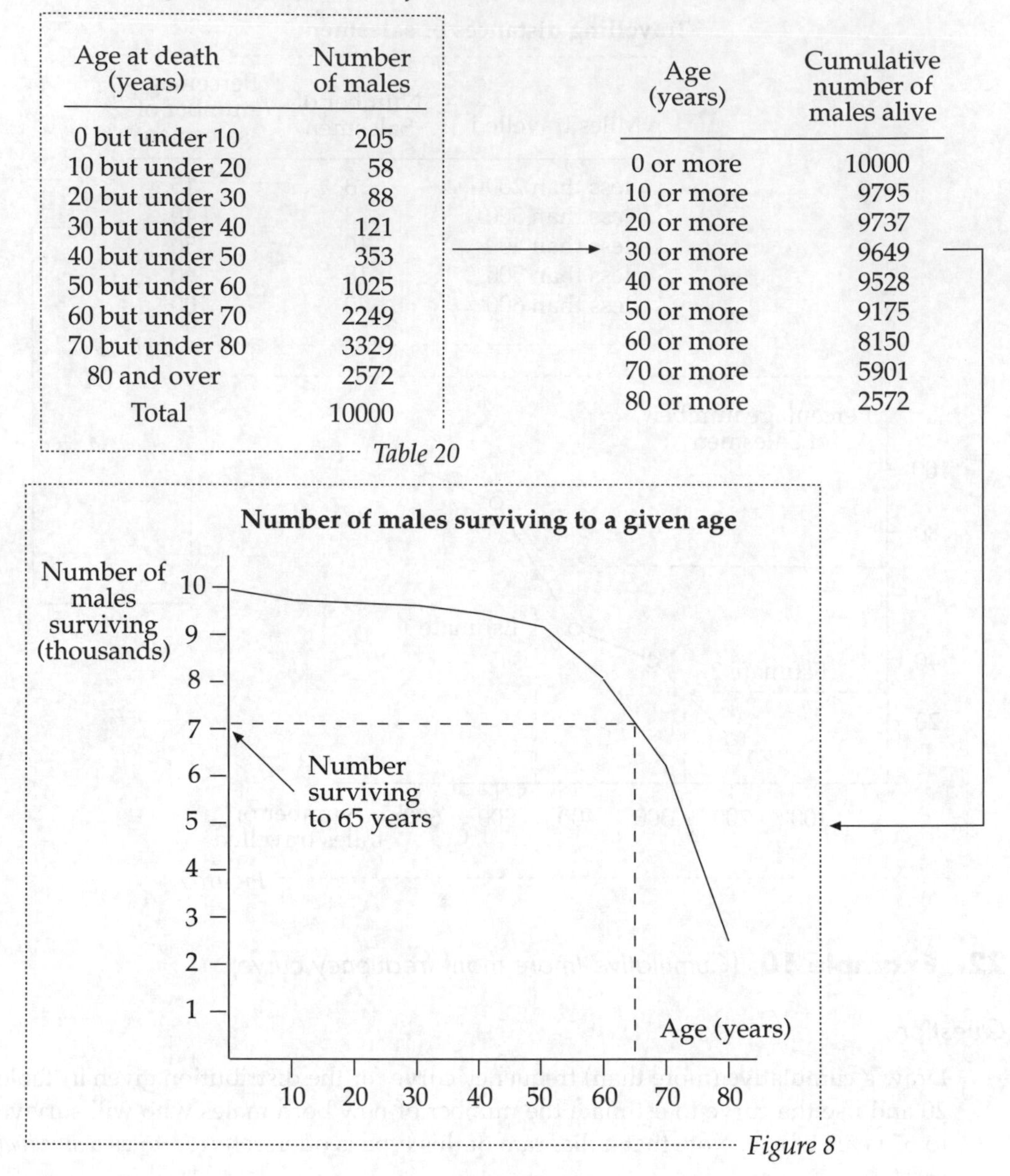

| Age at death (years) | Number of males |
|---|---|
| 0 but under 10 | 205 |
| 10 but under 20 | 58 |
| 20 but under 30 | 88 |
| 30 but under 40 | 121 |
| 40 but under 50 | 353 |
| 50 but under 60 | 1025 |
| 60 but under 70 | 2249 |
| 70 but under 80 | 3329 |
| 80 and over | 2572 |
| Total | 10000 |

*Table 20*

| Age (years) | Cumulative number of males alive |
|---|---|
| 0 or more | 10000 |
| 10 or more | 9795 |
| 20 or more | 9737 |
| 30 or more | 9649 |
| 40 or more | 9528 |
| 50 or more | 9175 |
| 60 or more | 8150 |
| 70 or more | 5901 |
| 80 or more | 2572 |



*Figure 8*

From the graph in Figure 8, the number of males who survive to age 65 is estimated as 7100. Also, since 5901 males survived to age 70 (from the cumulative frequency distribution), the proportion of 65 year-olds who survived to age 70 can be calculated as 5901/7100 = 0.83. That is, 83% of 65 year-olds are expected to survive to age 70.

## 23. Lorenz curves

A *Lorenz curve* is a similar type of graph to a cumulative percentage 'less than' frequency curve. The important difference however is that, where the cumulative 'less than' curve plots cumulative percentage frequency against upper class boundaries, a Lorenz curve plots cumulative percentage frequency against *cumulative percentage class totals*. A *class total* for any frequency distribution class is the total value of all the items belonging to the class.

In examinations, *class totals* are normally given if a Lorenz curve is asked for. If they are not given however, they can be estimated using: class total = class mid-point × class frequency.

The procedure for drawing a Lorenz curve, given a frequency distribution, is shown below.

**Procedure for drawing a Lorenz curve**

STEP 1 Calculate cumulative percentage frequency (for each class).

STEP 2 Calculate cumulative percentage class totals (for each class), using either:

i. given class totals, or

ii. estimated class totals (as described above).

STEP 3 Plot cumulative percentage frequency ($y$-axis) against cumulative percentage class totals ($x$-axis) as a set of points.

STEP 4 Join the points, giving the Lorenz curve required.

Example 10 demonstrates the use of this procedure.

## 24. Example 11 (Drawing a *Lorenz curve*)

The data in Table 21 overleaf give the values of properties handled by a dealer over a six-month period. Notice that the frequencies already total 100 and are thus, in effect, percentages.

Also notice that no class totals have been given, so they need to be estimated.

Table 22 overleaf shows the calculations necessary to identify plotted points for the Lorenz curve, and this is plotted in Figure 9 overleaf.

**Notes** on Table 22:

1. Column (1) – obtained from the given frequency distribution classes.
2. Column (2) – given.
3. Column (3) – accumulated column (2) values.
4. Column (4) – column (1) × column (2).
5. Column (5) – column (4) values as a percentage of column (4) total.
6. Column (6) – accumulated column (5) values.
7. The data pairs from columns (3) and (6) have been plotted and joined to form the Lorenz curve in Figure 9.

In Figure 9, notice the *line of equal distribution* which is always shown when a Lorenz curve is drawn. It is explained fully in the following section.

As with the cumulative polygons and curves considered previously, Lorenz curves can be used for estimation purposes. The estimate shown in Figure 9 reveals that the first 40% of properties account for only 30% of total property value.

*Table 21, Table 22 and Figure 9 Construction of Lorenz curve for property values and numbers*

| Value of property (£000) | Number of properties |
|---|---|
| 10 and less than 15 | 2 |
| 15 and less than 20 | 6 |
| 20 and less than 25 | 14 |
| 25 and less than 30 | 21 |
| 30 and less than 35 | 33 |
| 35 and less than 40 | 19 |
| 40 and less than 45 | 5 |
| Total | 100 |

*Table 21*

| Class mid-points | Frequency (%) | Cumulative % frequency | Class total (estimate) | Class total (%) | Cumulative % class total |
|---|---|---|---|---|---|
| (1) | (2) | (3) | (4)=(1)x(2) | (5) | (6) |
| 12.5 | 2 | 2 | 25.0 | 1 | 1 |
| 17.5 | 6 | 8 | 105.0 | 3 | 4 |
| 22.5 | 14 | 22 | 315.0 | 10 | 14 |
| 27.5 | 21 | 43 | 577.5 | 19 | 33 |
| 32.5 | 33 | 76 | 1072.5 | 36 | 69 |
| 37.5 | 19 | 95 | 712.5 | 24 | 93 |
| 42.5 | 5 | 100 | 212.5 | 7 | 100 |

*Table 22*

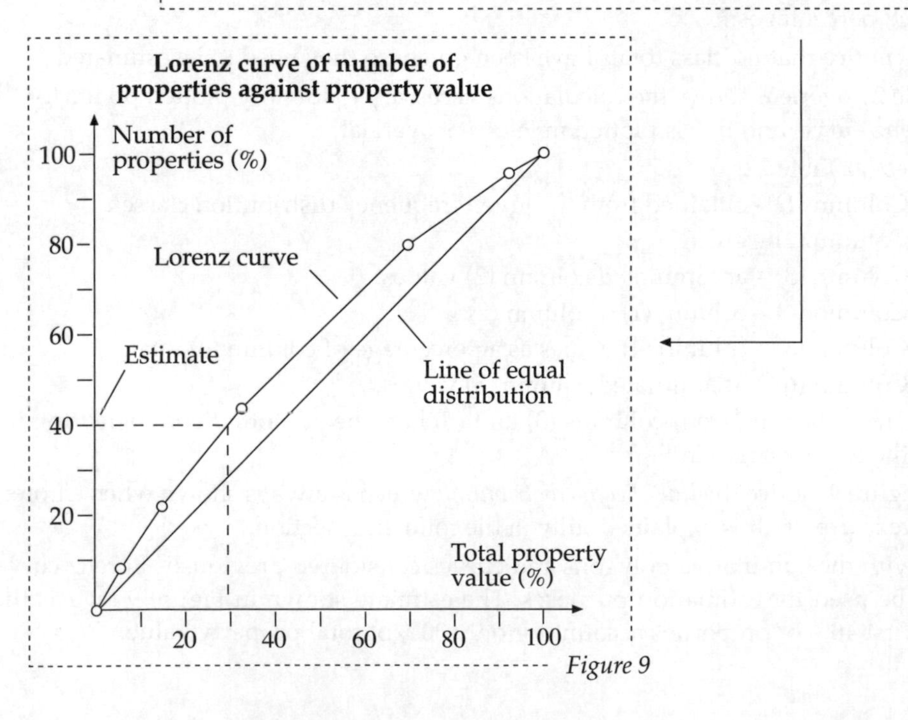


*Figure 9*

## 25. Interpretation and characteristics of Lorenz curves

The previous two sections were concerned with the construction of a Lorenz curve. Before stating the main purpose of the curve, the following needs to be understood. Consider an office in which the five employees are all paid the same salary. In this situation we would have total income earned shared out equally between the employees. It could be represented diagrammatically by plotting:

| | | | | | |
|---|---|---|---|---|---|
| % of employees | 20 | 40 | 60 | 80 | 100 |
| % of total income | 20 | 40 | 60 | 80 | 100 |

which would result in a straight line. This is known as the line of equal distribution (shown in Figure 9).

> **Aim of Lorenz Curve**
>
> The aim of a *Lorenz curve* is to show how the total value of the measurements of some economic variable is shared out among the subjects or items involved.

This aim can be realised by *comparing* a Lorenz curve with the line of equal distribution (LED). The further away the Lorenz curve is from the LED, the less equally the commodity involved is distributed. (There are quantitative techniques available for measuring inequality of distribution, but they are outside the scope of this manual.)

Figure 9 showed the Lorenz curve fairly close to the LED, demonstrating the fact that property values are shared out fairly equally between the properties.

Standard situations where Lorenz curves are used (and often quoted) are distributions of incomes (both before and after tax), personal wealth, turnover of companies, GNP of countries and similar monetary data. Example 12 shows the calculation of a Lorenz curve for a distribution of personal wealth.

## 26. Example 12 (Drawing a *Lorenz curve*)

### *Question*

The distribution in Table 23 overleaf gives the personal wealth of a certain cross-section of the population of Great Britain for a particular year. Draw a Lorenz curve to illustrate this data and use it to estimate:

a) the percentage of total personal wealth that the least wealthy 30% of persons have at their disposal;

b) the percentage of the most wealthy who command one half of all personal wealth.

### *Answer*

Notice that class totals, in this case total personal wealth, are given. Thus they do not need to be estimated (as was necessary in Example 11). For convenience, the following abbreviations have been used in the calculations that are shown in Table 24: $f$ = frequency = number of persons; $F$ = cumulative $f$; $v$ = values of class totals = estimate of total personal wealth; $V$ = cumulative $v$.

*Table 23 Personal wealth distribution*

| *Personal wealth (£)* | *Number of persons (000,000) (f)* | *Total personal wealth (£ 000m) (v)* |
|---|---|---|
| 0–2000 | 19 | 2.4 |
| 2000–5000 | 26 | 7.8 |
| 5000–10000 | 74 | 55.5 |
| 10000–15000 | 41 | 49.2 |
| 15000–20000 | 16 | 25.7 |
| 20000–25000 | 8 | 16.8 |
| 25000–50000 | 5 | 15.0 |
| 50000 and over | 1 | 6.3 |
| Total | 190 | 178.7 |

*Table 24 Lorenz chart workings table*

| *f* | *F* | *F%* | *v* | *V* | *V%* |
|---|---|---|---|---|---|
| 19 | 19 | 10 | 2.4 | 2.4 | 1 |
| 26 | 45 | 24 | 7.8 | 10.2 | 6 |
| 74 | 119 | 63 | 55.5 | 65.7 | 37 |
| 41 | 160 | 84 | 49.2 | 114.9 | 64 |
| 16 | 176 | 93 | 25.7 | 140.6 | 79 |
| 8 | 184 | 97 | 16.8 | 157.4 | 88 |
| 5 | 189 | 99 | 15.0 | 172.4 | 96 |
| 1 | 190 | 100 | 6.3 | 178.7 | 100 |

*Figure 10 Lorenz curve*

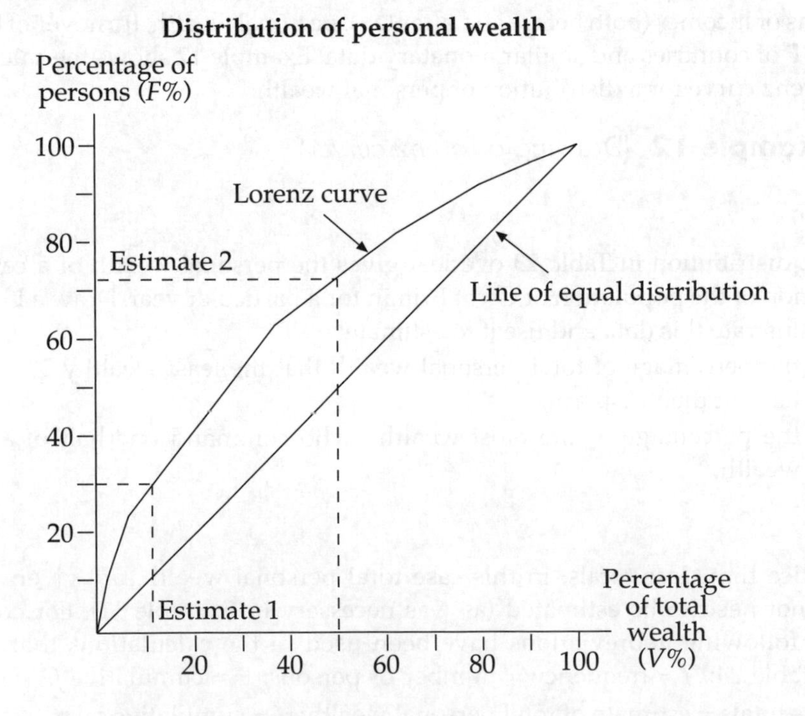

Figure 10 shows the required Lorenz curve as the plots of cumulative percentage frequency ($F\%$) against cumulative percentage personal wealth ($V\%$).

a) From the graph, estimate 1 gives: when $F\%=30$, $V\%=10$. That is, the least wealthy 30% of persons have only 10% of the total personal wealth at their disposal.

b) From the graph, estimate 2 gives: when $V\%=50$, $F\%=73$. That is, 50% of all personal wealth is at the disposal of the lower 73%, or the most wealthy 27%.

## 27. Summary

a) Raw data are data that have not been worked on.

b) A data array consists of raw data that have been arranged into size order.

c) A simple frequency distribution is a list of data values, each showing its frequency of occurrence. They are normally used to describe discrete data.

d) A grouped frequency distribution splits data values into groups, each one showing its frequency of occurrence. They can be used to describe discrete or continuous data.

e) Frequency distributions summarize data, giving an overall perspective. However, grouped distributions do not show individual data values.

f) i. Class limits and boundaries mark the practical and theoretical extent of a frequency distribution class.

   ii. A class width is the numerical difference between class boundaries. A class mid-point marks the centre of the class.

g) Rules for forming frequency distributions:

   i. All data values to be contained in one and only one class.

   ii. Classes listed in size order.

   Good practice when describing frequency distributions:

   i. There should be between 5 and 15 classes.

   ii. Class descriptions should be logical and natural for the data.

   iii. Equal class widths are preferable.

   iv. Express frequencies as percentages when comparing distributions having the same class structure.

h) A histogram is the main form of pictorial representation of a frequency distribution. It consists of a set of joined, vertical bars, each bar representing a single class.

i) Frequency polygons are alternatives to histograms. They are line diagrams, each point plotted representing a single class. They are preferred to histograms for comparing frequency distributions pictorially.

j) Cumulative frequency distributions tabulate data by how many values lie below ('less than') or above ('more than') graded points. The frequencies are often expressed in percentage terms. They can be represented pictorially using either polygons or curves.

k) Lorenz curves show how the total value of the measurements of some economic variable is shared out among the subjects or items involved. The curve is obtained by plotting cumulative percentage frequency against cumulative percentage class totals and compared with a line of equal distribution.

## 28. Points to note

a) Continuous data that has been subjected to systematic errors must be paid special attention to when frequency distribution classes are drawn up. For example, if ages have been measured in years, they will be recorded as 28, 41, 32, etc. However, an age of 28 really means the interval '28 and up to but excluding 29', and so on. Thus if frequency distribution classes need to be constructed, the form of classes should be:

| *like this* | *NOT like this* |
|---|---|
| 10 and up to 20 | 10 and up to 19 |
| 20 and up to 30 | 20 and up to 29 |
| 30 and up to 40 | 30 and up to 39 |
| ... | ... |
| etc | etc |
| Form (1) | Form (2) |

However, in an examination, if the classes for this type of data are given in terms of form (2), they must be treated as if they were in form (1).

b) When a histogram is used to represent a frequency distribution having one or more classes of different width to the rest, the *area* of the bar (and not its height) must represent class frequency. This can be ensured if the following procedure is carried out:

1. Choose a standard class width (normally the one that occurs the most).
2. A class having '2 times' the standard width has its bar drawn to $\frac{1}{2}$ times' its class frequency; a class having '$\frac{1}{2}$ times' the standard width has its bar drawn to '2 times' its class frequency; a class having '3 times' the standard width has its bar drawn to '$\frac{1}{3}$ times' its class frequency; ... and so on.

As an example, the histogram for the distribution in Table 25 is drawn in Figure 11.

## 29. Student self review questions

1. What is a data array? [3]
2. For what type of data is a simple frequency distribution used normally? [4]
3. Under what circumstances is a grouped (rather than a simple) frequency distribution used? [7]
4. What is gained and lost when a set of raw data is represented by a grouped frequency distribution? [7]
5. How do frequency distribution class limits differ to class boundaries? [9]
6. Name some of the rules and practices that should be followed when constructing grouped frequency distributions. [10]
7. What is a relative frequency distribution? [10]

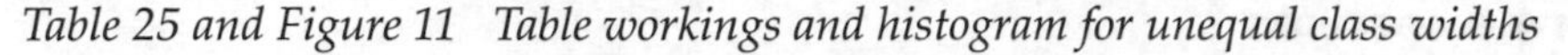

*Table 25 and Figure 11 Table workings and histogram for unequal class widths*

| Class width | Class | Frequency | Bar height |
|---|---|---|---|
| 5 | 0 and up to 5 | 1 (× 2) | = 2 |
| 5 | 5 and up to 10 | 3 (× 2) | = 6 |
| 10 | 10 and up to 20 | 11 | 11 |
| 10 | 20 and up to 30 | 25 | 25 |
| 10 | 30 and up to 40 | 41 | 41 |
| 10 | 40 and up to 50 | 36 | 36 |
| 10 | 50 and up to 60 | 16 | 16 |
| 30 | 60 and up to 90 | 15 (× 1/3) | = 5 |
| 60 | 90 and up to 150 | 9 (× 1/6) | = 1.5 |

*Table 25*

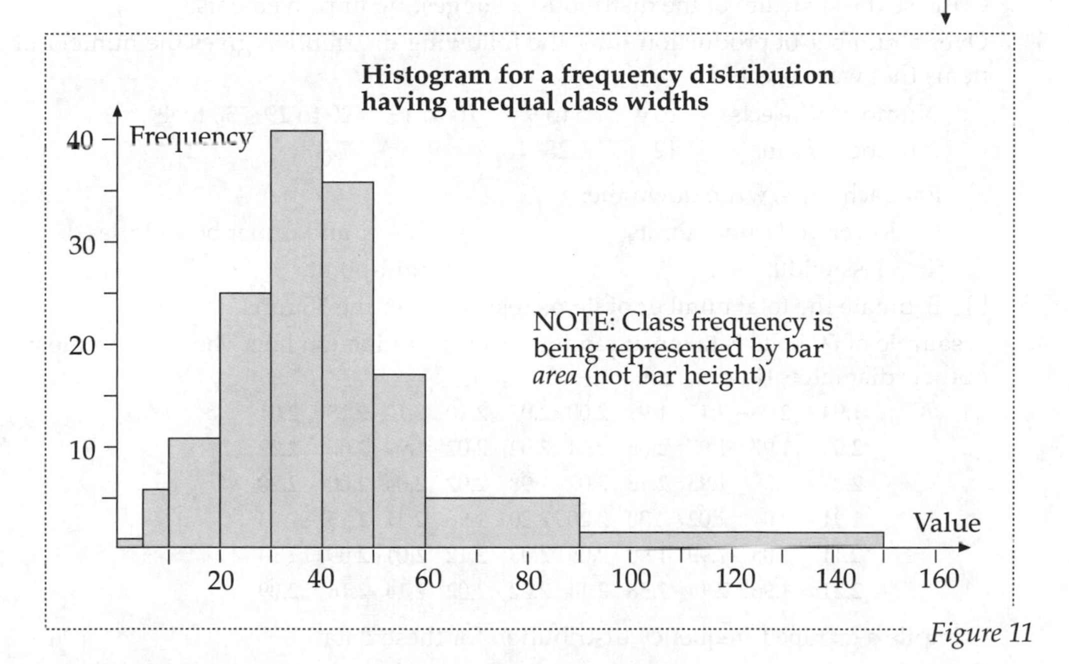


*Figure 11*

8. What is a histogram and how is it constructed? [13]
9. Under what circumstances might a frequency polygon (or curve) be used in preference to a histogram? [16]
10. What is the difference between a 'less than' and 'more than' cumulative frequency distribution? [19]
11. Describe precisely what points need to be plotted in order to obtain a cumulative percentage (less than) frequency polygon. [20]
12. What does a Lorenz curve aim to show? [25]
13. What is the 'line of equal distribution' with respect to a Lorenz curve? [25]

## 30. Student exercises

1. A large furniture removal company charts the availability of its vans on a daily basis. The following data give the number of vans not available for all or part of a day, each working day, over a sixty day period.

| | | | | | | | | | | | | | | | | | | | |
|---|---|---|---|---|---|---|---|---|---|---|---|---|---|---|---|---|---|---|---|
| 1 | 1 | 4 | 2 | 0 | 1 | 0 | 0 | 2 | 1 | 1 | 1 | 0 | 0 | 2 | 3 | 5 | 2 | 2 | 1 |
| 3 | 6 | 3 | 1 | 0 | 1 | 1 | 1 | 0 | 2 | 2 | 1 | 3 | 2 | 1 | 0 | 1 | 3 | 5 | 5 |
| 8 | 3 | 1 | 1 | 3 | 2 | 2 | 0 | 1 | 3 | 1 | 1 | 0 | 0 | 0 | 1 | 3 | 2 | 4 | 1 |

Compile a simple frequency distribution, paying attention to labelling. Comment on the results.

2. The following frequency distribution shows the number of components produced on a particular production line each day, over a sixty day period.

| Number of components | 500 to 520 | 520 to 540 | 560 to 580 | 540 to 560 |
|---|---|---|---|---|
| Number of days | 6 | 47 | 4 | 3 |

Criticise the structure of the distribution, suggesting improvements.

3. Over a number of production runs, the following distribution gives the number of items that were rejected.

| Number of rejects | 0 to 4 | 5 to 9 | 10 to 19 | 20 to 29 | 30 to 49 |
|---|---|---|---|---|---|
| Number of runs | 12 | 28 | 9 | 7 | 2 |

a) For each class, write down the:
   i. lower and upper limits
   ii. lower and upper boundaries
   iii. class width
   iv. mid-point.

b) Estimate the total number of items rejected over the 58 runs.

4. A sample of 60 bolts is taken at random from a production line. The measurements of their diameters (in mm) are:

| | | | | | | | | | |
|---|---|---|---|---|---|---|---|---|---|
| 1.94 | 2.06 | 2.15 | 1.99 | 2.00 | 2.07 | 2.10 | 2.12 | 2.18 | 2.01 |
| 2.03 | 1.97 | 1.97 | 2.06 | 2.04 | 2.02 | 2.02 | 1.99 | 2.00 | 2.02 |
| 2.05 | 2.09 | 1.95 | 2.16 | 2.07 | 1.98 | 2.02 | 2.04 | 2.00 | 2.20 |
| 1.91 | 2.04 | 2.02 | 2.30 | 2.20 | 2.20 | 1.96 | 2.11 | 2.15 | 2.23 |
| 2.11 | 2.08 | 1.99 | 1.90 | 2.05 | 2.03 | 2.12 | 2.01 | 2.09 | 2.14 |
| 2.21 | 1.96 | 2.14 | 2.58 | 2.14 | 2.02 | 2.02 | 2.14 | 2.16 | 2.09 |

Compile a grouped frequency distribution for these data.

5. Draw a histogram to represent the sales of men's shoes in a department store over a particular period:

| Size | 5 | 6 | 7 | 8 | 9 | 10 | 11 | 12 |
|---|---|---|---|---|---|---|---|---|
| Sales | 3 | 6 | 17 | 20 | 28 | 14 | 8 | 4 |

6. The time spent by cars in a car park in one day is given by the following distribution.

a) Draw a histogram for these data.

b) Estimate the daily total number of hours spent by all cars in the park during the day.

c) Given that there are 400 parking spaces in the car park and it is open from 8.00 a.m. to 8.00 p.m., estimate its percentage utilization.

| *Time parked (hrs)* | *Number of cars* |
|---|---|
| up to 1 | 452 |
| 1 and up to 2 | 737 |
| 2 and up to 3 | 646 |
| 3 and up to 4 | 121 |
| 4 and up to 5 | 44 |
| 5 and up to 6 | 37 |
| 6 and up to 7 | 24 |
| 7 and up to 8 | 16 |
| 8 and up to 9 | 9 |
| 9 and up to 10 | 2 |

7. The following data shows the income of females by highest educational qualification.

| *Range of weekly income(£)* | *Highest qualification* | |
|---|---|---|
| | *Degree or equivalent* | *"O"level or CSE Grade 1* |
| 40 and under 60 | 5 | 15 |
| 60 and under 80 | 7 | 30 |
| 80 and under 100 | 7 | 28 |
| 100 and under 120 | 18 | 14 |
| 120 and under 140 | 23 | 7 |
| 140 and under 160 | 14 | 3 |
| 160 and under 180 | 10 | 2 |
| 180 and under 200 | 16 | 1 |

Compare the two distributions using frequency polygons.

8. In a test to determine the working life of a type of electric light bulb, one hundred bulbs were selected at random from production and simultaneously connected to a power source. The following data show the number (of the original one hundred) still working at the end of successive periods of 100 hours, all bulbs having failed within 1000 hours.

| Elapsed time (hrs) | 100 | 200 | 300 | 400 | 500 | 600 | 700 | 800 | 900 |
|---|---|---|---|---|---|---|---|---|---|
| Number working | 99 | 98 | 90 | 82 | 70 | 45 | 26 | 12 | 3 |

Graph these data using a cumulative frequency polygon and use it to estimate:

a) the percentage of bulbs that lasted more than 750 hours;

b) the percentage of bulbs that did not last 350 hours;

c) the minimum guaranteed life of a bulb that the company could quote in order that only 5% of customers would have cause for complaint.

9. Transform the distribution given in exercise 8 into a grouped frequency distribution and represent the data pictorially with a histogram.

10. The distribution in Table 26 gives the annual mortgage repayments of a sample of home owners in a certain area. Draw up a cumulative percentage (less than) frequency distribution table and represent the data with a suitable graph. Estimate the percentage of these home owners who repay:

i. less than £1000 per annum

ii. more than £1500 per annum.

11. Using the data of the table below, estimate the total value of the repayments made by the members of each of the eight classes of the distribution.
    a) Use these same data to draw up a table of cumulative percentage number of home owners against cumulative percentage repayments.
    b) Use this table to construct a Lorenz curve.
    c) Use the Lorenz curve to estimate what percentage of the given sample pay the first 20% of total repayments.

| *Repayment (£000)* | *Number of home owners* |
|---|---|
| under 0.4 | 8 |
| 0.4 and under 0.8 | 60 |
| 0.8 and under 1.2 | 100 |
| 1.2 and under 1.6 | 108 |
| 1.6 and under 2.0 | 72 |
| 2.0 and under 2.4 | 32 |
| 2.4 and under 2.8 | 12 |
| 2.8 and under 3.2 | 8 |

12. For a particular industry, firms are classified into groups according to their number of employees. The data below gives, for each group (listed in ascending number of employees), the number of firms and their combined net output (£ million). Draw a Lorenz curve (cumulative % number of firms against cumulative % net output) in respect of these data. What percentage of the total net output are the largest 20% of firms responsible for?

| *Employee group* | *Number of firms* | *Net output* |
|---|---|---|
| 1 | 185 | 24 |
| 2 | 153 | 57 |
| 3 | 70 | 26 |
| 4 | 31 | 34 |
| 5 | 18 | 28 |
| 6 | 11 | 49 |

13. Draw a Lorenz curve showing the distribution of income before tax based on the following data.

| *Income class (£)* | *Number of incomes* | *Total income before tax for group (£000)* |
|---|---|---|
| up to 3000 | 8 | 20 |
| 3000 and up to 4000 | 24 | 91 |
| 4000 and up to 6000 | 30 | 165 |
| 6000 and up to 8000 | 52 | 380 |
| 8000 and up to 10000 | 50 | 450 |
| 10000 and up to 15000 | 19 | 219 |
| 15000 and up to 20000 | 11 | 187 |
| 20000 and up to 30000 | 6 | 132 |
| 30000 and over | 5 | 205 |

# 5 General charts and graphs

## 1. Introduction

The types of diagram described in this chapter include pictograms, various types of bar charts, pie charts, line and strata graphs, Z-charts, Gantt charts and semi-logarithmic graphs. Since all the above graphs and charts display either non-numeric frequency distributions or time series (in one form or another), the chapter begins by describing these types of data structure.

## 2. Non-numeric frequency distributions

Chapter 4 described the form, construction and pictorial display of numeric frequency distributions. Non-numeric frequency distributions take on a similar form to their numeric counterparts, except that the groups (or classes) describe qualitative (i.e. non-numeric) characteristics of the data. For example:

a) Table 1 shows the non-managerial workforce employed at a factory.

b) Table 2 shows the policies issued by an insurance brokers in one week.

*Table 1 Non-managerial workforce employed at a factory*

| *Job description* | *Number employed* |
|---|---|
| Labourers | 21 |
| Mechanics | 38 |
| Fitters | 9 |
| Clerks | 12 |
| Draughtsmen | 4 |

*Table 2 Policies issued by an insurance brokers in one week*

| Type of policy | Life | Disability | Household | Commercial | Other |
|---|---|---|---|---|---|
| Number issued | 10 | 2 | 25 | 3 | 8 |

## 3. Time series

The name given to data describing the values of some variable over successive time periods is a *time series*. For example, the data given in Tables 3 and 4 are typical.

*Table 3*

| Year | 1 | 2 | 3 | 4 | 5 | 6 | 7 | 8 |
|---|---|---|---|---|---|---|---|---|
| Production of company (million tonnes) | 140 | 183 | 162 | 160 | 181 | 214 | 215 | 227 |

*Table 4*

| Month | Jan | Feb | Mar | Apr | May | Jun |
|---|---|---|---|---|---|---|
| Sales of a product | 12 | 12 | 15 | 21 | 18 | 20 |

Combinations of non-numeric frequency distributions and time series often occur with business statistics. For example, Table 5 shows the breakdown of holiday locations booked through a travel agent. This type of data is sometimes known as a component time series, since the various classifications (in the above case, holiday destinations) can be thought of as the components of a meaningful total; in the above case, total number of holidays booked through the travel agent.

1981 1982 1983 1984 1985 1986

*Table 5 Holiday locations booked through a travel agent*

| | | | | | | |
|---|---|---|---|---|---|---|
| Europe | 385 | 350 | 326 | 341 | 286 | 297 |
| America | 86 | 78 | 124 | 112 | 95 | 137 |
| Middle East/Africa | 40 | 56 | 87 | 88 | 84 | 102 |
| Other | 112 | 65 | 156 | 143 | 112 | 161 |
| Total | 623 | 549 | 693 | 684 | 577 | 697 |

The data in Table 6, although similar in form to that of Table 5, have classifications for each month which *cannot* be added to form meaningful totals. They are sometimes called *multiple time series*, since the data given for each type of food is a separate time series.

*Table 6 Prices of selected dairy foods (pence)*

| | Mar | Apr | May | Jun | Jul | Aug | Sep | Oct |
|---|---|---|---|---|---|---|---|---|
| Eggs (per dozen) | 41 | 41 | 41 | 43 | 40 | 42 | 43 | 43 |
| Cheese (per lb) | 86 | 87 | 91 | 91 | 91 | 96 | 96 | 102 |
| Milk (per pint) | 31 | 30 | 30 | 30 | 32 | 32 | 34 | 35 |

## 4. Purpose of statistical diagrams

Statistical diagrams are generally drawn in order to:

i. present data in an attractive and colourful way;

ii. enable a general perspective of the data to be shown without excessive detail.

Diagrams can be used to replace listings or tabulation of data and often used, for example, when the intended audience is not very sophisticated. On the other hand, a long and detailed business report can be complemented by a liberal scattering of diagrams, which will help to 'break it up' and thus make it more palatable.

A checklist of points to be noted when drawing diagrams is given at the end of the chapter, in section 23.

## 5. Types of charts and graphs

The types of diagrams (i.e. charts and graphs) described in this chapter can be conveniently classed under three headings as follows.

a) *Diagrams to display non-numeric frequency distributions.*
   i. Pictograms.
   ii. Simple bar charts.
   iii. Pie charts.

These are described beginning at section 8.

b) *Diagrams to display time series.*
   i. Line diagrams.
   ii. Simple bar charts.

Line diagrams are described in section 11.

c) *Miscellaneous diagrams.*

These diagrams are generally used to display various combinations of multiple non-numeric frequency distributions or time series.

i. Component, percentage and multiple bar charts.
ii. Multiple pie charts.
iii. Strata charts.
iv. Z-charts.
v. Gantt charts.
vi. Semi logarithmic graphs.

These are described beginning at section 12.

## 6. Pictograms

A *pictogram* is a chart which represents the magnitude of numeric values by using only simple descriptive pictures. A picture (or symbol) is selected that easily identifies the data pictorially. It is then duplicated in proportion to the class frequency, for each class represented. Figure 1 shows a pictogram representing the data of Table 1.

*Figure 1 Pictogram representing the workforce employed at a factory*

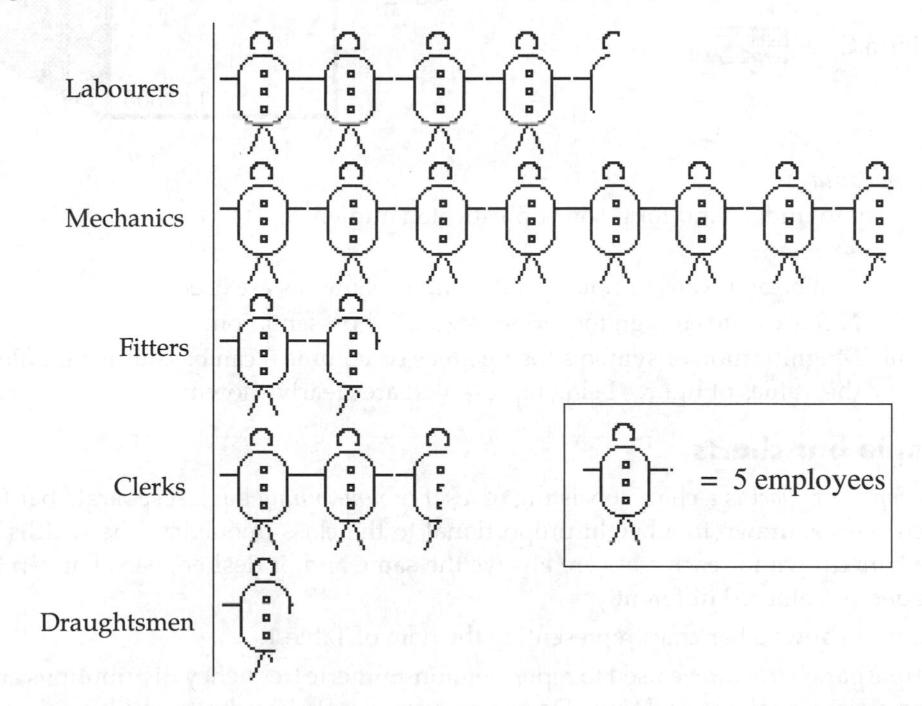


Pictograms are normally used for displaying a small number of classes, generally with non-numeric frequency distributions. However, they can be used for representing time series.

## 7. Some features of pictograms

a) Pictograms are sometimes referred to as *ideograms*.
b) The symbols are normally duplicated horizontally and, for the sake of accuracy, the numeric values being represented are sometimes shown. A scaled axis can be included.
c) An alternative method to duplicating the symbols used, is to magnify them. For example, Figure 2 represents different numbers of lorries by the *area* of the lorry

symbol and Figure 3 represents an increase in sales of detergent (in kg) by the *volume* of the two detergent boxes.

*Figure 2 Number of lorries in fleet* *Figure 3 Detergent sales (kg)*

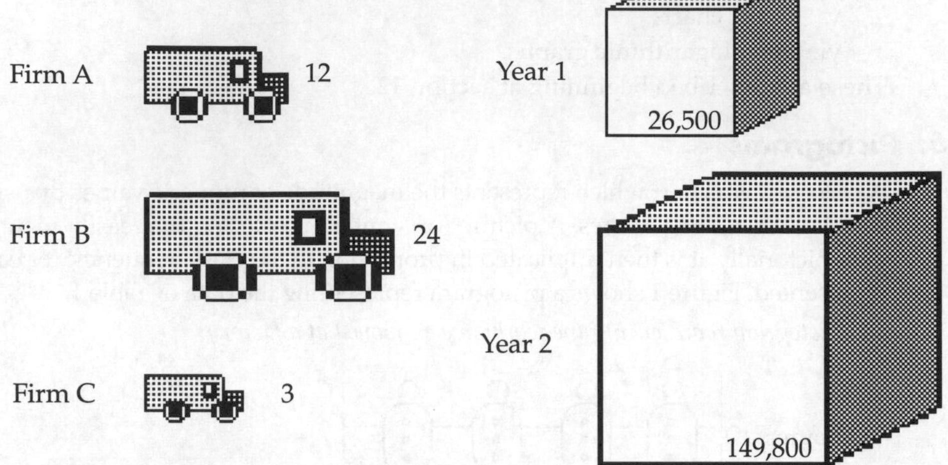


d) *Advantage*
Easy to understand for a non-sophisticated audience.

e) *Disadvantages*
 i. Can be awkward to construct if complex symbols are used.
 ii. Not accurate enough for serious statistical presentation.
 iii. Magnification of symbols (using areas or volumes) can be confusing unless the values of figures being represented are clearly shown.

## 8. Simple bar charts

A *simple bar chart* is a chart consisting of a set of *non-joining* bars. A separate bar for each class is drawn to a height proportional to the class frequency. The widths of the bars drawn for each class are always the same and, if desired, each bar can be shaded or coloured differently.

Figure 4 shows a bar chart representing the data of Table 1.

Simple bar charts can be used to represent non-numeric frequency distributions and time series equally well. [Note: Do not confuse simple bar charts and histograms. Histograms represent numeric data with joined bars. Simple bar charts represent non-numeric data (or time series) and have their bars separated from each other.]

## 9. Some features of simple bar charts

a) They can be drawn with vertical or horizontal bars, but must show a scaled frequency axis.

b) They are easily adapted to take account of both positive and negative values. For example, Figure 5 shows the Balance of Payments of a particular country.

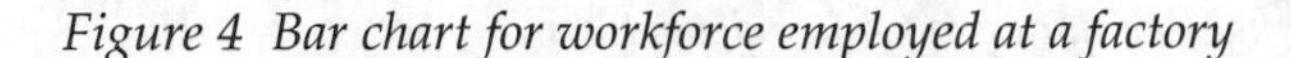
*Figure 4 Bar chart for workforce employed at a factory*

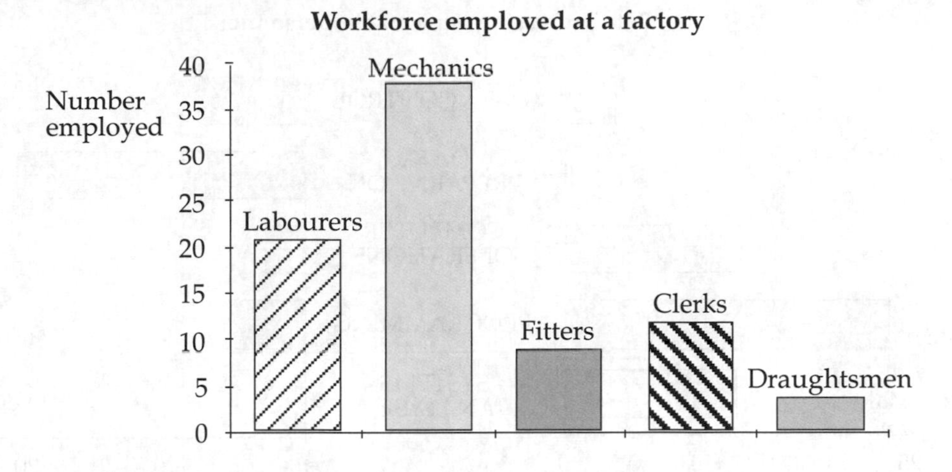


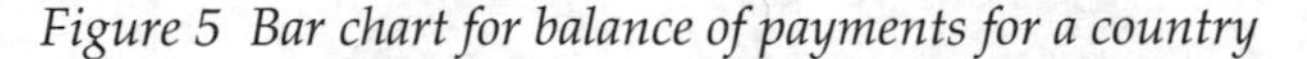
*Figure 5 Bar chart for balance of payments for a country*

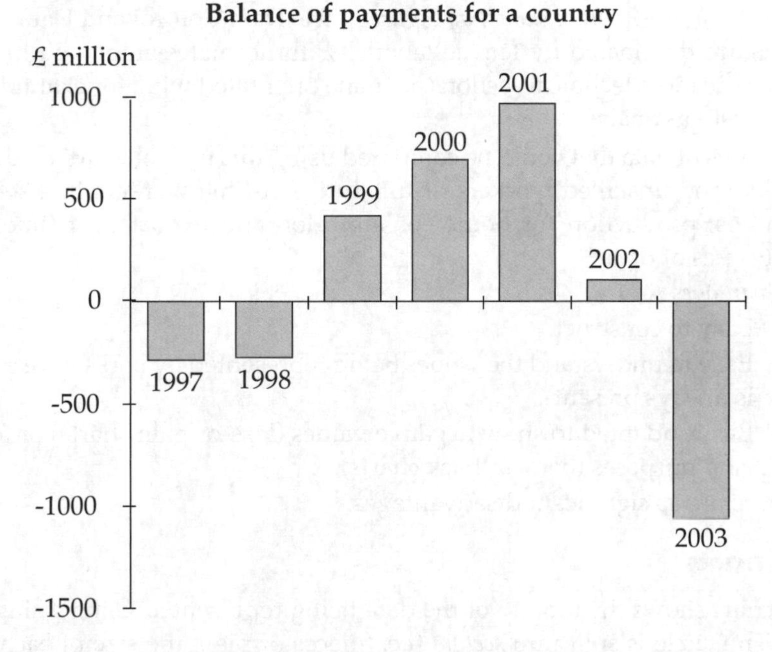


*Comments on the situation shown by Figure 5:*

Starting with a deficit of £300 million in 1997 and 1998, the balance of payments rose to a surplus of nearly £500 million in 1999, rising to a peak of £1000 million in 2001. This was virtually wiped out in 2002 and the following year followed the same pattern to show a deficit of £1000 million. This chart is sometimes known as a loss or gain bar chart.

c) Two bar charts can be placed back-to-back for comparison purposes. For example, Figure 6 shows the workforce of a computer department by sex.

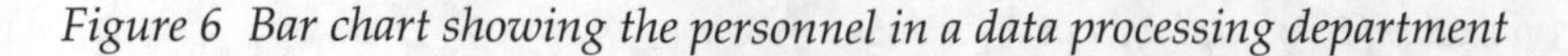

*Figure 6 Bar chart showing the personnel in a data processing department*

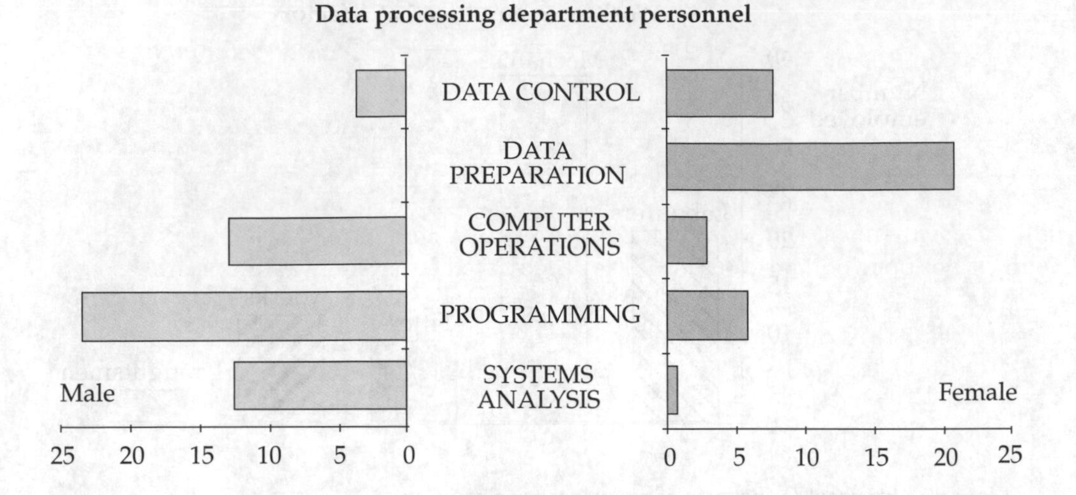


*Comments on the situation shown by Figure 6*: The Data Control and Data Preparation sections are dominated by female labour (with no males employed in the latter), whereas the more technical sections are male orientated with, for example, only one female systems analyst.

Other types of data that could be compared using this type of chart are the numbers of skilled and unskilled workers or full and part-time workers in a set of similar companies; production for home consumption and export over time and other similar types of data.

d) *Advantages*

i. Easy to construct.

ii. Easy to understand the values being represented by bars (since a scaled axis is always present).

iii. Easily adapted to show negative values (loss-or-gain charts) or for comparison purposes (back-to-back charts).

e) There are *no* significant disadvantages.

## 10. Pie charts

A pie chart shows the totality of the data being represented using a single circle (a 'pie'). The circle is split into *sectors* (i.e. 'pieces of pie'), the size of each one being drawn in proportion to the class frequency. Each sector can be shaded or coloured differently if desired.

Figure 7 shows a pie chart drawn for the data of Table 1. Pie charts are always used to represent non-numeric frequency distributions and are at their most effective where the classes need to be compared in *relative* terms.

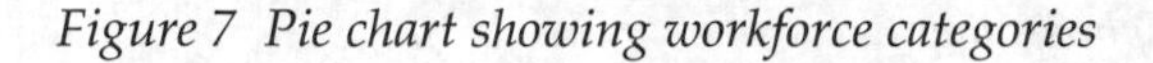

*Figure 7 Pie chart showing workforce categories*

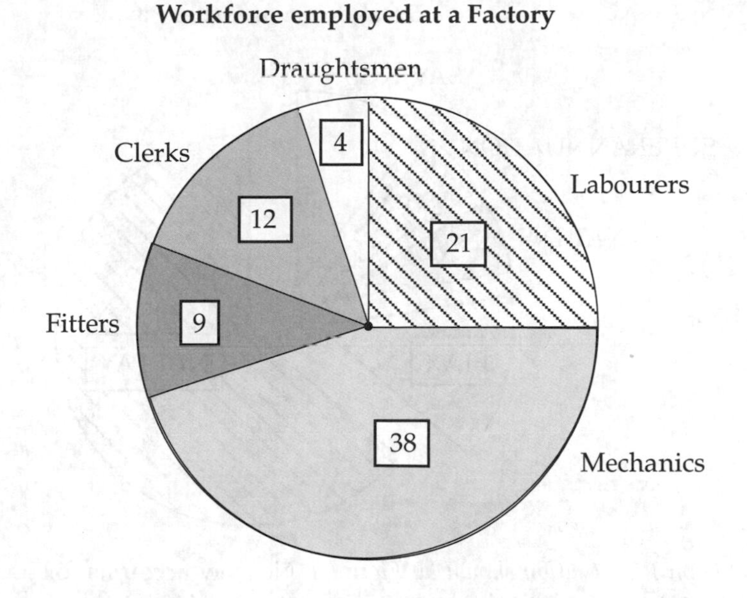


## 11. Some features of pie charts

a) Pie charts are sometimes referred to as *circular diagrams* or *divided circles*.

b) In order to construct a pie chart, the size of each sector in degrees needs to be calculated. The procedure is:

   i. Calculate the proportion of the total that each frequency represents.

   ii. Multiply each proportion by 360, giving the sizes of the relevant sectors (in degrees) that need to be drawn.

This procedure is demonstrated in Table 7, using the data of Table 1.

c) '*Exploding*' the different sectors is an effective way of presenting a pie chart. Figure 8 shows the breakdown of an employee's monthly pay, with each sector exploded.

*Table 7 Procedure for construction of pie chart*

| *Job description* | *Number employed* | *Proportion* | *Sector size (degrees)* |
|---|---|---|---|
| | (a) | (b)=(a)÷84 | (c)=(b)×360 |
| Labourers | 21 | 0.2500 | 90 |
| Mechanics | 38 | 0.4524 | 163 |
| Fitters | 9 | 0.1071 | 39 |
| Clerks | 12 | 0.1429 | 51 |
| Draughtsmen | 4 | 0.0476 | 17 |
| Total | 84 | | |

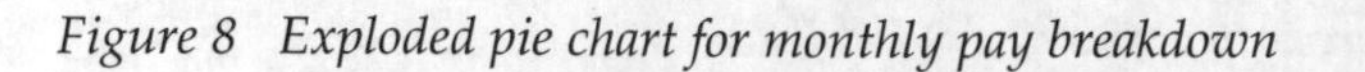

*Figure 8  Exploded pie chart for monthly pay breakdown*

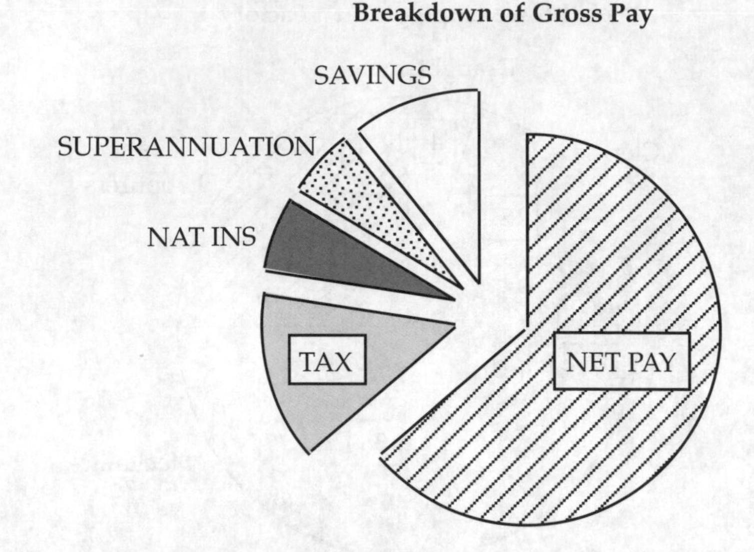


*Comments on the situation shown by Figure 8*: Net pay accounts for only about two-thirds of total pay, with the highest single 'stoppage' being tax.

d) *Advantages*
   i. A dramatic and appealing way of presenting data.
   ii. Good for comparing classes in relative terms.

e) *Disadvantages*
   i. Compilation is laborious. Circles should not be drawn by hand and sectors should be drawn using a protractor. However, without a protractor (once the size of each sector has been determined), their physical size within the circle can be intelligently guessed at.
   ii. Can be untidy if there are many classes (say, 8 or more) and different shadings or colourings are being used.

## 12. Line diagrams

A *line diagram* plots the values of a time series as a sequence of points joined by *straight lines*. The time points are always represented along the horizontal axis, values of the variable along the vertical axis. Figure 9(a) shows a line diagram representation of the data of Table 3. Since bar charts are often drawn for time series, Figure 9(b) shows a bar chart, also for the data of Table 3, for comparison with the line diagram.

*Comment on the situation shown by Figures 9 (a) and (b)*: A jump in year 2's production was followed by smaller falls in the next two years. From year 4 onwards, production has been steadily increasing to stand at about 230 million tonnes in year 8.

*Figure 9 (a) Line diagram showing annual production of a company*

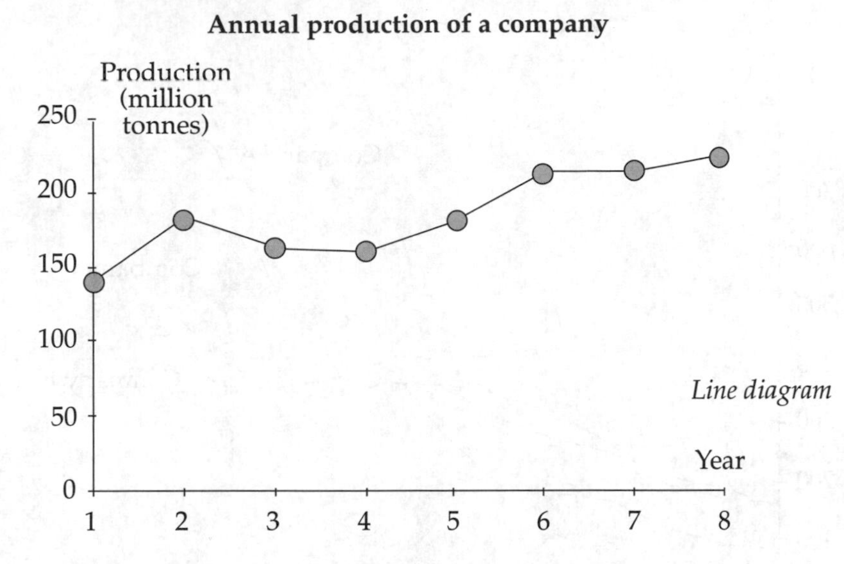


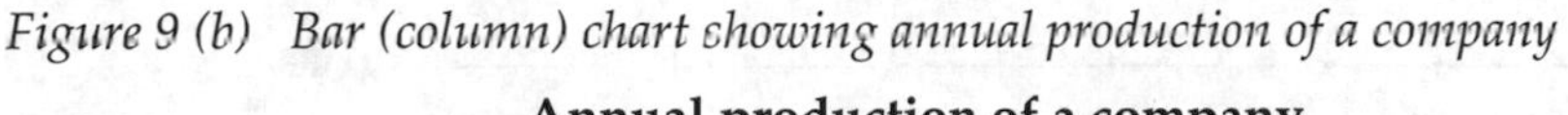
*Figure 9 (b) Bar (column) chart showing annual production of a company*

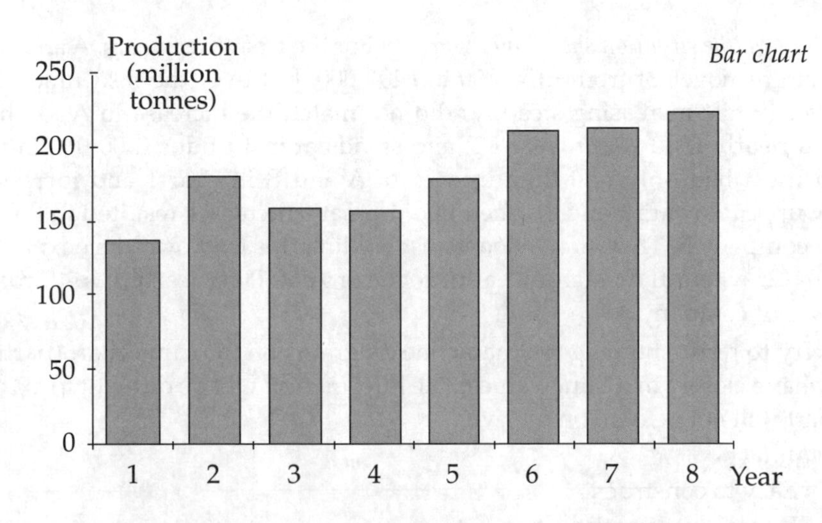


## 13. Some features of line diagrams

a) Line diagrams are technically known as *historigrams*.

b) For comparison purposes, several line diagrams can be drawn on the same set of axes (sometimes known as *multiple line diagrams*) representing either multiple or component time series. For example, Figure 10 shows the turnovers of three competing companies over a number of years.

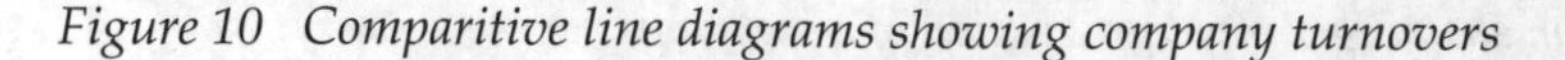

*Figure 10 Comparitive line diagrams showing company turnovers*

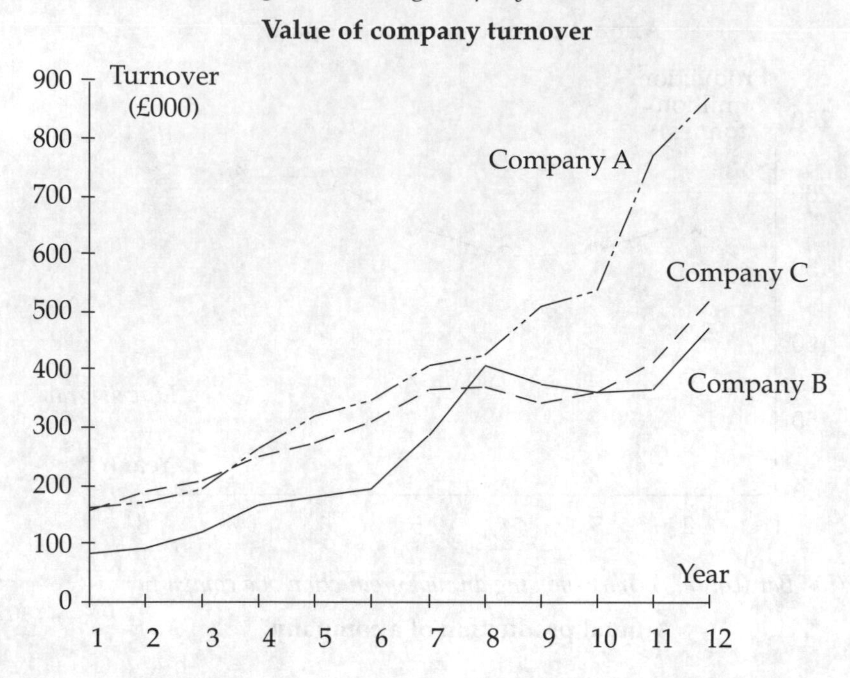


*Comments on the situation shown by Figure 10*: For the first three years, A and B shared the same turnover of from £150 000 to £200 000, but over the next nine years B's turnover (while increasing steadily) did not match the increase in A's (which has reached nearly £900 000 in year 12) and stands at just under £500 000 in year 12. Company C had only half the turnover of A and B in year 1 and increased this slowly up until years 7 and 8, when two dramatic increases resulted in them overtaking company B. They succeeded in just holding this lead over the next four years to year 12, where they stand at a turnover of £550 000, nearly two-thirds of the turnover of company A.

**Note:** Try to resist showing too many line diagrams on the same chart (particularly if they have closely matching values); the information will be difficult to extract and the chart will not be easy on the eye.

c) *Advantages*

  i. Easy to construct.
  ii. Easy to understand.
  iii. A sense of continuity is given by a line diagram which is not present in a bar chart.
  iv. The perfect medium to enable direct comparisons.

d) *Disadvantages*

  i. Might be confusing if too many diagrams with closely associated values are compared together.
  ii. Where several diagrams are displayed, there is no provision for total figures.

## 14. Component, percentage and multiple bar charts

These charts are used as extensions of simple bar charts, where another dimension of the data is given. For example, where a simple bar chart might show the production of a company by year, one of the above would be used if each year's production was split into, say, exports and home consumption i.e. a component time series.

*Component bar charts* have each bar representing a class and split up into constituent parts (components). Within each bar, components are always stacked in the same order.

*Percentage bar charts* have each bar representing a class but all drawn to the same height, representing 100% (of the total). The constituent parts of each class are then calculated as percentages of the total and shown within the bar accordingly. Within each bar, components are stacked in the same order.

*Multiple bar charts* have a set of bars for each class, each bar representing a single constituent part of the total. Within each set, the bars are physically joined and always arranged in the same sequence. Sets of bars should be separated.

For all three charts, the components are normally shaded and a legend (key) should be shown at the side of the chart. The following example shows the data of Table 5 represented in turn by each of the above charts.

## 15. Example 1 (Drawing *component*, *percentage* and *multiple* bar charts)

The data of Table 5 is reproduced below, with each figure calculated as a percentage of the yearly total, shown in brackets.

*Holiday locations booked through a travel agent*

| | 1981 | 1982 | 1983 | 1984 | 1985 | 1986 |
|---|---|---|---|---|---|---|
| Europe | 385 (62) | 350 (64) | 326 (47) | 341 (50) | 286 (50) | 297 (42) |
| America | 86 (14) | 78 (14) | 124 (18) | 112 (16) | 95 (16) | 137 (20) |
| Middle East/Africa | 40 (6) | 56 (10) | 87 (13) | 88 (13) | 84 (15) | 102 (15) |
| Other | 112 (18) | 65 (12) | 156 (22) | 143 (21) | 112 (19) | 161 (23) |
| Total | 623 | 549 | 693 | 684 | 577 | 697 |

Figure 11 overleaf shows (a) component, (b) percentage and (c) multiple bar charts representing this tabulated data.

*Comments on the situation shown by Figures 11 (a), (b) and (c)*: The number of bookings varies between 550 and 700 per year with all but about 20% of them taken in Europe, America or the Middle East/Africa. Europe was the most popular holiday location of the four categories given, for the years 1981 to 1986. However, over the six-year period, the number of bookings to Europe has been slowly decreasing and, where it accounted for 60% of all holidays booked in 1981, it now accounts for only just over 40%. Bookings to the Middle East and Africa have slowly increased from below 50 in 1981 to 100 in 1986, which accounted for about 15% of all bookings. Bookings to America have only marginally increased over the period and stood at about 140, or 20%, in 1986.

*Figure 11 Comparison of component, percentage and multiple bar charts for holiday data*

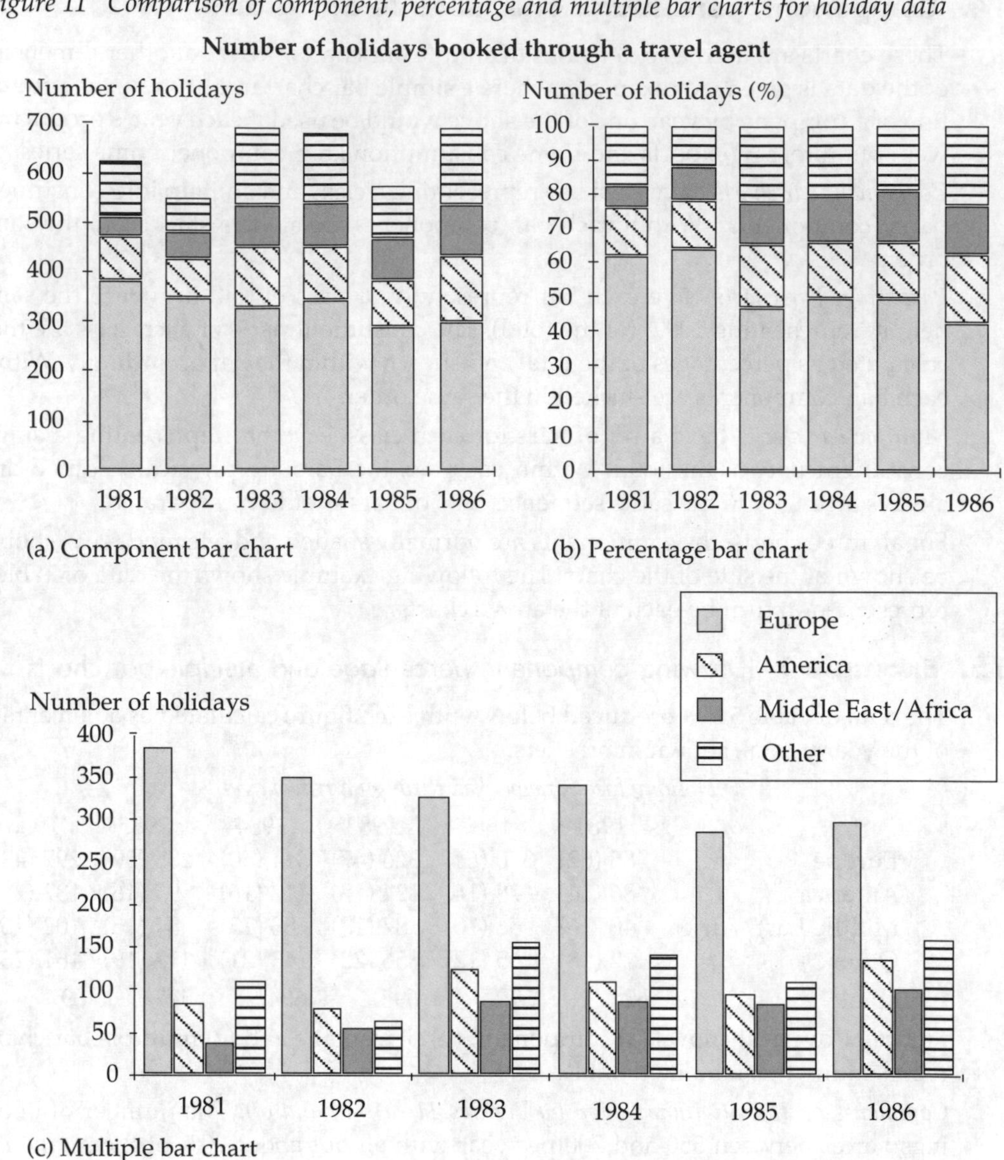


## 16. Comparison of component, percentage and multiple bar charts

In the following lists, '+' signifies a good point or an advantage and '–' signifies a drawback or disadvantage.

a) *Component bar charts.*
   i. Used particularly where class totals need to be represented (+). In Figure 9(a), the height of each bar shows the total number of holidays booked for that year.
   ii. Not easy to compare components relatively across classes owing to actual values being displayed (–).
b) *Percentage bar charts.*
   i. Used where relative comparisons between components are important (+).
   ii. Actual figures (including class totals) are lost (–).
c) *Multiple bar charts.*
   i. Good for comparing components both within and across classes in actual terms, since each bar is drawn from a fixed base (+).
   ii. Class totals are not easy to assimilate (–).
   iii. Can be unwieldy if there are a large number of classes (–).

## 17. Multiple pie charts

Multiple pie charts can be used as alternatives to percentage bar charts; that is, a pie chart (360 degrees) replaces a bar (100%) for each class or year. For example, Table 8 represents the skills classification of the workforces at two factories.

(Note that the degrees figure in Table 8 can be obtained by multiplying each percentage by 3.6.)

Figure 12 overleaf shows multiple pie charts for the data of Table 8.

*Table 8 Multiple pie charts calculations*

| | Factory A | | | Factory B | | |
|---|---|---|---|---|---|---|
| | Number of workers | % | Degrees | Number of workers | % | Degrees |
| Unskilled | 23 | (20) | [72] | 110 | (34) | [122] |
| Semi-skilled | 26 | (22) | [79] | 68 | (21) | [76] |
| Skilled | 67 | (58) | [209] | 144 | (45) | [162] |
| Total | 116 | (100) | [360] | 322 | (100) | [360] |

*Comments on the situation shown by Figure 12*: At both factories, about 20% of the workforce is semi-skilled. However, whereas unskilled workers account for only 20% of the workforce of factory A, they constitute about 35% of factory B's workforce.

a) The *advantage* of using multiple pie charts as opposed to a percentage bar chart is mainly visual impact; they are generally felt to be more attractive.
b) However, their construction is more involved and this is considered as a *major disadvantage*. (Most people prefer to work out percentages and draw straight line bars than calculate degrees of sectors and draw circles.)

*Figure 12 Multiple pie charts for skills classification of the workforce at two factories*

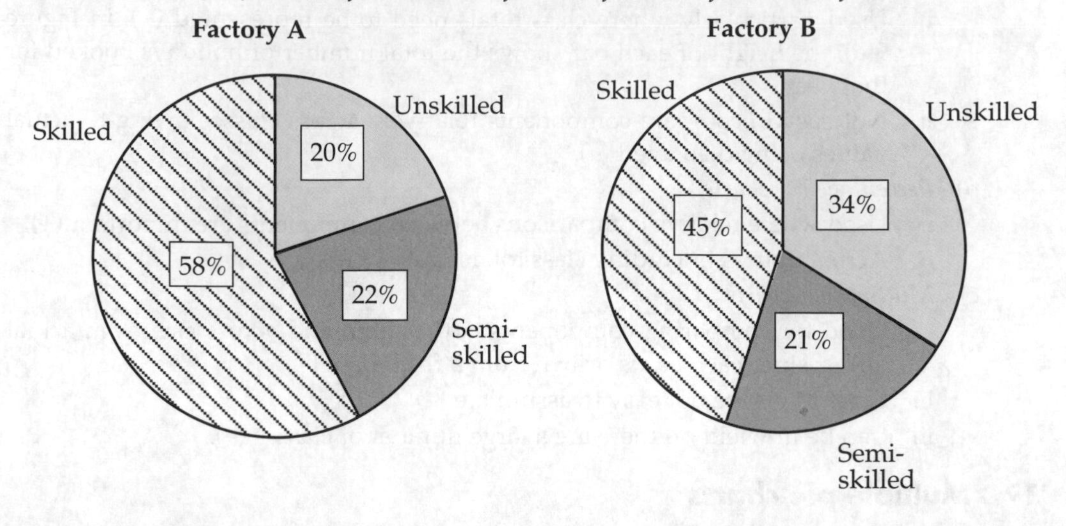


## 18. Proportional pie charts

Sometimes, the presentation of multiple pie charts can be taken a stage further than that shown in Table 8 and Figure 12. This is done by making the areas of the circles proportional to the class totals. Thus, if the total frequency of one class was twice that of another, the area of its representative circle would be twice as large. These are called *proportional pie charts*. The following procedure describes how to draw a pair of proportional pie charts, to represent two classes having totals $T_1$ and $T_2$ respectively. At the same time, the technique is demonstrated, using the data given in the previous section.

**Constructing proportional pie charts.**

STEP 1 Determine the two class totals.

For the previous data, $T_1 = 116$ (total number of workers at factory A) and $T_2 = 322$ (total number of workers at factory B).

STEP 2 Draw the first circle using ANY convenient radius.

STEP 3 Draw the second circle according to the following:

$$\text{Radius of circle 2} = (\text{radius of circle 1}) \times \sqrt{\frac{T_2}{T_1}}$$

where: $T_1$ = total for class 1

and: $T_2$ = total for class 2.

Thus, for the previous data, radius 2 $= \text{radius } 1 \times \sqrt{\frac{322}{116}}$

$= (1.7) \times \text{radius } 1$

STEP 4 Calculate the size of the sectors for both circles in the usual way.

Figure 13 shows the data of Table 8 represented by proportional pie charts, where the radius of the first circle has been multiplied by 1.7 to obtain the radius of the second circle.

*Figure 13 Multiple pie charts for skills classifications*

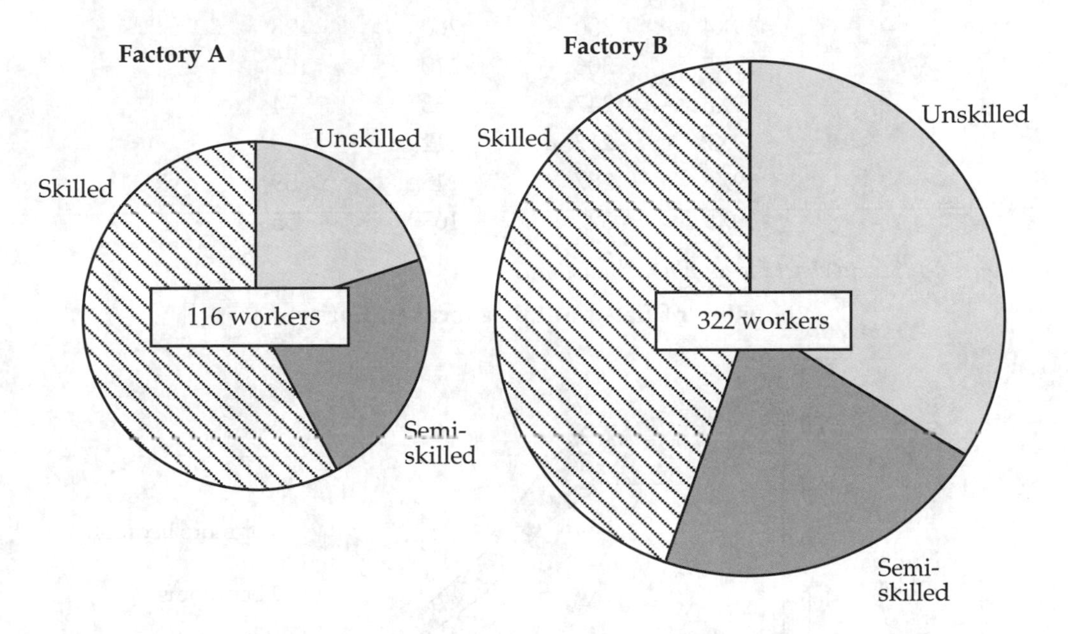


The obvious impact of proportional pie charts is often outweighed by the considerable effort (both calculation and drawing) that needs to be made in its construction. This is particularly significant if there are more than two classes involved.

## 19. Strata charts

*Strata charts* (otherwise known as *cumulative line diagrams or area graphs*) are used to represent *component time series*. As the middle of the above three titles implies, the separate line diagrams for the components are stacked (in a similar way to component bar charts).

Figure 14 shows a strata chart of the number of flats built (by type) for a local authority, for the data given in the accompanying table.

*Comments on the situation shown by Figure 14*: Total building decreased from nearly 800 flats in 19X1 to a 'low' of under 400 in 19X7, but over the next three years (up to 19X9) recovered nearly half to stand at about 550. Over the whole period, the building of both 2 and 3 or more bedroom flats have decreased by about half whereas 1 bedroom flats, despite a small slump in 19X7, have remained at just over 300 per year.

| | *Classification of flat* | | |
|---|---|---|---|
| | 1-bed | 2-bed | 3+bed |
| 19X1 | 333 | 326 | 116 |
| 19X2 | 319 | 290 | 115 |
| 19X3 | 308 | 283 | 123 |
| 19X4 | 330 | 260 | 128 |
| 19X5 | 306 | 210 | 102 |
| 19X6 | 247 | 153 | 74 |
| 19X7 | 217 | 112 | 49 |
| 19X8 | 287 | 143 | 59 |
| 19X9 | 322 | 168 | 55 |

*Figure 14 Strata chart for number of flats built*

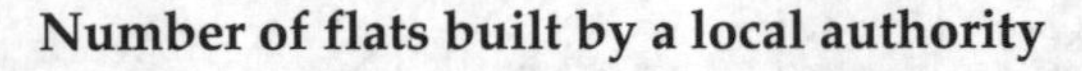


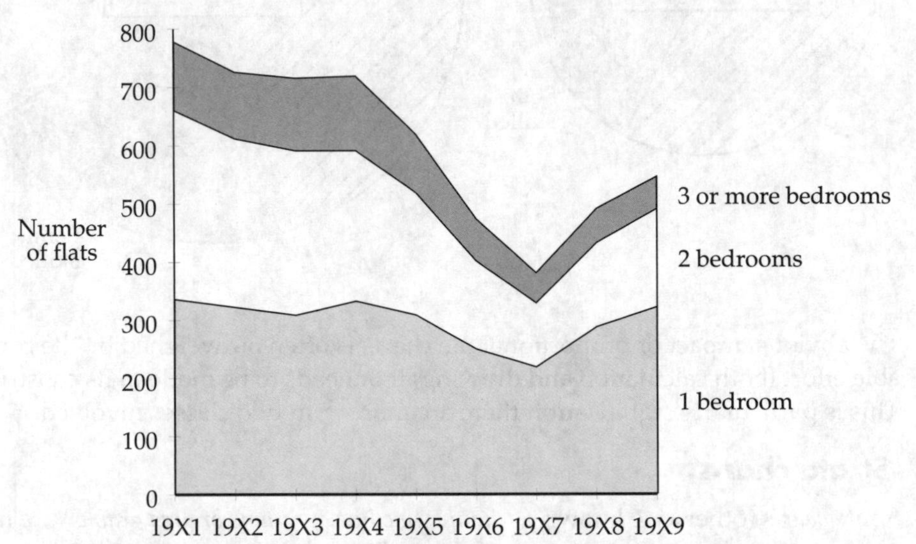


a) It is important to shade the various sections of the chart, as shown in Figure 14, since this emphasises the cumulative nature of the data. Compare this with the case of a multiple line diagram, where actual data values for each variable are plotted and no shading would be used.

b) The *advantages* of a strata chart, *compared with a component bar chart*, are:
   i. the impression of continuity that the lines give (appropriate for a time series);
   ii. easier to construct.

c) The *advantages* of a strata chart, *compared with a multiple line diagram*, are:
   i. a total is shown for each year;
   ii. as many components as desired can be plotted without confusion.

d) The *disadvantages* of using a strata chart, *compared with a multiple line diagram*, are:
   i. 'cross-over points' (i.e. where the value of one component overtakes the value of another) are not identified on a strata chart;
   ii. individual component comparisons are not as easy to make.

## 20. Z-charts

A *Z-chart* is the name given to a chart which portrays time series data in the form of a combination of three separate line diagrams, described as follows.

a) The first line diagram drawn describes the *actual time series values*. Normally, the time series consists of monthly measurements over one complete year, i.e. Jan, Feb, ..., through to Dec.

b) The second line diagram drawn describes the *accumulated time series values*. For monthly measurements, the first plot will coincide with the first plot of the diagram in (a), i.e. Jan's figure (which is the point corresponding to the bottom left-hand join of the Zed). The second plot will be Jan+Feb; the third, Jan+Feb+Mar and so on. This diagram is useful for charting monthly progress towards an annual total. The more removed from a straight line it is, the more variation there has been in the actual monthly figures.

*Table 9 Calculations for a Z-chart construction*

| | Jan | Feb | Mar | Apr | May | Jun | Jul | Aug | Sep | Oct | Nov | Dec |
|---|---|---|---|---|---|---|---|---|---|---|---|---|
| Actual year 1 | 150 | 154 | 183 | 162 | 181 | 149 | 130 | 152 | 186 | 199 | 193 | 168 |
| Actual year 2 | 162 | 163 | 171 | 158 | 175 | 145 | 121 | 138 | 172 | 175 | 163 | 152 |
| Cumulative Year 2 | 162 | 325 | 496 | 654 | 829 | 974 | 1095 | 1233 | 1405 | 1580 | 1743 | 1895 |
| Moving Totals Year 2 | 2019 | 2028 | 2016 | 2012 | 2006 | 2002 | 1993 | 1979 | 1965 | 1941 | 1911 | 1895 |

154 + 183 + ........ + 162

2016 + 158 - 162

2019 + 163 - 154

2028 + 171 - 183

c) The third line diagram is drawn such that each point describes the current month's figure plus the previous eleven month's figures, to form a *twelve-month total*. Collectively, the twelve values so obtained are called *moving totals* (see Table 9 above). Note that, in order to calculate moving totals for a particular year, the previous year's figures must be known. The first point plotted (at Jan, this year) will be the sum of the figures from Feb, last year, to Jan, this year.

The second point plotted will be the sum of the figures from Mar, last year, to Feb, this year; and so on. The last point moving total plotted will coincide with the last accumulated value of the diagram in b) (which is the point corresponding to the top right-hand join of the Zed). This particular diagram is useful for determining the long term underlying trend of the data.

Example 2 shows the construction of a Z-chart.

## 21. Example 2 (Construction of a *Z-chart*)

The data in Table 9 give the monthly production figures of a manufactured component and shows the calculations necessary for constructing a Z-chart.

Figure 15 shows the Z-chart, comprising actual, cumulative and moving totals for year 2.

*Figure 15 Z-chart of monthly production figures*

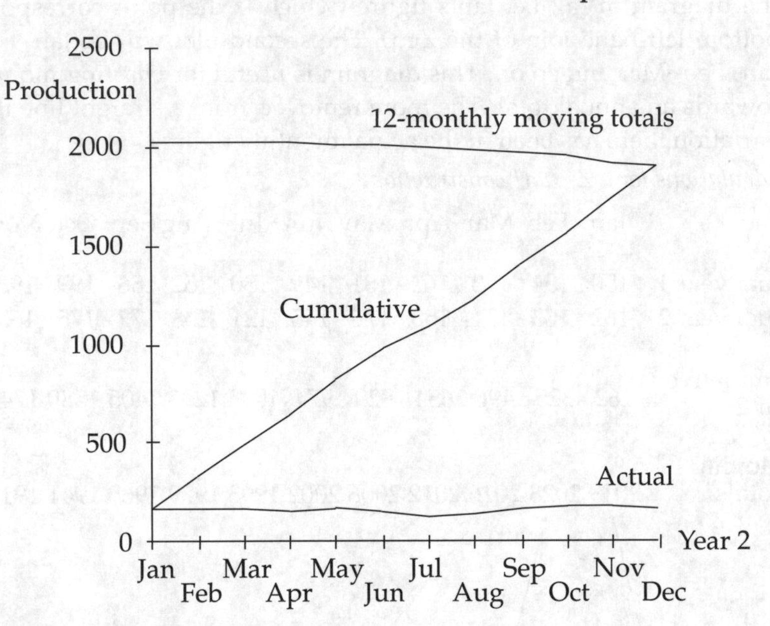


*Comment on the situation shown by Figure 15*: Production in year 2 was relatively steady with a slight drop in the summer months. The long term trend shows a drop in overall production.

## 22. Gantt charts

A method of charting the progress of some project against a defined plan is affected by means of a *Gantt chart*. A number of scaled natural time periods (days, weeks or months) are identified, within each of which three bars can be drawn. One bar shows the planned achievement and the other two show actual achievement and cumulative achievement (to date).

As an example, Table 10 shows the planned achievement of five weeks production for a project, together with the actual achievement up to the end of week three.

*Table 10 Planned and actual details for a project*

| Week number | Planned | Cumulative Planned | Cumulative Actual | Actual |
|---|---|---|---|---|
| 1 | 28 | 28 | 23 | 23 |
| 2 | 32 | 60 | 58 | 35 |
| 3 | 30 | 90 | 98 | 40 |
| 4 | 25 | 115 | | |
| 5 | 40 | 155 | | |

Figure 16 shows the Gantt chart drawn up for the data of Table 10. The comments shown on the chart are for information purposes and may or may not be included on an actual chart.

*Figure 16 Gantt chart for production figures project*

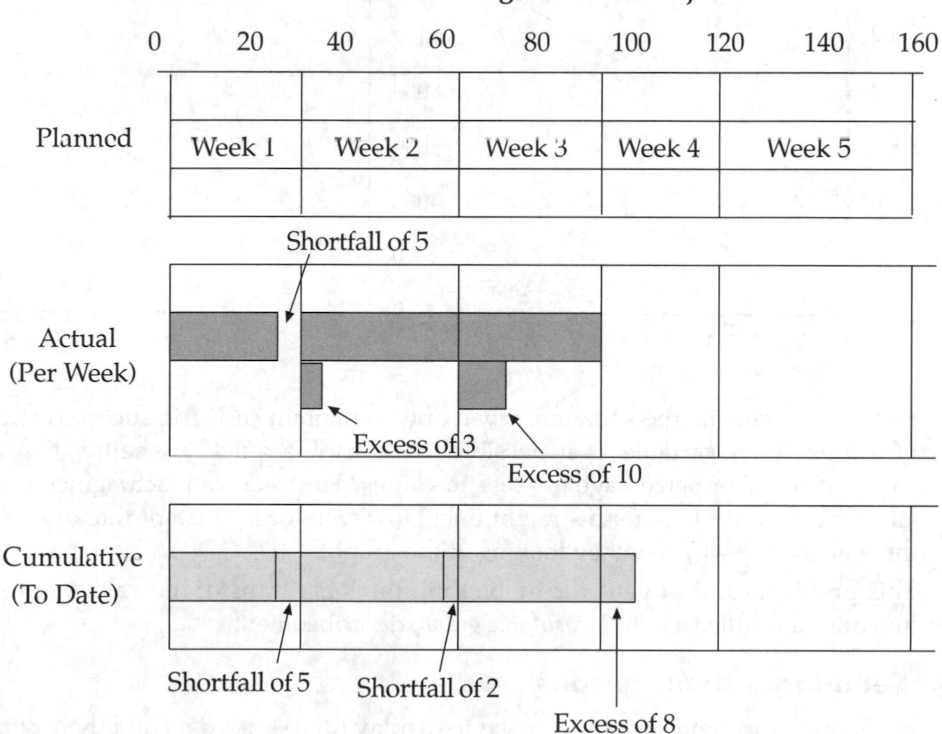


*Comment on the situation shown by Figure 16*: Despite a shortfall in actual production in week 1, weeks 2 and 3 both showed an excess of production which resulted in an overall excess of 8 units by the end of week 3.

## 23. Actual and percentage increases in time series

In Business, it is sometimes necessary to show clearly, using a graphical method, whether some time series variable is increasing in actual or percentage terms. The difference between these two can have a marked effect on successive values of the variable.

For example, suppose the yearly demand for a new technological product was estimated as 2000 in year 1. Table 11 shows the difference between an actual increase of 500 per year and a relative increase of 25% per year and Figure 17 shows these values plotted using line diagrams.

*Table 11 Comparison of constant actual and relative increases*

| Year | 1 | 2 | 3 | 4 | 5 | 6 | 7 | 8 | 9 |
|---|---|---|---|---|---|---|---|---|---|
| Situation 1: Actual increase of 500 per year | 2000 | 2500 | 3000 | 3500 | 4000 | 4500 | 5000 | 5500 | 6000 |
| Situation 2: Increase of 25% per year | 2000 | 2500 | 3125 | 3906 | 4883 | 6104 | 7629 | 9537 | 11921 |

*Figure 17 Line diagrams for the data of Table 11*

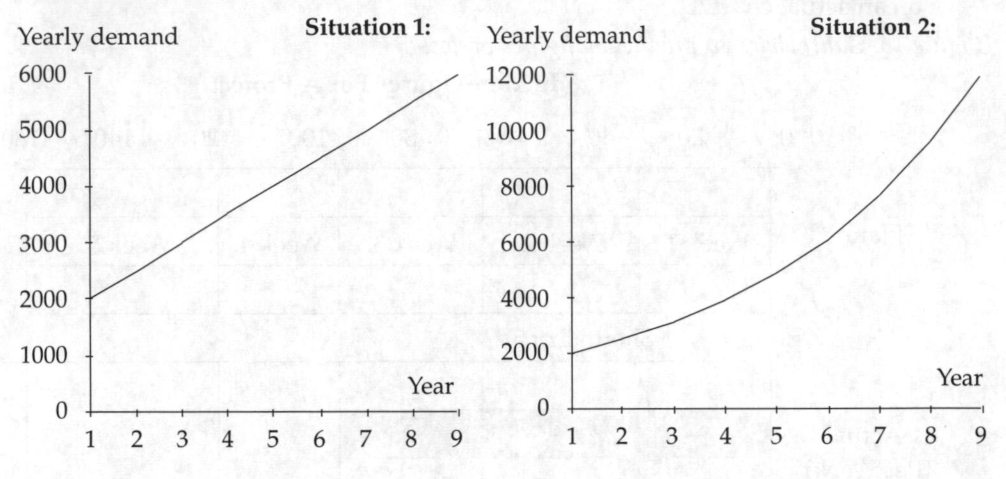


However, reversing the situation, given only a diagram showing successive values of a time series variable, is it possible to determine quickly whether there is a constant actual or percentage increase in values? For a constant actual increase, the answer is yes. We look for a straight line. However, for a constant rate of increase, there is no easy way to tell by looking at the graph.

This problem can be overcome by plotting the logarithms of the values to form a line diagram called a *semi-logarithmic graph*, described next.

## 24. Semi-logarithmic graphs

a) Semi-logarithmic graphs are used to display time series data and their purpose is to show whether the *rate of increase/decrease* in the values of the variable involved is constant or not. They are constructed by plotting the logarithms of the given values against their respective time points and, if a straight line results, the values are increasing or decreasing at a constant rate.

As an example, the data given in Table 11 (situation 2) will be plotted using a semi-logarithmic graph. The layout of the calculations is shown below.

| Year | 1 | 2 | 3 | 4 | 5 | 6 | 7 | 8 | 9 |
|---|---|---|---|---|---|---|---|---|---|
| Est'd production | 2000 | 2500 | 3125 | 3906 | 4883 | 6104 | 7629 | 9537 | 11921 |
| Log (base 10) | 3.301 | 3.398 | 3.495 | 3.592 | 3.689 | 3.786 | 3.882 | 3.979 | 4.076 |

The above logs are plotted against year in Figure 18. Notice that the points form a perfect straight line (which was expected, since the values were calculated using a constant 25% increase).

*Figure 18 Semi-logarithmic chart for estimated sales*

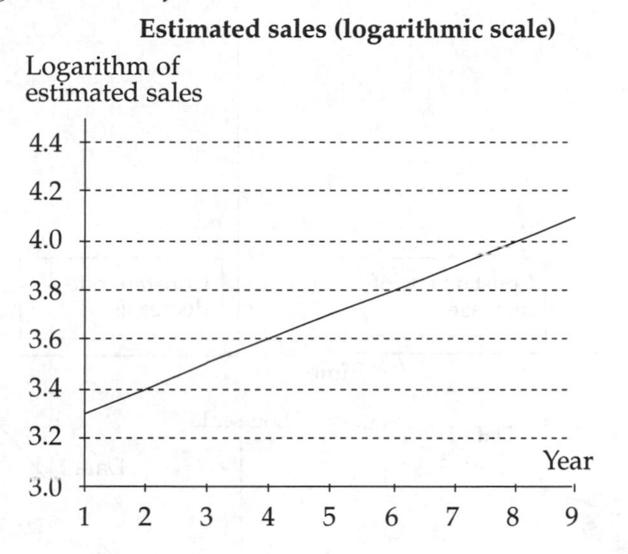


b) The larger the rate of increase of time series data, so the steeper the (semi-logarithmic) line will be. This fact enables the rates of increase of two or more time series to be compared on the same set of axes. Note that when comparing time series data in this way, it is ONLY the steepness (or inclination) of the lines that is of any interest AND NOT their positions. Figure 19 opposite shows the significance of some standard shapes of semi-logarithmic graphs.

## 25. Checklist for the construction of diagrams

The following points should be regarded as standards when constructing graphs and should be remembered.

a) All diagrams should be neat and attractive to look at. Always use graph paper and a ruler.

b) Diagrams should be easy to read, without excessive detail.

c) Always try to locate the diagram centrally on the paper, using as much of the paper as possible.

d) A general title must always be given which describes what is being portrayed but it should be as brief as possible and to the point. A long title, crammed with information, *will not* be read. (Important qualifications of any words in the title can be described in a footnote if necessary.)

e) Axes, if used, should be clearly labelled, giving the units of the data and a note of any break of scale.

f) Shading or colouring, if used, must be lightly done as it may detract from the presentation.

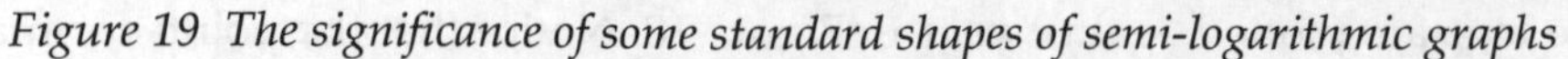

*Figure 19 The significance of some standard shapes of semi-logarithmic graphs*

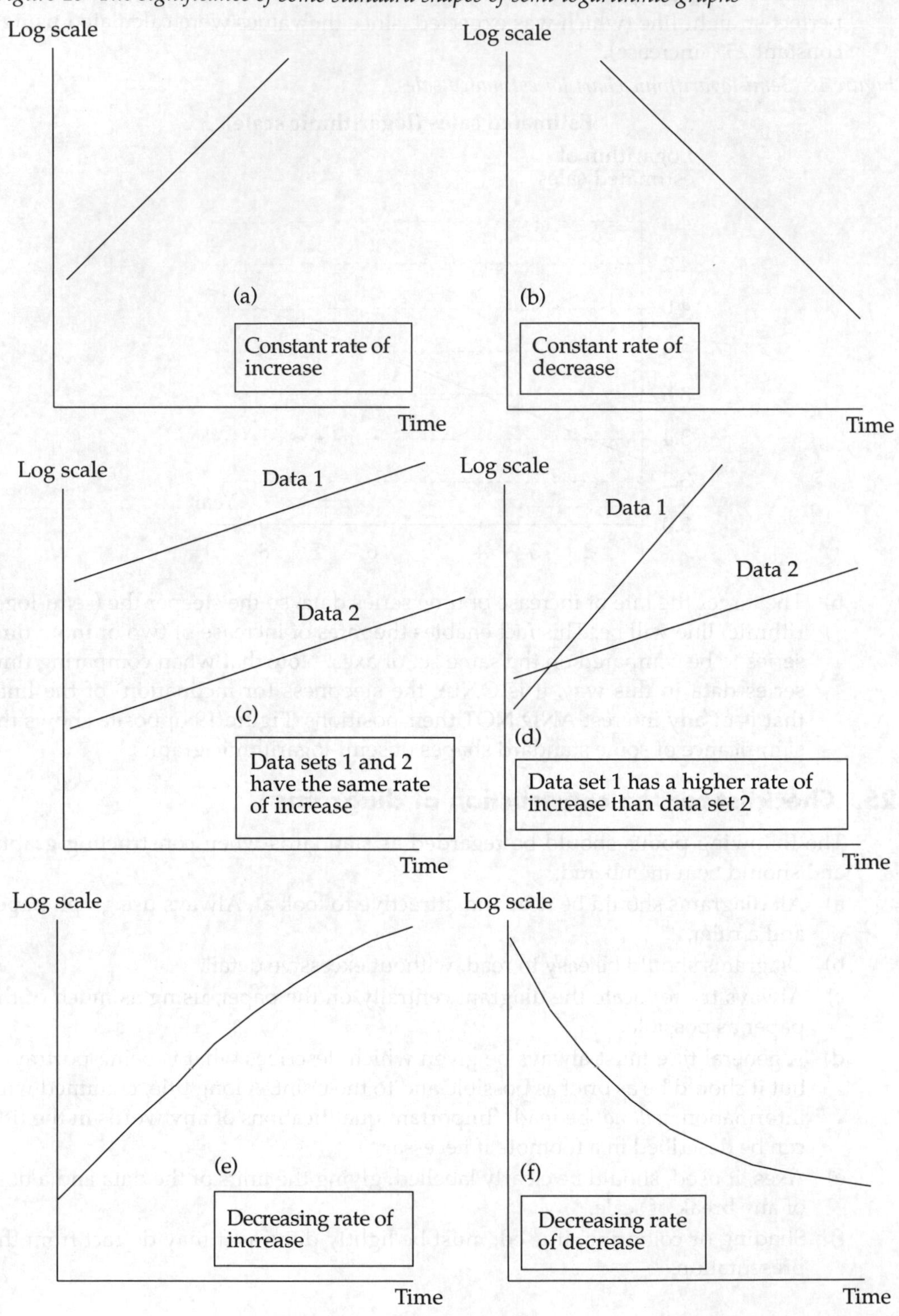

g) If two or more line diagrams appear together, distinguish between them clearly by labelling, colouring or dotting/dashing. If charts are coloured or shaded, label them clearly or provide a separate key.

## 26. Summary

a) Non-numeric frequency distributions describe data by their quality. Time series consist of measurements of some variable over time. Both of these data structures, or a combination of the two, are represented by the charts and graphs described in this section.

b) Pictograms are charts which represent different magnitudes by repeating or varying the size of symbols/pictures which are easily identifiable by a non-specialist audience. Their merit is that they are easy to understand; their disadvantages include possible misrepresentation, using varying sizes of symbol.

c) Simple bar charts use the heights of standard-width bars to represent magnitudes. Variations are 'loss or gain' charts (showing both positive and negative values) and 'back-to-back' charts (which are able to compare characteristics such as male and female or skilled and unskilled). Their merits include ease of construction and the ability to represent numeric values accurately.

d) Pie charts represent the totals of a set of classes using a circle, with individual sector sizes representing the magnitudes of the component classes. Their main merits are impact and the ability to compare classes in relative terms, but their compilation is laborious.

e) Line diagrams are used for time series alone and consist of plotted points, drawn to a height corresponding to class frequency and joined with straight lines. Multiple line diagrams can be drawn on the same set of axes for comparison purposes. They are more suitable for representing time series than bar charts, and are easy to understand and construct. However, too many line diagrams (with closely associated values) can confuse.

f) Component, percentage and multiple charts are extensions of simple bar charts which represent both classes and the components that make them up. Component charts show a bar for each class, subdivided into components and representing actual figures. Percentage charts show also a bar per class, but all drawn to the same height (100%) comprising components in percentage terms. Multiple charts show a bar per component, grouped together to form separate classes.

g) Multiple pie charts can be used to compare two or more sets of classes of components, including time series. Proportional pie charts is the name given to this type of representation when the circle areas are drawn in proportion to class totals. Their advantage is impact but their complexity generally outweighs this advantage.

h) Strata charts are cumulative line diagrams and represent time series split into components forming a natural total. They are easier to construct than the equivalent component bar charts and give a good representation of continuity. However, they do not show cross-over points.

i) Z-charts are three-in-one line diagrams representing (in turn) actual figures, cumulative figures and moving totals. They are normally used to portray monthly time series over a single year.
j) Gantt charts are used to record the progress of some time-orientated project. Three bars show, respectively, planned, actual (per time period) and cumulative (to date).
k) Semi-logarithmic graphs plot the logarithms of the values of a time series and if the graph results in a straight line, this shows that the data is increasing (or decreasing) at a constant rate. They can be used to compare two or more time series with respect to their magnitude of rate of change.
l) When constructing graphs: make them neat and easy to understand; give a general title and label the axes clearly; show breaks of scale distinctly; do not shade or colour too heavily; label multiple lines and give a key for colouring and shading.

## 27. Points to note

a) There is often no 'correct' graph to use for a given set of data. Some graphs are usually more appropriate than others in terms of bringing out various attributes of the data that is of interest; for example, relative rather than actual comparisons.
b) Misrepresentation of data, either by accident or design, is caused by not following accepted standards (the most important of these are listed in section 23). Common misrepresentations often result from not showing breaks of scale, faulty or missing scaling on a numeric axis and confusion over linear, area and volume pictograms.

## 28. Student self review questions

1. What is a time series? Explain the difference between multiple and component time series. [3]
2. What are the main purposes of statistical diagrams and in what circumstances can they be used? [4]
3. Which two types of diagrams could represent a time series of total numbers employed in some industry? [5]
4. What are the disadvantages of using pictograms? [7]
5. How can a simple bar chart be adapted to represent both positive and negative values? [9]
6. Give some examples of the type of data which could be represented by 'back-to-back' bar charts. [9]
7. What type of data would a single pie chart be used to represent? What are the main disadvantages of this form of representation? [10,11]
8. Why are line diagrams more suited to represent time series than simple bar charts? [13(c)]
9. What factor would decide whether a component or a percentage bar chart is chosen for a set of data? [16(a,b)]

10. What is the particular advantage that a multiple bar chart has over component and percentage bar charts? [16(c)]
11. Why are percentage bar charts generally favoured over multiple pie charts? [17(a,b)]
12. What are proportional pie charts? [18]
13. What is a strata chart and why should the separate strata be shaded? [19(a)]
14. What would be the deciding factors when choosing between a multiple line diagram and a strata chart? [19(c,d)]
15. What are the three separate line diagram components that go to make up a Z-chart? [20]
16. What is a Gantt chart? [22]
17. What is the purpose of a semi-logarithmic graph and what is the significance of the steepness of the line? [24]
18. Name some of the standards for constructing graphs. [25]

## 29. Student exercises

1. Represent the following data pictorially by means of both a bar and pie chart.

*Policies issued by an insurance broker in one week*

| Type of policy | Life | Disability | Household | Commercial | Other |
|---|---|---|---|---|---|
| Number issued | 10 | 2 | 25 | 3 | 8 |

2. Draw a line diagram to display the following data, which relates to reported accidents in a certain industry.

| Year | 1 | 2 | 3 | 4 | 5 | 6 | 7 | 8 | 9 | 10 | 11 |
|---|---|---|---|---|---|---|---|---|---|---|---|
| Number of accidents | 23 | 17 | 25 | 31 | 15 | 19 | 26 | 11 | 9 | 16 | 10 |

3. Represent the following data diagrammatically in an appropriate way. (Break the vertical scale, starting at 3500.)

| Shift number | 121 | 122 | 123 | 124 | 125 | 126 | 127 | 128 | 129 |
|---|---|---|---|---|---|---|---|---|---|
| Number of finished products | 3821 | 3657 | 4122 | 3966 | 3729 | 4103 | 3802 | 3755 | 4202 |

Comment on the presentation.

4. Represent the following data pictorially in TWO completely different ways, saying which one you prefer and why.

*Prices of selected dairy foods (pence)*

| | Mar | Apr | May | Jun | Jul | Aug | Sep | Oct |
|---|---|---|---|---|---|---|---|---|
| Eggs (per dozen) | 41 | 41 | 41 | 43 | 40 | 42 | 43 | 43 |
| Cheese (per lb) | 86 | 87 | 91 | 91 | 91 | 96 | 96 | 102 |
| Milk (per pint) | 31 | 30 | 30 | 30 | 32 | 32 | 34 | 35 |

5. Display the data (on the following page) using:
   a) a component bar chart;
   b) a percentage bar chart.

   Which of the two charts is more appropriate?

| Costs | Project A | B | C | D | E |
|---|---|---|---|---|---|
| Set-up | 26500 | 32000 | 18500 | 19000 | 24000 |
| Running | 41500 | 38500 | 19000 | 28000 | 18500 |
| Overhead | 8500 | 6500 | 3000 | 6000 | 9000 |
| Labour | 12000 | 18500 | 16000 | 11500 | 14000 |

6. Represent the following data pictorially using:
   a) a component bar chart;
   b) a strata chart.

*Amount of cereals produced (000 tonnes) in a particular region*

| | 1977 | 1978 | 1979 | 1980 | 1981 | 1982 | 1983 | 1984 | 1985 | 1986 |
|---|---|---|---|---|---|---|---|---|---|---|
| Wheat | 390 | 347 | 336 | 424 | 482 | 478 | 500 | 613 | 449 | 474 |
| Barley | 922 | 827 | 866 | 753 | 856 | 924 | 901 | 913 | 851 | 765 |
| Oats | 139 | 122 | 131 | 122 | 136 | 125 | 108 | 96 | 80 | 76 |
| Other | 15 | 17 | 25 | 27 | 23 | 24 | 22 | 16 | 14 | 10 |

7. Represent the following data using proportional pie charts.

*Monthly sales (£000) for a national supermarket chain*

| | Month 1 | Month 2 | Month 3 |
|---|---|---|---|
| Hardware | 26340 | 35500 | 38760 |
| Greengrocery | 42450 | 44780 | 43710 |
| Grocery | 86900 | 126880 | 141700 |
| Other | 32860 | 51190 | 74320 |

8. Use a strata chart to display the following data:

*Turnover (£m) of a large furniture chain*

| | 19X1 | 19X2 | 19X3 | 19X4 | 19X5 | 19X6 | 19X7 | 19X8 | 19X9 |
|---|---|---|---|---|---|---|---|---|---|
| Beds | 0.8 | 0.8 | 0.9 | 0.8 | 0.7 | 0.5 | 0.6 | 0.9 | 1.2 |
| Suites | 2.4 | 2.5 | 2.7 | 2.7 | 3.1 | 3.0 | 3.1 | 3.2 | 3.9 |
| Dining | 0.6 | 0.3 | 0.3 | 1.2 | 1.4 | 0.7 | 0.6 | 1.1 | 1.0 |
| Misc | 1.5 | 2.2 | 1.4 | 1.1 | 1.7 | 2.3 | 2.1 | 1.8 | 2.7 |

9. The following data give the monthly sales, in thousands of gallons, of a petrol service station over a two-year period.

| | Jan | Feb | Mar | Apr | May | Jun | Jul | Aug | Sep | Oct | Nov | Dec |
|---|---|---|---|---|---|---|---|---|---|---|---|---|
| Year 1 | 11.4 | 11.8 | 11.5 | 12.0 | 11.9 | 11.9 | 12.1 | 12.4 | 12.0 | 11.8 | 10.4 | 10.8 |
| Year 2 | 11.0 | 11.9 | 11.6 | 11.7 | 11.5 | 11.8 | 11.7 | 12.2 | 11.8 | 11.4 | 9.2 | 10.6 |

Represent these figures diagrammatically using a Z-chart.

10. A computer firm planned the production of a new model as 12000 in each of the first two months, 15000 in month 3, 20000 in each of the next two months and 10000 in month 6. Actual production and distribution to dealers in the first four months is given in the following table.

| Month | 1 | 2 | 3 | 4 |
|---|---|---|---|---|
| Production | 13500 | 14200 | 14800 | 13500 |
| Distribution | 11000 | 12750 | 14500 | 15000 |

Draw Gantt charts to show:

a) planned against actual production;
b) production against distribution.

11. Sales of gas (thousands of therms) in a particular region

| Year | 1 | 2 | 3 | 4 | 5 | 6 | 7 | 8 | 9 | 10 | 11 |
|---|---|---|---|---|---|---|---|---|---|---|---|
| Domestic | 2382 | 2812 | 3194 | 3522 | 3910 | 4489 | 4796 | 5360 | 5869 | 6174 | 6569 |
| Industrial | 894 | 956 | 1081 | 1472 | 2720 | 4278 | 4827 | 6011 | 5870 | 6258 | 6399 |

Plot the graphs of domestic and industrial sales on a semi-logarithmic scale and comment on the results shown.

# Examination questions

1. Sampling methods are widely used for the collection of statistical data in industry and business. Explain FOUR of the following, illustrating your answers with practical examples:
   i. simple random sampling;
   ii. stratification;
   iii. quota sampling;
   iv. sample frame;
   v. cluster sampling;
   vi. systematic sampling.

*CIMA*

2. a) Discuss the relative advantages and disadvantages of the postal questionnaire and the personal interview as a means of collecting data.
   b) Compare simple random sampling and quota sampling as methods of selecting a representative sample from a population.

*ICSA*

3. a) Real consumers' expenditure in 1984 – component categories (Seasonally adjusted, £000 million, 1980 prices):

| *Durable goods* | *Foods* | *Alcohol and tobacco* | *Clothing and footwear* | *Energy products* | *Other goods* | *Rent and rates* | *Other services* |
|---|---|---|---|---|---|---|---|
| 16 | 22 | 14 | 11 | 11 | 16 | 17 | 37 |

*(Source: Economic Trends, August 1985)*

You are required to:

i. draw an appropriate diagram to illustrate the relative shares of real consumers' expenditure in 1984;
ii. state one item / category of expenditure which falls under the heading Other goods, and one item which falls under the heading Other services.

b) Comparative profit before tax (1980=100) of six scotch whisky companies

| *Company* | *1980* | *1981* | *1982* | *1983* | *1984* |
|---|---|---|---|---|---|
| Bells | 100 | 116 | 160 | 183 | 208 |
| Distillers | 100 | 93 | 98 | 110 | 104 |
| Highland | 100 | 87 | 100 | 122 | 144 |
| Invergordon | 100 | 87 | 91 | 85 | not available |
| Macallan | 100 | 106 | 120 | 161 | 174 |
| MacDonald | 100 | 107 | 146 | 170 | 171 |

*(Source: Datastream, September 1985)*

You are required to draw an appropriate graph to compare the profit performance of the six companies.

*CIMA*

4. a) Define and give an example of each of the following statistical terms:
   i. absolute error  ii. relative error
   iii. compensating error  iv. biased error.

   b) A jobbing engineer has quoted £4,000 to his customer for the production of a special purpose machine. The material is expected to cost £2,600 to the nearest £100. The production time is estimated at 150 hours to the nearest 10 and the wage rate is £4 per hour but this might rise by 10%.
   Required: Calculate the estimated profit and state the error limits.

*AAT*

5. The production of each manufacturing department of your company is monitored weekly to establish productivity bonuses paid to the members of that department. 250 items have to be produced each week before a bonus will be paid. The production in one department over a forty week period is shown below:

382 367 364 365 371 370 372 364 355 347
354 359 359 360 357 362 364 365 371 365
361 380 382 394 396 398 402 406 437 456
469 466 459 454 460 457 452 451 445 446

Required:

a) Form a frequency distribution of five groups for the number of items produced per week.

b) Construct the ogive or cumulative frequency diagram for the frequency distribution established in a).

*ACCA (50%)*

6. The following table shows the number of insurance policies, by class of business (numbers expressed in thousands), issued by an insurance company during the years 1978-82.

| | Numbers in years | | | | |
|---|---|---|---|---|---|
| Policy type | 1978 | 1979 | 1980 | 1981 | 1982 |
| Life | 24 | 27 | 32 | 31 | 33 |
| Motor | 42 | 37 | 31 | 29 | 26 |
| Household | 10 | 14 | 21 | 28 | 35 |
| Other | 7 | 5 | 8 | 7 | 4 |

a) Carefully draw a suitable chart to illustrate the data.

b) What are the advantages and disadvantages of your form of representation?

c) Comment on the company's progress.

What are the limitations of the given information in interpreting the progress of the company?

*CII*

7. The following figures relating to the distribution of identified personal wealth in Great Britain were taken from the Annual Abstract of Statistics 1976 (Table 366). Present the information in the form of a Lorenz curve and comment on the results.

| Ranges of net wealth | | 1967 | | 1974 | |
|---|---|---|---|---|---|
| Over £ | Not over £ | Number of cases Thousands | £ thousand million | Number of cases Thousands | £ thousand million |
| - | 1,000 | 5,398 | 2.8 | 3,410 | 2.0 |
| 1,000 | 3,000 | 5,273 | 9.8 | 4,775 | 8.6 |
| 3,000 | 5,000 | 2,966 | 11.6 | 2,223 | 8.7 |
| 5,000 | 10,000 | 2,177 | 15.3 | 4,131 | 30.3 |
| 10,000 | 15,000 | 620 | 7.6 | 2,166 | 26.5 |
| 15,000 | 20,000 | 270 | 4.8 | 757 | 13.3 |
| 20,000 | 25,000 | 157 | 3.4 | 415 | 9.6 |
| 25,000 | 50,000 | 279 | 9.9 | 640 | 21.7 |
| 50,000 | 100,000 | 109 | 7.5 | 229 | 15.4 |
| 100,000 | 200,000 | 37 | 5.1 | 65 | 9.2 |
| 200,000 | - | 14 | 5.8 | 26 | 11.8 |
| | Total | 17,300 | 83.6 | 18,837 | 157.1 |

*CIMA*

8. Your company is in the course of preparing its published accounts and the Chairman has requested that the assets of the Company be compared in a component bar chart for the last five years. The data for this task are contained in the following table.

| | (£'000s) | | | | |
|---|---|---|---|---|---|
| Asset | 1978 | 1979 | 1980 | 1981 | 1982 |
| Property | 59 | 59 | 65 | 70 | 74 |
| Plant and Machinery | 176 | 179 | 195 | 210 | 200 |
| Stock and work-in-progress | 409 | 409 | 448 | 516 | 479 |
| Debtors | 330 | 313 | 384 | 374 | 479 |
| Cash | 7 | 60 | 29 | 74 | 74 |

Required:

i. Construct the necessary component bar charts.

ii. Comment upon the movement in the assets over the five year period.

*ACCA (35%)*

9. a) Show by means of sketches four different methods of presenting statistical data diagrammatically. For each method briefly outline the relative advantages and disadvantages.

   b) Using an appropriate method represent the following data diagrammatically.

*Marks & Spencer plc*

*Shareholders owning more than 100,000 shares*

| Type of owner | Number of shareholders |
|---|---|
| Insurance companies | 94 |
| Banks and nominee companies | 231 |
| Pension funds | 122 |
| Individuals | 189 |
| Others | 146 |
| Total | 782 |

*(Source: Company accounts)*

*ICSA*

10. A company is preparing future production plans for a new product. Research findings suggest that next year the company could make and sell 10,000 units (± 20%) at a price of £50 (± 10%), depending on size of order, weather, quality of supply, discounts etc.

The variable costs of production for next year, given these data, are also uncertain but have been estimated as follows:

| *Type* | *Costs* £ | *Margin of error* |
|---|---|---|
| Materials | 150,000 | ± 2% |
| Wages | 100,000 | ± 5% |
| Marketing | 50,000 | ± 10% |
| Miscellaneous | 50,000 | ± 10% |

You are required to find the range of possible error in next year's revenue, cost of production, contribution and contribution per unit, in each case stating your answer both in absolute (actual) and in relative terms.

$$\left[\text{Relative error (\%)} = 100 \times \frac{\text{Maximum error}}{\text{Estimated total}}\right]$$

# Part 2 Statistical measures

This part of the book deals with the basic analysis of univariate data (data obtained from measuring just one attribute). Statistical measures is the name given to describe this type of analysis, the measures themselves being split into various groups.

*Measures of location*, commonly called averages, are the most well known measures of numeric data. They are single values, intended as representatives, which can neatly characterise a whole group. When Trade Unions are negotiating pay rises for their members, they often use 'the average wage' as a standard on which to base increases; a business might use the average value of an order to forecast next year's turnover; an insurance company would use the average age of death to calculate the cost of a life assurance policy.

There are different types of averages to suit different situations and requirements and these are covered in chapters 6 to 8. Chapter 6 deals with the most commonly used, the arithmetic mean, while chapters 7 and 8 describe others of differing degrees of importance.

*Measures of dispersion* describe how spread out a set (or distribution) of values is. Averages alone are generally not enough to characterise numeric data in a meaningful way, since they only measure location and take no account of data spread. For example, suppose two separate haulage firms carried on average the same weight of load on their trips. A measure of dispersion could be used on both sets of values to see which firm's loads varied the most, information that could be important to an insurance company for instance.

*Measures of skewness* show how evenly a set of items is distributed, and these, together with various measures of dispersion, are dealt with in chapters 9 to 11.

# 6 Arithmetic mean

## 1. Introduction

This chapter describes the most commonly used average, the arithmetic mean. It is initially defined in words with an accompanying simple example, then some important notation for describing sets of data is given. Techniques for calculating the mean for (discrete) sets of data and frequency distributions are then demonstrated and the place of so-called 'weighted' means is shown.

## 2. Definition of the arithmetic mean

The **arithmetic mean** of a set of values is defined as 'the sum of the values' divided by 'the number of values'.

That is, arithmetic mean $= \dfrac{\text{the sum of all the values}}{\text{the number of values}}$

The arithmetic mean is normally abbreviated to just the 'mean'.

## 3. Example 1 (*Arithmetic mean* for a *set*)

a) If a firm received orders worth

£151, £52, and £280

for three consecutive months, their mean average value of orders per month would be calculated as:

$$\frac{£(151+52+280)}{3} = \frac{£483}{3} = £161$$

b) The mean of the values 12, 8, 25, 26 and 10 is calculated as:

$$\frac{12+8+25+26+10}{5} = \frac{81}{5} = 16.2$$

Note that the mean of a set of values is not necessarily the same as any of the original values in the set, as demonstrated in both a) and b) above.

## 4. Notation for general values

When dealing with a set of items that have known values, the values themselves can be written down and manipulated. However, if it is necessary to consider the values of a set of items in general terms, particularly for use in a formula, the following notation is used:

$x_1, x_2, x_3, \ldots, x_n,$

where $n$ is the number of items in the set. This notation is really just a compact way of saying: 'the 1st $x$-value, the 2nd $x$-value, ... and so on.

For example, if a salesman completes 4, 5, 12, 8 and 2 sales in consecutive weeks, this would correspond to: $x_1=4$ (i.e. the 1st $x$-value is 4), $x_2=5$, $x_3=12$, $x_4=8$ and $x_5=2$. Here, variable $x$ is 'number of sales' and $x_1$ is the 1st sale, $x_2$ is the 2nd sale, ... and so on. In this case, $n=5$. Note that there is nothing particularly sacrosanct about the letter '$x$'; it is simply the letter that is normally used.

## 5. Notation for the sum of values

Adding the values of sets of general items together occurs in so many different formulae in Statistics that it has its own notation, enabling sums to be written very compactly.

$x_1 + x_2 + x_3 + ... + x_n$ is written as $\sum x$

$\sum$ is the Greek symbol for capital 'S' (for Sum) and $\sum x$ can be simply translated as 'add up all the $x$-values under consideration'.

For the sales example given in section 4 above, we have:

$\sum x = 4+5+12+8+2 = 31.$

This notation is now used to give a compact formula for the mean.

## 6. Formula for the mean of a set of values

The *Arithmetic Mean* is the most commonly used average and is commonly known as the 'mean', which it will generally be called from hereon in the manual.

Using the notation of the previous section, the mean of a set of values $x_1, x_2, ..., x_n$ is calculated as follows.

**Mean for a set**

$$\bar{x} = \frac{x_1 + x_1 + ... + x_n}{n}$$

$$= \frac{\sum x}{n}$$

## 7. Example 2 (*Mean for a set*)

To calculate the mean for the set: 43,75,50,51,51,47,50,47,40,48

Here, $n = 10$ and $\sum x = 502$.

Therefore: $\bar{x} = \frac{\sum x}{n} = \frac{502}{10} = 50.2$

## 8. The mean of a simple frequency distribution

Large sets of data will normally be arranged into a frequency distribution, and thus the formula for the mean given in section 7 is not quite appropriate, since no account is taken of frequencies.

Consider the following simple (discrete) frequency distribution.

| $x$ | 10 | 12 | 13 | 14 | 16 | 19 |
|---|---|---|---|---|---|---|
| $f$ | 2 | 8 | 17 | 5 | 1 | 1 |

The definition of the mean (given in section 3) can still be used as is now demonstrated.

The total of all the values = the sub-total of the '10's (= 10×2 = 20)
+ the sub-total of the '12's (= 12×8 = 96)
+ the sub-total of the '13's (= 13×17 = 221)
+ the sub-total of the '14's (= 14×5 = 70)
+ the sub-total of the '16's (= 16×1 = 16)
+ the sub-total of the '19's (= 19×1 = 19)
= 452

Notice that in order to get the sub-totals 20, 96, 221, ... etc, $x$ is being multiplied by $f$ each time. In other words, the total of all the values is just $\sum fx$ (the sum of all the $fx$ values), which in this case is 452. Also, since there are 2 '10's, 8 '12's, 17 '13's, ... etc, the number of values included in the distribution is 2+8+17+5+1+1 = 34 = $\sum f$ (the sum of the frequencies).

Thus, the mean $\overline{x} = \frac{\sum x}{n} = \frac{452}{34} = 13.3$ (2D)

Note that the above formula is a slight adaptation of the formula for the mean of a set given in section 6.

## 9. Example 3 (*Mean for a simple frequency distribution*)

To calculate the mean of the following distribution:

| Number of vehicles serviceable ($x$) | 0 | 1 | 2 | 3 | 4 | 5 |
|---|---|---|---|---|---|---|
| Number of days ($f$) | 2 | 5 | 11 | 4 | 4 | 1 |

the normal layout for calculations is:

| $x$ | $f$ | $fx$ |
|---|---|---|
| 0 | 2 | 0 |
| 1 | 5 | 5 |
| 2 | 11 | 22 |
| 3 | 4 | 12 |
| 4 | 4 | 16 |
| 5 | 1 | 5 |
| Total | 27 | 60 |

Thus: $\overline{x} = \frac{\sum x}{n} = \frac{60}{27} = 2.2$ (1D)

Hence, the mean number of vehicles serviceable is 2.2 (1D).

## 10. The mean of a grouped frequency distribution

One of the disadvantages of arranging discrete data into the form of a grouped frequency distribution is the fact that individual values of items are lost. This is particularly inconvenient when a mean needs to be calculated since, clearly, it is impossible to find the total of the values of the items, which means, in effect, that it is impossible to calculate the mean exactly. However, it is possible to *estimate* it.

This is done by:

a) using the group (or class) mid-points as representative x-values,
b) estimating the total of the values in each group using $x$ times $f$ (the group frequency),
c) adding these totals together to form an estimate of the total of all values (i.e. $\sum fx$),
d) dividing by the total number of items ($\sum f$).

Notice that this gives an estimate of the mean as $\bar{x} = \frac{\sum fx}{\sum f}$ which is exactly the same formula as for a simple frequency distribution.

In the case of a continuous frequency distribution, *the mean cannot exist* since individual continuous values themselves are not determinable. Again, as in the case of discrete grouped data, the mean can only be estimated, the same formula as above being used.

The following formula summarises the work of the previous three sections.

## 11. Formula for the mean of a frequency distribution

The mean for a frequency distribution is calculated using the following formula.

**Mean for a frequency distribution**

$$\text{Mean, } \bar{x} = \frac{\sum fx}{\sum f}$$

**Note**: For a grouped frequency distribution, $x$ is the class mid-point.

## 12. Example 4 (*Mean for a grouped discrete frequency distribution*)

*Question*

The following data relates to the number of successful sales made by the salesmen employed by a large microcomputer firm in a particular quarter.

| Number of sales | 0-4 | 5-9 | 10-14 | 15-19 | 20-24 | 25-29 |
|---|---|---|---|---|---|---|
| Number of salesmen | 1 | 14 | 23 | 21 | 15 | 6 |

Calculate the mean number of sales.

*Answer*

The standard layout and calculations are shown in the following.

| Number of sales | Number of salesmen ($f$) | Class mid point ($x$) | ($fx$) |
|---|---|---|---|
| 0 to 4 | 1 | 2 | 2 |
| 5 to 9 | 14 | 7 | 98 |
| 10 to 14 | 23 | 12 | 276 |
| 15 to 19 | 21 | 17 | 357 |
| 20 to 24 | 15 | 22 | 330 |
| 25 to 29 | 6 | 27 | 162 |
| Totals | 80 | | 1225 |

Here: $\sum fx = 1225$

and $\sum f = 80$

$\therefore$ mean number of sales, $\bar{x} = \dfrac{\sum fx}{\sum f} = \dfrac{1225}{80} = 15.3$ (1D)

## 13. Notes on example 4

a) *Calculating* $\sum fx$. Use the 'accumulating memory' facility on your calculator for both generating and automatically accumulating values of *fx*. This will obviate the need for 're-keying' values already produced by the calculator and will ensure the least number of keystrokes for all the information required in the table. The procedure on the calculator is as follows.

| | |
|---|---|
| 1 × 2 M+ | Multiply 1 by 2 and press the 'add to memory' (M+) key; then write down the displayed value (2) in the table. |
| 14 × 7 M+ | and write down value (98) |
| ... | |
| ... etc | |
| 6 × 27 M+ | and write down value (162) |
| RM | Pressing the 'recall memory' (RM) key finally will display $\sum fx$, which can also now be entered into the table. |

b) *Validation of mean*. After calculating the mean it is always wise to check on the reasonableness of the value obtained. Since the mean is a measure of location, it should be roughly centrally located. In this case, 15.3 is clearly acceptable. An examiner will always give you credit for noticing that a calculated value is unreasonable (even if you have no time or capability of spotting the actual error!)

## 14. Example 5 (*Mean of a grouped continuous frequency distribution*)

*Question*

A machine produces circular bolts and, as a quality control test, 250 were selected randomly and the diameter of their heads measured. Find the mean of the following resulting diameters.

| *Diameter of head (cm)* | *Number of components* | *Diameter of head (cm)* | *Number of components* |
|---|---|---|---|
| 0.9747–0.9749 | 2 | 0.9765–0.9767 | 49 |
| 0.9750–0.9752 | 6 | 0.9768–0.9770 | 25 |
| 0.9753–0.9755 | 8 | 0.9771–0.9773 | 18 |
| 0.9756–0.9758 | 15 | 0.9774–0.9776 | 12 |
| 0.9759–0.9761 | 42 | 0.9777–0.9779 | 4 |
| 0.9762–0.9764 | 68 | 0.9780–0.9782 | 1 |

*Answer*

Since there are many classes and the data is fairly 'unwieldy', the given classes have not been repeated in the following table of calculations. This would be perfectly acceptable in an examination.

| Mid points $(x)$ | $(f)$ | $(fx)$ |
|---|---|---|
| 0.9748 | 2 | 1.9496 |
| 0.9751 | 6 | 5.8506 |
| 0.9754 | 8 | 7.8032 |
| 0.9757 | 15 | 14.6355 |
| 0.9760 | 42 | 40.9920 |
| 0.9763 | 68 | 66.3884 |
| 0.9766 | 49 | 47.8534 |
| 0.9769 | 25 | 24.4225 |
| 0.9772 | 18 | 17.5896 |
| 0.9775 | 12 | 11.7300 |
| 0.9778 | 4 | 3.9112 |
| 0.9781 | 1 | 0.9781 |
| Total | 250 | 244.1041 |

Here: $\sum fx = 244.1041$

and: $\sum f = 250$

Thus: $$\bar{x} = \frac{\sum fx}{\sum f} = \frac{244.1041}{250} = 0.97642 \text{ (5D)}$$

That is, the mean diameter of head = 0.97642 cms (5D).

## 15. Notes on example 5

a) The mean value obtained (0.97642) is located roughly in the centre of the values and thus is reasonable.

b) It is normal practice when calculating the mean to quote the answer to *at least one more place of decimals* than the original data. Thus, since the data was given to 4 places of decimals, the mean given to 5 decimal places is quite in order.

c) The usefulness of the result with regard to the production of the bolts is worth mentioning here. The mean of 0.97642 ins could be used as a specification to buyers or as a standard for comparison with other machines or for future production.

## 16. Manipulation of the mean formula

There are some circumstances where the simple manipulation of the formula:

$$\bar{x} = \frac{\sum x}{n}$$

to become

$$\sum x = n\bar{x}$$

can help solve common problems that arise with numbers involving means.

For example, if the mean weekly wage of 86 employees has been calculated as £172.45 and employee number 87 earned £158.80, what is the mean wage of all 87 employees?

To calculate the mean of the wages of all 87 employees, it is necessary to find their total wage and then divide by 87. But the total of the first 86 can be found using the above formula re-arrangement.

i.e. for the first 86 wages, $\sum x = n\bar{x}$

$= £86(172.45)$

$= £14830.70$

Therefore, total of all 87 wages (in £) $= 14830.70 +$ wage of number 87

$= 14830.70 + 158.80$

$= 14989.50.$

Hence, the mean of all 87 wages is £14989.50÷87 = £172.29.

## 17. Weighted means

Another common problem arises where the means of a number of groups need to be combined to form a grand mean. For example, suppose a company splits its home sales area into three regions, each having a sales representative. Over a particular period, representative A averages £86.42/sale from 24 sales, representative B, £112.91 from 37 sales and representative C, £104.22 from 25 sales. It is required to find the average value per sale overall. That is, the average of all sales completed by the three representatives.

The problem can be solved in a similar way to that of the previous section, where the total value of sales for each representative is calculated then divided by total number of sales. However, this is best thought of in terms of a calculation of a *weighted mean* where the means themselves are thought of as $x$-values and their constituent numbers as (frequency) $f$-values.

In this situation, the formula $\frac{\sum fx}{\sum f}$ is replaced by the more relevant $\frac{\sum n\bar{x}}{\sum n}$

For the data given above:

$$\text{mean} = \frac{24(86.42 + 37(112.91) + 25(104.22)}{24 + 37 + 25}$$

$$= 102.99$$

i.e. average value of all sales = £102.99

## 18. Significance of the mean

a) Generally understood as the standard average.

b) Technically, it is considered as the 'mathematical average', since its basic definition is given in arithmetical terms. This is not true of its two rivals, the mean and median, which are covered in the following two chapters.

c) Regarded as truly representative of the data, since all values are taken into account in its calculation.

d) The mean does not necessarily take a value that is the same as one of the original values (whereas often other averages do).
e) Not thought suitable for data sets that have extreme values at one end, since these are taken into account and can result in an average that is not really representative and therefore not usable in practice. This is seen as the major disadvantage of the mean. Consider the following weekly wages (in £) of a set of employees in a small workshop: 158, 138, 141, 148, 148, 146, 157 and 252 (the latter wage being that for the workshop manager). The mean is easily calculated as £161, which is representative of neither the seven smaller values or the single extreme larger value.

## 19. Summary

a) The arithmetic mean is the most well known example of a measure of location, or average, which aims to represent a set of items numerically.
b) The special notation, $x_1$, $x_2$, $x_3$, ... etc is used as a method of describing the individual values of the items in a group in general terms, without specifying their actual values.
c) The summation operator, $\sum$, is used to represent the addition of a set of values in general terms.
d) The mean for a set of values is found by dividing the sum of the values by their number.
e) The mean is the most popular average, being well understood and taking all items into account. Its main disadvantage is the fact that it takes extreme values too much into account and can be considered unrepresentative where such values occur.

## 20. Points to note

a) Calculators should be noiseless and cordless (for obvious reasons).
b) When there are open-ended classes in a frequency distribution and a mean needs to be calculated, it is necessary to fix a nominal limit to the class in question in order to determine a mid-point. There is no particular rule for doing this other than taking note of what the data are describing and then fixing a limit which seems reasonable. For example, if a set of classes of data describing the ages of employees had the two groups 'under 21' and '55 and over', these could be translated as '18 to 21' and '55 to 65' for the practical purpose of obtaining the two mid-points 19.5 and 60 respectively.
c) As a general rule, the mean is quoted to at least one more place of decimals than the original data. Thus the mean of data given in whole numbers would be quoted to at least 1D and the mean of data given to 4D would be quoted to at least 5D.
d) *The sum of the deviations from the mean is ZERO*. This is a particular feature of the mean that is useful in further statistical work (it is referred to in passing in chapter 11). This result can easily be demonstrated. For example, consider the set 2, 4, 6 and 8. The mean is 5, and subtracting it from each of the values gives:

$$2 - 5 = -3; 4 - 5 = -1; 6 - 5 = 1 \text{ and } 8 - 5 = 3.$$

When these four values are added, the result is zero, as stated above.

## 21. Student self review questions

1. What is a measure of location? [Part intro]
2. How is the arithmetic mean defined? [2]
3. Why is the special notation $x_1, x_2, ..., x_n$ used? [4]
4. What does $\sum fx$ mean? [8]
5. Why is the formula for the arithmetic mean of a frequency distribution different to that for the mean of a set? [8]
6. How is it that the mean of a grouped frequency distribution cannot be calculated exactly? [10]
7. How can a calculator's accumulating memory be used to advantage when calculating the mean for a frequency distribution? [13(a)]
8. In what situation would a weighted mean be used? [17]
9. Why is the mean considered to be the 'mathematical average'? [18(b)]
10. What is the main disadvantage of the mean? [18(e)]

## 22. Student exercises

1. Find the arithmetic mean of the following sets:
   a) 84, 92, 73, 67, 88, 74, 91, 74
   b) 0.53, 0.46, 0.50, 0.49, 0.52, 0.53, 0.44, 0.55, 0.54
2. Find the mean of the following frequency distributions:

   a)

| $x$ | 18.5 | 19.5 | 20.5 |
|---|---|---|---|
| $f$ | 5 | 12 | 20 |

   b)

| $x$ | 1 | 2 | 3 | 4 | 5 | 6 |
|---|---|---|---|---|---|---|
| $f$ | 2 | 8 | 24 | 52 | 31 | 11 |

3. *MULTI-CHOICE.* In a random selection of 20 invoices, the following numbers of errors were found:

| Number of errors | 0 | 1 | 2 | 3 | 4 | 5 | 6+ |
|---|---|---|---|---|---|---|---|
| Number of invoices | 6 | 3 | 4 | 4 | 2 | 1 | 0 |

   The expected value of the number of errors per invoice is:
   a) 1.8 b) 2.0 c) 2.1 d) 3.0
4. A firm recorded the number of orders received for each of 58 successive weeks to give the following distribution:

| Number of orders received | 10–14 | 15–19 | 20–24 | 25–29 | 30–34 | 35–39 |
|---|---|---|---|---|---|---|
| Number of weeks | 3 | 7 | 15 | 20 | 9 | 4 |

   Calculate the mean weekly number of orders received.
5. The ages of a company's employees are tabulated below:

| Age in years | 20 and under 25 | 25 and under 30 | 30 and under 35 | 35 and under 40 | 40 and under 45 | 45 and under 50 |
|---|---|---|---|---|---|---|
| Number of employees | 2 | 14 | 29 | 43 | 33 | 9 |

   Calculate the mean employee age in years.
6. A quality control section of a cannery inspected the contents of 130 randomly selected tins of cooked spaghetti from output. As part of their measurements, the following net weights (in grams) were tabulated:

| *Weight (in grams)* | *Number of tins* |
|---|---|
| under 424.9 | 1 |
| 424.900–424.925 | 1 |
| 424.925–424.950 | 6 |
| 424.950–424.975 | 18 |
| 424.975–425.000 | 33 |
| 425.000–425.025 | 46 |
| 425.025–425.050 | 14 |
| 425.050–425.075 | 5 |
| 425.075–425.100 | 5 |
| 425.1 and over | 1 |

Calculate the mean net weight of the contents of the tins and say whether you think that the consumer is getting reasonable value if the label on the tin advertises the contents as 425gms.

7. The following is an extract from a business report.

   '... Over the past 15 months, the number of orders received has averaged 24 per month with the best three months averaging 35. The lowest months saw only 14, 14, 16 and 22 orders respectively...'.

   a) Find the average number of orders that were received in the middle 8 months.
   b) If the target over 16 months is an average of 25, how many orders must be received in month 16 to achieve this?

8. During the 1984-85 session, a college ran 70 different classes of which 44 were 'science', with a mean class size of 15.2, and 26 were 'arts', with a mean class size of 19.2.

   The frequency distribution of class sizes is given below

   No student belonged to more than one class.

   a) Calculate the mean class size of the college.
   b) Suppose now that no class of 12 students or less had been allowed to run. Calculate what the mean class size for the college would have been if the students in such classes:
      i. had been transferred to the other classes;
      ii. had not been admitted to the college.
   c) The number of students enrolling in 1986-87 on science and arts courses is expected to rise by 20% and to fall by 10% respectively, compared with 1984-85. Calculate the maximum number of classes the college should run if the mean class size is to be not less than 20.

   *Frequency distribution of class sizes*

| *Size of class (no of students)* | *Number of science classes* | *Number of art classes* |
|---|---|---|
| 1–6 | 4 | 0 |
| 7–12 | 15 | 3 |
| 13–18 | 11 | 10 |
| 19–24 | 8 | 8 |
| 25–30 | 5 | 4 |
| 31–36 | 1 | 1 |

9. *MULTI-CHOICE*. Which one of the following is a false statement.
   a) The mean is not necessarily one of the original values
   b) The mean is affected by extreme values
   c) In a grouped frequency distribution, the mean can only be estimated.
   d) Sometimes the mean cannot be calculated for a large set of values.

# 7 Median

## 1. Introduction

The *median* is generally considered as an alternative average to the mean. This chapter defines the median and shows how to find its value for a set and for simple and grouped frequency distributions. In the case of a grouped frequency distribution, two equivalent methods are demonstrated, one using a formula, the other a graphical method. Also described are the standard situations where the median is most effectively used.

## 2. Definition of the median

Suppose a machine produces 5, 3, 5, 21 and 2 defective items each day over a five-day period. The mean number of defectives per day would be calculated as:

$$\text{mean} = \frac{5+3+5+21+2}{236} = \frac{36}{5} = 7.2$$

An objection to using 7.2 as an average here is that it is unrepresentative, both of the four lower values (5, 3, 5 and 2) and the largest value (21). As already mentioned in the previous chapter, the mean takes extreme items into account and thus is sometimes not very useful as a practical average. In cases such as these, the median is used. This is found by placing the values in size order and picking the *middle value* as the average. The above five values, in order of size, can be written as:

$$2, 3, \underline{5}, 5, 21$$

and the median (the middle value, underlined) is seen to be 5, which is more useful as a working average.

> **The median**
>
> The median of a set of data is the value of that item which lies exactly half way along the set (arranged into size order).

**Note 1**. When a set of data contains an even number of items, there is no unique middle or central value. The convention in this situation is to use the *mean* of the middle two items to give a (practical) median.

**Note 2**. For a set with an odd number ($n$) of items, the median can be precisely identified as the value of the $\frac{n+1}{2}$th item. Thus in a size-ordered set of 15 items, the median would be the $\frac{15+1}{2}$ th = the 8th item along.

## 3. Example 1 (*Median* of a set of values)

a) The median of 43,75,48,51,51,47,50 is determined by size-ordering the set as: 43,47,48,50,51,51,75 and then: median = middle item = 50.

b) The median of 2,4,6,1,2,3,3,2 is found by size-ordering the set as: 1,2,2,2,3,3,4,6 (noticing that there is an even number of items) which gives:
median = mean of middle two = $\frac{2+3}{2}$ = 2.5

## 4. Median for a simple frequency distribution

Where there is a large number of discrete items in a data set, but the range of values is limited, a simple frequency distribution will probably have been compiled. For example, if records had been kept of the number of vehicles not available for hire on each of 80 consecutive days for a large taxi fleet, the results might appear as follows.

| Number of vehicles unavailable | 0 | 1 | 2 | 3 | 4 | 5 | 6 |
|---|---|---|---|---|---|---|---|
| Number of days | 15 | 24 | 18 | 12 | 8 | 2 | 1 |

It should be realised that, whereas a grouped distribution 'loses' the individual values of the items, there is no real difference between the above (simple frequency distribution) structure and a long listing of all the items. Indeed, the above data is already ordered, so that all that is necessary in order to determine the median is to identify the middle item.

The following section describes the procedure to be followed, using the data above.

## 5. Procedure for calculating the median

To calculate the median for a simple (discrete) frequency distribution, the following procedure should be followed.

STEP 1 Calculate the value of $\frac{\Sigma f+1}{2}$ (identifying the central item)

STEP 2 Form a $F$ (cumulative frequency) column

STEP 3 Find that $F$ value which first exceeds $\frac{\Sigma f+1}{2}$

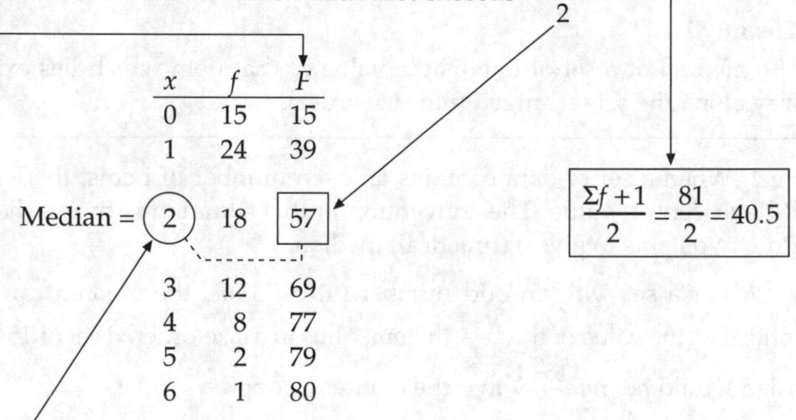


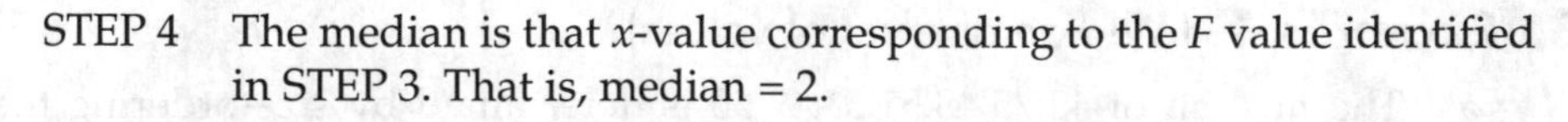

STEP 4 The median is that $x$-value corresponding to the $F$ value identified in STEP 3. That is, median = 2.

**Note**: Sometimes $\sum f$ is replaced by $N$ for convenience.

## 6. Example 2 (The *median* for a *simple discrete frequency distribution*)

*Question*

Calculate the median for the following distribution of delivery times of orders sent out from a firm.

| Delivery time (days) | 0 | 1 | 2 | 3 | 4 | 5 | 6 | 7 | 8 | 9 | 10 | 11 |
|---|---|---|---|---|---|---|---|---|---|---|---|---|
| Number of orders | 4 | 8 | 11 | 12 | 21 | 15 | 10 | 4 | 2 | 2 | 1 | 1 |

*Answer*

STEP 1 The median is the $\frac{N+1}{2}$th

$= \frac{91+1}{2}$th

= 46th item

STEP 2 The $F$ column is shown in the following table:

| Delivery time (days) ($x$) | Number of orders ($f$) | Cum $f$ ($F$) |
|---|---|---|
| 0 | 4 | 4 |
| 1 | 8 | 12 |
| 2 | 11 | 23 |
| 3 | 12 | 35 |
| 4 | 21 | 56 |
| 5 | 15 | 71 |
| 6 | 10 | 81 |
| 7 | 4 | 85 |
| 8 | 2 | 87 |
| 9 | 2 | 89 |
| 10 | 1 | 90 |
| 11 | 1 | 91 |

STEP 3 The first $F$ value to exceed 46 is $F = 56$.

STEP 4 The median is thus 4 (days).

## 7. Median for a grouped frequency distribution

As mentioned in the previous chapter, the penalty paid for grouping values is the loss of their individual identities and thus there is no way that a median can be calculated exactly in this situation. However, there are two methods commonly employed for estimating the median:

a) using an interpolation formula

b) by graphical interpolation.

*Interpolation* in this context is a simple mathematical technique which estimates an unknown value by utilizing immediately surrounding known values.

## 8. Estimating the median by formula

Given a grouped frequency distribution, the best that can be done is to identify the class or group that contains the median item. From there, using cumulative frequencies and the fact that the median must lie exactly one half of the way along the distribution, a formula has been derived that will pinpoint a 'theoretical' value for the median within the class. The following sections describe the procedure involved and gives a full, worked example of its use.

## 9. Procedure for estimating the median by formula

The procedure for estimating the median (by formula) for a grouped frequency distribution is:

STEP 1 Form a cumulative frequency ($F$) column.

STEP 2 Find the value of $\frac{N}{2}$ (where $N = \sum f$).

STEP 3 Find that $F$ value that first exceeds, which identifies the median class $M$.

STEP 4 Calculate the median using the following interpolation formula:

**Median interpolation formula**

$$\text{Median} = L_M + \left( \frac{\frac{N}{2} - F_{M-1}}{f_M} \right) . c_M$$

where:

$L_M$ = lower bound of the median class

$F_{M-1}$ = cumulative frequency of class immediately prior to the median class

$f_M$ = actual frequency of median class

$c_M$ = median class width.

## 10. Example 3 (The *median* for a *grouped frequency distribution* by *formula*)

*Question*

Estimate the median (using the interpolation formula) for the following data, which represents the ages of a set of 130 representatives who took part in a statistical survey.

| Age in years | 20 and under 25 | 25 and under 30 | 30 and under 35 | 35 and under 40 | 40 and under 45 | 45 and under 50 |
|---|---|---|---|---|---|---|
| Number of representatives | 2 | 14 | 29 | 43 | 33 | 9 |

*Answer*

STEP 1

| *Age (years)* | *Number of representatives* (*f*) | (*F*) |
|---|---|---|
| 20 and under 25 | 2 | 2 |
| 25 and under 30 | 14 | 16 |
| 30 and under 35 | 29 | 45 |
| 35 and under 40 | 43 | 88 |
| 40 and under 45 | 33 | 121 |
| 45 and under 50 | 9 | 130 |

STEP 2 $\frac{N}{2} = \frac{130}{2} = 65$

STEP 3 The median class is the class that has the first *F* greater than 65. Here, it is 35 to 40.

STEP 4 The median can now be estimated using the interpolation formula.
$L_M = 35; F_{M-1} = 43; c_M = 5$

$$\text{Thus, median} = L_M + \left(\frac{\frac{N}{2} - F_{M-1}}{f_M}\right).c_M$$

$$= 35 + \left(\frac{65 - 45}{43}\right) \times 5$$

$$= 37.33 \text{ years (2D)}$$

i.e. median = 37.33 years (2D)

## 11. Estimating the median graphically

This particular method can be thought of as the graphical equivalent of the previous interpolation formula (although in fact, as will be explained, there is a slight difference between them). A percentage cumulative frequency curve (or ogive) is drawn and the value of the variable that corresponds to the 50% point (i.e. half way along the distribution) is read off and gives the median estimate. The technique is summarised next, using a fully worked example.

## 12. Procedure for estimating the median graphically

The procedure for estimating the median (graphically) for a grouped frequency distribution is:

STEP 1 Form a cumulative (percentage) frequency distribution.

STEP 2 Draw up a cumulative frequency curve by plotting class *upper bounds* against cumulative percentage frequency and join the points with a smooth curve.

STEP 3 Read off the 50% point to give the median.

## 13. **Example 4** (Estimation of the *median* for a *grouped frequency distribution* using the *graphical method*)

The table on the left below gives the distribution of advertising expenditure for a number of companies in a month.

The table on the right shows the accompanying cumulative (%) frequency distribution and the corresponding cumulative (%) frequency curve is displayed in Figure 1.

The 50% point on the curve gives the median as approximately £1550.

*Figure 1 Cumulative frequency curve construction*

| *Expenditure (£)* | *Number of companies* |
|---|---|
| Less than 500 | 210 |
| 500 and up to 1000 | 184 |
| 1000 and up to 1500 | 232 |
| 1500 and up to 2000 | 348 |
| 2000 and up to 2500 | 177 |
| 2500 and up to 3000 | 83 |
| 3000 and up to 3500 | 48 |
| 3500 and up to 4000 | 12 |
| 4000 and over | 9 |

| *Upper bound* | *F* | *F%* |
|---|---|---|
| 500 | 210 | 16.1 |
| 1000 | 394 | 30.2 |
| 1500 | 626 | 48.0 |
| 2000 | 974 | 74.8 |
| 2500 | 1151 | 88.3 |
| 3000 | 1234 | 94.7 |
| 3500 | 1282 | 98.4 |
| 4000 | 1294 | 99.3 |
| 5000 | 1303 | 100 |

**Cumulative frequency curve of companies' expenditure**

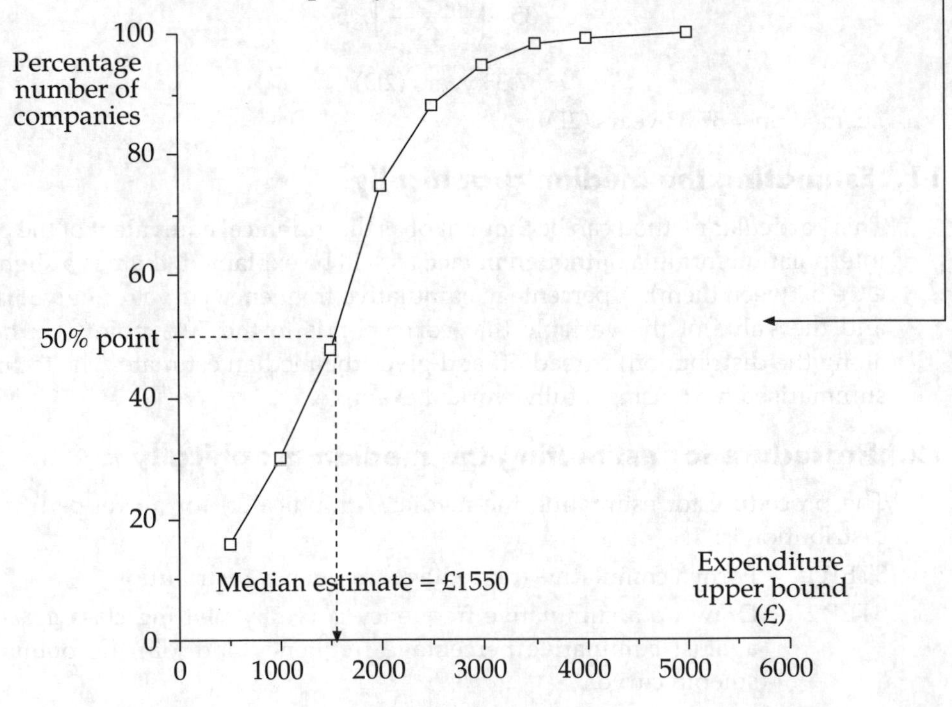

## 14. Comparison of formula and graphical techniques

When using a graphical technique to estimate the median, if a cumulative frequency *polygon* (i.e. straight lines join the points) is used rather than a cumulative frequency curve, the result would be identical to that obtained using the formula. That is, the formula method is the algebraic equivalent of estimating the median using a cumulative frequency polygon. However, since a curve (if it is drawn reasonably well) will fit the points plotted more naturally, it should be clear that the standard graphical technique, using a curve, is to be preferred.

Summary of characteristics.

a) By *formula*.
   i. Uses *linear* interpolation.
   ii. A more mechanical approach.
   iii. Can be used in computing routines.
b) By *smooth curve graph*.
   i. No formula to be used or remembered.
   ii. Uses (more accurate) *non-linear* interpolation, but graph needs to be drawn reasonably accurately.
   iii. Curve can be used for further types of estimates.

## 15. Median for a simple (continuous) frequency distribution

Occasionally, continuous data will be measured to a particular value rather than naturally allocated to true continuous groups. For example, during a work study exercise, the times taken by 46 workers to complete a particular job were measured (to the nearest minute) to give the following:

| Number of minutes | 11 | 12 | 13 | 14 | 15 | 16 | 17 | 18 | 19 |
|---|---|---|---|---|---|---|---|---|---|
| Number of workers | 2 | 6 | 18 | 12 | 5 | 0 | 1 | 1 | 1 |

Notice that, although at first sight the data might appear discrete, it is strictly continuous. In order to calculate the median, the values given for number of minutes must be translated as true continuous groups rather than discrete values. Thus, the first group ranges from 10.5 to 11.5, the second from 11.5 to 12.5 and so on. The median can only be estimated using either the formula or the graphical technique; it *should not* be calculated using the technique described in section 4. The calculation of the median for the above data is left as an exercise.

## 16. Characteristics of the median

a) It is an appropriate alternative to the mean when extreme values are present at one or both ends of a set or distribution.
   For the following set of employee weekly wages (in £): 158, 138, 141, 148, 148, 146, 157 and 252, the mean was previously calculated as £161 which is clearly unrepresentative. The median value, once the set has been sorted into ascending order, is calculated as £148, a more reasonable and understandable representative measure. The extreme value of £252 has effectively been discounted.
b) Can be used when certain end values of a set or distribution are difficult, expensive or impossible to obtain, particularly appropriate to 'life' data. In extreme

cases, the only numeric values that need to be determined are the middle one or two.

For example, 100 light bulbs can be tested for average (median) length of life by waiting only until the 50th 'goes out'. The length of time that this particular bulb has stayed alight is the value of the median. That is, it is not necessary to carry on the test for the rest of the bulbs.

c) Can be used with non-numeric data if desired, providing the measurements can be naturally ordered. (The mean cannot be calculated if the data values are non-numeric.)

For example, the median size of garment can be determined if all garments are measured in such terms as 'extra large', 'large', 'medium', etc.

d) Will often assume a value equal to one of the original items, which is considered as an advantage over the mean.

e) The main disadvantage of the median is that it is difficult to handle theoretically in more advanced statistical work, so its use is restricted to analysis at a basic level.

## 17. Summary

a) The median is an alternative average to the mean and is particularly useful where:
   i. a set or distribution has extreme values present and
   ii. values at the end of a set or distribution are not known.

b) The median is defined as the middle item in a size-ordered set or distribution.

c) For a grouped frequency distribution, the median can be estimated using either:
   i. the linear interpolation formula or
   ii. a cumulative frequency curve.

d) The median can be used when a distribution is skewed or when end values are not known. Its main disadvantage is the difficulty of handling it theoretically and it is not used in further statistical analysis.

## 18. Points to note

a) The median item can be identified as the $\frac{n+1}{2}$ th item along an odd-numbered set or a simple discrete frequency distribution, but when calculating (estimating) its value in a grouped frequency distribution, it is correct to consider it as the $\frac{\Sigma f+1}{2}$ th $= \frac{N}{2}$ th theoretical item.

b) The graphical estimation method is generally considered superior to the formula estimation method as long as a smooth cumulative frequency curve is drawn, due to its (superior) non-linear interpolation effect.

c) When estimating the median graphically, it is necessary to plot cumulative frequency against the mathematical upper bound of the given classes of values. For example, if the classes were described as: 10–19, 20–29, 30–39, 40–49, etc, the correct upper bounds to plot would be: 19.5, 29.5, 39.5, 49.5, etc. This consideration also applies to the use of the interpolation formula, where it is necessary to evaluate $L$, the lower bound of the median class. Thus, if the median class for the above set was 40–49, the correct lower bound, $L$, would be 39.5.

d) Sometimes it is convenient to express life data in the form of a cumulative 'more than' (as opposed to 'less than') distribution. If a cumulative frequency curve is drawn in this situation, the median can still be determined by finding the 50% point.

## 19. Student self review questions

1. How is the median defined? [2]
2. If a set has an even number of items, how can the median be determined? [2]
3. Describe briefly how to estimate the median of a grouped frequency distribution graphically. [11]
4. What is the graphical equivalent of the interpolation formula? [14]
5. On balance, why is the graphical method preferred to the formula method for estimating the median? [14]
6. Name two separate conditions under which the median rather than the mean would be chosen as a measure of location and explain why. [16]
7. What is the main disadvantage of the median? [16]

## 20. Student exercises

1. *MULTI-CHOICE.* Over 11 successive workdays, an employee of a company registered the following number of complete hours worked:

   35, 36, 36, 36, 40, 38, 40, 37, 35, 42, 43

   The median number of hours worked was:

   a) 36 b) 37 c) 38 d) 39

2. Find the median of the following sets of data:

   a) 2.52, 3.96, 3.28, 9.20, 3.75

   b) 84, 91, 72, 68, 87, 78, 78, 78, 82, 79

3. The following figures were obtained by sampling the output of bags of walnuts which were ready to be distributed to a national chain of supermarkets.

   | Number of walnuts | 19 | 20 | 21 | 22 | 23 |
   |---|---|---|---|---|---|
   | Number of bags | 2 | 11 | 29 | 36 | 10 |

   a) Find the median number of walnuts per bag.

   b) Why is the median a suitable average for the above data?

   c) State why the mean would probably be chosen as more suitable if the bags had been measured by weight of walnuts rather than by number.

4. Use the interpolation formula to estimate the median of the following data, which relate to the IQ of a special group of an organisation's employees.

   | IQ | 98–106 | 107–115 | 116–124 | 125–133 | 134–142 | 143–151 | 152–160 |
   |---|---|---|---|---|---|---|---|
   | Number of employees | 3 | 5 | 9 | 12 | 5 | 4 | 2 |

   Have you any criticism of the structure of the frequency distribution?

5. The following figures relate to the length of time spent by cars in a particular car park during one day.

| Time parked (hrs) | Up to 1 | 1–2 | 2–3 | 3–4 | 4–5 | 5–6 | 6–9 | 9–12 |
|---|---|---|---|---|---|---|---|---|
| Number of cars | 450 | 730 | 640 | 120 | 40 | 30 | 20 | 20 |

   Estimate the median parking time (to 2D).

6. The following data show the number out of one hundred electric light bulbs (simultaneously connected to a power source), still working at the end of successive periods of 100 hours.

| Elapsed time (hrs) | 0 | 100 | 200 | 300 | 400 | 500 | 600 | 700 | 800 | 900 | 1000 |
|---|---|---|---|---|---|---|---|---|---|---|---|
| Number working | 100 | 99 | 98 | 90 | 82 | 70 | 45 | 26 | 12 | 3 | 0 |

   By drawing a cumulative (more than) frequency curve, estimate the median life. [Hint: see section 18(d).]

# 8 Mode and other measures of location

## 1. Introduction

Although the mean and median will be the averages used in most circumstances, there are situations in which other averages are particularly appropriate. Whereas the mean can be said to find the *centre of gravity* and the median the *middle* of a set of items, the *mode* identifies the *most popular* item and is described in the following sections for sets and frequency distributions. Two other specialised averages described are the geometric mean and the harmonic mean, used to average percentages and rates respectively.

## 2. The mode

Sometimes a set of data is obtained where it is appropriate to measure a representative (average) value in terms of 'popularity'. For example, if a shop sold television sets, the answer to the question 'what price does the average television set sell at?' is probably best given as the price of the best-selling television. This value is the *mode*. In this type of instance, the mode would be more representative of the data than, for instance, the mean or median.

Other types of data for which the mode is sometimes used as the most appropriate average are shoe or clothes sizes, number of defectives found on production runs or size of company (by number of employees).

**The mode**

The *mode* of a set of data is that value which occurs most often or, equivalently, has the largest frequency.

## 3. Example 1

a) The mode of the set 2,1,3,3,1,1,2,4 is 1, since this value occurs most often.

b) The mode of the following simple discrete frequency distribution

| x | 4 | 5 | 6 | 7 | 8 | 9 | 10 |
|---|---|---|---|---|---|---|---|
| f | 2 | 5 | 21 | 18 | 9 | 2 | 1 |

is 6, since this value has the largest frequency (of 21).

## 4. The mode for grouped data

For a grouped frequency distribution, the mode (in line with the mean and median) cannot be determined exactly and so must be estimated. The technique used is one of *interpolation*, similar to that used to estimate the median of a frequency distribution. There are two methods that can be used to estimate the mode:

a) using an interpolation formula

b) graphically, using a histogram.

The procedure for estimating the mode using an interpolation formula is set out in section 5, while section 7 shows how the mode can be estimated graphically.

## 5. Mode of a grouped frequency distribution by formula

An estimate of the mode for a grouped frequency distribution can be obtained using the following procedure:

STEP 1 Determine the modal class (that class which has the largest frequency).

STEP 2 Calculate $D_1$ = difference between the largest frequency and the frequency *immediately preceding* it.

STEP 3 Calculate $D_2$ = difference between the largest frequency and the frequency *immediately following* it.

STEP 4 Use the following interpolation formula:

**Interpolation formula for the mode**

$$\text{Mode} = L + \left(\frac{D_1}{D_1 + D_2}\right).C$$

where: $L$ = lower bound of modal class

C = modal class width

and: $D_1$, $D_2$ are as described above in STEPS 2 and 3.

Example 2 demonstrates the use of this procedure.

## 6. Example 2 (Estimation of the *mode* of a *frequency distribution* using the *interpolation formula*)

*Question*

Estimate the mode of the following distribution of ages.

| Age (years) | 20–25 | 25–30 | 30–35 | 35–40 | 40–45 | 45–50 |
|---|---|---|---|---|---|---|
| Number of employees | 2 | 14 | 29 | 43 | 33 | 9 |

*Answer*

The table below shows the standard layout of the data, with the steps in the procedure clearly specified.

| | *Age (years)* | *Number of employees* | |
|---|---|---|---|
| STEP 1 | 20 and under 25 | 2 | |
| Modal class | 25 and under 30 | 14 | STEPS 2 and 3 |
| | 30 and under 35 | 29 | $D_1 = 43 - 29 = 14$ |
| → | 35 and under 40 | 43 | $D_2 = 43 - 33 = 10$ |
| | 40 and under 45 | 33 | |
| | 45 and under 50 | 9 | |

STEP 4

The lower class bound of the modal class, $L = 35$

The class width of the modal class, $C = 5$ (from 35 to 40)

Thus: mode $= L + \left(\frac{D_1}{D_1 + D_2}\right).C$

$$= 35 + \left(\frac{14}{14+10}\right).5$$

i.e. mode $= 37.92$ years (2D)

## 7. Graphical estimation of the mode

The graphical equivalent of the above interpolation formula is to construct three histogram bars, representing the class with the highest frequency and the ones on either side of it, and to draw two lines, as shown in Figure 1. The mode estimate is the $x$-value corresponding to the intersection of the lines.

*Figure 1 Technique for estimating the mode graphically*

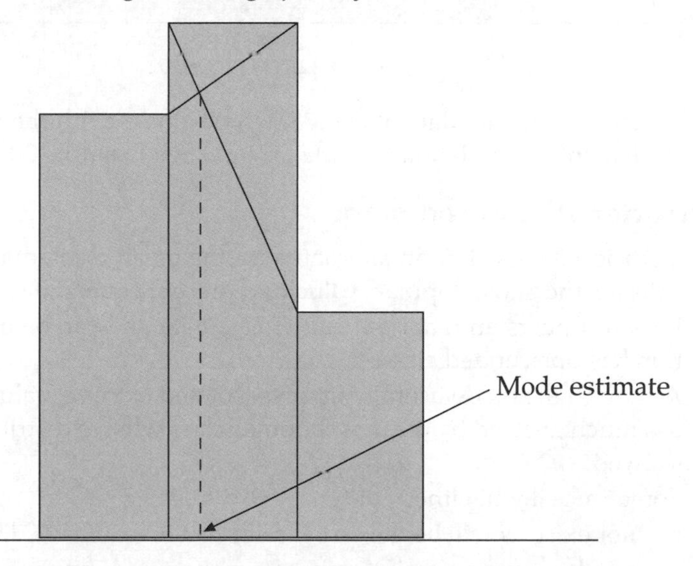


## 8. Example 3 (Estimation of the *mode* of a *frequency distribution* using the *graphical formula*)

The following extract of data demonstrates the use of the above graphical technique to estimate the mode for the data of Example 2. The histogram bars in Figure 2 represent the following three classes and frequencies:

| | |
|---|---|
| ... | |
| 30 and under 35 | 29 |
| 35 and under 40 | 43 |
| 40 and under 45 | 33 |
| ... | |

*Figure 2 Estimating the modal age graphically*

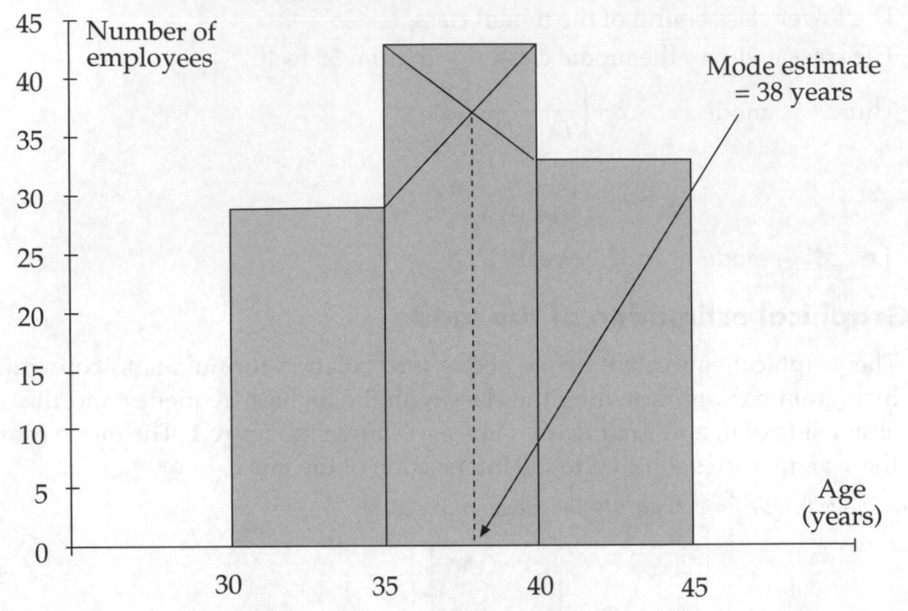


The estimate of the modal value is 38, which agrees with the value obtained using the equivalent interpolation formula (37.92). (See Example 2.)

## 9. Characteristics of the mode

a) Occasionally used as an alternative to the mean or median when the situation calls for the 'most popular' value to represent some data.

b) Easy to understand, not difficult to calculate and can be used when a distribution has open ended classes.

c) Although the mode usefully ignores isolated extreme values, it is thought to be too much affected by the most popular class when a distribution is significantly skewed.

d) Sometimes it will either:
   i. not exist, as will be the case if a set of items all have different values, or
   ii. not be unique, as is the case when two or more values occur equally frequently or a distribution has more than one 'hump' at the same height.

e) Unlike the mean and median, the mode has no 'natural' measure of dispersion to twin with, which is a particular disadvantage in most cases where further analysis is required.

f) Like the median, the mode is not used in advanced statistical work.

## 10. Graphical comparisons of mean, median and mode

Frequency curves of distributions may be relatively symmetric, but more often are skewed to some extent. Typical examples of this are distributions of wages, company turnover or times to component failure and it is of some interest to know the approximate relative positions of the three main averages, the mean, median and mode. Figure 3 shows these positions for moderately left-skewed, symmetric and moderately right-skewed distributions.

*Figure 3 Graphical positions of mean, median and mode*

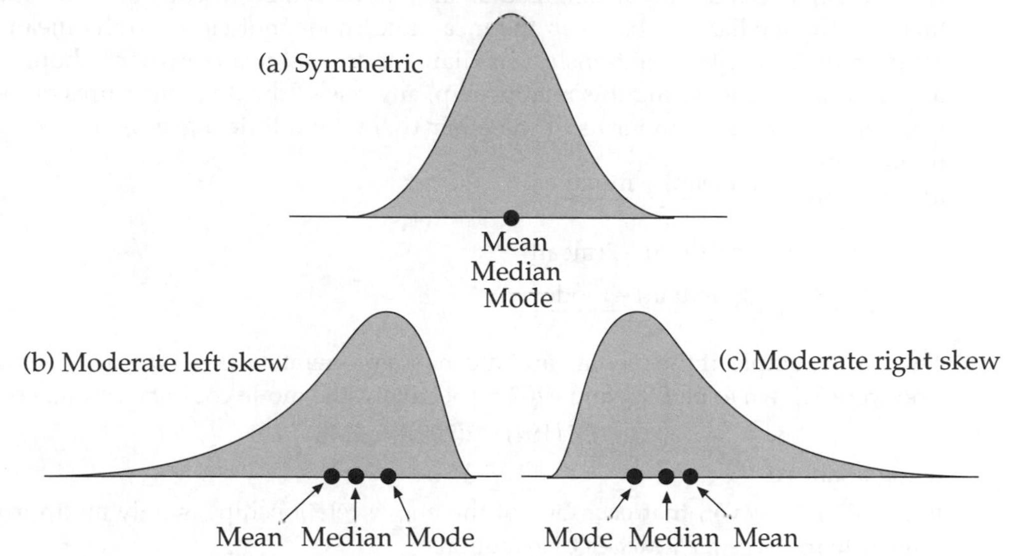


A useful aid in remembering the positions of the three averages in Figure 3 is to use the following characteristics of the three measures:

a) The mode is the item that occurs most frequently and so it must lie under the main 'hump'.

b) The mean is the average that is most affected by extremes and so it must lie towards the 'tail' of the distribution (except, of course, for a symmetric distribution).

c) The median is the middle item and it also lies in the middle of the other two averages (but slightly closer to the mean by a factor of 2 to 1 approximately).

## 11. Example 4 (*Comparison* of mean, median and mode)

Distributions of incomes are usually right-skewed; that is, although there will be a clustering of values in a particular area, there will always be a number of them that lie in the upper regions of the scale with very few (if any) lying in the lower regions. For example, the following data relates to Personal Incomes.

| Total income (£) | 675– | 1000– | 1500– | 2000– | 3000– | 4000– | 6000– | 8000– | 10000+ |
|---|---|---|---|---|---|---|---|---|---|
| Number of persons (000) | 1230 | 2500 | 2660 | 5190 | 4320 | 4470 | 970 | 300 | 330 |

Broadly speaking, the clustering of incomes is from £1000 to £5000 but there is a significant tail from £6000 upwards. This distribution is moderately right-skewed with the mean taking the largest and the mode the smallest value. For this type of distribution, the median would be the most appropriate average.

## 12. Relationships between mean, median and mode

For moderately skewed distributions [see diagrams (a) and (c) in Figure 3], simple relationships between the mean, median and mode can be worked out. Given the fact that the median lies between the mean and mode, but closer to the mean by a factor of 2 to 1, the relationship median–mode = 2(mean–median) should be approximately true. Using this relationship, any one of the three measures of location can be expressed in terms of the other two with a little algebraic re-arrangement. Namely:

a) $\text{median} = \dfrac{2(\text{mean}) + \text{mode}}{3}$

b) $\text{mode} = 3(\text{median}) - 2(\text{mean})$

c) $\text{mean} = \dfrac{3(\text{median}) - \text{mode}}{2}$

Thus, for example, if the median and mean of a moderately skewed wage distribution were known to be £184 and £202 respectively, the mode could be estimated as:

$$£[3(184) - 2(202)] = £148$$

using b) above.

Remember however, that each one of the above relationships is only an approximate rule for *moderately skewed* distributions.

## 13. Proportional increases and multipliers

The idea of a *proportional multiplier* needs to be introduced at this stage.

Consider increasing some value, £200 say, by 20%. The most efficient way to do this is to multiply the value by 120%. That is, resultant value = $£300 \times \dfrac{120}{100} = £300 \times 1.2$ = £360. Note that the 120% comes from (100+20)% or, equivalently, the 1.2 is made up of 1 + 0.2.

Thus, for example:

| | | | | |
|---|---|---|---|---|
| to add | 35% | multiply by | 135% | or 1.35 |
| to add | 50% | multiply by | 150% | or 1.50 |
| to add | 87% | multiply by | 187% | or 1.87 |

In the above, 1.35, 1.50 and 1.87 are called *proportional multipliers*. Using proportions rather than percentages simplifies the arithmetic involved in calculations.

## 14. Formula for the geometric mean

The *geometric mean* is a specialised measure, used to average *proportional increases*.

In calculating the geometric mean, there are three steps to follow.

STEP 1 Express the proportional increases ($p$, say) as proportional multipliers $(1+p)$

For example, suppose a small firm had been growing over a four year period, with its average number of employees per year given as

84, 97, 116 and 129.

The proportional increases from each year to the next are:

$$\frac{97-84}{84} = 0.155; \quad \frac{116-97}{97} = 0.196; \quad \frac{129-126}{126} = 0.112;$$

The proportional multipliers are therefore: 1.155, 1.196 and 1.112

STEP 2 Calculate the geometric mean multiplier using:

$$\text{Geometric mean multiplier} = \sqrt[n]{(1+p_1)(1+p_2)...(1+p_n)}$$

For the previous example we have:

$$\text{gm multiplier} = \sqrt[3]{(1.155)(1.196)(1.112)}$$

$$= \sqrt[3]{1.5361}$$

$$= 1.154 \text{ (3D)}$$

STEP 3 Subtract 1 from the gm multiplier to obtain the average proportional increase.

For the above example, average proportional increase = 1.154 – 1
= 0.154
= 15.4%

That is, the (geometric) mean rise in employees per year is 15.4%

Note that (using the original data) a 15.4% increase applied 3 times successively to 84 will give a result of 129. That is, $84 \times 1.154^3 = 129$.

The geometric mean can be used to average proportional increases in wages or goods, such as percentages or index numbers. Because of the way it is defined, it takes little account of extremes and is occasionally used as an alternative to the arithmetic mean.

The Financial Times (FT) Index is the most well known example of the practical use of the geometric mean. It is calculated as the geometric mean of a set of selected share values.

## 15. Example 5 (Calculation of the *geometric mean*)

If it is known that the price of a commodity has risen by 6%, 13%, 11% and 15% in each of four successive years, then the geometric mean rise can be calculated as follows.

STEP 1 The four proportional multipliers are 1.06, 1.13, 1.11 and 1.15

STEP 2 The geometric mean is given by

$$\text{gm} = \sqrt[4]{1.06 \times 1.13 \times 1.11 \times 1.15}$$

$$= \sqrt[4]{1.5290}$$

$$= 1.112 \text{ (3D)}$$

STEP 3 The (geometric) average rise = 1.112 – 1 = 0.112
= 11.2%

This value of 11.2% can be translated as the constant increase necessary each year to produce the final year price, given the starting year price.

## 16. Note on calculation of the geometric mean

There are several ways of calculating the geometric mean; for example, with a calculator, using logarithms or with a command line in the computer language BASIC (on a micro computer).

Using a *calculator*. Unless there are only two items, the calculator must have a special '$x^y$' key. If it has, the program for the above example is:

$1.06 \times 1.13 \times 1.11 \times 1.15 = x^y\ 0.25 = $ (Ans 1.112)

(Mathematical note: $\sqrt[4]{x} = x^{\frac{1}{4}} = x^{0.25}$)

## 17. The harmonic mean

The *harmonic mean* is another specialized measure of location used only in particular circumstances; namely when the data consists of a *set of rates*, such as prices (£/kilo), speeds (mph) or productivity (output/manhour). It is defined as the reciprocal of the mean of the reciprocals of the item values. Symbolically, the harmonic mean (hm) of $x_1, x_2, \ldots, x_n$, is given by:

**Harmonic Mean**

$$\text{hm} = \frac{n}{\sum \frac{1}{x}}$$

As a simple example, the harmonic mean of 2, 4 and 6 is given by:

$$\text{hm} = \frac{3}{\frac{1}{2}+\frac{1}{4}+\frac{1}{6}} = \frac{3}{0.5 + 0.25 + 0.17}$$

$$= 3.27 \text{ (2D)}$$

Notice that the harmonic mean in this example is *less* than the arithmetic mean (which is easily calculated as 4).

## 18. Use of harmonic mean compared with arithmetic mean

Care must be taken when considering averages of rates. Depending on the basis of measurement, rates can be averaged using either the harmonic or the arithmetic mean. Remembering that a rate is always expressed in terms of the ratio of two units (e.g. price/kg or hours worked per man or miles/hr), the criterion for choosing which average is appropriate can be stated as follows.

a) if the rates are being averaged over *constant numerator units*, the *harmonic mean* should be used,

BUT

b) if the rates are being averaged over *constant denominator units*, the *arithmetic mean* should be used.

Suppose we wished to average the productivity (in items/day) of two production lines which are both producing the same items. Then, if the productivity for each line was measured over, say, one week (i.e. the *same time*, thus making the denominator units constant for both lines), the arithmetic mean would be used to average the two rates. However, if the productivity for each line was measured over, say, the production of 5000 items (i.e. the *same quantity*, thus making the numerator units constant for both lines), the harmonic mean would be used to average the two rates.

Example 6 shows the calculation of the harmonic and arithmetic means in a practical situation, demonstrating the difference between cases a) and b) above.

## 19. Example 6 (*Comparison* of the *harmonic* and *arithmetic means*)

A firm has two types of lorry in its fleet.

a) If two lorries, one of each type, were tested over a distance of 200 miles to yield the two rates 14 and 18 mpg, then the numerator unit (miles or distance) is a uniform amount for both and so the harmonic mean is the appropriate average to use.

Here, average consumption = hm = $\frac{2}{\frac{1}{14}+\frac{1}{18}}$

= 15.75 mpg

b) If now the two lorries are each filled with 10 gallons of fuel and tested until the fuel runs out to yield the two rates 14 and 18 mpg, then the denominator unit (gallons or fuel units) is now a uniform amount for both and so the arithmetic mean is the appropriate average to use.

Here, average consumption = am = $\frac{14+18}{2}$

= 16 mpg

The interpretation of the difference between the two averages is that the arithmetic mean of 16 mpg is based on consumption using 10 gallons of fuel, while the harmonic mean of 15.75 mpg is based on consumption over 200 miles.

## 20. Characteristics of the geometric and harmonic means

a) For *all sets of data*, the following relationship is always true:

| Arithmetic mean > geometric mean > harmonic mean. |
|---|

As a demonstration of this relationship, using the rates in the above example:

am = 16 mpg, hm = 15.75 mpg and the gm = $1-\sqrt{1.14\times1.18} = 1-1.1598 = 0.1598$ = 15.98 mpg.

That is, the hm is the smallest and the am the largest.

b) Both of these means must be seen as special measures, used only in particular circumstances and not to be compared or contrasted with the other three standard averages (the mean, median and mode) except for the special case noted in section 19 above.

c) Both means have the advantage of taking little account of extreme values compared with, for instance, the arithmetic mean which is affected (often severely) by extremes.

For example, suppose the highest of one of five shares doubled in price overnight. Table 1 shows the effects that this change would have on both the geometric and arithmetic means.

*Table 1 Comparison of arithmetic and geometric means*

| | *Share prices* | | | | | | | *gm* | *am* |
|---|---|---|---|---|---|---|---|---|---|
| Yesterday | 142 | 181 | 270 | 180 | 95 | 164 | 174 | 164 | 174 |
| Today | 142 | 181 | 540 | 180 | 95 | 188 | 228 | 188 | 228 |

It is clear that the arithmetic mean is severely affected by such extremes, whereas the geometric mean can cope with the extremes adequately.

## 21. Summary

a) The mode is defined as that value of a set which occurs most often.

b) The mode is found by inspection for a set of values, and for a frequency distribution can be estimated using either:

i. the linear interpolation formula or

ii. a histogram.

c) The relative positions of the mean, median and mode are fairly well defined in symmetric and moderately skewed distributions. Given any two, it is possible to estimate the third.

d) The geometric mean is a specialized measure of location, generally used to average proportional increases and is defined as the $n$-th root of the product of $n$ values.

e) The harmonic mean is another specialized measure of location, generally used to average rates or ratios and defined as the reciprocal of the mean of the reciprocals of the given values.

## 22. Points to note

a) The mode may not be uniquely defined (unlike the mean and median) as it is possible to obtain two or more modes for a given set. For example, the set 1,3,4,4,2,6,3 has *two modes*, 3 and 4.

b) The modal value of a set may not exist, as is the case with the set 1,3,4,5,7,10, which has no one number that occurs more frequently than another.

c) The relationships between the mean, median and mode, described in sections 10, 11 and 12, are only approximate but will NOT hold for excessively skewed data.

## 23. Student self review questions

1. How is the mode defined? [2]
2. Why is the mode not used extensively in statistical analysis? [9]

3. Under what conditions may any one of the mean, median or mode be estimated, given the values of the other two? [12]
4. Write down the definition of the geometric mean and the type of values that it can be used to average. [14]
5. Write down the definition of the harmonic mean and the type of values that it can be used to average. [17]

## 24. Student exercises

1. Determine the value of the mode for the following sets of data:
   a) 10,11,10,12,11,10,11,11,11,12,13,11,12
   b) 2,1,1,2,3,2,3,3,4,6,4,1,2,3
   c)

| x | 14 | 15 | 16 | 17 | 18 | 19 | 20 |
|---|---|---|---|---|---|---|---|
| f | 14 | 26 | 18 | 9 | 2 | 1 | 1 |

2. Calculate a modal value for the following data of age at commitment of crime of 500 male criminals.

| Age (years) | Under16 | 16–17 | 18 | 19–20 | 21–27 | 28–36 |
|---|---|---|---|---|---|---|
| Number of men | 8 | 70 | 95 | 133 | 161 | 33 |

   Comment on the suitability of the mode to represent this data.
3. State the mode of the following distribution and comment on its use in this situation.

| Number of children in family | 0 | 1 | 2 | 3 | 4 | 5 | 6 or more |
|---|---|---|---|---|---|---|---|
| Number of families | 11 | 47 | 28 | 9 | 4 | 1 | 1 |

4. *MULTI-CHOICE*. For a right-skewed distribution, which one of the following statements is false.
   a) The mode is less than or equal to the mean.
   b) The median is less than or equal to the mean.
   c) The mean is less than or equal to the median.
   d) The mode is less than or equal to the median.
5. The following data relates to Personal Incomes.

| Total income (£) | 675– | 1000– | 1500– | 2000– | 3000– | 4000– | 6000– | 8000– | 10000+ |
|---|---|---|---|---|---|---|---|---|---|
| No. of persons (000) | 1230 | 2500 | 2660 | 5190 | 4320 | 3470 | 970 | 300 | 130 |

   Calculate estimates of the value of:
   a) the mean
   b) the median, using the appropriate interpolation formula and
   c) the mode, using the appropriate interpolation formula.

   Calculate the value of:
   d) 2(mean–mode)
   e) 3(median–mode)

   and give the reason why these two values should be approximately equal for this distribution.
6. Calculate the arithmetic mean, geometric mean and harmonic mean of the three numbers 2, 3 and 4.

7. Each of four large supermarkets, A, B, C and D, underwent a time and motion study to investigate cash-point service and movement of customers. The measurements began at the same time in each supermarket and ended in each after exactly 100 customer visits. As part of the results, the flow through A, B, C and D was measured as 80, 68, 81 and 74 customers per hour respectively. Find the average number of customers per hour per supermarket while the study measurements were being carried out.
8. The price of a particular model of car was £4760, £4983, £5104, £5421 and £5500, over five successive years. Use the geometric mean to calculate the average yearly percentage increase in price.
9. A group of workers have received 5.8%, 8.5% and 3.2% wage increases over the last three years. What percentage do they need this year in order to average 6% over the whole period?
10. In a car factory there are three lines working a scheduled seven hour day, each producing different models. Line A produced 44 cars per day, line B 68 per day and line C produced 9 cars per hour. Find the average daily rate of production:
    a) if each line worked steadily for 2 hours
    b) while each line produced 10 cars.
11. The cost of sand, small pebbles and cement is £12, £24 and £60 per yard respectively. A mix for heavy duty paths consists of equal parts of sand, small pebbles and cement. Another mix is used as mortar for bricklaying and consists of 5 parts of sand to 1 of cement. Find the average price per yard of the relevant constituent components for:
    a) a mortar mix
    b) a heavy duty path mix.

# 9 Measures of dispersion and skewness

## 1. Introduction

This chapter introduces dispersion and skewness in general terms and introduces the range and mean deviation as particular measures of dispersion. The two most important measures of dispersion, the standard deviation and quartile deviation, and related measures of skewness are covered separately in chapters 11 and 12.

## 2. Comparison of location, dispersion and skewness

*Dispersion* is the statistical name for the spread or variability of data, while *skewness* describes how non-symmetric (or 'lopsided') the data is.

a) Distributions 1 and 2 (in Figure 1), although spread out in the same way, are located differently. The two distributions might describe the daily rates of semi-skilled(1) and skilled (2) workers employed on a large contract.

*Figure 1 Distributions (of wages) with equal dispersion but different location*

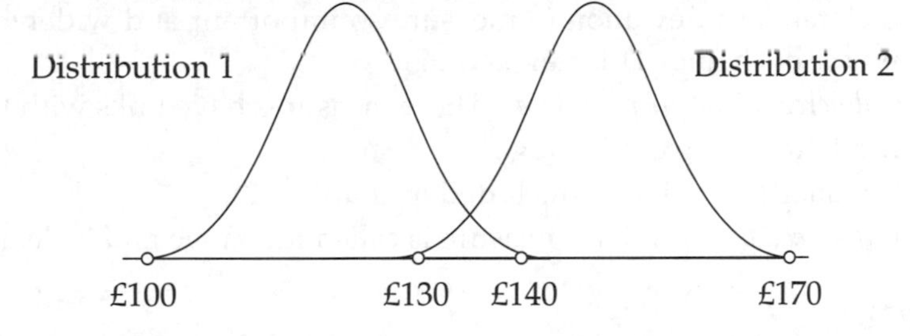


*Figure 2 Distributions (of sales) with equal location but different dispersion*

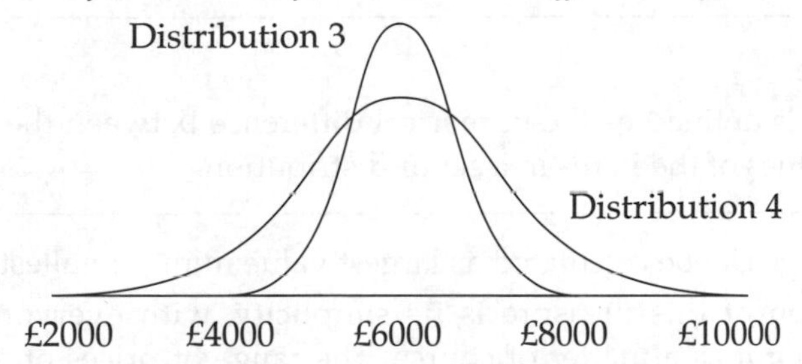


b) Distributions 3 and 4 (in Figure 2) are located in the same place but dispersed differently. These might be the values of monthly sales for two companies, where company 3 can be described as being more consistent (or less variable) than company 4.

c) Distributions 5 and 6 (in Figure 3) have equal dispersion but are skewed (and also located) differently. These might be the distributions of the ages of employees in two different companies

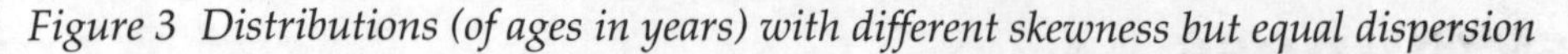

*Figure 3 Distributions (of ages in years) with different skewness but equal dispersion*

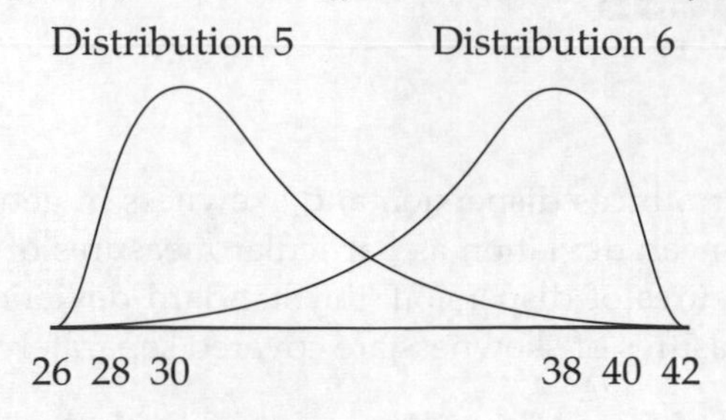


## 3. Measures of dispersion

Measures of dispersion describe how spread out or scattered a set or distribution of numeric data is. There are different bases on which the spread of data can be measured.

a) *Spread about the mean*. This is concerned with measuring the distance between the items and their common mean. There are two measures of this type used:
   i. the mean deviation
   ii. the standard deviation (a measure so important and widely used that the whole of chapter 11 is devoted to it).

b) *Central percentage spread of items*. These measures have links with the median.
   i. the 10 to 90 percentile range
   ii. the quartile deviation (included in chapter 12).

c) *Overall spread of items*. This measure is called the range and is dealt with next.

## 4. The range

The *range* is the simplest measure of dispersion available in statistical analysis.

> **The range**
>
> The *range* is defined as the numerical difference between the smallest and largest values of the items in a set or distribution.

Thus the range can be calculated as largest value minus smallest value.

The attraction of this measure is its simplicity, with everyone (at one time or another) using it as a matter of course. The range of prices of different models of the same car; the range of times of delivery of different items ordered; the range of manpower available on a production line at different times. All these examples provide relevant information on the dispersion of a set of values, albeit of a basic kind.

## 5. Example 1 (Determination of the *range*)

a) The daily number of rejected items detected from the separate output of two industrial machines over fourteen days were:

| | |
|---|---|
| Machine 1: | 4, 7, 1, 2, 2, 6, 2, 3, 0, 4, 5, 3, 7, 4 |
| Machine 2: | 3, 2, 2, 3, 3, 2, 4, 1, 1, 3, 2, 4, 2, 2 |

The range of values for machine 1 is 7 – 0 = 7 and for machine 2 is 4 – 1 = 3. Thus the daily production of rejects is more variable for machine 1.

b) For the two sets of data in Figure 2, distribution 3 has a range of £8000–£4000=£4000 and distribution 4 has range £10000–£2000=£8000.

c) The two distributions in Figure 3 have equal ranges of 14.

## 6. Characteristics of the range

a) The range is a simple concept and easy to calculate.

b) The major disadvantage of the range is the fact that it only takes two values into account (the smallest and largest) and is thus only too obviously affected by extreme values.

c) The range has no natural partner in a measure of location and is not used in further advanced statistical work.

d) One of its common practical uses is for quality control purposes. Small samples of output are taken at regular intervals and the sample mean and range are calculated and recorded on separate charts. The chart for the ranges of samples enables a check to be kept on the variability of production in a quick and easy way.

## 7. The mean deviation

**The mean deviation**

The *mean deviation* is a measure of dispersion that gives the average absolute difference (i.e. ignoring 'minus' signs) between each item and the mean.

The mean deviation is a much more representative measure than the range since all item values are taken into account in its calculation.

As an example, suppose an assembly line produced 3, 10, 5 and 2 defective products on four successive runs. The mean number of defectives = $\frac{3+10+5+2}{4} = \frac{20}{4} = 5$.

The absolute differences between each value and the mean (5) are respectively:

3 – 5 = –2 (or 2, ignoring the 'minus' sign)

10 – 5 = 5

5 – 5 = 0

2 – 5 = –3 (or 3, ignoring the 'minus' sign).

The mean deviation can now be calculated as the average of the above absolute differences. That is:

$$\text{mean deviation} = \frac{2+5+0+3}{4} = \frac{10}{4} = 2.5$$

Depending on whether the data consists of a set of items or a complete distribution, different formulae need to be used in the calculation of this measure. These are shown in the following section.

## 8. Formulae for the calculation of the mean deviation

The *mean deviation* (md) is calculated through the use of the following formulae.

**Mean deviation**

For a set: $md = \frac{\sum |x - \bar{x}|}{n}$

For a frequency distribution: $md = \frac{\sum f|x - \bar{x}|}{\sum f}$

**Note**: *The modulus symbol* | ..... | means 'the absolute value of' and simply ignores the sign of the expression inside it.

For example: $|-6| = |6| = 6$

and: $|2 - 5| = |-3| = 3$

## 9. Example 2 (The *mean deviation* for a *set*)

*Question*

Calculate the mean deviation of 43,75,48,39,51,47,50,47.

*Answer*

First determine the mean as: $\frac{400}{8} = 50$, and then:

$$md = \frac{\sum |x - \bar{x}|}{n}$$

$$= \frac{|43-50| + |75-50| + |48-50| + |39-50| + |51-50| + |47-50| + |50-50| + |47-50|}{8}$$

$$= \frac{7+25+2+11+1+3+0+3}{8}$$

$$= 6.5$$

In other words, each value in the set is, on average, 6.5 units away from the common mean.

## 10. Example 3 (*Mean deviation* for a *frequency distribution*)

*Question*

The data in Table 1 following relates to the number of successful sales made by the salesmen employed by a large microcomputer firm in a particular quarter. Calculate the mean and the mean deviation of the number of sales.

*Answer*

The standard layout and calculations are shown in Table 2. The mean is calculated first, then used to find the mean deviation.

*Table 1 Number of sales made by salesmen*

| Number of sales | 0–4 | 5–9 | 10–14 | 15–19 | 20–24 | 25–29 |
|---|---|---|---|---|---|---|
| Number of salesmen | 1 | 14 | 23 | 21 | 15 | 6 |

*Table 2 Layout of calculations*

| *Number of sales* | *Number of salesmen* ($f$) | *Mid-point* ($x$) | ($fx$) | $\lvert x-\bar{x}\rvert$ | $f\lvert x-\bar{x}\rvert$ |
|---|---|---|---|---|---|
| 0 to 4 | 1 | 2 | 2 | 13.3 | 13.3 |
| 5 to 9 | 14 | 7 | 98 | 8.3 | 116.2 |
| 10 to 14 | 23 | 12 | 276 | 3.3 | 75.9 |
| 15 to 19 | 21 | 17 | 357 | 1.7 | 35.7 |
| 20 to 24 | 15 | 22 | 330 | 6.7 | 100.5 |
| 25 to 29 | 6 | 27 | 162 | 11.7 | 70.2 |
| Totals | 80 | | 1225 | | 411.8 |

$$\text{Mean number of sales, } \bar{x} = \frac{1225}{80} = 15.3 \text{ (1D)}$$

$$\text{For the '0–4' group: } |x-\bar{x}| = |2-15.3| = |-13.3| = 13.3$$

$$\text{For the '5–9' group: } |x-\bar{x}| = |7-15.3| = |-8.3| = 8.3$$

... and so on.

$$\text{Thus, \textit{mean deviation}, md} = \frac{\sum f|x-\bar{x}|}{\sum f} = \frac{411.8}{80} = 5.1 \text{ sales}$$

## 11. Characteristics of the mean deviation

a) The mean deviation can be regarded as a good representative measure of dispersion that is not difficult to understand. It is useful for comparing the variability between distributions of like nature.

b) Its practical disadvantage is that it can be complicated and awkward to calculate if the mean is anything other than a whole number.

c) Because of the modulus sign, the mean deviation is virtually impossible to handle theoretically and thus is not used in more advanced analysis.

## 12. Other measures of dispersion

As mentioned earlier, the two most important measures of dispersion, the *standard deviation* and *quartile deviation*, are dealt with separately in the following two chapters.

## 13. Skewness

Remember that *skewness* is concerned with how non-symmetric or 'lop-sided' a distribution is. The direction of skew is determined by the position of the 'long tail' of the distribution; thus if the long tail is to the left, then the distribution is said to be left or negatively skewed. Wage (or salary) distributions normally have right skew, due to the fact that there will inevitably be a small proportion of the wages (or salaries) that are unrepresentatively high.

Chapter 8, section 10 described an approximate relationship between the mean, median and mode for moderately skewed distributions and it is this relationship that enables a simple measure of skewness to be derived. If the distribution is skewed to the right, then the mean is greater in value than the mode (and vice versa). Thus the numerical difference between the values of the mean and mode will be an indicator of the degree of skewness.

Generally, measures of skewness are given in terms of measures of location and dispersion, the two most commonly used involving the two measures of dispersion dealt with in the following chapters. Thus their precise forms are left until then.

## 14. Summary

a) Measures of dispersion can be linked to the mean, median or the spread of all items.
b) The range is the simplest measure of dispersion and is defined as the numerical difference between the smallest and largest items.
c) The range is used to measure the dispersion of small samples in industrial quality control, but is not used extensively in situations where more advanced statistical analysis is required.
d) The mean deviation is a measure of dispersion that measures the average distance between each item and the mean of a set or distribution.
e) Skewness describes the extent of non-symmetry of a distribution.

## 15. Point to note

Sometimes the mean deviation is called the mean deviation *from the mean*, since it is deviations from the mean that are averaged. It is possible for example to calculate the mean deviation *from the median* (or any other value) if required.

## 16. Student self review questions

1. Why are measures of location alone not enough to characterise data in a meaningful way? [2]
2. On what bases can dispersion be measured? [3]
3. What is the significant use of the range in an industrial situation? [6]
4. Why is the range not used extensively in statistical analysis? [6]

5. How does the mean deviation measure dispersion? [7]
6. What is the major disadvantage of the mean deviation? [11]
7. What is skewness? [13]

## 17. Student exercises

1. Find the range and the mean deviation of the following sets of values:
   84, 92, 73, 67, 88, 74, 91, 74.
2. Weekly withdrawals (to the nearest £) from petty cash for a six-week period before and after a new claim scheme was introduced were recorded as follows:

   Before new scheme: 145, 124, 84, 78, 124, 138

   After new scheme: 63, 128, 85, 57, 136, 122

   Show, using the mean and mean deviation, that although the new scheme has cut down the weekly claims, the variation in the claims has increased.
3. For the distribution:

| x | 1 | 2 | 3 | 4 | 5 | 6 |
|---|---|---|---|---|---|---|
| f | 2 | 8 | 24 | 52 | 31 | 11 |

   calculate:
   a) the mean
   b) the mean deviation from the mean
   c) the median
   d) the mean deviation from the median.
4. A firm recorded the number of orders received for each of 58 successive weeks to give the following distribution:

| Number of orders received | 10–14 | 15–19 | 20–24 | 25–29 | 30–34 | 35–39 |
|---|---|---|---|---|---|---|
| Number of weeks | 3 | 7 | 15 | 20 | 9 | 4 |

   Calculate the mean deviation from the mean, given that the mean number of orders is 25.2 (1D)

# 10 Standard deviation

## 1. Introduction

Recall from the previous chapter (section 11) that the mean deviation cannot be used extensively because of the awkward modulus sign. Instead, we use an adaptation of it to form the most commonly used measure of dispersion, the standard deviation. It is the ideal partner for the mean, as will be demonstrated in this chapter.

The standard deviation is first described for a set, then special computational formulae are given for both a set and a frequency distribution. Finally, two special measures based on the standard deviation are given; the first, a relative measure of dispersion called the coefficient of variation, the second, Pearson's measure of skewness.

## 2. Calculating the standard deviation

A procedure for calculating the standard deviation is now described and, at the same time, demonstrated using the set of values, 2, 4, 6 and 8.

STEP 1 Calculate the mean.

$$\bar{x} = \frac{2+4+6+8}{4} = 5$$

STEP 2 Find the sum of the squares of deviations of items from the mean.

$$\begin{aligned}(2-5)^2 + (4-5)^2 + (6-5)^2 + (8-5)^2 &= (-3)^2 + (-1)^2 + (1)^2 + (3)^2 \\ &= 9 + 1 + 1 + 9 \\ &= 20\end{aligned}$$

STEP 3 Divide this sum by the number of items and take the square root.

$$\sqrt{\frac{20}{4}} = \sqrt{5} = 2.24 \text{ (2D)}$$

This value so obtained, 2.24, is the standard deviation.

This procedure can be summarised in general terms as follows.

> **Standard deviation for a set of values**
>
> $$s = \sqrt{\frac{\sum (x-\bar{x})^2}{n}}$$

In words, the standard deviation can be defined as 'the root of the mean of the squares of deviations from the common mean' of a set of values.

**Note**: If the mean is not a whole number, the calculations could involve some awkward, decimal-bound work. An example of this follows.

## 3. Example 1 (*Standard deviation* for a *set*)

*Question*

Find the standard deviation of 6, 11, 14, 10, 8, 11 and 9.

*Answer*

The tabular layout shown below in Table 1 is used so as to be able to carry out the required calculations in a systematic manner.

The mean is calculated first and then used to obtain the values in the second two columns.

$$\bar{x} = \frac{69}{7} = 9.857 \text{ (3D)}; \ \sum (x-\bar{x})^2 = 38.854 \text{ (3D)}$$

$$s = \sqrt{\frac{\sum (x-\bar{x})^2}{n}} = \sqrt{\frac{38.854}{7}} = 2.360 \text{ (3D)}$$

*Table 1 Layout of calculations*

| $x$ | $x-\bar{x}$ | $(x-\bar{x})^2$ |
|---|---|---|
| 6 | −3.857 | 14.876 |
| 11 | 1.143 | 1.306 |
| 14 | 4.143 | 17.164 |
| 10 | 0.143 | 0.020 |
| 8 | −1.857 | 3.448 |
| 11 | 1.143 | 1.306 |
| 9 | −0.857 | 0.734 |
| 69 | | 38.854 (3D) |

## 4. The link between the standard deviation and the mean

At this point it is worth putting the standard deviation in its correct context.

Given a set of data, there would be little point in calculating a measure of dispersion as sophisticated as the standard deviation without reference to its natural partner, the mean. This is because the mean needs to be calculated anyway in order to obtain the value of the standard deviation, as can be seen quite clearly from Table 1 and the accompanying calculations.

Hence, from now on, the mean and standard deviation will be calculated and referred to as a linked pair.

## 5. Adaptation of the standard deviation formula

As mentioned in section 2 earlier, particular problems can occur when calculating the standard deviation if the mean is not a whole number. These are due to the fact that each item must have the mean (with its usual 'awkward' value) subtracted from it. In order to overcome these problems, the formula given in section 2 can be re-arranged into a more suitable form. The adapted (computational) formula avoids having to subtract the mean from each item.

It is described in section 6 and is followed with an example of its use.

## 6. Computational formula for the standard deviation of a set of values

The use of the computational formula is now shown, working with the same set of values (2, 4, 6 and 8) as in section 2.

STEP 1 Sum the squares of all values.

$2^2 + 4^2 + 6^2 + 8^2 = 120$

STEP 2 Divide this sum by the number of values.

$\frac{120}{4} = 30$

STEP 3 Subtract the square of the mean.

(The mean is $\frac{2+4+6+8}{4} = 5$, the square being 25.)

We now have: $30 - 25 = 5$.

STEP 4 Take the square root, giving the standard deviation.

Thus the standard deviation is: $\sqrt{5} = 2.24$ (2D)

Notice that the same value has been obtained here as in section 2.

The generalized formula for this procedure is now given.

**Computational formula for the standard deviation of a set**

$$s = \sqrt{\frac{\sum x^2}{n} - \overline{x}^2}$$

**Note**: The formula given will always yield the same value for the standard deviation as the formula in section 2.

However, the previous computational formula is generally preferred since it involves less awkward arithmetic.

## 7. Example 2 (*Standard deviation* for a *set* using the *computational formula*)

*Question*

Calculate the mean and standard deviation of the values

43,75,48,51,51,47,50,47,40,48

*Answer*

The standard layout is shown on the opposite page and the calculations are as follows:

$n = 10$; $\sum x = 500$; $\sum x^2 = 25802$. $\overline{x} = \frac{\sum x}{n} = \frac{500}{10} = 50$

$$s = \sqrt{\frac{\sum x^2}{n} - \overline{x}^2}$$

$$= \sqrt{\frac{25802}{10} - 50^2}$$

$$= 8.96 \text{ (2D)}$$

| $x$ | $x^2$ |
|---|---|
| 43 | 1849 |
| 75 | 5625 |
| 48 | 2304 |
| 51 | 2601 |
| 51 | 2601 |
| 47 | 2209 |
| 50 | 2500 |
| 47 | 2209 |
| 40 | 1600 |
| 48 | 2304 |
| 500 | 25802 |

## 8. Notes on example 2

a) *Calculating* $\sum x^2$ Use both the automatic squaring and 'accumulating memory' facility on your calculator to generate these values. This will ensure the least effort for the information needed. The procedure on the calculator is as follows:

| | |
|---|---|
| 43 × M+ | Squaring 43 and adding it to memory<br>(M+ is the 'add to memory' key)<br>Now write down the $x^2$ value displayed |
| 75 × M+ | Write down the $x^2$ value |
| 48 × M+ | Write down the $x^2$ value |
| ... | |
| ... | and so on, down to: |
| 48 × M+ | Write down the $x^2$ value |

Finally, press the 'memory recall' key to display $\sum x^2$

b) *Calculating the standard deviation*. The expression $\frac{25802}{10} - 50^2$ (in example 2) can be evaluated on any standard calculator, without needing to write down any intermediate results. The procedure to follow is always the same. Evaluate the mean, square it and transfer to memory. Then evaluate the left hand side and subtract the contents of memory, finally taking the square root.

In this case, the steps on the calculator are:

| | |
|---|---|
| 50 × | Squaring 50 (With some calculators it is necessary to key 50 × ×) |
| M + | Adding (2500) to memory |
| 25802÷10 | Evaluating left hand side |
| – RM | Subtract the contents of memory (RM is the 'recall memory' key) |
| = | Perform the subtraction, giving 80.2 (DO NOT forget this step!) |
| SQR | Square root, to give the result. |

## 9. Standard deviation for a frequency distribution

For large sets of data, a frequency distribution is normally compiled and the computational formula for the standard deviation for a set needs to be duly adapted. The adapted formula is given in section 10 and is followed by a worked example.

## 10. Formula for the standard deviation of a frequency distribution

The following standard deviation formula has been adapted from the formula for a set and can be used for both simple discrete and grouped distributions.

**Computational formula for the standard deviation of a frequency distribution**

$$\sqrt{\frac{\sum fx^2}{\sum f} - \left(\frac{\sum fx}{\sum f}\right)^2}$$

**Note**: For a *grouped* frequency distribution, $x$ is the class mid-point.

## 11. Example 3 (The *standard deviation* for a *frequency distribution*)

*Question*

The data in Table 2 below relates to the number of successful sales made by the salesmen employed by a large microcomputer firm in a particular quarter. Calculate the mean and standard deviation of the number of sales.

*Answer*

The standard layout and calculations are shown in Table 3 and the subsequent text.

*Table 2 Successful sales made*

| *Number of sales* | *Number of salesmen* |
|---|---|
| 0 to 4 | 1 |
| 5 to 9 | 14 |
| 10 to 14 | 23 |
| 15 to 19 | 21 |
| 20 to 24 | 15 |
| 25 to 29 | 6 |

*Table 3 Table of calculations*

| *Number of sales* | *Number of salesmen* (*f*) | *Mid-point* (*x*) | (*fx*) | (*fx*²) |
|---|---|---|---|---|
| 0 to 4 | 1 | 2 | 2 | 4 |
| 5 to 9 | 14 | 7 | 98 | 686 |
| 10 to 14 | 23 | 12 | 276 | 3312 |
| 15 to 19 | 21 | 17 | 357 | 6069 |
| 20 to 24 | 15 | 22 | 330 | 7260 |
| 25 to 29 | 6 | 27 | 162 | 4374 |
| Totals | 80 | 87 | 1225 | 21705 |

$$\text{Mean, } \bar{x} = \frac{1225}{80}$$

$= 15.3$ sales (1D)

$$\text{Standard deviation, } s = \sqrt{\frac{\sum fx^2}{\sum f} - \left(\frac{\sum fx}{\sum f}\right)^2}$$

$$= \sqrt{\frac{21705}{80} - \left(\frac{1225}{80}\right)^2}$$

= 6.1 sales (1D)

The next spreadsheet shows a layout for these calculations. The formulae necessary are shown on the immediately following spreadsheet.

Ch10-eg3

| | A | B | C | D | E | F | G | H | I | J |
|---|---|---|---|---|---|---|---|---|---|---|
| 1 | | | | | | | | | | |
| 2 | Σf = | 80 | | | | | | | | |
| 3 | Σfx = | 1225 | | | | | | | | |
| 4 | Σfx2 = | 21705 | | | | | | | | |
| 5 | mean = | **15.31** | | | | | | | | |
| 6 | sd = | **6.07** | | | | | | | | |
| 7 | | | | | | | | | | |
| 8 | Totals | 80 | 87 | 1225 | 21705 | | | | | |
| 9 | | | | | | | | | | |
| 10 | | f | x | fx | fx2 | | | | | |
| 11 | | 1 | 2 | 2 | 4 | | | | | |
| 12 | | 14 | 7 | 98 | 686 | | | | | |
| 13 | | 23 | 12 | 276 | 3312 | | | | | |
| 14 | | 21 | 17 | 357 | 6069 | | | | | |
| 15 | | 15 | 22 | 330 | 7260 | | | | | |
| 16 | | 6 | 27 | 162 | 4374 | | | | | |
| 17 | | | | | | | | | | |
| 18 | | | | | | | | | | |
| 19 | | | | | | | | | | |
| 20 | | | | | | | | | | |
| 21 | | | | | | | | | | |
| 22 | | | | | | | | | | |

Sheet1 Sheet2

Ch10-eg3

| | A | B | C | D | E | F |
|---|---|---|---|---|---|---|
| 1 | | | | | | |
| 2 | Σf = | =B8 | | | | |
| 3 | Σfx = | =D8 | | | | |
| 4 | Σfx2 = | =E8 | | | | |
| 5 | mean = | **=B3/B2** | | | | |
| 6 | sd = | **=SQRT(B4/B2-(B3/B2)^2)** | | | | |
| 7 | | | | | | |
| 8 | Totals | =SUM(B11:B20) | =SUM(C11:C20) | =SUM(D11:D20) | =SUM(E11:E20) | |
| 9 | | | | | | |
| 10 | | f | x | fx | fx2 | |
| 11 | | 1 | 2 | =B11*C11 | =D11*C11 | |
| 12 | | 14 | 7 | =B12*C12 | =D12*C12 | |
| 13 | | 23 | 12 | =B13*C13 | =D13*C13 | |
| 14 | | 21 | 17 | =B14*C14 | =D14*C14 | |
| 15 | | 15 | 22 | =B15*C15 | =D15*C15 | |
| 16 | | 6 | 27 | =B16*C16 | =D16*C16 | |
| 17 | | | | | | |
| 18 | | | | | | |
| 19 | | | | | | |
| 20 | | | | | | |
| 21 | | | | | | |
| 22 | | | | | | |

Sheet1 Sheet2

## 12. Notes on Example 3

a) *Take care when evaluating $fx^2$.*

'$fx^2$' means $f$ times $x$ times $x$ (and *not* $fx$ times $fx$). Thus, values of $fx^2$ can be calculated using $fx$ times $x$.

b) *Use of a calculator.*

When calculating (and at the same time accumulating) values of $fx$ and $fx^2$ , use the type of technique described in the previous example. It will ensure the least number of keystrokes for all the information required in the table. For example, in the case of $fx$ values, the calculator procedure is:

1 × 2 M+ and write down value (2)

14 × 7 M+ and write down value (98)

... etc, down to:

6 × 27 M+ and write down value (162)

RM gives the value of $\sum fx$ (1225)

c) *Validation of calculated measures.*

Whenever the mean and standard deviation are calculated, it is always wise to check on the reasonableness of the results.

i. Since the mean is a measure of location, it should be roughly centrally located. In this case, 15.3 is acceptable.

ii. The standard deviation should be approximately 'one-sixth of the range for roughly symmetric distributions. For moderately skewed distributions, it will be slightly larger. In this case, the distribution is not very skew and dividing the range (29–0=29) by 6 gives approximately 5, which agrees very well with the calculated value of 6.1.

An examiner will always give students credit for noticing that a calculated value is unreasonable (even if they have no time to correct their error).

## 13. Characteristics of the standard deviation

a) The standard deviation is the natural partner to the arithmetic mean in the following respects:

i. 'By definition'. The standard deviation is defined in terms of the mean.

ii. 'Further statistical work'. In further statistical analysis there is a need to deal with one of the most commonly occurring natural distributions, called the Normal distribution, which can only be specified in terms of both the mean and standard deviation.

b) It can be regarded as truly representative of the data, since all data values are taken into account in its calculation.

c) For distributions that are not too skewed:

i. virtually all of the items should lie within three standard deviations of the mean.

i.e. range = 6 × standard deviation (approximately).

ii. 95% of the items should lie within two standard deviations of the mean.

iii. 50% of the items should lie within 0.67 standard deviations of the mean.

## 14. The coefficient of variation

It is sometimes necessary to compare two different distributions with regard to variability. For example, if two machines were engaged in the production of identical components, it would be of considerable value to compare the variation of certain critical dimensions of their output. However, the standard deviation is used as a measure for comparison only when the units in the distributions are the same and the respective means are roughly comparable.

In the majority of cases where distributions need to be compared with respect to variability, the following measure, known as the *coefficient of variation*, is much more appropriate and is considered as the standard measure of relative variation.

**Coefficient of variation**

$$s = \frac{\text{standard deviation}}{\text{mean}} \times 100\%$$

In words, the coefficient of variation calculates the standard deviation as a percentage of the mean. Since the standard deviation is being divided by the mean, the actual units of measurement cancel each other out, leaving the measure unit-free and thus very useful for relative comparison.

## 15. Example 4 (Calculation of the *coefficient of variation*)

Over a period of three months the daily number of components produced by two comparable machines was measured, giving the following statistics.

Machine A: Mean=242.8; sd=20.5

Machine B: Mean=281.3; sd=23.0

The coefficient of variation for machine A $= \frac{20.5}{242.8} \times 100\%$

$= 8.4\%$

The coefficient of variation for machine B $= \frac{23.0}{281.3} \times 100\%$

$= 8.2\%$

Thus, although the standard deviation for machine B is higher in absolute terms, the dispersion for machine A is higher in *relative terms*.

## 16. Pearson's measure of skewness

Skewness was described in the previous chapter and it was shown that the degree of skewness could be measured by the difference between the mean and the mode. However, for most practical purposes, it is usual to require a measure of skewness to be unit-free (i.e. a coefficient) and the following expression, known as *Pearson's measure of skewness* (Psk) is of this type.

**Pearson's measure of skewness**

$$\text{Psk} = \frac{\text{mean} - \text{mode}}{\text{standard deviation}}$$

$$= \frac{3\times(\text{mean} - \text{median})}{\text{standard deviation}}$$

Thus the skewness of two different sets of employee's remuneration can be compared if, perhaps, one is given in terms of weekly wages and the other in terms of annual salary.

Note that:

Psk $< 0$ shows there is left or negative skew.

Psk $= 0$ signifies no skew (mean=mode for a symmetric distribution).

Psk $> 0$ means there is right or positive skew.

The greater the value of Psk (positive or negative), the more the distribution is skewed.

## 17. Example 5 (Determination of *Pearson's measure of skewness*)

*Question*

The following data relates to the number of successful sales made by the salesmen employed by a large microcomputer firm in a particular quarter.

| Number of sales | 0–4 | 5–9 | 10–14 | 15–19 | 20–24 | 25–29 |
|---|---|---|---|---|---|---|
| Number of salesmen | 1 | 14 | 23 | 21 | 15 | 6 |

The mean and standard deviation (earlier calculated) are 15.3 and 6.1 respectively. Estimate the value of the mode and thus calculate Pearson's measure of skewness.

*Answer*

The interpolation formula (given in chapter 8, section 5) is used to estimate the mode.

The modal class is 10–14 (since it has the largest frequency) with a lower class bound of 9.5. The differences between the largest frequency and those either side are: $D_1 = 23 - 14 = 9$; $D_2 = 23 - 21 = 2$ and the modal class width is 5.

Thus, mode = $9.5 + \left(\frac{9}{9+2}\right) \times 5 = 13.6$ (1D)

Pearson's measure of skewness, Psk $= \frac{\text{mean} - \text{mode}}{\text{standard deviation}}$

$$= \frac{15.3 - 13.6}{6.1}$$

$$= +0.28 \text{ (2D)}$$

This value of Psk demonstrates a small degree of right skew, which can be confirmed by inspecting the given frequency distribution table.

## 18. The derivation of the standard deviation

In chapter 9, it was noted that the mean deviation, although an ideal measure of dispersion, is awkward to handle both practically and theoretically. It is for this reason that the standard deviation was derived as a practical and theoretical alternative. The steps involved in the transition from mean deviation to standard deviation are now described.

a) Mean deviation = $\dfrac{\sum|x-\bar{x}|}{n}$

The objection to the above is the awkward modulus sign. We remove it, to give:

b) $\dfrac{\sum(x-\bar{x})}{n}$

However this expression is always zero (from Arithmetic Mean, chapter 6 section 20), since the deviations from the mean cancel each other out. The expression in (a) conveniently disregards all the minus signs! But there is another way of disregarding minus signs, by squaring. Thus:

c) $\dfrac{\sum(x-\bar{x})^2}{n}$

This expression is known as the *variance*. It is quite useful as a measure of dispersion and, indeed, is used for numerous purposes in more advanced statistical analysis. For practical purposes however it has one drawback, it is measured in square units. For example, if the original units were, say, in £, then the mean would be in *£ but the variance would be in £*$^2$! This particular inconvenience can be overcome, of course, by taking the square root of the above expression, to give:

d) $\sqrt{\dfrac{\sum(x-\bar{x})^2}{n}}$

## 19. Summary

a) The standard deviation of a set of values is defined as 'the root of the mean of deviations from the mean'.

b) This most important measure of dispersion is always paired with the arithmetic mean because:

   i. it is defined in terms of the mean
   ii. it is relatively easy to handle theoretically
   iii. both measures are needed when dealing with Normal distributions.

c) The standard deviation is usually calculated using the 'computational formula'.

d) For distributions that are approximately 'normal', the standard deviation should cover approximately one-sixth of the range.

e) The coefficient of variation is an alternative (unit-free) measure to the standard deviation when comparing distributions. It is found by calculating the standard deviation as a percentage of the mean.

f) Pearson's measure of skewness is a coefficient which gives the difference between the mean and mode as a proportion of the standard deviation.

## 20. Points to note

a) When calculating the standard deviation, care must be taken to use the actual value of the mean and not an approximate (rounded) value. Rounded values are often acceptable when quoting answers but should not be used in intermediate calculations.

b) Intelligent use of a calculator is very important in obtaining tables of calculations. Remember to accumulate automatically $x^2$, $fx$ and $fx^2$ at the same time as entering their values. This both saves time and cuts down the total amount of keying necessary.

c) Due to the fact that, for moderately skewed distributions:

$$\text{mean} - \text{mode} = 3(\text{mean} - \text{median})$$

the numerator '(mean – mode)' in Pearson's measure of skewness (given in section 16) can be replaced by '3(mean – median)'. This is particularly useful if (for the given set of data) either:

i. only the mean and median are known, or
ii. the mode is particularly unrepresentative.

## 21. Student self review questions

1. What characteristic of the mean deviation precludes it from being the natural partner to the mean? [1]
2. How is the standard deviation defined? [2]
3. What is the practical advantage in using the computational formula for calculating the standard deviation? [5]
4. Having just calculated the standard deviation of a frequency distribution, how can it be quickly validated? [12(c)]
5. 'The standard deviation is the natural partner to the mean'. Explain why this is so. [13(a)]
6. What percentage of an approximately symmetric distribution lies within two standard deviations of the mean? [13(c)ii.]
7. What is the coefficient of variation and how is it used? [14]
8. How is Pearson's measure of skewness calculated and how does it measure skewness? [16]
9. What is the variance and why is it not used for practical purposes as a measure of dispersion? [18(c)]

## 22. Student exercises

1. Calculate (to 2D) the standard deviation of the set: 2,4,2,3,6 and 1.
2. *MULTI-CHOICE*. A manager has obtained the values of petty cash amounts and found the mean to be £25 with standard deviation £5. All the values he has used now need to be expressed as a percentage of the maximum petty cash amount of £100. What (percentage) value would the standard deviation now take?

   a) √ 20%    b) √ 50%    c) 20%    d) 100%

3. a) Find the mean, mode, range and standard deviation of the following simple frequency distribution:

| x | 2 | 4 | 6 | 8 | 10 |
|---|---|---|---|---|---|
| f | 1 | 6 | 18 | 10 | 5 |

   b) Calculate Pearson's measure of skewness.

   c) Validate the standard deviation (using the range) and explain why it is larger in value than might be expected.

4. *MULTI-CHOICE*. A bus company has a mean average number of complaints of 12 with standard deviation 3. The coefficient of variation is:

   a) 0.25%  b) 4%  c) 25%  d) 400%

5. A large firm tabulated the ages of its employees to give the following:

| Age in years | Percentage of employees | Age in years | Percentage of employees |
|---|---|---|---|
| 15 up to 20 | 1.3 | 40 up to 45 | 7.8 |
| 20 up to 25 | 8.8 | 45 up to 50 | 9.2 |
| 25 up to 30 | 19.0 | 50 up to 55 | 7.4 |
| 30 up to 35 | 22.6 | 55 up to 60 | 2.3 |
| 35 up to 40 | 19.4 | 60 up to 65 | 2.2 |

   a) Estimate the mean and standard deviation age.

   b) Given that the above distribution is not too skewed, between what ages should the central 95% of employees lie?

6. A sample of 50 ball bearings was taken from the production of a machine and their diameters (in cms) were measured to give the following distribution.

| Diameter (cm) | Number of ball bearings |
|---|---|
| 0.160–0.162 | 1 |
| 0.162–0.164 | 3 |
| 0.164–0.166 | 9 |
| 0.166–0.168 | 20 |
| 0.168–0.170 | 14 |
| 0.170–0.172 | 2 |
| 0.172–0.174 | 1 |

   a) Calculate the mean and standard deviation of diameters.

   A second sample of 100 ball bearings was taken from another machine, giving a mean diameter of 0.16500 cms with standard deviation 0.00300 cms.

   b) Compare the dispersion of the diameters from the two machines using the coefficient of variation.

   c) Calculate the mean of the combined sample of 150 ball bearings. It is subsequently discovered that the measuring device used in the second sample was faulty and consistently underestimated the diameters by 0.003 cms.

   d) What should the true mean and standard deviation of the diameters in sample 2 be and would this affect the comparison made in b) above?

# 11 Quantiles and the quartile deviation

## 1. Introduction

This chapter introduces a particular type of measure known as a quantile (sometimes called a *fractile*), which collectively split a set or distribution of items up into equal portions.

The most important example of a quantile is called a quartile, and these are described together with their use in calculating measures of dispersion and skewness. Other types of quantiles described are deciles and percentiles.

## 2. Quantiles and their use

A **quantile** is the name given to an item that lies at some proportional way along a size-ordered set or distribution.

They are normally considered in groups, the members of which split up the data set into equal portions.

a) The median, as we already know, is the middle item of a size-ordered set. Thus, by definition, it can be said to split a set (or frequency distribution) into two equal portions. In other words, it is a particular example of a quantile.

b) Quantiles are best suited to types of business data that:

   i. are particularly susceptible to extremes: wages of employees; turnover of companies; value of customer orders; in fact, any distribution that has at least a moderate amount of skew.

   ii. have distributions that have either open-ended classes or data that are difficult, expensive or impossible to obtain at extremes.

Remember that this is just the type of data that was described in chapter 7 as being best suited for analysis using the median.

c) A particular problem at this stage is the fact that no measure of dispersion so far introduced can be paired naturally with the median, in order to represent data of the type described above.

The range is a simple measure of dispersion which stands alone.

Both the mean deviation and standard deviation are defined in terms of the mean, and thus unsuitable.

Thus it is necessary to find measures of dispersion and skewness, based on the idea of splitting size-ordered data into equal portions, that will naturally partner the median. Such a measure, the quartile deviation, is developed in the following sections.

## 3. Quartiles

A (size-ordered) set of data can be split up into four equal portions. The *three* values that do this, lying respectively one-quarter, one-half and three-quarters of the way along the set, are known as 'quartiles'.

For example, the set 7,4,5,3,3,9,8 can be size-ordered and the quartiles identified as follows:

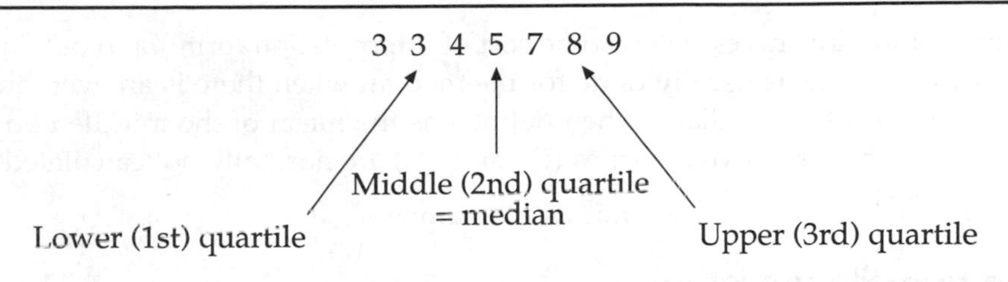


For this particular set of data, the quartiles are easily identified as 3, 5 and 8 respectively and are shown above. (Note that the middle quartile is the median.)
To summarise:

> **The quartiles of a distribution**
> The three *quartiles* of an ordered set or frequency distribution are those values that lie one-quarter, one-half and three-quarters of the way along the group, and are respectively called the lower, middle and upper quartiles ($Q_1$, $Q_2$ and $Q_3$)
> The median is the middle quartile ($Q_2$).

## 4. Identification of the quartiles for a set

For an ordered set of data, just as the median can be identified as the value of the $\frac{n+1}{2}$ th item, the other two quartiles can be identified as follows:

$$Q_1 \text{ is the value of the } \frac{n+1}{4} \text{ th item}$$

$$Q_3 \text{ is the value of the } \frac{3(n+1)}{4} \text{ th item}$$

Note that although the median is, by definition, the middle quartile, the term 'quartiles' is often used to describe only the lower and upper quartiles, $Q_1$ and $Q_3$ respectively.

## 5. Example 1 (Calculation of the *quartiles* for a *set*)

The quartiles of the set

43,75,48,51,51,47,50

are found first by size ordering as:

43,47,48,50,51,51,75

and then:

$Q_1$ is the value of the $\frac{7+1}{4}$ th = 2nd item, which is 47.

$Q_3$ is the value of the $\frac{3(7+1)}{4}$ th = 6th item, which is 51.

Notice that if there had been, say, one more item in the set, the values of $\frac{n+1}{4}$ and $\frac{3(n+1)}{4}$ would not have been whole numbers.

This would have necessitated some sort of interpolation formula to obtain (untypical) values. This is usually done for the median when there is an even number of items in a set (the median is then defined as the mean of the middle two values). However, the other two quartiles (*Q*1 and *Q*3) are normally not calculated in cases where $\frac{n+1}{4}$ and $\frac{3(n+1)}{4}$ are not whole numbers.

## 6. The quartile deviation

The identification of the quartiles enables a measure of dispersion to be defined. This is known as the *quartile deviation* and is defined as half the range of the middle 50% of items (i.e. the difference between the lower and upper quartiles divided by two). It is thus sometimes referred to as the '*semi-interquartile range*'.

For the set of values,

7,4,5,3,3,9,8

introduced in section 2, we found that: $Q_1$=3 and $Q_3$=8.

Therefore, the quartile deviation is calculated as: $\frac{8-3}{2} = 2.5$.

To summarise:

**Quartile deviation (semi-interquartile range)**

$$\text{qd} = \frac{Q_3 - Q_1}{2}$$

## 7. The place of the quartile deviation

It was stressed earlier that, because of the importance of the median, there is a need for a measure of dispersion to pair with it. This measure is the quartile deviation. $Q_3 - Q_1$ gives what is called the *interquartile range*, which is the range covered by the central 50% of items, and dividing by two gives (what can only be described as) the average distance between the median and the quartiles. Thus, approximately, it can be considered that approximately 50% of all items lie within one quartile deviation either side of the median.

Generally, from now on, the median and quartile deviation will be calculated and referred to as a linked pair.

## 8. Example 2 (The *quartile deviation* for a *set*)

For the set

43,75,48,51,51,47,50

the quartiles were found (in Example 1) as: $Q_1$=47 and $Q_3$=51.

$$\text{Thus, the quartile deviation, qd} = \frac{Q_3 - Q_1}{2} = \frac{51 - 47}{2} = 2.$$

## 9. Identifying the quartiles of a frequency distribution

a) The quartiles have the property that they split a distribution into four equal segments, which means effectively that the area under the frequency curve is divided into four equal parts. See Figure 1.

*Figure 1 The place of the median and quartiles in a distribution*

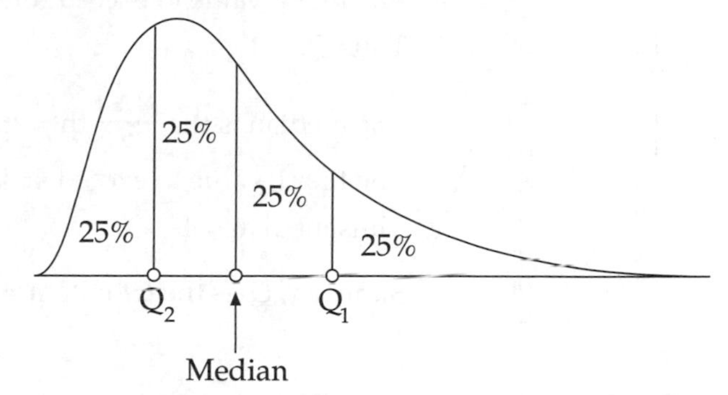


b) When calculating the median and quartiles for a *simple discrete frequency distribution* (consisting of discrete values which are not grouped), the technique used is essentially the same as that for a set. However, the form of the data requires the same approach as that used for the calculation of the median (chapter 7, sections 5 and 6). Example 3, which follows, demonstrates this technique for the calculation of the median and quartiles.

c) For a *grouped frequency distribution*, the quartiles can only be estimated (as was the case for the mean, standard deviation and median). The technique normally used is graphical. The procedure is described in section 11 and demonstrated in Example 4.

## 10. Example 3 (The median and *quartile deviation* for a *simple frequency distribution*)

*Question*

Calculate the median and quartile deviation for the following distribution:

| Delivery time (days) | 0 | 1 | 2 | 3 | 4 | 5 | 6 | 7 | 8 | 9 | 10 | 11 |
|---|---|---|---|---|---|---|---|---|---|---|---|---|
| Number of orders | 4 | 8 | 11 | 12 | 21 | 15 | 10 | 4 | 2 | 2 | 1 | 1 |

*Answer*

See the table for calculations overleaf.

So that, median $= 4$

$$\text{quartile deviation} = \frac{Q_3 - Q_1}{2} = \frac{5-2}{2} = 1.5$$

| Delivery time ($x$) | Number of orders ($f$) | Cum $f$ ($F$) |
|---|---|---|
| 0 | 4 | 4 |
| 1 | 8 | 12 |
| 2 | 11 | 23 * |
| 3 | 12 | 35 |
| 4 | 21 | 56 * |
| 5 | 15 | 71 * |
| 6 | 10 | 81 |
| 7 | 4 | 85 |
| 8 | 2 | 87 |
| 9 | 2 | 89 |
| 10 | 1 | 90 |
| 11 | 1 | 91 |

$Q_1$ is the $\frac{N+1}{4}$th = $\frac{92}{4}$th = 23rd item

The first $F$ value to exceed (or =) 23 is 23

Thus $Q_1 = 2$

The median is the $\frac{N+1}{2}$th = 46th item

The first $F$ value to exceed 46 is 56

Thus median = 4

Similarly, $Q_3$ is the 69th item = 5

## 11. Graphical estimation of quartiles

The procedure for estimating the median and quartile deviation for a grouped frequency distribution, which is an extension of the procedure given for the median in chapter 7, section 12, is:

STEP 1 Form a cumulative (percentage) frequency distribution (using the given frequency distribution).

STEP 2 Draw a cumulative frequency curve by plotting class *upper bounds* against cumulative percentage frequency and join the points with a smooth curve.

STEP 3 Read off the 25%, 50% and 75% points to give $Q_1$, the median and $Q_3$ respectively.

STEP 4 The quartile deviation is calculated in the usual way.

The following example demonstrates this technique.

## 12. Example 4 (Estimation of the median and *quartile deviation* using the *graphical method*)

*Question*

Estimate the median and quartile deviation for the distribution of advertising expenditure for a number of companies in a month, given in the table at the top of the following page.

*Answer*

Figure 2 shows the layout for calculations and the cumulative frequency curve drawn and used for the necessary estimating measures.

From the graph: Median = £1500; $Q_1$ = £800; $Q_3$ = £2000

Therefore, quartile deviation = £$\frac{2000-800}{2}$ = £600

*Figure 2 Cumulative frequency curve with layout for calculations*

| Expenditure (£) | Number of companies | Upper bound | $F$ | $F$% |
|---|---|---|---|---|
| Less than 500 | 210 | 500 | 210 | 16.1 |
| 500 and up to 1000 | 184 | 1000 | 394 | 30.2 |
| 1000 and up to 1500 | 232 | 1500 | 626 | 48.0 |
| 1500 and up to 2000 | 348 | 2000 | 974 | 74.8 |
| 2000 and up to 2500 | 177 | 2500 | 1151 | 88.3 |
| 2500 and up to 3000 | 83 | 3000 | 1234 | 94.7 |
| 3000 and up to 3500 | 48 | 3500 | 1282 | 98.4 |
| 3500 and up to 4000 | 12 | 4000 | 1294 | 99.3 |
| 4000 and over | 9 | 5000 * | 1303 | 100.0 |

* Nominal upper bound

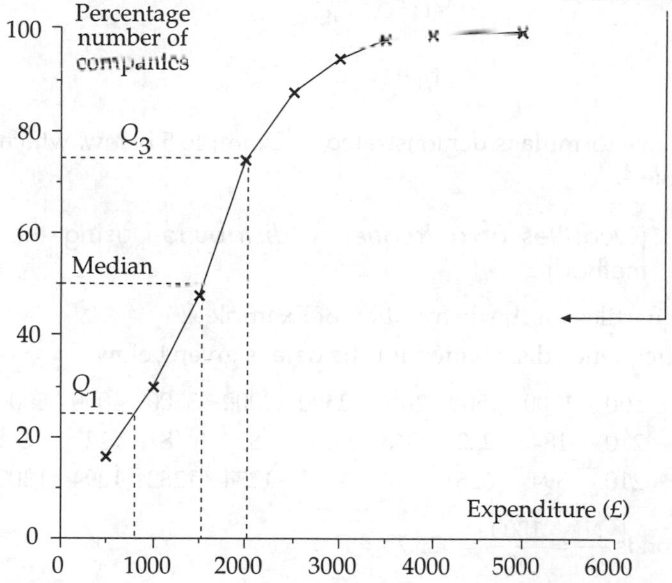

Percentage number of companies
100
80
60
40
20
0
$Q_3$
Median
$Q_1$
Expenditure (£)
0 1000 2000 3000 4000 5000 6000

## 13. General interpolation formula for quantiles

a) There are two methods of estimating the median value of a grouped frequency distribution; the graphical and interpolation methods, although it should be noted that the graphical method is generally preferred.

The median interpolation formula can be generalized to include the other two quartiles (and indeed any of the other quantiles described in following sections).

b) The *generalized interpolation formula* to estimate the value of some defined quantile, $Q$, is given as follows:

**Formula for a quantile**

$$Q = L_Q + \left[\frac{P_Q - F_{Q-1}}{f_Q}\right].c_Q$$

where: $L_Q$ = lower bound of the quantile class
$P_Q$ = position of quantile in distribution
$F_{Q-1}$ = cumulative frequency of class prior to the quantile class
$f_Q$ = actual frequency of quantile class
$c_Q$ = quantile class width

c) It should be noted that the above formula can be used for the median as well as other quantiles. In the case of the median,

$$P_Q = \frac{N}{2}$$

and for the other two quartiles,

$$Q_1 : P_Q = \frac{N}{4}$$

$$Q_3 : P_Q = \frac{3N}{4}$$

The use of the above formula is demonstrated in Example 5 below, which uses the data from Example 4.

## 14. Example 5 (*Quartiles* of a *frequency distribution* using the *formula* method)

To calculate the quartiles for the distribution of Example 4.
The cumulative frequency distribution for the data is given below.

| Upper bound | 500 | 1000 | 1500 | 2000 | 2500 | 3000 | 3500 | 4000 | 5000 |
|---|---|---|---|---|---|---|---|---|---|
| $f$ | 210 | 184 | 232 | 348 | 177 | 83 | 48 | 12 | 9 |
| Cum $f$ $(F)$ | 210 | 394 | 626 | 974 | 1151 | 1234 | 1282 | 1294 | 1303 |

For $Q_1$: Position is $\frac{N}{4} = \frac{1303}{4} = 325.75 = P_Q$.

So $Q_1$ lies in class 500–1000.

$L_Q$=500; $F_{Q-1}$ = 210; $f_Q$=184; $c_Q$=500

$$\text{Hence, } Q_1 = 500 + \left[\frac{325.75 - 210}{184}\right].500$$

= 814.5 (compared with 800 from graph)

For $Q_3$: Position is $3N \div 4 = 3(1303) \div 4 = 977.25 = P_Q$.

So $Q_3$ lies in class 2000–2500.

$L_Q$=2000; $F_{Q-1}$=974; $f_Q$=1151; $c_Q$=500

$$\text{Hence, } Q_3 = 2000 + \left[\frac{977.25 - 974}{177}\right].500$$

= 2009.2 (compared with 2000 from graph)

Notice that the values obtained above for $Q_1$ and $Q_3$ (814.5 and 2009.2) are very close to those obtained using a cumulative frequency curve (800 and 2000) in Example 4.

## 15. The quartile coefficient of dispersion

Just as it was necessary to be able to compare distributions in respect of variation involving the mean and standard deviation, so there is a need for such a comparison involving the median and quartiles. Such a measure is the *quartile coefficient of dispersion* (qcd), which measures the quartile deviation as a percentage of the median (in the same way that the coefficient of variation measures the standard deviation as a percentage of the mean). Thus the quartile coefficient of dispersion is a *relative* measure of variation.

**Quartile coefficient of dispersion**

$$\text{qcd} = \frac{\text{quartile deviation}}{\text{median}} \times 100\%$$

Example 6 demonstrates the use of this formula.

## 16. Quartile measure of skewness

For a symmetric distribution, the median ($Q_2$) lies exactly halfway between the other two quartiles. If a distribution is skewed to the right (positive skew), the median is pulled closer to $Q_1$ (or pulled closer to $Q_3$ for negative skew) and this relationship enables the following coefficient to be derived for measuring skewness.

**Quartile measure of skewness**

$$\text{qsk} = \frac{Q_1 + Q_3 - 2Q_2}{Q_3 - Q_1}$$

Note that: $\text{qsk} < 0$ shows there is left or negative skew,
$\text{qsk} = 0$ signifies no skew
$\text{qsk} > 0$ means there is right or positive skew.

Example 6 demonstrates the use of this formula.

## 17. Example 6 (*Skewness* and *relative dispersion* for a *frequency distribution*)

To calculate a measure of skewness (qsk) and relative dispersion (qcd) for the advertising expenditure data of Example 4.

The three quartiles were estimated as $Q_1$ = £800; median ($Q_2$) = £1500; $Q_3$ = £2000 and quartile deviation = £600.

$$\text{qsk} = \frac{Q_1 + Q_3 - 2Q_2}{Q_3 - Q_1}$$

$$= \frac{800 + 2000 - 2(1500)}{2000 - 800}$$

$$= -0.17 \text{ (2D)}$$

which signifies some degree of negative skew.

$$\text{qcd} = \frac{\text{quartile deviation}}{\text{median}} \times 100\%$$

$$= \frac{600}{1500} \times 100\%$$

$$= 40\%$$

This value could, for instance, be compared with the expenditure for the same or different companies in a later month.

## 18. Some other quantiles

a) *Deciles*. These are the *nine* points of a distribution that divide it up into TEN equal parts. They are normally denoted as $D_1$, $D_2$, ... $D_9$. Note that the fifth decile, $D_5$, is the median.

Sometimes the deciles are used in circumstances where 'blocks of 10%' are convenient, and particularly where extremes need to be discounted. For example, it is not uncommon to see the first and ninth deciles of national wage and income distributions used as practical low and high values, since the two ten percentiles are thought of as unrepresentative.

b) *Percentiles*. These are the *ninety nine* points of a distribution that divide it up into *one hundred* equal parts. They are normally denoted as $P_1$, $P_2$, ... $P_{99}$. Note that the 50th percentile, $P_{50}$, is the median.

c) *The 10 to 90 percentile range*. This particular measure of dispersion is defined as $P_{90}-P_{10}$, that is, the value obtained by subtracting $P_{10}$ from $P_{90}$. It is useful as a practical range, cutting out the ten percent extremes at the two ends of a distribution.

## 19. Summary

a) A quantile is the value of an item which lies at a particular place along an ordered set or distribution, the most well known being the median.

b) The three quartiles split a distribution up into four equal parts and can be identified as the numerical value of the $(n+1)/4$th, the $n/2$th (the median) and the $3(n+1)/4$th items.

c) The quartile deviation is the measure of dispersion that is paired with the median and is defined as the range covered by the first and third quartiles divided by two.

d) The quartiles of a frequency distribution can be estimated using either:

   i. a generalized linear interpolation formula or

   ii. a cumulative frequency curve.

e) The quartile coefficient of dispersion describes relative variation by measuring the quartile deviation as a percentage of the median.

f) The quartile measure of skewness enables a measure of skew to be calculated in terms of the quartiles.

g) Deciles and percentiles split a distribution into ten and one hundred equal parts respectively.

## 20. Points to note

a) Since, for an approximately symmetric distribution:
   i. 50% of items lie within 1 QD of the median (section 7), and
   ii. 50% of items lie within 0.67 SDs of the mean (chapter 10, section 13 (c) iii),
   it follows that, for an approximately symmetric distribution, QD = 0.67□SD.

b) Of the many quantiles that could be defined, the quartiles are obviously the most important. Deciles and percentiles are used occasionally in special circumstances, especially with wage or salary distributions.

## 21. Student self review questions

1. What are quantiles and for what type of data are they most commonly used? [1,2]
2. How can the quartiles of an ordered set or distribution be identified? [4]
3. How is the quartile deviation defined and why is it the natural measure of dispersion to pair with the median? [6,7]
4. Define the quartile coefficient of dispersion. [15]
5. What relationship between the three quartiles does the quartile measure of skewness use in its measurement of skew? [16]
6. What measure of dispersion is based on the percentiles? [18(c)]

## 22. Student exercises

1. *MULTI-CHOICE*. Which one of the following is a unit-free measure of variation.
   a) Coefficient of variation
   b) Mean deviation
   c) Quartile deviation
   d) Standard deviation
2. Find the median and the quartile deviation for the following set of values:

   32,30,24,24,36,33,29,29,30,28,30,32,32,34,28

3. An observer visited an 'off-licence' on 20 occasions to note the number of people in the queue. The results were as follows:

| Number of people in queue | 0 | 1 | 2 | 3 | 4 or more |
|---|---|---|---|---|---|
| Number of queues | 6 | 8 | 3 | 2 | 1 |

   Find the median and quartile deviation of the queue length.

4. The following data relates to the ages of policy holders in a particular insurance scheme.

| Age | 10–19 | 20–29 | 39–39 | 40–49 | 50–59 | 60–69 | 70–79 | 80–89 | 90+ |
|---|---|---|---|---|---|---|---|---|---|
| Number of policy holders | 3016 | 6894 | 9229 | 5714 | 3575 | 1492 | 170 | 9 | 1 |

   a) Find the median and quartile deviation of ages.
   b) Using an appropriate formula, measure the skew of the distribution.
   (**Note**: Pay careful attention to the class bounds and widths.)

5. *MULTI-CHOICE*. The diagram below shows an ogive for a sample of the values of 800 invoices.

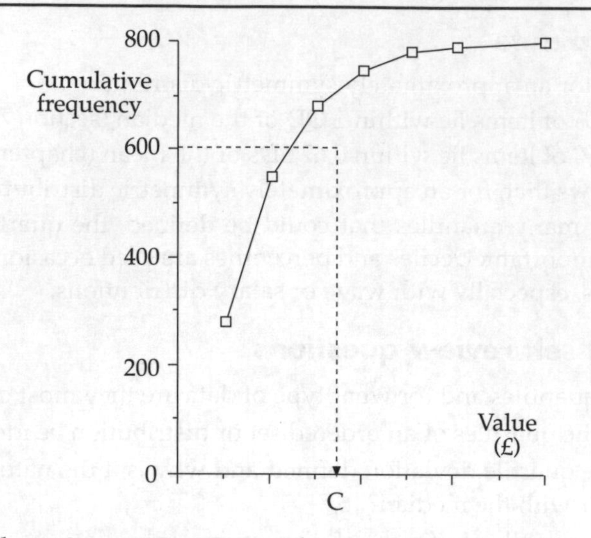


The point C on the Value axis represents the value

a) which 75% of the sample take

b) below which 75% of the sample values lie

c) which 25% of the sample take

d) above which 75% of the sample values lie

6. The life of electric light bulbs is tested by lighting 100 and observing at noon each day the numbers still burning or burnt out. The results of the observations are given in the following table.

| Time in days since lighting | 6 | 7 | 8 | 9 | 10 | 11 | 12 | 13 | 14 | 15 |
|---|---|---|---|---|---|---|---|---|---|---|
| Number still burning | 100 | 98 | 93 | 87 | 73 | 47 | 29 | 16 | 6 | 0 |

a) By plotting a smooth cumulative frequency curve, estimate the median and the 7th decile.

b) Explain what the value of the 7th decile shows.

**Notes:**

1. Either a 'more than' (as given) or a 'less than' curve can be drawn, but be careful with the calculation of the 7th decile.
2. Remember that a cumulative frequency curve is drawn using *class boundaries*.

7. One way that a company classifies its bad debts is by how many days overdue the particular account is. A new accountant decided to approximate the data by dividing the number of days overdue by seven and then rounding (to the nearest week). The following table is the result.

| Number of weeks overdue ($x$) | 1 | 2 | 3 | 4 | 5 | 6 | 7 |
|---|---|---|---|---|---|---|---|
| Number of accounts ($f$) | 2 | 5 | 12 | 23 | 18 | 7 | 1 |

By treating the data as continuous, estimate the median number of weeks overdue of a bad debt. [Hint: Express each $x$-value as a class of values.]

# **Examination example** (with worked solution)

*Question*

a) Give a *specific* business, commercial or industrial example of when:
   i. the median would be used in preference to the arithmetic mean;
   ii. the mode would be used in preference to the median;
   iii. the arithmetic mean would be used in preference to any other average.

b) A company has ten sales territories with approximately the same number of sales people working in each territory. Last month the sales orders (£000) achieved were as follows:

| Area | A | B | C | D | E | F | G | H | I | J |
|---|---|---|---|---|---|---|---|---|---|---|
| Sales | 150 | 130 | 140 | 150 | 140 | 300 | 110 | 120 | 140 | 120 |

For these sales data calculate the following:

i. arithmetic mean  ii. mode  iii. median
iv. lower quartile  v. upper quartile  vi. quartile deviation
vii. standard deviation  viii. mean deviation.

State clearly the most appropriate average and the most appropriate measure of dispersion for these data.

*Answer*

a) i. Whenever a distribution is significantly skewed or data is difficult or expensive to measure, the median will be the most appropriate measure of location. For example, salaries of employees, turnover of a large set of companies, time to destruction in tests of components.

ii. The mode is the most useful measure of location when the 'most common' or the 'most popular' item is required. For example, number of customers in a queue, number of defects in a sample, sales of shirts by neck sizes.

iii. The mean would always be chosen in symmetric distributions or where further statistical calculations or analysis might be required. For example, number of items produced per day on a large assembly line, number of orders received per month for a firm.

b)

| $x$ | $x^2$ | $x$ (in size order) |
|---|---|---|
| 150 | 22500 | 110 |
| 130 | 16900 | 120 |
| 140 | 19600 | 120 |
| 150 | 22500 | 130 |
| 140 | 19600 | 140 |
| 300 | 90000 | 140 |
| 110 | 12100 | 140 |
| 120 | 14400 | 150 |
| 140 | 19600 | 150 |
| 120 | 14400 | 300 |
| 1500 | 251600 | |

i. mean $= 1500 \div 10 = 150$

ii. mode $= 140$ (most common)

iii. median $= \frac{140 + 140}{2} = 140$

iv. lower quartile = 2.75th item = 120

v. upper quartile = 8.25th item = 150

vi. quartile deviation $= \frac{150 + 120}{2} = 15$

vii. standard deviation $= \sqrt{\frac{251600}{10} - 150^2} = 56$ (1D)

viii. mean deviation $= \frac{\sum |x - 150|}{10}$

$$= \frac{0+20+10+0+150+40+30+10+30}{10}$$

$$= \frac{300}{10}$$

$$= 30$$

Since there is a single (very) extreme value present, the median and (its natural partner) the quartile deviation would be the most appropriate measures to use, since they are not really affected by the extreme value. The mean and sd (and the md) are severely affected and thus most unrepresentative as measures.

*CIMA*

# Examination questions

1. As part of a marketing exercise your firm has collected the following data on the population of a small town.

| *Age* | *No of persons* |
|---|---|
| 0 and less than 5 | 39 |
| 5 and less than 15 | 91 |
| 15 and less than 30 | 122 |
| 30 and less than 45 | 99 |
| 45 and less than 65 | 130 |
| 65 and less than 75 | 50 |
| 75 and over | 28 |

a) Find the mean age of the population.
b) Find the standard deviation of the population's age.
c) Display the data using a histogram.
d) Draw a percentage less than ogive of the data.
e) What is the median age of the population?

*ACCA*

2. 16-Year-Old Entrants: Distribution by Length of Training Received, England and Wales 1978.

| Length of training (in weeks) | Male | | | Female | | |
|---|---|---|---|---|---|---|
| | All % | Apprentices % | Others % | All % | Apprentices % | Others % |
| No training | 36.4 | - | 57.7 | 48.2 | - | 52.2 |
| 1 to 2 | 2.6 | - | 4.2 | 4.4 | - | 4.7 |
| 3 to 8 | 7.5 | - | 11.9 | 19.1 | - | 20.7 |
| 9 to 26 | 8.0 | - | 12.7 | 13.0 | - | 14.1 |
| 27 to 52 | 3.3 | - | 5.2 | 3.5 | - | 3.8 |
| 53 to 104 | 6.0 | 8.1 | 4.7 | 4.6 | 21.8 | 3.2 |
| 105 or more | 36.2 | 91.9 | 3.6 | 7.2 | 78.2 | 1.3 |
| All entrants | 100 | 100 | 100 | 100 | 100 | 100 |
| All entrants sample size | 19793 | 7316 | 12477 | 14411 | 1110 | 13301 |

*(Source: Employment Gazette, December 1980, 16-year-olds entering employment in 1978, Table 4.)*

The table above shows the length of training received by a sample of school leavers in their first employment in England and Wales in 1978.

a) *Of those who received some training,* what percentage of
i. males ii. females
received at least 105 weeks of training?
b) Draw ogives (cumulative 'less than' frequency) on the same graph for both female and male *non-apprentices* who receive training.

From the ogives estimate the median length of training received by

i. male and

ii. female

non-apprentices.

c) What do you consider to be *three* of the main points shown by the data in the table?

*CIMA*

3. An analysis of access time to a computer disc system was made during the running of a particular computer program, which utilised disc file handling facilities. The results of the 120 access times were as follows:

| Access time in milliseconds | Frequency |
|---|---|
| 30 but less than 35 | 17 |
| 35 but less than 40 | 24 |
| 40 but less than 45 | 19 |
| 45 but less than 50 | 28 |
| 50 but less than 55 | 19 |
| 55 but less than 60 | 13 |

Required:

i. Determine the mean access time for this program.

ii. Determine the standard deviation of the access time for this program.

iii. Interpret for your superior, who is not familiar with grouped data, what the results of parts (i) and (ii) mean.

*ACCA*

4. A machine produces the following number of rejects in each successive period of five minutes:

| | | | | | | | | | |
|---|---|---|---|---|---|---|---|---|---|
| 16 | 21 | 26 | 24 | 11 | 24 | 17 | 25 | 26 | 13 |
| 27 | 24 | 26 | 3 | 27 | 23 | 24 | 15 | 22 | 22 |
| 12 | 22 | 29 | 21 | 18 | 22 | 28 | 25 | 7 | 17 |
| 22 | 28 | 19 | 23 | 23 | 22 | 3 | 19 | 13 | 31 |
| 23 | 28 | 24 | 9 | 20 | 33 | 30 | 23 | 20 | 8 |

a) Construct a frequency distribution from these data, using seven class intervals of equal width.

b) Using the frequency distribution, calculate an appropriate measure of: (i) average; (ii) dispersion.

c) Briefly explain the meaning of your calculated measures.

*CIMA*

5. Your company manufactures components for use in the production of motor vehicles. The number of components produced each day over a forty day period is tabulated below.

| | | | | | | | | | |
|---|---|---|---|---|---|---|---|---|---|
| 553 | 526 | 521 | 528 | 538 | 523 | 538 | 546 | 524 | 544 |
| 532 | 554 | 517 | 549 | 512 | 528 | 523 | 510 | 555 | 545 |
| 524 | 512 | 525 | 543 | 532 | 533 | 519 | 521 | 536 | 534 |
| 541 | 535 | 531 | 551 | 535 | 519 | 530 | 549 | 518 | 531 |

a) Group the data into five classes.
b) Draw the histogram of the frequency distribution that you have obtained in (a).
c) Establish the value of the mode of the frequency distribution from the histogram.
d) Establish the value of the mean of the distribution.
e) Establish the value of the standard deviation of the distribution.
f) Describe briefly the shape of the frequency distribution, using the values you obtain in (c), (d) and (e).

*ACCA*

6. *Building Society mortgages*

| Income of borrowers | Percentage of all mortgages |
|---|---|
| Under £30,000 | 5 |
| £30,000 – £39,999 | 2 |
| £40,000 – £44,999 | 3 |
| £45,000 – £49,999 | 5 |
| £50,000 – £59,999 | 10 |
| £60,000 – £69,999 | 15 |
| £70,000 – £99,999 | 18 |
| £100,000 – £149,999 | 21 |
| £150,000 and over | 4 |

Calculate suitable measures of average and dispersion and comment upon your results. Discuss in general terms the value of descriptive statistics for management purposes.

*R&V*

7. What is the best 'average', if any, to use in each of the following situations? Justify each of your answers.
   a) To find the average percentage increase per annum when the price of a share has doubled in a period of four years.
   b) To establish a typical wage to be used by an employer in wage negotiations for a small company of 300 employees, a few of whom are very highly paid specialists.
   c) To calculate the average speed for the entire trip when on the outward journey from A to B the average speed was 30mph while on the return journey on the selfsame route the average speed was 60mph.
   d) To determine the height to construct a bridge (not a drawbridge) where the distribution of the heights of all ships which would pass under is known and is skewed to the right.
   e) To ascertain the average annual income of all workers when it is known that the mean annual income of skilled workers is £4500 while the mean annual income of unskilled workers is £3500.
   f) To state the amount to be paid to each employee when the company introducing a profit sharing scheme requires that each employee receives the same amount.

*CIMA*

8. Income of females in 1981 by highest educational qualification.

| Range of weekly income (£) | Highest educational qualification | |
|---|---|---|
| | Degree or equivalent | O level or CSE grade 1 |
| | (percentages) | |
| 40 and under 60 | 5 | 15 |
| 60 .. .. .. .. 80 | 7 | 30 |
| 80 .. .. .. .. 100 | 7 | 28 |
| 100 .. .. .. .. 120 | 18 | 14 |
| 120 .. .. .. .. 140 | 23 | 7 |
| 140 .. .. .. .. 160 | 14 | 3 |
| 160 .. .. .. .. 180 | 10 | 2 |
| 180 .. .. .. .. 220 | 16 | 1 |
| Number interviewed | 104 | 553 |

i. For the *first* distribution, i.e. those with degree or equivalent qualifications, calculate the mean and standard deviation.

ii. For the *second* distribution, i.e. those with O level or CSE grade 1, the mean income was £88.00 with a standard deviation of £29.33. Using appropriate measures, compare the two distributions.

iii. Comment on your results.

*ICSA*

# Part 3 Regression and correlation

The analysis of bivariate numeric data is concerned with statistical measures of regression and correlation, which can be thought of as the bivariate equivalents of 'location' and 'dispersion'. Generally, regression locates bivariate data in terms of a mathematical relationship, able to be graphed as a line or curve, while correlation describes the nature of the spread of the items about the line or curve. The basic aim of the overall analysis is to discover the extent to which one variable is connected to the other.

Chapter 12 begins with a basic revision of linear functions and graphs.

All the standard linear regression and correlation techniques available are covered in the subsequent sections, including validation checks on results and useful calculator procedures.

Chapter 13 describes the various standard techniques for obtaining regression lines, including fitting by eye and the methods of semi-averages and least squares. The relationship between regression and correlation is dealt with in chapter 14, together with two measures of a coefficient of correlation, product moment and rank.

# 12 Linear functions and graphs

## 1. Introduction

This chapter is intended mainly as revision and covers linear (straight line) equations and their graphs. It deals specifically with:

a) Plotting a straight line, given its associated equation.
b) Identifying the equation of a given line graph.
c) The meaning of a 'gradient' and its determination.

## 2. Linear functions and graphs

If variable $y$ is expressed in terms of variable $x$ in the form $y=3+2x$, then the relationship between $y$ and $x$ can be represented in graphical form as a *straight line*, and because of this, $3+2x$ is sometimes called a *linear* (of a line) *function* of $x$.

[*Special note*. Those students who understand the above statement and feel confident in this area can quickly skip through the rest of the chapter.]

Other examples where $y$ is a linear function of $x$ are: $y=5-3x$, $y=-3+x$ and $y=9x+12$. In general, a linear function of $x$ takes the form $y=a+bx$, where $a$ and $b$ can be any numerical values, positive or negative.

If these relationships are plotted on a graph, the result is always a straight line. Section 3 gives the procedure for plotting a line, given a linear function.

> **A linear function**
>
> The function $y=a+bx$ is called a *linear* function, since, when the relationship is plotted on a graph, a straight line is obtained.
>
> $a$ is the $y$-intercept; $b$ is the gradient (or slope).

For example:

if $y=12+4x$, then the gradient is 4 and the $y$-intercept is 12;

if $y=6x-3$, then the gradient is 6 and the $y$-intercept is $-3$.

## 3. Plotting a straight line

Plotting a straight line is quite straightforward. Consider the line whose equation is $y=2+7x$. When $x=1$, the value of $y$ can be found by substituting $x=1$ into the given equation, i.e. $2+7(1) = 9$. Therefore the point defined by $x=1$ and $y=9$ must lie on the line. Any other $x$-value will also determine a point which lies on the line. For example $x=4$ gives $y = 2+7(4) = 30$. Thus the point at $x=4$ and $y=30$ defines another point on the line. These two points can be plotted. Any other points that are found can also be plotted and they can all be joined to form the line required.

Given a linear function in the form $y=a+bx$, the procedure for plotting the corresponding straight line is now described.

STEP 1 Select a few convenient values of $x$ (normally three).

STEP 2 For each $x$-value calculate the corresponding value of $y$, using the given function.

STEP 3 Draw a $y$ (vertical) axis and an $x$ (horizontal) axis, suitably scaled to accommodate all values of $x$ and $y$ obtained in (b).

STEP 4 Plot the pairs of $x$ and $y$ values as points.

STEP 5 Join the points to obtain the required straight line.

Note that only two points are absolutely necessary to define a straight line but if three are plotted, it will be obvious if any error has been made in the calculations (since the points will not all lie on a line!).

## 4. Example 1 (Plotting a *straight line*)

To plot the straight line corresponding to $y = 2x + 3$.

Choosing $x = 1$, 2 and 3 (for pure convenience), the calculation of the corresponding $y$-values is shown as follows.

For $x = 1$: $y = 2(1) + 3 = 5$

$x = 2$: $y = 2(2) + 3 = 7$

$x = 3$: $y = 2(3) + 3 = 9$.

These three points are plotted on the graph at Figure 1. Joining the points gives the required line.

*Figure 1 Plotting a line using three points*

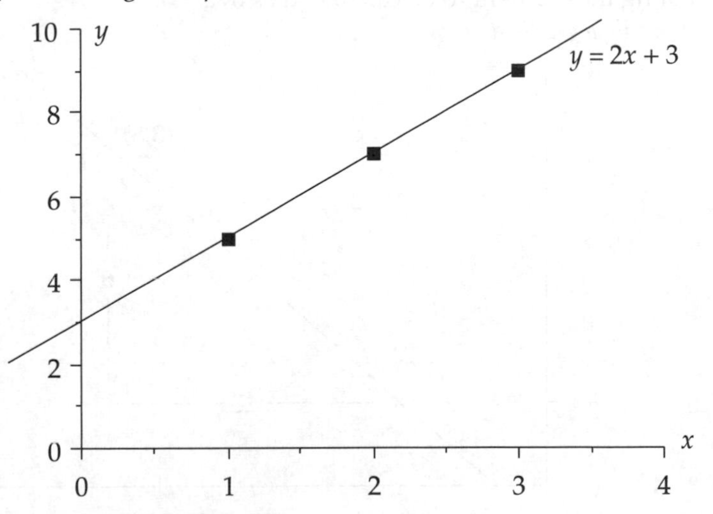


## 5. The gradient and y-intercept

a) The *y-intercept* of any linear function $y = a + bx$ is the point on the $y$-axis where the line crosses. In Example 1, the straight line crossed the $y$-axis at 3, which is the value of $a$ in the function $y = 2x + 3$.

b) The gradient of a linear function determines the slope of the line. In numerical terms, the gradient gives the rate of change of $y$ for each unit increase in $x$. In Example 1 above, three $x,y$ values were identified and notice that $y$ increased by 2 (the value of the gradient) for each unit increase in $x$. Thus if $y=4–3x$, for every unit increase in $x$, $y$ decreases by 3, the value of the gradient.

## 6. Determining the value of the gradient

If a line is given or shown on a graph without its equation, the procedure for calculating its gradient is shown below.

STEP 1 Identify any two points on the line.

STEP 2 Calculate:

i. the vertical numerical difference between the points, and

ii. the horizontal numerical difference between the points.

STEP 3 Calculate:

$$\text{gradient} = \frac{\text{vertical numerical difference}}{\text{horizontal numerical difference}}$$

The following example demonstrates this technique.

## 7. Example 2 (Finding a *gradient*)

a) To find the gradient of the line joining the point $x$=1, $y$=2 to the point $x$=3, $y$=8 [which in future will be written in the form (1,2) and (3,8)].
The difference in the $x$-values (horizontal difference) is 3–1 = 2; the difference in the $y$-values (vertical difference) is 8–2 = 6. Figure 2(a) shows these differences, enabling the gradient to be calculated as 6÷2 = 3.

*Figure 2(a) Finding a gradient*

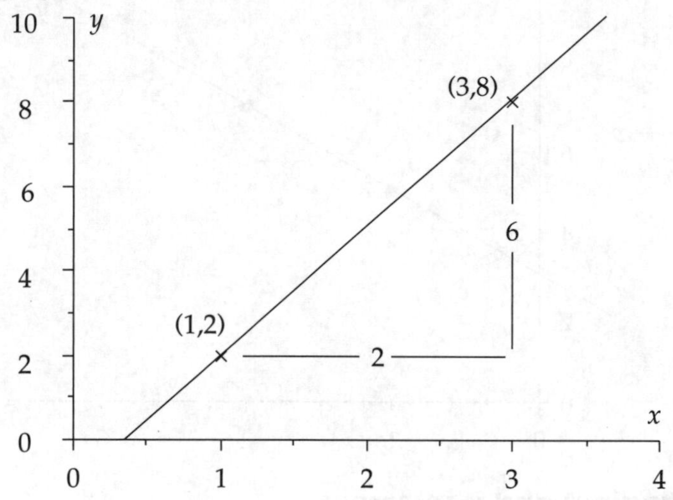


Note that horizontal and vertical differences defined here must always be consistent. In the example above, the differences were obtained by taking values of the first point from the second for both $x$ and $y$. The following example demonstrates a negative gradient.

b) To find the gradient of the line joining the points (0,1) and (12,–2).
Subtracting the values of the first point from the second gives:
difference in the $x$-values is 12–0 = 12; difference in the $y$-values is –2–1 = –3
See Figure 2(b).

Therefore, the value of the gradient is $\frac{-3}{12} = -0.25$.

*Figure 2(b) Finding a gradient*

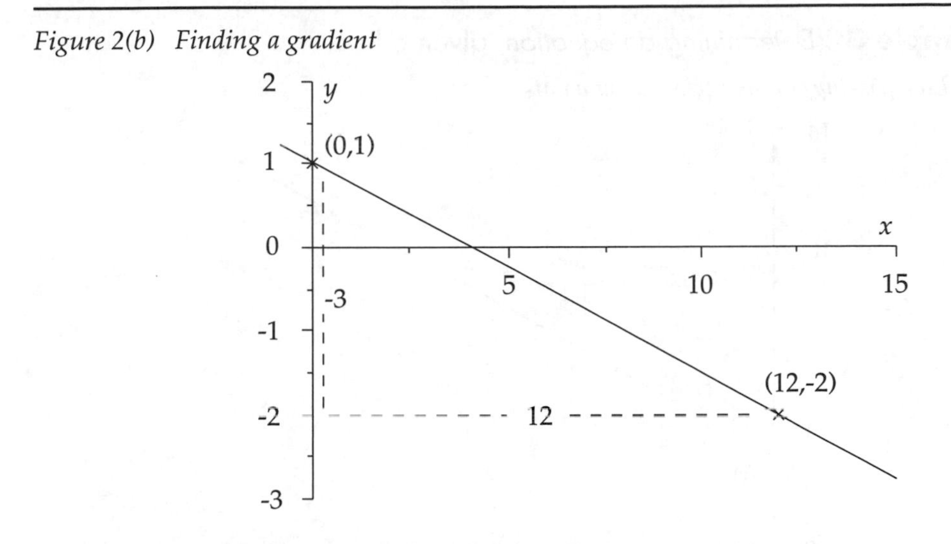


## 8. Determining the equation of a given line

In this section, we are concerned with determining the equation of a straight line, given its graph.

If the equation of the line is put in the standard form $y = a+bx$, then the problem is simply to find the values of $a$ and $b$.

Suppose it could be seen that a line passed through the two points (1,10) and (3,24). The gradient of the line is determined as the difference in the $y$-coordinates divided by the difference in the $x$-coordinates.

i.e. $b = \dfrac{24-10}{4-1} = \dfrac{14}{2} = 7.$

Thus the line can now be written as $y = a + 7x$ (since the gradient, $b$ =7).

But, since this relationship satisfies all $x$,$y$ points on the line, it must satisfy, for example, $x$=1 and $y$=10 (one particular point that we *know* lies on the line).

Substituting these two values into $y = a + 7x$ gives: $10 = a + 7(1)$. Therefore, $a = 3$.

Thus, the line has the equation $y = 3 + 7x$.

This procedure is now generalised.

To find the equation ($y = a + bx$) of a line given only its graph.

STEP 1 The value of the gradient, $b$, can be determined by identifying two points that lie on the line and proceeding as described in section 6.

STEP 2 The value of the $y$-intercept, $a$, can be found either:

i. by inspection – seeing where the line cuts the $y$-axis, or
ii. by substituting the values of $x$ and $y$ (of a point through which the line passes) into the equation.

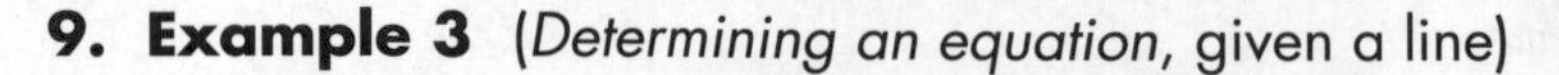

## 9. Example 3 (*Determining an equation*, given a line)

*Figure 3 Line passing through two known points*

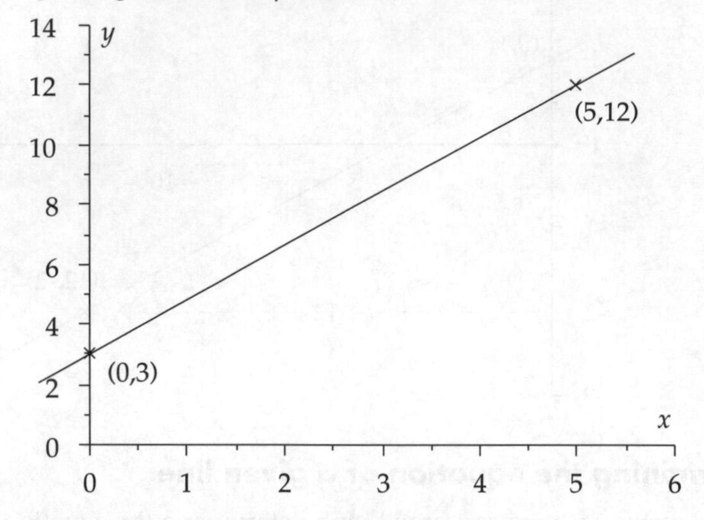


We wish to find the equation of the line in Figure 3. Let the line be $y = a + bx$.

The line can be seen to pass through the two points (0,3) and (5,12).

Therefore, gradient $= b = \dfrac{12-3}{5-0} = 1.8$

Thus the line has equation: $y = a + 1.8x$

The line also crosses the $y$-axis at $y$=3. Therefore $a$=3.

The linear equation required is thus $y = 3 + 1.8x$.

## 10. Example 4

To find the linear equation of the line that passes through the two points (18,6) and (14,54).

Let the required equation be $y = a + bx$.

Gradient $= b = \dfrac{54-6}{14-18} = \dfrac{48}{-4} = -12$.

Therefore $y = a - 12x$.

Also, since (18,6) lies on the line, we must have that: $6 = a - 12(18) = a - 216$. Therefore $a = 222$.

So that, $y = 222 - 12x$ is the required equation.

## 11. Summary

a) A linear function takes the form: $y = a + bx$, where:
   i. $a$ is the $y$-intercept, and
   ii. $b$ is the gradient.

b) If a linear function is plotted on a graph, a straight line results.

c) Plotting a straight line, given its linear function, involves substituting three convenient values of $x$ and obtaining the corresponding values of $y$. When plotted, the points obtained can be joined to form the required line.

d) i. The $y$-intercept of a linear function is the value of $y$ where the line crosses the $y$-axis.

ii. The gradient of a linear function is the rate that $y$ increases (changes) for each unit increase (change) in $x$.

e) Determining the gradient of the line joining two given points can be accomplished by evaluating the ratio of the vertical divided by the horizontal numerical differences between the two points.

## 12. Point to note

Sometimes a linear equation may not be given in the form of $y=a+bx$, but can be put into this form if necessary by a little re-arrangement.

For example:

i. $y + 3 = 3x + 10$ can be re-arranged to give $y = 3x + 7$

ii. $4y = 12x - 8$ can be re-arranged to give $y = 3x - 2$

iii. $2x = 2y + 14$ can be re-arranged to give $y = x - 7$

## 13. Student self review questions

1. What is a linear function and how is it used? [2]
2. How is the gradient of a line defined? [5(b)]
3. How can the gradient of a line be determined, given only two points through which the line passes? [6]

## 14. Student exercises

1. *MULTI-CHOICE*. The following formula is used in inventory control:

$$E = \sqrt{\frac{2DO}{H}}$$

Rearrange the formula to give H in terms of D, O and E.

a) $H = \frac{E^2}{2DO}$ b) $H = \frac{2DO}{E^2}$ c) $H = \sqrt{\frac{2DO}{E}}$ d) $H = \sqrt{\frac{E}{2DO}}$

2. Write down the gradients and $y$-intercepts of the following linear functions:

a) $y = 13 + 4x$

b) $y = 3x - 12$

c) $2y = 4x - 3$

d) $x = 4y - 2$

3. Plot the graphs corresponding to the following linear equations:

a) $y = 3x + 7$

b) $y = 12x - 5$

c) $2y = 4x - 3$

d) $x = 4y - 2$

4. Find the gradients of the lines joining the following pairs of points:

a) (1,2) and (2,3)

b) (3,4) and (2,–2)

c) (50,5) and (100,10)

5. Determine the equations of the lines that pass through the points:
   a) (1,2) and (2,3)
   b) (4,2) and (3,3)
   c) (–2,–2) and (6,2)
6. A straight line meets the $x$-axis at $x$=2 and the $y$-axis at $y$=–4. Determine its equation.
7. *MULTI-CHOICE.* The gradient of the line represented by the following relationship: $y + 2x - 4 = 0$ is:
   a) –2 b) +2 c) –4 d) +4
8. *MULTI-CHOICE.* A straight line passes through the point $x$=12, $y$=2 and crosses the $x$-axis at $x$=8. Which one of the following describes it?
   a) $y = 4 - \frac{x}{2}$ b) $y = 2 - \frac{x}{4}$ c) $y = \frac{x}{2} - 4$ d) $y = \frac{x}{4} - 2$

# 13 Regression techniques

## 1. Introduction

Regression is a technique used to describe a relationship between two variables in mathematical terms. This chapter describes:

a) the uses of a regression relationship, and

b) three common methods of obtaining a regression relationship.

## 2. Regression and its uses

Suppose a large firm (firm A) are interested in measuring the relationship between annual expenditure on advertising and their overall annual turnover. If a relationship between the two (i.e. a regression relationship) was found, the firm might reasonably expect the following question to be able to be answered: 'If we increase our annual advertising expenditure to £30000, what effect will this have on annual turnover?' This question can be answered and, in fact, demonstrates the most important use of regression; that is, the ability to estimate the value of one of the variables, given a value of the other.

Regression relationships are also useful for comparison purposes. For example, our firm might want to compare their own regression relationship between advertising expenditure and turnover with that of another firm, to see whether their advertising is generally more or less effective.

To summarise:

> **Regression**
>
> *Regression* is concerned with obtaining a mathematical equation which describes the relationship between two variables.
>
> The equation can be used for comparison or estimation purposes.

**Note**: For the purposes of the syllabuses covered in this manual, only *linear* regression equations are considered.

## 3. Example 1 (The *use* of a given *regression relationship*)

Suppose the firm (firm A in the preceding section) fed the values of turnover, $y$, and advertising expenditure, $x$, (both in £000) for the past eight years, into a computer and obtained the regression relationship $y = 26.7 + 8.5x$. How can this functional relationship be interpreted and used?

a) First, it acts as a model. That is, in the absence of any other information, the regression line describes the approximate relationship between advertising expenditure and turnover that might be expected from the firm. It could also be used for comparison with other firms of similar size, circumstances and business activity.

See Figure 1.

*Figure 1 Regression lines representing advertising expenditure*

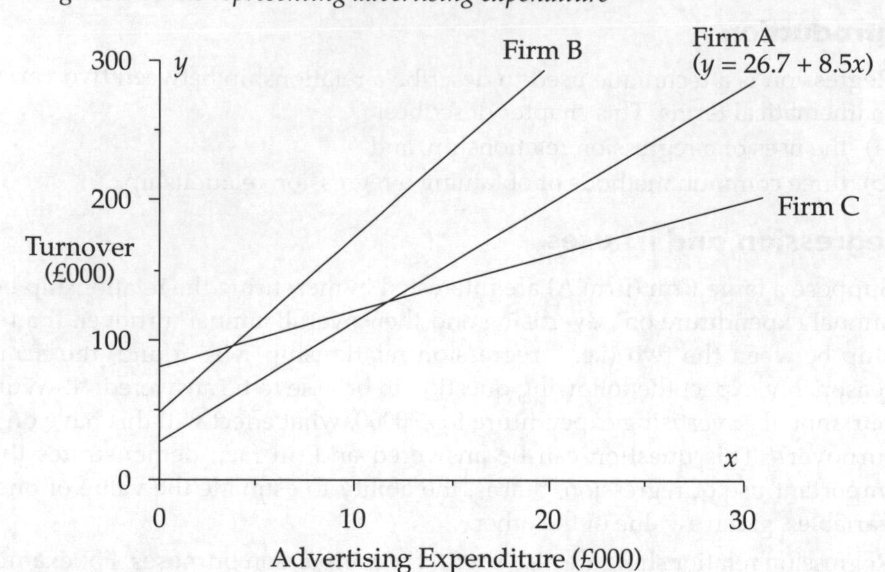


The other two regression lines shown are for similar firms B and C. Since firm B's line is always above firm A's line (in terms of turnover), it could be argued that firm B gets better value from its advertising. Similarly it seems that firm C, on average, gets slightly worse value than A from its advertising expenditure.

b) Second, the regression line can be used to estimate the value of turnover, given a particular level of advertising expenditure.

For example, suppose next year's advertising expenditure was planned as £24000. Then, by substituting $x = 24(£000)$ into $y = 26.7 + 8.5x$, we have:

$$\begin{aligned} y_{\text{est}} &= 26.7 + (8.5)(24) \\ &= 230.7(£000) \\ &= £230700 \end{aligned}$$

That is, using the regression model, the turnover can be estimated as £230700, if advertising expenditure was £24000.

## 4. Standard methods of obtaining a regression line

The process of obtaining a linear regression relationship for a given set of (bivariate) data is often referred to as *fitting* a regression line.

There are three methods commonly used to fit a regression line to a given set of bivariate data.

a) *Inspection*. This method is the simplest and consists of plotting a scatter diagram for the relevant data and then drawing in the line that most suitably fits the data. The main disadvantage of this method is that different people would probably draw different lines using the same data. It sometimes helps to plot the *mean point* of the data (i.e. the mean of the $x$'s and $y$'s respectively) and ensure the regression line passes through this. Example 2 demonstrates this technique.

b) *Semi-averages*. The technique consists of splitting the data into two equal groups, plotting the mean point for each group and joining these two points with a straight line. This technique, together with a worked example, is shown in sections 7 and 8.

c) *Least squares*. This is considered to be the standard method of obtaining a regression line. The derivation of the technique is mathematical and while examination boards do not require students to derive the formulae used in obtaining the least squares line, they do expect students to have an idea of its basis. The least squares technique, together with a fully worked example, is dealt with in sections 10, 11 and 12.

## 5. Example 2 (Fitting, identification and interpretation of a regression line *by inspection*)

### *Question*

The figures show the output (in thousands of tons) and the expenditure on energy (£) for a firm over ten monthly periods.

| | | | | | | | | | | |
|---|---|---|---|---|---|---|---|---|---|---|
| Output ($x$) | 20 | 22 | 25 | 26 | 21 | 23 | 28 | 20 | 25 | 29 |
| Expenditure ($y$) | 106 | 138 | 158 | 172 | 120 | 142 | 184 | 102 | 164 | 192 |

You are required to:

a) Draw a scatter diagram for the data.

b) Calculate the mean point of the data and plot it on the diagram.

c) Fit a regression line (by inspection) which passes through the mean point.

d) Find the equation of the regression line.

e) Estimate the energy expenditure if the following month's output is planned at 27000 tons.

### *Answer*

a) The scatter diagram is shown as part of Figure 2.

b) The mean point for the data is calculated as $\bar{x} = \frac{239}{10} = 23.9$; $\bar{y} = \frac{1478}{10} = 147.8$

c) By inspection, a regression line has been fitted (passing through the mean point) and is shown at Figure 2 overleaf.

d) Labelling the regression line as $y = a + bx$, the gradient, $b$, has been estimated (see the graph) as $\frac{52}{4} = 10.4$. So that the line can be written as $y = a + 10.4x$.

Also, since the line passes through the mean point (23.9,147.8), substituting for $x$ =23.9 and $y$ =147.8 in the above line gives $147.8 = a + (10.4)(23.9)$.

Hence, $a = -100.76$.

Therefore the regression line is $y = -100.76 + 10.4x$.

e) If the following month's output is planned as 27000 tons, then the energy expenditure could be estimated by substituting $x$ =27 into the above regression model to obtain:

$$y = -100.76 + (10.4)(27) = 180.04$$

i.e. estimated energy expenditure (at 27000 tons production) = £180 (approx).

*Figure 2 Output and expenditure on energy over ten months*

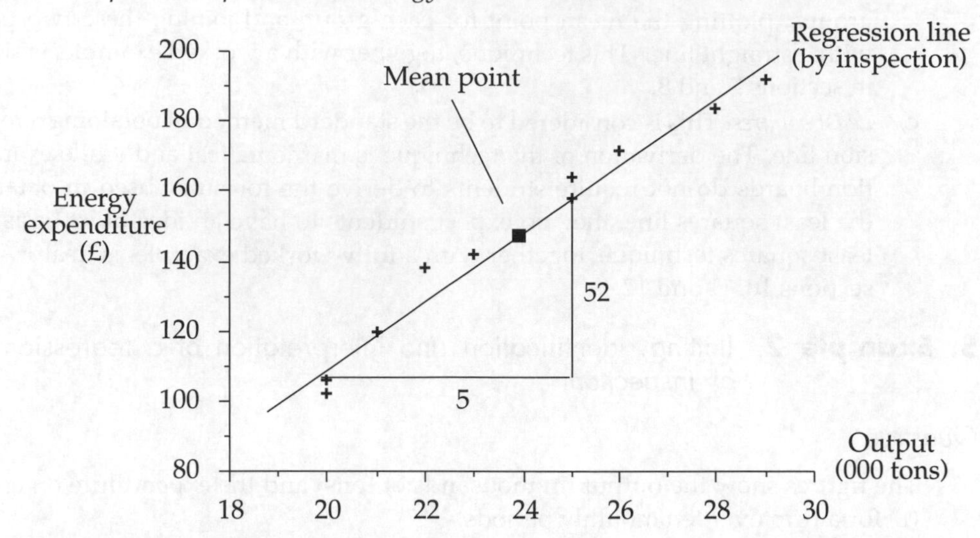


## 6. The 'y on x' regression line

As already mentioned, an important use of regression lines is for estimating the value of one variable given a value of the other. Before proceeding to the other two methods of obtaining a regression line, it is necessary to discuss briefly two distinct ways in which regression can be used for estimation purposes.

For any set of bivariate data, there are *two* regression lines which can be obtained.

a) The *y on x regression line* is the name given to that regression line which is used for estimating $y$ given a value of $x$.

b) The *x on y regression line* is the name given to that regression line which is used for estimating $x$ given a value of $y$.

These two regression lines are quite distinct.

Suppose, for example, a Sales Manager recorded the expense claims ($y$) against the number of visits ($x$) made by his salesmen for a defined period. If one particular salesman forgot to submit an expense claim, the manager could estimate it using the number of visits made. In this case, the 'claims' ($y$) on 'visits' ($x$) regression line would need to be determined and used.

For the syllabuses that this manual covers, only the $y$ on $x$ regression line is dealt with.

Note that the previous method of inspection gives a general regression line. It does not specify particularly which variable can be estimated and in this sense is thought of only as an approximate method.

## 7. The method of semi-averages

The method of semi-averages was briefly described in section 4.

The procedure for obtaining the $y$ on $x$ regression line using this method is described as follows.

STEP 1 Sort the (bivariate) data into size order by $x$-value.

STEP 2 Split the data up into two equal groups, a lower half and an upper half (if there is an odd number of items, *ignore* the central one).

STEP 3 Calculate the *mean point* for each group.

STEP 4 Plot the above mean points on a graph within suitably scaled axes and join them with a straight line. This is the required $y$ on $x$ regression line.

This method is considered superior to the method of inspection. Example 3 demonstrates its use.

## 8. Example 3 (Obtaining and interpreting a regression line using the method of *semi-averages*)

### *Question*

Twelve administrative trainees in a company took an aptitude test in two parts, one designed to test ability to do appropriate calculations and the other designed to test skill in interpreting results. Their scores were as follows:

| Trainee | A | B | C | D | E | F | G | H | J | K | L | M |
|---|---|---|---|---|---|---|---|---|---|---|---|---|
| Calculation score | 23 | 56 | 74 | 29 | 82 | 45 | 36 | 51 | 60 | 55 | 52 | 88 |
| Interpretation score | 16 | 38 | 65 | 39 | 32 | 51 | 11 | 19 | 47 | 54 | 43 | 50 |

a) Obtain the interpretation on calculation regression line, using the method of semi-averages, and plot it, together with the original data, on a scatter diagram.

b) Trainee N obtained 72 in the calculation test but was absent for the interpretation test. Use the regression line to estimate trainee N's interpretation score.

### *Answer*

The situation involves obtaining a regression line and using it to estimate an Interpretation score *given* a *Calculation score*. Thus, labelling Interpretation as $y$ and Calculation as $x$, the $y$ on $x$ regression line is the one required.

a) In Table 1, the data has been sorted into size order by $x$-value and split into two halves as required in the procedure described in the previous section.

Mean value 1 is (39.3,29.8) and mean value 2 is (69.2,47.7). These are both plotted, along with the original data in Figure 3.

*Table 1 Calculations for semi-average method*

| | *Lower half of data* | | *Upper half of data* | |
|---|---|---|---|---|
| | *Calculation* ($x$) | *Interpretation* ($y$) | *Calculation* ($x$) | *Interpretation* ($y$) |
| | 23 | 16 | 55 | 54 |
| | 29 | 39 | 56 | 38 |
| | 36 | 11 | 60 | 47 |
| | 45 | 51 | 74 | 65 |
| | 51 | 19 | 82 | 32 |
| | 52 | 43 | 88 | 50 |
| Totals | 236 | 179 | 415 | 286 |
| Averages | 39.3 | 29.8 | 69.2 | 47.7 |

b) It can be seen from Figure 3 that trainee N's Interpretation score is estimated as approximately 50.

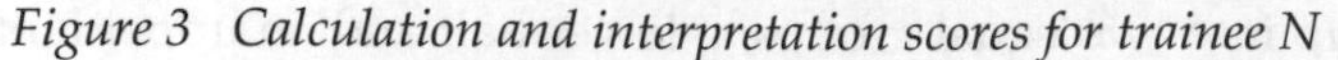

*Figure 3 Calculation and interpretation scores for trainee N*

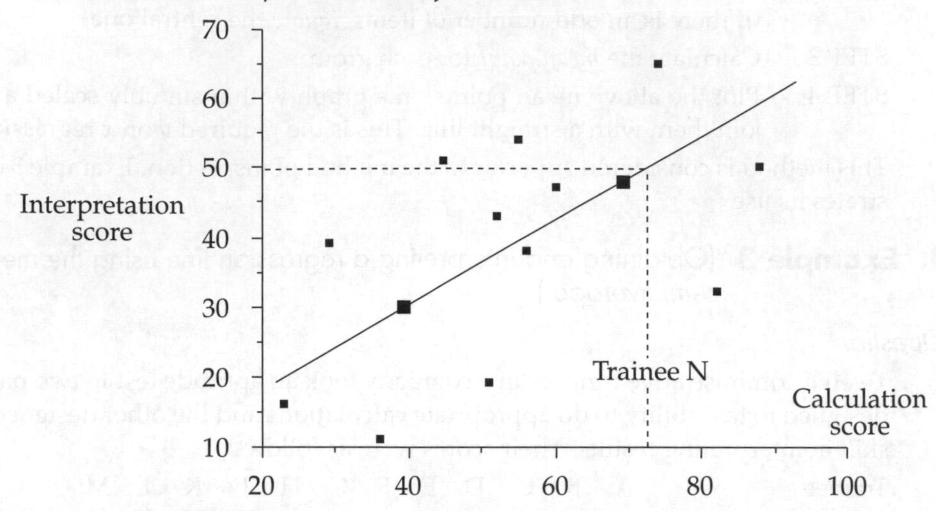


## 9. The background and basis of least squares regression

A major drawback of the semi-averages technique for obtaining a regression line is the fact that it relies on only two points, both means of the two respective data groups. If there are extreme values present, either or both of the means are easily distorted, thus so is the regression line. As mentioned earlier, the least squares method of obtaining a regression line has a mathematical base which involves all values and is thus considered to be superior. While its technical derivation is not required to be known by students, a general idea of its nature is.

Consider a scatter diagram of a set of bivariate data. Of all the regression lines that could be drawn to represent the data, the *least squares regression line of y on x* is that line for which the sum of squares of the *vertical deviations* of all the points from the line is least.

See Figure 4.

## 10. Formula for obtaining the *y* on *x* least squares regression line

The least squares method of obtaining the $y$ on $x$ regression line has already been outlined in section 9. If, as usual, we label the line as $y = a + bx$, the values of $a$ and $b$ can be obtained using the following formulae.

**Least squares regression formulae**

If the least squares regression line of $y$ on $x$ is given by $y = a + bx$, then

$$b = \frac{n\sum xy - \sum x \sum y}{n\sum x^2 - \left(\sum x\right)^2}; \quad a = \frac{\sum y}{n} - b\frac{\sum x}{n}$$

**Note**: There are alternative (but equivalent) formulae for determining the values of $a$ and $b$. Section 20 d) gives one of these alternatives.

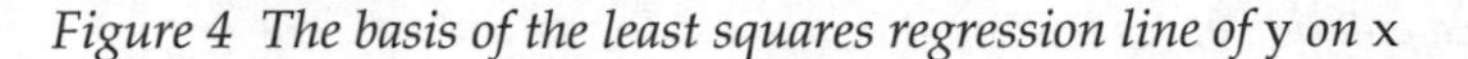

*Figure 4 The basis of the least squares regression line of* y *on* x

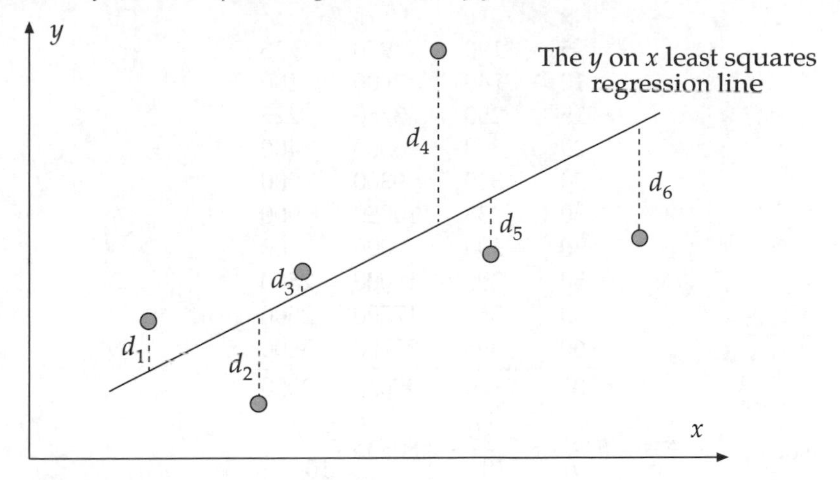


In Figure 4 the line shown is the least squares regression line of $y$ on $x$. $d_1$, $d_2$, ... etc are the numerical differences between each plotted value and the regression line.

**Note:** No other line which can be drawn through the mean point for the data has a value of $d_1^2+d_2^2+ .... +d_6^2$ smaller than the least squares regression line of $y$ on $x$, shown in the above diagram.

Example 4 demonstrates the use of the previous formulae and the standard layout for calculations.

## 11. Example 4 (Finding the *least squares* regression line for a given set of data)

*Question*

The data in Table 2 relates the weekly maintenance cost (£) to the age (in months) of ten machines of similar type in a manufacturing company. Find the least squares regression line of maintenance cost on age and use this to predict the maintenance cost for a machine of this type which is 40 months old.

*Table 2 The age and weekly maintenance cost of 10 machines*

| Machine | 1 | 2 | 3 | 4 | 5 | 6 | 7 | 8 | 8 | 10 |
|---|---|---|---|---|---|---|---|---|---|---|
| Age ($x$) | 5 | 10 | 15 | 20 | 30 | 30 | 30 | 50 | 50 | 60 |
| Cost ($y$) | 190 | 240 | 250 | 300 | 310 | 335 | 300 | 300 | 350 | 395 |

*Answer*

Notice that the regression line of $y$ on $x$ is being asked for, and if we express it in the form $y = a + bx$, the values of $a$ and $b$ can be found using the formulae given in the previous section. Table 3 shows the standard layout for calculations.

Here: $n$=10; $\Sigma x$=300; $\Sigma y$=2970; $\Sigma xy$=97650; $\Sigma x^2$=12050

Thus: $b = \dfrac{n\sum xy - \sum x \sum y}{n\sum x^2 - \left(\sum x\right)^2} = \dfrac{10 \times 97650 - 300 \times 2970}{10 \times 12050 - 300^2}$

$= \dfrac{85500}{30500} = 2.8033$ (4D)

*Table 3 Standard layout for calculation of regression line*

| $x$ | $y$ | $xy$ | $x^2$ |
|---|---|---|---|
| 5 | 190 | 950 | 25 |
| 10 | 240 | 2400 | 100 |
| 15 | 250 | 3750 | 225 |
| 20 | 300 | 6000 | 400 |
| 30 | 310 | 9300 | 900 |
| 30 | 335 | 10050 | 900 |
| 30 | 300 | 9000 | 900 |
| 50 | 300 | 15000 | 2500 |
| 50 | 350 | 17500 | 2500 |
| 60 | 395 | 23700 | 3600 |
| 300 | 2970 | 97650 | 12050 |

Also: $$a = \frac{\sum y}{n} - b\frac{\sum x}{n} = \frac{2970}{10} - 2.8033 \times \frac{300}{10}$$

$$= 212.90 \text{ (2D)}$$

The least squares regression line of cost on age is thus $y = 212.90 + 2.8033x$.
This can now be used to estimate the maintenance cost associated with an age of 40 months. Substituting $x = 40$ into the above regression line gives:

estimated maintenance cost = $y = 212.90 + (2.8033)(40) =$ £325.

## 12. Notes on Example 4

As demonstrated in previous chapters, the efficient use of a calculator saves time when calculating sums of multiples and squares of variables. The particular techniques that were used then can also be used when calculating sums of multiples and squares of variables involved with regression line calculations.

Using the data of Example 4:

a) *When calculating* $\sum xy$ (and writing intermediate products down), use the following program:

5 × 190 M+ (write down 950)
10 × 240 M+ (write down 2400)
15 × 250 M+ (write down 3750)
etc...

b) *When calculating* $\sum x^2$ (and intermediate squares):

5 × M+ (write down 25)
10 × M+ (write down 100)
15 × M+ (write down 225)
etc...

## 13. Dependent and independent values

Bivariate data always involves two distinct variables and in the majority of cases one variable will depend naturally on the other. The *independent variable* is the one that is chosen freely or occurs naturally. The other, the *dependent variable* occurs as a consequence of the value of the independent variable. The following cases demonstrate this idea.

a) Number of items produced (independent) and total cost of production (dependent). The total cost naturally depends on the number of items produced.

b) Cost per item and number of items produced. This is a well known relationship in business. The cost per item is normally dependent on the level of production.

c) Turnover (dependent) and year (independent). In this case the independent variable occurs naturally and the turnover can only be expressed in terms of some time period, such as a year.

In the above three examples, and in all cases where there is a clear case of a dependent and an independent variable, it is usual to label the dependent variable as $y$ and the independent variable as $x$. Thus it would be meaningful to want to estimate $y$ given $x$ (and not $x$ given $y$) which is the reason for the standard use of the regression line of $y$ on $x$.

Sometimes the relationship between a dependent and an independent variable is called a *causal relationship*, since it can be argued that the value of one variable (the dependent variable) has been 'caused' by the value of the other (the independent variable).

## 14. Non-causal relationships

Occasionally there are circumstances where a set of bivariate data will yield no obvious, clear-cut case of an independent and dependent variable. A classic example of this is a set of people measured for both height and weight. Does height generally depend on weight or weight on height? The true answer is, of course, neither! Both height and weight depend on a complex set of variables (i.e. diet, exercise routine, gene structure, etc...) that are inextricably linked with the physical structure of each person.

Data of this type are normally described as *non-causal*, since the value of any one of the variables is not directly attributable to the value of the other.

Some other examples of non-causal relationships are given as follows.

a) Gross turnover and number of employees (for a set of companies).

b) Annual salary and size of household (for a set of employees).

c) Number of days unserviceable and mileage (for a set of company trucks).

Sometimes it is debatable whether a relationship is causal or not, arguments being able to be put for both cases. However, in examinations, unless there is a clear-cut case of causality, a definite lead will be given in the question. Example 5 demonstrates this type of situation.

**15. Example 5** (Calculation of a least squares line, where a *causal relationship* is not obvious)

*Question*

The following figures relate to the carriage of Goods by Road in Great Britain.

| *Year* | *Goods carried (thousand million tonne kilometres)* | *Number of goods vehicles registered (thousands)* |
|---|---|---|
| 1968 | 79 | 1640 |
| 1969 | 83 | 1640 |
| 1970 | 85 | 1630 |
| 1971 | 86 | 1632 |
| 1972 | 88 | 1660 |
| 1973 | 90 | 1736 |
| 1974 | 90 | 1778 |
| 1975 | 95 | 1791 |
| 1976 | 96 | 1773 |
| 1977 | 98 | 1712 |

Labelling 'number of goods vehicles' as the *dependent* variable:

a) Calculate an appropriate regression line.

b) Use this line to estimate:

  i. the number of goods vehicles that would be expected if the goods carried amounted to 80 (thousand million tonne kilometres), and

  ii. the amount of goods carried when the number of goods vehicles is 1800 (thousand).

*Answer*

Notice that it is not absolutely clear which of the two variables is the naturally dependent one. However, we have been told that the variable 'number of vehicles' should be taken as the dependent one and thus this will be labelled as y and 'goods carried' as $x$.

a) The appropriate regression line is the $y$ on $x$ line and, as usual, it will be labelled as $y = a + bx$.

The layout for the calculations of $a$ and $b$ is identical to that demonstrated in Example 4 and the product totals are:

$n = 10; \Sigma x = 890; \Sigma y = 16992; \Sigma xy = 1515051; \Sigma x^2 = 79540.$

Using the formulae of section 10:

$$b = \frac{n\sum xy - \sum x \sum y}{n\sum x^2 - \left(\sum x\right)^2} = \frac{10 \times 1515051 - 890 \times 16992}{10 \times 79540 - 890^2}$$

$$= \frac{27630}{3300}$$

$$= 8.3727 \text{ (4D)}$$

$$a = \frac{\sum y}{n} - b\frac{\sum x}{n} = \frac{16992}{10} - 8.3727 \times \frac{890}{10}$$
$$= 954.0297 \text{ (4D)}$$

The $y$ on $x$ regression line is thus $y = 954.03 + 8.37x$ (2D).

b) i. Using this line, when $x$=80, $y = 954.03 + (8.37)(80)$
$= 1623.$

In other words, when the amount of goods carried is 80 thousand million tonne kilometres, the number of goods vehicles is estimated as 1 623 000.

ii. Using the same line, and substituting $y$ =1800,

gives: $1800 = 954.03 + 8.37x$

which yields: $x = \dfrac{1800 - 954.03}{8.37}$
$= 101.$

Thus, when the number of goods vehicles is 1 800 000, the goods carried is estimated as 101 thousand million tonne kilometres.

The following spreadsheet shows a layout for the calculations in part a). The formulae necessary are shown on the spreadsheet overleaf.

Ch13-eg5

| | A | B | C | D | E | F | G | H | I | J |
|---|---|---|---|---|---|---|---|---|---|---|
| 1 | | | | | | | | | | |
| 2 | n = | 10 | | | | | | | | |
| 3 | b (num) = | 27630 | | | | | | | | |
| 4 | b (denom) = | 3300 | | | | | | | | |
| 5 | b = | **8.37** | | | | | | | | |
| 6 | a = | **954.03** | | | | | | | | |
| 7 | | | | | | | | | | |
| 8 | Totals | 890 | 16992 | 1515051 | 79540 | | | | | |
| 9 | | | | | | | | | | |
| 10 | | x | y | xy | x^2 | | | | | |
| 11 | | 79 | 1640 | 129560 | 6241 | | | | | |
| 12 | | 83 | 1640 | 136120 | 6889 | | | | | |
| 13 | | 85 | 1630 | 138550 | 7225 | | | | | |
| 14 | | 86 | 1632 | 140352 | 7396 | | | | | |
| 15 | | 88 | 1660 | 146080 | 7744 | | | | | |
| 16 | | 90 | 1736 | 156240 | 8100 | | | | | |
| 17 | | 90 | 1778 | 160020 | 8100 | | | | | |
| 18 | | 95 | 1791 | 170145 | 9025 | | | | | |
| 19 | | 96 | 1773 | 170208 | 9216 | | | | | |
| 20 | | 98 | 1712 | 167776 | 9604 | | | | | |
| 21 | | | | | | | | | | |
| 22 | | | | | | | | | | |

Sheet1 / Sheet2

## 16. The least squares line and the mean point

A particular property of the least squares regression line of $y$ on $x$ is given as follows:

**Regression line and mean point**

For any set of bivariate data, a least squares regression line always passes through the mean point ($x$, $y$) of the data.

Ch13-eg5

| | A | B | C | D | E | F |
|---|---|---|---|---|---|---|
| 1 | | | | | | |
| 2 | n = | =COUNT(B11:B27) | | | | |
| 3 | b (num) = | =B2*D8-B8*C8 | | | | |
| 4 | b (denom) = | =B2*E8-B8^2 | | | | |
| 5 | b = | **=B3/B4** | | | | |
| 6 | a = | **=C8/B2-B5*B8/B2** | | | | |
| 7 | | | | | | |
| 8 | Totals | =SUM(B11:B20) | =SUM(C11:C20) | =SUM(D11:D20) | =SUM(E11:E20) | |
| 9 | | | | | | |
| 10 | | x | y | xy | x^2 | |
| 11 | | 79 | 1640 | =B11*C11 | =B11^2 | |
| 12 | | 83 | 1640 | =B12*C12 | =B12^2 | |
| 13 | | 85 | 1630 | =B13*C13 | =B13^2 | |
| 14 | | 86 | 1632 | =B14*C14 | =B14^2 | |
| 15 | | 88 | 1660 | =B15*C15 | =B15^2 | |
| 16 | | 90 | 1736 | =B16*C16 | =B16^2 | |
| 17 | | 90 | 1778 | =B17*C17 | =B17^2 | |
| 18 | | 95 | 1791 | =B18*C18 | =B18^2 | |
| 19 | | 96 | 1773 | =B19*C19 | =B19^2 | |
| 20 | | 98 | 1712 | =B20*C20 | =B20^2 | |
| 21 | | | | | | |
| 22 | | | | | | |

Sheet1 **Sheet2**

This property can be demonstrated for the data of Example 5. If the mean point lies on the regression line, then substituting $x = \bar{x}$ and $y = \bar{y}$ into the equation of the line should satisfy it.

Now, the mean point of the data is given by:

$$\bar{x} = \frac{\sum x}{n} = \frac{890}{10} = 89$$

$$\bar{y} = \frac{\sum y}{n} = \frac{16992}{10} = 1699.2$$

Substituting this point into the regression line $y = 954.03 + 8.37x$ gives:

$1699.2 = 954.03 + (8.37)(89)$

$= 1698.96$ (slight error due to rounding).

This shows that the mean value (89, 1699.2) *lies on the regression line.*

This is demonstrated diagrammatically in Figure 5, where the original bivariate data are shown together with the regression line.

The regression line, $y = 954.03 + 8.37x$, has been plotted by calculating two points through which it passes as follows:

when $x = 80$, $y = 954.03 + (8.37)(80) = 1624$

when $x = 100$, $y = 954.03 + (8.37)(100) = 1791$.

These two points have been plotted and the regression line drawn through them. Notice that the line *does* pass through the mean point.

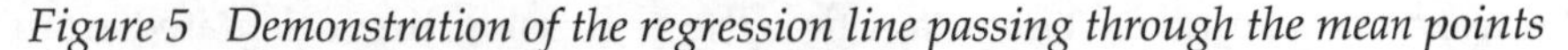

*Figure 5 Demonstration of the regression line passing through the mean points*

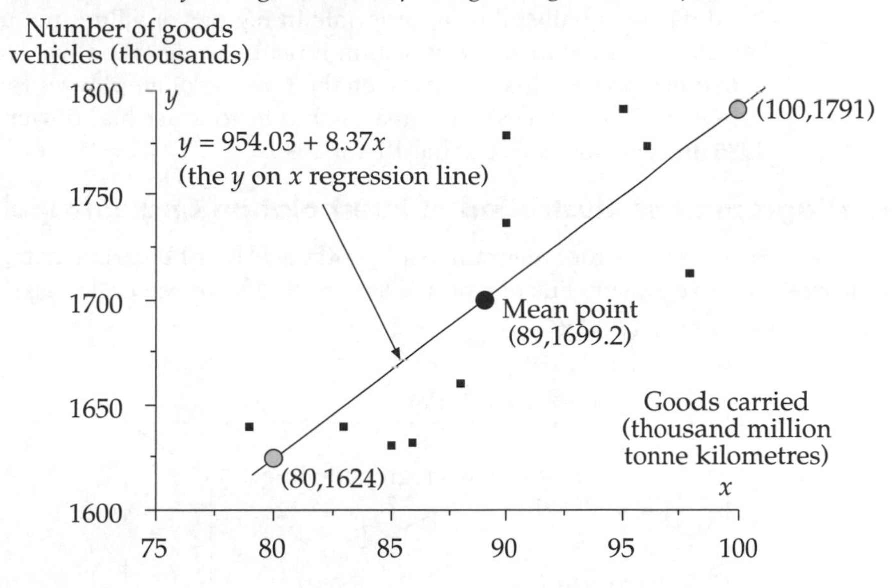


## 17. Interpolation and extrapolation

Estimation using a regression line can be classified in two distinct ways.

a) *Interpolation.*

   i. This is the name given to estimation carried out within the range of values given for the independent variable. As an example, consider the data for goods by road in the previous Example 5. The values used for $x$, to calculate the $y$ on $x$ line, ranged from 79 to 98 and so any estimate for $y$ based on a value of $x$ within this range is known as an interpolated estimate. The estimation carried out in part (b) (i) was interpolation, since the value of $x$ used was $x = 80$.

   ii. Interpolated estimates can usually be regarded with a degree of confidence.

b) *Extrapolation.*

   i. This is the name given to estimation that is based on values of the independent variable in a region that has not been considered in the calculation of the appropriate regression line. Thus, for the goods by road data, where the $x$ values ranged from 79 to 98, estimating a value for $y$ given $x = 120$, would be an example of extrapolation.

   ii. Extrapolation is most commonly used for forecasting, using values of a variable described over time, otherwise known as Time Series (and discussed later in chapters 15 to 17). For example, a firm's total number of employees ($y$) described over a number of years ($x$) and treated as bivariate data might have a $y$ on $x$ regression line calculated. One of its most obvious uses would be for forecasting levels of manpower for future years, which could be obtained by extrapolating the data (using the calculated regression line).

iii. Because one can never be certain that the regression line calculated from the data given will still be appropriate in regions of values not used in the calculation of the line, extrapolation must be undertaken with care. In the above manpower illustration, given the number of employees for the years between 1973 and 1985, it is reasonable to forecast manpower levels for 1986 and also for 1987, but hardly for 1995.

## 18. Diagrammatic illustration of interpolation and extrapolation

Figure 6 shows the scatter diagram of a hypothetical set of bivariate data, where the least squares regression line of $y$ on $x$ is assumed to have been calculated.

*Figure 6 A regression line and its use*

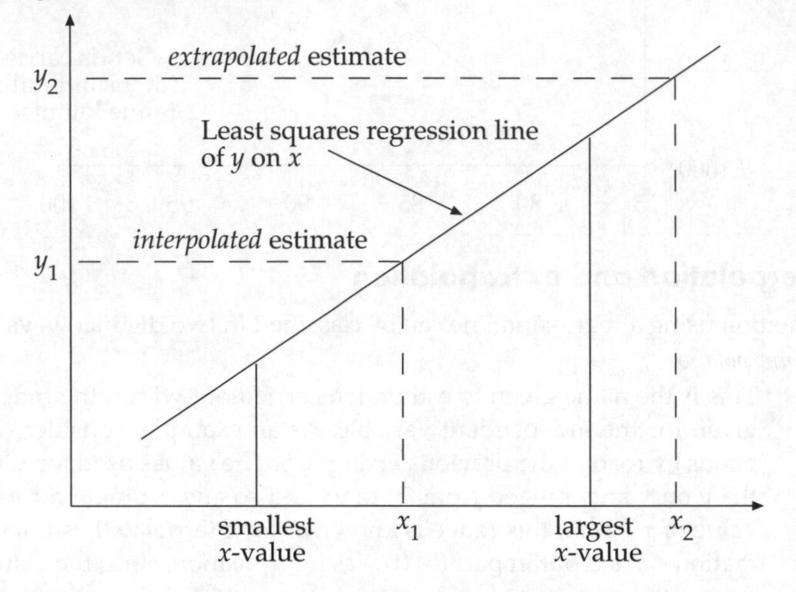


In Figure 6, $y_1$ is an interpolated estimate, since the $x$-value on which it is based lies *within the range of the given data x-values*. However, $y_2$ is an extrapolated estimate, since the $x$-value on which it is based lies *outside the range of given data x-values*.

## 19. Summary

a) Regression is concerned with producing a mathematical function which describes the relationship between two variables. It is normally used for estimation purposes.

b) There are three common methods for determining a regression line for a set of bivariate data.

   i. Inspection method,
   ii. method of Semi-averages and
   iii. method of Least Squares.

c) The method of least squares is the standard technique for obtaining a regression line.

d) For any set of bivariate data, the $y$ on $x$ regression line is used for estimating a value of $y$ given a value of $x$.

e) With respect to bivariate data, an independent variable is one that occurs naturally or is specially chosen in order to investigate another (dependent) variable. Relationships involving the dependence of one variable on the other are sometimes known as causal relationships.

f) For any set of bivariate data, the least squares regression line passes through mean point of the data.

g) Interpolation involves estimating a value of the dependent variable given a value of the independent variable within the range of the data used to calculate the regression line and can be carried out with some confidence. Estimation outside this range is known as extrapolation and the results obtained should be treated with caution.

## 20. Points to note

a) In most examination questions the bivariate variables involved will be labelled (usually $x$ and $y$) and the regression line of $y$ on $x$ will be asked for. Where this is not the case, it is usual to label the independent variable as $x$ and the dependent variable as $y$ and thus the $y$ on $x$ regression line will be appropriate.

b) Although only linear regression has been considered in this chapter, students *should be aware* that other forms of regression are sometimes appropriate and can be calculated. Figure 7 shows examples of bivariate data which display a distinct non-linear pattern, sometimes called *curvilinear regression*.

c) If the regression line $y = a + bx$ is used to describe the relationship between variables $x$ and $y$, the implication is that only variable $x$ affects the way variable $y$ behaves. However, this may not be the case, since there might be other factors or variables involved.

For example (and see exercise 7 in the following section), suppose that a regression line of overtime worked on orders received per week was being considered by a particular company. One could argue that although the number of orders received is the most important factor in determining overtime levels, other contributory factors could be: employee availability, machine readiness, material stock, complexity of order and so on. In more advanced statistical analysis this is dealt with by developing a *multiple regression model* such as, for instance: $y = a + bx + cz + dw$, where $x$, $z$ and $w$ are three numeric variables representing contributory factors which are thought to affect the value of $y$.

d) As mentioned in section 10, there are different but equivalent formulae for calculating the values of a and b in the least squares regression line $y = a + bx$. One alternative is:

**Alternative least squares regression formulae**

$$b = \frac{\sum xy - n\overline{x}\overline{y}}{\sum x^2 - n\overline{x}^2}; \quad a = \overline{y} - b\overline{x}$$

where the least squares regression line of $y$ on $x$ is $y = a + bx$.

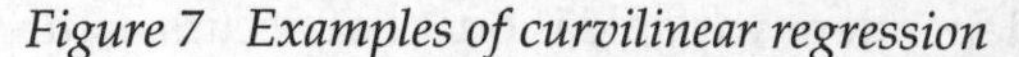

*Figure 7 Examples of curvilinear regression*

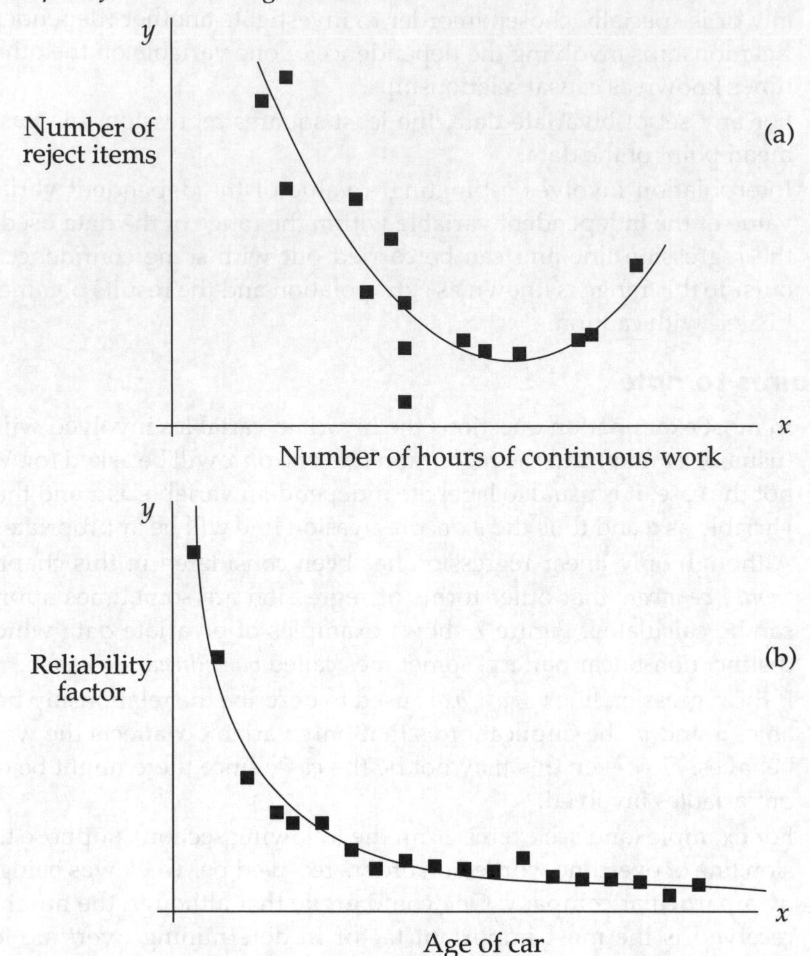


However, note that the above formula can cause difficulties since there is a temptation to use rounded (and thus inaccurate) values for $\overline{x}$ and $\overline{y}$. If you are required to use these alternative formulae, take care to calculate the mean values accurately using the normal formulae:

$$\overline{x} = \frac{\sum x}{n} \text{ and } \overline{y} = \frac{\sum y}{n}$$

rather than using rounded values.

## 21. Student self review questions

1. What is regression and what purpose does it serve? [1,2]
2. What is the 'method of inspection' for obtaining a regression line and what place does it have compared with other methods? [4]
3. What is the meaning of the expression '$y$ on $x$' regression line? [6]
4. Although the method of semi-averages is superior to the method of inspection for obtaining a regression line, why is it not used extensively? [9]

5. On what basis is the method of least squares derived? [9]
6. In the context of regression, what is an independent variable? [13]
7. What is meant by a 'causal' relationship? Give some examples of causal and non-causal relationships between variables. [13,14]
8. What is the essential difference between interpolation and extrapolation and what reliance can be placed on estimates obtained using each of them? [17]
9. Give two different, unconnected, reasons why a linear regression line such as $y = a + bx$ might not be appropriate to describe fully the movements and/or variation in values of $y$. [20(c,d)]

## 22. Student exercises

1. For the following data:

| $x$ | 1 | 3 | 4 | 6 | 8 | 9 | 11 | 14 |
|---|---|---|---|---|---|---|---|---|
| $y$ | 1 | 2 | 4 | 4 | 5 | 7 | 8 | 9 |

   a) Draw a scatter diagram.
   b) By inspection, fit a straight line to the data (ensuring it passes through the mean value).
   c) Estimate a value for $y$ corresponding to $x = 10$.
2. Rework the data of Exercise 1 using the method of semi-averages.
3. The following data were obtained by recording the number of days taken to complete 30 building contracts, where $x$ =maximum and $y$ =actual.

| $x$ | 42 | 57 | 27 | 45 | 76 | 88 | 20 | 67 | 46 | 52 | 43 | 73 | 39 | 66 | 53 |
|---|---|---|---|---|---|---|---|---|---|---|---|---|---|---|---|
| $y$ | 35 | 50 | 47 | 45 | 81 | 75 | 20 | 56 | 55 | 46 | 31 | 71 | 32 | 72 | 52 |

| $x$ | 75 | 50 | 36 | 47 | 62 | 82 | 64 | 41 | 48 | 54 | 73 | 32 | 56 | 39 | 34 |
|---|---|---|---|---|---|---|---|---|---|---|---|---|---|---|---|
| $y$ | 56 | 40 | 24 | 36 | 51 | 53 | 41 | 33 | 42 | 46 | 67 | 22 | 62 | 26 | 28 |

   Using the method of semi-averages, find the regression line of $y$ on $x$. Estimate the number of days that a contract which was allocated a maximum of 55 days would be expected to take in reality.
4. Obtain the least squares regression line of $y$ on $x$ for the following data:

| $x$ | 10 | 12 | 8 | 13 | 12 | 11 | 11 |
|---|---|---|---|---|---|---|---|
| $y$ | 27 | 41 | 27 | 32 | 34 | 26 | 38 |

5. *MULTIPLE-CHOICE.* Select the equation below which properly describes the regression line in Figure 8 that has been fitted to the plotted points by eye.
   a) $y = 6 - 6x$  b) $y = 6x + 6$  c) $y = x + 6$  d) $y = 6 - x$
6. A large company's Sales Manager has tabulated the price (£) against engine capacity (c.c.) for 10 models of car available for salesmen as follows:

| Price | 4900 | 5200 | 6160 | 7980 | 7930 | 3190 | 3190 | 5160 | 4050 | 7150 |
|---|---|---|---|---|---|---|---|---|---|---|
| Capacity | 1000 | 1270 | 1750 | 2230 | 1990 | 600 | 650 | 1500 | 1450 | 1650 |

   Obtain the least squares regression line of price on engine capacity.

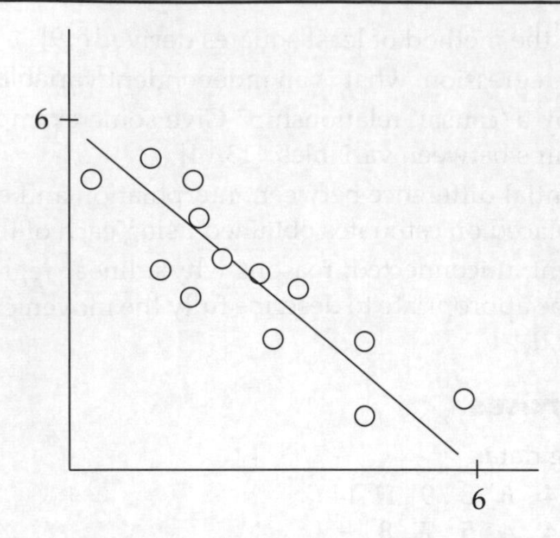


7. As part of an investigation into levels of overtime working, a company decides to tabulate the number of orders received weekly and compare this with the total weekly overtime worked to give the following:

| Week number | 1 | 2 | 3 | 4 | 5 | 6 | 7 | 8 | 9 | 10 |
|---|---|---|---|---|---|---|---|---|---|---|
| Orders received | 83 | 22 | 107 | 55 | 48 | 92 | 135 | 32 | 67 | 122 |
| Total overtime | 38 | 9 | 42 | 18 | 11 | 30 | 48 | 10 | 29 | 51 |

a) Use the method of least squares to obtain a regression line that will predict the level of total overtime necessary for 100 orders.

b) Given that normal weekly hours for an employee are currently 35, state in words what criterion is necessary for the company to consider taking on a new employee.

# 14 Correlation techniques

## 1. Introduction

Correlation is a technique used to measure the strength of the relationship between two variables. This chapter describes the general nature and purpose of correlation and gives two techniques for measuring correlation, known as product moment and rank respectively. Also covered is the connection between regression and correlation and how they can be linked through the coefficient of determination. Finally, correlation and causality are discussed, introducing the idea of 'spurious' correlation.

## 2. The purpose of correlation

The purpose of regression analysis is to identify a relationship for a given set of bivariate data. What it does not do however, is to give any indication of how good this relationship might be.

This is where correlation comes in. It provides a measure of how well a least squares regression line 'fits' the given set of data. The better the correlation, the closer the data points are to the regression line and hence the more confidence one would have in using the regression line for estimation.

As an illustration, suppose that a Data Preparation manager receives statistics on operator performance with regard to the number of errors made in keying in data from batches of documents. The two scatter diagrams shown in Figure 1 have had regression lines superimposed and show the performance of two particular operators.

*Figure 1 Two scatter diagrams showing a similar relationship but different 'spreads'*

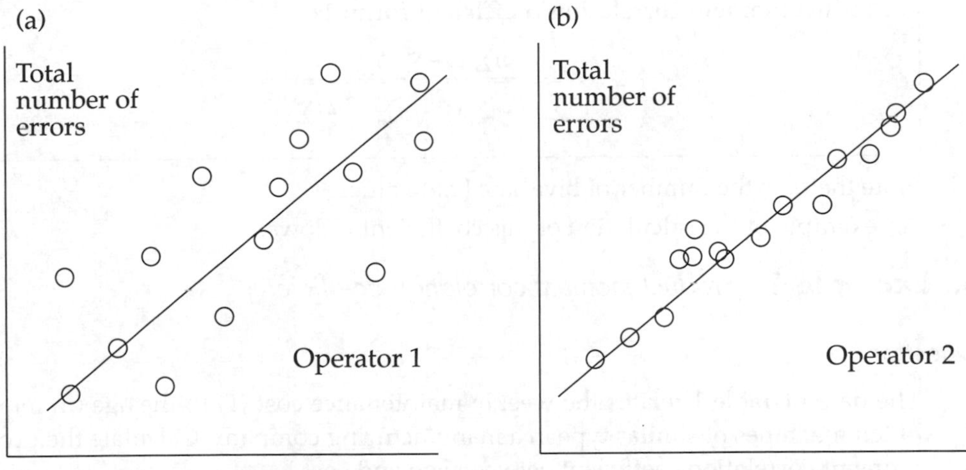


The striking difference between the two sets of data is the scatter of the points. Operator 2's data is much more consistent, giving a better fit to the regression line and thus a higher degree of correlation. We would be much more confident in estimating the number of errors that would be made by operator 2 than operator 1 if they were both about to key in the data from a number of batches.

## 3. The definition and meaning of correlation

Correlation can be defined as follows.

> **Correlation**
> *Correlation* is concerned with describing the strength of the relationship between two variables by measuring the degree of 'scatter' of the data values.

The less scattered (or variable) the data values are, the stronger the correlation is said to be. The following section describes the way in which the strength of correlation is measured.

## 4. The correlation coefficient

We need a way of measuring the strength of the correlation between two variables. This is achieved through a *correlation coefficient*, normally represented by symbol $r$. It is a number which lies between –1 and +1 (inclusive). That is: $-1 <= r <= +1$.

A value of $r = 0$ signifies that there is no correlation present, while the further away from 0 (towards –1 or +1) $r$ is, the stronger the correlation.

## 5. The product moment correlation coefficient

The standard measure of correlation that has the features described in section 4 is called the *product moment* correlation coefficient and, given a set of bivariate $(x,y)$ data, it is calculated using the following formula.

> **Product moment correlation coefficient formula**
>
> $$\mathrm{r} = \frac{n\sum xy - \sum x \sum y}{\sqrt{n\sum x^2 - (\sum x)^2}\,\sqrt{n\sum y^2 - (\sum y)^2}}$$

Note that $n$ is the number of bivariate $(x,y)$ values.

An example of the calculation of this coefficient follows.

## 6. Example 1 (*Product moment correlation coefficient*)

*Question*

The data of Table 1 relates the weekly maintenance cost (£) to the age (in months) of ten machines of similar type in a manufacturing company. Calculate the product moment correlation coefficient between age and cost.

*Answer*

The normal layout of the calculations is shown in Table 2.

*Table 1 Maintenance cost and age of ten machines*

| Machine | 1 | 2 | 3 | 4 | 5 | 6 | 7 | 8 | 9 | 10 |
|---|---|---|---|---|---|---|---|---|---|---|
| Age ($x$) | 5 | 10 | 15 | 20 | 30 | 30 | 30 | 50 | 50 | 60 |
| Cost ($y$) | 190 | 240 | 250 | 300 | 310 | 335 | 300 | 300 | 350 | 395 |

*Table 2 Normal layout for calculation of the product moment correlation coefficient*

| $x$ | $y$ | $xy$ | $x^2$ | $y^2$ |
|---|---|---|---|---|
| 5 | 190 | 950 | 25 | 36100 |
| 10 | 240 | 2400 | 100 | 57600 |
| 15 | 250 | 3750 | 225 | 62500 |
| 20 | 300 | 6000 | 400 | 90000 |
| 30 | 310 | 9300 | 900 | 96100 |
| 30 | 335 | 10050 | 900 | 112225 |
| 30 | 300 | 9000 | 900 | 90000 |
| 50 | 300 | 15000 | 2500 | 90000 |
| 50 | 350 | 17500 | 2500 | 122500 |
| 60 | 395 | 23700 | 3600 | 156025 |
| 300 | 2970 | 97650 | 12050 | 913050 |

From Table 2: $n=10$; $\sum x=300$; $\sum y=2970$; $\sum xy=97650$; $\sum x^2=12050$; $\sum y^2=913050$.

Thus: $r = \dfrac{n\sum xy - \sum x \sum y}{\sqrt{n\sum x^2 - (\sum x)^2}\sqrt{n\sum y^2 - (\sum y)^2}}$

$= \dfrac{10 \times 97650 - 300 \times 2970}{\sqrt{10 \times 12050 - 300^2}\sqrt{10 \times 913050 - 2970^2}}$

$= \dfrac{85500}{174.64 \times 556.42}$

i.e. $r = 0.880$ (3D)

The result shows a strong measure of correlation between machine maintenance cost and machine age.

## 7. Positive correlation

Correlation can exist in such a way that increases in the value of one variable tend to be associated with *increases* in the value of the other. This is known as *positive* (or *direct*) correlation. In this case, the correlation coefficient, $r$, will take a value between 0 and +1, with $r=+1$ signifying 'perfect' positive correlation.

Some examples of bivariate data which would be expected to be positively correlated are:

a) Age of employee and salary. (With age would come qualifications and experience, both of which would be reflected in a higher salary.)

b) Number of calls made by salesman and number of sales obtained. (The more calls a salesman makes, the more sales would be likely.)

c) Age of insured person and amount of premium. (The older a person is, the greater is the chance that they will fall ill or die.)

d) Machine maintenance cost and age. (Example 1 showed $r=+0.88$, a high degree of positive correlation for the particular data given.)

e) Number of vehicles licensed and road deaths.

The scatter diagram in Figure 2 illustrates the situation that might be expected for example (c).

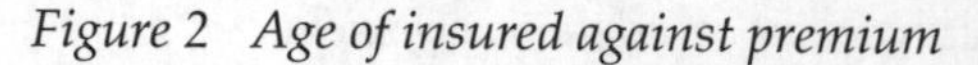

*Figure 2 Age of insured against premium*

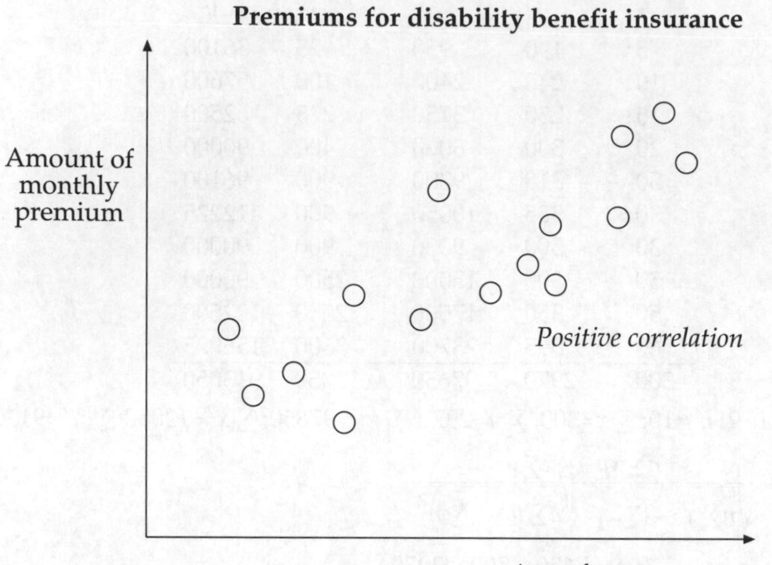


Notice that, generally speaking, the monthly premium would increase with age, with of course a certain amount of variation due to the particular state of health and past medical history of the insured. In other words, the two variables 'age' and 'amount of premium' are positively correlated.

## 8. Negative correlation

Correlation also exists when increases in the value of one variable tend to be associated with decreases in the value of the other (and vice versa). In this type of case the correlation is said to be *negative* (or *inverse*). In this case, the correlation coefficient, $r$, will take a value between 0 and –1, with $r = -1$ signifying 'perfect' negative correlation.

Some examples of bivariate data which one would expect to be negatively correlated are:

a) Number of weeks of experience and number of errors made. (As one becomes more experienced in performing a particular task, so less errors would be made.)

b) Age at death and age at retirement. (Actuarial records show that, within certain limits, the older a person is at retirement, the earlier they will be expected to die.)

c) Amount of goods sold and average cost per good.

Figure 3 shows the scatter diagram that might be expected from example a) above.

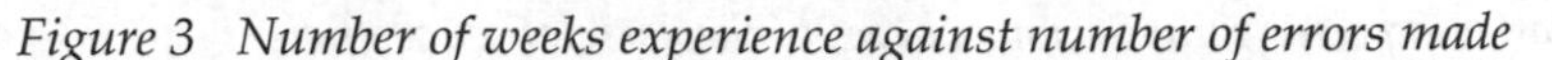

*Figure 3 Number of weeks experience against number of errors made*

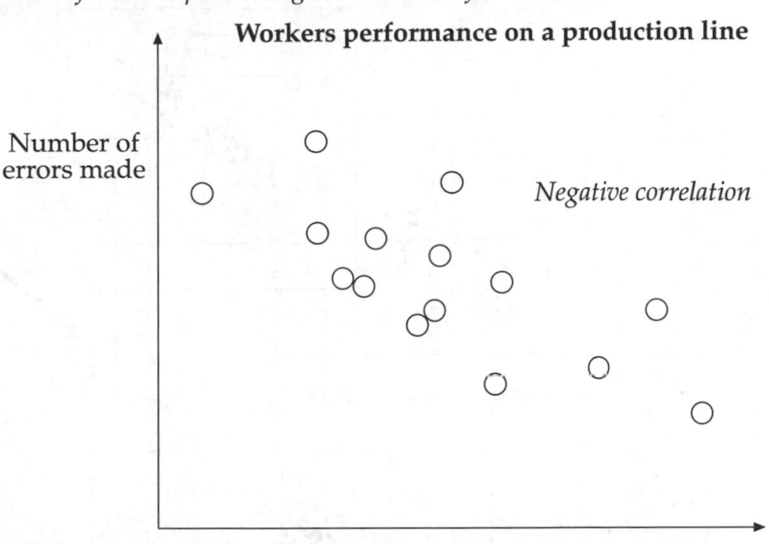


It would normally be the case that the more experience a worker had, the fewer errors would be made. There would obviously be a certain amount of variation due to particular workers' skills and learning ability.

## 9. Diagrammatic demonstration of correlation

The pictures in Figure 4 show a number of scatter diagrams, each with an approximate correlation coefficient.

Notice that when there is no correlation ($r$ =0) the points are scattered in a haphazard fashion but as the correlation increases in strength (in both directions, positive and negative) so the elliptical boundary containing the points narrows, which means there is much less scatter. When correlation is perfect, that is $r = -1$ or $r = +1$, so the points form a straight line. This straight line is the *least squares regression line of $y$ on $x$*.

## 10. Example 2 (*Product moment correlation coefficient*)

*Question*

The following data, obtained from claims drawn on life assurance policies for a particular category of employment, relates age at official retirement to age at death for nine males.

| | | | | | | | | | |
|---|---|---|---|---|---|---|---|---|---|
| Age at retirement | 57 | 62 | 60 | 57 | 65 | 60 | 58 | 62 | 56 |
| Age at death | 71 | 70 | 66 | 70 | 69 | 67 | 69 | 63 | 70 |

Calculate the product moment coefficient of correlation between age at retirement and age at death.

*Figure 4 Patterns of scatter expected with various values of the correlation coefficient*

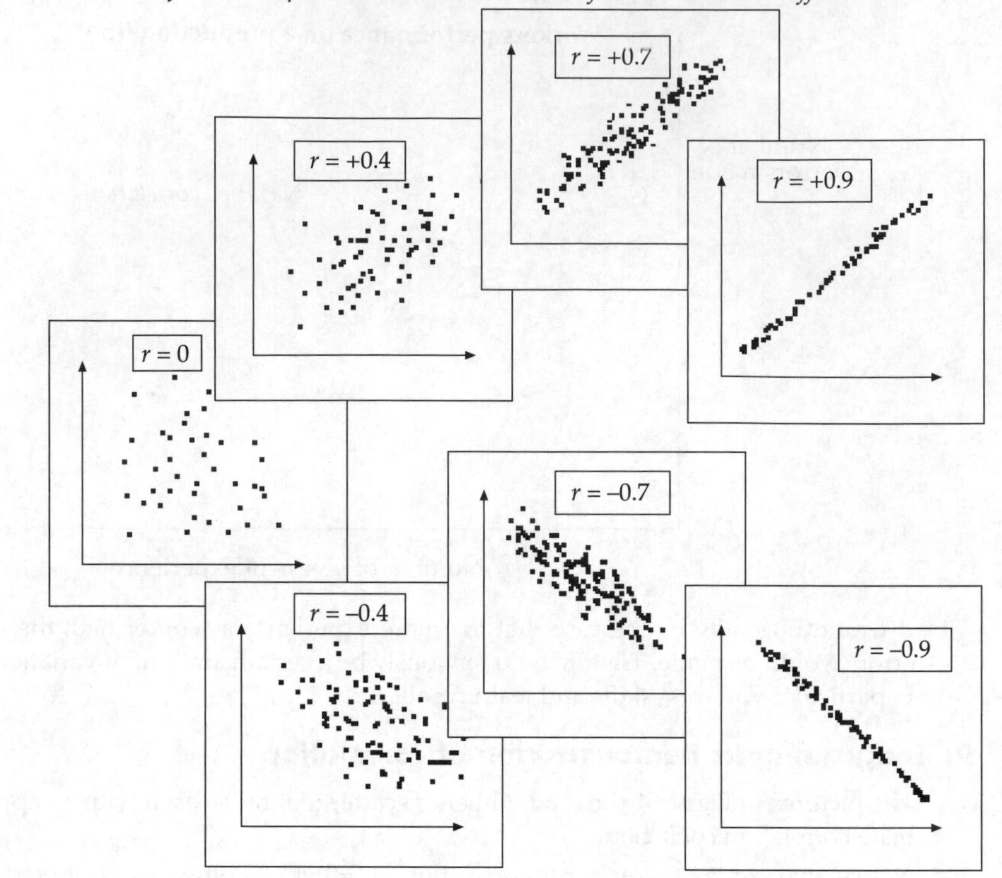


## Answer

Labelling age at retirement as variable $x$ and age at death as variable $y$, the following table shows the standard layout of the calculations.

| $x$ | $y$ | $xy$ | $x^2$ | $y^2$ |
|---|---|---|---|---|
| 57 | 71 | 4047 | 3249 | 5041 |
| 62 | 70 | 4340 | 3844 | 4900 |
| 60 | 66 | 3960 | 3600 | 4356 |
| 57 | 70 | 3990 | 3249 | 4900 |
| 65 | 69 | 4485 | 4225 | 4761 |
| 60 | 67 | 4020 | 3600 | 4489 |
| 58 | 69 | 4002 | 3364 | 4761 |
| 62 | 63 | 3906 | 3844 | 3969 |
| 56 | 70 | 3920 | 3136 | 4900 |
| 537 | 615 | 36670 | 32111 | 42077 |

Thus: $n = 9$; $\sum x = 537$; $\sum y = 615$; $\sum xy = 36670$; $\sum x^2 = 32111$; $\sum y^2 = 42077$.

Using the formula for the product moment coefficient in section 5 gives the following calculations.

$$r = \frac{n\sum xy - \sum x \sum y}{\sqrt{n\sum x^2 - (\sum x)^2}\sqrt{n\sum y^2 - (\sum y)^2}}$$

$$= \frac{9 \times 36670 - 537 \times 615}{\sqrt{9 \times 32111 - 537^2}\sqrt{9 \times 42077 - 615^2}}$$

$$= \frac{225}{174.64 \times 556.42}$$

i.e. $r = -0.414$ (3D)

This result shows only moderate negative correlation.

## 11. Correlation, regression and variation

Correlation was defined earlier as measuring the degree of scatter or variation that bivariate data exhibit. A particular use of the correlation coefficient is to determine the appropriateness of a corresponding *linear* least squares regression line. That is, given the regression line in the form $y = a + bx$ and a set of observed data, how well do the given variations in $x$ alone account for the observed variations in $y$? If, for example, the turnover of a company was being regressed against advertising expenditure, it would be of considerable value to know just how much of the variability of turnover was due to factors other than advertising; such as, for example, market share or product life.

The following sections explain how the $y$-variations can be split up and the particular way in which the correlation coefficient, $r$, is used to account for variation.

## 12. Explained and unexplained variation

Suppose figures for turnover ($y$) and advertising expenditure ($x$) were available for a company over a number of months. The turnover ($y$) values are obviously going to vary from one month to the next, and the best estimate of a typical value of turnover (given *no information* about advertising expenditure) is the *mean value* of turnover over the period. The variation between individual $y$ values and their mean is known as the *total variation*.

However, a better estimate of $y$ (given a specific value of $x$) can be obtained from the $y$ on x regression line. The differences between the values of $y$ given by the $y$ on $x$ regression line and the mean of the $y$'s is known as *explained variation* ('explained' by the $x$-variations).

The numerical difference between total and explained variation is thus called *unexplained variation*, since it must have been caused by factors (variables) not taken into account by the regression model.

## 13. The coefficient of determination

The *coefficient of determination* is the ratio of explained variation to total variation and is obtained by *squaring the value of r* (the product moment correlation coefficient).

In words, the coefficient of determination gives the proportion of all the variation (in the $y$-values) that is explained (by the variation in the $x$-values).

Suppose that for the example of turnover ($y$) measured against advertising expenditure ($x$), given in section 11, the correlation coefficient is calculated as $r = 0.76$.

Then: coefficient of determination, $r^2 = (0.76)^2$

$= 0.58$ (2D).

This means that *only 58% of the variation in turnover* is due to advertising expenditure.

Put another way, 42% of variation in turnover is due to factors other than advertising expenditure; perhaps product quality, changing trends or productivity.

**Coefficient of determination**

$$\text{cd} = \frac{\text{explained variation in all items}}{\text{total variation in all items}} = r^2$$

where $r$ is the product moment correlation coefficient.

Notice that, since $-1 < r < +1$, it follows that $0 < r^2 < +1$.

## 14. Example 3 (Calculation and interpretation of the *coefficient of determination*)

In Example 1, the correlation coefficient between maintenance cost and age of a set of ten machines was $r = 0.88$(2D).

Thus, the coefficient of determination $= r^2 = 0.882 = 0.77$(2D).

That is, only 77% of the variation in the maintenance costs is accounted for by the variation in the machine ages. The other 23% of the variation comes, perhaps, from amount of machine use and/or operator experience.

## 15. Spearman's rank correlation coefficient

An alternative method of measuring correlation, based on the *ranks* of the sizes of item values, is available and known as *rank correlation.*

The measure of rank correlation most commonly used is known as *Spearman's* rank correlation coefficient and the procedure for obtaining it is given as follows.

STEP 1 Rank the $x$ values (to give $r_x$ values).

STEP 2 Rank the $y$ values (to give $r_y$ values).

Note that the rankings of the $x$-values are performed quite independently of the rankings of the $y$-values and ranking is normally performed in ascending order (although this is not essential).

An example follows.

| $x$ | $r_x$ | $y$ | $r_y$ |
|---|---|---|---|
| 23 | 2 | 245 | 4 |
| 35 | 3 | 236 | 2 |
| 18 | 1 | 238 | 3 |
| 36 | 4 | 232 | 1 |
| 41 | 5 | 250 | 6 |
| 43 | 6 | 247 | 5 |
| 48 | 7 | 252 | 7 |

STEP 3 For each pair of ranks, calculate $d^2 = (r_x - r_y)^2$
Using the data above:
the first pair of ranks is 2 and 4, giving $d^2 = (2–4)^2 = 4$;
the second pair of ranks is 3 and 2, giving $d^2 = (3–2)^2 = 1$;
... and so on, giving the other $d^2$ values as 4, 9, 1, 1 and 0.

STEP 4 Calculate $\sum d^2$
Using the working data:
$\sum d^2 = 4 + 1 + 4 + 9 + 1 + 1 + 0 = 20.$

STEP 5 The value of the rank correlation coefficient can then be found using the following formula:

**Rank correlation coefficient**

$$r' = 1 - \frac{6\sum d^2}{n(n^2 - 1)}$$

where $n$ is the number of bivariate pairs.

Using the working data above: $r' = 1 - \frac{6 \times 20}{7 \times 48} = 0.643$ (3D)

## 16. Rank correlation and its uses

Rank correlation is used in the following circumstances.

a) *As an approximation to the product moment correlation coefficient*. Rank correlation is a quick and easy technique and so is sometimes used as an approximation to product moment correlation. This is particularly appropriate if the values of numeric bivariate data are difficult to obtain physically or involve great expense and yet can be ranked in size order.

b) *To measure correlation between non-numeric variables*. If one or both of the variables involved is non-numeric, the product moment correlation coefficient *cannot be calculated*. However, as long as the non-numeric values can be ranked in some natural way, rank correlation can be used. For instance, status or importance can usually be ranked; a set of companies can be ranked by industrial classification; in competitions, ranking is naturally performed by placing competitors as first, second, third and so on.

## 17. Example 3 (*Spearman's rank* correlation coefficient)

*Question*

The data of Table 3 show the average rent and rates (£ per square foot) for a selection of areas. Calculate Spearman's rank correlation coefficient to assess whether there is any correlation between rates and rent.

*Table 3 Average rent and rates for a selection of areas*

| | | | | | | | | | | |
|---|---|---|---|---|---|---|---|---|---|---|
| Rates ($x$) | 1.68 | 1.46 | 1.57 | 13.37 | 3.18 | 1.95 | 1.07 | 1.71 | 1.22 | 6.46 |
| Rent ($y$) | 3.81 | 4.19 | 4.87 | 22.85 | 6.47 | 6.48 | 2.66 | 6.49 | 5.33 | 15.23 |

*Answer*

The standard layout for calculations is shown in Table 4, where the steps detailed in the previous section have been carried out in sequence.

*Table 4 Standard layout for calculations*

| Rates | Rank of $x$ | Rent | Rank of $y$ | |
|---|---|---|---|---|
| $x$ | $r_x$ | $y$ | $r_y$ | $d^2$ |
| 1.68 | 5 | 3.81 | 2 | 9 |
| 1.46 | 3 | 4.19 | 3 | 0 |
| 1.57 | 4 | 4.87 | 4 | 0 |
| 13.37 | 10 | 22.85 | 10 | 0 |
| 3.18 | 8 | 6.47 | 6 | 4 |
| 1.95 | 7 | 6.48 | 7 | 0 |
| 1.07 | 1 | 2.66 | 1 | 0 |
| 1.71 | 6 | 6.49 | 8 | 4 |
| 1.22 | 2 | 5.33 | 5 | 9 |
| 6.46 | 9 | 15.23 | 9 | 0 |
| | | | | Total = 26 |

From the table: $n = 10$ and $\sum d^2 = 26$.

Thus: $$r' = 1 - \frac{6\sum d^2}{n(n^2-1)} = \frac{6\times 26}{10\times 99} = 0.842 \text{ (3D)}$$

This result demonstrates relatively high positive correlation. Thus high rates tend to be paired with high rents and vice versa.

## 18. Notes on the rank correlation procedure

a) Clearly, if rankings are already given for one or both sets of bivariate values, steps 1 and 2 in the procedure (of section 15) would not be necessary.

b) Ranks are usually allocated in ascending order; rank 1 to the smallest item, rank 2 to the next largest and so on, although it is perfectly feasible to allocate in descending order. However, whichever method is selected *must be used on both variables.*

c) If one or more groups of data items have the same value (known as *tied values*), the ranks that would have been allocated separately must be averaged and this average rank given to each item with this equal value. For example, the four numbers 14, 26, 26 and 28 would be allocated ranks 1, 2.5, 2.5 and 4 respectively (since two items have value 26, they each must be allocated the average of ranks 2 and 3. i.e. 2.5). The numbers 9, 8, 9, 9, 5, 10 and 12 would be allocated ranks 4, 2, 4, 4, 1, 6 and 7 respectively (since three items share the value 9, they must each be ranked as the average of 3, 4 and 5. i.e. rank 4).

d) Given a set of numeric bivariate data, both rank and product moment coefficients can be calculated and in general slightly different results will be obtained. *It should be understood* that the rank coefficient here is an *approximation* to the product moment coefficient.

## 19. Example 4 (*Rank* correlation coefficient for data with *tied values*)

### *Question*

The following data relate to the number of vehicles owned and road deaths for the populations of 12 countries.

| | | | | | | | | | | | | |
|---|---|---|---|---|---|---|---|---|---|---|---|---|
| Vehicles per 100 population | 30 | 31 | 32 | 30 | 46 | 30 | 19 | 35 | 40 | 46 | 57 | 30 |
| Road deaths per 100,000 population | 30 | 14 | 30 | 23 | 32 | 26 | 20 | 21 | 23 | 30 | 35 | 26 |

Calculate Spearman's rank correlation coefficient and comment on the result.

### *Answer*

Labelling vehicles as $x$ and deaths as $y$, the following table shows the results and layout of calculations.

| | Rank of $x$ | | Rank of $y$ | |
|---|---|---|---|---|
| $x$ | $r_x$ | $y$ | $r_y$ | $d^2$ |
| 30 | 3.5 | 30 | 9 | 30.25 |
| 31 | 6 | 14 | 1 | 25 |
| 32 | 7 | 30 | 9 | 4 |
| 30 | 3.5 | 23 | 4.5 | 1 |
| 46 | 10.5 | 32 | 11 | 0.25 |
| 30 | 3.5 | 26 | 6.5 | 9 |
| 19 | 1 | 20 | 2 | 1 |
| 35 | 8 | 21 | 3 | 25 |
| 40 | 9 | 23 | 4.5 | 20.25 |
| 46 | 10.5 | 30 | 9 | 2.25 |
| 57 | 12 | 35 | 12 | 0 |
| 30 | 3.5 | 26 | 6.5 | 9 |

In the previous table, successive values of $d^2$ have been evaluated as:

$(9 - 3.5)^2 = (5.5)^2 = 30.25;$

$(1 - 6)^2 = (5)^2 = 25;$

$(9 - 7)^2 = (2)^2 = 4;$

... and so on.

This results in $\sum d^2 = 127.00$ with $n = 12$.

$$r' = 1 - \frac{6\sum d^2}{n(n^2-1)}$$
$$= 1 - \frac{6\times127}{12\times143}$$
$$= 0.56 \text{ (2D)}$$

The result shows a moderate degree of positive correlation indicating that there is some correlation between vehicles owned and number of road deaths, but the relationship is not strong.

## 20. Comparison of rank and product moment correlation

Listed below are some significant features of each method, with '+' and '–' signifying whether the feature can be thought of as an advantage or disadvantage respectively.

a) *Product moment coefficient.*
   1. The standard measure of correlation. (+)
   2. Data must be numeric. (–)
   3. The calculations can be awkward. (–) This is particularly significant if there are many digits in each value of the original data.

b) *Rank coefficient.*
   1. Only an approximation to the product moment coefficient. (–)
   2. Easier to use with less involved calculations. (+)
   3. Can be used with non-numeric data. (+)
   4. Can be insensitive to small changes in actual values. (–) This is easily seen using the data values 12.3, 12.4 and 23, say, where the allocated ranks would be 1, 2 and 3. No account is taken of the small difference between the first two values compared with the large difference between the second and third values.

## 21. Correlation and causality

a) It will be recalled that a *causal relationship* is said to exist between two variables when the value of one is directly attributable to the other. Another way of thinking of this concept is that there is a distinct 'cause and effect' relationship between the two variables. Further simple examples of this type of situation are:

   Age of machine (cause) v Maintenance cost (effect);

   Number of TV licenses (cause) v Number of cinema attendances (effect);

   Age of insured person (cause) v Cost of life assurance premium (effect);

   In cases such as these it is usual to expect some degree of correlation, but it need not necessarily be strong.

b) Correlation might exist between *two variables* (and it could be strong), yet no causal relationship exist. This is sometimes known as *spurious correlation*. Two examples are:

   i. Yearly profit of company and rateable value of premises. Both of these variables are likely to increase if the size of the firm increases, but the two variables are not directly associated. That is, there is no causal relationship present.

ii. Average hours worked per week and percentage of fibre in diet. Clearly, the results of automation, general affluence and social awareness are cutting down the hours we all need to work, while general medical awareness is causing diets to become healthier. The correlation in this case, based, say, on yearly measurements, will obviously be high and the spurious deduction one would not wish to make is that 'healthy eating causes laziness'!

c) The conclusions that can be drawn from a) and b) above are:

i. Correlation *does not necessarily imply* causality;

whereas: ii. Causality *normally implies* correlation.

## 22. Summary

a) Correlation is concerned with describing how well two variables are associated by measuring the degree of 'scatter' of the data values.

b) There are two types of correlation that can exist:

i. Positive (or direct). This is where increases in one variable are associated with increases in the other.

ii. Negative (or inverse). This is where increases in one variable are associated with decreases in the other.

c) A quantitative measure of correlation is given by the (product moment) correlation coefficient, $r$, which satisfies $-1 <= r <= +1$, and:

i. $r = -1$ signifies perfect negative correlation

ii. $r = 0$ signifies no correlation

iii. $r = +1$ signifies perfect positive correlation.

iv. The product moment correlation coefficient is the standard measure of correlation for numeric data. It cannot be used with non-numeric data.

d) The product moment correlation coefficient is the standard measure of correlation for numeric data. It cannot be calculated for numeric data.

e) The coefficient of determination, $r^2$, is used to indicate the proportion of the total variation in the dependent variable ($y$) that is due to variations in the independent variable ($x$).

f) Spearman's rank correlation coefficient can be used:

i. as an approximation to the product moment coefficient,

ii. with non-numeric data that can be ranked.

g) Correlation does not necessarily imply causality.

## 23. Points to note

a) The type of correlation (i.e. positive or negative) can always be seen clearly from a scatter diagram as long as the lengths of the two axes are the same and each covers the range of values over which each variable is distributed. For example, suppose the $x$-values ranged from 20 to 40 and the $y$-values ranged from 300 to 1200. In this case a range of 20 (i.e. 40–20) on the $x$ axis should cover exactly the same physical distance as a range of 900 (1200–300) on the $y$ axis. In other words, always represent a scatter diagram inside an approximate square.

b) The correlation coefficient as described in this chapter is only concerned with *linear relationships* involving *two variables*. Thus a low value (either positive or negative) of $r$ signifies only that there is no significant (linear, two-variable) association present on the evidence of the data given. However, there still could be correlation:
   i. of the curvilinear type (for example $y = ax^2 + bx + c$), or
   ii. of the multiple linear type (for example $y = ax + bz + cw + d$).

c) Rank correlation is only considered as an approximation to product moment correlation when the two variables involved are numeric. If, however, one or both of the variables are non-numeric, rank is the only type of correlation that can be defined.

e) An alternative formula for the calculation of the product moment correlation coefficient which is sometimes used is given as follows.

**Alternative product moment correlation coefficient formula**

$$r = \frac{\sum xy - n\bar{x}\bar{y}}{\sqrt{\sum y^2 - n\bar{y}^2}\sqrt{\sum x^2 - n\bar{x}^2}}$$

Note however that care must be taken in the use of the above formula. Approximations, particularly for the two means, should *not* be used.

## 24. Student self review questions

1. How is correlation defined and how is it connected with regression? [1,2,3]
2. What is a correlation coefficient and between what limits must it lie? [4]
3. What are the two distinct types of correlation that can exist between two variables and give two examples of each? [7,8]
4. Diagrammatically illustrate examples of:
   i. moderate positive correlation,
   ii. no correlation, and
   iii. strong negative correlation. [9]
5. What is the coefficient of determination? [13]
6. What purpose does a rank correlation coefficient serve? [16]
7. What is a causal relationship and why is an understanding of it important in the interpretation of the value of a correlation coefficient? [21]

## 25. Student exercises

1. Calculate the product moment correlation coefficient for the following data:

| $x$ | 1 | 3 | 4 | 6 | 8 | 9 | 11 | 14 |
|---|---|---|---|---|---|---|---|---|
| $y$ | 1 | 2 | 4 | 4 | 5 | 7 | 8 | 9 |

2. *MULTI-CHOICE*. The rank correlation coefficient between the ages and the resale values of lorries in a large haulage fleet is 'minus 1'. Which one of the options below best describes the situation.

a) perfect correlation exists b) no correlation exists
c) weak negative correlation exists d) an error has been made in calculations

3. *MULTI-CHOICE*. Which one of the options below best describes the correlation shown in the diagram.

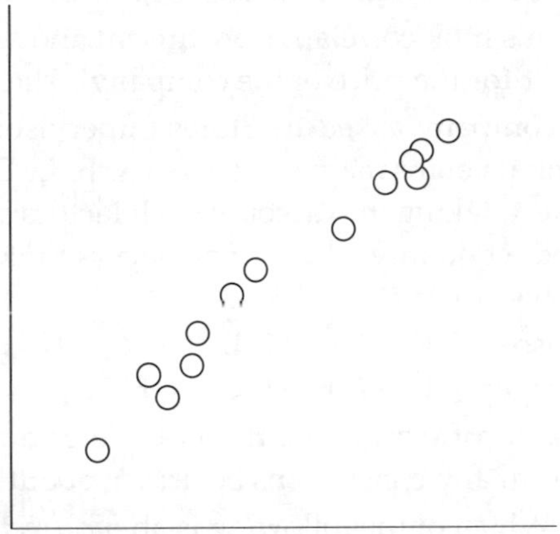

a) Strong positive correlation b) Moderate negative correlation
c) Strong negative correlation d) No correlation

4. The following data relate to length of public roads (000km) and number of goods vehicles with current licenses (000) for eight successive years.

| Length of roads | 316 | 318 | 320 | 322 | 325 | 327 | 329 | 331 |
|---|---|---|---|---|---|---|---|---|
| Number of licences | 1692 | 1640 | 1640 | 1630 | 1632 | 1660 | 1736 | 1778 |

Calculate the product moment correlation coefficient.

5. The following figures give (in units of £10m) the turnover and profit before taxation for a firm.

| Turnover | 106 | 125 | 147 | 167 | 187 | 220 |
|---|---|---|---|---|---|---|
| Profit | 10 | 12 | 16 | 17 | 18 | 22 |

Calculate the coefficient of determination for this data and comment on the result.

6. A cost accountant has derived the total cost (£000) against output (000) of standard size boxes from a factory over a period of ten weeks, yielding the following data.

| Output | 20 | 2 | 4 | 23 | 18 | 14 | 10 | 8 | 13 | 8 |
|---|---|---|---|---|---|---|---|---|---|---|
| Cost | 60 | 25 | 26 | 66 | 49 | 48 | 35 | 18 | 40 | 33 |

a) Calculate the product moment coefficient of correlation.
b) Interpret the result with a view to future extrapolation.

7. The following data shows median regional incomes for men aged 21 years and over in full-time employment and average regional house purchase prices for a particular year for the twelve major regions of the United Kingdom.

| Median income (£) | 57 | 54 | 54 | 51 | 63 | 56 | 52 | 56 | 55 | 55 | 56 | 50 |
|---|---|---|---|---|---|---|---|---|---|---|---|---|
| House purchase price (£000) | 10 | 9 | 10 | 12 | 15 | 15 | 12 | 11 | 10 | 10 | 11 | 10 |

Calculate:
a) The product moment correlation coefficient
b) Spearman's rank correlation coefficient.

8. On ten different days (picked at random) the following values were obtained for the price of a share for a particular company together with the value of the FT Index on that day.

| Share price (p) | 77 | 46 | 80 | 76 | 65 | 71 | 60 | 75 | 76 | 88 |
|---|---|---|---|---|---|---|---|---|---|---|
| FT Index | 319 | 315 | 387 | 339 | 383 | 340 | 340 | 356 | 358 | 398 |

Calculate Spearman's rank correlation coefficient and say whether the FT Index is a reasonable indicator for the price of the company's share.

9. As an exercise, a company asked its Stores Supervisor and Purchase Manager to independently rank its eight main suppliers (A, B, C, D, E, F, G and H) in order of value to the company, taking into account such factors as reliability, volume, special discounts and product quality. The two managers ranked the suppliers in order of preference as follows:

| Stores supervisor | E | C | G | H | B | D | A | F |
|---|---|---|---|---|---|---|---|---|
| Purchase manager | E | G | B | D | C | A | H | F |

Use Spearman's rank correlation coefficient to determine the amount of agreement between the two. Can any conclusions be drawn about the suppliers?

10. *MULTI-CHOICE*. Which of the following is an *incorrect* statement.
    a) The product moment correlation coefficient can be used with non-numeric data.
    b) The product moment correlation coefficient is the definitive measure of correlation.
    c) The rank correlation coefficient can be used as an approximation to the product moment correlation coefficient.
    d) The rank correlation coefficient can be used with non-numeric data.

# Examination examples (with worked solutions)

*Question 1*

A national consumer protection society investigated seven brands of paint to determine their quality relative to price. The society's conclusions were ranked according to the following table:

| Brand | T | U | V | W | X | Y | Z |
|---|---|---|---|---|---|---|---|
| Price/litre | 1.92 | 1.58 | 1.35 | 1.60 | 2.05 | 1.39 | 1.77 |
| Quality ranking | 2 | 6 | 7 | 4 | 3 | 5 | 1 |

Using Spearman's rank correlation coefficient, determine whether the consumer generally gets value for money.

*CIMA*

*Answer*

The quality value ($y$ say) has already been ranked in descending order of quality and it seems reasonable to do the same for price ($x$ say).

| | | | | | | | |
|---|---|---|---|---|---|---|---|
| $r_x$ | 2 | 5 | 7 | 4 | 1 | 6 | 3 |
| $r_y$ | 2 | 6 | 7 | 4 | 3 | 5 | 1 |
| $d^2$ | 0 | 1 | 0 | 0 | 4 | 1 | 4 |

With $n$ =7 and $\sum d^2$=10, the calculations for the rank coefficient are:

$$r' = 1 - \frac{6\sum d^2}{n(n^2-1)}$$

$$= \frac{6\times10}{7\times48}$$

$$= 0.821 \text{ (3D)}$$

A coefficient of this value shows a high degree of positive correlation which means that in general the consumer gets value for money.

*Question 2*

*Marks and Spencer plc*

| Year | 1977 | 1978 | 1979 | 1980 | 1981 | 1982 |
|---|---|---|---|---|---|---|
| Turnover (£10m) | 106 | 125 | 147 | 167 | 187 | 220 |
| Profit before taxation (£10m) | 10 | 12 | 16 | 17 | 18 | 22 |

i. Plot a scatter diagram showing the relationship between profit before taxation and turnover.

ii. Calculate the least squares regression line of profit before taxation on turnover.

iii. Comment generally on your results.

*ICSA*

*Answer*

i.

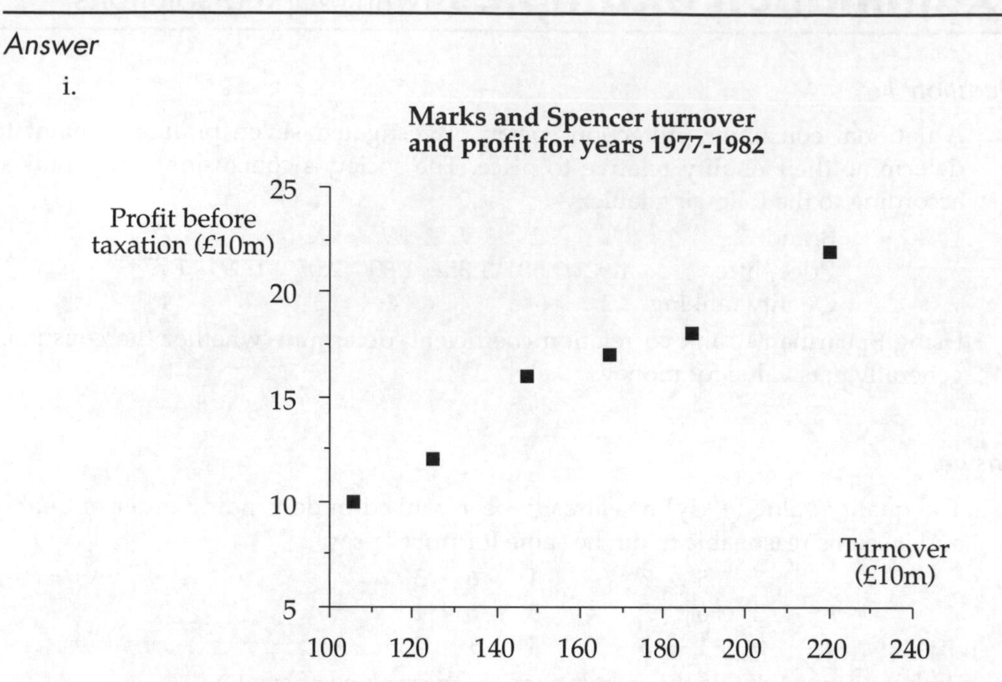


ii. For this data, the independent variable is turnover ($x$ say) and the dependent variable is profit before taxation ($y$ say). The table of values and calculations are shown below.

| Year | Turnover | Profit | | |
|---|---|---|---|---|
| | $x$ | $y$ | $xy$ | $x^2$ |
| 1977 | 106 | 10 | 1060 | 11236 |
| 1978 | 125 | 12 | 1500 | 15625 |
| 1979 | 147 | 16 | 2352 | 21609 |
| 1980 | 167 | 17 | 2839 | 27889 |
| 1981 | 187 | 18 | 3366 | 34969 |
| 1982 | 220 | 22 | 4840 | 48400 |
| | 952 | 95 | 15957 | 159728 |

For the calculations, $n = 6$; $\Sigma x = 952$; $\Sigma y = 95$; $\Sigma xy = 15957$; $\Sigma x^2 = 159728$.

The regression line required is $y$ on $x$, in the form $y = a + bx$.

$$\text{Thus: } b = \frac{n\sum xy - \sum x \sum y}{\sqrt{n\sum x^2 - \left(\sum x\right)^2}\sqrt{n\sum y^2 - \left(\sum y\right)^2}} = \frac{6\times 15957 - 952\times 95}{6\times 159728 - 952^2}$$

$$= \frac{5302}{52064} = 0.1018 \text{ (4D)}$$

$$a = \frac{\sum y}{n} - b\frac{\sum x}{n} = \frac{95}{6} - 0.1018\times\frac{952}{6} = -0.3193 \text{ (4D)}$$

The required regression line is thus: $y = -0.3193 + 0.1018x$

iii. The scatter diagram shows good correlation between the variables and, notwithstanding the small sample size ($n$=6), so the regression line will be useful for extrapolating profit from turnover for 1983.

# Examination questions

1. A cost accountant has derived the following data on the weekly output of standard size boxes from a factory.

| Week | Output ($x$) (thousands) | Total cost ($y$) (£000) |
|---|---|---|
| 1 | 20 | 60 |
| 2 | 2 | 25 |
| 3 | 4 | 26 |
| 4 | 23 | 66 |
| 5 | 18 | 49 |
| 6 | 14 | 48 |
| 7 | 10 | 35 |
| 8 | 8 | 18 |
| 9 | 13 | 40 |
| 10 | 8 | 33 |

$\Sigma x = 120$; $\Sigma y = 400$; $\Sigma x^2 = 1866$; $\Sigma y^2 = 18200$; $\Sigma xy = 5704$;

a) Plot a scatter diagram of the data.

b) Which weekly outputs, if any, appear to be different from the rest of the data?

c) State the co-ordinates of a point which must lie on a regression line fitted to the data above.

d) Find the least squares regression of total cost on output, and plot the line on the graph.

e) What is the fixed cost of the factory?

f) In a given week it is planned to produce 25000 standard size boxes. Use your regression equation to estimate the total cost of producing this quantity.

*CIMA*

2. Your company is planning a take-over of a small UK chain of multiple stores whose main competitors are co-operatives. As part of the preliminary work, you have been asked to investigate relationships between turnover, number of stores and region. Data for nine regions on the number of stores and turnover (£m) of multiples and co-operatives in 1985 is given below:

**Multiples**

| Region: | A | B | C | D | E | F | G | H | I |
|---|---|---|---|---|---|---|---|---|---|
| Stores ($X$): | 952 | 253 | 360 | 484 | 593 | 639 | 498 | 371 | 416 |
| Turnover ($Y$): | 3657 | 819 | 1250 | 1302 | 1861 | 1635 | 1452 | 717 | 1179 |

**Co-operatives**

| Region: | A | B | C | D | E | F | G | H | I |
|---|---|---|---|---|---|---|---|---|---|
| Stores ($X$): | 379 | 322 | 210 | 366 | 575 | 451 | 498 | 257 | 550 |
| Turnover ($Y$): | 260 | 236 | 194 | 308 | 445 | 427 | 286 | 130 | 335 |

*(Source: Market Research Society Yearbook, 1986 amended)*

The least squares regressions of turnover on stores have been calculated via MINITAB (a computer program). They are as follows:

$$Y = -508.50 + 4.04X, \quad r = 0.95 \text{ [Multiples]}$$

$$Y = 22.73 + 0.67X, \quad r = 0.83 \text{ [Co-operatives]}$$

a) On the same graph, plot scatter diagrams of $Y$ against $X$, including the regression lines, and interpret your results.

b) For a region with 500 stores of each type, predict the turnover for Multiples and for Co-operatives, and comment on the likely accuracy of these predictions.

c) By any method you consider appropriate, compare turnover per store between Multiples and Co-operatives.

*CIMA*

3. a) 'Correlation does not prove causation.' Discuss this statement.

b) Cinema admissions and colour television licences:

| Year | Cinema admissions (millions) | Colour T.V. licences (millions) |
|---|---|---|
| 1973 | 134 | 5.0 |
| 1974 | 138 | 6.8 |
| 1975 | 116 | 8.3 |
| 1976 | 104 | 9.6 |
| 1977 | 103 | 10.7 |
| 1978 | 126 | 12.0 |
| 1979 | 112 | 12.7 |
| 1980 | 96 | 12.9 |

*(Source: Annual abstract)*

Calculate the product moment correlation coefficient between the number of cinema admissions and the number of colour television licences issued. Briefly comment on your results.

*ICSA*

4. A cost accountant has derived the following data on the running costs (£ hundreds) and distance travelled (thousands of miles) by twenty of a company's fleet of new cars used by its computer salesmen last year. Ten of the cars are type $F$ and ten are type $L$.

| | Car $F$ | | Car $L$ | |
|---|---|---|---|---|
| | Distance travelled | Running costs | Distance travelled | Running costs |
| | $x$ | $y$ | $x$ | $y$ |
| | 4.0 | 5.3 | 3.5 | 6.9 |
| | 4.6 | 6.7 | 4.6 | 7.6 |
| | 5.9 | 7.5 | 5.3 | 7.9 |
| | 6.7 | 8.8 | 6.0 | 8.3 |
| | 8.0 | 8.0 | 7.2 | 8.8 |
| | 8.9 | 9.1 | 8.4 | 9.2 |
| | 8.9 | 10.5 | 10.1 | 9.6 |
| | 10.1 | 10.0 | 11.1 | 10.3 |
| | 10.8 | 11.7 | 11.5 | 10.1 |
| | 12.1 | 12.4 | 12.3 | 11.3 |
| Mean | 8.0 | 9.0 | 8.0 | 9.0 |

a) The least squares regression lines were calculated using a standard computer package, as follows:

Car $F$: $y = 2.650 + 0.794x$

Car $L$: $y = 5.585 + 0.427x$

Plot the two scatter diagrams and regression lines on the same graph, distinguishing clearly between the two sets of points.

b) Explain the meaning of the four regression coefficients for these data.

c) This year the company is expanding into a new region in which travelling distances are expected to be 50% higher than those in the example above. Given that the car type has to be *F* or *L*, which should the company choose for this region and why?

d) What will be the expected total running costs for 5 of these cars in this new region next year, if costs per mile are 10% higher than those in the example above?

*CIMA*

5. A sample of eight employees is taken from the production department of a light engineering factory. The data which follow relate to the number of weeks experience in the wiring of components, and the number of components which were rejected as unsatisfactory last week.

| Employee | A | B | C | D | E | F | G | H |
|---|---|---|---|---|---|---|---|---|
| Weeks of experience ($x$) | 4 | 5 | 7 | 9 | 10 | 11 | 12 | 14 |
| Number of rejects ($y$) | 21 | 22 | 15 | 18 | 14 | 14 | 11 | 13 |

$\Sigma x = 72$; $\Sigma y = 128$; $\Sigma xy = 1069$; $\Sigma x^2 = 732$; $\Sigma y^2 = 2156$;

a) Draw a scatter diagram of the data.

b) Calculate a coefficient of correlation for these data and interpret its value.

c) Find the least squares regression equation of rejects on experience. Predict the number of rejects you would expect from an employee with one week of experience.

*CIMA*

6. The following data gives the actual sales of a company in each of 8 regions of a country together with the forecast of sales by two different methods.

| Region | Actual sales | Forecast 1 | Forecast 2 |
|---|---|---|---|
| A | 15 | 13 | 16 |
| B | 19 | 25 | 19 |
| C | 30 | 23 | 26 |
| D | 12 | 26 | 14 |
| E | 58 | 48 | 65 |
| F | 10 | 15 | 19 |
| G | 23 | 28 | 27 |
| H | 17 | 10 | 22 |

i. Calculate the *rank* correlation coefficient between:

1. Actual sales and forecast 1;  2. Actual sales and forecast 2.

ii. Which forecasting method would you recommend next year?

*ICSA*

7. a) Define and explain the links between regression analysis and correlation analysis.

b) You are engaged as a managing accounting assistant by a television company which is investigating the relationship between cinema attendance and television viewing in Great Britain. As part of the preliminary statistical work you are asked to do the following:

i. Draw a scatter diagram of the data below and explain its meaning.

ii. Calculate a coefficient of *rank* correlation for the data and discuss briefly the hypothesis that colour television is emptying the cinemas in Great Britain.

| Year | Total number of colour television licences at year end (millions) | Total number of cinema admissions (millions) |
|---|---|---|
| 1971 | 1.3 | 176 |
| 1972 | 2.8 | 157 |
| 1973 | 5.0 | 134 |
| 1974 | 6.8 | 138 |
| 1975 | 8.3 | 116 |
| 1976 | 9.6 | 104 |
| 1977 | 10.7 | 104 |
| 1978 | 12.0 | 126 |
| 1979 | 12.7 | 112 |
| 1980 | 13.5 | 96 |
| 1981 | 14.1 | 88 |

*CIMA*

8. Prior to privatisation, the most recent annual sales and profit data (£ million) for distribution companies within the Central Electricity Board (England and Wales) were:

| Distribution company | Sales (£m) X | Profit (£m) Y |
|---|---|---|
| Norweb | 1,129 | 32.1 |
| Manweb | 808 | 26.6 |
| Midland | 1,181 | 38.4 |
| South Wales | 551 | 10.3 |
| South West | 687 | 30.0 |
| Southern | 1,134 | 65.4 |
| Seeboard | 912 | 27.2 |
| LEB | 1,050 | 39.9 |
| Eastern | 1,497 | 58.9 |
| East Midlands | 1,165 | 52.1 |
| Yorkshire | 1,140 | 49.3 |
| North East | 740 | 31.9 |

*[Source: The Times, Monday 16 April 1990]*

$\Sigma X = 11{,}994$; $\Sigma Y = 462.1$; $\Sigma XY = 498{,}912.2$; $\Sigma X^2 = 12{,}763{,}470$; $\Sigma Y^2 = 20{,}459.35$

You are required

a) to find the regression of profit on sales, to plot it on a scatter diagram and to predict profit for a similar company with sales of £1,000 million;

b) to interpret your analysis.

*CIMA*

# Part 4 Time series analysis

Statistical data that are described over time are known as a time series. What is required of such data is an understanding of the structure within which the data originate and the nature of the variation in both the short and long term. Why are sales higher on one day than another or in one quarter than another? Are purchases increasing or decreasing generally?

Time series analysis enables the structure of the data to be understood, trends to be identified and forecasts made.

The framework within which time series values are analysed is called a time series model. For the purposes of this book, two basic models are considered – additive and multiplicative. They are described in chapter 15, together with the main form of pictorial representation of time series data, the historigram.

The techniques involved in the various methods available for extracting a trend are shown in chapter 16, while chapter 17 is concerned with two significant applications of the analysis, seasonal adjustment and forecasting.

# 15 Time series model

## 1. Introduction

This chapter defines a time series and describes the structure (called the time series model) within which time series' movements can be explained and understood. The various components that go to make up each time series value are then discussed and, finally, brief mention is made of graphical techniques.

## 2. Definition of a time series

A time series is the name given to the values of some statistical variables measured over a uniform set of time points. Any business, large or small, will need to keep records of such things as sales, purchases, value of stock held and VAT and these could be recorded daily, weekly, monthly, quarterly or yearly. These are examples of time series.

> **Time series**
> A *time series* is a name given to numerical data that is described over a uniform set of time points.

Time series occur naturally in all spheres of business activity as demonstrated in the following example.

## 3. Example 1 (Situations in which *time series* occur naturally)

a) Annual turnover (in £m) of a firm for ten successive years.
b) Numbers unemployed (in thousands) for each quarter of four successive years.
c) Total monthly sales (£000) for a small business for three successive years.
d) Daily takings for a supermarket over a two month period.
e) Number of registered journeys for a Home Removals firm (see table below)

| | Qtr 1 | Qtr 2 | Qtr 3 | Qtr 4 |
|---|---|---|---|---|
| Year 1 | 73 | 90 | 121 | 98 |
| Year 2 | 69 | 92 | 145 | 107 |
| Year 3 | 86 | 111 | 157 | 122 |
| Year 4 | 88 | 109 | 159 | 131 |

## 4. Time series cycle

Normally, time series data exhibits a general pattern which broadly repeats, called a cycle. Sales of domestic electricity always have a distinct four-quarterly cycle; monthly sales for a business will exhibit some natural 12-monthly cycle; daily takings for a supermarket will display a definite 6-daily cycle. The cycle for the Home Removals data in 2(e) above can be seen to be 4-quarterly.

## 5. Time series models

Business records, and in particular certain time series of sales and purchases, need to be kept by law. Of course they are also used to help control current (and plan future) business activities. To use time series effectively for such purposes, the data have to be organised and analysed. In order to explain the movements of time series data, models can be constructed which describe how various components combine to form individual data values.

As an example, a Sales Manager could set up the following model to explain the expense claims of his sales force each week:

$$y = f + t$$

where: $y$ = total expenses for week,
$f$ = fixed expenses (meals, insurance etc), and
$t$ = travelling expenses (petrol, car maintenance, incidentals etc)

## 6. Time series analysis

It is the evaluation and extraction of the components of a model that 'break down' a particular series into understandable and explainable portions and enables:

a) trends to be identified,
b) extraneous factors to be eliminated and
c) forecasts to be made.

The understanding, description and use of these processes is known as *time series analysis*.

## 7. Standard time series models

Depending on the nature, complexity and extent of the analysis required, there are various types of model that can be used to describe time series data. However, for the purposes of this book, two main models will be referred to. They are known as the simple additive and multiplicative models.

The components that go to make up each value of a time series are described in the following definitions.

**The time series additive model**

$$y = t + s + r$$

where: $y$ is a given time series value
$t$ is the trend component
$s$ is the seasonal component
$r$ is the residual component.

**The time series multiplicative model**

$$y = t \times S \times R$$

where: $y$ is a given time series value
$t$ is the trend component
$S$ is the seasonal component
$R$ is the residual component.

Put another way, given a set of time series data, every single given ($y$) value can be expressed as the sum or product of three components. It is the evaluation and interpretation of these components that is the main aim of the overall analysis.

Note that although the trend component will be constant no matter which of the two models are used, the values of the seasonal and residual components will depend on which model is being used. In other words, given a set of data to which both models are being applied, both trend values would be identical whereas the respective seasonal and residual components would be quite different.

## 8. Description of time series components

a) *Trend.* The underlying, long-term tendency of the data. Various techniques for extracting a trend from a given time series are discussed in chapter 16.

b) *Seasonal variation.* These are short-term cyclic fluctuations in the data about the trend which take their name from the standard business quarters of the year. Note however that the word 'season' in this context can have many different meanings. For example:

   i. daily 'seasons' over a weekly cycle for sales in a supermarket,

   ii. monthly 'seasons' over a yearly cycle for purchases of a company,

   iii. quarterly 'seasons' over a yearly cycle for sales of electricity in the domestic sector.

   Techniques for obtaining and analysing seasonal variation are discussed in chapter 17.

c) *Residual variation.* These include other factors not explained by a) and b) above. This variation normally consists of two components:

   i. *Random factors.* These are disturbances due to 'everyday' unpredictable influences, such as weather conditions, illness, transport breakdowns, and so on. The *evaluation* of this component is not usually required in examinations, but its interpretation should be known and is discussed briefly in chapter 17.

   ii. *Long-term cyclic factor.* This can be thought of (if it exists) as due to underlying economic causes outside the scope of the immediate environment. Examples are standard trade cycles or minor recessions. This particular type of variation is not discussed further in this book since it requires techniques that are beyond the scope of the relevant syllabuses.

## 9. Example 2 (General comments on a given time series)

*Question*

Comment on the following data, which relates to visitors (in hundreds) to a hotel over a period of three years. Do not use any quantitative techniques or analyses.

| | Qtr 1 | Qtr 2 | Qtr 3 | Qtr 4 |
|---|---|---|---|---|
| Year 1 | 57 | 85 | 97 | 73 |
| Year 2 | 64 | 96 | 107 | 89 |
| Year 3 | 76 | 102 | 115 | 95 |

*Answer*

The data displays a distinct 4-quarterly cycle over the three year period, with the underlying trend showing a steady increase overall, as well as in each particular quarter.

It shows a significant seasonal effect with (not unexpectedly) the cycle peak in the summer quarter and a trough in the winter quarter. Increases are significantly less in the second and third quarters from year 2 to year 3, which may be due to an upper capacity limit in accommodation for those periods or some other random factor.

There is not enough data to identify any possible long-term cyclic factors.

## 10. Graphing a time series

a) The standard graph for a time series is a line diagram, known technically as a historigram. It is obtained by plotting the time series values (on the vertical axis) against time (on the horizontal axis) as single points which are joined by *straight line segments*.

b) Historigrams can be shown on their own but it is quite common to see both a historigram and the graph of associated derived data, such as a trend, plotted together on the same chart.

## 11. Example 3 (Drawing a historigram)

The data shown in section 3 e) are displayed in Figure 1.

*Figure 1 Historigram of numbers of journeys per year*

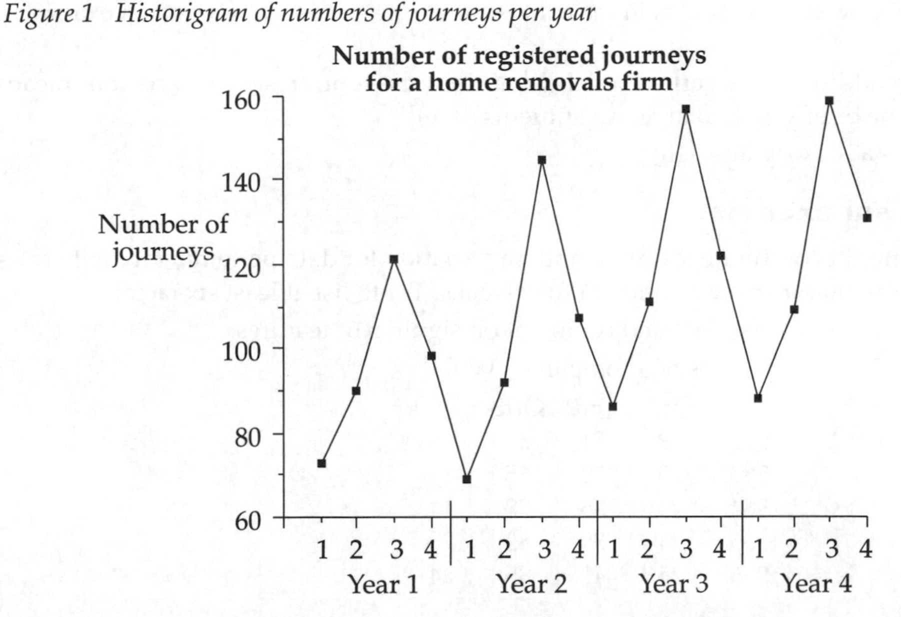


The data display a highly seasonal yearly cycle. The general upward trend appears not to be so marked in the second half of the data.

## 12. Summary

a) A time series is a set of data that is described over a uniform set of time points.
b) Cycles are general patterns that repeat and occur in most types of time series.
c) Time series models are used to gain an understanding of the factors that affect time series.
d) The time series additive model describes the way that the trend, seasonal and residual components independently make up each time series value.
e) A historigram is the standard way of displaying a time series diagrammatically.

## 13. Points to note

a) There are many types of model used to analyse time series, depending mainly on the extent of the analysis required. The additive and multiplicative models are the simplest of these and assume that the component factors are independent of each other.
b) Historigrams are always constructed with the time variable along the horizontal ($x$) axis and the series values along the vertical ($y$) axis.

## 14. Student self review questions

1. What is a time series? [2]
2. What are the aims of time series analysis? [6]
3. Describe the simple additive time series model and name its components. [7]
4. Describe what a 'season' is in the context of a time series and give some examples. [8(b)]
5. For an additive time series model, what does the term 'residual variation' mean? Describe briefly its two main constituents. [8(c)]
6. What is a historigram? [10]

## 15. Student exercises

1. What might contribute towards random variation for data pertaining to daily sales in a supermarket over a period of four weeks. Try to list at least six factors.
2. Graph the following data and comment on significant features.

Sales of a company (£000)

| | Qtr1 | Qtr2 | Qtr3 | Qtr4 |
|---|---|---|---|---|
| 1982 | 19 | 31 | 62 | 9 |
| 1983 | 20 | 32 | 65 | 17 |
| 1984 | 24 | 36 | 78 | 14 |
| 1985 | 24 | 39 | 83 | 20 |
| 1986 | 25 | 42 | 85 | 24 |

# 16 Time series trend

## 1. Introduction

This chapter describes the significance of trend values and the three most common methods of extracting a trend from a given time series. Each method is demonstrated using a common time series and the results compared graphically. Significant features of the three techniques are listed, including their advantages and disadvantages.

## 2. The significance of trend values

It will be recalled from the previous chapter that the object of finding the time series trend is to enable the underlying tendency of the data to be highlighted. Thus, a business sales trend will normally show whether sales are moving up or down (or remaining static) in the long term.

The trend can also be thought of as the core component of the additive time series model about which the two other components, seasonal ($s$) and residual ($r$) variation, fluctuate. This component is found by identifying separate trend ($t$) values, each corresponding to a time point. In other words, at each time point of the series, a value of $t$ can be obtained which forms one of the components that go to make up the observed value of $y$ as described in section 7 of the previous chapter. The following section summarises three different ways of obtaining trend values for a given time series.

## 3. Techniques for extracting the trend

There are three techniques that can be used to extract a trend from a set of time series values.

a) *Semi-averages*. This is the simplest technique, involving the calculation of two ($x$, $y$) averages which, when plotted on a chart as two separate points and joined up, form a straight line. A similar method was introduced in chapter 13, to find a regression line.

b) *Least squares regression*. This method, also introduced in chapter 13, similarly results in a straight line.

c) *Moving averages*. This is the most commonly used method for identifying a trend and involves the calculation of a set of averages. The trend, when obtained and charted, consists of straight line segments.

## 4. The method of semi-averages

The method of semi-averages for obtaining a trend for a time series is now demonstrated with a simple example.

Suppose the following sales (£00, to nearest £10) were recorded for a firm and it is required to obtain a semi-average trend.

| | Week 1 | | | | | Week 2 | | | | |
|---|---|---|---|---|---|---|---|---|---|---|
| | Mon | Tue | Wed | Thu | Fri | Mon | Tue | Wed | Thu | Fri |
| Sales ($y$) | 250 | 320 | 340 | 520 | 410 | 260 | 380 | 410 | 670 | 420 |

Note that the data is time-ordered, which is normal and natural for a time series.

The procedure for obtaining a trend using the method of semi-averages is:

STEP 1 *Split the data into a lower and an upper group.*

For the data given:

the lower group is 250, 320, 340, 520 and 410;

the upper group is 260, 380, 410, 670 and 420.

STEP 2 *Find the mean value of each group.*

The mean of the lower group (L) is $\frac{1840}{5} = 368$.

The mean of the upper group (U) is $\frac{2140}{5} = 428$.

STEP 3 *Plot, on a graph, each mean against an appropriate time point.*

'An appropriate time point' can always be taken as the median time point of the respective group. Thus L would be plotted against Wednesday of week 1 and U against Wednesday of week 2.

STEP 4 *The line joining the two plotted points is the required trend.*

Note that it is important that the two groups in question have an equal number of data values. If the given data, however, contains an odd number of data values, the middle value can be ignored (for the purposes of obtaining the trend line).

Once a trend line has been obtained, the trend values corresponding to each time point can be read off from the graph.

A fully worked example follows.

## 5. Working data (used for this chapter)

The following set of data will be referred to throughout the chapter in order to demonstrate the calculations involved in using each of the three methods for obtaining a time series trend.

*UK outward passenger movements by sea*

| | Year 1 | | | | Year 2 | | | | Year 3 | | | |
|---|---|---|---|---|---|---|---|---|---|---|---|---|
| Quarter | 1 | 2 | 3 | 4 | 1 | 2 | 3 | 4 | 1 | 2 | 3 | 4 |
| Number of passengers (millions) | 2.2 | 5.0 | 7.9 | 3.2 | 2.9 | 5.2 | 8.2 | 3.8 | 3.2 | 5.8 | 9.1 | 4.1 |

## 6. Example 1 (Calculating a time series trend using *semi-averages*)

*Question*

Using the working data, given in section 5:

a) Use the method of semi-averages to obtain and plot a trend line.

b) Draw up a table showing the original data ($y$) values against the trend ($t$) values (obtained from the graph).

*Answer*

a) The data has been split up into lower and upper groups, each one being totalled and then averaged.

| | | | | | |
|---|---|---|---|---|---|
| Year 1 | Q1 | 2.2 | Year 2 | Q3 | 8.2 |
| | Q2 | 5.0 | | Q4 | 3.8 |
| | Q3 | 7.9 | Year 3 | Q1 | 3.2 |
| | Q4 | 3.2 | | Q2 | 5.8 |
| Year 2 | Q1 | 2.9 | | Q3 | 9.1 |
| | Q2 | 2.2 | | Q4 | 4.1 |
| | Total | 26.4 | | Total | 34.2 |
| | Mean (L) | 4.4 | | Mean (U) | 5.7 |

In this situation, both L and U must be plotted against a hypothetical point between the middle two time points in their respective sets. That is, L is plotted at a time point between Year 1 Q3 and Year 1 Q4 and L is plotted corresponding to a point between Year 3 Q1 and Year 3 Q2.

In Figure 1, the two means have been plotted and joined by a straight line to form the trend line.

*Figure 1 Trend line formed using semi-averages*

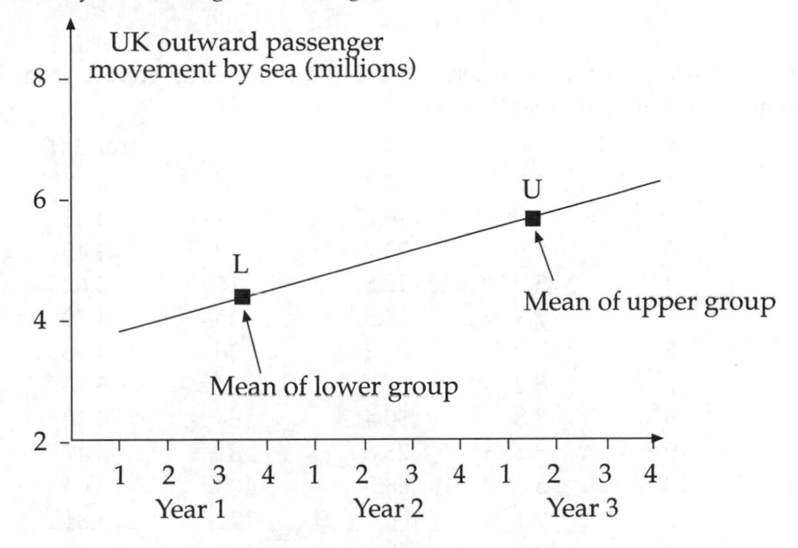


b) The trend values have been read from the graph and are tabulated below, together with the original data values.

| | Year 1 | | | | Year 2 | | | | Year 3 | | | |
|---|---|---|---|---|---|---|---|---|---|---|---|---|
| Quarter | 1 | 2 | 3 | 4 | 1 | 2 | 3 | 4 | 1 | 2 | 3 | 4 |
| Data ($y$) | 2.2 | 5.0 | 7.9 | 3.2 | 2.9 | 5.2 | 8.2 | 3.8 | 3.2 | 5.8 | 9.1 | 4.1 |
| Trend ($t$) | 3.9 | 4.1 | 4.3 | 4.5 | 4.7 | 4.9 | 5.2 | 5.4 | 5.6 | 5.8 | 6.0 | 6.2 |

## 7. The method of least squares regression

The technique of least squares regression was explained and demonstrated for bivariate data in chapter 13. In order to use this method to obtain a trend line for a time series, it is necessary to consider the time series data as bivariate. The procedure is given as follows.

STEP 1 Take the physical time points as values (coded as 1, 2, 3, etc if necessary) of the independent variable $x$.

STEP 2 Take the data values themselves as values of the dependent variable $y$.

STEP 3 Calculate the least squares regression line of $y$ on $x$, $y=a+bx$.

STEP 4 Translate the regression line as $t=a+bx$, where any given value of time point $x$ will yield a *corresponding value of the trend, t*.

An example of the use of this technique follows.

## 8. Example 2 (Determining a time series trend using *least squares*)

*Question*

For the working data of section 5, calculate, using least squares regression, a trend component for each time point given.

*Answer*

Put $y$ = number of passengers and $x$ = time point, coded from 1 to 10. i.e. 1=Year1(Qtr1) and 10=Year3(Qtr2)

| $x$ | $y$ | $xy$ | $x^2$ | trend ($t$) |
|---|---|---|---|---|
| 1 | 2.2 | 2.2 | 1 | 4.11 |
| 2 | 5.0 | 10.0 | 4 | 4.28 |
| 3 | 7.9 | 23.7 | 9 | 4.45 |
| 4 | 3.2 | 12.8 | 16 | 4.62 |
| 5 | 2.9 | 14.5 | 25 | 4.79 |
| 6 | 5.2 | 31.2 | 36 | 4.96 |
| 7 | 8.2 | 57.4 | 49 | 5.13 |
| 8 | 3.8 | 30.4 | 64 | 5.30 |
| 9 | 3.2 | 28.8 | 81 | 5.47 |
| 10 | 5.8 | 58.0 | 100 | 5.64 |
| 11 | 9.1 | 100.1 | 121 | 5.81 |
| 12 | 4.1 | 49.2 | 144 | 5.98 |
| 78 | 60.6 | 418.3 | 650 | |

From the table: $\sum x = 78$; $\sum y = 60.6$; $\sum xy = 418.3$; $\sum x^2 = 650$; $n = 12$.

Putting the regression line as $y=a+bx$, $a$ and $b$ are now calculated.

Thus: $$b = \frac{n\sum xy - \sum x \sum y}{n\sum x^2 - (\sum x)^2} = \frac{12 \times 418.3 - 78 \times 60.6}{12 \times 650 - 78^2}$$

$$= \frac{292.8}{1716}$$

i.e. $b = 0.17$ (2D)

and: $a = \frac{\Sigma y}{n} - b\frac{\Sigma x}{n} = \frac{60.6}{12} - (0.17)\frac{78}{12}$

i.e. $a = 3.94$ (2D)

Thus, the regression line for the trend is $t = 3.94 + 0.17x$ (2D).

(*Remember* that once the regression line is determined, it will be used for calculating trend values. So the normal '$y$' has been replaced by '$t$'.)

The time point values ($x$=1,2,3 etc) can now be substituted into the above regression line to give the trend values required.

When $x$=1 (Year1 Qtr1), $t = 3.94 + 0.17(1)$ i.e. $t$=4.11 (2D)

When $x$=2 (Year2 Qtr2), $t = 3.94 + 0.17(2)$ i.e. $t$=4.28 (2D) ... etc.

These and other values of $t$ are tabulated in the previous table.

## 9. The method of moving averages

This method of obtaining a time series trend involves calculating a set of averages, each one corresponding to a trend ($t$) value for a time point of the series. These are known as *moving averages*, since each average is calculated by moving from one overlapping set of values to the next. The number of values in each set is always the same and is known as the *period* of the moving average.

To demonstrate the technique, a set of moving averages of period 5 has been calculated below for a set of values.

| | | | | | | | | | | |
|---|---|---|---|---|---|---|---|---|---|---|
| Original values: | 12 | 10 | 11 | 11 | 9 | 11 | 10 | 10 | 11 | 10 |
| Moving totals: | | | 53 | 52 | 52 | 51 | 51 | 52 | | |
| Moving averages: | | | 10.6 | 10.4 | 10.4 | 10.2 | 10.2 | 10.4 | | |

The first total, 53, is formed from adding the first 5 items;

i.e. 53 = 12 + 10 + 11 + 11 + 9.

Similarly, the second total, 52 = 10 + 11 + 11 + 9 + 11, and so on. The averages are then obtained by dividing each total by 5.

Notice that the totals and averages are written down *in line with* the middle value of the set being worked on. These averages are the trend ($t$) values required.

It should also be noticed that there are no trend values corresponding to the first and last two original values. This is always the case with moving averages and is a disadvantage of this particular method of obtaining a trend.

## 10. Summary of the moving average technique

**Moving average**

Moving averages (of period $n$) for the values of a time series are arithmetic means of successive and overlapping values, taken $n$ at a time.

The (moving) average values calculated form the required trend components ($t$) for the given series.

The following points should be noted when considering a moving average trend.

a) The period of the moving average *must* coincide with the length of the natural cycle of the series. Some examples follow.
   i. Moving averages for the trend of numbers unemployed for the quarters of the year must have a period of 4.
   ii. Total monthly sales of a business for a number of years would be described by a moving average trend of period 12.
   iii. A moving average trend of period 6 would be appropriate to describe the daily takings for a supermarket (open six days per week) over a number of months.

b) Each moving average trend value calculated *must* correspond with an appropriate time point. This can always be determined as the median of the time points for the values being averaged. For moving averages with an odd-numbered period, 3, 5, 7, etc, the relevant time point is that corresponding to the 2nd, 3rd, 4th, etc value. See the example in the previous section, where the moving averages had a period of 5 and thus each average obtained was set against the 3rd value of the respective set being averaged.

However, when the moving averages have an even-numbered period (2, 4, 6, 8, etc), there is no obvious and natural time point corresponding to each calculated average. The following section describes the technique known as 'centering', which is used in these circumstances.

## 11. Moving average centering

When calculating moving averages with an even period (i.e. 4, 6 or 8), the resulting moving average would seem to have to be placed in between two corresponding time points. As an example, the following data has a 4-period moving average calculated and shows its placing.

| Time point | 1 | | 2 | | 3 | | 4 | | 5 | | 6 | | 7 | | 8 | | 9 | | 10 |
|---|---|---|---|---|---|---|---|---|---|---|---|---|---|---|---|---|---|---|---|
| Data value | 9 | | 14 | | 17 | | 12 | | 10 | | 14 | | 19 | | 15 | | 10 | | 16 |
| Totals (of 4) | | | | 52 | | 53 | | 53 | | 55 | | 58 | | 58 | | 60 | | | |
| Averages (of 4) | | | | 13.00 | | 13.25 | | 13.25 | | 13.75 | | 14.50 | | 14.50 | | 15.00 | | | |

The placing of these averages as described above would not be satisfactory when the averages are being used to represent a trend, since the trend values need to coincide with particular time points. A method known as *centering* is used in this type of situation, where the calculated averages are themselves averaged in successive overlapping pairs. This ensures that each calculated (trend) value 'lines up' with a time point.

This technique is now shown for the previous data.

| Time point | 2 | | 3 | | 4 | | 5 | | 6 | | 7 | | 8 | | 9 |
|---|---|---|---|---|---|---|---|---|---|---|---|---|---|---|---|
| Averages (of 4) | | 13.00 | | 13.25 | | 13.25 | | 13.75 | | 14.50 | | 14.50 | | 15.00 | |
| Averages (of 2) | | | 13.125 | | 13.250 | | 13.500 | | 14.125 | | 14.500 | | 14.750 | | |

A worked example follows which uses this technique.

## 12. Example 3 (calculating trend values using *moving average centering*)

*Question*

Calculate trend values for the working data of section 5, using moving averages with an appropriate period. Plot a graph of the original data with the trend superimposed.

*Answer*

The cycle of the data is clearly 4-quarterly and we thus need a (centered) 4-quarterly moving average trend, using the technique described in section 11 above. Table 1 demonstrates the standard columnar layout of the calculations.

*Table 1 Standard layout of calculations*

| Qtr | | Original data ($y$) | Moving totals of 4 | Moving average | Centered moving average ($t$) |
|---|---|---|---|---|---|
| Year 1 | 1 | 2.2 | | | |
| | 2 | 5.0 | | | |
| | | | 18.3 | 4.575 | |
| | 3 | 7.9 | | | 4.66 |
| | | | 19.0 | 4.750 | |
| | 4 | 3.2 | | | 4.78 |
| | | | 19.2 | 4.800 | |
| Year 2 | 1 | 2.9 | | | 4.84 |
| | | | 19.5 | 4.875 | |
| | 2 | 5.2 | | | 4.95 |
| | | | 20.1 | 5.025 | |
| | 3 | 8.2 | | | 5.06 |
| | | | 20.4 | 5.100 | |
| | 4 | 3.8 | | | 5.18 |
| | | | 21.0 | 5.250 | |
| Year 3 | 1 | 3.2 | | | 5.36 |
| | | | 21.9 | 5.475 | |
| | 2 | 5.8 | | | 5.51 |
| | | | 22.2 | 5.550 | |
| | 3 | 9.1 | | | |
| | 4 | 4.1 | | | |

Notice that the two starting and ending time points do not have a trend value. As mentioned previously, this type of omission *will always occur* with a moving average trend.

Figure 2 overleaf shows a graph of the original data with the trend values superimposed.

## 13. Comparison of techniques for trend

For the working data given in section 4, all three methods of obtaining a trend have now been demonstrated. The method of semi-averages (Example 1), least squares (Example 2) and moving averages (Example 3). Figure 3 overleaf shows the graphs for comparison.

The three sets of trend values are quite distinct and this underlines the fact that there is no unique set of trend values for a time series. Each method will yield a different trend, as has been evidenced.

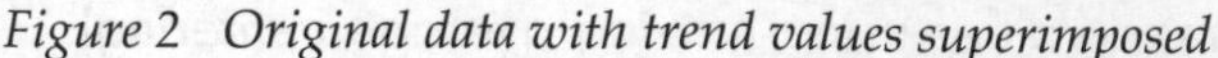

*Figure 2 Original data with trend values superimposed*

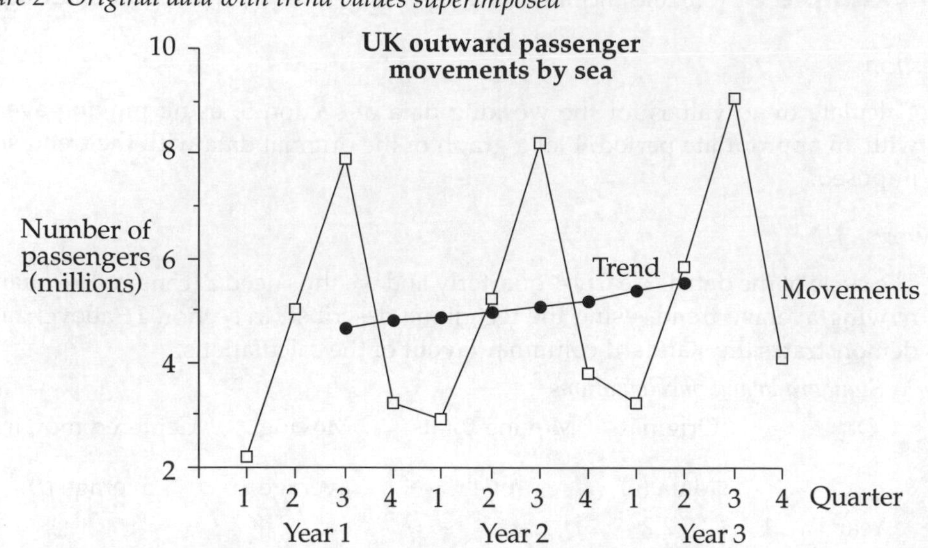


Significant features of each method are now summarised.

a) *Semi-averages*. Although simple to apply, the fact that only two plotted points are used in its construction leads to the general feeling that it is unrepresentative. It also assumes that a strictly linear trend is appropriate to the data.

b) *Least squares*. Although mathematically representative of the data, it assumes that a linear trend is appropriate. It is generally thought unsuitable for highly 'seasonal' data.

c) *Moving averages*. The most widely used technique for obtaining a trend. If the period of the averages is chosen appropriately, it will show the true nature of the trend, whether linear or non-linear. One disadvantage is the fact that no trend values are obtained for the beginning and end time points of a series.

*Figure 3 Comparison of techniques*

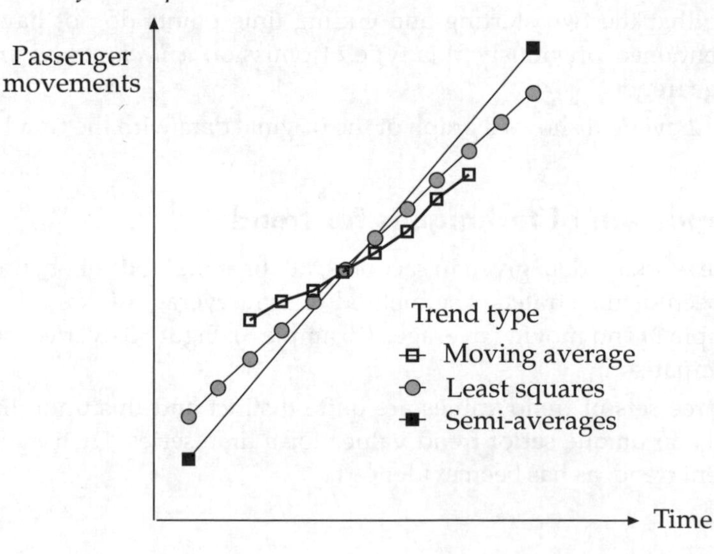

## 14. Summary

a) Three common techniques for identifying trend components are:
   i. semi-averages
   ii. least squares regression
   iii. moving averages.

b) For time series that have a significant seasonal effect, the moving average technique is generally preferred.

c) When moving averages are used for identifying trend components, the period of the average must coincide with the cycle of the data being analysed. This is done in order to remove possible cyclical fluctuations.

d) Even-period moving averages must be centred in order that their values coincide with actual time points.

## 15. Point to note

There is no unique set of trend values for a given time series. The particular method chosen needs to take into account the nature of the data and the use to which trend values will be put.

## 16. Student self review questions

1. What are the three most common techniques for obtaining a time series trend? [3]
2. Describe the stages involved in obtaining a time series trend using the method of semi-averages. [4]
3. What are moving averages and what is the 'period' of a moving average? [9]
4. What is moving average centering and why is it employed? [11]
5. What is the main advantage and disadvantage of (a) semi-averages and (b) moving averages, as methods of obtaining a time series trend? [13]

## 17. Student exercises

1. Calculate a set of trend values (to 1D) using the method of semi-averages, for the following data:

   16, 12, 15, 14, 18, 12, 14, 13, 18, 13.

2. *MULTI-CHOICE.* Sales of a particular product are recorded weekly for a period of 48 weeks. The 4-point (centred) moving averages are plotted on a graph. How many points are plotted?

   a) 44 b) 45 c) 46 d) 48

3. Calculate a set of moving averages of period: (a) 3 (b) 5 for the following time series data:

   8, 11, 10, 21, 4, 9, 12, 10, 23, 5, 10, 13, 11, 26, 6.

   Which set of moving averages is the correct one to use for obtaining a trend for the series?

4. Draw a historigram for the data described in question 3 above, superimposing the correct trend values.

5. The number of houses (in thousands) built each year between 1953 and 1969 (inclusive) are given as:

| Year | 1 | 2 | 3 | 4 | 5 | 6 | 7 | 8 | 9 |
|---|---|---|---|---|---|---|---|---|---|
| Number of houses | 319 | 348 | 317 | 308 | 308 | 329 | 332 | 354 | 378 |

| Year | 10 | 11 | 12 | 13 | 14 | 15 | 16 | 17 |
|---|---|---|---|---|---|---|---|---|
| Number of houses | 364 | 358 | 383 | 391 | 396 | 415 | 426 | 378 |

Assuming a seven-year cycle, eliminate the cyclical movement by producing a moving average trend and plot this, together with the original data on the same chart.

6. The following figures relate to Rate receipts (in £m) for a Local Authority.

| | Year 1 | Year 2 | Year 3 |
|---|---|---|---|
| Qtr 1 | 2.8 | 3.0 | 3.0 |
| Qtr 2 | 4.2 | 4.2 | 4.7 |
| Qtr 3 | 3.0 | 3.5 | 3.6 |
| Qtr 4 | 4.6 | 5.0 | 5.3 |

Plot a historigram for the data, together with a least squares regression trend

# 17 Seasonal variation and forecasting

## 1. Introduction

This first part of the chapter describes the nature of seasonal variation in a time series and how it can be calculated. Forecasting, or the ability to estimate future values of a given time series using seasonal variation, is dealt with in the latter half of the chapter.

## 2. The nature of seasonal variation

*Seasonal* (or short-term cyclic) *variation* is present in many time series. Winter sportswear will sell well in autumn and winter and badly in spring and summer; supermarket sales are higher at the end of the week than at the beginning; unemployment figures are always seasonally inflated at Easter and in the early summer owing to school leavers.

When values are obtained to describe seasonal variation, they are sometimes known as seasonal values or factors and are expressed as *deviations* (i.e. '+' or '–') from the underlying trend. They show, on average, by how much a particular season will tend to increase or decrease the underlying trend. Thus we would expect the seasonal variation for winter sportswear to be positive in autumn and winter and negative in spring and summer.

**Seasonal variation**

*Seasonal variation* components give an average effect on the trend which is solely attributable to the 'season' itself. They are expressed in terms of deviations from (additive model) or percentages of (multiplicative model) the trend.

The use of seasonal variation figures are of great importance to organisations operating in environments where a seasonal factor is significant. For example, a regional Electricity Board needs to know the average increase in demand expected in the winter months in order to be able to meet this demand.The following two sections describe and demonstrate the technique for calculating seasonal variation.

## 3. Technique for calculating seasonal variation

a) *Additive model*

Given the original time series ($y$) values, together with the trend ($t$) values, the procedure for calculating the seasonal variation is given as follows.

STEP 1 Calculate, for each time point, the value of $y–t$ (the difference between the original value and the trend).

STEP 2 For *each season* in turn, find the average (arithmetic mean) of the $y–t$ values.

STEP 3 If the total of the averages differs from zero, adjust one or more of them so that their total is zero.

The values so obtained are the appropriate seasonal variation values; i.e.the '$s$' figures in the additive model $y = t + s + r$.

b) *Multiplicative model*

Given the original time series ($y$) values, together with the trend ($t$) values, the procedure for calculating the seasonal variation is given as follows.

STEP 1 Calculate, for each time point, the value of $(y-t)/t$ (the difference between the original value and the trend expressed as a proportion of the trend).

STEP 2 For each season in turn, find the arithmetic mean of the above proportional changes.

[Note that this should strictly involve calculating the geometric mean of 1 + proportional change values. In practise however this is felt to be too complex!]

STEP 3 If the total of the averages differs from zero, adjust one or more of them so that their total is zero.

The values so obtained are the appropriate seasonal variation values; i.e.the '$S$' figures in the multiplicative model $y = t \times S \times R$.

## 4. Example 1A (Calculating seasonal variation figures using the *additive model*)

The sales of a company ($y$,in £000) are given below, together with a previously calculated trend ($t$). The subsequent calculations to find the seasonal variation are shown, laid out in a standardised way.

STEP 1

| | $y$ | $t$ | $y-t$ |
|---|---|---|---|
| Year 1 Qtr 1 | 20 | 23 | −3 |
| 2 | 15 | 29 | −14 |
| 3 | 60 | 34 | 26 |
| 4 | 30 | 39 | −9 |
| Year 2 Qtr 1 | 35 | 45 | −10 |
| 2 | 25 | 50 | −25 |
| 3 | 100 | 55 | 45 |
| 4 | 50 | 61 | −11 |

STEP 2

| Deviations ($y-t$) | Q1 | Q2 | Q3 | Q4 | Sum |
|---|---|---|---|---|---|
| Year 1 | −3 | −14 | 26 | −9 | |
| Year 2 | −10 | −25 | 45 | −11 | |
| Totals | −13 | −39 | 71 | −20 | |
| Averages | −6.5 | −19.5 | 35.5 | −10.0 | −0.5 |

STEP 3

Since the averages sum to −0.5 (and not zero), it is necessary to adjust one or more of them accordingly. In this case, since the difference is so small, only one will be adjusted. In order to make the smallest percentage error, the largest value (35.5) is changed to 36.0. This adjustment is shown in the following table:

| | Q1 | Q2 | Q3 | Q4 | |
|---|---|---|---|---|---|
| Initial $s$ values | −6.5 | −19.5 | 35.5 | −10.0 | |
| Adjustment | 0 | 0 | +0.5 | 0 | |
| Adjusted $s$ values | −6.5 | −19.5 | 36.0 | −10.0 | (Sum = 0) |

The interpretation of the figures is that the average seasonal effect for quarter 1, for instance, is to *deflate* the trend by 6.5 (£000) and that for quarter 3 is to *inflate* the trend by 36 (£000).

## 5. Example 1B (Calculating seasonal variation figures using the *multiplicative model*)

The sales of a company ($y$,in £000) are given below, together with a previously calculated trend ($t$). The subsequent calculations to find the seasonal variation are shown, laid out in a standardised way.

Step 1

| | $y$ | $t$ | $\frac{y-t}{t}$ | $S=1+\frac{y-t}{t}$ |
|---|---|---|---|---|
| Year 1 Qtr 1 | 20 | 23 | –0.13 | 0.87 |
| 2 | 15 | 29 | –0.48 | 0.52 |
| 3 | 60 | 34 | 0.76 | 1.76 |
| 4 | 30 | 39 | –0.23 | 0.77 |
| Year 2 Qtr 1 | 35 | 45 | –0.22 | 0.78 |
| 2 | 25 | 50 | –0.50 | 0.50 |
| 3 | 100 | 55 | 0.82 | 1.82 |
| 4 | 50 | 61 | –0.18 | 0.82 |

Step 2

Deviations $\left(1+\frac{y-t}{t}\right)$

| | Q1 | Q2 | Q3 | Q4 | Sum |
|---|---|---|---|---|---|
| Year 1 | 0.87 | 0.52 | 1.76 | 0.77 | |
| Year 2 | 0.78 | 0.50 | 1.82 | 0.82 | |
| G. Means | 0.82 | 0.51 | 1.79 | 0.79 | 3.91 |

STEP 3

Since the averages sum to 3.91 (and not 4), it is necessary to add 0.09 to one or more of them accordingly. In this case, as in Example 1A, since the difference is so small, only one will be adjusted. In order to make the smallest percentage error, the largest value (1.79) is changed to 1.88. This adjustment is shown in the following table:

| | Q1 | Q2 | Q3 | Q4 | |
|---|---|---|---|---|---|
| Initial $S$ values | 0.82 | 0.51 | 1.79 | 0.79 | |
| Adjustment | 0 | 0 | +0.9 | 0 | |
| Adjusted $S$ values | 0.82 | 0.51 | 1.88 | 0.79 | (Sum = 4.00) |

The interpretation of the figures is that the average seasonal effect for quarter 1, for instance, is to deflate the trend by 18% (since 0.82 is 0.18 less than 1) and that for quarter 3 is to inflate the trend by 88%.

## 6. Seasonally adjusted time series

One particular and important use of seasonal values is to *seasonally adjust* the original data. The effect of seasonal adjustment is to smooth away seasonal fluctuations, leaving a clear view of what might be expected 'had seasons not existed'.

The technique is similar for both models but is shown separately for clarity.

*Additive model*:

The adjustment is performed by subtracting the appropriate seasonal figure from each of the original time series values and represented algebraically by $y$–$s$.

As an example, the data of Examples 1A and 1B are seasonally adjusted below.

| | | $y$ | $s$ | $y$–$s$ | |
|---|---|---|---|---|---|
| Year 1 | Qtr 1 | 20 | –6.5 | 20–(–6.5)=26.5 | |
| | 2 | 15 | –19.5 | 15–(–19.5)=34.5 | |
| | 3 | 60 | 36.0 | 60–36.0=24.0 | |
| | 4 | 30 | –10.0 | 30–(–10.0)=40.0 | Seasonally |
| Year 2 | Qtr 1 | 35 | –6.5 | 35–(–6.5)=41.5 | adjusted values |
| | 2 | 25 | –19.5 | 25–(–19.5)=44.5 | |
| | 3 | 100 | 36.0 | 100–36.0=64.0 | |
| | 4 | 50 | –10.0 | 50–(–10.0)=60.0 | |

*Multiplicative model*:

The adjustment is performed by dividing each of the original time series values by $S$ and is represented algebraically by $y/S$.

As an example, the data of Example 1 are again seasonally adjusted below.

| | | $y$ | $S$ | $\frac{y}{S}$ | |
|---|---|---|---|---|---|
| Year 1 | Qtr 1 | 20 | 0.82 | 20/0.82=24.3 | |
| | 2 | 15 | 0.51 | 15/0.51=29.5 | |
| | 3 | 60 | 1.88 | 60/1.88=31.9 | |
| | 4 | 30 | 0.79 | 30/0.79=37.8 | Seasonally |
| Year 2 | Qtr 1 | 35 | 0.82 | 35/0.82=42.6 | adjusted values |
| | 2 | 25 | 0.51 | 25/0.51=49.2 | |
| | 3 | 100 | 1.88 | 100/1.88=53.2 | |
| | 4 | 50 | 0.79 | 50/0.79=63.0 | |

To summarise:

**Seasonal adjustment**

Seasonally adjusted time series data are obtained by subtraction (additive model) or division (multiplicative model) as follows:

*Additive model*: seasonally adjusted value = $y - s$

*Multiplicative model*: seasonally adjusted value = $y/S$.

The importance of seasonal adjustments is reflected in the fact that the majority of economic time series data published by the Central Statistical Office is presented both in terms of 'actual' and 'seasonally adjusted' figures.

## 7. Example 2 (*Seasonal adjustment* of a time series)

*Question*

The following data gives UK outward passenger movements (in millions) by sea, together with a 4-quarterly moving average trend (calculated previously in chapter 16, Example 3). Find the values of the seasonal variation for each of the four quarters (using an additive model) and hence obtain seasonally adjusted outward passenger movements. Plot the results on a graph.

| | Year 1 | | | | Year 2 | | | | Year 3 | | | |
|---|---|---|---|---|---|---|---|---|---|---|---|---|
| Quarter | 1 | 2 | 3 | 4 | 1 | 2 | 3 | 4 | 1 | 2 | 3 | 4 |
| Number of passengers ($y$) | 2.2 | 5.0 | 7.9 | 3.2 | 2.9 | 5.2 | 8.2 | 3.8 | 3.2 | 5.8 | 9.1 | 4.1 |
| Trend ($t$) | | | 4.66 | 4.78 | 4.84 | 4.95 | 5.06 | 5.18 | 5.36 | 5.51 | | |

*Answer*

The deviations are calculated and displayed in column 5, and the calculations for the seasonal variation are shown in the lower table and the results, together with the seasonally adjusted data, have been added in the last 2 columns.

| | Original data ($y$) | Centered moving average ($t$) | Deviations ($y$–$t$) | Seasonal variation ($s$) | Seasonally adjusted data ($y$–$s$) |
|---|---|---|---|---|---|
| Year 1 Qtr 1 | 2.2 | | | –2.03 | 4.23 |
| 2 | 5.0 | | | 0.28 | 4.72 |
| 3 | 7.9 | 4.66 | 3.24 | 3.21 | 4.69 |
| 4 | 3.2 | 4.78 | –1.58 | –1.46 | 4.66 |
| Year 2 Qtr 1 | 2.9 | 4.84 | –1.94 | –2.03 | 4.93 |
| 2 | 5.2 | 4.95 | 0.25 | 0.28 | 4.92 |
| 3 | 8.2 | 5.06 | 3.14 | 3.21 | 4.99 |
| 4 | 3.8 | 5.18 | –1.38 | –1.46 | 5.26 |
| Year 3 Qtr 1 | 3.2 | 5.36 | –2.16 | –2.03 | 5.23 |
| 2 | 5.8 | 5.51 | 0.29 | 0.28 | 5.52 |
| 3 | 9.1 | | | 3.21 | 5.89 |
| 4 | 4.1 | | | –1.46 | 5.56 |

| | Q1 | Q2 | Q3 | Q4 | Sum |
|---|---|---|---|---|---|
| Year 1 | | | 3.24 | –1.58 | |
| Year 2 | –1.94 | 0.25 | 3.14 | –1.38 | |
| Year 3 | –2.16 | 0.29 | | | |
| Totals | –4.10 | 0.54 | 6.38 | –2.96 | |
| Averages | –2.05 | 0.27 | 3.19 | –1.48 | –0.07 |
| Adjustments | +0.02 | +0.01 | +0.02 | +0.02 | +0.07 |
| Adjusted averages | –2.03 | 0.28 | 3.21 | –1.46 | 0.00 |

The required graph is plotted in Figure 1.

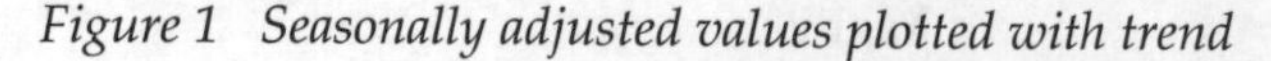

*Figure 1 Seasonally adjusted values plotted with trend*

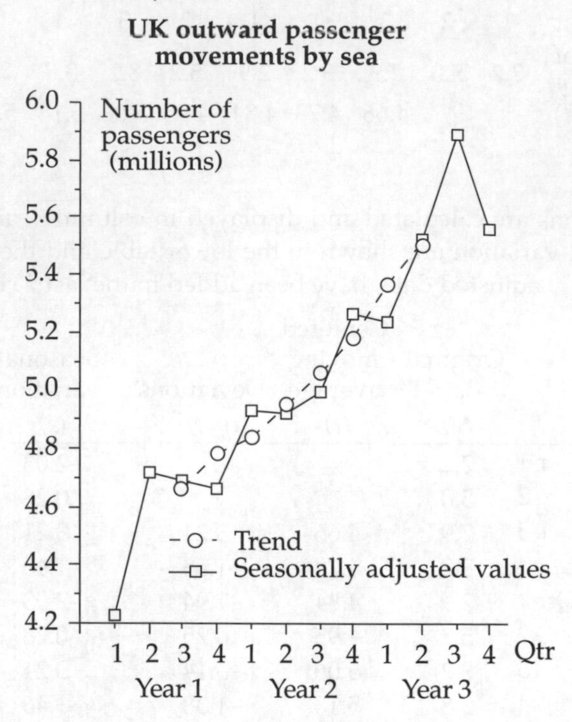


## 8. Notes on Example 2

1. It is usual to show the calculation of the seasonal values in rectangular form as demonstrated above.
2. Notice in the lower half of the previous table that the sum of the averages was –0.07 and thus they need to be adjusted by +0.07. However, rather than adding all of this to just one of the averages, it was divided up into the four parts +0.02, +0.02, +0.02 and +0.01, each being added to a separate average. This is generally regarded as a fairer way to adjust.
3. Even though the moving average trend values are missing at the beginning and end time points, the seasonal values calculated can still be used at these points and thus seasonal adjustment can be performed for all original data items.

## 9. Forecasting

a) A particular use of time series analysis is in *forecasting*, sometimes called *projecting* the time series. Clearly, business life would be much easier if monthly sales for the next year were known or the number of transport breakdowns next month could be determined. However, no-one can predict the future; the best that can be done is to estimate the most likely future values, given the analysis of previous years' sales or last month's breakdowns.

b) Forecasting can be performed at different levels, depending on the use to which it will be put. Simple guessing, based on previous figures, is occasionally adequate. However, where there is a large investment at stake (in plant, stock and manpower for example), structured forecasting is essential.

c) Any forecasts made, however technical or structured, should be treated with caution, since the analysis is based on past data and there could be unknown factors present in the future. However, it is often reasonable to assume that patterns that have been identified in the analysis of past data will be broadly continued, at least into the short-term future.

## 10. Technique for forecasting

Forecasting a value for a future time point involves the following steps.

STEP 1 Estimate a trend value for the time point. There are a number of ways of estimating future trend values and some of these are described in section 12.

STEP 2 Identify the seasonal variation value appropriate to the time point. Seasonal variation values are calculated in the manner already described in section 5.

STEP 3 Add (or multiply, depending on the model) these two values together, giving the required forecast.

**Time series forecasting** can be attempted using the simple additive or multiplicative model in the following adapted form:

*Additive:* $y_{est} = t_{est} + s$

*Multiplicative:* $y_{est} = t_{est} \times S$

where: $y_{est}$ = estimated data value

$t_{est}$ = projected trend value

*s and S* = appropriate seasonal variation value.

Notice that there is no provision for residual variation in the above forecasting models.

## 11. Example 3 (Time series *forecasting*)

Forecast the values for the four quarters of year 4, given the following information which has been calculated from a time series. Assume that the trend in year 4 will follow the same pattern as in years 1 to 3 and an additive model is appropriate.

| | Year 1 | | | | Year 2 | | | | Year 3 | | | |
|---|---|---|---|---|---|---|---|---|---|---|---|---|
| Quarter | 1 | 2 | 3 | 4 | 1 | 2 | 3 | 4 | 1 | 2 | 3 | 4 |
| Trend ($t$) | 42 | 44 | 46 | 48 | 50 | 52 | 54 | 56 | 58 | 60 | 62 | 64 |

$s_1$ = seasonal factor for quarter 1 = –15; $s_2 = -8$; $s_3 = +6$; $s_4 = +17$

STEP 1 Estimate trend values for the relevant time points. Note that, in this case, the trend values increase by exactly 2 per quarter.

Trend for year 4, quarter 1 = $t_{4,1} = 66$.

Similarly, $t_{4,2} = 68$, $t_{4,3} = 70$ and $t_{4,4} = 72$.

STEP 2 Identify the appropriate seasonal factors.The seasonal factors for year 4 are taken as the given seasonal factors. That is, seasonal factor for year 4, quarter 1 = $s_1 = -15$ etc.

STEP 3 Add the trend estimates to the seasonal factors, giving the required forecasts.

Forecast for year 4, quarter 1 = $t_{4,1} + s_1$ = 66–15 = 51;
Forecast for year 4, quarter 2 = $t_{4,2} + s_2$ = 68–8 = 60;
Forecast for year 4, quarter 3 = $t_{4,3} + s_3$ = 70+6 = 76;
Forecast for year 4, quarter 4 = $t_{4,4} + s_4$ = 72+17 = 89.

## 12. Projecting the trend

Projecting the trend for the data of Example 3 was straightforward since the given trend values increased uniformly, thereby displaying a distinct linear pattern. In general, trend values will not conform conveniently in this way.

There are a number of techniques available for projecting the trend, depending on the method used in obtaining the trend values themselves. The most common are now listed.

a) *Linear trend.* Whether the method of least squares or semi-averages has been used, the projection involves simply extending the trend line already calculated.

b) *Moving average trend.* There is no one universal method. Three common means of projecting are listed below.

   i. 'By eye' (or inspection) from the graph. This involves adding a projection freehand in a manner that seems most appropriate. This might seem fairly arbitrary, but remember that any form of projection (no matter how technical) is still only an estimate. This particular method can be employed when the calculated trend values are distinctly 'non-linear'.

   ii. Using the method of semi-averages on the calculated trend values to obtain a linear projection of the trend. This method can be employed with 'fluctuating linear' trend values.

   iii. Using the first and last of the calculated trend values to obtain a linear projection of the trend. This method can be employed with fairly 'steady linear' trend values.

## 13. Example 4 (Time series *forecasting*)

*Question*

Forecast the four quarterly values for year 4 for the following data, which relates to UK outward passenger movements by sea (in millions). The trend (calculated previously in chapter 16, example 3) and the seasonal variation components (using the multiplicative model) are given below.

| | Year 1 | | | | Year 2 | | | | Year 3 | | | |
|---|---|---|---|---|---|---|---|---|---|---|---|---|
| Quarter | 1 | 2 | 3 | 4 | 1 | 2 | 3 | 4 | 1 | 2 | 3 | 4 |
| Number of passengers ($y$) | 2.2 | 5.0 | 7.9 | 3.2 | 2.9 | 5.2 | 8.2 | 3.8 | 3.2 | 5.8 | 9.1 | 4.1 |
| Trend ($t$) | | | 4.66 | 4.78 | 4.84 | 4.95 | 5.06 | 5.18 | 5.36 | 5.51 | | |

Seasonal variation ($S$): Qtr1 = 0.60; Qtr2 = 1.05; Qtr3 = 1.65; Qtr4 = 0.70

Plot the original values, trend and forecasts on a single chart.

*Answer*

*STEP 1*

Estimate trend values for the relevant time points. Since there is a fairly steady increase in the trend values, demonstrating an approximate linear relationship, method iii (from section 12 (b)) is appropriate for projecting the trend.

Range of trend values = 5.51 – 4.66 = 0.85

Therefore, average change per time period = 0.85÷7 = 0.12 (approx).

[Note that since there are 8 trend values, there are correspondingly only 7 'jumps' from the first to the last. Hence the divisor of 7 in the above calculation.]

The last trend value given is 5.51 for Year 3 Quarter 2 and this must be used as the base value to which is added the appropriate number of multiples of 0.12.

Thus, the trend estimates are:

$t$(Year4 Qtr1) = 5.51 + 3(0.12) = 5.87; $t$(Year4 Qtr2) = 5.51 + 4(0.12) = 5.99;
$t$(Year4 Qtr3) = 5.51 + 5(0.12) = 6.11; $t$(Year4 Qtr4) = 5.51 + 6(0.12) = 6.23.

*STEP 2*

Identify the appropriate seasonal factors.

These values are given in the question as: $S_1$=0.60; $S_2$=1.05; $S_3$=1.65; $S_4$=0.70.

*STEP 3*

Multiply the trend estimates by the respective $S$ values, giving the required forecasts.

$y$(Year4 Qtr1) = 5.87 × 0.60 = 3.52; $y$(Year4 Qtr2) = 5.99 × 1.05 = 6.29;
$y$(Year4 Qtr3) = 6.11 × 1.65 = 10.08; $y$(Year4 Qtr4) = 6.23 × 0.70 = 4.36.

These values are plotted in Figure 2, along with the original data and trend.

*Figure 2 Forecast (projected) values plotted with trend and original values*

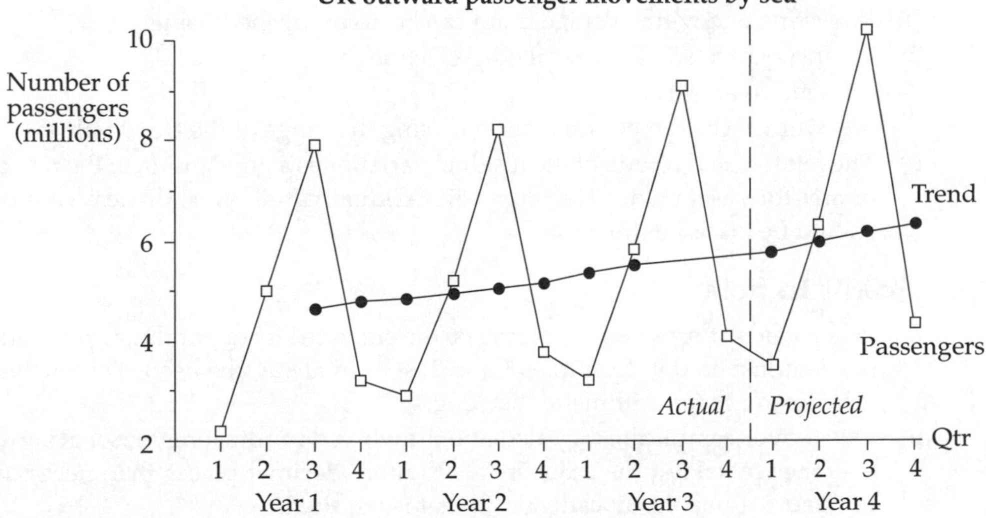

## 14. Forecasting and residual variation

Residual variation was described in chapter 18 as that variation which takes into account everything else other than trend and seasonal factors. In the main it consists of small random fluctuations which, although not controllable, have little effect. If the residual variation is relatively large however, it will make forecasts less dependable, since they effectively ignore residual elements.

Thus, being able to identify a residual element in a time series will normally be a pointer to how reliable any projection will be.

## 15. Summary

a) Seasonal factors are of importance to management as a control factor wherever seasonal effects are significant.

b) Seasonal factors:
   i. are individually expressed as deviations from (additive model) or percentages of (multiplicative model) the trend;
   ii. should collectively sum to either 0 (additive model) or 4 (multiplicative model).

c) Seasonally adjusted values are calculated by
   i. subtracting seasonal factors from trend values (additive model); or
   ii. dividing trend value by seasonal factor (multiplicative model).

d) Seasonally adjusted values are used to eliminate the effect of seasonal variation.

e) Time series forecasting involves
   i. adding the appropriate seasonal factors to calculated trend projections (additive model);
   ii. multiplying the calculated trend projections by the appropriate seasonal factor (multiplicative model).

f) Projecting a moving average trend can be accomplished using:
   i. inspection (sometimes called projection 'by eye');
   ii. semi-averages; or
   iii. average change per time period, using the range of the trend values.

g) The relative magnitude of the residual variation is a good guide to the reliability of any forecasts made. The larger the residual variation is, the less confidence should be placed in forecasts.

## 16. Points to note

a) When calculating seasonal averages, the standard form of tabular presentation (as demonstrated in Examples 1 and 2) should always be used. This enables an easy reference for arithmetic checking.

b) Never over-complicate the calculations involved in adjusting seasonal variation averages by taking the adjustments to more decimal places than necessary. A fair and reasonable allocation is all that is required.

c) Seasonally adjusted values are sometimes referred to as *deseasonalised data.*

d) It is normal to project or forecast one further cycle of a given time series provided that the data consists of at least three cycles. The reliability of further projections will depend on knowledge of the situation and the experience and ability of the investigator.

e) A range of error can be calculated for projected values. The further forward the projection, the greater the calculated range of error will be. The technique however is beyond the scope of this text.

f) Residual variation ($r$ or $R$ values) can be calculated as long as trend and seasonal values have already been identified.
This can be accomplished by a simple rearrangement of
 the time series additive model: $y = t + s + r$ to read: $r = y - t - s$;
 the time series multiplicative model: $y = t \times S \times R$ to read: $R = \frac{y}{t \times S}$.
In other words, residual variation at any time point is what is left when trend and seasonal values are eliminated from the original data value.

## 17. Student self review questions

1. What is seasonal variation? [2]
2. Given a time series with trend figures already calculated, describe *in words only* (as concisely as possible) the method for calculating seasonal variation values using the additive model. [3]
3. How is time series data seasonally adjusted? [6]
4. How might the management of an organisation use seasonal variation figures and seasonally adjusted data? [2,6]
5. Why must forecasts be treated with caution? [9]
6. How are the two time series analysis models adapted in order to produce forecasts? [10]
7. When projecting a moving average trend, what basis would make the choice of one of the following appropriate?
   a) Projecting 'by eye'
   b) Using the method of semi-averages.
   c) Using the average change in trend per period from the range. [12(b)]
8. How does residual variation affect a time series projection? [14]

## 18. Student exercises

1. The following data describes the sales of components for a particular firm:

| | Quarters | | | |
|---|---|---|---|---|
| | 1 | 2 | 3 | 4 |
| Year 1 | | | | 130 |
| Year 2 | 140 | 160 | 90 | 140 |
| Year 3 | 160 | 170 | 120 | 170 |
| Year 4 | 180 | 200 | 130 | |

Seasonally adjust these sales, using:
a) an additive model  b) a multiplicative model

2. Using the data of STUDENT EXERCISE 2, CHAPTER 16, together with the calculated moving average trend of period 5, calculate (to 1D):
   a) seasonal variation (for five 'seasons', counting the first time point as season 1 etc), and hence
   b) seasonally adjust the original data and
   c) forecast the values for the next full cycle of the data.

   Assume an additive model.
3. The following figures (measured to the nearest 10) relate to the number of holidays booked at an Austrian resort hotel together with the centred four-quarterly moving average trend.

| Bookings | Q1 | Q2 | Q3 | Q4 | Moving average trend | Q1 | Q2 | Q3 | Q4 |
|---|---|---|---|---|---|---|---|---|---|
| 1983 | 220 | 260 | 260 | 350 | 1983 | | | 270.0 | 265.0 |
| 1984 | 200 | 240 | 240 | 330 | 1984 | 260.0 | 255.0 | 255.0 | 262.5 |
| 1985 | 220 | 280 | 310 | 390 | 1985 | 276.25 | 292.5 | 303.75 | |
| 1986 | 250 | | | | | | | | |

   Calculate, using a multiplicative model:
   a) the seasonal variation for each quarter
   b) the deaseasonalised figures for 1985 Q3 and Q4 and 1986 Q1
   c) forecasts for the remaining three quarters of 1986 (take the three trend projections as 304, 308 and 312 respectively).
4. The data below relates to Rate receipts (in £m) for a Local Authority with a corresponding trend value in brackets.

| | 1982 | 1983 | 1984 |
|---|---|---|---|
| Quarter 1 | 2.8 (3.3) | 3.0 (3.7) | 3.0 (4.2) |
| Quarter 2 | 4.2 (3.4) | 4.2 (3.9) | 4.7 (4.3) |
| Quarter 3 | 3.0 (3.5) | 3.5 (4.0) | 3.6 (4.4) |
| Quarter 4 | 4.6 (3.6) | 5.0 (4.1) | 5.3 (4.5) |

   Assuming an additive model:
   a) calculate the seasonal variation
   b) estimate the receipts for 1985.

The following THREE questions (5, 6 and 7) are all based on the following data, where a *multiplicative model* should be assumed. (Note that the seasonal variation is given in two different but equivalent ways)

Quarterly sales (units) of Brand X, 2003

| | Q1 | Q2 | Q3 |
|---|---|---|---|
| Sales (units) | 1,600 | 4,400 | 1,680 |
| Seasonal variation (% change) | –20% | +100% | –30% |
| Seasonal variation (proportion) | 0.8 | 2.0 | 0.7 |

5. *MULTI-CHOICE*. The trend value for Q1 sales (units) is:

   a) 1,280  b) 1,920  c) 2,000  d) none of these
6. *MULTI-CHOICE*. The seasonal variation for Q4 in 2003 is

   a) -50% or 0.5  b) 0% or 1  c) +50% or 1.5  d) none of these

7. *MULTI-CHOICE.* The forecast for the fourth quarter's sales (units), Q4, in 2003, assuming the trend pattern continues, is closest to:
   a) 1,300   b) 2,300   c) 3,800   d) 5,200

8. The following data describes personal savings as a percentage of earned income for a particular region of the country.

| | 1980 | 1981 | 1982 |
|---|---|---|---|
| Quarter 1 | 10.1 | 12.6 | 11.9 |
| Quarter 2 | 8.6 | 7.6 | 8.7 |
| Quarter 3 | 8.0 | 7.6 | 8.3 |
| Quarter 4 | 5.8 | 6.2 | 7.2 |

   Use both additive and multiplicative models to seasonally adjust the above percentages and forecast the percentage savings for quarter 1 of 1983. Comment on the results.

# **Examination example** (with worked solution)

*Question*

Analyse the following supermarket sales data and present the results in graphical form, including a forecast for the daily sales in week 5.

*Supermarket sales (£000) for a particular period*

| | Week 1 | Week 2 | Week 3 | Week 4 |
|---|---|---|---|---|
| Monday | 22 | 22 | 24 | 26 |
| Tuesday | 36 | 34 | 38 | 38 |
| Wednesday | 40 | 42 | 43 | 45 |
| Thursday | 48 | 49 | 49 | 50 |
| Friday | 61 | 58 | 62 | 64 |
| Saturday | 58 | 59 | 58 | 58 |

*Answer*

The Main Table below shows the calculations laid out in standard form with supplementary Tables 1 and 2. Table 3 shows the calculations involved in the projection for week 5.

*Main table*

| | Sales (*y*) | Totals of 6 | Average | Centered average (*t*) | (*y-t*) | Seasonal factor (*s*) | Seasonal adjustment (*y-s*) |
|---|---|---|---|---|---|---|---|
| Week 1 Mon | 22 | | | | | -21.36 | 43.36 |
| Tue | 36 | | | | | -8.82 | 44.82 |
| Wed | 40 | 265 | 44.17 | | | -2.24 | 42.24 |
| Thu | 48 | 265 | 44.17 | 44.17 | 3.83 | 3.87 | 44.13 |
| Fri | 61 | 263 | 43.83 | 44.00 | 17.00 | 15.37 | 45.63 |
| Sat | 58 | 265 | 44.17 | 44.00 | 14.00 | 13.18 | 44.82 |
| Week 2 Mon | 22 | 266 | 44.33 | 44.25 | -22.25 | -21.36 | 43.36 |
| Tue | 34 | 263 | 43.83 | 44.08 | -10.08 | -8.82 | 42.82 |
| Wed | 42 | 264 | 44.00 | 43.92 | -1.92 | -2.24 | 44.24 |
| Thu | 49 | 266 | 44.33 | 44.17 | 4.83 | 3.87 | 45.13 |
| Fri | 58 | 270 | 45.00 | 44.67 | 13.33 | 15.37 | 42.63 |
| Sat | 59 | 271 | 45.17 | 45.09 | 13.91 | 13.18 | 45.82 |
| Week 3 Mon | 24 | 271 | 45.17 | 45.17 | -21.17 | -21.36 | 45.36 |
| Tue | 38 | 275 | 45.83 | 45.50 | -7.50 | -8.82 | 46.82 |
| Wed | 43 | 274 | 45.67 | 45.75 | -2.75 | -2.24 | 45.24 |
| Thu | 49 | 276 | 46.00 | 45.84 | 3.16 | 3.87 | 45.13 |
| Fri | 62 | 276 | 46.00 | 46.00 | 16.00 | 15.37 | 46.63 |
| Sat | 58 | 278 | 46.33 | 46.17 | 11.83 | 13.18 | 44.82 |
| Week 4 Mon | 26 | 279 | 46.50 | 46.42 | -20.42 | -21.36 | 47.36 |
| Tue | 38 | 281 | 46.83 | 46.67 | -8.67 | -8.82 | 46.82 |
| Wed | 45 | 281 | 46.83 | 46.83 | -1.83 | -2.24 | 47.24 |
| Thu | 50 | | | | | 3.87 | 46.13 |
| Fri | 64 | | | | | 15.37 | 48.63 |
| Sat | 58 | | | | | 13.18 | 44.82 |

*Table 1*

**Calculations for seasonal variation**

| | Mon | Tue | Wed | Thu | Fri | Sat |
|---|---|---|---|---|---|---|
| Week 1 | | | | 3.83 | 17.00 | 14.00 |
| Week 2 | -22.25 | -10.08 | -1.92 | 4.83 | 13.33 | 13.91 |
| Week 3 | -21.17 | -7.50 | -2.75 | 3.16 | 16.00 | 11.83 |
| Week 4 | -20.42 | -8.67 | -1.83 | | | |
| Total | -63.84 | -26.25 | -6.50 | 11.82 | 46.33 | 39.74 |
| Average | -21.28 | -8.75 | -2.17 | 3.94 | 15.44 | 13.25 |
| Adjustment | -0.08 | -0.07 | -0.07 | -0.07 | -0.07 | -0.07 |
| Adjusted average | -21.36 | -8.82 | -2.24 | 3.87 | 15.37 | 13.18 |

*Table 2*

**Seasonally adjusted sales**

| | Mon | Tue | Wed | Thu | Fri | Sat |
|---|---|---|---|---|---|---|
| Week 1 | 43.36 | 44.82 | 42.24 | 44.13 | 45.63 | 44.82 |
| Week 2 | 43.36 | 42.82 | 44.24 | 45.13 | 42.63 | 45.82 |
| Week 3 | 45.36 | 46.82 | 45.24 | 45.13 | 46.63 | 44.82 |
| Week 4 | 47.36 | 46.82 | 47.24 | 46.13 | 48.63 | 44.82 |

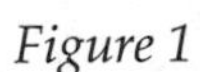

*Figure 1*

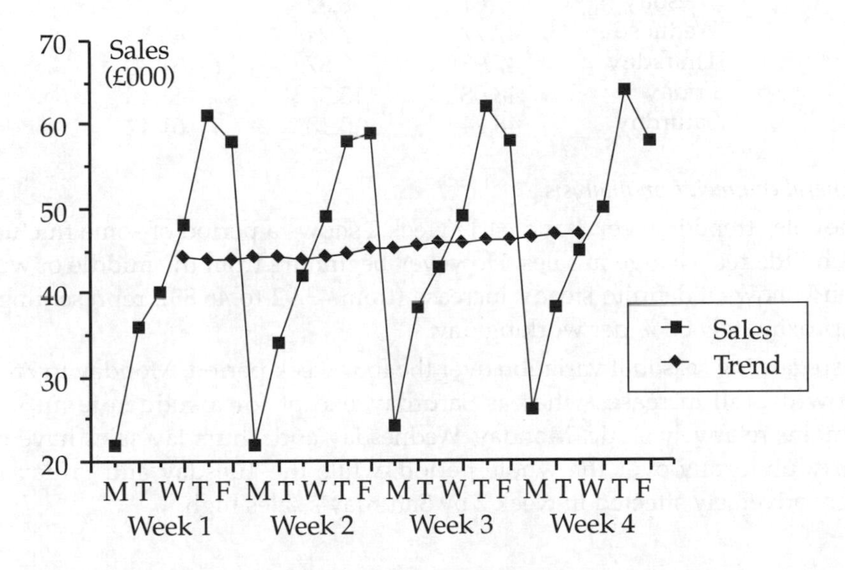

*Figure 2*

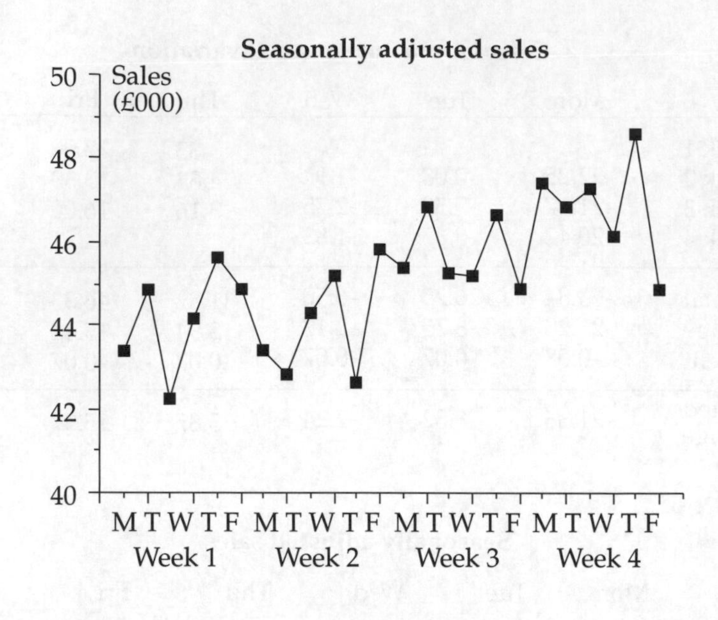


*Table 3*

**Projection for week 5**

| | $t_{est}$ | $s$ | $t_{est} + s$ |
|---|---|---|---|
| Monday | 47.46 | -21.36 | 26.88 |
| Tuesday | 47.61 | -8.82 | 38.79 |
| Wednesday | 47.77 | -2.24 | 45.53 |
| Thursday | 47.93 | 3.87 | 51.80 |
| Friday | 48.08 | 15.37 | 63.45 |
| Saturday | 48.24 | 13.18 | 61.42 |

*General comments on analysis.*

The sales trend in week 1 and early week 2 shows a period of some fluctuation but with little real change in sales. However, beginning from the middle of week 2, the trend shows a definite steady increase (from 43.92 to 46.83), representing 6.5% or approximately 0.5% per working day.

Disregarding seasonal variation over the four week period, Monday to Friday sales show overall increases whereas Saturday, except for a sudden jump in week 2, remains relatively static. Monday, Wednesday and Thursday sales have increased fairly uniformly over the whole period, while the Tuesday and Friday increases were adversely affected in week 2 by Saturday's sales high.

# Examination questions

1. The quarterly electricity account for your company is tabulated as follows:

Electricity account (£)

| | Quarter | | | |
|---|---|---|---|---|
| Year | 1 | 2 | 3 | 4 |
| 1982 | | 662 | 712 | 790 |
| 1983 | 686 | 718 | 821 | 846 |
| 1984 | 743 | 782 | 827 | 876 |
| 1985 | 805 | 842 | 876 | |

Required:

a) Assuming the additive model, establish the centered trend values for the data using a method of moving averages.

b) i. Plot the original data and the trend values together on a properly labelled graph.

ii. Draw the trend line on the graph.

c) If the seasonal variations are -56 for quarter 1, -39 for quarter 2, 45 for quarter 3 and 51 for quarter 4, deseasonalise the original data.

d) By extending the trend line, establish forecasts for the electricity account values for quarter 4, 1985 and quarter 1, 1986.

*ACCA*

2. The following information has been supplied by the Sales department. (Sales are in units.)

| | Quarter | | | |
|---|---|---|---|---|
| | 1 | 2 | 3 | 4 |
| Year | | | | |
| 1973 | 100 | 125 | 127 | 102 |
| 1974 | 104 | 128 | 130 | 107 |
| 1975 | 110 | 131 | 133 | 107 |
| 1976 | 109 | 132 | | |

You are required to:

a) calculate a four quarterly moving average of the above series;

b) calculate the sales corrected for seasonal movements;

c) plot the actual sales and the sales corrected for seasonal movements on a single graph; and

d) comment on your findings.

*CIMA*

3. The daily output of your company over a four week period is shown in the table below.

*Number of units of output*

| | Monday | Tuesday | Wednesday | Thursday | Friday |
|---|---|---|---|---|---|
| Week 1 | 187 | 203 | 208 | 207 | 217 |
| Week 2 | 207 | 208 | 210 | 206 | 212 |
| Week 3 | 202 | 210 | 212 | 205 | 214 |
| Week 4 | 208 | 215 | 217 | 217 | 213 |

Required:

a) Using the additive time series model, establish the five-period moving average trend of output.

b) Display on a graph the actual data together with the trend figures.

c) Establish the daily deviations from the trend and use these to determine the average daily variations.

d) Forecast the daily output for the first two days of Week 5 to the nearest unit of production.

e) Comment upon the accuracy of the forecast that you have made.

*ACCA*

4. a) Briefly explain the components which make up a typical time series.

b) Unemployed school leavers in the United Kingdom (figures in thousands) is tabulated below.

| Year | January | April | July | October |
|---|---|---|---|---|
| 1979 | 22 | 12 | 110 | 31 |
| 1980 | 21 | 26 | 150 | 70 |
| 1981 | 50 | 36 | 146 | 110 |

*Source: Employment Gazette*

i. By means of a moving average, find the trend and the seasonal adjustments.

ii. Give the data for January and April 1981 seasonally adjusted.

*ICSA*

# Part 5 Index numbers

Index numbers provide a standardised way of comparing the values, over time, of commodities such as prices, volume of output and wages. They are used extensively, in various forms, in Business, Commerce and Government.

Chapter 18 describes index relatives, the simplest form of index number, and some of the ways that they can be presented and manipulated.

Chapter 19 covers composite index numbers, which describe the change over time of groups or classes of commodities that have something in common. The two forms of composite index covered are the weighted average of relatives and the weighted aggregate.

Chapter 20 is devoted to a description of some of the most significant officially published indices. These include the Index of Retail Prices, Producer Price Indices, Index of Average Earnings and Index of Output of the Production Industries.

# 18 Index relatives

## 1. Introduction

This chapter introduces index numbers and describes the most simple form; the index relative. Relatives are defined, calculated as time series and compared (using a base-changing technique). Finally, time series deflation is described, which is a method of calculating an index of the real values of a time series.

## 2. Definition of an index number

> An **index number** measures the percentage change in the value of some economic commodity *over a period of time.*
> It is always expressed in terms of a base of 100.

'Economic commodity' is a term of convenience, used to describe anything measurable which has some economic relevance. For example: price, quantity, wage, productivity, expenditure, and so on.
Examples of typical index number values are:
125 (an increase of 25%), 90 (a decrease of 10%), 300 (an increase of 200%).

## 3. Simple index number construction

a) Suppose that the price of standard boxes of ball-point pens was 60p in January and rose to 63p in April. We can calculate as follows:

$$\text{percentage increase} = \frac{63-60}{60} = \frac{100}{20} = 5$$

In other words, the price of ball-point pens rose by 5% from January to April. To put this into index number form, the 5% increase is added to the base of 100, giving 105. This is then described as follows:

*the price index of ball-point pens in April was 105 (January=100).*

Note that any increase must always be related to some time period, otherwise it is meaningless. Index numbers are no exception, hence the (January=100) in the above statement, which:

i. gives the starting point (January) over which the increase in price is being measured;
ii. emphasises the base value (100) of the index number.

b) If the productivity of a firm (measured in units of production per man per day) decreased by 3% over the period from 1983 to 1985, this percentage would be *subtracted* from 100 to give an index number of 97. Thus we would say:

*'the productivity index for the company in 1985 was 97 (1983=100)'.*

## 4. Some notation

a) It is convenient, particularly when giving formulae for certain types of index numbers, to be able to refer to an economic commodity at some general time point.

Prices and quantities (since they are commonly quoted indices) have their own special letters, $p$ and $q$ respectively. In order to bring in the idea of time, the following standard convention is used.

**Index number notation**

$p_0$ = price at *base* time point

$p_n$ = price at *some other* time point

$q_0$ = quantity at *base* time point

$q_n$ = quantity at *some other* time point.

In the example in 3(a) above, time point 0 was January and time point n was April, with p0=60 and pn=63.

b) It is also convenient on occasions to label index numbers themselves in a compact way. There is no standard form for this but, for example (from section 3 b)), the following is sometimes used:

$$I_{1985}(1983{=}100) = 97$$

or

$$I_{1985/1983} = 97$$

which is translated as:

*'the index for 1985, based on 1983 (as 100), is 97'.*

## 5. Index relatives

An *index relative* (sometimes just called a *relative*) is the name given to an index number which measures the change in a single distinct commodity. A price relative was calculated in section 3 a) and a productivity relative was found in section 3 b). However, there is a more direct way of calculating relatives than that demonstrated in section 3.

The following shows the method of calculating a price and quantity relative.

**Price and quantity relatives**

$$\text{Price relative: } I_P = \frac{p_n}{p_0} \times 100$$

$$\text{Quantity relative: } I_Q = \frac{q_n}{q_0} \times 100$$

Expenditure and productivity relatives can be calculated in a similar fashion.

## 6. Example 1 (Calculation of *price* and *quantity relatives*)

The following table gives details of prices and quantities sold of two particular items in a department store over two years.

| Item | 1984 Price | 1984 Number sold | 1985 Price | 1985 Number sold |
|---|---|---|---|---|
| | $p_0$ | $q_0$ | $p_n$ | $q_n$ |
| Video recorder PX21 | £438 | 37 | £462 | 18 |
| 27-inch television X8 | £322 | 26 | £384 | 45 |

We wish to find price and quantity relatives for 1985 (1984=100) for both items.

Year 0 = 1984 and year $n$ = 1985.

*For the video*:

$$\text{Price relative} = \text{I(VP)}_{85/84} = \frac{p_n}{p_0} \times 100 = \frac{462}{438} \times 100$$

$$= 105.5$$

$$\text{Quantity relative} = \text{I(VQ)}_{85/84} = \frac{q_n}{q_0} \times 100 = \frac{18}{37} \times 100$$

$$= 48.6$$

*For the television*:

$$\text{Price relative} = \text{I(TP)}_{85/84} = \frac{p_n}{p_0} \times 100 = \frac{384}{322} \times 100$$

$$= 119.3$$

$$\text{Quantity relative} = \text{I(TQ)}_{85/84} = \frac{q_n}{q_0} \times 100 = \frac{45}{26} \times 100$$

$$= 173.1$$

The above calculations and presentation demonstrates typical index number notation. Notice the use of the bracketed VP, VQ, etc used for clarity.

Thus it can be seen that an index number is a compact way of describing percentage changes over time.

## 7. Time series of relatives

It is often necessary to see how the values of an index relative change over time. Given the values of some commodity over time (i.e. a time series), there are two distinct ways in which relatives can be calculated.

a) *Fixed base relatives*. Here, each relative is calculated based on the *same fixed* time point. This approach can only be used when the basic nature of the commodity is unchanged over the whole period. That is, fixed base relatives are used for comparing 'like with like'. For example, the price of Canadian cheddar cheese in a supermarket over six monthly periods or weekly family expenditure on entertainment.

b) *Chain base relatives*. In this case, each relative is calculated with respect to the *immediately preceding* time point. This approach can be used with any set of commodity values, but must be used when the basic nature of the commodity changes over the whole time period. For example, a company might wish to construct a monthly index of total petrol costs for the standard model of car that its salesmen use. However, the model is likely to change yearly with, for instance, different tyres or 'lean-burn' engines being fitted as standard. Both of these features would affect petrol consumption and thus, also, the petrol cost index. Therefore, in this case, a chain base relative should be used.

Example 2 demonstrates the use of the two techniques for the values of a commodity over time.

## 8. Example 2 (*Fixed* and *chain base* set of relatives for a given time series)

The data in Table 1 relate to the production of beer (thousands of hectolitres) in the United Kingdom for the first six months of a year.

Table 2 shows the calculation of both fixed and chain base relatives, together with some descriptive calculations.

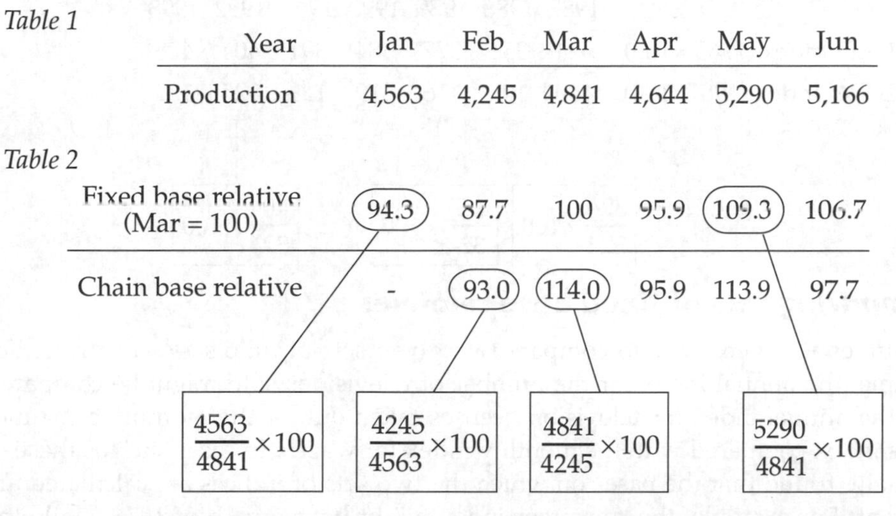

*Table 1*

| Year | Jan | Feb | Mar | Apr | May | Jun |
|---|---|---|---|---|---|---|
| Production | 4,563 | 4,245 | 4,841 | 4,644 | 5,290 | 5,166 |

*Table 2*

| | | | | | | |
|---|---|---|---|---|---|---|
| Fixed base relative (Mar = 100) | 94.3 | 87.7 | 100 | 95.9 | 109.3 | 106.7 |
| Chain base relative | - | 93.0 | 114.0 | 95.9 | 113.9 | 97.7 |



In Table 2, the fixed base relatives have been calculated by dividing each month's production by the March production (4841) and multiplying by 100. They enable each month's production to be compared with the March production. Thus, for example, May's production (relative=109.3) was 9.3% up on March.

The chain base relatives in Table 2 have been calculated by dividing each month's production by the previous month's production and multiplying by 100. They enable changes from month to month to be highlighted. Thus, for example, February's production (chain relative=93.0) was 7% down on January, March's production (chain relative=114.0) was 14% up on February, and so on.

## 9. Changing the base of fixed-base relatives

Given a time series of relatives, it is sometimes necessary to change the base. One of the reasons for doing this might be that the original base time point is too far in the past to be relevant today and a more recent one is needed. For example, the following set has a base of 1965, which would probably now be considered out-of-date.

| | 1987 | 1988 | 1989 | 1990 | 1991 | 1992 | 1993 |
|---|---|---|---|---|---|---|---|
| Index (1965=100) | 324 | 351 | 377 | 384 | 391 | 404 | 428 |

The procedure for changing the base of a time series of relatives is essentially the same as that for calculating a set of relatives for a given time series of values. However, the procedure is given below and demonstrated, using the above set of relatives:

STEP 1 Choose the required new base time point and thus identify the corresponding relative.

We will choose 1987 as the base year, with a corresponding relative of 324.

STEP 2 *Divide* each relative in the set by the value of the relative identified above and *multiply the result* by 100.

Thus, each index relative given needs to be divided by 324 and multiplied by 100. Table 3 shows the new index numbers.

*Table 3*

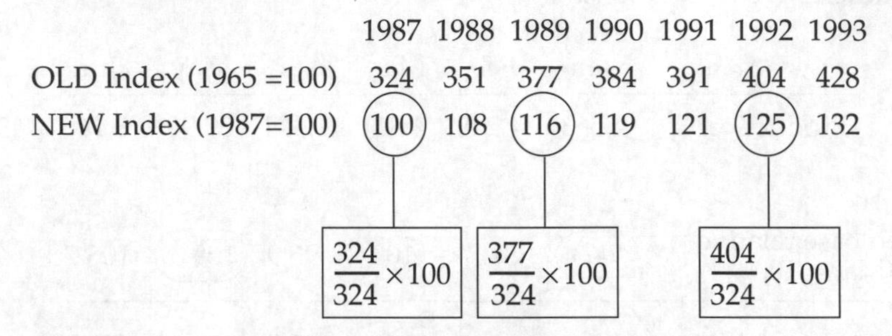

| | 1987 | 1988 | 1989 | 1990 | 1991 | 1992 | 1993 |
|---|---|---|---|---|---|---|---|
| OLD Index (1965 =100) | 324 | 351 | 377 | 384 | 391 | 404 | 428 |
| NEW Index (1987=100) | 100 | 108 | 116 | 119 | 121 | 125 | 132 |



## 10. Comparing sets of fixed base relatives

Sometimes it is necessary to compare two given sets of time series relatives. For example, the annual index for the number of televisions sold might be compared with the annual index for television licenses taken out, or the monthly consumer prices index compared with the monthly index for wages. In cases such as these, it is usually found that the bases on which the two sets of indices are calculated are different. For example, the consumer index might have a base of 1974, while the wage index has a base of 1983. This can make comparisons difficult because the two sets of index relatives will be of different magnitudes. As an illustration, consider the data of Table 4.

| Year | 1986 | 1987 | 1988 | 1989 | 1990 | 1991 | 1992 |
|---|---|---|---|---|---|---|---|
| No. of TV sets sold (1988=100) | 61 | 88 | 100 | 135 | 165 | 192 | 210 |
| No. of TV licences taken out (1970=100) | 210 | 230 | 250 | 300 | 360 | 410 | 500 |

Comparing the indices given above is not easy. Many percentage increases will have to be calculated before any worthwhile comparisons can be made. This type of problem can be overcome by changing the base of one set of indices to match the base of the other. The following example shows the calculations necessary.

## 11. Example 3 (Time series comparison by *changing the base* of one of the sets)

*Question*

Compare the figures given in Table 4 by changing the base of one of the sets and comment on the results.

*Answer*

The base of the television licence relatives will be changed to coincide with the base of the televisions sold relatives. The following table shows the new figures.

*Table 4*

| Year | 1986 | 1987 | 1988 | 1989 | 1990 | 1991 | 1992 |
|---|---|---|---|---|---|---|---|
| Number of television sets sold (1988=100) | 61 | 88 | 100 | 135 | 165 | 192 | 210 |
| Number of television licences taken out (1970=100) | 210 | 230 | 250 | 300 | 360 | 410 | 500 |
| Number of television licences taken out (1988=100) | 84 | 92 | 100 | 120 | 144 | 164 | 200 |

$$92 = \frac{230}{250} \times 100 \qquad 144 = \frac{360}{250} \times 100$$

The two sets of relatives are now much easier to compare. Before 1988 and up to 1991, sales of television sets increased at a much faster rate. However, over the last year, the number of television licenses taken out increased dramatically, showing the same percentage increase (over 1988) as the sales of television sets (possibly due to detector van publicity).

## 12. Actual and real values of a commodity

In times of significant inflation, the actual value of some commodity is not the best guide of its *'real' value* (or *worth*).

The worth of any commodity can only be measured relative to the value of some associated commodity. In other words, some relevant 'indicator' is necessary against which to judge value.

For example, suppose that the annual rent of some business premises last year was £2000 and this year it has been increased by 10% to £2200. Clearly the actual cost is higher. However, if we are now given the information (as an indicator) that the cost of business premises in the region as a whole has risen by 15% over the past year, we can rightly argue that the real cost of the given premises *has decreased*.

On the other hand, if business turnover for the premises (as an alternative indicator) has only increased by 5%, we might consider that the real cost of the premises *has increased*. Thus, depending on the particular indicator chosen, the real value of a commodity can change.

Two standard national indicators are the rate of inflation (normally represented by the Retail Prices Index) and the Index of Output of the Production Industries. (Both of these indices are described in chapter 20.)

The following section describes a method of constructing a series of relatives to measure the real value of some commodity over time. This is known as time series deflation.

## 13. Time series deflation

*Time series deflation* is a technique used to obtain a set of index relatives that measure the changes in the real value of some commodity with respect to some given indicator.

*Table 5*

| Month | 1 | 2 | 3 | 4 | 5 | 6 | 7 | 8 |
|---|---|---|---|---|---|---|---|---|
| Average daily wage (£) | 17.60 | 18.10 | 18.90 | 19.60 | 20.25 | 20.30 | 20.60 | 21.40 |
| Retail price index | 106.1 | 107.9 | 112.0 | 113.1 | 116.0 | 117.4 | 119.5 | 119.7 |

The procedure for calculating each index relative is given below, using the data of Table 5 to demonstrate calculating the real wage index for month 7 (month 1 = 100) as an example.

STEP 1 Choose a base for the index of real values of the series.

In this case, month 1 has been chosen.

Then, *for each time point* of the series:

STEP 2 Find the ratio of the current value to the base value.

For month 7, this gives: $\frac{20.60}{17.60} = 1.17$

This step expresses the increase in the actual value as a multiple.

STEP 3 Multiply by the ratio of the base indicator to the current indicator (notice that these two values are in reverse order compared with the two in the previous step).

For month 7, this gives: $1.17 \times \frac{106.1}{119.5} = 1.039$, 'deflating' the above wage multiple.

STEP 4 Multiply by 100.

For month 7, this gives: $1.039 \times 100 = 103.9$.

This step changes the multiple of the previous step into an index.

The above steps can be summed up both in symbols and words as follows.

**Real Value Index (RVI)**

Given a time series ($x$-values) and some indicator index series ($I$-values) for comparison, the *real value index* for period $n$ is given by:

$$\text{RVI} = \frac{\text{current value}}{\text{base value}} \times \frac{\text{base indicator}}{\text{current indicator}} \times 100$$

$$= \frac{x_n}{x_0} \times \frac{I_0}{I_n} \times 100$$

The following example duplicates the data of Table 5 and shows the real wage index relatives, the calculations (using the above steps) being demonstrated for selected values.

## 14. Example 4 (*Index relatives of real values*)

Table 6 following shows the values of the real wage index relative for the data of Table 5.

*Table 6*

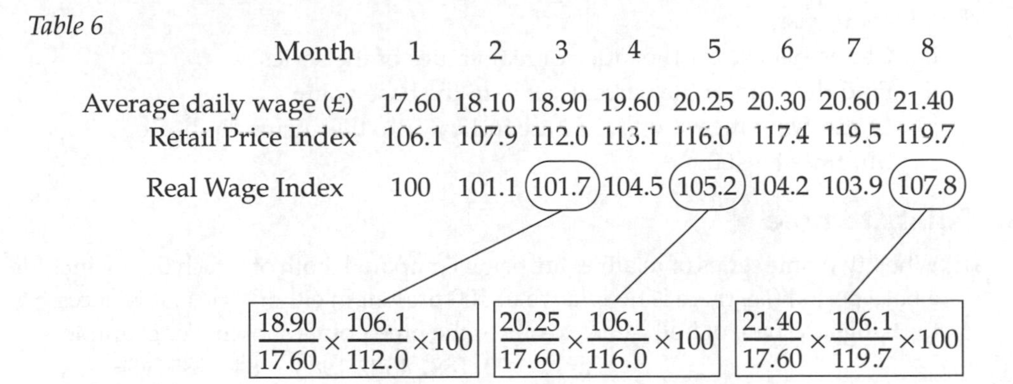

| Month | 1 | 2 | 3 | 4 | 5 | 6 | 7 | 8 |
|---|---|---|---|---|---|---|---|---|
| Average daily wage (£) | 17.60 | 18.10 | 18.90 | 19.60 | 20.25 | 20.30 | 20.60 | 21.40 |
| Retail Price Index | 106.1 | 107.9 | 112.0 | 113.1 | 116.0 | 117.4 | 119.5 | 119.7 |
| Real Wage Index | 100 | 101.1 | 101.7 | 104.5 | 105.2 | 104.2 | 103.9 | 107.8 |

$$\frac{18.90}{17.60} \times \frac{106.1}{112.0} \times 100 \qquad \frac{20.25}{17.60} \times \frac{106.1}{116.0} \times 100 \qquad \frac{21.40}{17.60} \times \frac{106.1}{119.7} \times 100$$

The real wage index shows that the real value of the average weekly wage has increased by 7.8% over the nine-month period. In real terms, wages increased steadily with larger than usual increases in months 4 and 8 and small decreases in months 6 and 7.

## 15. Summary

a) An index number measures the percentage change in the value of some economic commodity over a period of time. It is always expressed in terms of a base of 100.

b) Some notation used in the calculation of index numbers is:

$p_0$ = price at base time point; $p_n$ = price at some other time point;

$q_0$ = quantity at base time point; $q_n$ = quantity at some other time point.

c) An index relative is the name given to an index number which measures the change in a single distinct commodity.

d) A price relative can be calculated as the ratio of the current price to the base price multiplied by one hundred. Quantity, expenditure and productivity relatives are calculated in a similar manner.

e) Fixed base relatives are found by calculating relatives for each value of a time series based on the same fixed time point. Chain base relatives are found by calculating relatives for each value of a time series based on the immediately preceding time point.

f) In order to compare two time series of relatives, each series should have the same base time point.

g) To change the base of a given time series of relatives:

i. select the set which is to have its base changed,

ii. divide each relative in this set by the value of the relative which matches 100 in the other set and multiply the result by 100.

h) The real value of some commodity can only be measured in terms of some 'indicator'. Standard indicators are the Retail Price Index or the Index of Output of the Production Industries.

j) Time series deflation is a technique used to obtain a set of index relatives that measure the changes in the real value of some commodity with respect to some given indicator.

The procedure is:

i. Choose a base for the index of real values of the series.
ii. Find the ratio of the current value to the base value.
iii. Multiply by the ratio of the base indicator to the current indicator.
iv. Multiply by 100.

## 16. Points to note

a) When two time series of relatives are being compared, both of which do not include a base period (i.e. there is no relative of 100 present in either series), it is necessary to change the base of both series to one of the time points shown. For example:

| Year | 1980 | 1881 | 1982 | 1983 | 1984 | 1985 | 1986 |
|---|---|---|---|---|---|---|---|
| Productivity relative (1976=100) | 127 | 124 | 134 | 134 | 146 | 141 | 146 |
| Wage relative (1970=100) | 184 | 195 | 211 | 220 | 245 | 253 | 286 |

In this case, both series would need a base change to the same year (between 1980 and 1986).

b) *Chain linking* is the name sometimes given to the process of converting from a chain base to a fixed base set of index numbers. The following table and calculations demonstrate the technique.

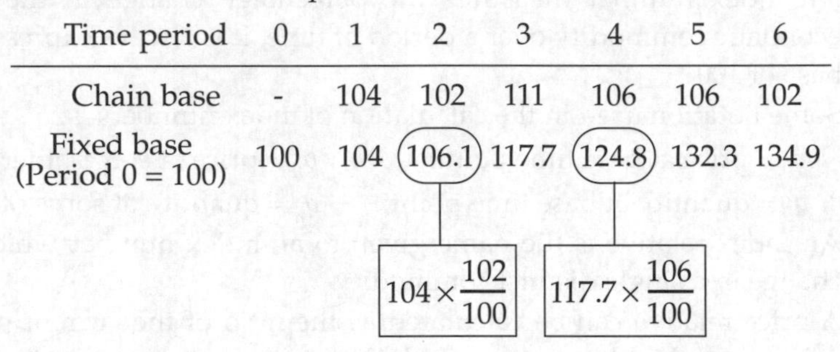

| Time period | 0 | 1 | 2 | 3 | 4 | 5 | 6 |
|---|---|---|---|---|---|---|---|
| Chain base | - | 104 | 102 | 111 | 106 | 106 | 102 |
| Fixed base (Period 0 = 100) | 100 | 104 | 106.1 | 117.7 | 124.8 | 132.3 | 134.9 |

$104 \times \frac{102}{100}$ $\quad$ $117.7 \times \frac{106}{100}$

The fixed base index must show the same percentage increase as the chain base index each year. Thus, since from period 1 to 2 there was a 2% increase (signified by the chain base relative of 102), the fixed base index for period 1 (104) must be increased by 2%, as shown above. Similarly for the other time points.

c) Sometimes 'points' increases are quoted for index numbers as opposed to percentage increases. A *points increase* in an index number is the numeric difference in its value as measured from one time point to another. For example, if an index number has risen from 245 to 257, then:

i. the points increase = 257 – 245 = 12, but
ii. the percentage increase $\frac{257 - 245}{245} \times 100 = 4.90$.

**Note**: The distinction between a 'points' difference and a 'percentage' difference is an important one which is often misunderstood and misused.

d) Often the base time point of a series is chosen as the first time point, but this does not have to be the case. Any time point of the series can be chosen as a standard (i.e. base) for comparison purposes.

## 17. Student self review questions

1. What is an index number? [2]
2. Explain what $I_{1985/1983}$ means. [4]
3. What is a price relative? [5]
4. What is the difference between 'fixed base' and 'chain base' relatives? [7]
5. Under what conditions would chain base, rather than fixed base, relatives be more suitable for describing a commodity over time? [7]
6. Give two reasons for changing the base of a time series of relatives. [9,10]
7. What is meant by the 'real value' of a commodity and what place does an 'indicator' variable have in its measurement? [12]
8. Name two standard indicator index number series. [12]
9. What is time series deflation? [13]

## 18. Student exercises

1. *MULTI-CHOICE.* An index number increases each year by 5% of its value in the previous year. If its value in 2002 was 140, its value in 2005 is closest to:
   a) 160 b) 161 c) 162 d) 163
2. The average price of Brand X video cassettes this year was £3.33, which represented a decrease of 10% over last year's average price. The number bought (at these prices) last year was 2500, but increased by 750 this year. Calculate price, quantity and *expenditure* relatives for these cassettes for this year (based on last year).
3. The following data relate to the production of cars from a particular assembly line over a number of months.

| | Mar | Apr | May | Jun | Jul | Aug | Sep | Oct |
|---|---|---|---|---|---|---|---|---|
| Production | 142 | 126 | 128 | 104 | 108 | 146 | 158 | 137 |

   Calculate sets of productivity relatives (to 1D) with: (a) Mar=100 (b) May=100 (c) Aug=100.
4. Butter stocks (thousand tonnes) in the United Kingdom in a particular year

| Mar | Apr | May | Jun | Jul | Aug | Sep | Oct | Nov |
|---|---|---|---|---|---|---|---|---|
| 216.9 | 225.1 | 234.6 | 237.2 | 235.2 | 230.1 | 224.4 | 226.1 | 220.2 |

   Calculate (to 1D) a set of:
   a) fixed base relatives (Mar=100); b) chain base relatives.
   Comment on the results.
5. Quantity of main crop potatoes harvested for a region (000 tonnes).

| | 19X1 | 19X2 | 19X3 | 19X4 | 19X5 | 19X6 | 19X7 | 19X8 | 19X9 |
|---|---|---|---|---|---|---|---|---|---|
| Index (19X3=100) | 97 | 102 | 100 | 112 | 128 | 121 | 145 | 149 | 152 |

   a) Convert the above information to show a set of chain base relatives (to 1D).
   b) Given that the amount of potatoes harvested in 19X4 was 587,000 tonnes, calculate the amount of potatoes harvested (to the nearest 1000 tonnes) each year from 19X6 to 19X9.

6. The yearly index for the production of an important product for a firm is contrasted with a national production index for the same type of product.

| | 19X0 | 19X1 | 19X2 | 19X3 | 19X4 | 19X5 | 19X6 | 19X7 |
|---|---|---|---|---|---|---|---|---|
| Production index for firm (19X2=100) | 101 | 96 | 100 | 107 | 98 | 98 | 103 | 107 |
| National production index (19X0=100) | 384 | 382 | 427 | 445 | 416 | 410 | 427 | 444 |

Compare the firm's production record with national production by changing the base of the national index.

7. Compare the following series, using the same fixed base, and comment on the results.

*Average earnings index numbers*

| | Feb | Mar | Apr | May | Jun | Jul | Aug | Sep | Oct | Nov |
|---|---|---|---|---|---|---|---|---|---|---|
| Whole economy | 164.6 | 168.1 | 169.4 | 169.4 | 171.9 | 173.7 | 173.4 | 176.1 | 173.9 | 176.8 |
| Coal and Coke | 78.2 | 122.5 | 137.9 | 139.5 | 148.0 | 149.5 | 150.7 | 152.9 | 153.6 | 159.3 |

8. The figures below compare the fuel costs of a small garage with a national price index.

| Time point | 1 | 2 | 3 | 4 | 5 | 6 | 7 |
|---|---|---|---|---|---|---|---|
| Cost of fuel (£000) | 34.1 | 34.8 | 33.6 | 33.6 | 33.4 | 33.1 | 33.4 |
| Producer (fuel) price index | 169.8 | 173.9 | 163.8 | 151.1 | 148.9 | 147.4 | 147.4 |

Produce an index (time point 1 = 100), to 1D, of the real cost of fuel to the garage by deflating the given fuel costs by the Producer (Fuel) Price Index.

9. The data below show the gross income of a particular category of family compared with the Retail Price Index over a seven year period.

| | 19X5 | 19X6 | 19X7 | 19X8 | 19X9 | 19Y0 | 19Y1 |
|---|---|---|---|---|---|---|---|
| Family income (£000) | 6,989 | 8,105 | 8,416 | 10,037 | 11,475 | 13,443 | 16,140 |
| Retail price index | 134.8 | 157.1 | 182.0 | 197.1 | 223.5 | 263.7 | 295.0 |

Calculate:

a) an index of real gross income (19X5=100),

b) a chain base index of real gross income, using the Retail Price Index as an indicator.

*The next two questions (10 and 11) are based on the following data*

| | 2000 | 2001 | 2002 | 2003 |
|---|---|---|---|---|
| Weekly money wages index (2000=100) | 100 | 105 | 110 | 115 |
| Index of inflation (1992=100) | 180 | 190 | 200 | 210 |

10. *MULTI-CHOICE*. Consider the following two statements. (i) Inflation has increased by more than money wages (ii) Money wages have increased by 5% each year, year on year. Which one of the following is true?

a) (i) only b) (ii) only c) Both (i) and (ii) d) Neither (i) nor (ii)

11. *MULTI-CHOICE*. 'Real wages' are money wages that have been adjusted for inflation (or deflated). Over the period 2000 to 2003, real wages have approximately:

a) remained unchanged b) decreased by 1.43%

c) decreased by 1.67% d) increased by 1.45%

# 19 Composite index numbers

## 1. Introduction

This chapter defines a composite index number and the idea of weighting factors. Two types of composite index numbers are described, average of relatives and aggregates, and some factors in the construction of index numbers are discussed.

## 2. Composite index numbers

A *composite index number* is an index number which is obtained by combining the information from a set of economic commodities (in this context, conveniently called *components*) of like kind. Some examples are now given.

a) An index number of housing costs might consider components such as mortgage payments or rent, rates, repairs, insurance, and so on.

b) The official Retail Prices Index (generally regarded as a national consumer price index) considers components such as food, alcoholic drink, fuel and light, transport and vehicles, and so on.

c) The official Index of the Output of the Production Industries considers components as industrial groups such as metal, coal and coke, food, drink and tobacco, printing and publishing, and so on.

## 3. Weighting of components

A composite index number normally cannot be calculated unless each component is *weighted*. A weighting factor can be thought of as an indicator of the importance of the component with respect to the type of index being calculated.

a) *The need for weights*.

   i. Consider an electrical retail outlet that sells only two types of four-way 'gang' sockets, standard (£11.50) and super (£15.00). An attempt to represent the price of gang sockets by, say, the average of these two prices would be unreasonable, since the outlet might only sell one super socket per week as opposed to perhaps three or four standard sockets per day. However, if each of the sockets was weighted by sales volume, a weighted mean of the two prices could be reasonably used to represent the price of the gang sockets.

   ii. Consider a particular problem in calculating a price to represent dairy products bought by a family. If milk is 38p per pint, butter 96p per pound and eggs 120p per dozen, again, an average of these three prices would be meaningless. However, if each price was weighted by, say, expenditure, this would indicate the relative importance of each of the prices to the family budget as a whole.

b) *Some examples of weighting factors used in practice.*

   i. A price index would normally have its components weighted either by quantity or expenditure.

   ii. A quantity (or volume) index usually has its components weighted either by price or expenditure.

   iii. A productivity (output per man) index could have each component (a distinct product) weighted by the number of men involved in its manufacture.

## 4. Types of composite index number

Given a set of economic commodities, their values (for two separate time points) and a set of weights, there are two alternative methods available for obtaining a composite index number.

a) *Weighted average of relatives*. This method involves calculating index relatives for each of the given components, then using the given weights to obtain a weighted average of the relatives.

b) *Weighted aggregates*. This method involves multiplying each component value by its corresponding weight and adding these products to form an aggregate. This is done for both time points. Finally, the two aggregates obtained are used to calculate the index.

A useful way of differentiating between the two is: the weighted average of relatives method calculates indices first and applies weights last; the weighted aggregate method applies weights first and calculates an index last.

The following sections deal with each of these two types.

## 5. Example data for calculating indices

The data of Table 1 give details of the mix, prices and standard quantities used in making up a large bag of mortar.

*Table 1 Mix components for a large bag of mortar*

| Components of mix | Price (£) Year 1 | Year 2 | Standard quantity |
|---|---|---|---|
| | $p_0$ | $p_n$ | $w$ |
| A | 1.50 | 3.00 | 8 |
| B | 3.40 | 4.25 | 3 |
| C | 10.40 | 8.84 | 1 |

This data will be used to demonstrate the calculations involved in finding the values of the two standard composite index numbers.

## 6. Weighted average of relatives

The steps involved in calculating a weighted average of relatives index number are given below and are demonstrated using the data of Table 1.

STEP 1 Calculate an index relative for each component.

Here: $I_A = \frac{p_n}{p_0} \times 100 = \frac{3.00}{1.50} \times 100 = 200.$

Similarly, $I_B = \frac{4.25}{3.40} \times 100 = 125$

and: $I_C = \frac{8.84}{10.40} \times 100 = 85$

STEP 2 Calculate a weighted average of relatives, using:

**Weighted average of relatives**

$$I_{AR} = \frac{\sum wI}{\sum w}$$

where, for each component: $w$ = weighting factor
$I$ = index relative.

Here: average $= \frac{8(200)+3(125)+1(85)}{12} = \frac{2060}{12} = 171.7$, which is the required overall index number.

The standard layout for calculations is shown in Table 2, using the data from Table 1.

*Table 2*

| Components of mix | Price (£) Year 1 | Price (£) Year 2 | Standard quantity | Price relative | |
|---|---|---|---|---|---|
| | $p_0$ | $p_n$ | $w$ | $I$ | $wI$ |
| A | 1.50 | 3.00 | 8 | 200 | 1600 |
| B | 3.40 | 4.25 | 3 | 125 | 375 |
| C | 10.40 | 8.84 | 1 | 85 | 85 |
| | | Totals ... | 12 | | 2060 |

Weighted average of relatives = IAR = $\frac{\sum wI}{\sum w} = \frac{2060}{12} = 171.7$, agreeing (of course) with the previous calculation.

## 7. Weighted aggregates

The steps involved in calculating a weighted aggregate index number are given below and are demonstrated using the data of Table 1.

STEP 1 Calculate the products of weight and base values and add these to form a (base) aggregate total.

For the data given: Base year aggregate $= 8(1.50) + 3(3.40) + 1(10.40)$
$= 32.60$

STEP 2 Calculate the products of weight and current values and add these to form a (current) aggregate total.

For the data given: Current year aggregate $= 8(3.00) + 3(4.25) + 1(8.84)$
$= 45.59$

STEP 3 Calculate an index for the current aggregate, compared with the base aggregate, which is the required weighted aggregate index.

For the data given:

$$\text{weighted aggregate index} = \frac{45.59}{32.60} \times 100 = 139.8$$

This procedure can be symbolized by the following formula:

**Weighted aggregate index**

$$I_{AG} = \frac{\sum wv_n}{\sum wv_0} \times 100$$

where, for each component:

$v_n$ = value of commodity at current time point
$v_0$ = value of commodity at base time point
$w$ = weighting factor.

The standard layout is shown in Table 3, using the data from Table 1.

*Table 3 Standard layout for calculations*

| Components of mix | Price (£) Year 1 | Year 2 | Standard quantity | Year 1 products | Year 2 products |
|---|---|---|---|---|---|
| | $p_0$ | $p_n$ | $w$ | $wp_0$ | $wp_n$ |
| A | 1.50 | 3.00 | 8 | 12.00 | 24.00 |
| B | 3.40 | 4.25 | 3 | 10.20 | 12.75 |
| C | 10.40 | 8.84 | 1 | 10.40 | 8.84 |
| | | | Totals... | 32.60 | 45.59 |

Notice here that the values ($v$) are in fact prices ($p$).

Thus, weighted aggregate price index $= I_{AG} = \frac{\sum wv_n}{\sum wv_0} \times 100$

$$= \frac{45.59}{32.60} \times 100$$

$$= 139.8 \text{ (1D)}$$

agreeing with the value obtained at the bottom of the previous page.

**Note**: It will be seen that the value obtained here is NOT the same as that obtained using the weighted average of relatives method (in the previous section). This is because they measure *different relative changes*. It would only be coincidence if the same index number was produced using both methods.

## 8. Example 1 (*Composite index numbers* for a given set of commodities)

*Question*

The data in Table 4 below relates to the volumes of production of four commodities A1, A2, A5 and C1 over two years X and Y.

Calculate:

a) a weighted average of relatives volume index
b) a weighted aggregate volume index

for year Y (year X = 100).

*Answer*

Tables 5 and 6 show the extra columns that need to be constructed to deal with the calculations necessary for the respective indices.

| Commodity | Standard weight $w$ | Volume Year $X$ $v_0$ | Volume Year $Y$ $v_n$ |
|---|---|---|---|
| A1 | 41 | 35 | 40 |
| A2 | 16 | 12 | 14 |
| A5 | 30 | 54 | 60 |
| C1 | 12 | 40 | 44 |
| Totals | 99 | | |

*Table 4*

| $I$ | $wI$ |
|---|---|
| 114.3 | 4685.7 |
| 116.7 | 1866.7 |
| 111.1 | 3333.3 |
| 110.0 | 1320.0 |
| | 11205.7 |

*Table 5*

| $wv_0$ | $wv_n$ |
|---|---|
| 1435 | 1640 |
| 192 | 224 |
| 1620 | 1800 |
| 480 | 528 |
| 3727 | 4192 |

*Table 6*

(a) Weighted average of relatives volume index:

$$I_{AR} = \frac{\Sigma wI}{\Sigma w} = \frac{11205.7}{99} = 113.2$$

(b) Weighted aggregate volume index:

$$I_{AG} = \frac{\Sigma wv_n}{\Sigma wv_0} \times 100 = \frac{4192}{3727} \times 100 = 112.5$$

## 9. Comparison of the two composite indices

a) No one method of the two (average of relatives and aggregate) is 'correct'. They are simply alternative ways of combining the information for a set of commodities into an index number.

b) The two separate methods given for calculating a composite index number will generally yield two different values for the index, given the same set of data. This obviously lays index numbers open to abuse by unscrupulous presenters of statistics, since they can present whichever index (higher or lower) suits them. The only way round this type of problem is:

   i. the technique used in calculating the index number should be quoted or publicized (this is always done with official index numbers);

   ii. if indices are being calculated over a set of time periods, the same method should be used for each time period.

c) The characteristics of the two methods which cause different values of the final index to occur are:

   i. The aggregate index uses the *magnitudes of the actual values* of the component commodities and so is affected by actual increases or decreases. Thus it can claim to be truly representative of the data, but has the disadvantage of being affected by extreme values.

ii. The average of relatives index uses the *value of the relatives* for each component commodity (and not the actual values) and so is affected by *relative* increases or decreases. Thus it can be used for smoothing out extreme values, but it could be claimed that this method is not truly representative of the given data.

## 10. Example 2 (*Comparison* and *interpretation*)

a) In the example of section 5, which has been worked with both methods, the results were: average of relatives index = 171.7 and aggregate index = 139.8. The aggregate index has taken the relatively large values of component C (which were the only ones to decrease) into significant account, thus deflating the final figure. On the other hand, the average of relatives index has been much more affected by the large value of the relative calculated for component C, which has resulted in an inflated final figure. This is an extreme case, but is useful as an example of the difference in index values that can be obtained using the two methods.

b) In example 1 (in the previous section), notice that the two indices obtained: average of relatives index = 113.2 and aggregate index = 112.5 have approximately the same value. This is because the relatives of the volumes of the four commodities given were similar in value and there were no significant extremes in the actual volumes given.

## 11. Laspeyres indices

A *Laspeyres index* is a special case of a weighted aggregate index which always uses *base time period weights*. It is most commonly associated with price and quantity indices, where:

a Laspeyres *price* index uses base time period *quantities* as weights;

a Laspeyres *quantity* index uses base time period *prices* as weights.

The formulae for calculating a Laspeyres index can be derived from the general formula for a weighted aggregate index given in section 7. The two formulae most commonly used are given below:

**Laspeyres index**

$$\text{Price index: } L_P = \frac{\sum q_0 p_n}{\sum q_0 p_0} \times 100$$

$$\text{Quantity index: } L_Q = \frac{\sum p_0 q_n}{\sum p_0 q_0} \times 100$$

## 12. Paasche indices

A Paasche index is a special case of a weighted aggregate index which always uses *current time period weights*. It is most commonly associated with price and quantity indices, where:

a Paasche *price* index uses current time period *quantities* as weights;

a Paasche *quantity* index uses current time period *prices* as weights.

As with a Laspeyres index, the formulae for calculating a Paasche index can be derived from the general formula for a weighted aggregate index given in section 7. The two formulae most commonly used are given below:

**Paasche index**

$$\text{Price index: } P_P = \frac{\sum q_n p_n}{\sum q_n p_0} \times 100$$

$$\text{Quantity index: } P_Q = \frac{\sum p_n q_n}{\sum p_n q_0} \times 100$$

## 13. Example 3 (*Paasche* and *Laspeyres* price indices)

*Question*

The following data relate to a set of commodities used in a particular process. Calculate Laspeyres and Paasche price indices for period 1.

| | | Base period | | Period 1 | |
|---|---|---|---|---|---|
| Commodity | Unit of purchase | Price (£) | Quantity (units) | Price (£) | Quantity (units) |
| A | 2 gallon drum | 36 | 100 | 40 | 95 |
| B | 1 tonne | 80 | 12 | 90 | 10 |
| C | 10 pounds | 45 | 16 | 41 | 18 |
| D | 100 metres | 5 | 1100 | 6 | 1200 |

*Answer*

The layout for the calculations is shown in Table 7.

*Table 7*

| | | | | | calculations for Laspeyres index | | calculations for Paasche index | |
|---|---|---|---|---|---|---|---|---|
| Commodity | $p_0$ | $q_0$ | $p_n$ | $q_n$ | $q_0p_n$ | $p_0q_0$ | $q_np_n$ | $q_np_0$ |
| A | 36 | 100 | 40 | 95 | 4000 | 3600 | 3800 | 3420 |
| B | 80 | 12 | 90 | 10 | 1080 | 960 | 900 | 800 |
| C | 45 | 16 | 41 | 18 | 656 | 720 | 738 | 810 |
| D | 5 | 1100 | 6 | 1200 | 6600 | 5500 | 7200 | 6000 |
| | | | | | 12336 | 10780 | 12638 | 11030 |

$$\text{Laspeyres price index, } L_p = \frac{\sum q_0 p_n}{\sum q_0 p_0} \times 100$$

$$= \frac{12336}{10780} \times 100$$

$$= 114.4 \text{ (1D)}$$

$$\text{Paasche price index,} \quad P_p = \frac{\sum q_n p_n}{\sum q_n p_0} \times 100$$

$$= \frac{12336}{10780} \times 100$$

$$= 114.6 \text{ (1D)}$$

Both indices are showing an increase in prices of about 14.5%, although notice that the two indices have slightly different values. This is to be expected, since the weights used for the commodities are different. In general, a greater difference than that shown above would not be unusual.

## 14. Comparison of Laspeyres and Paasche price indices

a) *What the indices are calculating.*
*Laspeyres*: compares base period expenditure with a hypothetical current period expenditure at base period quantities. *Paasche*: compares current period expenditure with a hypothetical base period expenditure at current period quantities.

b) *When prices are rising.*
The *Laspeyres* index tends to *over-estimate* price increases and the *Paasche* index tends to *under-estimate* price increases. This is thought of as a disadvantage for both types of index.

c) *Nature of the indices.*
The *Laspeyres* index is thought of as a 'pure' price index, since (from period to period) it is comparing like with like, which is an advantage. The *Paasche* index is not considered as 'pure', since (from period to period) the weights change and thus, strictly, like is not being compared with like. This is considered a disadvantage of the index.

d) *Weights used in calculations.*
Since the *Laspeyres* index uses *base* period quantities as weights, they can easily become out of date (disadvantage), while the *current* quantities that the *Paasche* index uses as weights are always up-to-date (advantage).

e) *Ease of calculations.*
The *Laspeyres* index needs only the base period quantities, no matter how many periods the index is being calculated for, which is a considerable advantage over the *Paasche* index, which needs new quantities for each time period. Quantities are normally more difficult to determine than prices.

Similar considerations apply to *Laspeyres* and *Paasche* quantity (or volume) indices. Generally, on balance, the *Laspeyres* index is favoured to the *Paasche* index.

## 15. Some considerations in the construction of index numbers

a) *Selection of component commodities.*

Which commodities to include in a composite index will of course depend on the purpose to which the final index will be put. A particular problem however, is changing patterns of trend or behaviour. Items that were once included (and significant) might no longer be fashionable or might not be used or produced any more. For example, writing ink or obsolete products. Also new items might appear, such as video recorders (relevant to a cost of living index) or Value Added Tax (relevant to a company costs index).

b) *Selection of weights.*

As mentioned earlier, weights are simply a device to reflect the importance of the component within the overall composite index, and normally prices are weighted by quantities, quantities by prices and productivity by number of workers involved. In order to determine weights, sampling of some sort might be necessary. The weights used for the items covered by the Retail Price Index are determined by a complementary survey, the Family Expenditure Survey, which is a continuous investigation into the expenditure of thousands of families.

c) *Collection of the data.*

This involves finding the values (i.e. prices, volumes, productivity etc) for each component, together with measuring the magnitude of the weights selected. Normally it is found that quantities are much more difficult to determine than prices, which is what makes a Laspeyres price index more popular than a Paasche price index. The more complex the structure of the index, the more cost will be involved in collecting the data.

d) *Selection of a base time period.*

Technically, it makes no difference what base is chosen for a time series of index numbers. The information that indices convey (percentage increases or decreases) is independent of which base point is chosen. However, for practical purposes, a base might be chosen that is not too far in the past and might correspond with new working practices, a new company organisation or even a new method of collecting data. For the comparison of two series of index numbers, the base dates for both should be identical.

## 16. Some uses of index numbers

Index numbers are used to reflect general economic conditions over a period of time. For example, the Retail Price Index measures changes in the cost of living; the Index of Industrial Production reflects changes in industrial output; the Financial Times Ordinary Share Index reflects the general state of the Stock Market.

In particular, they can be used by Government to decide on tax changes, subsidies to industries or regions or national retirement pension increases. Trades Unions often use national cost of living and production indices in wage negotiations or to compare the cost of living across national boundaries, regions or professions. Insurance companies use various cost indices to index-link house (building or contents) policies.

## 17. Some limitations of index numbers

a) Indices, by their nature, give only general indications of changes often over wide geographical areas. Thus, they generally will not cater for minority groups or professions or measure regional variations. For example, an individual's food bill for a particular month may have increased significantly, yet the Food Index (part of the Retail Price Index) for that month might show a decrease.

b) Weighting factors can become out of date, thus the index may not be comparing 'like with like'.

c) If samples have been used to obtain data for either values or weights, the information obtained might be biased, incomplete or false.

d) Index numbers can be misinterpreted by the uninformed layman. For example:

   i. Suppose a chain base price index for a number of months yielded the values: 113, 109, 108 and 105. This might be confused with a fixed base index and interpreted as a decrease in prices,whereas in fact prices are still rising (although the *rate of increase* in prices is decreasing!).

   ii. If a production index increases from, say, 400 to 410, this might be interpreted as a 10% increase in production. In fact, there has been a 10 points increase, which is a 10 out of 400 or 2.5% increase in reality.

## 18. Summary

a) A composite index number is one which is obtained by combining the information from a set of economic commodities of like kind.

b) A weighting factor is an indicator of the importance of each component commodity. Standard weights for:

   i. price indices are quantities, and for

   ii. quantity indices are prices.

c) There are two standard types of composite index number:

   i. Weighted average of price relatives;

   ii. Weighted aggregates.

d) A weighted average of relatives can be found in the following way:

   i. calculate an index relative ($I$) for each component;

   ii. calculate a weighted average using the appropriate formula.

   Remember: indices are calculated first, then weights are applied.

e) A weighted aggregate index can be found by using the appropriate formula. Remember: weights are applied first, then an index is calculated.

f) Both methods of obtaining a composite index should be considered as alternatives. The aggregate index uses the magnitudes of the actual values, while the average of relatives index uses the values of the relatives, to determine the index.

g) A Laspeyres index is a weighted aggregate index which uses base time period weights. In particular:

   i. a Laspeyres price index uses base period quantities as weights;

ii. a Laspeyres quantity index uses base period prices as weights.

*Advantages*: it is a 'pure' index (since weights are constant) and only one set of weights is needed for a time series of indices. *Disadvantages*: tends to over-estimate price increases in a period of rising prices and base period weights can become out of date.

h) A Paasche index is a weighted aggregate index which uses current time period weights. In particular:

i. A Paasche price index uses current time period quantities as weights;

ii. A Paasche quantity index uses current time period prices as weights.

*Advantage*: current period weights are always up to date. *Disadvantages*: tends to under-estimate price increases in a period of rising prices, not a 'pure' index and (for a price index) current period quantities are difficult to obtain.

j) Considerations in the construction of index numbers include:

i. selection of component commodities;

ii. selection of weights;

iii. collection of data;

iv. selection of a base time period.

k) Index numbers are used to reflect general economic conditions over a period of time by Government, Trades Unions, insurance companies etc.

l) Limitations of index numbers include:

i. they give only general indications of changes;

ii. weighting factors can become out of date;

iii. samples, if used, never give precise values;

iv. they can be misinterpreted by the uninformed layman.

## 19. Student self review questions

1. What is a composite index number? Give some examples. [2]
2. Why are weights needed with a composite index number? [3]
3. Give some examples of weighting factors. [3(b)]
4. What is the difference between a weighted average of relatives index number and a weighted aggregate index number? [4]
5. What factors contribute to the difference in values of the index obtained when both methods are used? [9]
6. Define Laspeyres and Paasche indices. [11,12]
7. What are the advantages and disadvantages of using Laspeyres and Paasche indices? [14]
8. What problems are met in the selection of commodities for a composite index number? [15(a)]
9. What considerations are involved in the selection of a base time period for an index? [15(d)]
10. Name some ways in which index numbers are used. [16]
11. Give some limitations of index numbers. [17]

## 20. Student exercises

1. *MULTI-CHOICE*. An index number is made up of two items, A and B, as follows:

| Subgroup | Weight | Index |
|---|---|---|
| A | 7 | 130 |
| B | 3 | X |
| All | 10 | 127 |

The index number for subgroup B (X) is closest to:

a) 120  b) 123  c) 125  d) 128

2. Calculate a) a weighted average of price relatives and b) a weighted aggregate price index, both for year 1 (based on year 1), for the following commodities.

| | | Price (£) | |
|---|---|---|---|
| Commodity | Weight | Year 1 | Year 2 |
| W | 5 | 215 | 210 |
| X | 12 | 250 | 275 |
| Y | 2 | 1100 | 1300 |
| Z | 8 | 950 | 950 |

3. For a particular process, a firm uses three basic raw materials. They wish to onstruct a volume (quantity) index, using an average of relatives method, for February and March (both based on January). They have also decided that the weighting factor will be the latest (March) prices. Using the table below, calculate the two indices required and briefly comment on the values.

| Raw material | March price (£) (standard units) | Quantities used Jan | Feb | Mar |
|---|---|---|---|---|
| A | 12.50 | 11 | 11 | 9 |
| B | 6.45 | 34 | 38 | 40 |
| C | 5.10 | 18 | 24 | 23 |

4. The following data relate to the pay of workers employed at a factory.

| | | Average number of hours worked per week | | |
|---|---|---|---|---|
| Type of worker | Rate of pay (£/hr) | June | November | Number of workers employed |
| Skilled | 6.50 | 37 | 38.5 | 26 |
| Semi-skilled | 5.10 | 38.5 | 39.5 | 14 |
| Unskilled | 3.50 | 40.5 | 40 | 42 |

Calculate a weighted aggregate index of average weekly pay for November (June=100), using the number of workers employed as a weighting factor.

5. A survey of household expenditure showed the following changes over the same week in each of three years for an average family.

| Item | Unit | Quantity purchased Year 1 | Price Year 1 | Year 2 | Year 3 |
|---|---|---|---|---|---|
| Sugar | 2lb | 4lb | 20 | 22 | 57 |
| Bread | loaf | 4 | 21 | 23 | 32 |
| Tea | 0.25lb | 0.5lb | 18 | 21 | 24 |
| Milk | pint | 20 | 13 | 13 | 22 |
| Butter | 0.5lb | 1.5lb | 21 | 32 | 41 |

a) Using year 1 as base, calculate index numbers for Year 2 and Year 3 using the Laspeyres method.

b) State, with reasons, whether a survey based on these items represents a reasonable assessment of changes in the cost of food over the three years.

6. The data below show area of land used (thousand hectares) and value of output (£m) of three types of farm crops for the United Kingdom.

| Type of | Area of land used | | | Value of output | | |
|---|---|---|---|---|---|---|
| crop | 19X0 | 19X1 | 19X2 | 19X0 | 19X1 | 19X2 |
| Wheat | 1441 | 1491 | 1663 | 1431 | 1482 | 1635 |
| Barley | 2330 | 2327 | 2222 | 440 | 438 | 531 |
| Oats | 148 | 144 | 129 | 400 | 453 | 449 |

a) Calculate the price (in £m) obtained per thousand hectares to 2D for each type of crop for the three years.

b) Thus calculate, using *value of output* as a weighting factor:

i. Laspeyres price indices for 19X1 and 19X2;

ii. Paasche price indices for 19X1 and 19X2.

7. The following data give the quantities and costs of materials for the four divisions of a company for two years. Calculate Paasche price (cost) and quantity index numbers for year 2 (year 1 = 100).

| | Quantity (tonnes) | | Cost (£) | |
|---|---|---|---|---|
| Division | Year 1 | Year 2 | Year 1 | Year 2 |
| A | 175 | 201 | 1540 | 1830 |
| B | 32 | 46 | 1270 | 1490 |
| C | 48 | 43 | 2760 | 2490 |
| D | 65 | 66 | 2190 | 2070 |

# 20 Special published indices

## 1. Introduction

This chapter describes some of the most important official index numbers. The price indices described are the Retail Prices Index (which includes the important Family Expenditure Survey), Purchasing Power of the Pound, the Tax and Price Index and Index numbers of Producer Prices. Indices of Average Earnings are also covered. Volume (or quantity) indices described are the Index of Output of the Production Industries and the Index of Retail Sales. Some indices described cover more than one section.

## 2. The Retail Prices Index

The *Retail Prices Index* (or *RPI*), is probably the best known of all the published indices.

a) It is published monthly (on a Tuesday near the middle of the month) by the Department of Employment and displayed (to different levels of complexity) in the following publications: Monthly Digest of Statistics, the Annual Abstract of Statistics, the Department of Employment Gazette and Economic Trends.

b) The RPI measures the annual changes, month by month, in the average level of prices of most goods purchased by the great majority of households in the United Kingdom. It takes account of practically all wage earners and most small and medium salary earners.

c) The items covered by the RPI are classified into 5 major groups and 14 subgroups. For example, the major group Food & Catering has Catering as a subgroup. Each group is sub-divided into sections. For example, Catering is sub-divided into 1) Restaurant meals 2) Canteen meals and 3) Take-aways and snacks. These subdivisions are often split up into further items.

d) Each month, an overall index is published, together with separate indices for each group, subgroup, section and individual item (of which there are approximately 350).

e) Each group (and further sections and specific items) are weighted according to expenditure by a 'typical family', the weights being updated annually for the year ended in the previous June. See following section.

f) The weights are obtained from a continuous investigation known as the Family Expenditure Survey (described in section 5).

g) As at the publication of this 6th edition, July 2003, the base date for the RPI was January 1987 and the latest 'All items' RPI was published for May 2003 at 181.5.

## 3. Main RPI groups and their weights

Table 1 shows the main groups of the RPI, their separate price indices (as at January 1986) and their weights for three different dates.

*Table 1 Retail Price Index subgroups*

| Main subgroups | Price index May 2003 (1987=100) | Weights 1962 | 1985 | 2003 |
|---|---|---|---|---|
| Food | 151.7 | 350 | 190 | 109 |
| Catering | 225.6 | | 45 | 51 |
| Alcoholic drink | 200.5 | 71 | 75 | 68 |
| Tobacco | 302.9 | 80 | 37 | 30 |
| Housing | 243.6 | 87 | 153 | 203 |
| Fuel and Light | 130.0 | 55 | 65 | 29 |
| Household Goods | 144.0 | 66 | 65 | 72 |
| Household Services | 172.7 | | | 61 |
| Clothing and Footwear | 101.8 | 106 | 75 | 51 |
| Personal Goods and Services | 198.0 | | | 41 |
| Motoring Expenditure | 181.8 | 68 | 156 | 146 |
| Fares and Other Costs | 210.7 | | | 20 |
| Leisure Goods | 103.2 | 59 | 77 | 48 |
| Leisure Services | 248.1 | | | 71 |
| Overall | 181.5 | 1000 | 1000 | 1000 |

*Notes on Table 1:*

a) Weights are always calculated to add to 1000.
b) 'Catering' was not included in the 1962 weightings.
c) Certain items of expenditure are not included in the RPI. These include:
   i. Income tax and National Insurance payments;
   ii. Insurance and pension payments;
   iii. Mortgage payments for house purchase (except for interest payments which are included);
   iv. Gambling, gifts, charity, etc.

It should be noted that the RPI is *not strictly a 'Cost of Living' index of basic essentials*. In practice it would be virtually impossible to decide on a basket of goods and services which everyone considered to be essential. Instead, the RPI gives us a measure of what we would need to spend in order to buy the same goods we bought in an earlier period.

## 4. **Example 1** (Comments on the data in Table 1)

a) The Retail Prices Index for May 2003 (1987=100) was 181.5. This represents an overall (or average) increase in prices of just over 80% from 1987 to 2003.
b) Food has been subject to below average price increases (151.7 index = 52% increase) and expenditure has continued to decrease significantly. Since food is a basic necessity of life, this is a good indication of our increasing affluence.
c) Tobacco has seen the highest increase in price (index = 302.9) with a definite downward trend in expenditure (as evidenced in the weights). The latter trend is probably due to both high price and health warnings.
d) Clothing and Footwear has had the lowest increase in price (index = 101.8), representing only a 2% increase in price over the previous 10 years, but this

group has still seen a downward trend in expenditure. Since there is no reason to suppose that we now buy fewer items of clothing and footwear, it probably means that these are much cheaper in real terms.

e) Housing and Motoring/Fares both show a similar upward trend in expenditure. However, where Motoring/Fares is only showing an average price increase, Housing shows the third highest (index = 243.6). Upward expenditure on transport clearly signifies our increasing mobility (in both work and recreation). Extra expenditure on housing probably reflects social and ecological factors as much as increase in price.

## 5. The Family Expenditure Survey

The *Family Expenditure Survey* (*FES*) is a continuous major investigation which, among other things, measures average consumption levels. These are used to obtain the (annually revised) weights for items included in the RPI.

a) The FES originated from the *Household Expenditure Survey*, a special investigation carried out in 1953, the results of which were used to structure the groups, sections and items in the RPI and determine their weights.

b) The FES involves a stratified random sample, spread over the course of a year, of about 10000 households. Each household is visited by an interviewer.

c) Each member of the household over the age of sixteen years is required to keep a detailed diary of all expenditure for a continuous 14-day period, which is checked and retained by the interviewer.

d) The interviewer also completes a *Household Schedule*, which contains information on longer term spending such as rent, council tax, carpets, cars, and so on. (An *Income Schedule* is also filled out for the members of the household.)

e) The published weights are calculated, not from a single year's FES data, but as an average of the previous three year's data. This ensures that large items of expenditure do not unfairly influence average patterns of spending.

## 6. Price collection and calculation of the RPI

a) Prices are collected by Department of Employment staff, based at 200 local offices, on the middle Tuesday of each month.

b) Different types of retail outlets, from village shops to large supermarkets, are visited. To ensure uniformity, the same ones are used each month and these will be the type of retail outlet used by households examined by the FES.

c) Price relatives are calculated (for each item covered by the RPI) for each retail outlet and averaged for a local area. Average relatives for all local areas are in turn averaged to obtain a national average of relatives (for each of the 350 items covered by the index).

d) Weights are then used to calculate composite indices using the average of relatives method for items within sections, sections within groups and, finally, groups. Thus the RPI is a weighted average of relatives of each group.

## 7. The Purchasing Power of the Pound (index)

a) The (internal) Purchasing Power of the Pound (PPP) is an index which, since 1962, has been based solely on the annual average of the RPI.

b) The philosophy behind the index is: when prices go up, the amount which can be purchased with a given sum of money goes down.

c) The index is described in terms of two particular years. If the purchasing power of the pound is taken to be 100p in the first year, the comparable purchasing power in a later year is calculated as:

$$100 \times \frac{\text{average price index for first year}}{\text{average price index for later year}}$$

d) For example, the PPP index for 2002 (1990=100) is 72. This can be interpreted as:

i. the pound (in 2002) is worth only 72% of its 1990 value, or
ii. 100p buys (in 2002) what would only have cost 72p in 1990.

## 8. The Tax and Price Index

The *Tax and Price Index* (*TPI*), published monthly, is another index which is linked to the Retail Prices Index. It was first published in January, 1978.

a) The TPI measures the increase in gross taxable income needed to compensate taxpayers for any increase in retail prices (as measured by the RPI). As at May 2003, the TPI (1987=100) was 165.2.

b) It is considered as a more comprehensive index than the RPI since, while the RPI measures changes in retail prices, the TPI additionally takes account of the changes in liability to direct taxes (including employees' national insurance contributions) facing a representative cross-section of taxpayers.

c) Some people would argue that the TPI is a better measure of the cost of living than the RPI since it takes direct taxes into account. However, whether or not this is acceptable depends on the meaning of the phrase 'cost of living' - it has different meanings to different people and circumstances. Another complicating factor is that the TPI (a relatively new index) is regarded suspiciously by some political opponents of the Government in office at the time of its introduction.

d) Data for the TPI is collected via the Survey of Personal Incomes, which consists of a stratified sample from the population of tax units (i.e. single people or married couples who pay tax).

e) Excluded from the survey are those units with incomes over a certain amount, since it is considered that they are not necessarily representative of the majority of taxpayers. The relevant percentage excluded amounts to only the top 4% of households (which are also excluded from the calculation of the RPI).

## 9. Example 2 (Comparison of the *TPI* and *RPI*)

The TPI for June 1985 (January 1978=100) was 191.7 [INDEX 1]
The RPI for June 1985 (January 1974=100) was 376.4 [INDEX 2]
The RPI for 1978 (1974=100) was 197.1 [INDEX 3]

*(Source of data: Monthly Digest of Statistics)*

Note that it is difficult to compare the first two indices, since their base dates are different. However, the information contained in INDEX 3 allows the RPI (INDEX 2) to be base-changed to coincide with the base of the TPI (INDEX 1), for a direct comparison.

$$\text{Thus: } RPI_{85/78} = \frac{RPI_{85/74}}{RPI_{78/74}} \times 100$$

$$= \frac{376.4}{197.1} \times 100$$

$$= 191.0 \text{ (1D)}$$

Therefore the TPI for June 1985 shows a slightly higher increase (91.7%) than the RPI for June 1985 (91.0%).

Note, however, that INDEX 3 is based on annual averages whereas the other two indices are based on actual months of the year. Hence the above base change will cause the resultant figure to be slightly in error.

## 10. Producer Price Indices (PPI)

The *Producer Price Indices (PPI)* is a monthly survey that measures the price changes of goods bought and sold by UK manufacturers. The PPI is conducted by the Office for National Statistics and provides a key measure of inflation, alongside other indicators such as the Retail Price Index (RPI).

They work on the 'basket of goods' concept. A wide collection of representative products is selected and the prices of these fixed sets of goods are collected each month. The movements in these prices are weighted to reflect the relative importance of the products in a chosen year (known as the base year, currently 1995). These are then aggregated for various sectors of industry to give the required indices.

Two important sets of these indices are described as follows.

a) The *output price indices* measure change in the prices of goods produced by UK manufacturers (these are often called 'factory gate prices'). For example, Output of Manufactured Products for May 2003 was 110.2 (1995=100).

b) The *input price indices* measure change in the prices of materials and fuels bought by UK manufacturers for processing. These are not limited to just those materials used in the final product, but also include what is required by the company in its normal day to day running. For example, Input for Materials and Fuel Purchased for May 2003 was 91.3 (1995=100).

## 11. Indices of Average Earnings

The *Indices of Average Earnings* are supplied by the Department of Employment monthly and measure the changes in average gross income. They are published for manual workers and all workers and given for 26 industry groups of the 1992 Standard Industrial Classification (at the time of writing), all manufacturing industries, production industries and the whole economy. Actual and seasonally adjusted indices are given for certain tables. The series as at 2003 are all based on 1995=100.

## 12. Index of Output of the Production Industries

The *Index of Output of the Production Industries* is often known as the *Index of Production (IOP).*

a) It provides a general measure of monthly changes in the volume of output of the production industries in the United Kingdom.
b) Energy, water supply and manufacturing are included in the index. However, agriculture, construction, distribution, transport, communications, finance and all other public and private services are excluded.
c) The index covers the production of intermediate, investment and consumer goods, both for home and export.
d) The index is calculated as a weighted average of 328 separate relatives, each of which describes the change in the volume of the output of a small sector of industry. The base for all the series currently presented is 1995 and the weight used for each small sector is based on its value added for that year. Information for weights is also taken from national accounts income data sources for energy and water supply industries.
e) The indices are classified to the 1992 revision to the Standard Industrial Classification (SIC). The production industries, which accounted for 26.8 per cent of gross domestic product in 1995, cover mining and quarrying (Section C of the SIC), manufacturing (Section D) and electricity, gas and water supply (Section E).
f) Many of the series presented are seasonally adjusted. This excludes any changes in production resulting from public and other holidays and from other seasonal factors. The adjustments are designed to eliminate normal month to month fluctuations and thus to show the trend more clearly.
g) Example 3 shows the main industrial groups, their SIC, weights and the group indices for 1985.

## 13. Example 3 (Main groups of the *index of production*)

Table 2 shows the weights and the Index of Output of the Production Industries for 1998, 2000 and May 2003 , together with their Standard Industrial Classification (SIC).

a) Notice that Chemicals & man-made fibres has the highest index of 117.7, representing an increase in output of 17.7% since 1995. The weighting factor of 89 (out of 1000) shows that it represents nearly 9% contribution to the overall index.
b) The index of 60 for Textiles, leather & clothing (representing a 40% decrease in output) is reflecting the continuing slow decline of this sector in our overall production.
c) The overall index for May 2003 (shown on the bottom row in italics as 99.1 can be confirmed by calculating a weighted average of the indices for the SIC groups C, D and E as follows:

$$\frac{95(96.9)+814(98.0)+91(111.0)}{1000}=99.1$$

So that, production, as measured by SIC groups C, D and E has fallen by just 0.9% since 1995.

*Table 2 Index of Production – the constituent SIC groups*

| Sector | SIC (1995) | Weight | Index (1995=100) | | |
|---|---|---|---|---|---|
| | | | 1998 | 2000 | May 2003 |
| Mining and Quarrying (inc Oil & gas extraction) | C | 95 | 104.3 | 99.6 | 96.9 |
| Oil & gas extraction | C1 | 80 | 107.5 | 102.1 | 99.5 |
| Manufacturing | D | 814 | 102.8 | 98.5 | 98.0 |
| Coke, refined petrol and nuclear fuels | DF | 17 | 88.3 | 81.0 | 72.0 |
| Chemicals & man-made fibres | DG | 89 | 104.0 | 117.5 | 117.7 |
| Basic metals & metal products | DJ | 94 | 99.2 | 88.4 | 85.4 |
| Engineering & allied industries | DK,DM | 251 | 110.4 | 103.0 | 104.1 |
| Food, drink & tobacco | DA | 106 | 101.5 | 102.4 | 101.3 |
| Textiles, leather & clothing | DB,DC | 45 | 89.0 | 63.6 | 60.0 |
| Other manufacturing | DD,DN | 211 | 99.7 | 96.7 | 96.5 |
| Electricity, Gas & Water Supply | E | 91 | 107.5 | 113.0 | 111.0 |
| *Production Industries* | *C,D,E* | *1000* | *103.4* | *99.9* | *99.1* |

## 14. Index of Retail Sales

The *Index Numbers of Retail Sales* are published monthly by the Business Statistics Office. They cover the retail trades (excluding the motor trades) in Great Britain and, as at August 2003, the base year is 1995. Index numbers are given for both volume and value of sales. For example, the All-Retailers (predominantly non-food) volume index for May 2003 was 138.2 (1995=100). The comparable value index was 142.0.

a) The indices are compiled on the type of business, rather than on a commodity basis and take account of the results of major enquiries into retailing.

b) A voluntary panel of about 3,200 small retailers (with turnover less than £2 million for 1982) and about 350 large retailers (who account for about 80% of sales of this sector) fill out statistical returns of both volume and value of sales. Mail order firms are included on the panel.

c) The indices are calculated on a chain base system, month by month, and then chain-linked back to the base year.

## 15. Other index numbers

Some other index numbers that are given in main publications, such as the Monthly Digest of Statistics and the Annual Abstract of Statistics are:

a) Index numbers of Output (at constant factor cost);
b) Index numbers of Expenditure (at 1980 prices, currently);
c) Volume Index of Sales of Manufactured Goods;
d) Indices of Labour costs;
e) (External Trade) Volume and Unit Value Index numbers;

and an important non-official publication:

f) The Financial Times Ordinary Share Index.

## 16. Summary

a) The Retail Prices Index (RPI) is published monthly and measures the percentage changes in the average level of prices of the commodities and services purchased by most households in the UK. It is classified into eleven main groups, each weighted by expenditure, the values of which are obtained through the Family Expenditure Survey.
b) The Family Expenditure Survey is a continuous investigation, involving over 10,000 households, which measures average consumption levels. Information is obtained through stratified random samples and each member of the household gives details of spending for a fourteen day period.
c) Prices for the RPI are collected through 200 local offices, which examine all types of retail outlet. The RPI is a weighted average of relatives index.
d) The Purchasing Power of the Pound (index) gives the percentage worth of a current pound compared with a pound in a previous period.
e) The Tax and Price Index measures the increase in gross taxable income needed to compensate taxpayers for any increases in retail prices (as measured by the RPI). It takes account of direct taxation.
f) The Producer Price Indices measure manufacturers prices for main industrial groupings as described by the Standard Industrial Classification (SIC). Weightings are mainly derived from the 1979 Purchases Inquiry.
g) The Indices of Average Earnings measure the changes in average gross income for manual and other workers across 26 industry groups.
h) The Index of Output of the Production Industries provides a general measure of monthly changes in the volume of output of the production industries in the UK. Certain services (such as agriculture and finance) are excluded from the index.
i) Index numbers of Retail Sales give both volume and value indices and are compiled on the type of business rather than on a commodity basis.

## 17. Student self review questions

1. What is the Retail Prices Index (RPI)? [2(b)]
2. Name at least *five* of the eleven main groups into which the RPI is divided. [3]
3. Name some of the items of expenditure that are not included in the calculation of the RPI. [3]
4. What connection does the Family Expenditure Survey (FES) have with the RPI? [5]
5. Explain briefly how information on expenditure is obtained from households in the FES. [5]
6. How are prices collected for the RPI? [6(a,b)]
7. Explain what the 'Purchasing Power of the Pound' is and how it is calculated. [7]
8. What does the Tax and Price Index (TPI) measure? [8(a)]
9. Compare the RPI and TPI. [8(b-e)]
10. Describe some aspects of the Index Numbers of Producer Prices. [10]
11. Describe some aspects of the Indices of Average Earnings. [11]
12. What items are not included in the Index of Output of the Production Industries? [12(b)]
13. What is the Index of Retail Sales and how are the data in its construction collected? [14]

# Examination questions

1. You have been requested by a UK based client to research into the area of salaries paid to their systems analysis team, and to prepare a report in order that a pay review may be carried out.

   The following table shows the salaries together with the retail price index (or consumer price index) for the years 1977-85.

| Year | Average salary £ | Retail Price Index* (1975=100) |
|---|---|---|
| 1977 | 9,500 | 135.1 |
| 1978 | 10,850 | 146.2 |
| 1979 | 13,140 | 165.8 |
| 1980 | 14,300 | 165.8 |
| 1981 | 14,930 | 218.8 |
| 1982 | 15,580 | 237.7 |
| 1983 | 16,200 | 248.6 |
| 1984 | 16,800 | 261.0 |
| 1985 | 17,500 | 276.8 |

** Source: Economic Trends (HMSO)*

*Required:*

i. What is the purpose of an index number?

ii. Tabulate the percentage increases on a year earlier for the average salary and the retail price index.

iii. Revalue the average salary for each year to its equivalent 1985 value using the retail price index.

iv. Using the results from (ii) and (iii) above, comment on the average salary of the systems analysts of your client.

*ACCA (55%)*

2. a) Briefly explain *two* commercial, industrial or business uses of index numbers.

   b) A cost accountant has derived the following information about basic weekly wage rates (W) and the number of people employed (E) in the factories of a large chemical company.

*Basic weekly rates (£'s) and number of employees (100's)*

| Technical group of employees | July 1979 W | July 1979 E | July 1980 W | July 1980 E | July 1981 W | July 1981 E |
|---|---|---|---|---|---|---|
| Q | 60 | 5 | 79 | 4 | 80 | 4 |
| R | 60 | 2 | 65 | 3 | 70 | 3 |
| S | 70 | 2 | 85 | 2 | 90 | 1 |
| T | 90 | 1 | 110 | 1 | 120 | 2 |

   i. Calculate a Laspeyres (base weights) all-items index number for the July 1980 basic weekly wage rates, with July 1979 = 100.

   ii. Calculate a Paasche (current weights) all-items index number for the July 1981 basic weekly wage rates, with July 1979 = 100.

iii. Briefly compare your index numbers for the company with the official government figures for the Chemical and Allied Industries which are given below.

Yearly annual averages:

| | 1979 | 1980 | 1981 |
|---|---|---|---|
| Weekly wage rates (July 1976 = 100) | 156.3 | 187.4 | 203.4 |

*(Source: Employment Gazette, November 1981)*

*CIMA*

3. Your company Manco plc is about to enter wage negotiations with its Production Department. In the following table are tabulated the average weekly earnings and hours worked of the full-time manual workers 21 years of age and over in the Department. Also included in the table is the General Index of Retail Prices (in some countries known as the Consumer Price Index) for the years 1974 to 1981.

| Year | Average weekly earnings | Average hours worked | Retail Price* Index (1975=100) |
|---|---|---|---|
| 1974 | 40.19 | 44 | 80.5 |
| 1975 | 52.65 | 45 | 100.0 |
| 1976 | 62.03 | 45 | 116.5 |
| 1977 | 70.20 | 46 | 135.0 |
| 1978 | 76.83 | 46 | 146.2 |
| 1979 | 91.90 | 46 | 165.8 |
| 1980 | 107.51 | 45 | 195.6 |
| 1981 | 121.95 | 43 | 218.9 |

* *Source: Economic Trends – HMSO*

Required:

a) Assuming that normal time for the Production Department has been 40 hours for the period covered, tabulate the year, average weekly earnings, average hours worked and add the normal weekly rate to the table. You may also assume that overtime is paid at time and a half i.e. one and a half times the basic hourly rate of pay.

b) Plot the figures for the Retail Price Index on a semi-logarithmic (ratio scale) graph, by plotting the logarithm of the Retail Price Index against time.

c) What characteristic does this graph demonstrate about the Retail Price Index?

d) Deflate the Normal Weekly Rate by the Retail Price Index and show this in tabulated form against the Normal Weekly Rate and the year.

e) What is the significance of the deflated series with regard to the wage negotiations?

*ACCA*

4. Index of industrial production in the United Kingdom:

| | Weight | July 1982 (1975=100) Index |
|---|---|---|
| Mining and quarrying* | 41 | 361 |
| Manufacturing | | |
| Food, drink and tobacco | 77 | 106 |
| Chemicals | 66 | 109 |
| Metal | 47 | 72 |
| Engineering | 298 | 86 |
| Textiles | 67 | 70 |
| Other manufacturing | 142 | 91 |
| Construction | 182 | 84 |
| Gas, electricity and water | 80 | 115 |

* including North Sea oil.

*(Source: Monthly Digest of Statistics)*

i. Calculate the index of industrial production for July 1982:
(1) all industries
(2) all industries except mining and quarrying
(3) manufacturing industries

ii. Comment on your results.

*ICSA*

5. The total weights of the Index of Retail Prices in both 1961 and 1981 were 1000. However, the distribution in the two years of the weights over the item headings differed as the following items show:

| | 1961 | 1981 |
|---|---|---|
| Food | 350 | 207 |
| Housing | 87 | 135 |
| Clothing | 106 | 81 |
| Transport | 68 | 152 |

Explain how such changes can occur and the effects they can have upon the index number.

*RVA*

# Part 6 Compounding, discounting and annuities

Chapter 21 describes simple and compound interest and common depreciation techniques. It also includes the mathematical progressions which form the bases of these techniques.

Chapter 22 deals with Investment Appraisal techniques based on discounted cash flows and chapter 23 is concerned with annuities (regular fixed payments), which are used as methods of loan repayment and asset depreciation.

# 21 Interest and depreciation

## 1. Introduction

This chapter begins by introducing special sequences of numbers known as arithmetic and geometric progressions and describes the form of each, together with formulae for calculating totals.

The differences between simple and compound interest are then explained, formulae for their calculation are given and some of their applications, such as APR, are described. The final part of the chapter deals with two common methods of depreciation, straight line and reducing balance.

## 2. Progressions and interest

Interest and depreciation are fundamental concepts in a business or commercial environment. Interest is earned on money invested and paid on money borrowed; the value of most physical assets will depreciate with time.

However, in order to cover interest and depreciation (and other topics in the manual which follow) in a meaningful way, it is necessary to understand what progressions are. The following few sections deal with these.

## 3. Arithmetic progressions

An arithmetic progression (ap for short) is a sequence of numbers, called *terms*, in which any term after the first can be obtained from its immediate predecessor by adding a fixed number, called the *common difference* ($d$).

For example, the sequence of numbers: 3, 7, 11, 15, 19, etc is an arithmetic progression, since the common difference between one term and the next is 4 (the value of $d$).

The *first term* of an arithmetic progression is usually denoted by $a$. Thus, the progression takes the general form:

$$a, a+d, a+2d, a+3d, \ldots, \text{etc.}$$

In various situations, it is convenient to be able to identify the value of a particular term and to add together a set of consecutive terms. The following formulae can be used for these purposes.

**Formulae for an arithmetic progression**

If an arithmetic progression has starting term $a$ and difference $d$, then:

a) the $n$-th term is:

$$T_n = a + (n-1)d$$

b) the sum of the first $n$ terms is:

$$S_n = \frac{n}{2}[2a + (n-1)d]$$

## 4. Example 1 (Identifying an AP and evaluating terms and sums)

a) i. Consider the progression 6, 11, 16, 21, 26, 31, ...
This is an ap, since each term is obtained by *adding* 5 to the previous term.
Thus, $a=6$ and $d=5$.

ii. The progression 54, 50, 46, 42, 38, ... is an ap, since each term is obtained by *adding* –4 to (i.e. subtracting 4 from) the previous term.
Thus, $a=54$ and $d=-4$.

b) To find the value of the 10th term and the sum of the first 12 terms of 6, 11, 16, 21, ...

First, identify the progression as an ap with: $a$ = starting term = 6 and $d$ = difference between terms = 5.

Now: $T_n = a + (n-1)d$

and with $n=10$: $T_{10} = 6 + (9)5 = 51$

That is, the 10th term of this ap has value 51.

Also: $S_n = \frac{n}{2}[2a + (n-1)d]$

and with $n=12$: $S_{12} = \frac{12}{2}[2(6) + 11(5)] = 6(12+55) = 402.$

In other words, the sum of the first 12 terms of this ap is 402.

## 5. Geometric progressions

A *geometric progression* (gp for short) is a sequence of numbers, called *terms*, in which any term after the first can be obtained from its immediate predecessor by multiplying by a fixed number, called the *common ratio* ($r$). The first term of a gp is usually denoted by $a$.

For example, the sequence of numbers: 3, 12, 48, 192, etc is a gp since each term is being multiplied by the same value (4 in this case) to obtain the next term in the sequence.

The progression takes the general form:

$$a, ar, ar^2, ar^3, \ldots \text{ etc}$$

In the example above, $a=3$ and $r=4$, giving the progression as:

$$3, \quad 3(4), \quad 3(4)^2, \quad 3(4)^3, \quad \ldots$$

or: $$3, \quad 12, \quad 48, \quad 192, \quad \ldots$$

As with arithmetic progressions, it is convenient to be able to identify the value of a particular term and to add together a set of consecutive terms.

The following formulae can be used for these purposes.

> **Formulae for a geometric progression**
> If a geometric progression has starting term $a$ and common ratio $r$, then:
>
> a) the $n$-th term is: $T_n = ar^{n-1}$
>
> b) the sum of the first $n$ terms is: $S_n = \dfrac{a(r^n - 1)}{r-1} = \dfrac{a(1-r^n)}{1-r}$
>
> c) If $|r|<1$, then the sum to infinity is: $S_\infty = \dfrac{a}{1-r}$

For the gp: 3, 12, 48, 192, ... etc (with $a$=3 and $r$ =4):

i. the 7th term, $T_7 = 3(4^{7-1}) = 3(4^6) = 12288$.

ii. the sum of the first 4 terms, $S_4 = \dfrac{3(4^{4-1})}{4-1} = 255$

(check: 3+12+48+192 = 255)

and for the gp: 4, 2, 1, 0.5, 0.25, ... etc (with $a$=4 and $r$=0.5):

iii. the sum to infinity, $S_\infty = \dfrac{a}{1-r} = \dfrac{4}{1-0.5} = 8$

## 6. Example 2 (Geometric progressions)

a) i. Consider the progression

4, 8, 16, 32, 64, 128, ...

This is a gp because each term is obtained by multiplying the previous term by 2.

Thus, $a$=4 and $r$=2.

ii. The progression 16, 4, 1, 0.25, ... is a gp, since each term is obtained by multiplying the previous term by 0.25 (or, equivalently, dividing the previous term by 4.)

Thus, $a$=16 and $r$ =0.25.

b) To find the 11th term and the sum of the first 20 terms of the gp 4, 8, 16, 32, 64, ...

First identify the series as a gp, with $a$=4 and $r$=2.

Then, using $T_n = ar^{n-1}$, we have:

$$T_1 = 4(2)^{11-1} = 4(2)^{10} = 4096$$

Also, using $S_n = \dfrac{a(r^n - 1)}{r-1}$, we have:

$$S_{20} = \frac{4\left(2^{20} - 1\right)}{2-1} = 4{,}194{,}300$$

c) To find the sum to infinity of the gp 16, 4, 1, 0.25, ... etc. Notice that $a$=16 and $r$=0.25 (i.e. $r$<1).

Thus the sum to infinity, $S_\infty = \frac{a}{1-r} = \frac{16}{1-0.25} = 21.33$ (2D).

As mentioned earlier, a knowledge of progressions is necessary before various topics which follow can be truly understood. Before dealing with interest techniques however, it is useful to specify some common business terms, which will be used from now on.

## 7. Some terms used in business calculations

a) *Principal amount* ($P$). This is the amount of money that is initially being considered. It might be an amount about to be invested or loaned or it may refer to the initial value or cost of plant or machinery. Thus if a company was considering a bank loan of say £20000, this would be referred to as the principal amount to be borrowed.

b) *Accrued amount* ($A$). This term is applied generally to a principal amount after some time has elapsed for which interest has been calculated and added. It is quite common to qualify $A$ precisely according to time elapsed. Thus $A^1$, $A^2$, etc would mean the amount accrued at the end of the first and second years and so on. The company referred to in (a) above might owe, say, an accrued amount of £22000 at the end of the first year and £24200 at the end of the second year (if no repayments had been made prior to this time).

c) *Rate of interest* ($i$). Interest is the name given to a proportionate amount of money which is added to some principal amount (invested or borrowed). It is normally denoted by symbol $i$ and expressed as a *percentage rate per annum*. For example if £100 is invested at interest rate 5% per annum (pa), it will accrue to £100 + (5% of £100) = £100 + £5 = £105 at the end of one year. Note however, that for calculation purposes, a percentage rate is best written *as a proportion*. Thus, an interest rate of 10% would be written as $i = 0.1$ and 12.5% as $i = 0.125$ and so on.

d) *Number of time periods* ($n$). The number of time periods over which amounts of money are being invested or borrowed is normally denoted by the symbol $n$. Although $n$ is usually a number of years, it could represent other time periods, such as a number of quarters or months.

## 8. Difference between simple and compound interest

When an amount of money is invested over a number of years, the interest earned can be dealt with in two ways.

a) *Simple interest*. This is where any interest earned is *NOT* added back to the principal amount invested.

For example, suppose that £200 is invested at 10% simple interest per annum. The following table shows the state of the investment, year by year:

| Year | Amount on which interest is calculated | Interest earned | Cumulative amount accrued |
|---|---|---|---|
| 1 | £200 | 10% of £200 = £20 | £220 |
| 2 | £200 | 10% of £200 = £20 | £240 |
| 3 | £200 | 10% of £200 = £20 | £260 |
| ... | | | |
| ... etc | | | |

b) *Compound interest*. This is where interest earned is *added back* to the previous amount accrued.

For example, suppose that £200 is invested at 10% compound interest. The following table shows the state of the investment, year by year:

| Year | Amount on which interest is calculated | Interest earned | Cumulative amount accrued |
|---|---|---|---|
| 1 | £200 | 10% of £200 = £20 | £220 |
| 2 | £220 | 10% of £220 = £22 | £242 |
| 3 | £242 | 10% of £242 = £24.20 | £266.20 |
| ... | | | |
| ... etc | | | |

The difference between the two methods can easily be seen by comparing the above two tables. Notice that the amount on which simple interest is calculated is always the same; namely, the original principal.

Note that whether a principal amount is being invested [as in a) and b)] or borrowed makes no difference to the considerations for interest.

## 9. Formula for amount accrued (simple interest)

If £1000 is invested at 5% simple interest (pa) then 5% of £1000 = £50 will be earned every year.

*In general terms*, if amount $P$ is invested at $100i\%$ simple interest (per time period), then the amount of interest earned per time period is given by $P.i$.

For the data given, $P = 1000$ (£) and $100i\% = 5\%$ (i.e. $i = 0.05$).

Therefore, the interest earned per year $= P.i = 1000(0.05) = £50$.

Also, if £3000 ($P = 3000$) is invested at a (simple) interest rate of 9.5% ($i = 0.095$), the interest earned in any year is $P.i = 3000(0.095) = £285$. Over a period of, say, 5 years, the accrued amount of the investment would be given by: £[3000 + 5(285)] = £4425. That is, the original principal plus 5 year's-worth of interest.

In general terms, if amount $P$ is invested at $100i\%$ simple interest, then the amount accrued over $n$ years is given by the following formula.

**Accrued amount formula (simple interest)**

$$A_n = P(1 + i.n)$$

where: $A_n$ = accrued amount at end of $n$-th year
$P$ = principal amount
$i$ = (proportional) interest rate per year
$n$ = number of years.

Generally, simple interest is of no great practical value in modern business and commercial situations, since in practice interest is always compounded.

## 10. Formula for amount accrued (compound interest)

Section 8 gave details of the year-by-year state of an investment of £200 at 10% compounded per annum (sometimes called an *investment schedule*). The following derives, in general terms, a formula to give the accrued amount of a principal at the end of any time period.

If principal $P$ is invested at a (compound) interest rate of $100i$% over $n$ time periods, then:

$A_1 = P + I$ — i.e. the total amount accrued at the end of the first time period is the amount of the original principal plus the interest earned ($I$) for this period.

So, $A_1 = P + P.i$ — the interest earned in one time period is $P.i$

i.e. $A_1 = P(1+i)$ — factorising

$A_2 = P(1+i) + I$ — amount accrued at end of second time period is the amount at the beginning of the period plus the interest earned ($I$)

i.e. $A_1 = P(1+i) + P(1+i)i$ — interest earned is accrued amount (at beginning of the period) times $i$

$= P(1+i)(1+i)$ — factorising

i.e. $A_2 = P(1+i)^2$

Similarly, $A_3 = P(1+i)^3$, $A_4 = P(1+i)4$ ..... and so on.

For example, if £5000 ($P$=5000) is invested at 9% p.a. ($i$=0.09), then:

Amount accrued after 1 year: $A_1 = P(1+i) = 5000(1.09) = £5450$.

Similarly: $A_2 = P(1+i)^2 = 5000(1.09)^2 = £5940.50$.

and $A_3 = P(1+i)^3 = 5000(1.09)^3 = £6475.15$.

... and so on.

In general terms, if amount $P$ is invested at $100i$% compound interest, then the amount accrued over $n$ years is given by the following formula.

> **Accrued amount formula (compound interest)**
>
> $$A_n = P(1 + i)^n$$
>
> where: $An$ = accrued amount at end of $n$-th year
> $P$ = principal amount
> $i$ = (proportional) interest rate per year
> $n$ = number of years.

## 11. Notes on previous formula

a) The above formula can be transposed to make $P$ the subject as follows:

$$P = \frac{A}{(1+i)^n}$$

So that, given an interest rate and a time period, a principal can be found if the accrued amount is known.

For example, if some principal amount is invested at 12% ($i = 0.12$) and amounts to £4917.25 (=$A$) after 3 years ($n$ =3), then $P$ (the principal amount) can be found using the above formula as:

$$P = \frac{4917.25}{1.12^3} = \frac{4917.25}{1.4049} = £3{,}500$$

b) The standard time period for the calculation of interest is usually a year. Hence, from this point on, the value of $n$ (the number of time periods) will be assumed to be a number of years *unless stated otherwise.*

## 12. Example 3 (using *accrued amount formula* to solve simple problems)

a) What will be the value of £450 compounded at 12% for 3 years?
Here, $P$=450; $i = \frac{12}{100}$ =0.12 and $n$ =3.
Therefore $A = P(1 + i)^n = 450(1+0.12)^3 = 450(1.12)^3 = £632.22$

b) A principal amount accrues to £8500 if it is compounded at 14.5% over 6 years. Find the value of this original amount.

Here it is necessary to use the inverted formula, since $P$ needs to be found.
With $A$=8500, $i$=0.145 and $n$ =6 (and using the formula in section 11(a)):

we have: $$P = \frac{A}{(1+i)^n} = \frac{8500}{(1+0.145)^6} = \frac{8500}{1.145^6} = 3772.12$$

That is, £3772.12 needs to be invested (at 14.5% over 6 years) in order to accrue to £8500.

## 13. Notes on calculations involving compound interest

a) *Using compound interest tables.*
Appendix 1 gives an example of how compound interest tables are set out. The rows marked 'C' (for compounding) give a multiplier, which is the exact value of $(1+i)^n$.

For example, if the interest rate was 13% ($i$=0.13) and the number of years invested was 7 (=$n$), then looking along the C row at 13% and the $n$ =7 column yields the value 2.3526.
This means that $(1+0.13)^7 = 2.3526$. Thus, if £2500 was invested at 13% for 7 years, the accrued amount could be calculated as £2500(2.3526) = £5881.50.
Similarly from tables, 11 years at 26% gives a multiplier of 12.708. Therefore, if £6000 was invested at 26% over 11 years, it will accrue to £6000(12.708) = £76,248.00.

b) *Calculator technique*
The *constant facility* on calculators can also be used when evaluating expressions such as $(1+i)^n$. For example, to calculate $(1.045)^4$, the following keys should be pressed:

$$1.045 \times = = =$$

The 'times' (×) sets up the constant facility, the first '=' generating $1.045^2$, the second '=' generating $1.045^3$, and so on.
For all values of $n$, the '=' needs to be pressed one less time than the value of $n$. Thus to generate the value of $1.065^{25}$, after keying in 1.065 followed by '×', the '=' key needs to be pressed 24 times!

c) *Determining i, given A, P and n.*
The accrued amount formula can also be arranged to give:

$$1+i = \sqrt[n]{\frac{A}{P}} = \left[\frac{A}{P}\right]^{\frac{1}{n}}$$

Calculators that have an '$x^y$' function can cope with this easily, but if this function is not available, logarithms must be used as follows:

i. Find the log of $\frac{A}{P}$.
ii. Divide the above value by $n$.
iii. Find the anti-log of the result (to give the value of $1+i$).

An example of this technique follows.

## 14. Example 4 (Finding an *interest rate*, given the principal, amount accrued and time)

### Question

Find the interest rate necessary for £20000 to accrue to £50000 in 12 years.

### Answer

Here, $P = 20000$; $A = 50000$; $n$ =12. To find the value of $i$.

Now, $= \frac{A}{P} = \frac{50000}{20000} = 2.5$ and, from 13 c) above: $1+\mathrm{i} = \sqrt[12]{2.5}$

(i) The log of 2.5 is 0.3979 (from log tables)
(ii) $\frac{0.3979}{12} = 0.0332$
(iii) The antilog of 0.0332 is 1.08 (2D, from log tables)

Thus $1 + i = 1.08$, giving $i = 0.08$.
The interest rate required is 8%.

## 15. Example 5 (Value of an *accrued amount* of a number of equal payments)

*Question*

A firm plans to invest an amount of money at the beginning of every year in order to accrue a sum of £100,000 at the end of a five year period. What is the value of the amount, if the investment rate is 14%?

*Answer*

Note that money is being invested as a *series* of equal payments and *not* as a single principal amount.
Let the amount invested at the beginning of each of the five years be $B$ (in £).
The first payment of $B$ accrues to $B(1.14)^5$ (since $B$ will have been invested for 5 years).
The second payment of $B$ accrues to $B(1.14)^4$ ... and so on.
But the sum of the separate accruals must add to 100,000. That is:

$$100{,}000 = B(1.14)^5 + B(1.14)^4 + B(1.14)^3 + B(1.14)^2 + B(1.14)$$

i.e. $$100{,}000 = B(1.14 + 1.14^2 + 1.14^3 + 1.14^4 + 1.14^5).$$

There are in general two ways available for calculating the value of the terms within the brackets:

a) by evaluating each term separately, which can be done using a calculator or compounding tables;
b) by using the fact that the terms form a gp and evaluating their sum using the formula of section 4(b). [Note that the terms *do* form a gp since each term is being multiplied by 1.14 to obtain the following term.]

Evaluating the terms in the brackets *using method a)* gives:

$1.14 + 1.2996 + 1.4815 + 1.6890 + 1.9254 = 7.5355$
(using compounding tables).

Evaluating the terms in the brackets *using method b)* gives:

a gp with $a$ = starting term = 1.14; $r$ = common ratio = 1.14 and $n$ = number of terms = 5

Now: $$S_5 = \frac{(1.14)[(1.14)^5 - 1]}{1.14 - 1}$$
$= 7.5355$ [agreeing with method a) above]

Thus: $100000 = B(7.5355)$,

giving: $$B = \frac{100000}{7.5355}$$
$= £13270.52$

In other words, the firm would need to invest £13,270.52 at the beginning of each year.

## 16. Nominal and effective interest rates

a) It is normal practice in business and commerce to express rates of interest as per annum figures, *even though the interest may be compounded over time periods of less than a year*. In this type of case, the given annual rate is called a *nominal rate*.

b) Compounding may be six-monthly, quarterly, monthly or daily. The actual annual rate of interest, called the *effective rate* or *actual percentage rate* (*APR*), will *always be greater* than the nominal rate.

For example, suppose the nominal rate of interest is 10%, but interest is being compounded six-monthly. This means that interest is being charged at 5% per six months. Thus, if £100 was to be invested, it would be worth £[100(1.05)] = £105 after the first six months. Then this £105 would be invested for a further six months at 5% and would be worth £[105(1.05)] = £110.25 at the end of a year. In other words, £100 has grown to £110.25 after one year, giving an *effective* interest rate, or APR, of 10.25% (and *not* 10%).

c) The standard method of determining the APR is to make the effective time period equal to the compounding period and actually compound over a period of a year. Examples of this follow.

## 17. Example 6 (Problems involving *effective* interest rates)

a) £100 is subject to interest at 12%, compounded quarterly. What is the accrued amount after one year?

The actual interest rate per quarter is $\frac{12\%}{4} = 3\% = 0.03$. Thus, £100 must be compounded at 0.03 for four quarters.

Therefore, the accrued amount is: $100(1.03)^4 = 112.55 =$ £112.55. (Note that this is not the same as £100 for one year at 12%, which would have accrued to only £112.)

b) A finance company advertises money at 22% nominal interest, but compounds monthly.

To find the APR, we can proceed as follows.

Proportional annual (nominal) interest rate $= \frac{22}{100} = 0.22$

Proportional *monthly* interest rate $= \frac{0.22}{12} = 0.0183$ (4D).

Thus, £1 invested would accrue (over 12 months) to £$1(1+0.0183)^{12} = (1.0183)^{12}$ $= 1.24$ (2D)

That is, £1 would accrue to £1.24 after one year.

Thus the effective rate (APR) is $1.24 - 1 = 0.24$. i.e. 24%.

**Note**: The discrepancy between the nominal rate and the APR gets larger as the number of years increases.

The following section gives a general formula for calculating APR.

## 18. Formula for calculating APR

**Formula to calculate APR**

Given a nominal annual rate of interest, the effective rate or actual percentage rate (APR) can be calculated as:

$$\text{APR} = \left[1 + \frac{i}{n}\right]^n - 1$$

where: $i$ = given nominal rate (as a proportion)

$n$ = number of equal compounding periods in one year.

Thus, for example:

a) 10% nominal, compounded quarterly, has APR = $(1.025)^4 - 1 = 0.1038 = 10.38\%$

b) 24% nominal, compounded monthly, has APR = $(1.02)^{12} - 1 = 0.2682 = 26.82\%$

## 19. Example 7 (To calculate principal amount and APR)

*Question*

A company will have to spend £300,000 on new plant in two years from now. Currently investment rates are at a nominal 10%.

a) What single sum should now be invested, if compounding is six-monthly?

b) What is the APR?

*Answer*

a) Since compounding is six-monthly, the investment (P, say) must accrue to a value of £300,000 after four *six-monthly* periods. Note also that the interest rate for each six-month period is (10/2)% = 5%.

Using the compounding (accrued amount) formula, $300000 = P(1+0.05)^4$

and re-arranging gives: $P = \dfrac{300000}{1.05^4} = 246810.75$

That is, the amount to be invested is £246,810.75.

b) Using the previous APR formula:

APR $= (1+0.05)^2 - 1 = (1.05)^2 - 1 = 0.1025 = 10.25\%$

## 20. Depreciation

*Depreciation* is an allowance made in estimates, valuations or balance sheets, normally for 'wear and tear'.

It is normal accounting practice to depreciate the values of certain assets. There are several different techniques available for calculating depreciation, two of which are:

a) *Straight line* (or Equal Instalment) depreciation and

b) *Reducing balance* depreciation.

These two methods can be thought of as the converse of the interest techniques dealt with so far in the chapter. That is, instead of adding value to some original principal amount (as with interest), value is taken away in order to reduce the original amount. Straight line depreciation is the converse of simple interest with amounts being subtracted (rather than added), while the reducing balance method is the converse of compound interest.

## 21. Straight line depreciation

Given that the value of an asset must be depreciated, this technique involves subtracting the same amount from the original book value each year. Thus, for example, if the value of a machine is to depreciate from £2500 to £500 over a period of five years, then the annual depreciation would be £(2500–500)÷5 = £400. The term 'straight line' comes from the fact that 2500 can be plotted against year 0 and 500 against year 5 on a graph, the two points joined by a straight line and then values for intermediate years can be read from the line. See Figure 1.

*Figure 1*

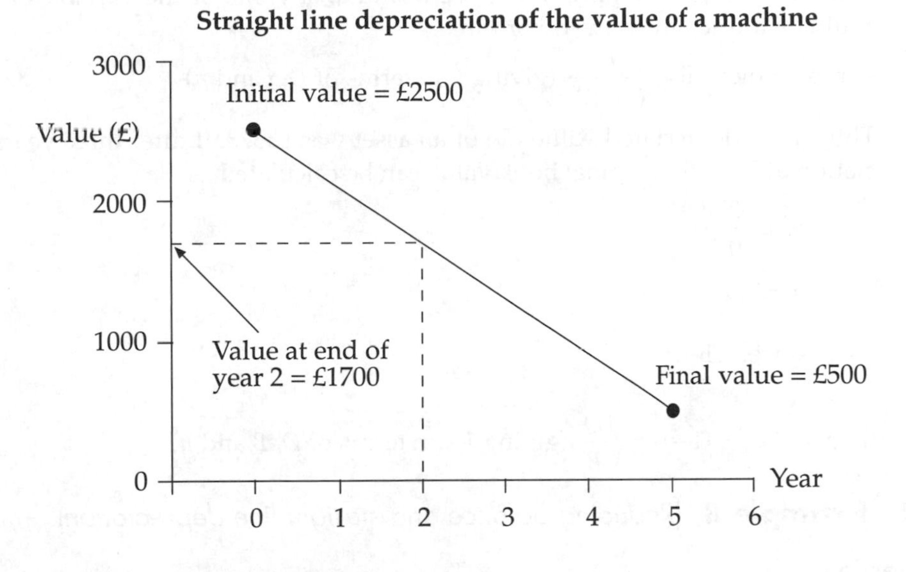


## 22. Reducing balance depreciation

We already know that to *increase* some value $P$ by $100i\%$, we need to calculate the quantity $P(1+i)$, and if this is done $n$ times, successively, the accrued value is given by $P(1+i)n$ (from section 10).

A similar argument is applied if we need to *decrease* (or equivalently depreciate) some value $B$ say by 100%. Here, we need to calculate $B(1–i)$. Notice that the multiplier is now $1–i$, rather than $1+i$. If the depreciation is carried out successively, $n$ times, the accrued value will be given as $B(1–i)^n$.

For example, £2550 depreciated by 15% is £2550(1–0.15) = £2550(0.85) = £2167.50.

Also, if £2550 was successively depreciated over four time periods by 15%, the final depreciated value = $£2550(1–0.15)^4$ = $£2550(0.85)^4$ = £1331.12.

*Reducing balance depreciation* is the name given to the technique of depreciating the book value of an asset by a constant percentage.

## 23. Formula for reducing balance depreciation

**Reducing balance depreciated value formula**

If book value $B$ is subject to reducing balance depreciation at rate $100i\%$ over $n$ time periods, the depreciated value at the end of the $n$-th time period is given by:

$$D = B(1-i)^n$$

where: $D$ = depreciated value at the end of the $n$-th time period
$B$ = original book value
$i$ = depreciation rate (as a proportion)
$n$ = number of time periods (normally years).

Note that by re-arranging the above formula, any one of the variables $B$, $i$ and $n$ could be found, given the other three.

For example: $B = \dfrac{D}{(1-i)^n}$ (giving $B$ in terms of $D$, $i$ and $n$)

Thus if the depreciated value ($D$) of an asset was £5378.91 after three years' depreciation at 25%, the original book value can be calculated as:

$$B = £\frac{5378.91}{0.75^3}$$

$$= £12{,}750.$$

Also, since: $(1-i)^n = \dfrac{D}{B}$

then: $(1-i) = \sqrt[n]{\dfrac{D}{B}}$ (giving $1-i$ in terms of $D$, $B$ and $n$)

## 24. Example 8 (*Reducing balance* and *straight line depreciation*)

*Question*

A mainframe computer whose cost is £220,000 will depreciate to a scrap value of £12000 in 5 years.

a) If the reducing balance method of depreciation is used, find the depreciation rate.

b) What is the book value of the computer at the end of the third year?

c) How much more would the book value be at the end of the third year if the straight line method of depreciation had been used?

*Answer*

Variables given are: $D = 12000$, $B = 220000$, $n = 5$ and $\dfrac{D}{B} = \dfrac{12000}{220000} = 0.0545$

a) Now $1-i = \sqrt[5]{0.0545}$

The right-hand side of the above expression can be calculated using either a calculator (which has an '$x^y$' function key) or using logarithms. For revision purposes, the method using logarithms now follows.

We have, $\log(0.0545) = \bar{2}.7364$

and: $\dfrac{\bar{2}.7364}{5} = \dfrac{(\bar{5}+3.7364)}{5} = \bar{1}.7473$

Finally, antilog $(\bar{1}.7473) = 0.5589$.

Therefore, $1-i = 0.5589$

and so $i = 0.4411$.

Thus the depreciation rate is 44.11%.

b) The book value at the end of year 3 is given by: $B(1-i)3 = 220000(0.5589)^3 = £38408$.

c) Using the straight line method, the annual depreciation is:

$$\frac{£(220000-12000)}{5} = £41600$$

Thus, after three years, the book value would be: £[220000 – 3(41600)] = £95200. So that the book value, using this method, would be £[95200–38408] = £56792 *more* than using the reducing balance method (at the end of the third year).

## 25. Summary

a) An arithmetic progression takes the form: $a, a+d, a+2d, a+3d,$ ... etc.
   i. The $n$th term, $T_n = a + (n-1)d$
   ii. Sum of first $n$ terms, $S_n = \frac{n}{2}[2a+(n-1)d]$

b) A geometric progression takes the form: $a, ar, ar^2, ar^3,$ ... etc.
   i. The $n$th term, $T_n = ar^{n-1}$
   ii. Sum of first $n$ terms, $S_n = \dfrac{a(r^n-1)}{r-1} = \dfrac{a(1-r^n)}{1-r}$
   iii. Sum to infinity, $S_\infty = \dfrac{a}{1-r}$

c) Simple interest is where the interest earned is not added back to the original principal for calculating future interest.
   Accrued amount after year n, $A_n = P(1 + i.n)$.

d) Compound interest is where the interest earned is added back to the original principal for calculating future interest.
   Accrued amount after year $n$, $A_n = P(1 + i)^n$.

e) Interest rates are usually quoted as nominal annual rates, even though the compounding period may be a proportion of a year.

Effective annual rate (or APR) can be calculated as:

$$\text{APR} = \left[1 + \frac{i}{n}\right]^n - 1$$

where $100i\%$ is the given nominal (annual) rate and $n$ is the number of compounding periods in a year.

f) Two common methods of depreciation are:

i. Straight line: where the same amount is deducted from the original book value each year.

ii. Reducing balance: where the same percentage is deducted from the original book value each year (the opposite of compound interest) and depreciated value at end of $n$-th year is given by:

$$D = B(1 - i)^n$$

## 26. Student self review questions

a) What is the essential difference between arithmetic and geometric progressions? [3,5]

b) If a geometric progression has starting term $a$ and common ratio $r$, what is the expression used to calculate the sum of the first $n$ terms? [5(b)]

c) Explain the terms $P$, $A$, $i$ and $n$, that are commonly used in interest calculations. [7]

d) How does simple interest differ from compound interest? [8]

e) State the formula for the amount accrued after compounding principal $P$ at rate $100i\%$ for $n$ years. [10]

f) What is an 'effective' rate of interest (or APR)? [16]

g) What is 'reducing balance' depreciation and how is it connected with compound interest? [22]

## 27. Student exercises

1. Find the 15th term and the sum of the first 15 terms of the progression:

   2, 5, 8, 11, ...

2. Find the sum of the progression:

   1.00, 1.04, 1.08, ..., 2.12, 2.16

3. Find the sum of the first 20 terms of the progression:

   48, 40, 32, 24, ...

4. Find the 8th term and the sum of the first 8 terms of:

   1, 3, 9, 27, ...

5. Find the sum (to 2D) of the first 15 terms of:

   $1, 1.03, (1.03)^2, (1.03)^3, \ldots$

6. Find the sum to infinity of the geometric progression:

   $14, 14(0.8), 14(0.8)^2, \ldots$

7. £1200 is invested at 12.5% simple interest. How much will have accrued after 3 years?

8. Find the amount of:
   a) £480 compounded at 14% over 5 years.
   b) £1240 compounded at 11.5% over 12 years.
9. After 3 years, what principal value will amount to £1100 at 8% compounded?
10. £10000 was compounded at 12.5% and amounted to £52015.80. How many years did it take?
11. What percentage compound interest rate (to 2D) will treble the value of an investment over a period of 5 years?
12. A firm borrows £6000 from the bank at 13% compounded semi-annually. If no repayments are made, how much is owed after 4 years?
13. *MULTI-CHOICE*. A loan company advertises that it charges 1% per month interest. What is its actual (or annual) percentage rate over the period of one year?
    a) 11.2% b) 12.0% c) 12.4% d) 12.7%
14. A loan company advertises money at 18% (nominal). What is the APR (to 2D), if compounding is (a) semi-annual (b) quarterly (c) monthly?
15. *MULTI-CHOICE*. The market value of a multi-national company falls from £46 billion to £5 billion in just 3 years. The average annual depreciation in company value is closest to:
    a) 45% b) 50% c) 60% d) 100%
16. A special council amenity costs £500,000 and its useful life is planned over 25 years. If the reducing balance depreciation method is used at a 10% rate, what is the book value of the amenity at the end of its life?
17. *MULTI-CHOICE*. A £20,000 new car depreciates in value by 20% ± 2% each year (year end). The car depreciates by the 'reducing balance method', which means that a constant percentage is applied each year to the written-down value. After 3 years, the car's value is most accurately estimated by:
    a) between £9,491 and £11,027 b) £10,240 ± £1,024
    c) £10,240 ± £205 d) between £12,168 and £13,448
18. A company buys a computer for £125,000 and houses it in a specially constructed suite at a cost of £20,000.
    a) If the computer depreciates at 25% (reducing balance) and the suite appreciates at 5% (compound), what is the book value of suite and computer after 5 years?
    b) Taking computer and suite together and using the reducing balance method, what is the overall depreciation rate?
19. A firm invests £4000 per year (at the end of every year) at 8% compound to meet a fixed commitment at a particular time. If the commitment is £15000 to be paid in exactly three years from now, what single sum *invested now* needs to be added in order to meet the commitment?

# 22 Present value and investment appraisal

## 1. Introduction

This chapter describes the technique of present value and how it can be applied to future cash flows in order to find their worth in today's money values. This in turn enables discounted cash flows to be calculated for investments, leading to descriptions of the main methods of investment appraisal.

## 2. Present value

Suppose money can be invested at 10%. Then £100 could be invested and be worth £110 in one year's time. Put another way, the value of £110 *in one year's time* is exactly the same as £100 *now* (if the investment rate is 10%). Similarly, £100 now has the same value as £100$(1.1)^2$ (= £121) in two years time, assuming the investment rate is 10%. This demonstrates the concept of the *present value* of a future sum.To state the above ideas more precisely, if the current investment rate is 10%, then:

the present value of £110 in one years time is $£\dfrac{110}{1.1} = £100$

similarly, the present value of £121 in two years time is $£\dfrac{121}{1.1^2} = £100$

and so on.

The investment rate, used in this context, is sometimes referred to as the *discount rate*.

In the following section, the above technique for calculating present value is generalized to give a formula and then followed by some examples.

## 3. Formula for present value

**Present value formula**

The present value of amount £A, payable in $n$ year's time, subject to a discount rate of $100i\%$ is given by:

$$P = \frac{A}{(1+i)^n}$$

where: $P$ = present value

$A$ = amount, payable in $n$ year's time

$i$ = discount rate (as a proportion)

$n$ = number of time periods (normally years).

The above formula is a simple re-arrangement of the formula given for calculating an accrued amount using compound interest (chapter 21, section 10).

## 4. Use of discounting tables

The quantity $\dfrac{1}{(1+i)^n}$ is sometimes known as the *present value factor* or *discount factor*.

Appendix 1 gives tables showing the values of the discount factor (these are the rows marked '*D*') for a wide range of values of $i$ and $n$. These are known as *discounting tables*.

As an example of their use, suppose the discount rate was 14% (i.e. $i$=0.14) and the number of years (or other periods of time) involved was $n$=4.

Looking along the *D* row for 14% and down the $n$=4 column yields the value 0.5921.

This can be translated as: $\frac{1}{(1+0.14)^4} = 0.5921$

Suppose we wanted to find the present value of £1500 in 6 years' time, subject to a discount rate of 19%.

The discount factor (from tables, with *D*=19% and *N*=6) is 0.3521.

Therefore, present value = £1500(0.3521) = £528.15.

## 5. Example 1 (*Present value*)

### *Question*

A department store advertises goods at £700 deposit and three further equal annual payments of £500. If the discount rate is 7.5%, calculate the present value of the goods.

### *Answer*

There are four payments to consider, but the first (£700) is payable *now* and thus its present value is the same as its face value. The other three payments of £500 have to be discounted over 1, 2 and 3 years respectively, using rate 7.5% = 0.075.

Therefore:

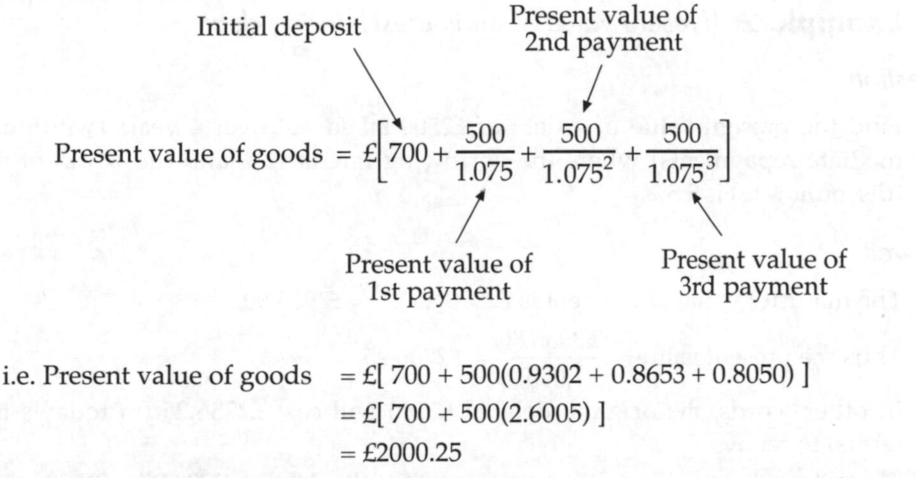


i.e. Present value of goods = £[ 700 + 500(0.9302 + 0.8653 + 0.8050) ]

= £[ 700 + 500(2.6005) ]

= £2000.25

Note that £2000.25 could be considered as the equivalent cash price of the goods.

## 6. Present value of an interest-bearing debt

When an amount of money is borrowed it will attract interest at an appropriate borrowing rate. If no intermediate payments are made, at the end of a period the debt, plus interest, must be repaid as a lump sum. The present value of amounts such as these are often required.

Given that the amount borrowed is $P$ at $100i\%$ compounded for $n$ years, the debt will amount to $P(1+i)^n$. If the investment (discount) rate is $100j\%$, the debt will have to be discounted at this rate back over the $n$ years. Thus, the present value of the interest-bearing debt is given by the following.

**Present value of an interest-bearing debt**

$$PV = \frac{P(1+i)^n}{(1+j)^n}$$

where: $i$ = proportional borrowing rate
$j$ = proportional discount rate
$P$ = original amount borrowed
$n$ = number of time periods (normally years).

Notice that, since the borrowing rate will always be greater than the investment rate, $i > j$.

So the ratio $\frac{P(1+i)^n}{(1+j)^n}$ will always be greater than 1.

In other words, PV (the present value of the debt) will always be *greater* than $P$ (the original amount borrowed).

## 7. Example 2 (*Present value* of an *interest-bearing debt*)

*Question*

Find the present value of a debt of £2500 taken out over 4 years (with no intermediate repayments) where the borrowing rate is 12% and the worth of money (discount rate) is 9.5%.

*Answer*

The maturity value of the debt is £2500$(1.12)^4$ = £3933.80.

Thus the present value = $\frac{£3933.80}{(1.095)^4}$ = £2736.25

In other words, the original debt of £2500 will cost £2736.25 (in today's money terms) to repay.

Thus, it can be considered that the *real cost* of the debt is £2736.25 – £2500 = £236.25 (in today's money terms).

## 8. Capital investments

For the purposes of this manual, a *capital investment* is a project which consists of:

a) An initial outlay of capital.

b) A set of estimated cash inflows and outflows over the life of the project.

c) Optionally, a resettlement figure, which might be caused by resale of plant or shares, or a cash settlement to clear any liabilities incurred.

There are many ways of appraising and comparing capital investments. Two methods, *both using discounting techniques*, covered in the following sections are:

i. discounted cash flow and net present value;

ii. internal rate of return.

## 9. Discounted cash flow

The *discounted cash flow* technique of investment appraisal involves calculating the sum of the present values of all cash flows associated with a project. This sum is known as the *net present value* (*NPV*) of the project. The cash flows are normally tabulated in net terms per year and the standard presentation is shown in the following example.

## 10. Example 3 (*Net present value*)

*Question*

A business project is being considered which has £12000 initial costs associated with revenues (i.e. inflows) over the following four years of £8000, £12000, £10000 and £6500 respectively. If the project costs (i.e. outflows) over the four years are estimated as £8500, £3000, £1500 and £1500 respectively and the discount rate is 18.5%, evaluate the project's NPV.

*Answer*

The project's cash flows are tabulated in Table 1 (following page) and the discounted cash flow table (Table 2) is shown, yielding the net present value of the project

## 11. Notes on Example 3

a) It is normal to include a 'discount factor' column, whether or not the values have been obtained using a calculator, given in an examination or read from tables.

b) The figures in the 'year' column are normally understood as 'end of year', in particular the first value, 0, being interpreted as 'now' (i.e. 'end of year 0').

c) Bracketed figures indicate negative amounts.

*Table 1 Business project cash flows*

| Year | Cash inflow (a) | Cash outflow (b) | Net cash flow (a) – (b) |
|---|---|---|---|
| 0 | – | 12000 | (12000) |
| 1 | 8000 | 8500 | (500) |
| 2 | 12000 | 3000 | 9000 |
| 3 | 10000 | 1500 | 8500 |
| 4 | 6500 | 1500 | 5000 |

*Table 2 Business project discounted cash flow table*

| Year | Net cash flow | Discount factor at 18.5% | Present value |
|---|---|---|---|
| 0 | (12000) | 1.0000 | (12000.00) |
| 1 | (500) | 0.8439 | (421.95) |
| 2 | 9000 | 0.7121 | 6408.90 |
| 3 | 8500 | 0.6010 | 5108.50 |
| 4 | 5000 | 0.5071 | 2535.50 |
| | | Net present value = | 1630.95 |

## 12. Interpretation of NPV

NPV can be interpreted *loosely* in the following way:

NPV > 0 ➔ project is in profit (i.e. worthwhile)
NPV = 0 ➔ project breaks even, and
NPV < 0 ➔ project makes a loss (i.e. not worthwhile).

However, it is important to understand fully the phrase 'project breaks even'. Consider the following example. Suppose we have £100 to invest in a project which returns a single inflow of £110 in a year's time, where the investment rate is 10%. If we put discount rate = investment rate = 10%, this will yield an NPV of zero. Thus *breaking even* here means that the return from the project is no better than a safe investment of the original capital (at 10%). Another way of interpreting this is that the project *earns* exactly 10%.

In other words, if the NPV of any project is negative, this means that the project *does not earn as much as the discount rate* (used in the calculations). Conversely, if the NPV is positive, this means that the project *earns more than the discount rate*.

**Interpretation of NPV**

NPV > 0 ➔ project earns *more* than the discount rate
NPV = 0 ➔ project earns *the same* as the discount rate
NPV < 0 ➔ project earns *less* than the discount rate.

Note that the discount rate is that rate which is used to discount the cash flows. Also, taxation and inflation have been ignored in the above interpretation.

## 13. Example 4 (*Net present value*)

*Question*

It is estimated that an investment in a new process will cause the following cash flow (in £):

| End year | 0 | 1 | 2 | 3 | 4 | 5 | 6 |
|---|---|---|---|---|---|---|---|
| Cash inflow | | | 15000 | 20000 | 20000 | 20000 | 20000 |
| Cash outflow | 60000 | 10000 | | | | | |

The firm wishes to earn at *least* 15% per annum on projects of this type. Calculate the Net Present Value of the project and comment on the course of action to be taken.

*Answer*

With respect to a discount rate of 15%, if the project earned exactly 15% per annum, the NPV would be zero.

Thus a positive NPV (obtained using a discount rate of 15%) would signify that the return on the project is higher than 15% and, on the criterion of present value, the firm should initiate the project. However, if the NPV is negative, the firm should not proceed with the project.

The discounted (at 15%) cash flow is shown in Table 3.

*Table 3 Discounted cash flow table at 15%*

| Year | Net cash flow | Discount factor at 15% | Present value |
|---|---|---|---|
| 0 | (60000) | 1.0000 | (60000.00) |
| 1 | (10000) | 0.8696 | (8696.00) |
| 2 | 15000 | 0.7561 | 11341.50 |
| 3 | 20000 | 0.6575 | 13150.00 |
| 4 | 20000 | 0.5718 | 11436.00 |
| 5 | 20000 | 0.4972 | 9944.00 |
| 6 | 20000 | 0.4323 | 8646.00 |
| | | Net present value = | (14178.50) |

Since the NPV is negative and comparatively large, the project clearly earns much less than 15% and hence would not be considered.

## 14. Internal rate of return (IRR)

The *Internal Rate of Return* (sometimes referred to as the '*Yield*') is an alternative method of investment appraisal to Net Present Value.

**Internal rate of return**

The Internal Rate of Return (IRR) of a project is the value of the discount rate that gives an NPV of zero.

Alternatively (and more to the point when projects are being compared) it can be translated as *the rate that a project earns.*

In Example 4, the IRR would work out at much less than 15%, since the NPV (calculated at a discount rate of 15%) was negative. Example 5 calculates this IRR.

There is no precise formula for calculating the IRR of a given project. However, it can be estimated (using a linear interpolation technique) either:

a) graphically, or
b) by formula.

Both of these techniques need the the NPV calculated using *two different discount rates* and these techniques are explained in the following two sections.

## 15. Graphical estimation of IRR

In order to estimate the IRR of a project graphically:

a) scale the vertical axis to include both NPVs;
b) scale the horizontal axis to include both discount rates;
c) plot the two points on the graph and join them with a straight line;
d) identify the *estimate of the IRR* where this line crosses the horizontal (discount rate) axis.

This technique is demonstrated in Figure 1, using the following project results:

discount rate 15% yields an NPV of £14000;
discount rate 17% yields an NPV of –£7000.

*Figure 1 Graphical estimation of IRR*

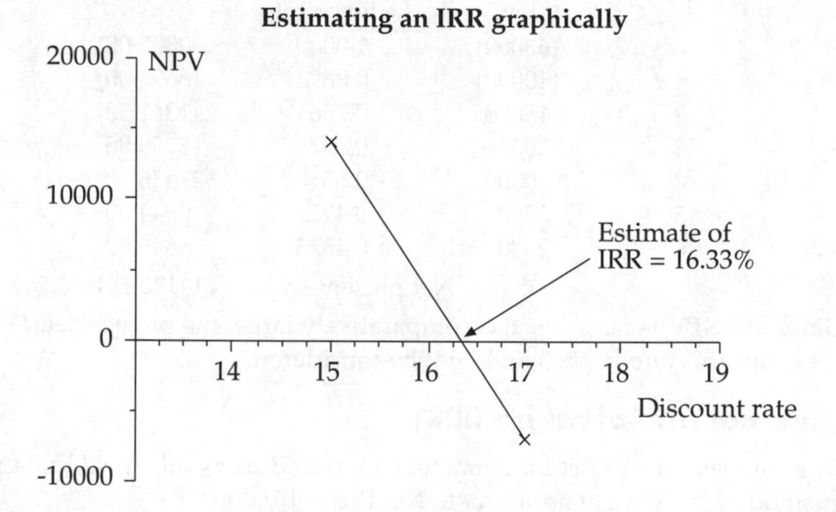


From Figure 1, the estimated value of the IRR for this project is seen to be 16.33%.

This graphical technique can be used with any two discount rates and their respective NPV values. (However, see section 22 for a note on the case of the NPVs being both positive or both negative.)

## 16. Estimation of IRR by formula

The exact formula equivalent of the graphical linear interpolation method is now given:

**Formula for estimating IRR**

$$\text{IRR} = \frac{N_1 I_2 - N_2 I_1}{N_1 - N_2}$$

where: discount rate $I_1$ gives NPV $N_1$
and: discount rate $I_2$ gives NPV $N_2$.

Using the example given in the previous section:
discount rate 15% yields an NPV of £14000 (i.e. $I_1$=0.15 and $N_1$=14000);
discount rate 17% yields an NPV of –£7000 (i.e. $I_2$=0.17 and $N_2$=–7000),
we have:

$$\text{IRR} = \frac{(14000)(0.17) - (-7000)(0.15)}{14000 - (-7000)}$$

$$= \frac{2380 + 1050}{21000}$$

$$= 0.1633$$

That is, an estimate of the IRR is 16.33% (agreeing with the result obtained in the previous section).

**Note**: The $I$ values substituted into the formula can be either proportions (as used above) or percentages.

Example 5, which follows, demonstrates the use of this formula from a discounted cash flow table.

## 17. Example 5 (*Internal rate of return*)

### *Question*

Using the cash flows of Example 4, (where the firm wishes to earn at least 15% p.a. on this type of project), calculate the NPV corresponding to each of the two discount rates 6% and 9%. Estimate the Internal Rate of Return for the project and interpret the result.

### *Answer*

When calculating the NPVs of a project using two separate discount rates, it is worth combining both sets of calculations into a single table. The dual discounted cash flow table for the project in question is shown in Table 4.

*Table 4 Discounted cash flow table*

| Year | Net flow | Discount factor at 6% | Present value | Discount factor at 9% | Present value |
|---|---|---|---|---|---|
| 0 | (60000) | 1.0000 | (60000) | 1.0000 | (60000.00) |
| 1 | (10000) | 0.9434 | (9434) | 0.9174 | (9174.00) |
| 2 | 15000 | 0.8900 | 13350 | 0.8417 | 12625.50 |
| 3 | 20000 | 0.8396 | 16792 | 0.7722 | 15444.00 |
| 4 | 20000 | 0.7921 | 15842 | 0.7084 | 14168.00 |
| 5 | 20000 | 0.7473 | 14946 | 0.6499 | 12998.00 |
| 6 | 20000 | 0.7050 | 14100 | 0.5963 | 11926.00 |
| | | | 5596 | | (2012.50) |

Using the formula to estimate IRR, we have:

$$I_1 = 6; N_1 = 5596 \text{ and } I_2 = 9; N_2 = -2012.5$$

and thus:

$$\begin{aligned} \text{IRR} &= \frac{N_1 I_2 - N_2 I_1}{N_1 - N_2} \\ &= \frac{(5596)(9) - (-2012.5)(6)}{5596 - (-2012.5)} \\ &= \frac{50634 + 12075}{7608.5} \\ &= 8.21\% \end{aligned}$$

The Internal Rate of Return is thus 8.21%, which is below the 15% required.

*Therefore the project should be rejected.*

(The value of 8.21% obtained above should be checked, using the graphical method, as an exercise.)

## 18. Comparison of projects

There is no universal formula for deciding which one of a number of projects is the best. Sometimes a choice will be obvious, and at other times not so clear, depending on a particular company's circumstances with regard to factors such as:

a) whether capital is available or whether it needs to be borrowed,

b) the ability to pay rates of interest,

c) general liquidity,

d) uncertainty of estimated flows.

However, given that the above factors are not markedly different for two or more similar projects, both NPV and IRR can be used to compare them.

The following section describes how this can be done.

## 19. Comparison of appraisal techniques

a) *NPV*.

It is normal to choose that project which has the largest NPV as the most profitable. The NPV technique is most suited to those projects which have a similar pattern of cash flows over the same length of time.

The *advantage* of using this method of appraisal is that it is practical and relevant in that it discounts net cash flows, and (if necessary) further adjustments can be made to take account of factors such as inflation and taxation. Another point in its favour is the fact that it gives results in (real) money terms, which is useful if the scale of the projects (i.e. how much money is involved) is important. A *disadvantage* is that it relies on the choice of a discount factor. In certain cases (for example, where negative cash flows occur during the life of a project), changing the discount rate might change a previous choice of project.

b) *IRR*.

Using this method, the project which has the greatest IRR is chosen. One *advantage* of this method of investment appraisal is that it does not depend on any external rates of interest (remember that the NPV method depends on the choice of a discount rate). A particular *disadvantage* though is the fact that the IRR returns a relative (percentage) value and thus does not differentiate between the scale of projects. That is, one project could involve cash flows in units of £100,000, the other in units of £100 (which might be of significance for some project comparisons). This would be taken no account of using the IRR method.

Note that in most cases where two or more similar projects are being ranked in order of preference, the methods of IRR and NPV will generally agree on the best project, but this is not a hard and fast rule.

## 20. Example 6 (*Comparison of investments*)

*Question*

The following two capital projects, involve the purchase, use and final disposal of two machines *A* and *B*.

| | Initial cost | Net cash flows | | | |
|---|---|---|---|---|---|
| | | Year 1 | Year 2 | Year 3 | Year 4 |
| Machine A | 50000 | 25500 | 24500 | 17000 | 14000 |
| Machine B | 45000 | 12500 | 15500 | 21000 | 38000 |

Note that year 4 includes scrap values of £5000 for machine *A* and £4000 for machine *B*.

Choose between the two projects using each of the following methods in turn:

a) Net Present Value – using a cost of capital of 22% (*and* 28%).

b) Internal Rate of Return – estimate its value using the results of a).

*Answer*

Tables 5 and 6 show the calculations for the required costs of capital.

*Table 5 Cash flows machine A*

| Year | Net flow | Discount factor at 22% | Present value | Discount factor at 28% | Present value |
|---|---|---|---|---|---|
| 0 | (50000) | 1.0000 | (50000.00) | 1.0000 | (50000.00) |
| 1 | 25500 | 0.8197 | 20902.35 | 0.7813 | 19923.15 |
| 2 | 24500 | 0.6719 | 16461.54 | 0.6104 | 14954.79 |
| 3 | 17000 | 0.5507 | 9361.90 | 0.4768 | 8105.59 |
| 4 | 14000 | 0.4514 | 6319.60 | 0.3725 | 5215.00 |
| | | | 3045.40 | | (1801.45) |

*Table 6 Cash flows machine B*

| Year | Net flow | Discount factor at 22% | Present value | Discount factor at 28% | Present value |
|---|---|---|---|---|---|
| 0 | (45000) | 1.0000 | (45000.00) | 1.0000 | (45000.00) |
| 1 | 12500 | 0.8197 | 10246.25 | 0.7813 | 9766.25 |
| 2 | 15500 | 0.6719 | 10414.45 | 0.6104 | 9461.19 |
| 3 | 21000 | 0.5507 | 11564.70 | 0.4768 | 10012.80 |
| 4 | 38000 | 0.4514 | 17153.19 | 0.3725 | 14155.00 |
| | | | 4378.60 | | (1604.75) |

a) At rate 22%, *machine B has the highest NPV and thus would be chosen as best*. (Note also that at rate 28%, machine *B* has the highest NPV.)

b) *For machine A*, $I_1=22$; $N_1=3045.40$; $I_2=28$; $N_2=-1801.45$

$$\text{and estimate of IRR} = \frac{(3045.4)(28) - (-1801.45)(22)}{3045.4 - (-1801.45)} = 25.8\%$$

*For machine B*, $I_1=22$; $N_1=4378.60$; $I_2=28$; $N_2=-1604.75$

$$\text{and estimate of IRR} = \frac{(4378.6)(28) - (-1604.75)(22)}{4378.6 - (-1604.75)} = 26.4\%$$

Thus, machine *B* has the highest rate of return, at 26.4%, and thus would be chosen as best, agreeing with the choice in a). Overall, machine *B* is clearly the best choice.

Of particular value, as noted earlier, is the ability to try out 'what if' scenarios.

The spreadsheets on the opposite page show (with formulae) the layout for the various NPV, and the two separate IRR, calculations of the above worked example. They are laid out slightly differently to the tables above, simply to facilitate formulae presentation. Clearly, as can be seen, this particular presentation is just as clear as the tables above.

Note that the direct calculations involved in the spreadsheet avoid the compounded errors inherent in the practical (via calculator) technique used above which are mainly due to the rounding in the discount factors. This results in the spreadsheet 'PVals' differing slightly to the 'Present Value' figures in Tables 5 and 6.

Ch22-eg6

| | A | B | C | D | E | F | G | H | I |
|---|---|---|---|---|---|---|---|---|---|
| 1 | | | | | | | | | |
| 2 | | IRR est A = | **25.8** | | | | | | |
| 3 | | | 22 | 28 | | | | | |
| 4 | Year | NetFlowA | PValA | PValA | | | | | |
| 5 | 0 | (50,000) | (50,000.00) | (50,000.00) | | | | | |
| 6 | 1 | 25,500 | 20,901.64 | 19,921.88 | | | | | |
| 7 | 2 | 24,500 | 16,460.63 | 14,953.61 | | | | | |
| 8 | 3 | 17,000 | 9,362.02 | 8,106.23 | | | | | |
| 9 | 4 | 14,000 | 6,319.59 | 5,215.41 | | | | | |
| 10 | Total | | 3,043.87 | (1,802.87) | | | | | |
| 11 | | | | | | | | | |
| 12 | | | | | | | | | |
| 13 | | IRR est B = | **26.4** | | | | | | |
| 14 | | | 22 | 28 | | | | | |
| 15 | Year | NetFlowB | PValB | PValB | | | | | |
| 16 | 0 | (45,000) | (45,000.00) | (45,000.00) | | | | | |
| 17 | 1 | 12,500 | 10,245.90 | [illegible],765.63 | | | | | |
| 18 | 2 | 15,500 | 10,413.87 | 9,460.45 | | | | | |
| 19 | 3 | 21,000 | 11,564.84 | 10,013.58 | | | | | |
| 20 | 4 | 38,000 | 17,153.17 | 14,156.10 | | | | | |
| 21 | Total | | 4,377.78 | (1,604.24) | | | | | |
| 22 | | | | | | | | | |

Sheet1 / Sheet1F / Sheet2 / Sheet3

Ch22-eg6

| | A | B | C | D |
|---|---|---|---|---|
| 1 | | | | |
| 2 | | IRR est A = | **=100*(C10*D3/100-D10*C3/100)/(C10-D10)** | |
| 3 | | | 22 | 28 |
| 4 | Year | NetFlowA | PValA | PValA |
| 5 | 0 | -50000 | =B5 | =B5 |
| 6 | 1 | 25500 | =B6/(1+C$3/100)^$A6 | =B6/(1+D$3/100)^$A6 |
| 7 | 2 | 24500 | =B7/(1+C$3/100)^$A7 | =B7/(1+D$3/100)^$A7 |
| 8 | 3 | 17000 | =B8/(1+C$3/100)^$A8 | =B8/(1+D$3/100)^$A8 |
| 9 | 4 | 14000 | =B9/(1+C$3/100)^$A9 | =B9/(1+D$3/100)^$A9 |
| 10 | Total | | =SUM(C5:C9) | =SUM(D5:D9) |
| 11 | | | | |
| 12 | | | | |
| 13 | | IRR est B = | **=100*(C21*D14/100-D21*C14/100)/(C21-D21)** | |
| 14 | | | =C3 | =D3 |
| 15 | Year | NetFlowB | PValB | PValB |
| 16 | 0 | -45000 | =B16 | =B16 |
| 17 | 1 | 12500 | =B17/(1+C$3/100)^$A17 | =B17/(1+D$3/100)^$A17 |
| 18 | 2 | 15500 | =B18/(1+C$3/100)^$A18 | =B18/(1+D$3/100)^$A18 |
| 19 | 3 | 21000 | =B19/(1+C$3/100)^$A19 | =B19/(1+D$3/100)^$A19 |
| 20 | 4 | 38000 | =B20/(1+C$3/100)^$A20 | =B20/(1+D$3/100)^$A20 |
| 21 | Total | | =SUM(C16:C20) | =SUM(D16:D20) |
| 22 | | | | |

Sheet1 / Sheet1F / Sheet2 / Sheet3

It is of some value to vary the two discount rates in order to pinpoint the IRRs more accurately.

That is, in order to home in on IRR A, we can change cells C3 and D3 by small increments while still keeping C10 positive and D10 negative.

For example, putting 25.5 into C3 yields NPV 117.98 in C10 and putting 25.8 into D3 yields NPV –119.61 in D10 as can be seen from the following spreadsheet.

Ch22-eg6

| | A | B | C | D | E | F | G | H | I |
|---|---|---|---|---|---|---|---|---|---|
| 1 | | | | | | | | | |
| 2 | | IRR est A = | **25.6** | | | | | | |
| 3 | | | 25.5 | 25.8 | | | | | |
| 4 | Year | NetFlowA | PValA | PValA | | | | | |
| 5 | 0 | (50,000) | (50,000.00) | (50,000.00) | | | | | |
| 6 | 1 | 25,500 | 20,318.73 | 20,270.27 | | | | | |
| 7 | 2 | 24,500 | 15,555.31 | 15,481.21 | | | | | |
| 8 | 3 | 17,000 | 8,600.38 | 8,539.00 | | | | | |
| 9 | 4 | 14,000 | 5,643.56 | 5,589.92 | | | | | |
| 10 | Total | | 117.98 | (119.61) | | | | | |
| 11 | | | | | | | | | |
| 12 | | | | | | | | | |
| 13 | | IRR est B = | **26.3** | | | | | | |
| 14 | | | 25.5 | 25.8 | | | | | |
| 15 | Year | NetFlowB | PValB | PValB | | | | | |
| 16 | 0 | (45,000) | (45,000.00) | (45,000.00) | | | | | |
| 17 | 1 | 12,500 | 9,960.16 | 9,936.41 | | | | | |
| 18 | 2 | 15,500 | 9,841.11 | 9,794.23 | | | | | |
| 19 | 3 | 21,000 | 10,624.00 | 10,548.18 | | | | | |
| 20 | 4 | 38,000 | 15,318.23 | 15,172.64 | | | | | |
| 21 | Total | | 743.51 | 451.45 | | | | | |
| 22 | | | | | | | | | |
| 23 | | | | | | | | | |

Sheet1 / Sheet1F / **Sheet2** / Sheet3

However, the interesting result is that the IRR B estimate in C13 has now changed to 26.3%, a better estimate than before and almost certainly correct to 1D.

Notice also that the IRR A estimate has changed to 25.6%. Is this the best estimate? With a little extra manipulation (ending with cell C3 at 26.2 and cell D3 at 26.4), the following spreadsheet can be seen to confirm 25.6%!

Ch22-eg6

| | A | B | C | D | E | F | G | H | I |
|---|---|---|---|---|---|---|---|---|---|
| 1 | | | | | | | | | |
| 2 | | IRR est A = | **25.6** | | | | | | |
| 3 | | | 26.2 | 26.4 | | | | | |
| 4 | Year | NetFlowA | PValA | PValA | | | | | |
| 5 | 0 | (50,000) | (50,000.00) | (50,000.00) | | | | | |
| 6 | 1 | 25,500 | 20,206.02 | 20,174.05 | | | | | |
| 7 | 2 | 24,500 | 15,383.22 | 15,334.58 | | | | | |
| 8 | 3 | 17,000 | 8,458.06 | 8,417.98 | | | | | |
| 9 | 4 | 14,000 | 5,519.38 | 5,484.53 | | | | | |
| 10 | Total | | (433.31) | (588.86) | | | | | |
| 11 | | | | | | | | | |
| 12 | | | | | | | | | |
| 13 | | IRR est B = | **26.3** | | | | | | |
| 14 | | | 26.2 | 26.4 | | | | | |
| 15 | Year | NetFlowB | PValB | PValB | | | | | |
| 16 | 0 | (45,000) | (45,000.00) | (45,000.00) | | | | | |
| 17 | 1 | 12,500 | 9,904.91 | 9,889.24 | | | | | |
| 18 | 2 | 15,500 | 9,732.24 | 9,701.47 | | | | | |
| 19 | 3 | 21,000 | 10,448.19 | 10,398.68 | | | | | |
| 20 | 4 | 38,000 | 14,981.19 | 14,886.59 | | | | | |
| 21 | Total | | 66.54 | (124.02) | | | | | |
| 22 | | | | | | | | | |
| 23 | | | | | | | | | |

Sheet1 / Sheet1F / Sheet2 / **Sheet3**

The above procedure for homing in on better and better estimates is not the most efficient way of arriving at answers. There are other (more advanced) 'iterative' techniques that will give more automated and faster results. However, it serves the purpose of demonstrating the 'what if' technique.

## 21. Summary

a) Present Value is a technique that enables a future cash flow to be represented in the equivalent of today's money terms. It is calculated by multiplying the flow by a discount factor.

b) The present value of an interest bearing debt is obtained by:
   i. compounding the debt over the appropriate time period, using the appropriate borrowing rate;
   ii. discounting (using the present value technique) back to today, using the appropriate investment rate.

c) The NPV of a project is found by discounting all cash flows and adding them. The interpretation is:

   NPV > 0 means that the project is profitable.
   NPV = 0 means that the project breaks even.
   NPV < 0 means that the project is not profitable.

d) The IRR of a project is that discount rate that causes the NPV to be zero. It can be interpreted as the rate (of interest) that a project earns.

e) Given the NPVs corresponding to two discount rates for a particular project, the IRR can be estimated, through linear interpolation, either graphically or by using a formula.

f) Investment appraisal, using discounting techniques, involves considering the worth of one or more capital projects using either Net Present Value (NPV) or Internal Rate of Return (IRR) techniques.

## 22. Points to note

a) When estimating the IRR graphically, if both NPV values are positive or both negative, the value of the IRR obtained will lie outside the range of the two given NPVs. Thus, ensure that the horizontal discount rate scale is adequate to 'capture' the estimate.

b) When estimating the IRR either graphically or by formula, a much better estimate is given if:
   i. one NPV is negative and the other positive, and
   ii. they are not too far removed from zero (in relative terms).

c) The present value techniques considered in this chapter have used a discount rate equated to an investment rate. However, sometimes, with capital projects, it is more relevant to equate the discount rate to a borrowing rate, since the capital necessary to fund the project might have to be found from the money market. For this reason, the discount rate used in these circumstances is often referred to as the *cost of capital*, or occasionally, the *worth of money*. For example, suppose £100 is borrowed (at 15%) to invest in a project which returns a single inflow of £115 in a year's time. If we put discount rate = borrowing rate =15%, this will yield an NPV of zero. Thus 'breaking even' in this context means that the return from the project is *just enough to pay the interest on the debt*. If the discount rate is equated to a borrowing rate in this way, the IRR of a project can be interpreted as the maximum borrowing rate that can be tolerated in order that the project be profitable.

## 23. Student self review questions

1. What is Present Value and how is it used? [2]
2. Write down how to calculate the present value of amount $A$ in $n$ year's time using a discount rate of $100i\%$. [3]
3. Why is it useful to know the present value of an interest-bearing debt? [6,7]
4. What methods are available for appraising and comparing capital investments? [8]
5. How is Net Present Value (NPV) calculated for a given project? [9]
6. What is the Internal Rate of Return (IRR) of a project? [14]
7. What methods are available to estimate an IRR? [15,16]
8. Name some factors which might be considered when choosing between NPV and IRR as methods of investment appraisal? [19]

## 24. Student exercises

1. Find the present value of:
   a) £1500 in 3 years at discount rate 15%
   b) £2000 in 5 years at discount rate 12%
   c) £5500 in 4 years if money is worth 11% compounded semi-annually.
2. A debtor can discharge his liabilities by paying £9000 now or £10000 in 2 years' time. If money is worth only 5% compounded quarterly, which is better?
3. What is the present value of a debt of £12000 taken out over 6 years at 14.5% interest, if the discount rate is 9.5%?
4. A retailer is considering buying 10 TV sets at £250 each, which he estimates he can clear (i.e. sell and obtain all monies) in three months. If he is considering borrowing the money at 16% and the discount rate is 12% (both compounded quarterly) and his overheads are 10% of costs, what is the minimum selling price of a set?
5. A second-hand car is being advertised at £1000 now and £1000 in one years time. The dealer offers it to you at £1800 cash. Which should you take and why?
6. The net cash flows on a project are estimated to be as follows:

| Year | 0 | 1 | 2 | 3 | 4 | 5 |
|---|---|---|---|---|---|---|
| Net cash flow (£) | –25000 | 8000 | 12000 | 9000 | 7000 | 7500 |

   Calculate the net present value of the investment, using a discount rate of 14%.
7. A machine that costs £100,000 is expected to have a life of 5 years and then a scrap value of £15000. If its expected net returns are year 1 £20000; year 2 £50000; year 3 £35000; year 4 £35000; year 5 £35000, and projects of this type are expected to return at least 18%, comment on whether the machine should be purchased.
8. Using the project in question 7, obtain the NPVs for the two discount rates 18% and 25% and thus estimate the Internal Rate of Return.
9. A firm is considering two separate capital projects with cash flows as follows:

| Year | 0 | 1 | 2 | 3 | 4 | 5 |
|---|---|---|---|---|---|---|
| Project 1 | (80000) | 18000 | 20000 | 25000 | 38000 | 45000 |
| Project 2 | (120000) | 30000 | 50000 | 50000 | 50000 | 15000 |

a) Using the NPV criterion and a discount rate of 15%, choose the project that is more profitable.
b) Find the NPVs using a discount rate of 20% and use the results to estimate the IRR for each project.
c) Verify that, using the IRR criterion, the decision in a) is reversed and attempt to explain why.

10. A company is thinking of borrowing £70000 to invest in a project which is expected to yield £20000 at the end of each of the next 6 years. If the cost of capital is 20%:
 a) Draw up a discounted cash flow table and hence calculate the NPV of the project.
 b) Interpret the above value of the NPV in the light of the situation.

# 23 Annuities

## 1. Introduction

This chapter describes various techniques associated with fixed payments (or receipts) over time, otherwise known as annuities. The invested value and present value of annuities are described, together with the amortization and sinking fund methods of debt repayment. Also shown is how annuities can be used to depreciate an asset.

## 2. Annuities and their uses

An *annuity* is a sequence of *fixed equal payments* (or receipts) made over uniform time intervals. Some common examples of annuities are: weekly wages or monthly salaries; insurance premiums; house-purchase mortgage payments; hire-purchase payments.

Annuities are used in all areas of business and commerce. Loans are normally repaid with an annuity; investment funds are set up to meet fixed future commitments (for example, asset replacement) by the payment of an annuity; perpetual annuities can be purchased with a (single) lump sum payment to enhance pensions.

## 3. Definition and types of annuity

An **annuity** is a sequence of fixed equal payments (or receipts) made over uniform time intervals.

a) Annuities may be paid:
   i. at the end of payment intervals (an *ordinary* annuity), or
   ii. at the beginning of payment intervals (a *due* annuity)

b) The term of an annuity may:
   i. begin and end on fixed dates (a *certain* annuity), or
   ii. depend on some event that cannot be fixed (a *contingent* annuity)

c) A *perpetual* annuity is one that carries on indefinitely.

## 4. Some aspects and examples of annuities

a) The most common form of an annuity is *certain* and *ordinary*. That is, the annuity will be paid at the end of the payment interval (or 'in arrears') and will begin and end on fixed dates. For example, most domestic hire purchase contracts will involve the payment of an initial deposit and then equal monthly payments, payable at the end of each month, up to a fixed date. Personal loans from banks or finance companies are paid off in a similar manner, but normally without the initial deposit.

b) Annuities that are being invested however are often *due*; that is, paid 'in advance' of the interval. For example, a savings scheme, paid as an annuity (with a bank, building society or insurance company) will not be deemed to have started until the first payment has been made.

c) Standard pension or superannuation schemes can be thought of as two-stage annuities. The first stage involves a due, certain annuity (i.e. regular *payments* into the fund up to retirement age); the second stage being the receipt of a contingent annuity (i.e. regular *receipts* until death).

## 5. Notes on calculations involving annuities

The rest of the chapter describes the standard situations in which annuities are involved. All the mathematical techniques which will be used in annuity calculations *have already been covered* in the previous two chapters. In particular, there are two previous formulae which it is ESSENTIAL to remember. These are:

a) Accrued amount (compound interest): $A_n = P(1 + i)^n$

b) Sum of the first $n$ terms of a gp: $S_n = \dfrac{a(r^n - 1)}{r - 1}$

There are NO OTHER formulae involved.

## 6. Accrued value of an invested annuity

Suppose an annuity of three annual payments of £12,000 is invested in a fund that pays 12%.

The first payment, made at the beginning of year 1, will be invested for three years and thus will amount to £12000(1.12)$^3$.

The second payment, made at the beginning of year 2, will be invested for only two years and will amount to £12000(1.12)$^2$.

The third payment, made at the beginning of year 3, will be invested for only one year and will amount to £12000(1.12).

Thus, the total value of the invested annuity can be expressed in the form:

$$\pounds12000(1.12) + \pounds12000(1.12)^2 + \pounds12000(1.12)^3$$
$$= \pounds12000\,[(1.12) + (1.12)^2 + (1.12)^3] \quad \ldots\ldots [1]$$

The terms inside the square bracket, in [1] above, can be evaluated using either a calculator or compounding tables.

Thus, value of investment: = £12000 [1.12 + 1.2544 + 1.4049]
= £12000(3.7793)
= £45,351.60

This method of calculating the maturity value of the fund is only practical with a relatively small number of time periods (normally expressed in years). In general, there are two other ways of tackling this type of problem:

a) *Using a (year-by-year) schedule*. This entails calculating the value of the fund on a yearly basis, taking into account new payments each year and all accumulating interest. This method is used normally when the value of the fund is needed on a yearly basis but is still only practical (particularly in an examination) for a relatively small number of years.

b) *Using a geometric progression*. The terms inside the square bracket at expression [1] above are evaluated using the formula for the sum of the first $n$ terms of a geometric progression. This method would normally be used when there are a large number of years.

Both of these methods are demonstrated in Examples 1 and 2.

## 7. Example 1 (The value of an *invested annuity* using a *schedule*)

What is the maturity value of a fund paying 9.5% into which 5 (advance) annual payments of £2400 are made? This is easily answered using a schedule, shown at Table 1, which will also give the yearly position of the fund.

*Table 1 Fund value*

| *Year* | *Payment* | *Total in fund* | *Interest* | *Amount in fund (year end)* |
|---|---|---|---|---|
| 1 | 2400 | 2400.00 | 228.00 | 2628.00 |
| 2 | 2400 | 5028.00 | 477.66 | 5505.66 |
| 3 | 2400 | 7905.66 | 751.04 | 8656.70 |
| 4 | 2400 | 11056.70 | 1050.39 | 12107.08 |
| 5 | 2400 | 14507.08 | 1378.17 | 15885.25 |

Thus, the maturity value of the fund is £15885.25

## 8. Example 2 (*Invested annuity* value using a *geometric progression*)

12 monthly payments of £100 are made into a Building Society account which pays a fixed nominal rate of 10.75%, compounded monthly. How much is the account worth at the end of the year? First, we calculate the proportional monthly investment rate as $\frac{0.1075}{12} = 0.00896$.

The *first* payment is invested for the whole 12 months and will accrue to £100$(1.00896)^{12}$

The *second* payment is invested for only 11 months and will accrue to £100$(1.00896)^{11}$

and so on ...

The *last* payment (made at the beginning of month 12) will only be invested for 1 month and will accrue to £100(1.00896)

Thus, the sum of all the accruals (i.e. the value of the fund) will be:

$$£100(1.00896) + £100(1.00896)^2 + ... + £100(1.00896)^{12}$$

$$= £100[(1.00896) + (1.00896)^2 + (1.00896)^3 + ... + (1.00896)^{12}]$$

But the terms inside the square bracket form a gp with $i = r = 1.00896$ (since each term is being multiplied by 1.00896 to obtain the next term).

Their sum is $S_{12} = \frac{(1.00896)\left((1.00896)^{12} - 1\right)}{1.00896 - 1} = 12.7223$ (sum of the terms of a gp).

Therefore, *the value of the fund is*: = £100(12.7223) = £1272.23.

## 9. Net present value of an annuity

If an annuity consists of payments £$A$ over $n$ years subject to a discount rate of 100$i$%, then (assuming an ordinary annuity) the *present values of the 1st, 2nd, 3rd,..., n-th payments* (at the end of the 1st, 2nd, 3rd, ... n-th years) is given by:

$$\frac{A}{1+i}, \frac{A}{(1+i)^2}, \frac{A}{(1+i)^3}, \quad \dots \quad \frac{A}{(1+i)^n},$$

Thus, the net present value of the annuity (adding all the above terms) is:

$$\begin{aligned} P &= \frac{A}{1+i} + \frac{A}{(1+i)^2} + \frac{A}{(1+i)^3} + \quad \dots \quad + \frac{A}{(1+i)^n} \\ &= A\left\{\frac{1}{1+i} + \frac{1}{(1+i)^2} + \frac{1}{(1+i)^3} + \quad \dots \quad + \frac{1}{(1+i)^n}\right\} \end{aligned}$$

The terms in the curly brackets can be calculated using either:

a) a calculator, or b) discounting tables.

Alternatively, if there are many terms to calculate within the curly brackets above, the following formula can be used.

**Net present value of an annuity**

The *net present value* of a regular annuity $A$, paid or received over $n$ years and subject to a discount rate of 100$i$%, is given by:

$$P = \frac{A}{i}\left[1 - \frac{1}{(1+i)^n}\right]$$

If the annuity is *in perpetuity*, the above formula is adjusted to:

$$P = \frac{A}{i}$$

## 10. Example 3 (*Net present value* of an *annuity*)

a) The net present value of £125, payable at the end of each of five years and subject to a discount rate of 8%, is:

$$\begin{aligned} P &= £\left\{\frac{125}{1.08} + \frac{125}{1.08^2} + \dots + \frac{125}{1.08^5}\right\} \\ &= £125\left\{\frac{1}{1.08} + \frac{1}{1.08^2} + \dots + \frac{1}{1.08^5}\right\} \\ &= £125\{0.9259 + 0.8573 + 0.7938 + 0.7350 + 0.6806\} \end{aligned}$$

(from discount tables or calculator)

$= £125(3.9926)$

$= £499.08$

b) The tenants of a rented house have their rent fixed at £1650 per annum *in advance* with immediate effect. They plan to stay in the property for 15 years. Find the total value of the payments (in today's terms), if the average discount rate is estimated at 10%. Notice that this is a *due* annuity.

The net present value of the annuity (rent) can be put into the form:

$$P = £1650 + \frac{£1650}{1.1} + \frac{£1650}{1.1^2} + \ldots + \frac{£1650}{1.1^{14}}$$

$$= £1650\left\{1 + \frac{1}{1.1} + \frac{1}{1.1^2} + \ldots + \frac{1}{1.1^{14}}\right\}$$

But the terms in the bracket form a gp with $r = \frac{1}{1.1} = 0.9091$ and $a$=1.
Note also that there are $n$=15 terms altogether.

Now, $S_n = \frac{a(r^n - 1)}{r - 1}$ giving $S_{15} = \frac{0.9091^{15} - 1}{0.9091 - 1} = 8.3671$ (in the bracket).

Thus, $P$ = £1650(8.3671) = £13805.72.
That is, the total present value of the payments is £13,805.72.

c) The present value of a pension of £12,000 per annum (in perpetuity) at interest rate 7.5% is calculated as:

$$P = \frac{A}{i} = \frac{12000}{0.075} = £160{,}000$$

## 11. Amortization annuity

If an amount of money is borrowed over a period of time, one way of repaying the debt is by paying an *amortization annuity*. This consists of a regular annuity (ordinary and certain) in which each payment accounts for both repayment of capital and interest. The debt is said to be *amortized* if this method is used. Many of the loans issued by banks and building societies for house purchase are of this type, where it is known as a repayment mortgage.

The standard problem is: given a borrowed amount £$P$ subject to interest at 100$i$%, what must the annual payment, $A$, be in order to amortize the debt in exactly $n$ years? The value of $A$ can be found if we consider:

$$P = \frac{A}{1+i} + \frac{A}{(1+i)^2} + \frac{A}{(1+i)^3} + \ldots + \frac{A}{(1+i)^n}$$

$$= \frac{A}{i}\left[1 - \frac{1}{(1+i)^n}\right]$$

and thus we have the following result:

**Amortization annuity payments**

The annual payment, $A$, necessary to amortize a debt, $P$, in exactly $n$ years is given by.

$$A = \frac{Pi}{\left[1 - \frac{1}{(1+i)^n}\right]}$$

## 12. Example 4 (*Amortization* of a debt)

*Question*

A company negotiates a loan of £200,000 over 15 years at 10.5% per annum. Calculate the annual payment necessary to amortize the debt.

*Answer*

Using the result given in the previous section, we need to equate the amount of the debt to the net present value of the annual repayments.

Here, $P = 200{,}000$; $i = 0.105$ (and $1+i = 1.105$); $n = 15$ and $A$ is the annual payment to be found.

*Method 1*

So, $200000 = A\left\{\dfrac{1}{1.105} + \dfrac{1}{1.105^2} + \dfrac{1}{1.105^3} + \ldots + \dfrac{1}{1.105^{15}}\right\}$

But the bracketed expression is a gp with $a = r = \dfrac{1}{1.105} = 0.9050.$

Therefore its sum to 15 terms is $S_{15} = \dfrac{a(r^n - 1)}{r - 1} = \dfrac{(0.905)\left[(0.905)^{15} - 1\right]}{0.905 - 1} = 7.3950$

Therefore, $200000 = A(7.395)$,

giving: $A = 27045.30.$

Thus the annual payment necessary to amortize the debt is £27,045.

*Method 2*

Here, we use the formula.

So, $$200000 = \frac{A}{0.105}\left[1 - \frac{1}{(1.105)^{15}}\right]$$

Rearranging gives: $$A = \frac{0.105(200000)}{\left[1 - \dfrac{1}{(1.105)^{15}}\right]}$$

yielding: $A = 27050.$

Thus the annual payment necessary to amortize the debt is £27,050.

The slight difference in the values obtained from the two different methods is due to rounding errors. The value derived from method 2 is the more accurate.

## 13. Amortization schedule

An *amortization schedule* is a specification, period by period (normally year by year) of the state of the debt. It is usual to show for each year:

a) Amount of debt outstanding at beginning of year,
b) Interest paid,
c) Annual payment, and, optionally,
d) Amount of principal repaid.

For examination purposes, a schedule would only be asked for if the period was relatively short; for example, up to 5 or 6 time periods.

## 14. Example 5 (*A schedule* for an amortized debt)

*Question*

A debt of £5000 with interest at 5% *compounded 6-monthly* is amortized by equal semi-annual payments over the next three years.

a) Find the value of each payment.
b) Construct an amortization schedule.

*Answer*

a) Making the standard time period 6 months, the interest rate is 2.5% with $n$=6 time periods. i.e. P=5000; $n$=6 and $i$=0.025 (1+$i$ = 1.025). We need to put the amount of the debt equal to the net present value of the repayment annuity, $A$.

$$\text{Thus: } 5000 = A\left\{\frac{1}{1.025}+\frac{1}{1.025^2}+\frac{1}{1.025^3}+\frac{1}{1.025^4}+\frac{1}{1.025^5}+\frac{1}{1.025^6}\right\}$$

$$= A[\ 0.97561 + 0.95181 + 0.92860 + 0.90595 + 0.88385 + 0.86230\ ]$$

$$= A[5.50812]$$

$$\text{giving, } A = \frac{5000}{5.50812} = £907.75$$

b) The amortization schedule is shown at Table 2.

Note that:

1. Principal repaid = Payment made – Interest paid
2. Outstanding debt (current year) = Outstanding debt (previous year) – Principal repaid (previous year)

*Table 2 Amortization schedule*

| *6-month period* | *Outstanding debt* | *Interest paid* | *Payment made* | *Principal repaid* |
|---|---|---|---|---|
| 1 | 5000.00 | 125.00 | 907.75 | 782.75 |
| 2 | 4217.25 | 105.43 | 907.75 | 802.32 |
| 3 | 3414.93 | 85.37 | 907.75 | 822.38 |
| 4 | 2592.55 | 64.81 | 907.75 | 842.94 |
| 5 | 1749.61 | 43.74 | 907.75 | 864.01 |
| 6 | 885.60 | 22.14 | 907.75 | 885.61 |
| Balance | (0.01) | | | |

## 15. Sinking funds

A **sinking fund** can be defined as an annuity invested in order to meet a known commitment at some future date.

Sinking funds are commonly used for the following purposes:

a) repayment of debts,
b) to provide funds to purchase a new asset when the existing one is fully depreciated.

## 16. Debt repayment using a sinking fund

Here, a debt is incurred over a fixed period of time, subject to a given interest rate. A sinking fund must be set up to mature to the outstanding amount of the debt.

For example, if £25000 is borrowed over three years at 12% compounded, the value of the outstanding debt at the end of the third year will be: $£25000(1.12)^3$ = £35123.20. If money can be invested at 9.5%, we need to find the value of the annuity, *A*, which must be paid into the fund in order that it matures to £35123.20. Assuming that payments into the fund *are in arrears*, we need:

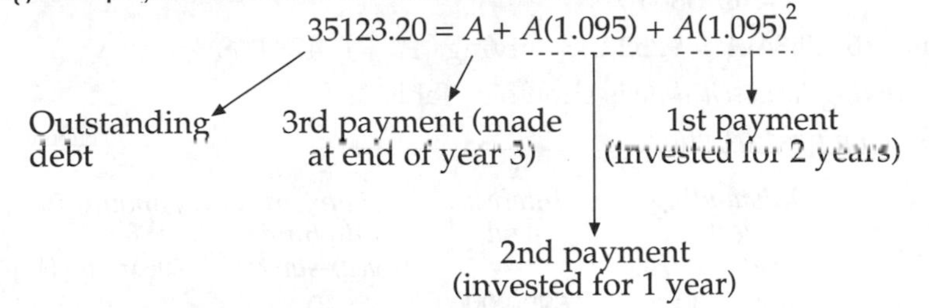


$$\text{i.e.} \quad 35123.20 = A.[1 + 1.095 + 1.095^2],$$

$$\text{giving} \quad 35123.20 = A.[3.2940].$$

$$\text{Therefore,} \quad A = \frac{35123.20}{3.2940} = 10662.78$$

That is, the annual payment into the sinking fund is £10,662.78 (which will produce, at 9.5%, £35,123.20 at the end of 3 years).

## 17. Sinking fund schedule

The standard presentation for the schedule is to show for each year:

a) for the debt:
   i. the outstanding amount
   ii. interest paid
b) for the fund:
   i. the regular payment (deposit)
   ii. interest earned
   iii. amount in fund.

## 18. Example 6 (*Sinking fund payment* and *schedule*)

*Question*

A company borrows £46000, which is compounded at 15%, to finance a new production line. The debt will be discharged at the end of the 5 years with regular annual payments into a sinking fund which pays 11.25%. Calculate the annual payment into the fund and construct a schedule, assuming that the first payment into the fund is made at the *end* of the first year.

*Answer*

The payments into the fund in this case form an ordinary annuity. At the end of 5 years, the debt amounts to £46000$(1.15)^5$ = £92522.43.

The *payments into the fund* will mature to:

$$A + A(1.1125) + A(1.1125)^2 + A(1.1125)^3 + A(1.1125)^4$$
$$= A[1 + 1.1125 + (1.1125)^2 + (1.1125)^3 + (1.1125)^4]$$
$$= A(1 + 1.1125 + 1.23766 + 1.37689 + 1.53179)$$
$$= (6.25884)A$$

Thus: $(6.25884)A = 92522.43$ giving: $A = £14782.68$

The *sinking fund schedule* is shown at Table 3.

*Table 3 Sinking fund schedule*

| Year | *Outstanding debt (year start)* | *Interest paid* | *Payment into fund (year start)* | *Amount in fund (year start)* | *Interest earned* |
|---|---|---|---|---|---|
| 1 | 46000.00 | 6900.00 | 0 | 0 | 0 |
| 2 | 52900.00 | 7935.00 | 14782.68 | 14782.68 | 1663.05 |
| 3 | 60835.00 | 9125.25 | 14782.68 | 31228.41 | 3513.20 |
| 4 | 69960.25 | 10494.04 | 14782.68 | 49524.29 | 5571.48 |
| 5 | 80454.29 | 12068.14 | 14782.68 | 69878.45 | 7861.33 |
| 6 | 92522.43 | | 14782.68 | 92522.46 | |
| Balance | (0.03) | | | | |

## 19. Sinking fund method of depreciation

a) One reason for depreciating an asset is to take proper account of its replacement. Thus one can consider the periodic book payments in respect of depreciation, the *depreciation charge,* as forming a pool that, at the end of the asset's useful life, will fund a replacement or alternative.

b) The *sinking fund method of depreciation* considers the depreciation charge payments as being available for investment into a fund (a *depreciation fund,* paying a market rate of interest) which will mature to some predetermined value. The book value of the respective asset at the end of any year can be determined by subtracting the *current amount in the fund* from the *original book value of the asset.*

c) A *depreciation schedule* can be set up showing, for each year:

  i. payment into fund (depreciation charge),
  ii. interest earned,
  iii. amount in fund,
  iv. current book value of asset.

## 20. Example 7 (Depreciation charge using a *sinking fund method* and preparation of *depreciation schedule*)

*Question*

A machine valued at £12500, with a 6 year life, is estimated to have a scrap value of £450. If the depreciation fund earns 8%:

a) use the sinking fund method *based on an ordinary annuity* to find the annual deposit into the fund (depreciation charge), and
b) prepare a depreciation schedule.
c) During what year does the value of the machine reach 50% of its original value?

*Answer*

The difference between the original value of the machine and its scrap value = £12500 – £450 = £12050, which must be the value of the depreciation fund after 6 years. Interest paid to fund = $i = 0.08$ with $n=6$.

a) We need to find $A$ such that:

$$A.[1 + 1.08 + 1.08^2 + 1.08^3 + 1.08^4 + 1.08^5] = 12050$$

This gives: $$\frac{A(1.08^{6-1})}{1.08-1} = 12050 \text{ [sum of a gp]}$$

Hence: $A.(7.36) = 12050$
giving $A = 1642.60$.
Thus, the annual depreciation charge is £1642.60.

b) The depreciation schedule follows.

*Table 4 Depreciation schedule*

| *Year* | *Depreciation charge* | *Amount in fund* | *Interest earned* | *Book value* |
|---|---|---|---|---|
| 0 | 0 | 0 | 0 | 12500.00 |
| 1 | 1642.60 | 1642.60 | 0 | 10857.40 |
| 2 | 1642.60 | 3416.61 | 131.41 | 9083.39 |
| 3 | 1642.60 | 5332.54 | 273.33 | 7167.46 |
| 4 | 1642.60 | 7401.74 | 426.60 | 5098.26 |
| 5 | 1642.60 | 9636.48 | 592.14 | 2863.52 |
| 6 | 1642.60 | 12050.00 | 770.92 | 450.00 |

c) 50% of the original value of the machine is £12500÷2 = £6250, and from the schedule, this value is first reached during the 4th year.

## 21. Summary

a) An annuity is a sequence of equal payments (or receipts) over uniform time intervals. Annuities can be:
   i. ordinary (paid at end of period)
   ii. due (paid at beginning of period).
b) Annuities normally begin and end on fixed dates (certain), but occasionally depend on an event that cannot be fixed (contingent).
c) Annuities sometimes carry on indefinitely (perpetual).
d) Annuities are commonly used:
   i. to repay debts
   ii. as investments to meet fixed known commitments (known as sinking funds).

e) An amortized debt is one that is repaid with an annuity that takes account of both interest and repayment of principal amount borrowed.

fs) A depreciation fund is an invested annuity which matures to cover the replacement cost of an asset. It is generally thought to be more realistic than straight line and reducing balance methods.

## 22. Student self review questions

1. What is an annuity? [1,3]
2. What is the difference between ordinary and due annuities? [3]
3. Explain what a contingent annuity is and give an example. [4]
4. What is an amortized debt? [11]
5. What is an amortization schedule and how is it used? [13]
6. What is a sinking fund and how is it used? [15]

## 23. Student exercises

1. Find the maturity value of an annuity of £500 invested at 8% compound over 5 years if the annuity is:
   a) ordinary b) due.
2. £25 per month is invested in a fund for a year and attracts interest at 10.5% compounded monthly. How much is the fund worth at the end of the year?
3. *MULTI-CHOICE*. An annual year-end income of £17,000 is required in perpetuity. Assuming a fixed rate of interest of 6% each year, and ignoring administration charges, the sum required now to purchase the annuity is closest to:
   a) £15,400 b) 28,300 c) 154,000 d) 283,000
4. What is the net present value of an ordinary annual annuity of £2500:
   a) over 5 years
   b) over 10 years
   c) indefinitely

   if the investment rate is 11.5%?
5. *MULTI-CHOICE*. A loan of £200,000 is to be paid back with 15 equal year-end payments. If the interest rate is fixed at 6%, the year-end payment will be closest to:
   a) £14,133 b) £20,593 c) £31,954 d) £83,400
6. A debt of £42,800 is amortized over 5 years. If the borrowing rate is 16.5%:
   a) How much are the annual payments?
   b) How much interest is paid on the loan each year?

7. £18,500 is borrowed for house purchase from a Building Society over 25 years at a fixed 11% interest and is to be amortized by equal annual payments.
   a) How much are the annual payments?
   b) If the annual payment is divided by 12 and charged monthly (in arrears), what is the Society's annual gain as a percentage of the original principal borrowed? (Assume the Society can invest at 9% compounded monthly).
8. A debt of £10,000 compounded at 15% is to be discharged over 4 years using a sinking fund method. Find the annual payment (based on an ordinary annuity) if the fund earns 11.5%. Draw up a schedule showing both the position of the debt and the fund each year.
9. A machine costing £15,000 has a life of 5 years, after which time its scrap value will be £550. Using a sinking fund method (where the fund will pay 5% interest on a due annuity basis), calculate the annual depreciation charge and prepare a depreciation schedule.
10. The sinking fund method (with interest at 8% and using an ordinary annuity) is used to depreciate an asset from its purchase price of £40,000 to a scrap value of £5,000 after 4 years. Find the annual depreciation charge and the book value of the asset at the end of the 3rd year.

# Examination examples (with worked solutions)

*Question 1*

A finance director estimates that his company will have to spend £0.25 million on new machinery in two years from now. Two alternative methods of providing the money are being considered, both assuming an annual rate of interest of 10%.

a) A single sum of money, £$A$, to be set aside and invested now, with interest compounded every six months. How much should this single sum be, and what is the *effective* annual rate of interest?

b) £$B$ to be put into a reserve fund every six months, *starting now*. If interest is compounded every six months, what should £$B$ be, in order that the £0.25 million will be available in two years from now?

*CIMA*

*Answer*

a) If £$A$ is invested now, it must be worth £0.25 million in two years time. Take the standard period as 6 months at rate 5%.

Thus: $250000 = A.(1.05)^4$,

giving $A = \dfrac{250000}{(1.05)^4} = £205{,}675.62.$

The *effective annual* rate is $(1.05)^2 - 1 = 0.1025 = 10.25\%$.

b) The standard time period is still 6 months and here £$B$ is the amount of a due annuity (since the first payment is now).

Therefore: $B(1.05) + B(1.05)^2 + B(1.05)^3 + B(1.05)^4 = 250000$

That is, $B[\ 1.05 + 1.052 + 1.053 + 1.054\ ] = 250000$

Giving: $B\dfrac{(1.05)(1.05^4 - 1)}{1.05 - 1} = 250000$

i.e. $B = \dfrac{(250000)(0.05)}{(1.05)(1.05^4 - 1)} = £55{,}240.93$

*Question 2*

a) A company has to repay a loan of £1 million in exactly four years from now. It is planned to set aside nine equal amounts of £$X$ every six months, first one now, each attracting interest at a nominal rate of 12% per annum, but compounded twice a year.

What should $X$ be?

b) A company plans to buy equipment for £100,000, half of which is due on delivery, with the balance due exactly one year later. The year-end cash inflows are expected to be £25,000 per annum, for five years. After exactly five years the equipment will be sold for £10,000.

If the company has to borrow at 14% per annum, analyse whether it is a worthwhile purchase.

*CIMA*

*Answer*

a) The amount set aside now will have value £X.

In 6 months' time, the total set aside will have value $X.r + X = X(r+1)$

In 12 months, the value will be $X(r+1).r + X + X(r^2 + r + 1)$

In 4 years' time, the value will be $X(r^8 + ... + r + 1) = X\frac{(r^9 - 1)}{(r-1)}$

But this must equal £1 million, and 12% 6-monthly gives $r = 1.06$,

and so: $X\frac{(1.06^9 - 1)}{(1.06 - 1)} = 1{,}000{,}000$

giving: $X = £87{,}022.24$

b)

| Time (years) | 0 | 1 | 2 | 3 | 4 | 5 |
|---|---|---|---|---|---|---|
| *PV factor* | *1* | *0.88* | *0.77* | *0.67* | *0.59* | *0.52* |
| Inflow | 0 | 25,000 | 25,000 | 25,000 | 25,000 | 35,000 |
| Outflow | 50,000 | 50,000 | 0 | 0 | 0 | 0 |
| Difference | (50,000) | (25,000) | 25,000 | 25,000 | 25,000 | 35,000 |
| Present value | (50,000) | (22,000) | 19,250 | 16,750 | 14,750 | 18,200 |

Adding up the bottom row of the above table shows that the net present value of the equipment purchase is (£3,050) and so the purchase is *not* worthwhile, i.e. is not earning 14%.

**Note**: Present Value Factors are from Table 11, "Mathematical Tables for Students". More accuracy can be obtained by calculating the factors from the formula stated in the table using a calculator.

*Question 3*

a) A machine costing £25,650 depreciates to a scrap value of £500 in ten years. You are required to calculate:

i. the annual percentage rate of depreciation if the reducing balance method of depreciation is to be used; and

ii. the book value at the end of the sixth year.

b) It is estimated that a mine will yield an annual net return (i.e. after all operating costs) of £50,000 for the next 15 years. At the end of this time the property will be valueless. Calculate the purchase price of the mine to yield a return of 12% per annum.

*CIMA*

*Answer*

a) i. £25650 depreciates to £500 in 10 years.

Thus, using the depreciation formula: $25650(1-i)^{10} = 500$.

Therefore: $1 - i = \sqrt[10]{\frac{500}{25650}} = 0.6745$

giving $i = 0.3255 = 32.55\%$

ii. At the end of the sixth year, the depreciated value is given by:

$25650(0.6745)^6 = £2415.34$

b) Note that there is no replacement fund mentioned here (and no separate fund rate), thus assume that this method is not considered. Hence, the purchase price must be equivalent to the present value of a £50000 annuity over 15 years at 12% discount rate.

$$\text{Thus:} \quad \text{purchase price} = \frac{50000}{1.12} + \frac{50000}{1.12^2} + \dots + \frac{50000}{1.12^{15}}$$

$$= \frac{50000}{1.12^{15}}\left[\frac{1.12^{15}-1}{0.12}\right]$$

$$\text{i.e:} \quad P = £340,543.20$$

# Examination questions

1. A mortgage of £40,000 is to be repaid by 80 equal quarterly instalments (in arrears) of £X. Interest of 4% is charged each quarter on the remaining part of the debt.
   a) Show mathematically that after six months the amount owed is:
   $$£(40{,}000R^2 - RX - X)$$
   where $R = 1.04$.
   b) Find £X, stating why your answer is reasonable.
   c) *Without* carrying out any calculations, briefly explain why the repayments on a mortgage of £80,000 would or would not be £2X.

   **Note**: The sum, S of a geometric progression of $n$ terms, with first term, $A$, and common ratio, $R$, is:
   $$S = \frac{A(R^n - 1)}{R - 1}$$

   *CIMA*

2. Your company uses a machine in its production department which costs £12000 at the beginning of 1983. The machine will be replaced after five years usage by a new machine at the end of 1988. During the five years of operation of the machine it is estimated that the net cash inflows at the beginning of each year will be as follows:

| Year | 1984 | 1985 | 1986 | 1987 | 1988 |
|---|---|---|---|---|---|
| Net cash inflow (£) | 6600 | 6000 | 4500 | (1000) | (2600) |

   *Required*:
   a) If the machine is being purchased with a five year loan, which is compounded annually at 15%, produce an amortization schedule for the five equal annual repayments of the loan.
   b) If the £12000 debt, which is compounded annually at 15%, is to be discharged in 1988 by a sinking fund method, under which equal annual deposits will be made into a fund paying 10% annually, produce the schedule for the sinking fund.
   c) Calculate the net present value of the net cash flows over the five years of operation of the machine at the 10% and 15% discount rates.
   d) Determine the Internal Rate of Return and comment on and compare the three sets of results ignoring taxation with a view to making payment for the machine.

   *ACCA*

3. a) At the beginning of each year a company sets aside £10000 out of its profits to form a reserve fund. This is invested at 10% per annum compound interest. What will be the value of the fund after four years?
   b) A new machine is expected to last for four years and to produce year end savings of £2000 per annum. What is its present value, allowing interest at 7% per annum?
   c) On 1st January 1977, £1000 was invested. It remained invested and, on 1st January of each successive year, £500 was added to it. What sum will have accumulated by 31st December 1981 if interest is compounded each year at 10% per annum?
   d) £20000 is borrowed from a building society, repayable over 20 years at 14% per annum compound interest. How much must be repaid each year?

   *CIMA*

4. A client of your company wishes to take out a mortgage on his property and has asked for a demonstration of the working of two models, one called the Repayment Model and the other called the Endowment Model.

   The mortgage required is for £10,000 over a five year period. In the Repayment Model, interest is charged on the mortgage at the rate of 12% per annum and is discharged by equal annual instalments at the end of each year.

   In the Endowment Model, interest is charged on the mortgage at the rate of 12% per annum and is discharged by one payment at the end of the period of the mortgage. This one payment is achieved by investing in an insurance fund which is growing at a rate of 15% per annum. The five equal annual insurance premiums are paid at the beginning of each year.

   *Required*:

   a) Determine the equal annual repayments for the Repayment Model.
   b) Demonstrate in tabular form the schedule of interest charges and repayments for the Repayment Model.
   c) Determine the annual insurance premium for the Endowment Model.
   d) Demonstrate in tabular form the schedule of interest charges and insurance premiums for the Endowment Model.

   [*Author note*: In the examination, an extract from present value (discounting) tables for both 12% and 15% is given here.]

   *ACCA (part)*

5. a) A company borrows £100,000 on the last day of the year and agrees to repay it by four equal amounts, the repayments being made at the end of each of the following four years. Compound interest at 12% per annum is payable. What is the amount of each repayment?
   b) A new machine will cost £50,000 and has an expected life of 5 years. Scrap value at the end of the fifth year will be £1000. Annual depreciation is to be calculated as a fixed percentage of its current book value. What should the percentage be?
   c) £1000 is invested at 12% per annum, with interest added every quarter-end. How long will it take for the investment to amount to £3000?

   *CIMA*

6. Your company is about to undertake a new project which requires it to invest in a specialized machine costing £75,000. The project will last for five years, after which it is estimated that the scrap value of the machine will be £1250, to be received at the end of the sixth year. It is necessary to inject money into the project at the end of each year to ensure that the machine is kept in proper working order for the next year of the project. The amount needed at the end of the first year is £1000 and it is estimated that this amount needs to be increased by 10% at the end of each succeeding year over the period of the project. The revenue produced from the project through the use of the machine is estimated to be £20000 at the end of the first year, and this will increase by 7.5% at the end of each succeeding year over the period of the project.

*Required*

a) Establish and tabulate the net cash flows for the project.

b) Establish the net present value of the project using a discount rate of 10%.

c) Establish the net present value of the project using a discount rate of 15%.

d) Establish the internal rate of return of the project by interpolation using a diagram.

e) Interpret the meaning of the internal rate of return that you have obtained in d).

[*Author note*: An extract from Net Present Value (discount) tables for 10% and 15% given here.]

*ACCA*

7. In the near future a company has to make a decision about its computer, C, which has a current market value of £15,000. There are three possibilities:

   i. sell C and buy a new computer costing £75,000;

   ii. overhaul and upgrade C;

   iii. continue with C as at present.

   Relevant data on these decisions are given below:

| *Decision* | *Initial outlay* | *Economic life* | *Re-sale value after 5 years* | *Annual service contract plus operating cost (payable annually in advance)* |
|---|---|---|---|---|
| | £ | years | £ | £ |
| (i) | 75,000 | 5 | 10,000 | 20,000 |
| (ii) | 25,000 | 5 | 10,000 | 27,000 |
| (iii) | 0 | 5 | 0 | 32,000 |

   Assume the appropriate rate of interest to be 12% and ignore taxation.

   *You are required,* using the concept of net present value, to find which decision would be in the best financial interest of the company, stating why, and including any reservations and assumptions.

*CIMA*

8. a) Your company has decided to set up a fund for its employees with an initial payment of £2750 which is compounded six monthly over a four year period at 3.5% per six months.

   Required:

   i. Calculate the size of the fund to two decimal places at the end of the four years.

   ii. Calculate the effective annual interest rate, to two decimal places.

   b) The company has purchased a piece of equipment for its production department at a cost of £37,500 on 1st April 1984. It is anticipated that this piece of equipment will be replaced after five years of use on 1st April 1989. The equipment is purchased with a five year loan, which is compounded annually at 12%.

*Required*:

i. Determine the size of the equal annual payments.

ii. Display a table which shows the amount outstanding and interest for each year of the loan.

c) If in b) the £37,500 debt is compounded annually at 12% and is discharged on 1st April 1989 by using a Sinking Fund method, under which five equal annual deposits are made starting on 1st April 1984 into the fund paying 8% annually,

i. determine the size of the equal annual deposits in the sinking fund, and

ii. display a table which demonstrates the growth of the loan and the sinking fund.

[*Author note*: In the examination, an extract from Net Present Value (discount) tables for 8% and 12% given here.]

*ACCA*

9. a) How much is it worth paying for an annuity of £3,000 a year for 10 years, payable twice a year in arrears, assuming an interest rate of 9% per annum and ignoring taxation?

b) A simple mortgage of £50,000 has to be repaid by equal quarterly instalments in arrears over 25 years at an interest rate of 10% a year. What is the amount of each quarterly instalment?

c) A savings scheme specifies payments of £300 a month for five years, with interest guaranteed to be compounded at a minimum rate of 1% a month. What sum will be guaranteed at the end of the five-year term? If inflation is assumed to occur at 5% a year what will be the 'real' value of this sum?

d) For each of (a), (b) and (c) above give a different reason, apart from taxation, why your answer may in practise be only approximate.

*CIMA*

10. A Health and Fitness Centre has to buy one of two types of machine, A or B. Machine A would cost £200,000, half of which would be due on delivery, the remainder a year later. Machine B would cost £240,000, with payment due in the same way as for machine A. Both machines last for 6 years and have an expected scrap value of 10% of their original cost price. Taking into account operating costs and maintenance, machine A would produce year-end net operational cash flows of £40,000 and machine B year-end net operational cash flows of £50,000. In both cases the relevant cost of capital is 10% each year.

*Required*:

a) Calculate the net present value of each machine.

b) Recommend which machine should be bought, giving your reasons and assumptions.

c) Estimate the internal rate of return for your recommended machine, using a graph or calculation, and state its meaning.

*CIMA*

# Part 7 Business equations and graphs

This part of the book deals with methods of graph plotting and of solving standard equations which can occur in Business and Accounting, and how these can be used in business applications.

Linear, quadratic and simple cubic functions and their graphs are covered in chapter 24 and chapters 25 and 26 deal respectively with the forms and methods of solution of linear and quadratic equations.

Chapter 27 introduces the elements of differentiation and integration which are enough to cope with simple applications such as cost, revenue and profit functions and equations which are specifically covered in chapter 28.

# 24 Functions and graphs

## 1. Introduction

A function is a way of describing certain types of quantitative relationships mathematically.

This chapter aims to give an understanding of the more common functional forms that occur naturally in business. It begins by reinforcing linear functions and extends these to quadratic, cubic and some other functions and their graphs.

## 2. Functions and equations

In many areas of business and commerce, functional relationships can enable structures to be understood, controlled and adapted. Assembly line production can be considered as a function of time or numbers of machines or both; sales revenue and production costs are normally functions of level of production; net present value is a function of discount rate.

> **Functions and equations**
>
> A *function* is an expression involving one or more variables.
>
> An *equation* specifies an exact relationship between two functions.

For example, the constituent parts of a manufactured product A, say, might be expressed specifically in terms of the *function* $10w + 2t + 3m$, where $w$ represents wood, $t$ for tin and $m$ miscellaneous material, where the numeric constants (10, 2 and 3) record the part weights.

This relationship could, for example, be used for simple costing purposes. In terms of an *equation*, the relationship might be put into the form:

$$A = 10w + 2t + 3m$$

and translated as: 'the value of $A$ is given by the value of the function $10w + 2t + 3m$'.

However, note that the terms 'expression', 'relationship', 'function' and 'equation' are often used interchangeably, particularly in courses where the mathematics involved is used for business applications rather than as a study in its own right.

## 3. Linear and quadratic equations

The following definitions enable two important types of equations to be recognised.

> **Form of a linear equation**
>
> A linear equation involves no powers of the variable involved greater than the first and takes the general form:
>
> $$y = ax + b$$
>
> where: $x$ is the variable
>
> $a$, $b$ are numeric coefficients.

The coefficients $a$ and $b$ can take any numeric value. For example:

$y = 15x + 10$ ($a$=15, $b$=10);
$y = 2.5x - 11.5$ ($a$=2.5, $b$=–11.5);
$y = 14 - x$ ($a$ = –1, $b$=14);
$y = 20$ ($a$ =0, $b$=20);
$y = 12x$ ($a$=12, $b$=0).

**Form of a quadratic equation**

A quadratic equation involves no powers of the variable involved greater than the second and takes the general form:

$$y = ax^2 + bx + c$$

where: $x$ is the variable
$a, b, c$ are numeric coefficients.

For example:

$y = 5x^2 + 3x - 10$ ($a$=5, $b$=3 and $c$=–10);
$y = 12 + 10x - 4x^2$ ($a$=–4, $b$=10 and $c$=12);
$y = 10x^2 + 20$ ($a$=10, $b$=0 and $c$=20).

The coefficients $a$, $b$ and $c$ can take *any* numeric value (with the exception that $a$ cannot equal zero).

## 4. Special form of a linear function

Linear functions were previously dealt with in chapter 12, where plotting lines and finding the equation of a line given its graph were both covered. In the previous case, the line was always considered in the general form $y=a+bx$. However, it is necessary to consider a slightly quicker method of plotting lines that are given in a particular way. Namely, $ax+by=c$. (Note that the methods mentioned previously can still be used here, but the following techniques are slightly more convenient when the line is in this special form.)

Consider, for example, the line $y = 100 - x$, which can be *rewritten* in the form $x + y = 100$.

Putting $x = 0$, gives $y = 100$. i.e. the line crosses the $y$-axis at 100.
Putting $y = 0$, gives $x = 100$. i.e. the line crosses the $x$-axis at 100.

Since we now know two points through which the line passes, the line can be drawn. This approach can be used for all lines in this special form.

In general:

The straight line equation $ax + by = c$ can be drawn by using the following two facts:

a) the line cuts the $x$-axis at $x = \frac{c}{a}$ b) the line cuts the $y$-axis at $y = \frac{c}{b}$ .

This is sometimes known as the *axis-intersection ratio method*.

Consider another line, $5x + 10y = 200$.

When $x=0$, $10y=200$; that is, $y=200 \div 10=20$. Thus, the line crosses the $y$-axis at $y=20$.

When $y=0$, $5x=200$; that is, $x=200 \div 5=40$. Thus the line crosses the $x$-axis at $x=40$.

The line can again be quickly drawn by joining these two points.

## 5. Example 1 (Plotting *straight lines* of the form $ax+by=c$)

To plot the two lines: $x+6y=120$ and $6x+3y=240$ on the same diagram.

For line 1, the $x$-axis (the line where $y=0$) is crossed at $x=120$ and the $y$-axis (line $x=0$) at $\frac{120}{6} = 20$.

For line 2, the $x$-axis is crossed at $x = 240 \div 6 = 40$ and the $y$-axis at $y = \frac{240}{3} = 80$. Both of these lines are plotted in Figure 1.

*Figure 1 Plotting two lines using axes crossing points*

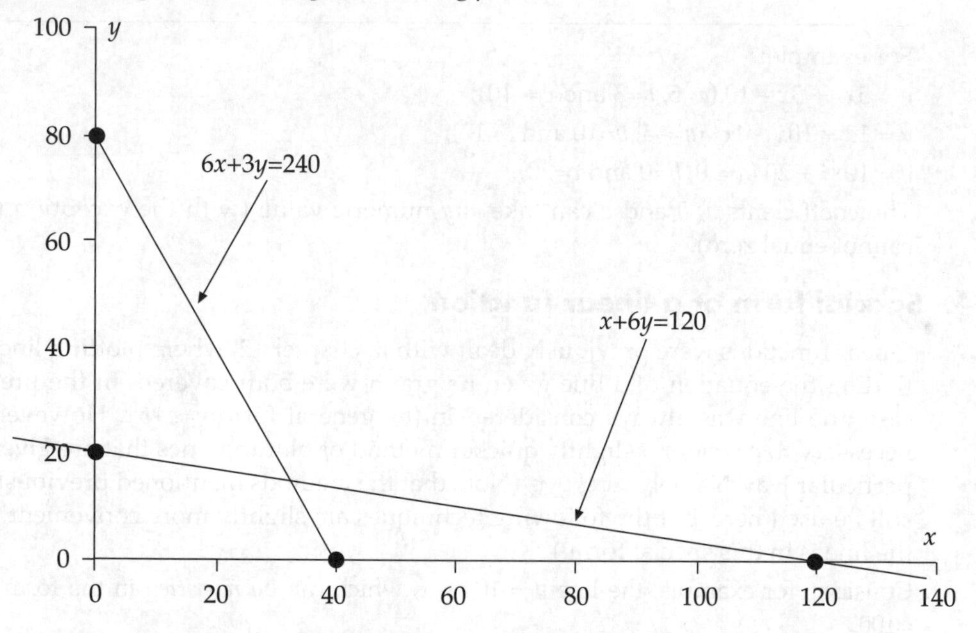


## 6. Quadratic functions and their graphs

a) A quadratic function when plotted on a graph is called a parabola and always takes on the same distinct shape, as shown in Figure 2.
Whether the basic form is 'valley' as in Figure 2(a) or a 'mountain' as in Figure 2(b) depends only on whether the value of $a$ (in $y = a.x^2 + b.x + c$) is positive or negative. Different values of $b$ and $c$ will only move or flatten (or elongate) the curve vertically or horizontally.

b) *Plotting a parabola* consists of:
   1) choosing at least four or five *relevant* $x$-values,
   2) substituting these into the given quadratic function to find the corresponding $y$-values,
   3) plotting these points on a graph, and
   4) joining the points with a *smooth curve*.

*Figure 2 The distinct shape of a parabola*

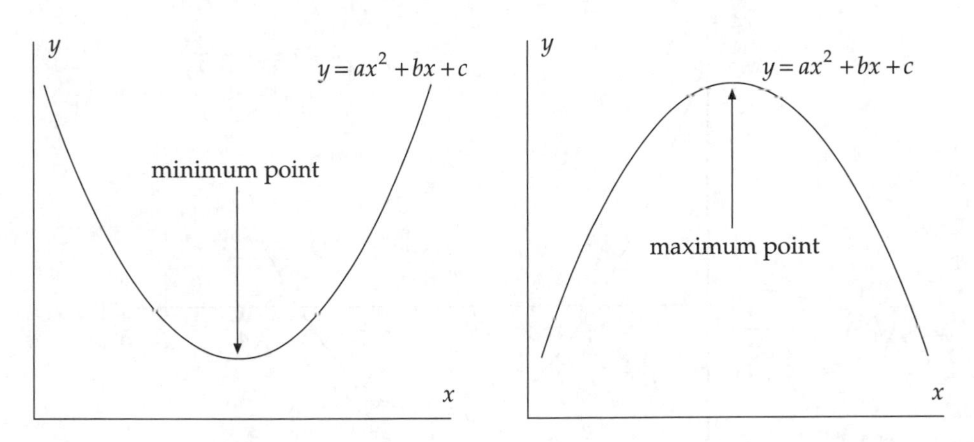


Note that a parabola is always symmetric about the vertical line that passes through the middle of it.

Because of the risk of incorrectly calculating values of $y$ for a quadratic function, it is always wise to deal with the calculations in the way shown in Examples 2 and 3.

## 7. Example 2 (Plotting a *parabola*)

*Question*

Plot the graph of the function $y = x^2 - 4.5x + 3.5$ between $x = 0$ and $x = 4$.

*Answer*

In this case, choosing values of $x$ from 0 to 4 in steps of 1 gives a reasonable number of points. When calculating values of a quadratic function, $y$, for a set of values of $x$, it is wise to break $y$ up into its three parts and calculate the value of each separately. Finally, adding the values of the parts will give the value of $y$. Thus, the function given has been split up into the three parts $x^2$, $-4.5x$ and 3.5 and is shown in Table 1.

*Table 1 Table layout for plotting*

| $x$ | 0 | 1 | 2 | 3 | 4 |
|---|---|---|---|---|---|
| $x^2$ | 0 | 1 | 4 | 9 | 16 |
| $-4.5x$ | 0 | −4.5 | −9 | −13.5 | −18 |
| 3.5 | 3.5 | 3.5 | 3.5 | 3.5 | 3.5 |
| $y=x^2-4.5x+3.5$ | 3.5 | 0 | −1.5 | −1.0 | 1.5 |

The pairs of $x,y$ values calculated in Table 1 [i.e. (0,3.5), (1,0) etc] are plotted as points and joined with a *smooth curve* in Figure 3.

*Figure 3 The plotted graph*

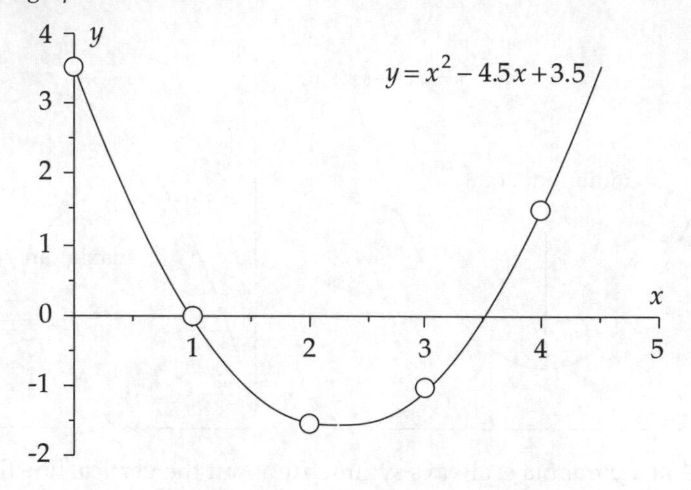


## 8. Example 3 (Plotting a *parabola* and its interpretation)

*Question*

A company's *total profit* (£000) over a particular period is given by the function $17x^2 - 12x - 5x^3$ where $x$ is the number of items produced (in hundreds). If it is known that the maximum production possible for the period is 300 items, plot the company's profit *per unit* curve.

*Answer*

If $17x^2 - 12x - 5x^3$ is the total profit, then $\frac{17x^2 - 12x - 5x^3}{x}$ will give the profit per unit = $y$, say.

That is, $y = 17x - 12 - 5x^2$.

Since it is known that the maximum value of $x$ is 3 (hundred), it seems reasonable to choose values of $x$ from 0 to 3 in steps of 0.5. The calculations necessary to obtain the plotted points are shown in Table 2.

*Table 2 Table layout for plotting*

| x | 0 | 0.5 | 1 | 1.5 | 2 | 2.5 | 3 |
|---|---|---|---|---|---|---|---|
| $17x$ | 0 | 8.5 | 17 | 25.5 | 34 | 42.5 | 51 |
| –12 | –12 | –12 | –12 | –12 | –12 | –12 | –12 |
| $-5x^2$ | 0 | –1.25 | –5 | –11.25 | –20 | –31.25 | –45 |
| $y=17x-12-5s^2$ | –12 | –4.75 | 0 | 2.25 | 2 | –0.75 | –6 |

The $(x,y)$ points in Table 2 have been plotted to obtain the curve shown in Figure 4.

*Figure 4 The plotted graph*

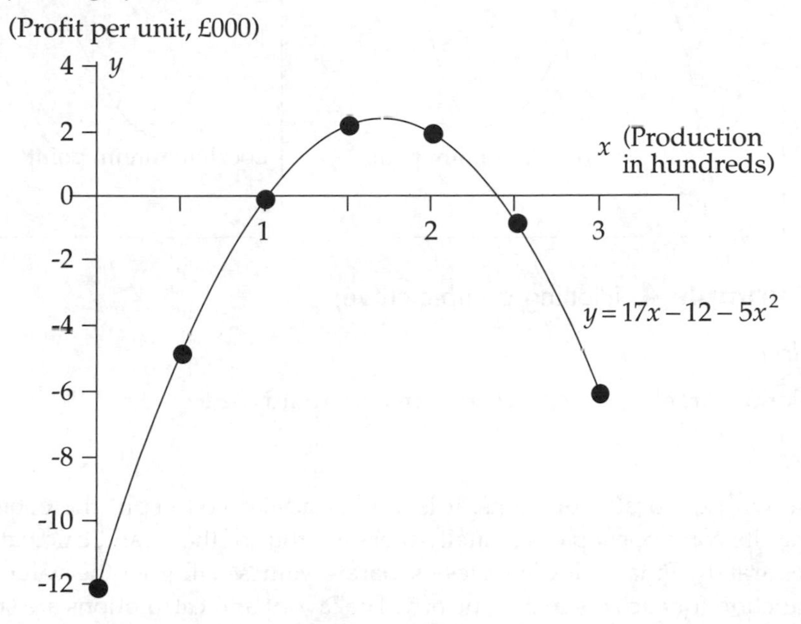


It is clearly profitable to produce within the range 100 to 240, the maximum profit obtainable being about £2,000 at a production level of about 170.

## 9. Cubic functions and equations

a) A cubic equation takes the general form:

$$y = ax^3 + bx^2 + cx + d$$

where $a$, $b$ and $c$ can take any numeric values (other than $a$=0). Note that a cubic function variable has a maximum power of 3.

b) There are certain distinct shapes for a cubic curve that can occur for particular values of $a$, $b$ and $c$. The two most common are shown in Figure 5.

c) The techniques for plotting cubic functions are really no different to those for quadratic functions. Example 4 demonstrates the technique.

*Figure 5 Two distinct shapes for a cubic curve*

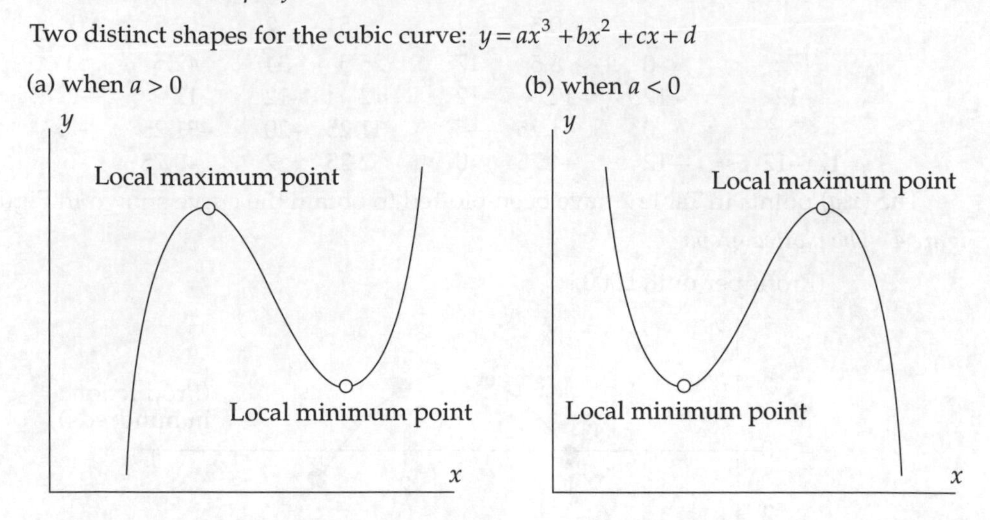


## 10. Example 4 (Plotting a *cubic* curve)

*Question*

Plot the graph of $y = 4x^3 - 12x^2 - 2x + 30$ from $x=-2$ to $x=4$.

*Answer*

As with quadratic functions, it is always advisable to split the cubic function up into its component parts (usually there are four of them) and calculate their values separately. Finally, adding these separate values will give the value of the whole function for each distinct value of $x$. The layout and calculations are shown in Table 3 and the graph in Figure 6.

*Table 3 Table layout for plotting*

| $x$ | –2 | –1 | 0 | 1 | 2 | 3 | 4 |
|---|---|---|---|---|---|---|---|
| $4x^3$ | –32 | –4 | 0 | 4 | 32 | 108 | 256 |
| $-12x^2$ | –48 | –12 | 0 | –12 | –48 | –108 | –192 |
| $-2x$ | 4 | 2 | 0 | –2 | –4 | –6 | –8 |
| $+30$ | 30 | 30 | 30 | 30 | 30 | 30 | 30 |
| $y=4x^3-12x^2-2x+30$ | –46 | 16 | 30 | 20 | 10 | 24 | 86 |

## 11. Polynomials and turning points

a) Linear, quadratic and cubic functions are particular examples of what are called *polynomials*. A polynomial is the most general form of expressing one variable in terms of another. For example:

i. $7x^6 - 3x^2 + 11x$ is a polynomial of *the 6-th order* (or *degree*), since 6 is the highest power of $x$.

ii. $6 - 4x + 3x^3 + 10x^4 - 3x^5$ is a *5-th degree* polynomial.

*Figure 6 The plotted graph*

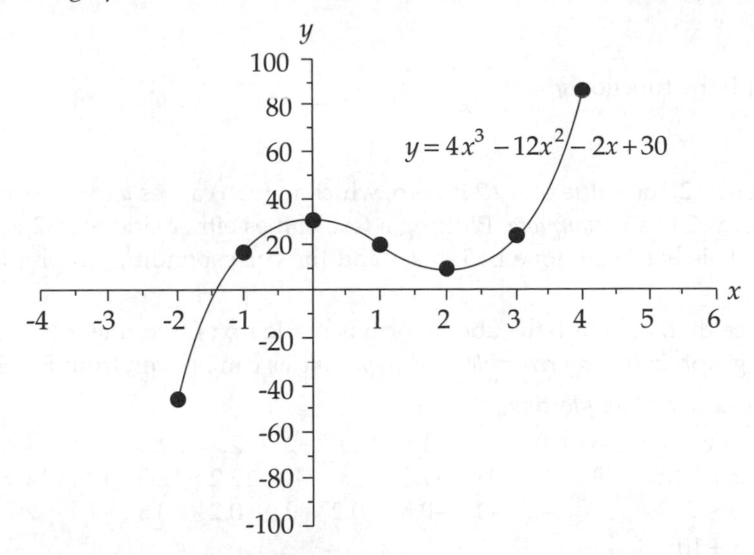


b) All polynomials (except linear functions) have points which are unique, the most important of these reaching either a peak (maximum points) or a trough (minimum points) with respect to points on either side of them. They are called *turning points*. Quadratic functions *always* have just one turning point, either a maximum or minimum. Cubic functions normally have two distinct turning points, one minimum and one maximum.

## 12. Ratio of two linear functions

a) There is another family of functions that is of some interest at the level at which this manual is aimed. They are functions of the form:

$$y = \frac{ax + b}{cx + d}$$

i.e. the ratio of two (in this case, linear) functions.
For example,

i) $y = \dfrac{3x + 10}{x - 3}$ and

ii) $y = \dfrac{4 + 3x}{2x + 4}$

b) The technique for plotting this type of function consists of choosing a reasonable number of relevant $x$-values and calculating the corresponding $y$-values from the given function. However, a little knowledge of the structure of graphs of this form helps a great deal in the plotting and can also help to explain (seemingly) strange plotted points. Because of the divisor (in the above case, $cx+d$), the curve has what are known as *asymptotes*. These are vertical and horizontal straight lines across which the curve can never pass. Example 5 demonstrates the technique for plotting functions of this type.

## 13. **Example 5** (Graph of the *ratio of two linear functions*)

*Question*

Graph the function $y = \dfrac{x+10}{x-2}$

*Answer*

When $x$=2, the value of $x-2$ is zero, which in turn causes $y$ to have an infinite value. Here, $x$=2 is an *asymptote*. Plotting a few values either side of $x$=2 will demonstrate this. This has been done in Table 4 and the corresponding graph plotted in Figure 7.

Notice that, although the above form is made up of two linear functions, the resultant graph *in no way resembles a straight line* as can be seen from Figure 7.

*Table 4 Table layout for plotting*

| $x$ | −1 | 0 | 1 | 1.5 | 1.8 | 2 | 2.2 | 2.5 | 3 | 4 | 5 |
|---|---|---|---|---|---|---|---|---|---|---|---|
| $x+10$ | 9 | 10 | 11 | 11.5 | 11.8 | 12 | 12.2 | 12.5 | 13 | 14 | 15 |
| $x-2$ | −3 | −2 | −1 | −0.5 | −0.2 | 0 | 0.2 | 0.5 | 1 | 2 | 3 |
| $\dfrac{x+10}{x-2}$ | −3 | −5 | −11 | −23.0 | −59.0 | $\infty$ | 61.0 | 25.0 | 13 | 7 | 5 |

*Figure 7 The plotted graph*

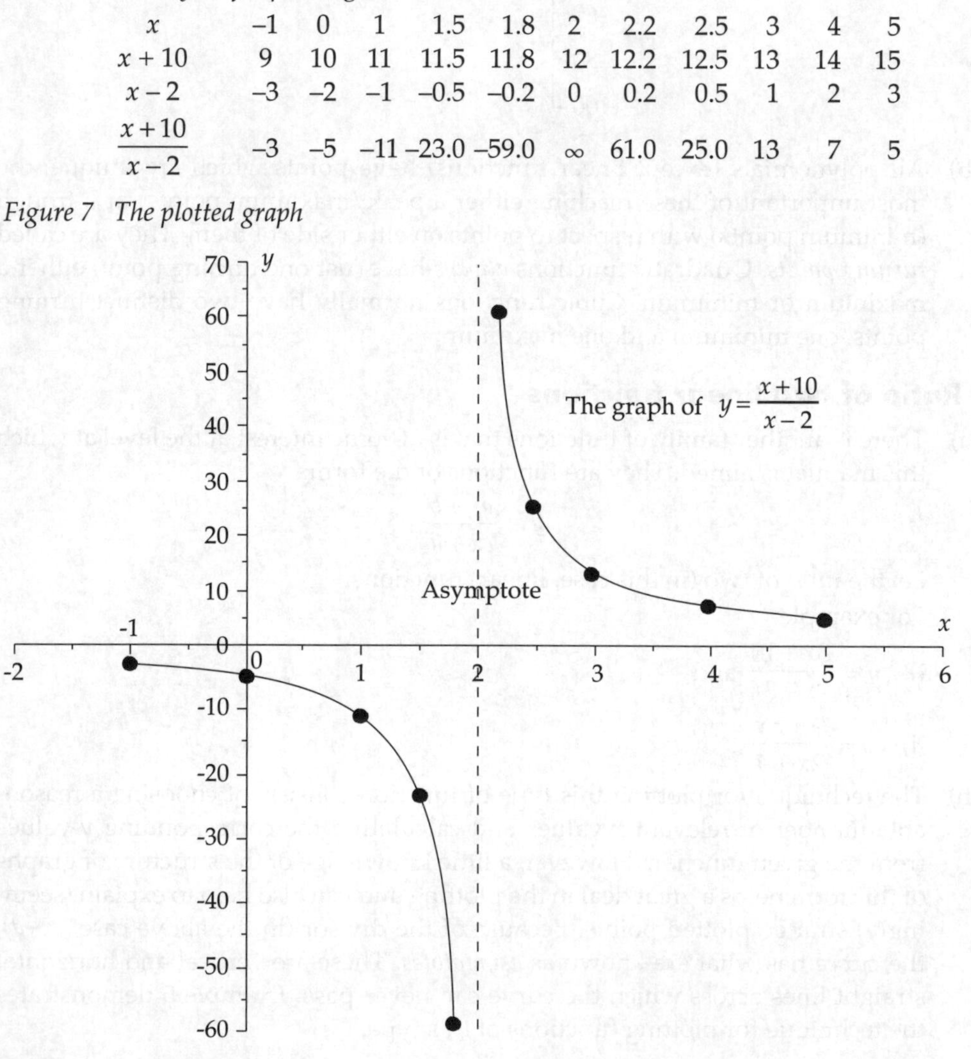

## 14. Points to note on figure 7

a) Notice that the closer $x$ gets to 2, the larger (both positive and negative) $y$ becomes and the closer the curve gets to the asymptote marked at $x$=2 (but will NEVER meet or cross it).

b) Consider what happens when $x$ gets larger in the positive direction. The fraction $(x+10)/(x-2)$ gets closer to 1. This also happens as $x$ gets larger in the negative direction. Thus there is another asymptote at $y$=1 (although this has not been shown in Figure 7).

c) When $y$=0, $x$+10=0, which gives $x$=–10. In other words, the graph crosses the $x$-axis at $x$=–10.

## 15. Example 6 (Plotting two functions and interpreting the results)

*Question*

Two machines, *A* and *B*, are being used to process certain items. The cost function for each machine is:

Machine *A*: $y = 15 + 3x$ Machine *B*: $y = 18 - x + x^2$

where: $y$ = cost of producing $x$ items (£)

and $x$ = number of items processed per hour (hundreds).

If the maximum speed at which both machines can run is 400 items/hour:

a) Plot the graphs of the two cost functions on the same diagram.

b) Use the graphs to find the range of production for which each item is produced more cheaply using:

i. machine *A*

ii. machine *B*.

c) Use the graphs to find the total cost during one hour:

i. of producing 150 items on machine *A*;

ii. of producing 350 items on machine *B*.

*Answer*

a) Tables 5 and 6 show the calculations for the graph plots.

*Table 5*

| $x$ | 0 | 1 | 2 | 3 | 4 |
|---|---|---|---|---|---|
| 15 | 15 | 15 | 15 | 15 | 15 |
| $3x$ | 0 | 3 | 6 | 9 | 12 |
| $15 + 3x$ | 15 | 18 | 21 | 24 | 27 |

*Table 6*

| x | 0 | 1 | 2 | 3 | 4 |
|---|---|---|---|---|---|
| 18 | 18 | 18 | 18 | 18 | 18 |
| –x | 0 | –1 | –2 | –3 | –4 |
| x2 | 0 | 1 | 4 | 9 | 16 |
| 18 –x + x2 | 18 | 18 | 20 | 24 | 30 |

Figure 8 shows the two graphs plotted.

*Figure 8 Graphs for machines A and B plotted for comparison*

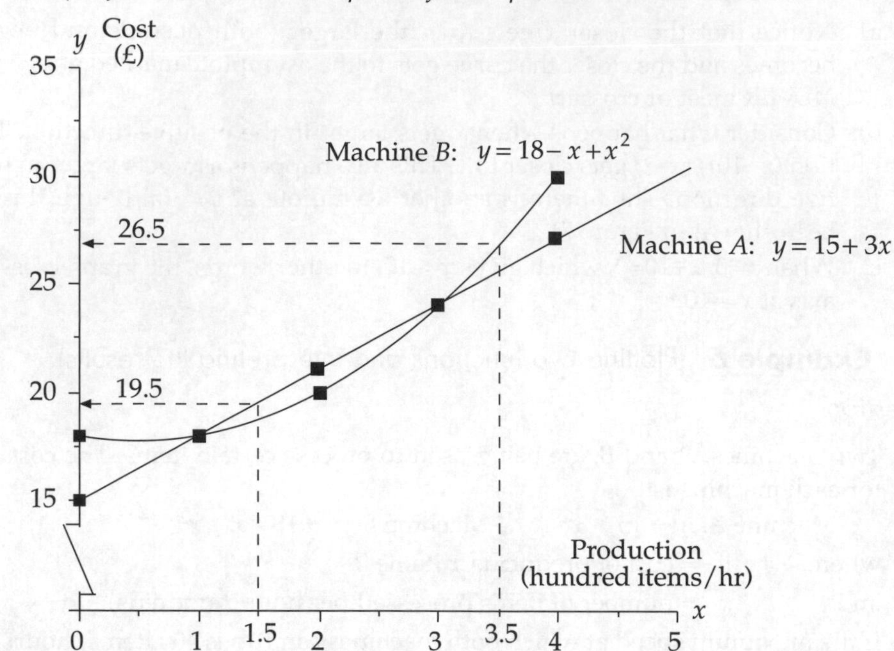


b) From Figure 8, the graphs cross at two places. Where $x$=1 and $x$=3. Thus, for (i), items are produced more cheaply on machine *A* if the rate of production is less than 100 items/hour or more than 300 items/hour. For (ii), items are produced more cheaply on machine *B* if the rate of production is between 100 and 300 items/hour.

c) i. On machine *A*, using the graph, the cost of producing 150 items per hour is £19.50. ii. On machine *B*, using the graph, the cost of producing 350 items per hour is £26.50.

## 16. Summary

a) A mathematical function is a specification of the relationship that exists between two or more variables. Functions specifying one variable ($y$) in terms of just one other ($x$) are the only ones considered in this manual.

b) A linear function takes the general form $y = ax + b$. The graph of a linear function is always a straight line.

c) A quadratic function takes the general form $y = ax^2 + bx + c$. The graph of a quadratic function is called a parabola and takes on a distinctive shape:

   i. 'U-shaped' (or 'valley'), if $a$>0, which includes a distinct turning point – a local minimum.

   ii. inverted 'U-shaped' (or 'mountain'), if $a$<0, which includes a distinct local maximum point where the curve turns.

d) A cubic function takes the general form $y = ax^3 + bx^2 + cx + d$. Generally, the graph will have two distinct turning points, one maximum and one minimum.

e) Plotting curves is accomplished by choosing a number (four to six) of relevant $x$-values, calculating the corresponding $y$-values from the given function, plotting these as points and joining the points to obtain a *smooth* curve. It is essential to perform the calculations for $y$ using a table that splits $y$ into manageable portions.

f) When plotting curves of the form $y = \dfrac{ax+b}{cx+d}$, notice should be taken of the position of the asymptotes (lines across which the curve never passes).

## 17. Points to note

a) If you are required to plot a parabola and no range of $x$-values is given, it is normal to include as much of the area around the turning point as possible. Also, if the $y$-values plotted have one change of sign, then make some more plots in order to include the second change of sign. Similar considerations apply to plotting cubic and other functions.

b) When plotting any graph, other than a straight line, always try to join the plotted points with a smooth curve (not straight line segments). This will ensure that you get a more accurate graph.

c) Plotting polynomials of a higher order than 3 is normally not asked for in examinations at this level, but if it is, the procedure is really no different to that for plotting quadratic and cubic functions.

## 18. Student self review questions

1. What general form does a quadratic function take? [3]
2. What is a parabola? [6]
3. What is a turning point? [11]
4. How many turning points does the graph of a cubic function have? [11]
5. What is an asymptote? [12]

## 19. Student exercises

1. MULTI-CHOICE. Evaluate the expression $\dfrac{20}{0.2} \times \dfrac{1}{r}$ when $r = 100$

   a) 1  b) 10  c) 100  d) 1000

2. Plot the two functions

   i) $y=14x-2$

   ii) $y=12-3x$

   on the same diagram and estimate the $x$ and $y$ values of the point at which they meet.

3. Write down the intersections with the $x$ and $y$-axes respectively for each of the following lines:

   a) $2x+y=200$

   b) $4x+3y=12$

   c) $10x - 5y=-20$

   d) $x=y + 12$

4. Plot the graph of $y=2x^2+5x-3$ between $x=-4$ and $x=2$. For what values of $x$ is the function $2x^2+5x-3$
   i) negative
   ii) positive?
5. By plotting the graphs of
   i) $y=14-3x$ and
   ii) $y=2x^2-12x+14$ between $x=-1$ and $x=5$, find the values of $x$ between which the function $2x^2-12x+14$ has a smaller value than the function $14-3x$.
6. Plot the graphs of
   i) $y=2x^2-4x-5$ and
   ii) $y=9+5x-x^2$

   between $x=-2$ and $x=4$ and find the $x$ values of their points of intersection.
7. a) Multiply $(x+1)(x-2)(x-3)$ out to obtain a cubic function of $x$.
   b) Plot the graph of this function between $x=-1$ and $x=4$ and find the points where the graph crosses the $x$-axis. [Compare these results with the form of the function given in (a)].
   c) Without plotting, where does the graph of $y=x(x-1.5)(x+12)$ cross the $x$-axis?
8. The profit in a week, $y$, (£000) from a certain production process can be expressed in the functional form $y = 9.5x - 7 - 3x^2$, where $x$ is the production in thousands. Plot the graph of the profit carefully between $x=1$ and $x=2.5$ in steps of 0.25 and use it to find:
   i. the 'break even points' (i.e. the level of production where profit = 0), and
   ii. the maximum profit possible and the associated level of production.
9. A haulage contractor has estimated that his business can support up to 6 lorries and the running costs (in £) per lorry per week are given by:

$$y = \frac{200}{x} + 110 + 5x$$

   where $x$ is the number of lorries.
   a) Plot the graph of the above function for relevant values of $x$.
   b) How does the function $200 + 110x + 5x^2$ relate to this situation?
   c) If the business currently runs 5 lorries, estimate their total annual running costs.

# 25 Linear equations

## 1. Introduction

Plotting the graph of a function and solving an equation are two complementary areas. This chapter explains and demonstrates the connection between the two, describing single linear equations and systems of two and three (simultaneous) linear equations. The solutions of these types of equations are covered both algebraically and graphically.

## 2. Types of linear equations

The types of linear equations that are covered in the following sections are classified as follows.

a) *Linear (in one variable).*

The general form of these equations is: $ax + b = 0$.

For example (i) $3x + 6 = 0$ (ii) $x - 2 = 0$ (iii) $6 - 3x = 0$.

Note that equations of this type can appear in other forms. For example:

$$\frac{2x-1}{3} = \frac{x+10}{4}$$

But this can be rearranged if necessary as:

| | | | |
|---|---|---|---|
| | $4(2x - 1)$ | $= 3(x + 10)$ | [by cross multiplication] |
| | $8x - 4$ | $= 3x + 30$ | [multiplying out the brackets] |
| giving finally: | $5x - 34$ | $= 0$ | (which is in the general form). |

Thus, $a$ and $b$ can take any numeric values.

The equation is 'solved' by finding the unique value of $x$ that satisfies the equation (i.e. makes the equation true).

Thus: $x - 2 = 0$ is *solved* by rearranging it as $x = 2$;

$3x + 6 = 0$ is *solved* by rearranging it as $3x = -6$ or $x = -2$; and so on.

b) *Simultaneous linear (in two variables).*

The general form of these equations is:

$$ax + by = c$$
$$dx + ey = f$$

where $a$, $b$, $c$, $d$, $e$ and $f$ can take any numeric values.

For example: (i) $3x - 2y = 12$, $2x + 7y = -4$ (ii) $12x + 5y = 20$, $-3x + 3y = 4$

If simultaneous linear equations are given in any other form, they should always be rearranged into the general form given above.

For example: $y = 12x - 3$ and $5x = 2y + 10$ can be rearranged as:

$$12x - y = 3$$
$$5x - 2y = 10.$$

The equations are *solved* by finding the unique values of $x$ and $y$ that satisfy *both* equations (simultaneously).

c) *Simultaneous linear (in three variables).*

These equations take the general form:

$$ax + by + cz = d$$
$$ex + fy + gz = h$$
$$jx + ky + lz = m$$

where $a, b, c, d, e, f, g, h, j, k, l$ and $m$ can take any numeric values.

For example:
$$3x - 2y + z = 10$$
$$10x + y - 5z = 5$$
$$2x + 3z = -3 \quad \text{(here, the coefficient of } y \text{ is zero)}$$

The equations are *solved* by finding the unique values of $x$, $y$ and $z$ that satisfy all three equations (simultaneously).

## 3. General methods of solution

In this chapter, two methods of solving the types of equation mentioned above will be covered.

a) *Analytical (algebraic) techniques.*

These involve manipulating the given equation(s) algebraically. This may take different forms and the most common will be described. This method will *always give exact solutions to equations.*

b) *Graphical techniques.*

These involve drawing one or more appropriate graphs and finding where they intersect each other or the $x$-axis. This type of method will sometimes give only approximate solutions to equations. Note that, although in 'real life' these techniques are normally alternatives, examiners are quite likely to ask for one or both to be demonstrated.

## 4. Solving linear equations algebraically

Solving this type of equation analytically is normally quite straightforward. For example, to solve $2x-8=0$, proceed as follows:

$$2x - 8 = 0$$

gives: $2x = 8$ (adding 8 to both sides).

Therefore: $x = 4$ (dividing both sides by 2).

Similarly, to solve $4x-3=33$, proceed:

$$4x - 3 = 33$$

then $4x = 36$

and finally $x = 9$.

Sometimes linear equations can take the form of algebraic fractions, which need a more detailed approach.

For example, to solve: $\dfrac{x+1}{1+2x} = \dfrac{1}{3}$,

the following approach can be taken:

$$\frac{x+1}{1+2x} = \frac{1}{3}$$

gives $3(x + 1) = 1(1 + 2x)$ (by 'cross multiplication').

| | | |
|---|---|---|
| Multiplying out gives: | $3x + 3 \quad = 1 + 2x$ | |
| and: | $x + 3 \quad = 1$ | (subtracting $2x$ from both sides). |
| Finally: | $x \quad = -2$ | (subtracting 3 from both sides). |

## 5. Solving linear equations graphically

Consider, as an example, the following two equations:

$y = 0$

and $y = 2x + 3$

The first represents a horizontal straight line, the $x$-axis. The second is a straight line with gradient 2 and $y$-intercept 3. Both of these are shown in Figure 1.

*Figure 1*

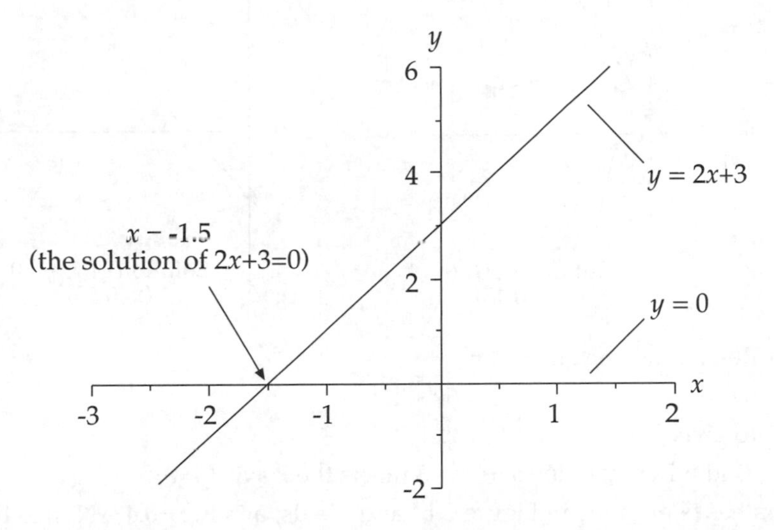


Now, since $y$ is common to both of the above straight lines, algebraically we can combine the two to give:

$$2x + 3 = 0 \quad \text{(step 1)}$$

This equation is easily solved to give:

$$x = -\frac{3}{2} \quad = -1.5 \quad \text{(step 2)}$$

The graphical equivalent of step 1 is to find the meeting point of the two lines; and the graphical equivalent of step 2 is reading off the corresponding $x$ value. In words, the graphical method of solving an equation is to plot the two functions (on either side of the '=' sign) on a graph and to find the $x$-value of their meeting point.

## 6. **Example 1** (Solving a linear equation *graphically*)

*Question*

Solve the linear equation $4 - x = 6x + 1$ graphically.

*Answer*

This can be approached in two different ways.

a) Plot the two lines $y = 4 - x$ and $y = 6x + 1$ and find the $x$ coordinate of their intersection. This has been done in Figure 2(a), yielding $x$=0.43.

*Figure 2*

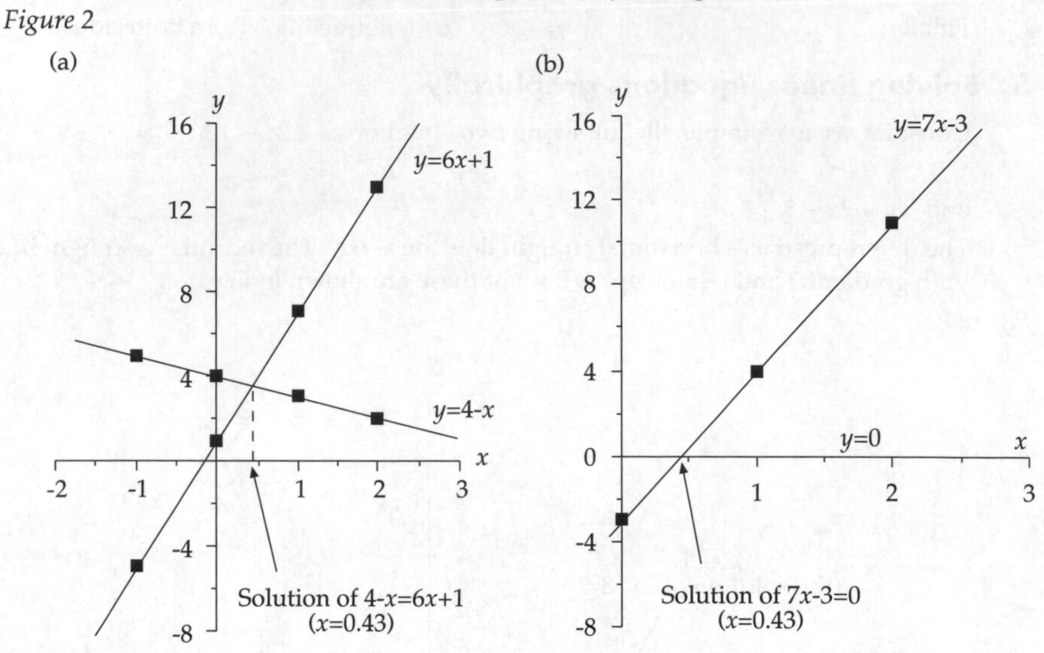


b) Re-arrange the equation:

$$4 - x = 6x + 1$$

to give: $$0 = 7x - 3$$

and find where the line $y = 7x - 3$ meets the $x$ axis ($y$=0).

This has been done in Figure 2(b) and yields, again, $x$=0.43. Notice that this particular method of solution requires far less work than the method demonstrated in (a).

## 7. The existence of solutions of equations

a) Consider the equation $x$+2$y$=6.
There are an infinite number of 'solutions' (i.e. values of $x$ and $y$) that satisfy it: $x$=0, $y$=3; $x$=1, $y$=2.5; $x$=2, $y$=2 and so on. In fact, if the relationship is graphed, all the points that lie on the resulting line *are solutions*.
That is, given a single equation, it is possible to 'solve' it only if there is just one variable involved.

b) When two equations are given, there must be two variables involved ($x$ and $y$ say), in order for a unique solution to exist which satisfies both equations. These are normally referred to as 2×2 (2 by 2) *simultaneous equations*. Only *linear* simultaneous equations are considered in this manual, and their solutions are demonstrated both algebraically and graphically.

c) Similarly equations that involve three variables must come in sets of three before a unique solution is possible. These are normally referred to as 3×3 (3 by 3) simultaneous linear equations and their solutions are only demonstrated using algebraic techniques.

## 8. Manipulating linear equations

In order to solve simultaneous linear equations, it is necessary to understand two simple rules for manipulating linear equations.

> **Rules for manipulating linear equations**
> 1. Any linear equation can be multiplied or divided throughout by any number without altering its truth or meaning
> 2. Any two linear equations can be added or subtracted (one from the other) algebraically to give a third, equally valid, equation.

For example:

i. if $x = 3$
then: $2x = 6$ (multiplying throughout by 2)
and: $-5x = -15$ (multiplying by –5)

ii. if $9x + 6y = 30$
then: $3x + 2y = 10$ (dividing throughout by 3)

iii. if $2x - y + 4z = 3$
then: $10x - 5y + 20z = 15$ (multiplying by 5).

iv. if: $2x - 3y = 5$ [1]
and: $x + 4y = 12$ [2]
then: $3x + y = 17$ (adding [1] and [2] together)
and: $x - 7y = -7$ (subtracting [2] from [1])

The reason for needing to be able to manipulate equations in the way demonstrated above is explained in the following section.

## 9. Procedure for solving 2x2 simultaneous equations

The procedure for solving 2x2 simultaneous equations algebraically is:

STEP 1 *Eliminate one of the variables* (using any of the techniques illustrated in section 8.)

STEP 2 *Solve* the resulting simple equation (to yield the value of the other variable).

STEP 3 *Substitute* this value back into one of the original equations, equation 1 say (to yield the value of the first variable).

STEP 4 *Check* the solutions (by substituting both values into original equation 2).

This technique is sometimes referred to as the *method of elimination*.

Some examples of this procedure follow.

## 10. Example 2 (Solving 2x2 simultaneous equations *algebraically*)

*Question*

Solve the simultaneous equations: $y = 3x + 5; 2y + 3x = 28$ for $x$ and $y$.

*Answer*

First, re-arrange the equations into the general form:

$$3x - y = -5 \quad [1]$$
$$3x + 2y = 28 \quad [2]$$

Since both equations already have a '$3x$' term, subtracting one from the other will eliminate $x$.

[1] – [2] gives: $-3y = -33$.
and hence: $y = 11$.
Substituting back into [1] with $y$=11 gives: $3x - 11 = -5$
which yields: $x = 2$.
The solutions to the equations are thus $x$=2 and $y$=11.
[Check in equation [2]: $3(2) + 2(11) = 6 + 22 = 28$. OK]

## 11. Example 3 (Solving 2x2 simultaneous equations *algebraically*)

*Question*

Solve, for $x$ and $y$, the equations:

$$7x - 3y = 41 \quad [1]$$
$$3x - y = 17 \quad [2]$$

*Answer*

Notice immediately that there are no $x$ and $y$ terms that are the same, but multiplying equation [2] by 3 will give both equations a '$-3y$' term.

Thus: $7x - 3y = 41$ [1]
and [2] × 3 gives: $9x - 3y = 51$ [3]
Now, [3] – [1] gives: $2x = 10$.
Therefore $x = 5$.
Substituting for $x$=5 into equation [1] gives: $35 - 3y = 41$.
Hence $y = -2$.
The solutions to the equations are $x$=5 and $y$=–2.
[Check in equation [2]: $3(5) - (-2) = 15 + 2 = 17$. OK.]

## 12. Example 4 (Solving 2x2 simultaneous equations *algebraically*)

*Question*

Find the values of $x$ and $y$ if: $4x-3y=18.5$ and $4y=7x-35.5$.

*Answer*

Re-arranging gives:

$$4x - 3y = 18.5 \quad [1]$$
$$-7x + 4y = -35.5 \quad [2]$$

There is no simple way to manipulate just one equation in this particular case. However, both equations can be made to contain the term '$12y$' as follows:

| | | |
|---|---|---|
| [1] × 4 gives: | $16x - 12y = 74.0$ | [3] |
| and [2] × 3 gives: | $-21x + 12y = -106.5$ | [4] |
| [3] + [4] gives: | $-5x = -32.5$ | |
| Therefore: | $x = 6.5$ | |

Substituting back into equation [1] for $x$=6.5 gives: $4(6.5) - 3y = 18.5$

| | |
|---|---|
| Therefore: | $26 - 3y = 18.5$ |
| which gives: | $3y = 7.5$ |
| and so | $y = 2.5$. |

The solutions to the equations are $x$=6.5 and $y$=2.5.

[Check in equation (2): $-7(6.5) + 4(2.5) = -45.5 + 10 = -35.5$. OK.]

## 13. Procedure for solving 2×2 simultaneous equations graphically

The procedure for solving 2×2 simultaneous equations graphically is:

STEP 1 *Graph* the two lines represented by the two given equations.

STEP 2 *Identify the x and y values* at the intersection of the lines.

These $x$ and $y$ values are the required solution.

## 14. Example 5 (Solving 2×2 simultaneous equations *graphically*)

*Question*

Solve the equations: $3x-y=2$ and $x-2y=-8$ graphically.

*Answer*

First, the two lines need to be plotted.

Now, $3x-y=2$ meets the $x$-axis (i.e. where $y$=0) at $x=\frac{2}{3}$ and the $y$-axis at $y=-2$.

$x-2y=-8$ meets the $x$-axis at $x=-8$ and the $y$-axis at $y=4$. ($x=-8$ is slightly impractical for plotting so choose, say, $x=4$, which yields $y=6$).

These two lines are drawn in Figure 3, where the solution of the equations is seen to be $x=2.4$ and $y=5.2$.

## 15. Example 6 (To construct 2×2 *simultaneous equations* and solve them *both algebraicly and graphically*)

*Question*

A company manufactures two products X and Y by means of two processes A and B. The maximum capacity of process A is 1750 hours and of process B 4000 hours. Each unit of product X requires 3 hours in A and 2 hours in B, while each unit of product Y requires 1 hour in A and 4 in B. Use the algebraic method to calculate how many units of products X and Y are produced if the maximum capacity available is utilized. Demonstrate the situation graphically.

*Figure 3*

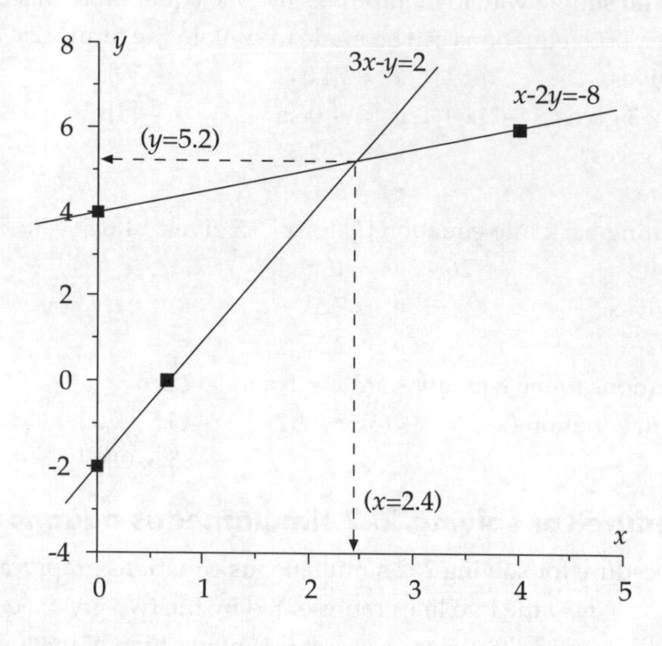


## Answer

Let $x$ be the number of units of product X, and $y$ be the number of units of product *Y*.

For process *A*, each of the $x$ items require 3 hours and thus the total time for all the X products is $3x$ hours. Each of the $y$ items require 1 hour, giving a total time of $1y$ (or just $y$) hours for all the *Y* products.

Thus, since the maximum capacity of process *A* is 1750 hours (and all this capacity *must be used*), we have that:

$$3x + y = 1750 \quad [1]$$

Similarly, for process *B*: $$2x + 4y = 4000 \quad [2]$$

*Algebraic solution*

[1] and [2] form a pair of simultaneous linear equations, which can be solved uniquely for $x$ and $y$.

[1] × 4 gives: $12x + 4y = 7000$ [3]

[3] – [2] gives: $10x = 3000$

i.e. $x = 300$.

Substitute in [1] for $x = 300$: $3(300) + y = 1750$

i.e. $y = 850$

That is: $x=300$ and $y=850$

[Check for $x=300$ and $y=850$ in (2): $2(300)+4(850) = 600+3400 = 4000$ (OK)]

Hence, 300 units of *X* and 850 units of *Y* should be produced if the maximum capacity available is utilized.

*Graphical solution*

The graphical interpretation of this procedure is first to plot the two lines:

$3x + y = 1750$ (cuts $x$-axis at $1750 \div 3 = 583$ and $y$-axis at 1750)

$2x + 4y = 4000$ (cuts $x$-axis at $4000 \div 2 = 2000$ and $y$-axis at $4000 \div 4 = 1000$)

and second to find their point of intersection giving the values of $x$ and $y$ required. Figure 4 below shows these values to be $x=300$ and $y=850$.

*Figure 4*

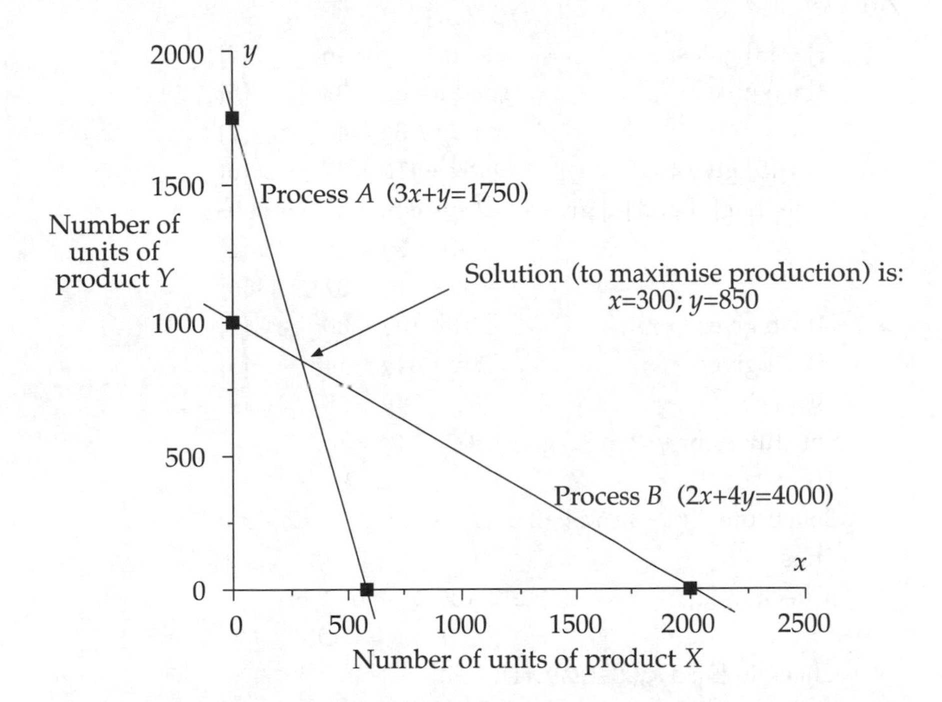


Note that the graphical solution of $x=300$ and $y=850$ seen in Figure 4 is of course the same as that obtained algebraically.

## 16. Procedure for solving 3×3 simultaneous equations algebraically

The procedure for solving 3×3 simultaneous equations algebraically is:

STEP 1 Using any two of the given equations, *eliminate one of the variables* (using the equation-manipulating techniques previously described) to obtain an equation in two variables.

STEP 2 Using another pair of equations, *eliminate the same variable* as in STEP 1, which will give a second equation in two variables.

STEP 3 *Solve* this 2×2 system of equations in the normal way.

STEP 4 *Substitute* into one of the three original equations to find the value of the third variable.

STEP 5 *Check* the solutions.

## 17. Example 7 (Solving 3x3 *simultaneous linear equations*)

*Question*

Solve the 3×3 system of equations:

$$3x - y + z = 5 \quad [1]$$
$$2x + 2y + 3z = 4 \quad [2]$$
$$x + 3y - z = 11 \quad [3]$$

*Answer*

[1] + [3] gives: $4x + 2y = 16$ [4]

[3] × 3 gives: $3x + 9y - 3z = 33$ [5]

$2x + 2y + 3z = 4$ [2]

[2] + [5] gives: $5x + 11y = 37$ [6]

Note that [4] and [6] give a 2×2 system, which is solved as:

$4x + 2y = 16$ [4]

$5x + 11y = 37$ [6]

[4] × 5 gives: $20x + 10y = 80$ [7]

[6] × 4 gives: $20x + 44y = 148$ [8]

[8] – [7]: $34y = 68$ i.e. $y=2$.

Substitute for $y=2$ in [4] gives: $4x + 2(2) = 16$.

Thus $x = 3$.

Substitute for $x=3$ and $y=2$ in [1] gives: $3(3) - 2 + z = 5$

Thus $z = -2$.

{Check in [2]: $2(3)+2(2)+3(-2) = 6+4-6$
$= 4$. OK.

Check in [3]: $3+3(2)-(-2) = 11$. OK.}

Thus solution is $x=3$, $y=2$ and $z=-2$.

## 18. Summary

a) A linear equation takes the form $ax + b = 0$ and has a unique solution.

b) Simultaneous linear equations can take one of two forms:

i. 2×2 systems: $ax + by = c$
$dx + ey = f$

where $a, b, c, d, e$ and $f$ can take any numeric values.

ii. 3×3 systems: $ax + by + cz = d$
$ex + fy + gz = h$
$jx + ky + lz = m$

where $a, b, c, d, e, f, g, h, j, k, l$ and $m$ can take any numeric values.

c) Equations can be solved analytically (i.e. algebraically) or graphically to give the same result, although some graphical solutions may only be approximate.

d) The following manipulations are usually necessary when solving simultaneous equations algebraically.
   i. Any linear equation can be multiplied or divided throughout by any number without altering its truth or meaning.
   ii. Any two linear equations can be added or subtracted algebraically to give a third valid equation.

e) The method of elimination for solving a 2×2 system involves eliminating one variable to find the value of the other and substituting back to find the value of the first. Always check solutions.

f) Solving a 2×2 system graphically involves plotting each of the given equations (lines) and identifying their point of intersection as the solution.

g) Only an algebraic solution is possible for a 3×3 system of linear equations, involving a similar (but extended) technique to that for a 2×2 system solution.

## 19. Points to note

a) The method of elimination is not the only way of solving 2×2 systems of equations algebraically. The *method of substitution* is another and a brief example follows.

To solve: $3x+2y=-3$ ...[1] and $2x-y=8.5$ ...[2]

From [2]: $y=2x-8.5$, and substituting for $y$ in [1] gives:

$3x + 2(2x-8.5) = -3$ giving $3x + 4x - 17 = -3$. i.e. $7x=14$ giving $x=2$.

Substituting for $x=2$ in [2] gives: $y = 2(2) - 8.5$. Thus $y=-4.5$.

However this method can involve some awkward algebra and the method of elimination is generally considered simpler.

b) *Ratios* of the values of variables can be given instead of clearly defined equations.

For example:

$$x:y = 2:3$$

is equivalent to:

i. $\frac{x}{y} = \frac{2}{3}$

or ii. $3x = 2y$

or iii. $3x - 2y = 0$

For example, to solve the system:

$$3x-2y = 11;$$
$$x:y = 5:2$$

i. we would translate $x:y = 5:2$ as $2x - 5y = 0$;

ii. then we would have the two equations: $3x-2y = 11$; $2x-5y = 0$ to solve in the usual manner.

c) *Three-part ratios* of the form $x:y:z$ = a:b:c can be made into only *two unique equations*.

For example $x:y:z = 2:4:5$ is equivalent to $4x=2y$ and $5y=4z$.

The third part of the ratio, $5x=2z$, is not unique, being obtainable from the other two.

For example, to solve: $3x–2y–z = 3; x:y:z = 1:2:4$

i. $x:y:z = 1:2:4$ can be translated to: $y = 2x$ (or $2x–y = 0$) and $z = 2y$ (or $2y–z = 0$)

ii. we would then have the three equations:

$$\begin{aligned} 3x - 2y - z &= 3; \\ 2x - y &= 0; \\ 2y - z &= 0 \end{aligned}$$

to solve in the usual manner.

## 20. Student self review questions

1. What is the difference between analytical and graphical solutions to equations? [3]
2. Explain in words how the equation $ax+b=0$ can be solved graphically. [5]
3. Why cannot the equation $y=ax+b$ be solved uniquely? [7(b)]
4. What is the 'method of elimination' for solving a 2×2 system of linear equations? [9]
5. How are 2×2 systems of linear equations solved graphically? [13]

## 21. Student exercises

1. *MULTI-CHOICE*. The straight lines $y=3x–2$ and $y=x+2$ intersect at which one of the following points:

   a) $x=2, y=4$ b) $x=0, y=–2$ c) $x=1, y=1$ d) $x=4, y=2$

2. Solve the following linear equations for $x$.

   a) $3x–6 = 12$ b) $2x+2 = 3x–11$ c) $5+4x = x+2$

   d) $1.2x–0.8 = 0.8x+1.2$ e) $5(x–2) = 15$ f) $3(x–1.5)+2(2x+3) = 5$

   g) $4(x–5) = 7–5(3–2x)$ h) $\frac{x}{5} = \frac{x+1}{3}$ i) $\frac{x}{5}+\frac{x}{3} = 2$

   j) $\frac{x+1}{3}+\frac{x-1}{2} = \frac{3}{2}$ k) $\frac{4}{x-2}+\frac{4}{3} = \frac{8}{3}$

3. Plot the graphs of $y=x+2$ and $y=4x–7$ on the same diagram and use them to solve the equation $x+2=4x–7$.
4. Draw the graph of $3x+4y=12$ and use it to find the value of $y$ that satisfies the equations $3x+4y=12$ and $x=2$.
5. Use the method of elimination to find the values of $x$ and $y$ that satisfy the following simultaneous equations:

   (a) $x+y=4$ (b) $3x+2y=19$ (c) $3x+2y=7$ (d) $x+3y=7$ (e) $7x–4y=37$

   $3x–y=8$ $3x–3y=–6$ $x+y=3$ $2x–2y=6$ $6x+3y=51$

6. Use a graphical method to solve the following simultaneous equations:

   (a) $3x+5y = 30$ (b) $10x+12y = 48$ (c) $x–y = 0$ (d) $y = 6x+1$

   $2x+2y = 16$ $5x+10y = 30$ $4x+6y = 40$ $3y = 1–2x$

7. A bill for £74 was paid with £5 and £1 notes, a total of 50 notes being used. How many £5 notes were there? (Solve this mentally first, then algebraically.)

8. An invoice clerk receives a bill for £37.50 for ten blank ledger books and three special filing trays. On phoning the supplier (after discovering the order had been written out incorrectly) the clerk agrees to return three of the ledger books in return for the supplier sending an extra filing tray. Given that there will be an extra £2.50 to pay and that there is a 10% discount for orders of 10 ledger books or more:
   a) derive linear equations in $x$ and $y$ to represent the components of the original and revised invoices, where $x$ is the price of a single ledger book and $y$ is the price of a filing tray, and
   b) solve the two equations simultaneously for $x$ and $y$.
9. A furniture factory manufactures two types of coffee table, $A$ and $B$. Each table goes through two distinct costing stages, assembly and finishing. The maximum capacity for assembly is 195 hours and for finishing, 165 hours. Each $A$ table requires 4 hours assembly and 3 hours finishing, while a $B$ table requires 1 hour for assembly and 2 hours for finishing. Calculate the number of $A$ and $B$ tables to be produced to ensure that the maximum capacity available is utilized.
10. Solve the following systems of linear equations for $x$, $y$ and $z$.

| (a) $3x+2y-z=-1$ | (b) $3x+y-z=10$ | (c) $y:z=2:1$ |
|---|---|---|
| $x+y+z=6$ | $x+2y+z=7$ | $10x+y=0$ |
| $3x+y+2z=15$ | $x-z=3$ | $5x+y+2z=15$ |

11. A company manufactures three products X, $Y$ and Z, each of which must go through three processes $A$, $B$ and $C$ for the following times:

| *Product* | *Time spent in process* | | |
|---|---|---|---|
| | *A* | *B* | *C* |
| X | 3 | 3 | 1 |
| Y | 3 | 2 | 3 |
| Z | 2 | 0 | 1 |

The maximum capacities of processes $A$, $B$ and $C$ are 130, 85 and 60 respectively. Calculate the number of units to be produced of products X, $Y$ and Z to ensure the utilization of maximum capacity.

# 26 Quadratic and cubic equations

## 1. Introduction

This chapter defines the general form of quadratic equations, their links with plotted parabolas and both algebraic and graphical methods of obtaining their solutions. The form of cubic equations is covered together with their solution using a graphical method only.

## 2. Quadratic equations

a) A *quadratic equation* (in one variable) can always be put into a general form.

**General form of a quadratic equation**

$$ax^2 + bx + c = 0$$

where $a$, $b$ and $c$ can take any numeric values.

b) For example:

i. $2x^2 + 3x - 10 = 0$ (here, $a$=2, $b$=3 and $c$=–10)

ii. $15 - 3x2 = 0$ (here, $a$=–3, $b$=0 and $c$=15)

iii. $0.4 - 11.6x + 2.9x^2$ (here, $a$=2.9, $b$=–11.6 and $c$=0.4

iv. $\frac{x}{4} - \frac{-3}{x} = 2$ can be re-arranged to give $\frac{x^2 - 12}{4x} = 2$

i.e $x^2 - 12 = 8x$ or, in the general form, $x^2 - 8x - 12 = 0$

c) A quadratic equation is *solved* by finding the values of $x$ that satisfy it. There are, at most, two solutions (or roots as they are sometimes called). Note that in a quadratic equation the variable has a maximum power of 2.

## 3. Example 1 (Identifying a *quadratic* equation)

$\frac{3}{x-3} = \frac{x}{4+2x}$ is a quadratic equation as can be seen from the following.

Cross multiplying gives: $3(4+2x) = x(x-3)$

multiplying out: $12 + 6x = x^2 - 3x$

and rearranging gives: $x^2 - 9x - 12 = 0$ (with $a$=1, $b$=–9 and $c$=–12).

## 4. Techniques for solving quadratic equations

As with linear equations, quadratic equations can be solved using either an algebraic approach or graphically.

a) *Algebraically*, there are two techniques available and both are covered in following sections:

i. by formula, and

ii. by factorising.

b) The *graphical approach* is similar to that for linear equations with some extensions.

Generally, a quadratic equation will yield two solutions, but students should be aware that in fact quadratic equations can have either two, one or no solutions. This is demonstrated graphically in section 10.

## 5. Solving quadratic equations by formula

**Quadratic equation formula**

The quadratic equation $ax^2 + bx + c = -0$ can be solved using the following formula:

$$x = \frac{-b \pm \sqrt{b^2 - 4ac}}{2a}$$

**Note:** If $b^2 < 4ac$ there are *no* solutions.
If $b^2 = 4ac$ there is *one* solution.
If $b^2 > 4ac$ there are *two* distinct solutions.

## 6. Example 2 (Solving a *quadratic* equation using the *formula*)

*Question*

Solve the equation $2x^2 - 11x + 22 = 10$.

*Answer*

First, put the equation into its general form, i.e. $2x^2 - 11x + 12 = 0$. Here then $a$=2, $b$=–11 and $c$=12.

So, $$x = \frac{-b \pm \sqrt{b^2 - 4ac}}{2a}$$

$$= \frac{11 \pm \sqrt{(-11)^2 - 4(2)(12)}}{2(2)}$$

$$= \frac{11 \pm \sqrt{25}}{4}$$

i.e. $$x = \frac{6}{4} \text{ or } \frac{16}{4}$$

So that, the solutions of $2x^2 - 11x + 22 = 10$ are $x$=1.5 and $x$=4.

## 7. Solving quadratic equations by factorization

Consider the relationship $ab = 0$. It can only mean that either $a$=0 or $b$=0 (or both $a$ and $b$ =0).

In the same vein, for example: if $(x–a)(x–b)=0$ then either $x–a=0$ or $x–b=0$. In other words $x=a$ or $x=b$.

So that, if a quadratic equation can be factorized (i.e. put into the form of the product of two linear factors), it can be solved.

Now, multiplying $(x-2)(x-3)$ out gives: $x.x + x.(-3) + (-2).x + (-2).(-3)$

$$= x^2 - 3x - 2x + 6$$
$$= x^2 - 5x + 6$$

Thus, $x^2-5x+6 = 0$ can be put into the form $(x-2)(x-3)=0$, which means that either $x-2=0$ or $x-3=0$.

In other words $x=2$ or $x=3$. Therefore, the solutions of the equation $x^2-5x+6 = 0$ are $x=2$ and $x=3$.

Similarly, since $(2x-5)(x+1)$ multiplies out to $2x^2-3x-5$, the equation $2x^2-3x-5=0$ can be factorized as $(2x-5)(x+1)=0$, giving the solutions $x=5\div2=2.5$ and $x=-1$.

**Note**. The factorization technique can only be used if the equation is in the form $ax^2+bx+c=0$.

For example, putting the equation $x^2-x-2=4$ into the form $(x+1)(x-2)=4$ *will not* help in solving it.

For students who are reasonably adept at factorising quadratic functions, this method provides a quick and easy way of solving quadratic equations. Otherwise, the formula method can always be used and will *always* give the solutions (if they exist).

## 8. Solving quadratic equations graphically

a) The technique employed is similar to that for solving linear equations graphically, described in the previous chapter.

The procedure for solving the quadratic equation $ax^2 + bx + c = 0$ graphically is:

STEP 1 Plot the graph of $y = ax^2 + bx + c$.

STEP 2 *Determine the x-values* of the points where the graph meets the $x$- axis $(y=0)$

These $x$-values give the required solution(s).

b) This method can be used with quadratic equations that are given originally in any form, since all that is needed is to convert to the general form $ax^2+bx+c=0$.

For example:

i. $x^2-5x-4=10$ can be put into the form $x^2-5x-14=0$

ii. $2x^2+10x-1=5x+3$ can be changed to $2x^2+5x-4=0$ ... and so on.

**Note**. If necessary (and sometimes examiners ask for this particularly), $x^2-5x-4=10$ can be solved by finding the $x$ values of the points where the parabola $y=x^2-5x-4$ meets the (horizontal) line $y=10$. (Plotting parabolas was covered in chapter 24.)

Also, $2x^2+10x-1=5x+3$ can be solved, if required, by identifying the points where the parabola $y=2x^2+10x-1$ meets the line $y=5x+3$.

## 9. Example 3 (Solving a *quadratic* equation *graphically)*

*Question*

Solve the equation $3x^2 - 8x - 5 = 0$ graphically.

*Answer*

This equation can be solved by plotting the graph of $y = 3x^2 - 8x - 5$ and finding where it crosses the $x$-axis.

[Often, in examination questions, a range of values for plotting a curve is given. However, if this is not given, using some knowledge of the shape of the curve together with some trial and error calculations is usually necessary. This type of approach is demonstrated now.]

Now, substituting $x$=0 gives $y$=–5. Also, the curve is a 'valley' shape (remember from chapter 27, this is because the coefficient of $x^2$ is positive). Both of these facts mean that the curve must meet the $x$-axis either side of the $y$-axis. So, initially, try values of $x$ between, say, –4 and +4, which are shown in Table 1. The sign of $y$ changes between $x$=–2 and 0, and again between $x$=2 and 4. So, adding three more relevant $x$-values around these areas (and ignoring those points corresponding to large values of $y$) gives the values in Table 2.

*Table 1*

| $x$ | –4 | –2 | 0 | 2 | 4 |
|---|---|---|---|---|---|
| $3x^2$ | 48 | 12 | 0 | 12 | 48 |
| $-8x$ | 32 | 16 | 0 | –16 | –32 |
| –5 | –5 | –5 | –5 | –5 | –5 |
| $y$ | 75 | 23 | –5 | –9 | 11 |

*Table 2*

| $x$ | –1 | 0 | 1 | 2 | 3 | 4 |
|---|---|---|---|---|---|---|
| $3x^2$ | 3 | 0 | 3 | 12 | 27 | 48 |
| $-8x$ | 8 | 0 | –8 | –16 | –24 | –32 |
| –5 | –5 | –5 | –5 | –5 | –5 | –5 |
| $y$ | 6 | –5 | –10 | –9 | –2 | 11 |

The $x$ and $y$ values from Table 2 are plotted and the corresponding graph drawn in Figure 1, where it can be seen that the graph crosses the $x$-axis at –0.5 and 3.2.

Thus, the solutions of the equation $3x^2 - 8x - 5 = 0$ are $x$=–0.5 and $x$=3.2 (1D).

## 10. Graphs and the solution of equations

Earlier, in section 5, referring to the quadratic equation $ax^2+bx+c=0$, the comparison of the values of $b^2$ and $4ac$ was shown to relate to how many roots (solutions) the equation had. This is demonstrated graphically in Figure 2.

In (a) the curve crosses the $x$-axis in two distinct points (two solutions), in (b) the curve just touches the $x$-axis (one solution only) and in (c) the curve does not meet the $x$-axis anywhere (no solutions).

*Figure 1 Solving $3x^2 - 8x - 5 = 0$ graphically*

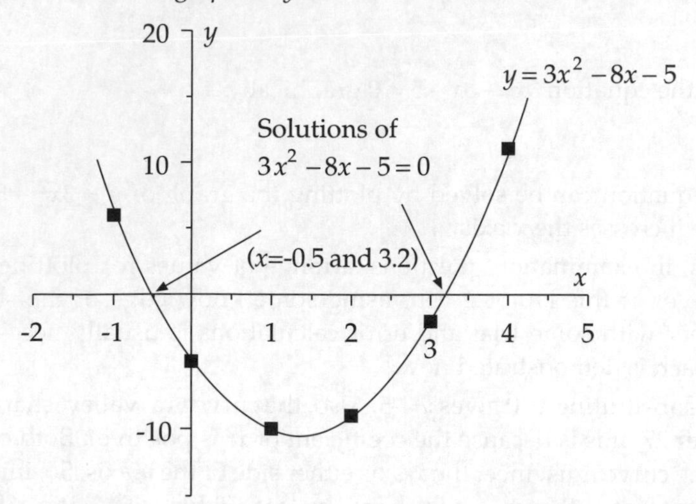


*Figure 2 Types of solutions to a general quadratic equation*

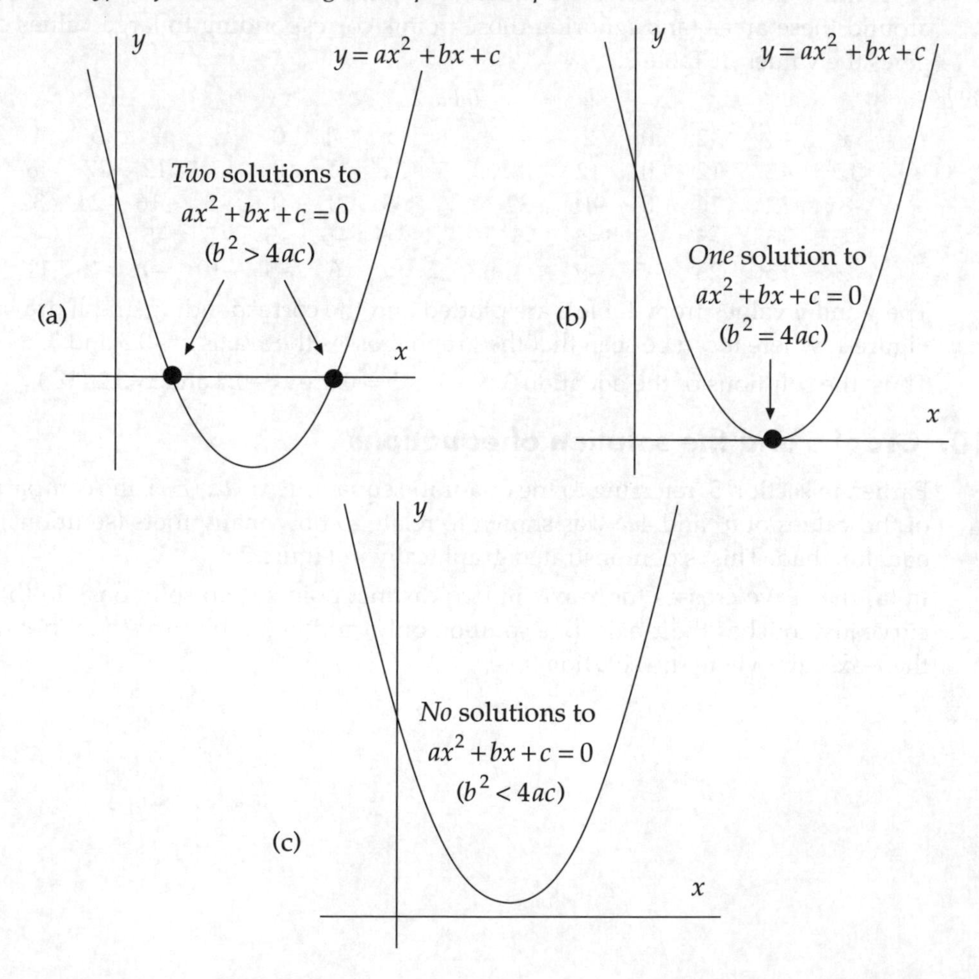

## 11. Example 4 (*Graphical* solution of a *quadratic* equation and interpretation)

*Question*

A company invests in a particular project and it has been estimated that after $x$ months of running, the cumulative profit (£000) from the project is given by the function $31.5x - 3x^2 - 60$, where $x$ represents time in months. The project can run for nine months at the most.

a) Draw a graph which represents the profit function.
b) Calculate the 'break even' time points for the project.
c) What is the initial cost of the project?
d) Use the graph to estimate the best time to end the project.

*Answer*

a) For the graph, put profit = $y = 31.5x - 3x^2 - 60$. Clearly, the lowest relevant value of $x$ is 0 and the highest 9 (since nine months is the maximum length of the project). Table 3 shows the calculations and Figure 3 the profit graph plotted.

*Table 3*

| $x$ | 0 | 2 | 4 | 6 | 7 | 8 | 9 |
|---|---|---|---|---|---|---|---|
| $31.5x$ | 0 | 63 | 126 | 189 | 220.5 | 252 | 283.5 |
| $-3x^2$ | 0 | –12 | –48 | –108 | 147 | –192 | –243 |
| –60 | –60 | –60 | –60 | –60 | –60 | –60 | –60 |
| $y$ | –60 | –9 | 18 | 21 | 13.5 | 0 | –19.5 |

*Figure 3*

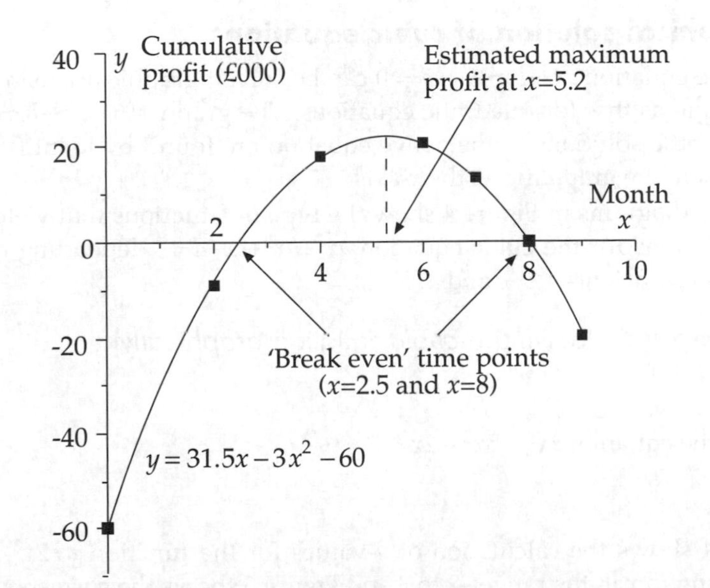


b) The 'break even' time points are those $x$ values that make the profit zero. These are found by solving the equation $31.5x–3x^2–60=0$, which, graphically, means identifying where the graph of $y=31.5x–3x^2–60$ meets the $x$-axis ($y=0$). From the graph (in Figure 3), these can be seen to be $x=2.5$ and $x=8$. Thus, the profit 'break even' times are after 2.5 and 8 months.

c) At time $x$=0, the profit (from the graph) can be seen to be –60 (£000), and since the project is not yet underway at this time, £60,000 must be the initial cost of the project.

d) Since the function given is cumulative profit, the best time to end the project would be when this function attains its maximum value. From the graph, this can be seen to be after (approximately) 5.2 months.

## 12. Cubic equations

a) A *cubic equation* (in one variable) can always be put into a general form.

> **General form of a cubic equation**
>
> $$ax^3 + bx^2 + cx + d = 0$$
>
> where $a$, $b$, $c$ and $d$ can take any numeric values.

b) For example:
   i. $2x^3 - 5x^2 + x - 20 = 0$ ($a$=2, $b$=–5, $c$=1 and $d$=–20)
   ii. $30 + 3x^2 - 0.5x^3 = 0$ ($a$=–0.5, $b$=3, $c$=0 and $d$=30)

c) A cubic equation is *solved* by finding the values of $x$ that satisfy it. There are, at most, three solutions (or roots as they are sometimes called). Note that in a cubic equation the variable has a maximum power of 3.

d) Various algebraic and graphical methods of solution are available, but for the purposes of this manual *only a graphical method is given*.

## 13. Graphical solution of cubic equations

a) The equation $ax^3+bx^2+cx+d = 0$ can be solved graphically using the same technique as that for quadratic equations. The graph of $y=ax^3+bx^2+cx+d$ is plotted and the solutions to the above equation are found by identifying the $x$ values where the graph meets the $x$-axis.

b) The diagrams in Figure 4 show the type of situations that yield either 1, 2 or 3 solutions for the cubic equation $ax^3+bx^2+cx+d=0$, depending on the values of the coefficients $a$, $b$, $c$ and $d$.

## 14. Example 5 (Solving a *cubic* equation *graphically*)

*Question*

Solve the equation $2x^3 - 5x^2 - 2x + 5 = 0$.

*Answer*

Table 4 shows the calculation of $y$-values for the function $y=2x^3-5x^2-2x+5$ corresponding to $x$ in the range –2 to 3 and Figure 5 shows the corresponding graph.

*Figure 4*

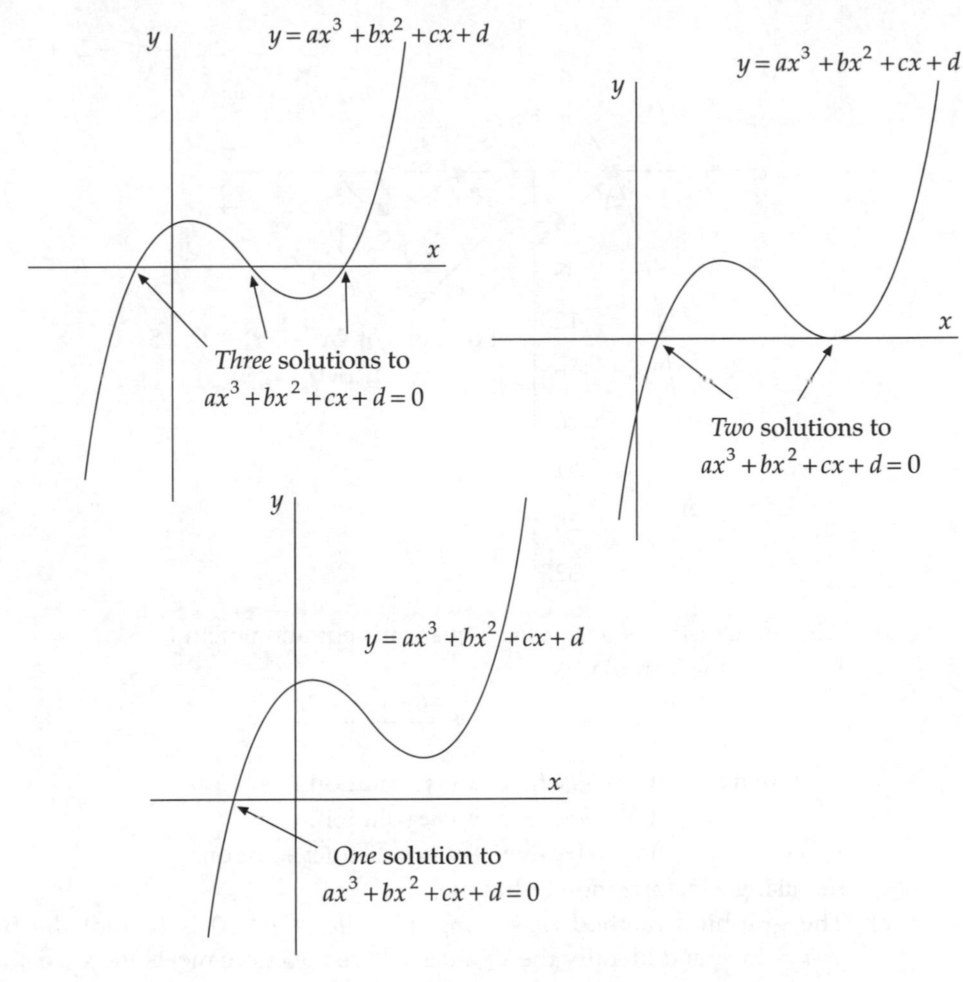


*Table 4*

| $x$ | −2 | −1 | 0 | 1 | 2 | 3 |
|---|---|---|---|---|---|---|
| $2x^3$ | −16 | −2 | 0 | 2 | 16 | 54 |
| $-5x^2$ | −20 | −5 | 0 | −5 | −20 | −45 |
| $-2x$ | 4 | 2 | 0 | −2 | −4 | −6 |
| 5 | 5 | 5 | 5 | 5 | 5 | 5 |
| $y$ | −27 | 0 | 5 | 0 | −3 | 8 |

The graph (in Figure 5 overleaf) crosses the $x$-axis at −1, 1 and 2.5. Therefore, the solutions of the given equation are $x$=−1, $x$=1 and $x$=2.5.

## 15. Summary

a) A quadratic equation takes the general form: $ax^2 + bx + c = 0$ where $a$, $b$ and $c$ can take any numeric values (except $a$=0).

*Figure 5*

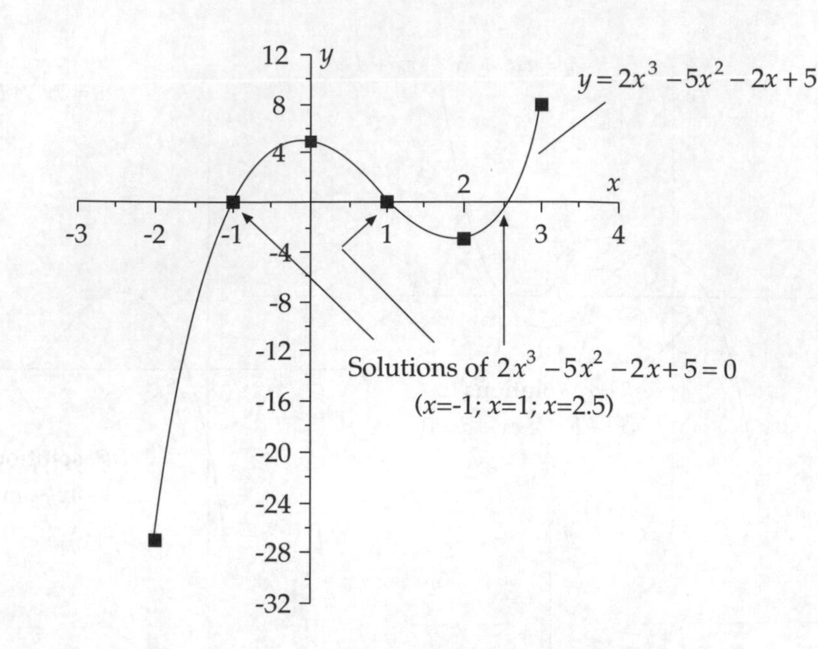


b) Algebraic methods of solving a quadratic equation include:

i. using the formula:

$$x = \frac{-b \pm \sqrt{b^2 - 4ac}}{2a}$$

where: if $b^2 < 4ac$ there are *no* solutions.
if $b^2 = 4ac$ there is *one* solution.
if $b^2 > 4ac$ there are *two* distinct solutions.

ii. using a factorization technique.

c) The graphical method of solving $ax^2 + bx + c = 0$ is to plot the function $y=ax^2+bx+c$ and identify the $x$ values where the curve meets the $x$-axis.

d) A cubic equation takes the general form: $ax^3 + bx^2 + cx + d = 0$ where $a$, $b$, $c$ and $d$ can take any numeric values (except $a$=0). The equation can have from 1 to 3 roots.

## 16. Student self review questions

1. What is the general form of a quadratic equation? [2]
2. Write down the formula that will give the solutions to the quadratic equation $ax^2 + bx + c = 0$. [5]
3. What are the conditions on $a$, $b$ and $c$ for there to be 2, 1 or no solutions to the quadratic equation $ax^2+bx+c=0$? [5]
4. Draw a quick sketch to represent the fact that a quadratic equation has no roots. [10]
5. How many roots does a cubic equation have? [12]

## 17. Student exercises

1. Solve the following quadratic equations:

   a) $(x-4)(x+3)=0$ b) $(3x+1)(5-2x)=0$
   
   c) $2x^2=8$ d) $4x(5x-2)=0$
   
   e) $(2x+1)(x-7)=17$ f) $x^2-7x+12=0$
   
   g) $2x^2-11x+14=0$ h) $x(2x-1)=10$

2. *MULTI-CHOICE*. Use the quadratic equation formula to show which one of the following options gives the solution to the equation $x^2-2x-24=0$.

   a) $\dfrac{-2\pm 10}{2}$ b) $\dfrac{-2\pm\sqrt{92}}{2}$ c) $\dfrac{-2\pm\sqrt{96}}{2}$ d) $\dfrac{2\pm 10}{2}$

3. Find the roots (to 2D if appropriate) of the following equations:

   a) $x^2-4x+3.75=0$ b) $x^2-7x+11=0$
   
   c) $x^2-7x+12=0$ d) $x^2-7x+13=0$
   
   e) $3x^2-2x-3=0$ f) $4-12x+3x^2=0$
   
   g) $0.9x^2-9.1x+19.4=0$ h) $4x^2+5x-1=2x+2$

4. a) Carefully plot the graph of $y=x^2-4.5x+4.5$ between $x=0$ and $x=8$.

   b) Use the above curve (in conjunction with any other plotted function if necessary) to estimate (to 1D) the solutions of the following equations:
   
      i. $x^2-4.5x+4.5=0$
      
      ii. $x^2-4.5x+4.5-3$
      
      iii. $x^2-4.5x+4.5=x-1$
      
      iv. $x^2-7.5x+2.5=0$

5. Solve (to 2D) the equations:

   a) $\dfrac{x-3}{x}=\dfrac{2x}{x-1}$ b) $\dfrac{x-3}{x}=\dfrac{2x}{x-1}+1$

6. Solve the equation $x^3-3x^2+2x=0$ (note that it can is easily factorised).

7. a) Plot the function $y=-25-18x+8x^2+2x^3$ carefully for values of $x$ between $-6$ and 3.

   b) Use the plotted curve to solve the equations:
   
      i. $2x^3+8x^2-18x-25=0$
      
      ii. $2x^3+8x^2-20x-45=0$

8. Solve the following equations graphically:

   a) $x^3-10.55x^2+36.4x-40.8=0$
   
   b) $4x^2(x-3)+3=x$
   
   c) $-20+8x+12x^2-3x^3=0$

9. Find the solution of $5x^3-10x^2-2x+12=0$ that lies between $-2$ and $+2$.

# 27 Differentiation and integration

## 1. Introduction

The chapter begins with an explanation of how differentiation and integration can be used in business situations. Then follows the rule for differentiation, its interpretation and its practical use in finding the maximum and minimum values of functions or turning points on curves. Finally, the rules of integration are given.

## 2. Differentiation and integration and their use

a) The general area of mathematics that includes differentiation and integration is known as *calculus*. The syllabuses that this manual covers deal with calculus *only at an elementary level* and require:
   i. familiarity with a few simple rules;
   ii. an understanding of some relevant applications.

   Contrary to (often popular) belief, it is NOT a difficult area of study, especially when (i) and (ii) above are borne in mind.

b) *Differentiation* can be used to find the maximum or minimum points of the curves of certain business functions. For example, cost, revenue and profit functions. If for a particular business process the profit is known to be a function of the level of production, the production level can be determined that will give maximum profit. Differentiation can also be used to measure 'rate of change', which is normally applied to cost and revenue functions to obtain *marginal* cost or revenue.

c) *Integration* can be regarded as the reverse process to differentiation. It can be used to identify specific revenue and cost functions, given marginal revenue and cost functions.

## 3. Simple functions

First, for pure convenience, a *simple function* of $x$ is defined to have the form $ax^b$, where $a$ and $b$ can take any numerical value, positive or negative.

For example: $3x^2$, $2x^3$ and $-12x^5$ are all simple functions of $x$.

Note that each of $\frac{1}{x}$ $(= x^{-1})$, $\frac{4}{x^2}$$(= 4x^{-2})$, and $\frac{-5}{x^3}$ $(= -5x^{-3})$ are also simple functions of $x$.

## 4. Rule and notation for differentiation

Differentiation can be thought of as a process which transforms one function into a different one. The new function is known as the *derivative* of the original one.

The use of the singular word 'rule' above is intentional. There is only one simple rule to learn in order to be able to differentiate any mathematical function. It is stated as follows:

**Derivative of a simple function**

The simple function $y = ax^b$ (where a and b are any numbers) can be *differentiated* to give the new function:

$$\frac{dy}{dx} = abx^{b-1}$$

where $\frac{dy}{dx}$ is translated as 'the derivative of $y$ with respect to $x$' and spoken as 'dee $y$ by dee $x$'.

*In particular*: if $y = ax$, then $\frac{dy}{dx} = a$, and if $y = a$ (a constant), then $\frac{dy}{dx} = 0$.

Thus, for example, if: $y= 5x^3$, $\frac{dy}{dx} = (5)(3)x^{3-1} = 15x^2$

If $y = 8x$, then $\frac{dy}{dx} = 8$; if $y = 5$, then $\frac{dy}{dx} = 0$.

## 5. Example 1 (*differentiation* of simple functions)

Using the above rule for differentiation:

a) If $y = 4x^2$, then $\frac{dy}{dx} = (4)(2).x^{2-1} = 8x^1 = 8x$

b) If $y = -12x^6$, then $\frac{dy}{dx} = -72x^5$

c) If $y = 3x^3$, then $\frac{dy}{dx} = 9x^2$

d) Also if $y = \frac{3}{x^2} = 3x^{-2}$, then $\frac{dy}{dx} = (3)(-2).x^{-2-1} = -6x^{-3} = \frac{-6}{x^3}$

e) Also if $y = 25x$, then $\frac{dy}{dx} = 25$

f) If $y = -3x$, then $\frac{dy}{dx} = -3$.

g) If $y = 15$, then $\frac{dy}{dx} = 0$ (in other words, the derivative of a constant, i.e. a number not involving $x$, is always zero).

## 6. An extension of differentiation

Obviously, the need is to be able to differentiate all functions of a variable $x$, not just simple ones as demonstrated above. The following statement, an extension of the previous rule, completes the picture and, for all practical purposes, allows all functions to be differentiated.

**A rule for derivatives**
The *derivative of the sum* of two (or more) simple functions is the *sum of the separate derivatives* of each simple function.

For example:

a) if $y = 4x^3 - 12x^2 + 3x + 12$, then $\dfrac{dy}{dx} = 12x^2 - 24x + 3$

b) if $y = \dfrac{5}{x^2} - \dfrac{12}{x} + 14 + 2x - 18x^2 + 14 + 2x - 18x^2 = 5x^{-2} - 12x^{-1} + 14 + 2x - 18x^2$,

then $\dfrac{dy}{dx} = -10x^{-3} + 12x^{-2} + 2 - 36x = \dfrac{-10}{x^3} + \dfrac{12}{x^2} + 2 - 36x$.

## 7. The practical interpretation of differentiation

The larger the value of the gradient of a straight line ($y = a + bx$ say), the greater the rate of change of $y$ with respect to $x$. The gradient (or slope) of a line is the way that a rate of change can be measured. For example, $y = 10 + 4x$ has a gradient of 4, which means that for each *unit increase* in $x$, *$y$ increases by 4 units*.

The gradient of the line above is just $\dfrac{dy}{dx} = 4 =$ the rate of change.

The derivative of any function measures its **rate of change.**

Notice that, for the example above, $y = 10 + 4x$, the rate of change was constant (=4), since this particular function is a straight line. In general however, rate of change is a *function of $x$*. That is, it varies from one point to the next.
Now consider the function $y = 2x^2 - 4x + 10$. The rate of change is $\dfrac{dy}{dx} = 4x - 4$, which can be calculated for any $x$-value.

For example: at $x = 1$, $\dfrac{dy}{dx} = 4(1) - 4 = 0$; at $x = 2$, $\dfrac{dy}{dx} = 4$; etc.

## 8. Second derivative of a function

Differentiation can be repeated as many times as necessary on any given function. Of particular use for the syllabuses covered by this manual is the *second derivative* of a function, defined as follows.

The **second derivative** of any function $y$ is written as

$$\frac{d^2y}{dx^2}$$

(spoken as 'dee two $y$ by dee $x$ squared') and obtained by differentiating $y$ twice.

For example if: $y = 4x^3 - 2x^2 + 10x - 12$,

then: $\frac{dy}{dx} = 12x^2 - 4x + 10$

and: $\frac{d^2y}{dx^2} = 24x - 4$

As another example, if $y = 6x^4 + 3x^3 - 22$,

then: $\frac{dy}{dx} = 24x^3 + 9x^2$

and: $\frac{d^2y}{dx^2} = 72x^2 + 18x$

The particular use to which the second derivative is put is explained in section 10.

## 9. The practical use of differentiation

Differentiation has an important practical use. That is to *determine the position of any turning points* of the curve defined by a given function. Of course, the graph of a given function can be drawn and the position of any turning points found approximately, but differentiation will always find them precisely.

The technique to be employed for identifying the turning points (maxima or minima) of a given function is described in the next section and followed by an example.

## 10. Procedure for identifying the turning points of a curve

It will already have been seen that the maximum (or minimum) value of a given function can be found by plotting the corresponding curve and reading off the value required at a relevant turning point (maximum or minimum). However, this is only an approximate method. Using differentiation techniques, turning points of curves can be identified exactly. The steps in the procedure are given below, and demonstrated using the function $y = 0.5x^2 - 8x + 60$, where $y$ is the cost (in £000) of manufacturing $x$ (hundred) items for some process.

The procedure for identifying the turning points on a curve is:

STEP 1 *Obtain* $\frac{dy}{dx}$ for the given function.

For the example given, we have $y = 0.5x^2 - 8x + 60$

Thus, $\frac{dy}{dx} = x - 8$

STEP 2 *Solve* the equation $\frac{dy}{dx} = 0$, which will give the $x$-coordinates of any turning points that exist.

Since $\frac{dy}{dx} = x–8$, the equation $x - 8 = 0$ needs to be solved.

Thus $x$=8 identifies some turning point.

STEP 3 *Evaluate* $\frac{d^2y}{dx^2}$ at each (and any) $x$-value found in STEP 2.

$\frac{d^2y}{dx^2} > 0$ signifies a *Minimum* point;

$\frac{d^2y}{dx^2} < 0$ signifies a *Maximum* point.

For the example given, $\frac{dy}{dx} = x - 8.$

and thus: $\frac{d^2y}{dx^2} = 1.$

The turning point at $x$=8 is therefore a *minimum*. Hence, 800 items must be manufactured in order to minimise total costs.

Note that the above procedure identifies *only x-coordinates* of any turning points. To find the corresponding $y$-coordinates, the $x$-values found need to be substituted into the original $y$ function.

For the example given, the minimum cost can be calculated by substituting $x$=8 into $y = 0.5x^2 - 8x + 60$ which gives:

minimum cost $= 0.5(8)^2 - 8(8) + 60 = 32 - 64 + 60 = 28$ (£000).

Thus, the minimum total cost of £28,000 is realised when 800 items are manufactured.

Two examples of the use of this technique follow.

## 11. Example 2 (To determine the *turning points* of a given function)

*Question*

Determine the coordinates and nature of any turning points on the curve represented by the function $y = x^3 - 7.5x^2 + 18x + 6$.

*Answer*

Now, $\frac{dy}{dx} = 3x^2 - 15x + 18,$

and putting: $\frac{dy}{dx} = 0$

gives: $3x^2 - 15x + 18 = 0$ [1]

Factorizing the expression on the left hand side of [1] gives:

$3(x-2)(x-3) = 0$ [2]

Solving the equation in [2], we have: $x$=2 and $x$=3.

**Note**: if you cannot see how the factors in [2] have been obtained, try using the quadratic equation formula (section 5) to solve the equation in [1]. This will also yield the solutions $x$=2 and $x$=3.

That is, turning points exist on the curve $y = x^3 - 7.5x^2 + 18x + 6$ at $x$=2 and $x$=3.

Also, $\frac{d^2y}{dx^2} = 6x - 15$

and when $x=2$, $\dfrac{d^2y}{dx^2} = -3$ (i.e. $< 0$), signifying a *maximum*.

when $x=3$, $\dfrac{d^2y}{dx^2} = +3$ (i.e. $> 0$), signifying a *minimum*.

Substituting $x=2$ into the original equation will give the $y$-coordinate of the maximum as:

$$y = (2)^3 - (7.5)(2)^2 + 18(2) + 6 = 20.$$

Hence, the maximum point for the curve is at $x=2$ and $y=20$ or (2, 20).

Similarly, substituting $x=3$ into the original equation will give the $y$-coordinate of the minimum as:

$$y = (3)^3 - (7.5)(3)^2 + 18(3) + 6 = 19.5.$$

Hence, the minimum point for the curve is at $x=3$ and $y=19.5$ or (3, 19.5).

## 12. Example 3 (To identify a *turning point* and *interpret* its significance)

*Question*

A food processing plant has a particular problem with the delivery and processing of perishable goods with a short life. All deliveries must be processed in a single day and, although there are a number of processing machines available, they are very expensive to run. A researcher has developed the function $y = 12x - 2a - ax^2$ to describe the profit ($y$, in £00) given the number of machines used ($x$) and the number of deliveries ($a$) in a day.

a) Show that the system is uneconomic if 4 deliveries are made in one day (i.e. $a=4$).

b) If three deliveries are made in one day, find the number of processing machines that should be used in order that the profit is maximised. In this case, what is the maximum profit?

*Answer*

a) Here, $a=4$, and substituting this into the given function, we have:

$$y = 12x - 2(4) - 4x^2.$$

i.e. $$y = 12x - 8 - 4x^2.$$

An uneconomic system will result if the profit contribution is not positive *no matter how many processing machines are used,* that is for all values of $x$.

For the given function: $\dfrac{dy}{dx} = 12 - 8x$

and: $\dfrac{d^2y}{dx^2} = -8$ (signifying a maximum, as we would expect).

Now, when $\dfrac{dy}{dx} = 0$, $12 - 8x = 0$.

i.e. $x = 1.5$.

Now, since the number of processing machines used *must be a whole number*, it is only necessary to show that the profit from using 1 or 2 machines is not positive.

When $x$=1, $y = 12(1) - 8 - 4(1)^2 = 0$ (£00)

and when $x$=2, $y = 12(2) - 8 - 4(2)^2 = 0$ (£00).

That is, in both cases there is no profit.

Thus the system is uneconomic if 4 deliveries are made.

b) Here, $a$=3.

Therefore, $y = 12x - 2(3) - 3x^2$

$= 12x - 6 - 3x^2$;

Also $\frac{dy}{dx} = 12–6x$

and $\frac{d^2y}{dx^2} = –6.$

When $\frac{dy}{dx} = 0$, $12 - 6x = 0.$

i.e. $x = 2.$

Hence, *2 processing machines should be used* in order to maximise profit contribution.

Also, maximum profit $= 12(2) - 6 - 3(2)^2$

$= 6$ (£00).

Thus, using two processing machines, the profit is £600.

## 13. Integration

As mentioned earlier, integration can be regarded as the opposite process to differentiation. For example, since differentiating $2x^3$ gives $6x^2$, integrating $6x^2$ *should* give $2x^3$. Well, almost!

A minor problem occurs because differentiating a constant will always give zero. So that, for example, all of the following functions:

$5x^2 + 6x - 12$;

$5x^2 + 6x + 2$;

$5x^2 + 6x + 50$

are differentiated to the *same function*, namely $10x$+6.

Therefore, the best that can be done is to integrate $10x$+6 to give $5x^2 +6x+C$, where $C$ is some arbitrary constant. Sometimes information is known about a particular situation that enables $C$ to be calculated. An example of this is given in Example 4 part e).

The rule for integrating a simple function of $x$, $ax^b$ say, is:

a) to increase the power of $x$ by 1 (thus $b$ goes to $b$+1, and

b) to divide the whole function by $b$+1.

For example, to integrate $5x^3$, the power 3 is increased by 1 to 4 and the whole function is then divided by 4.

Thus, the integral of $5x^3$ is $\frac{5}{4}x^4 + C$ (always remember the arbitrary constant C).

The previous statement can be expressed in mathematical form as:

$$\int 5x^3 dx = \frac{5}{4}x^4 + C$$

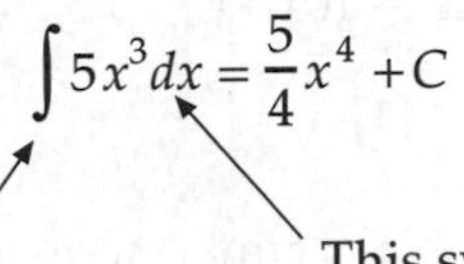

This symbol means 'the variable is $x$'

This symbol represents 'the integral of'

A formal statement of the rule follows.

## 14. Notation and rule for integration

The integral of the simple function $ax^b$ is given by:

$$\int ax^b dx = \frac{a}{b+1}x^{b+1} + C$$

where C is an arbitrary constant

In particular: $\int a\,dx = ax + C$

Thus, for example, $\int 4x^2\,dx = \frac{4}{2+1}x^{2+1} + C = \frac{4}{3}x^3 + C$

Also: $\int 10\,dx = 10x + C$

**Rule for integrals**
The *integral of a sum* of simple functions is the *sum of the separate integrals* of the simple functions.

## 15. Example 4 (Finding the *integrals* of functions)

a) $\int (3x^2 - 4x)\,dx = \frac{3}{2+1}x^{2+1} - \frac{4}{1+1}x^{1+1} + C$
$= x^3 - 2x^2 + C.$

b) $\int (8x^3 - 3x^2 + 6x - 10)\,dx = 2x^4 - x^3 + 3x^2 - 10x + C$

c) $\int (14 + 12x^2)\,dx = 14x + 4x^3 + C$

d) $\int (2.4x^2 + 8.6x - 3)\,dx = 0.8x^3 + 4.3x^2 - 3x + C$

e) If $\frac{dy}{dx} = 6x - 10$ and $y$=12 when $x$=0, to find y in terms of $x$.

Now, integrating $\frac{dy}{dx}$ will give $y$.

That is, $y = \int (6x-10)\,dx = 3x^2 - 10x + C$

$= 3x^2 - 10x + C$ [1]

But $y$=12 when $x$=0.

Therefore, substituting in [1], $12 = 3(0)^2 - 10(0) + C$,

giving $C = 12$.

Thus $y = 3x^2 - 10x + 12$.

Note that, in this case, with the information given, the arbitrary constant, C, was able to be evaluated.

## 16. Example 5 (Using *integration* and *differentiation* in a business situation)

*Question*

The total revenue obtained (in £000) from selling $x$ hundred items in a particular day is given by $R$, which is a function of variable $x$.

Given that $\frac{dR}{dx} = 20 - 4x$:

a) Determine the total revenue function $R$;

b) Find the number of items sold in one day that will maximise the total revenue and evaluate this total revenue.

*Answer*

a) We are given $\frac{dR}{dx} = 20 - 4x$. Integrating this must therefore give $R$.

That is, $\int \frac{dR}{dx}\,dx = \int (20-4x)\,dx$

$= 20x - 2x^2 + C.$

But when *no* items are sold (i.e. $x$=0), there will be *no* revenue (i.e. $R$=0). Thus, substituting $x$=0 into $R$ above gives C=0.

So that, $R$ = total revenue = $20x - 2x^2$.

b) The value of $x$ that maximises $R$ is found by solving the equation $\frac{dR}{dx} = 0$.

That is, where $20 - 4x = 0$.

This gives $x = 5$.

In other words, total revenue is maximised if 500 items are sold in a day.

The value of this total revenue is found by substituting $x$=5 into $R$.

This gives: $20(5) - 2(5)^2 = 100 - 50 = 50$.

Thus, the maximum total revenue (obtained by selling 500 items) is £50,000.

## 17. Summary

a) Differentiation can be used to find maxima or minima of business functions. The process can be thought of, in practical terms, as obtaining the rate of change of a function.

b) The derivative of the simple function $y = ax^b$ is given by $\frac{dy}{dx} = abx^{b-1}$

c) The derivative of the sum of two (or more) simple functions is the sum of the separate derivatives of the functions.

d) Differentiation can be repeated as many times as necessary on any given function. If $y$ is any function of $x$:

   i. $\frac{dy}{dx}$ is the *first* derivative

   ii. $\frac{d^2y}{dx^2}$ is the *second* derivative

e) Solving the equation $\frac{dy}{dx} = 0$, will identify the turning points of a function. The second derivative can then be used to indicate whether the turning point is a maximum or minimum.

f) Integration can be regarded as the opposite process to differentiation.
   The integral of $ax^b = \int ax^b dx = \frac{a}{b+1}x^{b+1} + C$

## 18. Point to note

Only the barest description of both differentiation and integration has been given in this chapter, enough to be able to use them in particular business functional situations. Demonstrations of this type of use were given in Examples 3 and 5 and these will be extended in the following chapter.

## 19. Student self review questions

1. Give examples of how differentiation is used. [2]
2. What does $\frac{dy}{dx}$ mean? [4]
3. Give a practical interpretation of $\frac{dy}{dx}$ in relation to a graph of $y$. [7]
4. How is the second derivative, $\frac{d^2y}{dx^2}$, obtained? [8]
5. Explain how the functions $\frac{dy}{dx}$ and $\frac{d^2y}{dx^2}$ are used in obtaining any turning points of some function $y$. [10]
6. What is the relationship between differentiation and integration? [13]

## 20. Student exercises

1. Find $\frac{dy}{dx}$, given the following functions:
   a) $y = 4x^2$
   b) $y = 3x^4 + 2x^2 - 10$
   c) $y = 12 - 10x + 30x^2$
   d) $y = 8x - 4 + \frac{2}{x}$ [Hint: express $\frac{2}{x}$ as $2x^{-1}$]
   e) $y = (4+x)(x-1)$ [Hint: multiply out the brackets first]
   f) $2y = 3x^3 + 10$ [Hint: divide both sides by 2 first]
   g) $xy = 2x^2 - 5x + 4$
2. Calculate $\frac{dy}{dx}$ and $\frac{d^2y}{dx^2}$ for the following functions of $x$:
   a) $y = 5x^2 - 2x + 4$
   b) $y = 12 - 10x + 6x^2 - 2x^3$
   c) $xy = 3x^4 + 4x^2$
3. Find the $x$-value of the single turning point on the graph of $y = 2x^2 - 8x$ and determine whether it is a minimum or maximum point.
4. Find the coordinates of the two turning points of the function $y = x^3 - 8x^2 + 5x + 3$.
5. Integrate the following functions with respect to $x$.
   a) $9x^2$
   b) $2x^3$
   c) $8x^3 - 3x^2 + 8x - 10$
   d) $\frac{2}{x^3}$
6. For a particular function, $\frac{dy}{dx} = 4x - 3$.

   If it is known that when $x=1$, $y=5$, find $y$ in terms of $x$.
7. The profit (in £00) from a daily production run is given by $P$, which is a function of the level of production, $x$ (in thousands).

   If $\frac{dP}{dx} = 11 - 2x$ and one profit break-even point is known to be a production of 3000, find:

   a) $P$ as a function of $x$;
   b) the other profit break-even point;
   c) the daily production run that gives the maximum profit;
   d) the value of the maximum daily profit.

# 28 Cost, revenue and profit functions

## 1. Introduction

Described in the chapter is the basic make-up of cost and revenue functions, together with how they relate to produce a functional expression for profit. Cost, revenue and profit functions are minimized and maximized using the techniques described in the previous chapter and also considered are marginal cost and revenue functions and their relationship in maximizing profit.

## 2. The basic profit equation

In any commercial environment, the (gross) profit can be considered as a simple function of the difference between the revenue obtainable from the sale of a number of products and the costs involved in their production. That is, Profit ($P$) = Revenue ($R$) – Costs ($C$).

To take a simple example, if a wholesaler can buy items at £6.50 each and sells them at £7.20 each, the profit per item is £7.20 – £6.50 = £0.70. The total profit realised from the handling of $x$ items can be expressed in a similar form:

Profit from $x$ items = Revenue from $x$ items – Cost of $x$ items

$$= £7.20x - £6.50x$$
$$= £0.70x.$$

## 3. The profit function

Normally, both cost and revenue are functions of $x$ (the number of items demanded or supplied) and so an expression for profit, also as a function of $x$, can be derived as the difference between the revenue and cost functions.

**The profit function**

$$P(x) = R(x) - C(x)$$

where: $x$ is the quantity of items demanded (supplied or produced)

$P(x)$ is the *profit* function in terms of $x$

$R(x)$ is the *revenue* function in terms of $x$

$C(x)$ is the *cost* function in terms of $x$.

**Note.** It is usual to assume that the supply and demand of items (or products) is identical unless specifically stated otherwise.

A *maximum profit point* for some process can be defined through normal calculus methods as described in the following.

**Maximum profit point**

The *maximum profit* point for some process can be found by solving the equation:

$$\frac{dP}{dx} = 0$$

for $x$, where:

$x$ is the quantity of items demanded (supplied or produced)

$P$ is the *profit* function in terms of $x$.

## 4. Example 1 *(Profit* function and *maximum profit point)*

*Question*

A manufacturer knows that if $x$ (hundred) products are demanded in a particular week: (i) the total cost function (£000) is $14 + 3x$, and (ii) the total revenue function (£000) is $19x - 2x^2$.

a) Derive the (total) profit function.
b) Find the profit break-even points.
c) Calculate the level of demand that maximizes profit (i.e. the maximum profit point) and the amount of profit obtained.

*Answer*

a) Here, the total cost function, $C(x) = 14 + 3x$ and the total revenue function, $R(x) = 19x - 2x^2$.

Therefore, total profit function, $P(x) = R(x) - C(x)$

$= 19x - 2x^2 - (14 + 3x)$

i.e. $P(x) = 16x - 2x^2 - 14$.

b) The profit break-even points are the levels of demand which make $P(x)=0$.

Now, $P(x) = 0$ when: $16x - 2x^2 - 14 = 0$

or $8x - x^2 - 7 = 0$ (dividing throughout by 2)

Solving this quadratic equation gives $x=1$ and $x=7$.

Thus, the profit break-even points are when the demand is 100 or 700 products.

c) Profit is maximized when $\frac{dP}{dx} = 0$.

Now, since $P = 16x - 2x^2 - 14$, $\frac{dP}{dx} = 16 - 4x$.

Putting, $\frac{dP}{dx} = 0$

gives $16 - 4x = 0$,

which yields $x = 4$.

Note that $\frac{d^2P}{dx^2} = -4$, $(< 0)$ showing that $x = 4$ gives the *maximum* profit.

Finally, when $x = 4$, $P = 16(4) - 2(4)^2 - 14$
$= 18.$

Hence, the maximum profit of £18000 is realised when the weekly demand is 400.

## 5. Cost and revenue functions

The following sections describe the standard make-up of cost and revenue functions. It is not always the case that the precise form of cost and revenue functions are known (or, more to the point, given in an examination) as in Example 1. Sometimes it is necessary to derive them from supplied information and the techniques employed for doing this are described fully.

Cost functions are described first, then price (sometimes called demand) functions are covered, together with how they relate to the formulation of revenue functions.

## 6. Cost functions

The costs involved in standard processes can normally be categorized as follows.

a) *Fixed (or set-up) costs*. This type of cost is normally associated with the purchase, rent or lease of equipment and fixed overheads. It can sometimes include transportation and manpower movement costs also. In general, it can be considered as all those costs that need to be borne before production can physically begin (and thus independent of the number of items to be produced).

b) *Variable costs*. These are the costs normally associated with the supply of the raw materials and overheads necessary to manufacture each product. Thus, they will depend on the number of items produced. For example, if $x$ is the number of items produced in a day and each item costs £2.50, then the daily variable cost is £$2.5x$.

c) *Special costs*. This optional cost factor is sometimes included in a total cost function and might cover costs relating to storage, maintenance or deterioration. The total costs involved are normally expressed in the form $cx^2$, where c is a relatively small value and $x$ the number of items under consideration. The effects of this type of cost would only be significant for large production runs or other large processes. Hence, a total cost function would take a general form as follows.

**Form of a cost function**

$$C(x) = a + bx + cx^2$$

where:
- $x$ is the quantity of items demanded (supplied or produced)
- $a$ is the fixed cost associated with the product
- $b$ is the variable cost per item
- $c$ is the (optional) special cost factor.

## 7. Example 2 (To formulate and interpret a production *cost* function)

*Question*

The variable costs associated with a certain process are £0.65 per item. The fixed costs per day have been calculated as £250 with special costs estimated as £$0.02x^2$, where $x$ is the size of the production run (i.e. number of items produced).

a) Derive a function to describe cost *per item* for a day's production.
b) Calculate the size of the daily run that will minimise cost per item.
c) Find the cost of a day's production for a run that minimises cost per item.

*Answer*

a) The total cost function, $TC = 250 + 0.65x + 0.02x^2$ for the production of $x$ items (using the information given in the question).

Therefore, the cost per item, $C = \dfrac{TC}{x}$

$$= \frac{250+0.65x+0.02x^2}{x}$$

$$= \frac{250}{x}+0.65+0.02x$$

b) The cost per item will be minimised when $\dfrac{dC}{dx} = 0$.

Now, C can be expressed in the form $250x^{-1} + 0.65 + 0.02x$.

Thus $\dfrac{dC}{dx} = -250x^{-2} + 0.02$

$$= \frac{-250}{x^2}+0.02.$$

The cost (per item) function turning point is found by solving the equation:

$$\frac{dC}{dx} = 0.$$

This gives: $\dfrac{-250}{x^2}+0.02 = 0.$

i.e. $x^2 = \dfrac{250}{0.02} = 12{,}500.$

Thus, $x = 112$ (nearest whole number).

That is, the size of the daily run that minimises cost per item is 112.

c) The total cost function is $TC = 250 + 0.65x + 0.02x^2$.
When the daily run is 112 (i.e. the size of run which minimises cost per item), The cost of a day's run which minimises cost per item is found by substituting $x$=112 into $TC$.
This gives $250 + 0.65(112) + 0.02(112)^2 = 573.68$.
That is, the cost of a day's run which minimises cost per item is £573.68.

## 8. Revenue functions

An expression for the revenue obtained from the output of some process is given as follows.

**Form of a revenue function**

$$R(x) = x.\,p_r(x)$$

where: $x$ is the quantity of items demanded (supplied or produced)

$p_r(x)$ is the fixed cost associated with the product.

It is quite usual in a business environment for item price to depend on the number of items in demand. That is, the more items that are in demand, the less the *price per unit* is. Hence, the reason that price (in the above box) is expressed in terms of $x$. It is usually known as the *demand function*, where it should be remembered that (unless otherwise stated) it is assumed that supply = demand. The following two sections outline its form and the technique that sometimes needs to be used to identify it.

## 9. Demand (price) function

Price (or demand) functions are used to vary the price of an item according to how many items are being considered. As mentioned in the previous section, the more the number of items, the less the price per item and vice versa. This is simply the standard business principle of the 'economy of scale', where it is generally more efficient to operate on as large a scale as can be coped with. Demand functions are normally *linear*.

**Form of a demand function**

$$p_r(x) = a + bx$$

where: $x$ is the quantity of items demanded (supplied or produced)

$a, b$ are coefficients that can take any numeric value

$p_r(x)$ is the price function.

So that, if $p_r(x) = 25–2x$,

then the revenue function $R(x) = x(25–2x) = 25x - 2x^2$.

## 10. Identifying the demand function

In cases where the (assumed linear) demand function needs to be identified specifically, certain information needs to be known about it. In examination questions, the information necessary is usually given in the form of two separate examples of price, together with their appropriate demand levels. For example, when 100 items are demanded, the price is £0.85 per item; when 500 items are demanded, the price falls to £0.68 per item.

The procedure for obtaining the demand function (assuming that two examples of price and their corresponding demand levels are given) is:

STEP 1 Put the demand function in the *form* $p_r = a + bx$, say.

STEP 2 *Substitute* the first price/demand pair into $p_r$ to obtain one equation in $a$ and $b$.

STEP 3 *Substitute* the second price/demand pair into $p_r$ to obtain another equation in $a$ and $b$.

STEP 4 *Solve* the two equations obtained, for $a$ and $b$.

Example 3 demonstrates this technique and Example 4 shows how a demand function is used to obtain a revenue function.

## 11. Example 3 (To identify a linear *demand function*)

*Question*

Given that the price of an item is £3.50 when 250 items are demanded, but when only 50 are demanded, the price rises to £5.50 per item, identify the linear demand function and calculate the price per item at a demand level of 115.

*Answer*

Putting $p_r = a + bx$ and:

a) substituting for $p_r$=3.5 and $x$=250 gives: $3.5 = a + 250b$ [1]

b) substituting for $p_r$=5.5 and $x$=50 gives: $5.5 = a + 50b$ [2]

Subtracting equation [2] from [1] gives: $200b = -2$.

Hence, $b = -0.01$.

Substituting for $b = -0.01$ in equation [1] gives: $3.5 = a + (250)(-0.01)$

$= a - 2.5$

i.e. $a = 6$ and $b = -0.01$

The demand function is thus $p_r = 6 - 0.01x$.

Substituting into the above demand function gives:

when $x$=115, $p_r = 6 - 0.01(115)$

$= 4.85$.

Therefore, at a demand level of 115, the price is £4.85/item.

## 12. Example 4 (Obtaining a *revenue* function from a given demand function)

*Question*

Given the demand function

$p_r = 10.4 - 1.3x$ (where $x$ is in hundreds),

find the level of production (i.e. the value of $x$) which maximises *total revenue*.

*Answer*

Now, revenue is found by multiplying the demand function ($p_r$) by the number of items demanded.

Thus, the revenue function is: $R = x.p_r$

$= x(10.4 - 1.3x)$

i.e. $R = 10.4x - 1.3x^2$.

To maximise revenue, we need to solve the equation $\frac{dR}{dx} = 0$.

Now, $\frac{dR}{dx} = 10.4 - 2.6x$. Also $\frac{d^2R}{dx^2} = -2.6$, signifying a maximum as expected.

So when $\frac{dR}{dx} = 0$,

$10.4 - 2.6x = 0$

which gives $2.6x = 10.4$.

Thus, $x = 4$.

That is, the level of production necessary to maximise revenue is 400 units.

## 13. Marginal cost and revenue functions

If for some process, the total cost function, $C$, and the revenue function, $R$, are identified, where $x$ is the level of activity:

a) $\frac{dC}{dx}$ is defined as the *marginal cost function* and can be interpreted as the extra cost incurred of producing another item at activity (demand) level $x$.

b) $\frac{dR}{dx}$ is defined as the *marginal revenue function* and can be interpreted as the extra revenue obtained from producing another item at activity level $x$.

c) These two functions can be used to give an alternative way of obtaining the maximum profit point (i.e. the value of $x$ that maximises profit) for the process, described as follows.

**Maximum profit point**

The *maximum profit point* for some process can be found by solving the equation:

$$\frac{dR}{dx} = \frac{dC}{dx}$$

where: $C$ is the total cost function

$R$ is the total revenue function.

The following example shows the use of this result.

## 14. Example 5 (*Cost, demand* and *profit* functions and their interpretation)

*Question*

A refrigerator manufacturer can sell all the refrigerators of a particular type that he can produce.

The total cost (£) of producing $q$ refrigerators per week is given by $300q + 2000$.

The demand function (£) is estimated as $500 - 2q$.

a) Derive the revenue function, $R$.
b) Obtain the total profit function.
c) How many units per week should be produced in order to maximise profit?
d) Show that the solution of the equation $\frac{dR}{dx} = \frac{dC}{dx}$ where $C$ represents the cost function, gives the same value for $q$ as in part c).
e) What is the maximum profit available?

*Answer*

a) The demand function is given as $p_r = 500 - 2q$.
Therefore, revenue function, $R = p_r . q = (500–2q)q$
i.e. $R = 500q - 2q^2$
b) The cost function, is given as $C = 300q + 2000$.
Therefore the profit function, $P = R - C = 500q - 2q^2 - (300q + 2000)$
This gives $P = 200q - 2q^2 - 2000$.
c) Now $\frac{dP}{dq} = 200 - 4q$ (and $\frac{d^2P}{dq^2} = -4$, signifying a maximum point).
Thus: $\frac{dP}{dq} = 0$
giving: $200 - 4q = 0$.
Therefore: $q = 50$.
That is, 50 units/week need to be produced in order to maximise profits.
d) $\frac{dR}{dq} = 500 - 4q$ and $\frac{dC}{dq} = 300$.
So that, if $\frac{dR}{dq} = \frac{dC}{dq}$
then: $500 - 4q = 300$.
i.e. $4q = 200$
So: $q = 50$ agreeing with the answer to part (c).
e) Substituting $q= 50$ into the profit function gives:
maximum profit $= 200(50) - 2(50)^2 - 2000$
$= 3000$.
That is, maximum profit available is £3000.

## 15. Summary

a) The basic profit equation describing some process is:

Profit ($P$) = Revenue ($R$) – Costs ($C$),

or if revenue and costs are given as functions,

$P(x) = R(x) - C(x)$, where $x$ is the quantity demanded.

b) A total cost function can normally be expressed in the form: $C(x) = a + bx + cx^2$, where $a$ is the fixed cost, $b$ is the variable cost and $c$ is the (optional) special cost factor.

c) A price (or demand) function normally takes the form: $p_r = a + bx$, where $x$ is the quantity demanded.

d) A revenue function is normally expressed in the form: $R = x.p_r$, where $x$ is the number of items supplied and $p_r$ is the demand function.

e) $\frac{dC}{dx}$ is the marginal cost function, and $\frac{dR}{dx}$ the marginal revenue function.

f) The solution of the equation $\frac{dC}{dx} = \frac{dR}{dx}$ defines the maximum profit point.

## 16. Points to note

a) The term 'process' can be translated very broadly. It can describe a complex manufacturing situation, where a large number of raw materials are put through a series of sub-processes and end up as a sophisticated product such as a motor car or washing machine. On the other hand, the term can be used to describe the operations of a wholesaler, who will buy in bulk from various manufacturers and channel through smaller amounts to retailers, without the physical goods changing their nature in any way.

b) It was mentioned earlier that, to all intents and purposes, supply equals demand and, in this context, the two words have been used interchangably in the text. In the rare instances where examiners have differentiated between the two, the question has involved algebraic manipulation and equation solving of a fairly routine nature. For example, given a (linear) supply function $S(x)$, say, and a (linear) demand function $D(x)$, say, the question is invariably 'for what level of production does supply equal demand?'. This is answered by solving the equation $S(x) = D(x)$.

## 17. Student self review questions

1. For a given process, what form does the standard profit function take? [3]
2. What is the difference between 'set-up' and 'variable' costs? [6]
3. What general form does a total cost function take? [6]
4. Under what conditions would revenue be described by a 'linear function of the quantity $x$'? [8]
5. What purpose does a price (or demand) function serve? [9]
6. What information does a marginal cost function give? [13]
7. What is the significance of the equation $\frac{dC}{dx} = \frac{dR}{dx}$? [13]

## 18. Student exercises

1. *MULTI-CHOICE*. In the function $C = 4 + 0.8Q$, $C$ is the total cost of sales (in £000) and $Q$ is the number of thousands of units sold. The total cost of sales for 2,000 units is closest to:

   a) £1,600  b) £5,600  c) £8,200  d) £12,400

2. *MULTI-CHOICE*. A product's weekly costs, TC, are less than or equal to £1,000. Weekly revenue, R, is a minimum of £1,200. Which one of the following statements is true?
   a) TC<£1000 and R>£1200 and R>TC
   b) TC ≥ £1000 and R ≤ £1200 and TC>R
   c) TC ≤ £1000 and R>£1200 and R<TC
   d) TC ≤ £1000 and R ≥ £1200 and R>TC
3. A process has a total cost function given by $C = 20+4x$ and a revenue function given by $R = 22x-4x^2$, where $x$ is the level of activity (in hundreds). $C$ and $R$ are both in units of £1000.
   a) Derive an expression for the (total) profit, $P$.
   b) Calculate the level of activity that maximises profit and the amount of profit at this level.
   c) What is the profit situation when 350 units are produced?
4. A manufacturing process costs £6500 to set up for one year's use. If items cost £85 each to produce and other costs amount to $3.5x^2$, where $x$ is the production in hundreds, find the level of production that will minimise the cost per item over the year. What will the total cost amount to at this level of production?
5. The price of a particular product is set at £12 per item for 10 items and falls to £8.50 per item when 30 are ordered. Derive the (assumed linear) demand function.
6. The demand function (£) for a particular process is given by $p_r = 1800-100q$, where $q$ is the quantity demanded in a month. What is the maximum amount of revenue obtainable in a month?
7. A production department has determined that the marginal profit (the differential of the total profit function) for a particular product is £0.50 and the marginal cost is £2, with a *fixed cost* of £2000.
   a) What is the selling price per unit?
   b) Derive expressions for (i) cost/unit and (ii) profit/unit.
   c) Calculate the profit break-even point.
   d) What level of production will give a profit of £0.10/unit?
   e) If the selling price is increased by 20%, find the new profit break-even point.
8. At a selling price of £3.80 per unit the expected sales of a particular product would be 10200, but would fall to 8400 if the selling price was £4.70. The total cost function (in £) for the product is $15000+1.8x$, where $x$ is the number of units.
   a) Derive the demand function, assuming it is linear.
   b) Derive an expression for total profit.
   c) Calculate the maximum profit and the level of sales which achieves it.
   d) What price is charged per unit at the maximum profit point?

# Examination examples (with worked solutions)

*Question 1*

a) Draw the graph of $y=x^2-6x+9$ from $x=1$ to $x=5$, and solve $x^2=6x-9$ graphically.

b) Two processes are necessary for the production of two separate products $x$ and $y$. The first process $A$ has a maximum capacity of 24000 hours and the second process $B$ has a maximum capacity of 21000 hours. Each unit of $x$ requires 4 hours processing in $A$ and 3 hours processing in $B$, while each unit of $y$ requires 2 hours processing in $A$ and 3 hours processing in $B$. The company seeks to make maximum use of the plant capacity. State how many units of $x$ and $y$ should be produced to ensure that the plant capacity is fully employed and there is no idle time in either process.

CIMA

*Answer*

a) The following table shows the calculations to determine $y$ for $x$ ranging from 1 to 5.

| $x$ | 1 | 2 | 3 | 4 | 5 |
|---|---|---|---|---|---|
| $x^2$ | 1 | 4 | 9 | 16 | 25 |
| $-6x$ | −6 | −12 | −18 | −24 | −30 |
| $+9$ | 9 | 9 | 9 | 9 | 9 |
| $y=x^2-6x+9$ | 4 | 1 | 0 | 1 | 4 |

The graph is drawn in Figure 1.

*Figure 1*

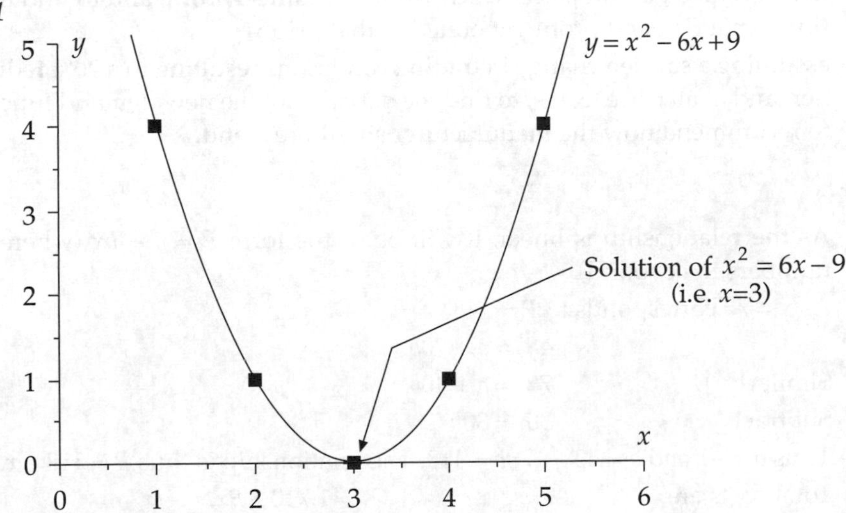


If $x^2=6x-9$ then $x^2-6x+9=0$. This equation is solved by finding the x-values where the graph of $y=x^2-6x+9$ meets the x-axis. From the graph in Figure 1, this can seen to be $x=3$ only.

b) Let $u$ be the number of units of $x$ produced and $v$ be the number of units of $y$ produced. The maximum capacity of process $A$ is 24000 hours and each one of the $u$ will take 4 hours, each one of the $v$ will take 2 hours.

Thus, for maximum use of process $A$: $4u + 2v = 24000$ [1]

Similarly, for process $B$: $3u + 3v = 21000$ [2]

In order to find the values of $u$ and $v$, equations (1) and (2) need to be solved simultaneously.

| | | |
|---|---|---|
| [1]×3 gives: | $12u + 6v = 72000$ | [3] |
| and [2]×2 gives: | $6u + 6v = 42000$ | [4] |
| Subtracting, ([3]–[4]) gives: | $6u = 30000.$ | |
| Therefore | $u = 5000.$ | |
| Substituting back into [1] for $u$=5000 gives | $4(5000) + 2v = 24000.$ | |
| i.e. | $v = 2000.$ | |

The solutions of the simultaneous equations are $u$=5000 and $v$=2000.

Thus for no idle time and maximum use, 5000 units of $x$ and 2000 units of $y$ should be produced.

### *Question 2*

A manufacturer of a new patented product has found that he can sell 70 units a week direct to the customer if the price is £48. In error, the price was recently advertised at £78 and , as a result, only 40 units were sold in a week. The manufacturers fixed costs of production are £1,710 a week and variable costs are £9 per unit. You are required

a) to show the equation of the demand function linking price ($P$) to quantity demanded ($X$), assuming it to be a straight line, is $P = 118 - X$;

b) to find where the manufacturer breaks even;

c) to recommend a unit price which would maximise profit, and to find the quantity demanded and profit generated at that price;

d) assuming a sudden change in trading conditions resulting in a 20% reduction in demand at all price levels, to find the equation of the new demand function and to recommend how the manufacturer should respond.

*CIMA*

### *Answer*

a) As the relationship is linear, it will be of the form $P = a + bx$ (where $a$, $b$ are numbers to be found).

As $x = 70$ corresponds to $P$ = £48

$$48 = a + 70b$$

Similarly $\quad 78 = a + 40b$

Subtract $\quad -30 = 30b$

Thus $b = -1$ and $a = 438 - 70b = 118$. The relationship is thus $P = 118 - x$.

b) Total costs are $\quad C = 1{,}710 + 9x.$

Revenue is given by $\quad R = \text{demand price} = x \,.\, P$

$$R = 118x - x^2$$

The break-even points occur where $\quad R = C$

$$118x - x^2 = 1{,}710 + 9x$$

$$x^2 - 109x + 1{,}710 = 0$$

$$(x - 19)(x - 90) = 0$$

$x = 19$ or $x = 90$ (or use quadratic formula).

Correspondingly: $P = 99$ or $P = 28$

Thus the manufacturer breaks even at prices of £28/unit (demand = 90 units) and £99/unit (demand =19 units).

c) The profit, $G$, is given by $i = R \quad C = -x^2 + 109x - 1{,}710$

Thus $\dfrac{dG}{dx} = -2x + 109$ and $\dfrac{d^2G}{dx^2} = -2$ which means that profit is a maximum where $G' = 0$.

This gives: demand $x = 54.5$, with price $P = £63.50$.

The maximum profit is $G = -54.52 + 109 \times 54.5 - 1{,}710 = £1{,}260.25$.

d) At a price of £48, the demand falls to 56, hence the demand equation changes to

$$48 = a + 56b.$$

Similarly $\quad 78 = a + 32b.$

Subtract: $\quad -30 = 24b$

This yields: $b = -1.25$, $a = 118$.

Thus $\quad P = 118 - 1.25x.$

The profit function now changes to

$$G = 118x - 1.25x^2 - 1{,}710 - 9x = -1.25x^2 + 109x - 1{,}710$$

$$\frac{dG}{dx} = -2.5x + 109$$

$\dfrac{dG}{dx} = 0$ now gives $x = 43.6$. The corresponding price is $P = £63.50$.

Hence the manufacturer should not amend his price.

# Examination questions

1. A small company produces specialised posters. The total cost is made up of three elements – materials, labour and administration – as follows:

| | |
|---|---|
| Materials | £0.50 per poster |
| Labour | £15 per hour |
| Administration | £10 per hundred, plus £50 |

The set-up time for printing takes 2 hours, and the posters are run off at the rate of 300 per hour.

You are required

a) to calculate the total cost of producing 1,000 posters;

b) to produce a simple formula for the total cost, £$C$, in terms of $N$, where $N$ is the numbers of posters produced, and to explain its meaning;

c) to find how many posters can be produced for (i) £500, and (ii) £$N$.

*CIMA*

2. The expression to determine the net present value £$P$ of a certain project is given by:

$$P = (-32r^2 + 884r - 5985)$$

where $r$ is the rate of interest.

Required:

a) Determine the internal rates of return of this project, by solving the appropriate quadratic equation.

b) Tabulate and plot the function $(-32r^2 + 884r - 5985)$ for the SIX integer values of $r$ which include the values of the internal rates of return in their range.

c) Using the graph of part (b) explain the significance of net present value as applied to this project.

d) Using the methods of differential calculus, determine and confirm the maximum net present value of the project.

*ACCA*

3. a) A cost accountant for a bus company has calculated that the number of passengers on a certain route increased by 30% when the fare was reduced by 10%. Find the percentage increase in revenue.

b) On another route, when the fare was reduced by $x$% the number of passengers increased by $2x$%. Show *algebraically* that the revenue was multiplied by: $(1 + 0.01x - 0.0002x^2)$.

c) Using a graph, or otherwise, find the value of $x$ which produces maximum revenue, and hence calculate the percentage increase in revenue which this would bring. Explain any assumptions you have made.

*CIMA*

4. a) A company manufactures a single product. Each unit is sold for £15.
The operating costs are:

| | £ |
|---|---|
| Fixed, per week | 800 |
| Prime, per unit produced | 5 |

The weekly maintenance cost $M$ is given by $M = 0.009x^2$ where $x$ is the weekly production in units.
You are required to:
   i. calculate the range of possible production in whole units, when no units are left unfinished at the end of the week, to provide a weekly profit of at least £200;
   ii. illustrate your answer to (i) above by drawing a graph of the profit equation.

b) Over a six year period, an investment depreciated in value from £32000 to £23500. If a reducing balance method of depreciation was used, determine the rate per annum to one decimal place.

*CIMA*

5. a) The total cost function is given by $C = x^2 + 16x + 39$, where $x$ units is the quantity produced and £$C$ the total cost.
Required:
   i. Write down the expression for the average cost per unit.
   ii. Sketch the average cost function against $x$, for values of $x$ between $x$=0 and $x$=8.

b) The demand function is given by $p = x^2 - 24x + 117$, where $x$ units is the quantity demanded and £$p$ the price per unit.
Required:
   i. Write down an expression for the total revenue for $x$ units of production.
   ii. Using the methods of differential calculus, establish the number of units of production and the price at which total revenue will be maximised.
   iii. If elasticity of demand is defined as: $\dfrac{1}{\left[\dfrac{x}{P} \times \dfrac{dP}{dx}\right]}$, determine the elasticity of demand for the quantity which maximises the total revenue.

*ACCA*

6. Your company manufactures large scale items. It has been shown that the marginal (or variable) cost, which is the gradient of the total cost curve, is $(92-2x)$ £thousands, where $x$ is the number of units of output per annum. The fixed costs are £800,000 per annum. It has also been shown that the marginal revenue, which is the gradient of the total revenue curve, is $(112-2x)$ £thousands.
Required:
a) Establish by integration the equation of the total cost curve.
b) Establish by integration the equation of the total revenue curve.
c) Establish the break-even situation for your company.

d) Determine the number of units of output that would:
   i. maximise the total revenue, and
   ii. maximise the total costs,

   together with the maximum total revenue and total costs.

e) Assuming that your company cannot manufacture more than 60 units of output per annum, what interpretation can be put on the results you obtain in d)? (A sketch of the total revenue and cost curve will be helpful.)

*ACCA*

7. a) A company subsidises a certain rail journey for some of its employees. When the price of the tickets is increased by £6, the number of tickets which the company can purchase for £2,850 is reduced by 36.

   You are required to find the percentage increase in the price of the ticket.

   b) A mathematically-minded street trader with no overheads has found that the weekly volume of sales of a toy are approximately $100/p2$, where £$p$ is the fixed price of the toy. The toy costs the trader 15 pence.

   You are required, by using a graph, or any method you consider appropriate, to find:

   i. the level of $p$ which maximises profit;
   ii. the level of the maximum profit;
   iii. the weekly volume of toys sold at this level.

*CIMA*

8. Your firm has recently started to give economic advice to your clients. Acting as a consultant you have estimated the demand curve of a client's firm to be

$$AR = 200 - 8x$$

where $AR$ is average revenue (£) and $x$ is output.

Investigation of the client firm's cost profile shows that marginal cost is given by

$$MC = x^2 - 28x + 211$$

where $MC$ is marginal cost (£). Further investigation has shown that the firm's costs when not producing output are £10.

Required:

a) If total cost is the integral of marginal cost find the equation of total cost.

b) If total revenue is average revenue multiplied by output find the equation of total revenue.

c) Profit is total revenue minus total cost. Using the methods of differentiation finding the turning point(s) of the firm's profit curve and say whether these point(s) are maxima or minima.

d) Marginal revenue is the first differential of total revenue. Find the equation of marginal revenue.

e) On the same axes sketch the marginal cost and marginal revenue curves.

*ACCA*

9. a) The cost of producing a quantity $Q$ of a product is given by an equation of the form

$$C = aQ^2 - bQ + c \qquad \text{where } a, b, c \text{ are constants.}$$

It is observed that if $Q = 10$, $C = 2900$, but if $Q = 40$, $C = 800$, or if $Q = 100$, $C = 2000$.

(i) Determine $a$, $b$ and $c$.

(ii) Plot your Cost Function for values of $Q$ between 0 and 120 and use the plot to determine the quantity that should be produced to minimise the total production cost.

(iii) The product is sold at £20 per unit. Write down the Revenue Function and plot it on the same graph as your Cost Function. Determine the range of production quantities over which the company can make a profit.

(iv) Estimate the minimum price at which the product must be sold to avoid a loss.

b) It is estimated that in 10 years time production costs will have increased by £4000 (ie £4000 is added to the Cost Function).

(i) Calculate the new Total Costs and find their Net Present Values assuming an interest rate of 10% per year. Plot the Net Present Value Function on your graph and determine the production quantities for which real costs are the same.

(ii) Estimate the minimum price that the product must be sold at in 10 years to avoid a loss.

# Part 8 Probability

Chapter 29 introduces the elementary theory of Sets; their basic make-up, their pictorial representation in Venn diagrams and methods of combining and enumerating them. This topic needs to be studied in order to fully understand the concept of probability.

Chapter 30 deals with the foundations of probability, introducing the idea of experiments and events. Probability is defined and the addition and multiplication rules are described.

Chapter 31 covers more advanced ideas in probability including expectation and conditional probability and their ability in solving problems in practical business situations.

# 29 Set theory and enumeration

## 1. Introduction

This chapter introduces some basic concepts in Set Theory, describing sets, elements, Venn diagrams and the union and intersection of sets. This is followed by a discussion of the problems of set enumeration.

## 2. Sets and elements

Sets of objects, numbers, departments, job descriptions ... etc are things that we all deal with every day of our lives. Mathematical Set Theory just puts a structure around this concept so that sets can be used or manipulated in a logical way. The type of notation used is a reasonable and simple one.

For example, suppose a company manufactured 5 different products *a*, *b*, *c*, *d*, and *e*. Mathematically, we might identify the whole set of products as *P*, say, and write:

$$P = \{a,b,c,d,e\}$$

which is translated as 'the set of company products, *P*, consists of the *members* (or *elements*) *a*, *b*, *c*, *d* and *e*'.

The elements of a set are usually put within braces (curly brackets) and the elements separated by commas, as shown for set *P* above.

> A **mathematical set** is a collection of distinct objects, normally referred to as *elements* or *members*.
>
> Sets are usually denoted by a capital letter and the elements by small letters.

## 3. Example 1 (Illustrations of *sets*)

a) The employees of a company working in the purchase department could be written as:

$P$ = {A.R.Jones, R.Wilson, Mrs E.Smith, A.Smythe, Miss F.Gait}

b) The warehouse locations of a large supermarket chain could be written as:

$W$ = {Edinburgh, Carlisle, Leeds, Nottingham, Wrexham, Coventry}

## 4. Further set concepts

a) *Subsets*. A subset of some set *A*, say, is a set which contains some of the elements of *A*. For example:

if $A = \{h,i,j,k,l\}$, then:

$X = \{i,j,l\}$ is a subset of $A$

$Y = \{h,l\}$ is a subset of $A$

$Z = \{i,j\}$ is a subset of $A$ and also a subset of $X$.

b) The *number* of a set. The number of a set *A*, written as $n[A]$, is defined as the number of elements that *A* contains.

For example:
if $A = \{a,b,c,d,e\}$, then $n[A]=5$ (since there are 5 elements in $A$);
if $D$ = {Sales, Purchasing, Inventory, Payroll}, then $n[D]=4$.

c) *Set equality*. Two sets are equal only if they have identical elements. Thus, if $A = \{x,y,z\}$ and $B = \{x,y,z\}$, then $A = B$.

d) The *universal* set. In some problems involving sets, it is necessary to consider one or more sets as belonging to some larger set that contains them. For example, if we were considering the set of skilled workers ($S$, say) on a production line, it might be convenient to consider the universal set ($U$, say) as all of the workers on the line. In other words, where a universal set has been defined, all the sets under consideration must necessarily be subsets of it.

e) The *complement* of a set. If $A$ is any set, with some universal set $U$ defined, the complement of $A$, normally written as $A'$, is defined as 'all those elements that are not contained in $A$ but are contained in $U$'. For the example of the workers on the production line (given in d) above), $S$ was specified as the set of skilled workers within the universal set of all workers on the line. Therefore, $S'$ would be all the workers that were not skilled. i.e. the set of unskilled workers.

## 5. Venn diagrams

A *Venn diagram* is a simple pictorial representation of a set. For example, if $M = \{a,b,c,d,e,f,g\}$, then we could represent this information in the form of a Venn diagram as in Figure 1.

*Figures 1 and 2 Venn diagrams – set M with elements and set A with subset D*

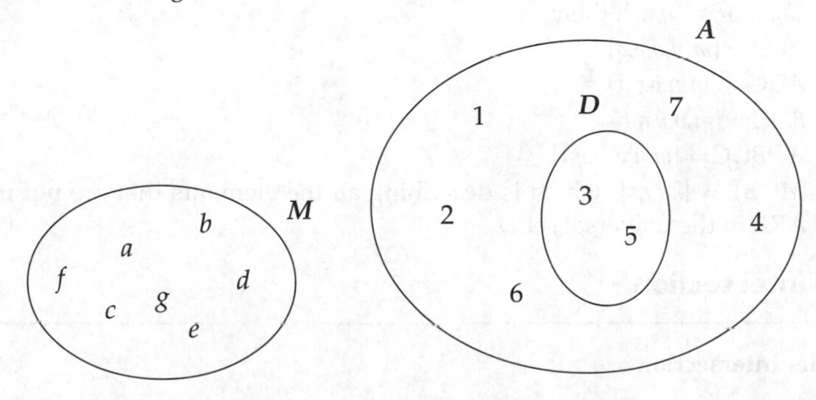


Venn diagrams are useful for demonstrating general relationships between sets. For example, if a firm maintains a fleet of 7 cars, we might write $A = \{1,2,3,4,5,6,7\}$ (each car being numbered for convenience). If also it was important to identify those cars of the fleet that were being used by the directors, we might have $D = \{3,5\}$. i.e. Cars 3 and 5 are director's cars. This situation could be represented in Venn diagram form as in Figure 2. This diagram nicely demonstrates the fact that $D$ is a subset of $A$, which normally means that $n[D]<n[A]$. In this case $n[D]=2$ and $n[A]=7$.

## 6. Operations on sets

In ordinary arithmetic and algebra there are four common operations that can be performed; namely, addition, subtraction, multiplication and division. With sets, however, just two operations are defined. These are set *union* and set *intersection*. Both of these operations are described, with examples, in the following sections.

## 7. Set union

**Set union**

The *union* of two sets $A$ and $B$ is written as $A \cup B$ and defined as that set which contains all the elements lying within *either* $A$ or $B$ or *both*.

For example, if $A = \{c,d,f,h,j\}$ and $B = \{d,m,c,f,n,p\}$, then the union of $A$ and $B$ is $A \cup B = \{c,d,f,h,j,m,n,p\}$, these being the elements that lie in either $A$ or $B$. So that any element of $A$ *must* be an element of $A \cup B$; similarly any element of $B$ *must also* be an element of $A \cup B$.

Set union for three or more sets is defined in an obvious way. That is, if $A$, $B$ and $C$ are any three sets, $A \cup B \cup C$ is the set containing all the elements lying within (i) any one of $A$, $B$ or $C$, (ii) any two of them or (iii) all three.

## 8. Example 2 (To demonstrate set *union*)

If $A = \{m,n,o,p\}$; $B = \{m,o,p,q\}$; $C = \{m,p,r\}$; and the universal set is defined as $U = \{k,l,m,n,o,p,q,r,s\}$, then:

a) $A \cup B = \{m,n,o,p,q\}$
b) $A \cup C = \{m,n,o,p,r\}$
c) $B \cup C = \{m,o,p,q,r\}$
d) $A \cup B \cup C = \{m,n,o,p,q,r\}$
e) $(A \cup B)' = \{k,l,r,s\}$, which is describing all the elements that are not in $A \cup B$ but ARE in the universal set $U$.

## 9. Set intersection

**Set intersection**

The *intersection* of two sets $A$ and $B$ is written as $A \cap B$ and defined as that set which contains all the elements lying within *both* $A$ or $B$.

For example, if $A = \{a,b,c,d,f,g,\}$ and $B = \{c,f,g,h,j\}$, then the intersection of $A$ and $B$ is $A \cap B = \{c,f,g\}$, since these are the elements that lie in both sets.

The intersection of three or more sets is a natural extension of the above. If $P$, $Q$ and $R$ are any three sets then $P \cap Q \cap R$ is the set containing all the elements that lie in *all three sets*.

Any combinations of union and intersection can be used with sets. For example, if $X$ and $Y$ are the sets specified above and $Z = \{d,f,g,j\}$, then: $(X \cap Y) \cup Z = \{c,f,g\} \cup \{d,f,g,j\} = \{c,d,f,g,j\}$, which can be described in words as 'the set of elements that are in either both of $X$ and $Y$ or in $Z$'.

## 10. Example 3 (To demonstrate set *intersection*)

If $A = \{m,n,o,p\}$; $B = \{m,o,p,q\}$; $C = \{n,q,r\}$; with a universal set defined as $\{k,l,m,n,o,p,q,r,s\}$, then:

a) $A \cap B = \{m,o,p\}$, since all these elements are in *both* sets.
   Similarly,

b) $A \cap C = \{n\}$

c) $B \cap C = \{q\}$.

d) $A \cap B \cap C$ has no elements, is sometimes called the *empty set* and can be written $A \cap B \cap C = \{\}$. Note $n[\{\}]=0$.

e) $(A \cap B)' = \{k,l,n,q,r,s\}$ is the complement of $A \cap B$ and is the set of all elements that are NOT in both $A$ and $B$.

f) $(A \cup B) \cap C = \{m,n,o,p,q\} \cap \{n,q,r) = \{n,q\}$ is the set of all elements that are in $A$ or $B$ AND ALSO in C.

## 11. Example 4 (The *union* and *intersection* of given sets)

*Question*

In a particular insurance Life office, employees Smith, Jones, Williamson and Brown have 'A' levels, with Smith and Brown also having a degree. Smith, Melville, Williamson, Tyler, Moore and Knight are associate members of the Chartered Insurance Institute (ACII) with Tyler and Moore having 'A' levels. Identifying set $A$ as those employees with 'A' levels, set $C$ as those employees who are ACII and set $D$ as graduates:

a) Specify the elements of sets $A$, $C$ and $D$.

b) Draw a Venn diagram representing sets $A$, $C$ and $D$, together with their known elements.

c) What special relationship exists between sets $A$ and $D$?

d) Specify the elements of the following sets and for each set, state in words what information is being conveyed.
   i. $A \cap C$ ii. $D \cup C$ iii. $D \cap C$

e) What would be a suitable universal set for this situation?

*Answer*

a) $A$ = {Smith, Jones, Williamson, Brown, Tyler, Moore};
   $C$ = {Smith, Melville, Williamson, Tyler, Moore, Knight}; $D$ = {Smith, Brown}

b) The Venn diagram is shown in Figure 3.

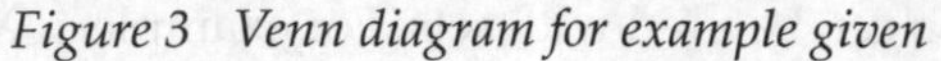
*Figure 3 Venn diagram for example given*

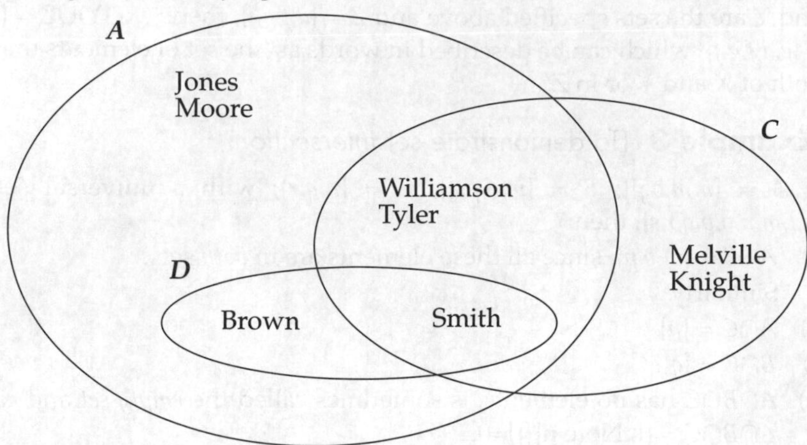


c) From the diagram, it can be seen that $D$ is a subset of $A$.

d) This information can be obtained either from the Venn diagram or from the sets listed in a) above.

   i. $A \cap C$ = {Williamson, Tyler, Smith}. This set gives the employees who have both 'A' levels and are ACII.

   ii. $D \cup C$ = {Brown, Smith, Williamson, Tyler, Melville, Knight}. This set gives the employees who are either graduates or ACII.

   iii. $D \cap C$ = {Smith}. This set gives the single employee who is both a graduate and ACII qualified.

e) A suitable universal set for this situation would be the set of all the employees working in the Life office.

## 12. Example 5 (Some useful results using set number, union and intersection)

*Question*

a) What does $n[A \cap B]=0$ imply about sets $A$ and $B$?

b) What is the largest value that $n[A \cap B]$ can take (in terms of $n[A]$ and/or $n[B]$) and what does this imply about sets $A$ and $B$?

c) What is implied about sets $A$ and $B$ if $n[A \cup B]=n[B]$?

d) What is the largest value that $n[A \cup B]$ can take (in terms of $n[A]$ and/or $n[B]$) and what does this imply about sets $A$ and $B$?

*Answer*

a) $n[A \cap B]=0$ means that there are no elements in the intersection of the two sets $A$ and $B$. That is, they do not intersect. This relationship is shown in Venn diagram form in Figure 4. Sometimes sets that have no intersection are known as *disjoint*.

*Figure 4 Venn diagram showing non-intersecting sets*

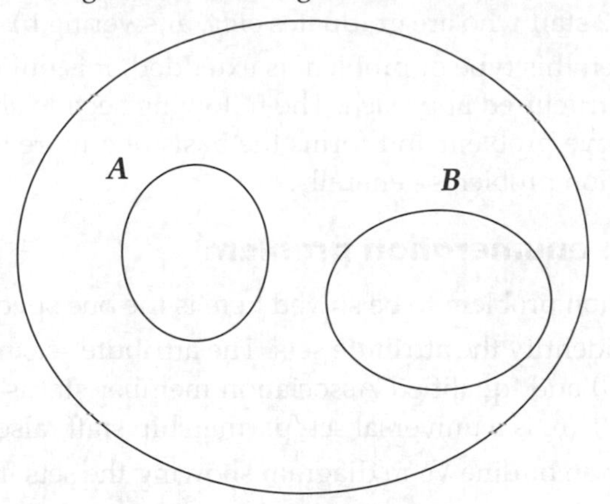


b) The more elements that two sets $A$ and $B$ have in common, the larger their intersection will be. The largest number of elements that can be contained in their intersection will thus be the smaller of the two numbers $n[A]$ and $n[B]$. In this case, one of the sets *must be a subset of the other.*
For example, $A = \{a,b,c,d\}$ and $B = \{a,b,c,d,e,f\}$. This gives $A \cap B = \{a,b,c,d\}$. Therefore, $n[A \cap B]=n[A]=4$.

c) In this case, it must be that set $A$ adds nothing to the union. The only way this can happen is if $A$ is a subset of $B$. For example, taking sets $A$ and $B$ as those in (b) above, where $A$ is a subset of $B$, we have: $A \cup B = \{a,b,c,d,e,f\}$, giving $n[A \cup B]=n[B]=6$.

d) The largest value $n[A \cup B]$ can take is $n[A]+n[B]$ and this can only happen when sets $A$ and $B$ are disjoint. For example, $A = \{a,b,c\}$ and $B = \{d,e,f,g\}$. Here, $A \cup B = \{a,b,c,d,e,f,g\}$ and $n[A \cup B] = 7 = n[A]+n[B]$. See also the result in a) above and the associated Venn diagram.

## 13. Set enumeration

The rest of the chapter is concerned with considering sets only in terms of the number of elements contained within the various areas defined by their union or intersection. Identifying the numbers of elements in these areas is known as *set enumeration.*

## 14. An enumeration problem

Suppose an accountancy partnership currently employs 16 staff. Given that 3 staff have no formal qualifications, and of the 7 staff who are graduates, 5 are also qualified Association members, it is possible to evaluate: a) the number of staff who are non-graduate, qualified members of the Association and b) the number of graduates who are not qualified members of the Association.

These two values can be calculated as follows.

i. Since 3 staff have no formal qualifications, there must be 16 – 3 = 13 staff who have at least one of the two qualifications.

ii. But there are 7 staff who are graduates, which means that 16 – 7 – 3 = 6 staff must be non-graduate, qualified members of the Association, which answers a).

iii. Also, since 5 staff are qualified Association members and graduates, there must be 7 – 5 = 2 staff who are graduates *only*, answering b).

However, when this type of problem is extended, it is quite difficult to solve using the above unstructured approach. The following section shows a procedure which tackles the above problem and forms the basis of a more logical approach of tackling enumeration problems generally.

## 15. Solving an enumeration problem

The enumeration problem to be solved here is the one specified in section 14.

First, we will identify the attribute sets. The attribute sets here are 'graduate status' (call this set *G*) and 'qualified Association member status' (call this set *A*). Notice however that there is a universal set 'partnership staff' also involved.

Next we draw an outline Venn diagram showing the sets involved. See Figure 5.

*Figure 5 Venn diagram showing numbers in groups within the accountancy partnership*

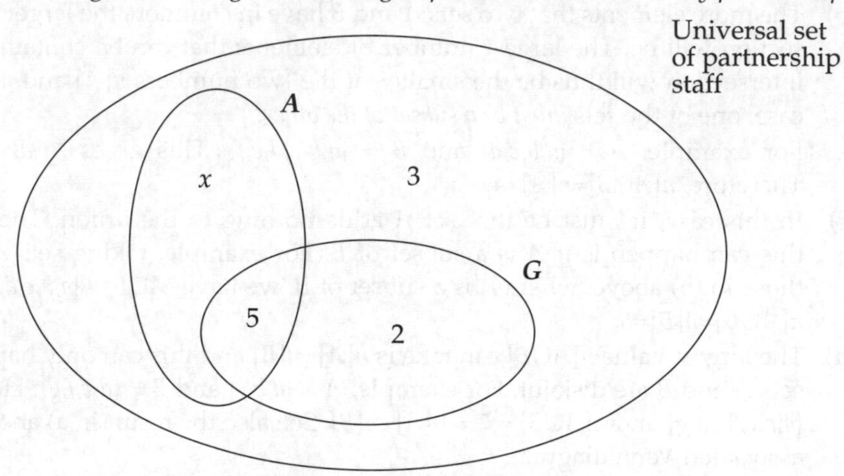


*Note*: The values (2, 3, 5 and *x*) shown are the number of elements contained in each of the four defined regions.

Now we can use the information given to fill in as much of the diagram as possible. Note that if two sets intersect within some defined universal set, there are four distinct areas defined; in this particular case: '*A* and *G*' (or more technically, $A \cap G$), '*A* alone', '*G* alone' and 'neither *A* nor *G*' (or more technically, $(A \cup G)'$).

Since there are 16 staff in total, the sum of the numbers in the four areas must be 16. Using the other information given:

i. 5 are graduates *and* Association qualified members, thus this number can be entered in the '*A* and *G*' area.

ii. Since there are 7 graduates altogether, there must be 7–5=2 graduates who are not qualified members of the Association. This number is entered in the '*G* alone' area.

iii. 3 of the staff had no qualifications at all and this number can be entered in the 'neither *A* nor *G*' area.

Finally we can evaluate the number of elements in any remaining unknown areas. Putting $x$ as the unknown number in the area '$A$ alone' that is required, we must have that:

$$3 + 5 + 2 + x = 16 \quad \text{(the total number of staff).}$$
$$\text{i.e. } x = 6.$$

That is, there are 6 staff who are qualified Association members but not graduates.

## 16. Defining the general enumeration problem

It is worthwhile at this stage to state clearly what the *general enumeration problem* is.

**The general enumeration problem**

Given:

a) overlapping attributes which can be represented as intersecting sets on a Venn diagram and thus defining a number of distinct areas;

b) the number of elements in a selection of areas

to find the number of elements in *each* distinct area.

**Note**, in particular, that there are:

a) 4 distinct areas (for two attribute sets), and

b) 8 distinct areas (for three attribute sets).

To put the particular problem given in section 14 into terms of the above definition, we had:

i. *Two* overlapping attributes, 'graduate status' and 'qualified members of Association' (overlapping in the sense that any member of staff could possess either none, one or both of the attributes), which were represented as intersecting sets $G$ and $A$ respectively.

ii. The 4 distinct areas on the Venn diagram were: '$A$ and $G$', '$A$ alone', '$G$ alone' and 'neither $A$ nor $G$'.

iii. Information was given that enabled a given area to be enumerated.

## 17. Solution of the general enumeration problem

The procedure for solving an enumeration problem is:

STEP 1 Identify the attribute sets.

STEP 2 Draw an outline Venn diagram.

STEP 3 Use the information given to fill in as much of the diagram as possible.

STEP 4 Evaluate the number of elements in any unknown areas.

This procedure is also relevant to either a 2-set or 3-set problem, the latter being described in sections 20 and 21.

## 18. Notation for the 2-set problem

Although the situation is quite straightforward for two attribute sets (as the example in section 14 showed), with no real difficulty in enumerating an unknown area, it is necessary to introduce a very convenient notation. This is particularly necessary when dealing with 3 attribute sets, but serves well as an introduction at this 2-set stage.

If the two attribute sets in question are labelled as $A$ and $B$, the Venn diagram in Figure 6 can be set up.

*Figure 6 Venn diagram illustrating notation for numbers of elements in various areas*

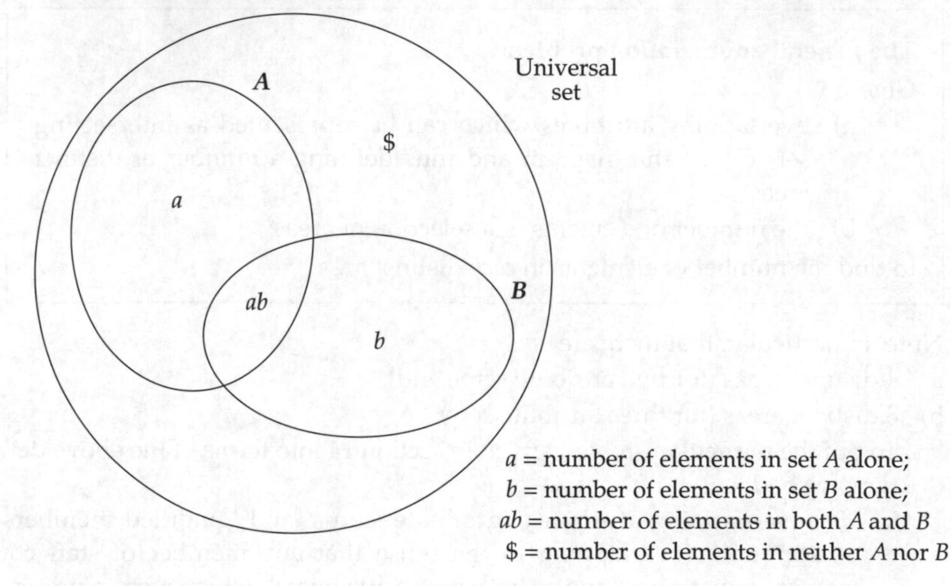


Thus: $n[A] = a + ab$; $n[B] = b + ab$; $n[A \cup B] = a + b + ab$, which can easily be seen from the above Venn diagram.

## 19. Example 6 (*A 2-set enumeration* problem)

*Question*

A survey was carried out by a local Chamber of Commerce, one of the aims being to discover to what extent computers were being used by firms in the area. 32 firms had both Stock Control and Payroll computerised, 65 firms had just one of these two functions computerised and 90 firms had a computerised payroll. If 22 firms had neither of these functions computerised, how many firms were included in the survey?

*Answer*

STEP 1 The two attributes are computerised payroll (with set $P$, say) and computerised stock control (with set $S$, say).

STEP 2 Using the standard notation, we have $p$ = number of firms with a computerised payroll only; $s$ = number of firms with a computerised stock control only; $ps$ = number of firms with both payroll and stock control computerised and \$ = number of firms with neither functions computerised. Figure 7 shows a Venn diagram describing the situation.

*Figure 7*

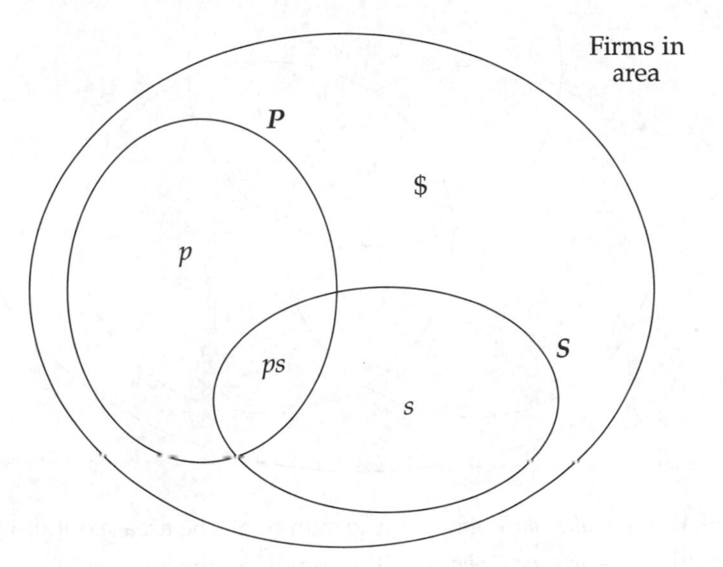


STEP 3 The following equations can be set up from the information given:
$ps = 32$ [1]; $p + s = 65$ [2]; $ps + p = 90$ [3]; \$ = 22 [4]

STEP 4 Substituting for $ps = 32$ in [3] gives $p = 58$.
Substituting for $p = 58$ in [2] gives $s = 7$.

Thus, the number of firms included in the survey = $p + s + ps$ + \$ = 58 + 7 + 32 + 22 = 119

## 20. Notation for the 3-set problem

The notation for the 3-set problem follows exactly the same lines as that for the 2-set problem. Here of course there is one extra attribute present, necessitating a Venn diagram involving three intersecting sets having 8 distinct areas.

Labelling the three sets in question *A*, *B* and *C*, the standard Venn diagram is shown in Figure 8,

where:

*a* = the number of elements in set *A* alone (i.e. not in *B* or *C*).

*ab* = the number of elements in just *A* and *B* (i.e. not in *C*).

*abc* = the number of elements in *A and B and C*

.... and so on.

\$ = the number of elements in *neither A nor B nor C*.

*Figure 8 Enumerating the areas involved with three intersecting sets*

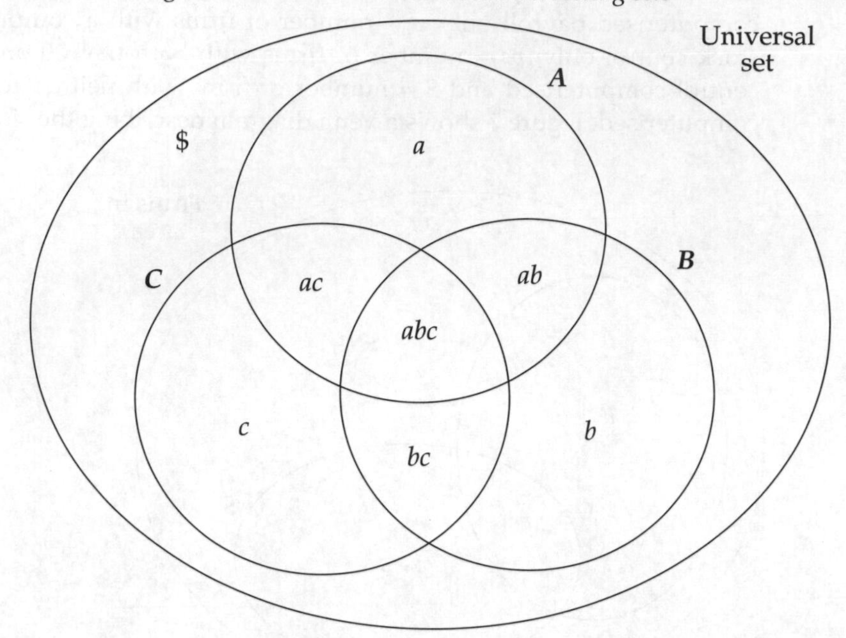


Also: $n[A] = a + ab + ac + abc$ (the sum of all the areas containing an '*a*').

$n[B] = b + ab + bc + abc$ (the sum of all the areas containing a '*b*').

$n[C] = c + ac + bc + abc$ (the sum of all the areas containing a '*c*').

$n[A \cup B]$ = the sum of all areas – $c$ – $ (this can be seen from the diagram).

Similarly for $n[A \cup C]$ and $n[B \cup C]$.

$n[A \cup B \cup C]$ = the sum of all areas – $.

The following example demonstrates the use of this notation to solve a 3-set problem.

## 21. Example 7 (A 3-set *enumeration* problem)

*Question*

Some stock items at a production plant may be classified into one or more of three categories: *P*, for perishable items; *S*, for special orders and *E*, for export, but 24 items are not classified. 6 items were classified only as *P*, 8 were *E* only and 4 were *S* only. Exactly 7 items were classified into just two special categories and no items were classified into all three. Given also that there were 14 export items and 9 special order items, find:

a) how many items were classified as perishable, and

b) how many different stock items were held.

*Answer*

STEP 1 The attribute sets have already been defined in the question as *E*, *P* and *S*.

STEP 2 The Venn diagram is shown in Figure 9, using the usual notation, where:
$p$ = the number of elements in set $P$ alone (i.e. not in $E$ or $S$).
$pe$ = the number of elements in just $P$ and $E$ (i.e. not in $S$).
$pes$ = the number of elements in $P$ and $E$ and $S$
... and so on.
\$ = the number of elements in neither $P$ nor $E$ nor $S$.

STEP 3 Writing down the information given symbolically gives: $p$=6, $e$=8, $s$=4, $eps$=0 and \$=24.

| | | | |
|---|---|---|---|
| number of items with 2 categories: | $ep + ps + es$ | = 7 | [1] |
| number of items for export: | $e + es + ep + eps$ | = 14 | [2] |
| number of special order items: | $s + es + ps + eps$ | = 9 | [3] |

*Figure 9*

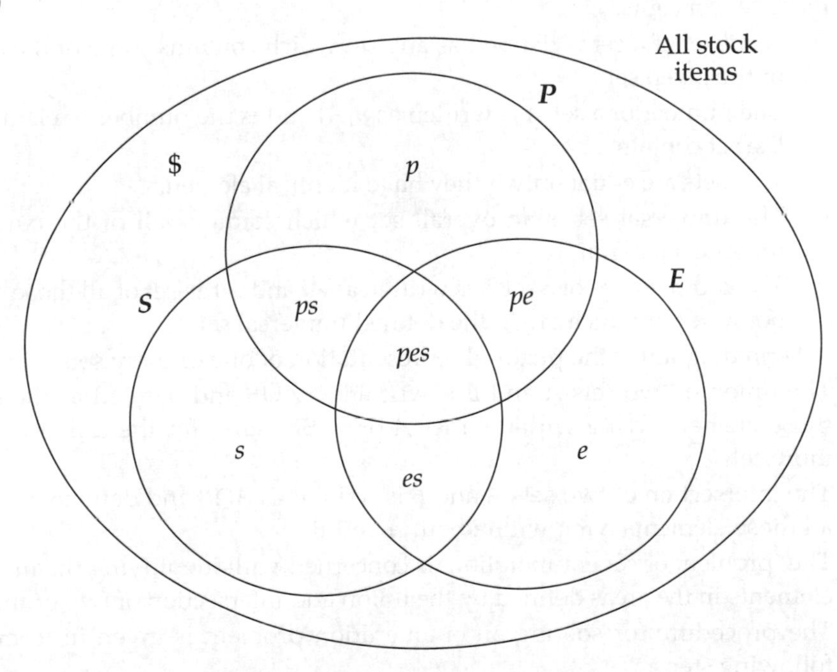


STEP 4 For (a), we need to find $n[P] = p + ep + ps + eps$
$= 6 + ep + ps + 0$ (since $p$=6 and $eps$=0).

To find $ep + ps$. Substituting the given individual area values into [1], [2] and [3] gives:

| | | |
|---|---|---|
| [2] as: | $es + ep = 6$ | [4] |
| [3] as: | $es + ps = 5$ | [5] |

and adding (4) and (5) gives: $ep + ps + 2es = 11$
But [1] is: $ep + ps + es = 7$.
Therefore: $es = 4$.
Substituting back into [1] gives: $ep + ps + 4 = 7$.
Therefore: $ep + ps = 3$.

Hence: $n[P] = 6 + 3 + 0 = 9$
$= \textit{number of perishable items}.$

For b), we need the number of stock items equal tothe sum of all the separate areas on the Venn diagram

$= \$ + p + s + e + ep + ps + es + eps.$

(But $\$=24$, $p=6$, $s=4$, $e=8$, eo+$ps$=3, es=4 and $eps$=0.)

Therefore: *number of stock items* = 24+6+4+8+3+4 = 49.

## 22. Summary

a) A set is a collection of distinct objects, called elements, which are normally enclosed within braces, and separated by commas. In general terms, sets are usually described by capital letters, elements by small letters.

b) Basic set concepts:
   i. A subset of a set is defined as any set which contains some of the elements of the given set.
   ii. The number of a set $A$ is written as $n[A]$ and is the number of elements that the set contains.
   iii. Two sets are equal only if they have identical elements.
   iv. The universal set is an overall set which contains all of the current sets under consideration.
   v. The complement of a set $A$ is written as $A'$ and is the set of all those elements not in $A$ but which are in the defined universal set.

c) A Venn diagram is the pictorial representation of one or more sets.

d) The union of two sets $A$ and $B$ is written as $A \cup B$ and defined as the set of all those elements lying within either $A$ or $B$. Similarly for the union of three or more sets.

e) The intersection of two sets $A$ and $B$ is written as $A \cap B$ and defined as the set of all those elements lying within both $A$ and $B$.

f) The 'problem of set enumeration' is concerned with identifying the numbers of elements in the areas defined by the union and intersection of two or more sets.

g) The procedure for solving an enumeration problem is given in terms of the following steps:

STEP 1 Identify the attribute sets.

STEP 2 Draw a Venn diagram.

STEP 3 Use the information given to fill in as much of the diagram as possible.

STEP 4 Evaluate the number of elements in any unknown areas.

## 23. Point to note

The notation *a,b, ab*, ... etc introduced in the general enumeration problem is simply a convenience, since defining the areas in proper set notation terms is often clumsy and unnecessary. For example, in the 2-set problem (with attribute sets $A$ and $B$ say) $a$ is replacing $A \cap B'$ and \$ is replacing $(A \cup B)'$ ... etc. In the 3-set problem (with attribute sets $A$, $B$ and $C$ say): $a$ is replacing $A \cap B' \cap C'$; $ab$ is replacing $A \cap B \cap C'$; \$ is replacing $(A \cup B \cup C)'$ and so on.

## 24. Student self review questions

1. What is a set and how is it normally described? [2]
2. What is a universal set? [4(d)]
3. Describe what the complement of a set is. [4(e)]
4. What is a Venn diagram? [5]
5. Write down the notation for, and the meaning of, the union of two sets *A* and *B*. [7]
6. Write down the notation for, and the meaning of, the intersection of two sets X and Y. [9]
7. What do you understand by the term 'disjoint sets'? [12(a)]
8. What is the 'general enumeration problem'? [16]
9. How many distinct areas are defined when three sets intersect within some defined universal set? [16(b)]
10. What does the symbol '*ab*' mean when used in the 3-set enumeration problem? [20]

## 25. Student exercises

1. If $A = \{a,b,c,d,e,f,g\}$, $B = \{e,f,g,h\}$, $C = \{f,g\}$, $D = \{f,h\}$, and $E = \{a,b,c,d\}$, state whether the following are true or false and, if false, try and correct them.
   a) *B* is a subset of *A*
   b) *A* is a subset of *C*
   c) *C* is a subset of *B*.
   d) If U is the universal set for *A*, *B*, *C*, *D* and *E*, the smallest value of $n[U]$ is 2.
   e) $C = D$
   f) $n[C] = n[D]$
   g) If the universal set is $\{a,b,c,d,e,f,g,h\}$, then *E* is the complement of set *B* (i.e. $E = B'$).
2. If $X = \{a,b,c,d,e\}$, $Y = \{c,d,e,f\}$ and $Z = \{a,c,d,e,g,h\}$ within a universal set of $\{a,b,c,d,e,f,g,h,i\}$, list the elements of the following sets:
   a) $X \cup Y$ b) $X \cup Z$ c) $X \cap Y$ d) $Y \cap Z$ e) $Y \cup Z$ f) $(X \cup Y) \cap Z$
   g) $(X \cap Y)'$ h) $Y' \cap Z'$ i) $Y' \cup Z'$
3. A company, which has 5 regular customers, stocks products *r, s, t, u, v, w, x* and *y*. Customer *A* buys products *r, s, t* and *v* only and this is represented in set form as $A = \{r,s,t,v\}$. Also, $B = \{r,t,v,w,x\}$, $C = \{r,t,x\}$, $D = \{r,v,w\}$ and $E = \{r,v,w,x\}$. Specify the elements of each of the following sets, giving its meaning in words.
   a) A universal set, *U*, for *A, B, C, D* and *E*
   b) $C \cup D$
   c) $E \cap B$
   d) $C'$
   e) $(A \cup C) \cap B'$
   f) $A \cap B \cap C \cap D \cap E$
   g) $(A \cup B \cup C \cup D \cup E)'$

4. $A$ and $B$ are two intersecting sets and $a$, $b$, $ab$ and \$ are the usual symbols for the number of elements contained in the four defined areas of the associated Venn diagram. Find the value of $ab$ and $a$ if:

   $n[A]=28$ $a+b=36$ \$=48 the number of elements in the universal set is 96.
5. A firm keeps accounts for 28 customers. 12 of these are defined as new customers and 14 are based locally. If just three new customers live locally, how many established customers do not live locally?
6. $A$, $B$ and $C$ are three intersecting sets and $a$, $b$, $ab$, ... etc are the usual symbols for the number of elements contained in the eight defined areas of the associated Venn diagram. Given that:

   $ac=15$ $b=65$ $c=51$ $abc+ab=15$ $a+b+ab=117$ $b+c+bc=128$ \$=0 $n$[Universal set]=200,

   calculate:
   a) $abc$
   b) $a+b+c$
   c) $n[A]$
   d) $a$
   e) $ab$.
7. A music shop specialises in organs and various other keyboard machines. Many modern machines have a MIDI (Musical Instrument Digital Interface, enabling a computer to be linked) and any machine can either be portable or fixed. The shop also stocks second-hand machines. Part of the manager's quarterly report, reflecting the stock position, reads: 'Currently, 31 machines are in stock of which 8 are second-hand. New MIDIs account for the majority of stock at 17, of which 12 are portable. There are 17 portable machines in total, 14 having MIDI. We currently have no fixed, second-hand MIDIs or non-MIDI, second-hand portables.'

   How many of the machines in stock:
   a) have MIDI?
   b) are new portables?
   c) that are new and fixed, do not have MIDI?
   d) are second-hand, portable MIDIs?

# 30 Introduction to probability

## 1. Introduction

This chapter begins by describing the structure of experiments and events, including mutually exclusive and independent events. Probability is then defined, including the addition and multiplication rules.

## 2. The concept of probability

Probability is a concept that most people understand naturally, since words such as 'chance', 'likelihood', 'possibility' and 'proportion' (and indeed probability itself) are used as part of everyday speech. For example, most of the following, which might be heard in any business situation, are in fact statements of probability:

a) 'There is a 30% chance that this job will not be finished in time'.

b) 'There is every likelihood that the business will make a profit next year'.

c) 'Nine times out of ten he arrives late for his appointments'.

d) There is no possibility of delivering the goods before Tuesday'.

Probability, in the statistical sense, simply puts a well defined structure around the concept of everyday probability, enabling a logical approach to problem solving to be followed. Before defining probability however, it is necessary to lay its foundations in the form of experiments and events.

## 3. Statistical experiments

A *statistical experiment* can be described as any situation, specially set up or occurring naturally, which can be performed, enacted or otherwise considered in order to gain useful information. Statistical experiments can be simple or complex, and repeated as many times as necessary in order to gain the information desired. Consider the following as examples.

a) In order to determine whether a production line is turning out products at a fast enough rate, a foreman could count the number of items produced in a five minute period, say, possibly repeating this procedure at regular intervals for consistency. This would be considered as a fairly simple experiment.

b) A questionnaire might be devised to put alternatives to, and ask the opinions of, the workforce employed by a large corporation concerning the possible introduction of various profit-sharing incentive schemes. This would be thought of as a complex experiment.

## 4. Outcomes of experiments

a) An *outcome set* for an experiment is a specification of all possible distinct results (i.e. *outcomes*) of the experiment when it is performed once. Put another way, when an experiment is performed, the result must be one and only one element of the outcome set. For example, the outcome set for the production line experiment of section 3(a) above would be {0,1,2,3,4,5, ... }.

b) *Equally likely outcomes* are defined when any one outcome of an experiment is no more likely to occur than any other when the experiment is performed. In practice, equally likely outcomes are not common, but the concept is important (in order to understand the nature of probability). For example:
   i. The outcomes for the production line experiment of section 3(a) above can in no way be considered as equally likely, since the outcomes '0' and '586', say, would be far less likely to occur than, say '13' or '40'.
   ii. A classic example of an equally likely outcome set is that obtained when an unbiased standard six-sided die (dice is plural!) is to be rolled. The outcome set for this experiment is {1,2,3,4,5,6} and as long as the die is in fact 'fair' (i.e. unbiased) then we would not expect outcome '1' to occur any more than '2', '3', '4', '5' or '6', when the die is rolled.

## 5. Statistical events

In any experiment, an *event* is defined as any subset of the given outcome set that is of interest. Thus an event consists of a set of outcomes, and the event is said to have occurred if the outcome of the experiment (when and if it is performed) is contained in the event set. For example:

i. For the production line experiment of section 3(a) above, we might define the event 'fast production' as 15 or more products in five minutes. That is, the event set is {15,16,17,18, ....}. If the experiment is performed (i.e. the number of products coming off the line in a particular 5-minute period is determined) and the outcome is 23, then the event WILL have occurred, since 23 is an element of the event set. That is, production will be regarded as 'fast' in this case.
ii. For the die-rolling experiment, the event 'even number' is the set {2,4,6}. If the die is rolled and a '5' occurs, then the event 'even number' will NOT have occurred since '5' is not an element of its event set.

In Set Theory terms, the outcome set for any experiment can be considered as the universal set for any events that may be defined, since (by definition) all events are subsets of the outcome set.

As many events as desired may be defined on an experimental outcome set. Some might overlap, others might be quite distinct (or disjoint). The following example demonstrates this point.

## 6. Example 1 (Some defined *events* of an experiment)

A set of nine people working in an accounts section are classified in three different ways:

a) by sex (M or F)
b) by Association qualification (Q = qualified; U = unqualified)
c) by status (F = full time; P = part time).

Person 1 is male, qualified and full time and is symbolised as 1:MQF. The other eight are: 2:MQF, 3:MUP, 4:MQF, 5:MUP, 6:FQF, 7:FQF, 8:FUF and 9:FUP. The defined experiment is 'choose a person at random'.

Three events are defined as follows: P = Part time; M = Male and F = female. A Venn diagram is constructed to demonstrate these events in Figure 1.

*Figure 1*

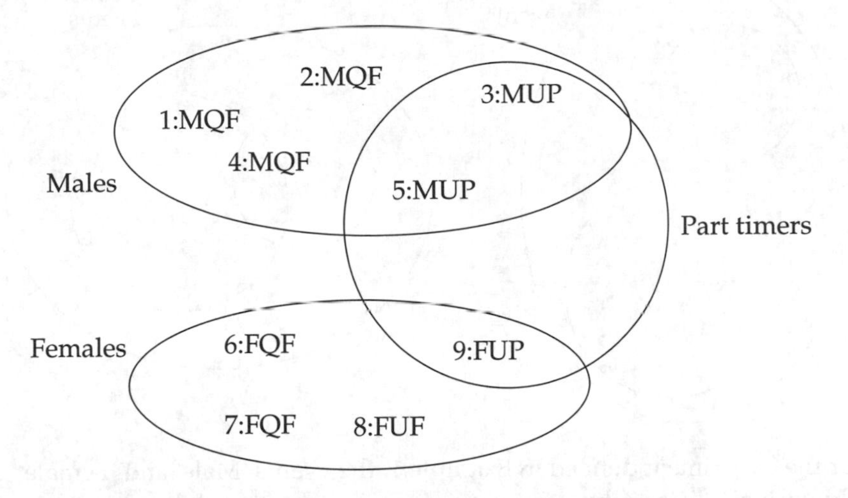


In Figure 1, notice that the events 'Male' and 'Female' do not overlap while, for example, 'Male' and 'Part time' do.

If the above experiment is performed and the outcome, say, is person 8, then the event 'Female' WILL have occurred (since this outcome is in the event set). The events 'Male' and 'Part time' will NOT have occurred (since person 8 is not contained in either of these two event sets).

If the experiment is performed and the outcome is person 3, then 'Male' and 'Part time' will BOTH have occurred, while 'Female' will not have occurred.

## 7. Definition of mutually exclusive events

**Mutually exclusive events**

Two events of the same experiment are said to be *mutually exclusive* if their respective event sets do not overlap.

Note that this is equivalent to saying that, when the experiment is performed, events that are mutually exclusive *cannot happen at the same time* (i.e. cannot occur together).

Figure 2 shows the general Venn diagram representation of two events of an experiment where, in (a), the events $A$ and $B$ are mutually exclusive (no overlap) and, in (b), the events $A$ and $B$ are NOT mutually exclusive (since there is a distinct overlap).

*Figure 2 Mutually exclusive and non-mutually exclusive events*

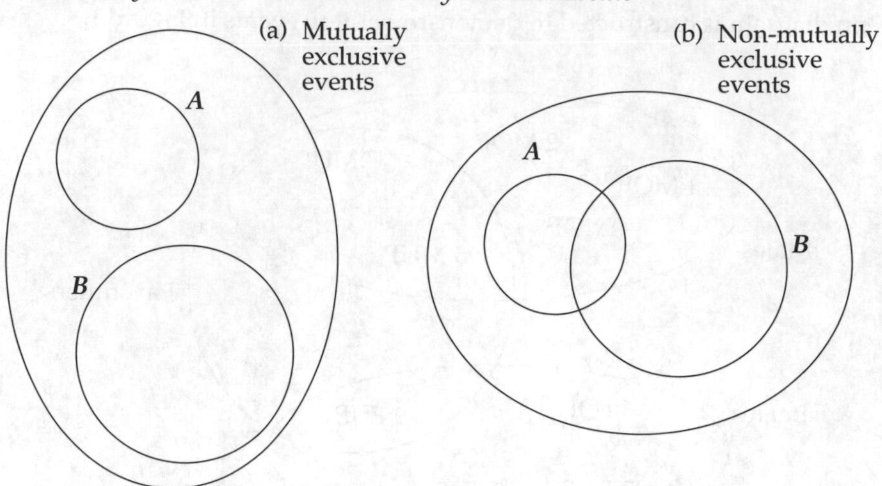


For the experiment defined in Example 1, the events 'Male' and 'Female' are mutually exclusive, since there is no overlap. The fact that they cannot happen at the same time should be clear since, when the experiment is performed, only one outcome can occur and it cannot be in both event sets.

The term 'mutually exclusive' is used to describe events of this kind because the occurrence of one necessarily excludes the occurrence of the other.

## 8. Example 2 (*mutually exclusive* events)

a) If the choice of one car from a large fleet of cars is considered (as an experiment), then some pairs of events can be defined as follows.

   i. 'Automatic' and 'Manual gears' would be mutually exclusive, since any car would have either automatic or manual, not both.

   ii. 'Over 6 months old' and 'New' would similarly be mutually exclusive.

   iii. 'Over 6 months old' and 'Rear wheel drive' would almost certainly NOT be mutually exclusive, since rear wheel drive cars are very common. (Note that it is not possible to be more certain than this unless precise details of each car is known. It could be that, for this particular fleet, no car was in both sets, but this would be exceptional.)

b) For the lengths of time that completed jobs have taken in a factory, the events 'On time' and 'Over 2 days late' would be mutually exclusive, since if a job was completed on time it could not possibly be over two days late as well!

c) When considering rolling an unbiased six-sided die, the two events 'Even number' and 'Odd number' would be mutually exclusive. Note however that the events 'Even number' and 'a number less than 4' are not mutually exclusive, since both events could happen simultaneously if the die was rolled and, for example, a '2' occurred.

## 9. Definition of independent events

**Independent events**
Two events are said to be *independent* if the occurrence (or not) of one of the events will in no way affect the occurrence (or not) of the other.
Alternatively, two events that are defined on two physically different experiments are said to be independent.

Both definitions given are necessary. Example 3 demonstrates both of these cases.

Note that two mutually exclusive events cannot be independent, since, by definition, the occurrence of one excludes the other (i.e. they are dependent in a particular way).

## 10. Example 3 (*independent* events)

a) The events 'a particular supplier will deliver on time' and 'a particular customer's account will not be paid within a specified time' would be independent events, because they are each based on physically different situations or experiments. The first experiment is 'delivery time of a particular supplier', the second is 'time to payment of a particular customer's account'.

b) The events 'shorthand typist' and 'brown eyes' would generally be considered as independent, since it is not likely that the occurrence of one would affect the other.

c) Suppose that two people had to be chosen randomly from 6 men and 3 women. The events 'a woman chosen as the first person' and 'a woman chosen as the second person' would NOT be independent, since if the first event did occur at the first selection, it would mean that there are less women to choose from for the second selection. In fact, any two events defined (as two successive selections of people) on this particular experiment will not be independent, since the selection of the first person changes the situation for the choice of the second (i.e. affects the choice of the second).

## 11. Theoretical and empirical probability

There are two separate ways of calculating probability, depending on whether some experiment is performed or not.

a) *Theoretical probability* is the name given to probability that is calculated *without the experiment being performed*, that is, using only information that is known about the physical situation.

b) *Empirical probability* is the name given to probability that is calculated *using the results of an experiment that has been performed a number of times*. Note that empirical probability is sometimes called *relative frequency* or *subjective* probability.

The following definitions and examples demonstrate the difference between these two types.

## 12. Definition of theoretical probability

**Theoretical probability**

If $E$ is some event of an experiment that has an equally likely outcome set ($U$), then the theoretical probability of event $E$ occurring when the experiment is performed is written as $p(E)$ and given by:

$$\text{p(E)} = \frac{\text{number of different ways that the event can occur}}{\text{number of different outcomes of the experiment}}$$

$$= \frac{n(E)}{n(U)}$$

where: $n(E)$ is the number of outcomes in event set $E$

$n(U)$ is the total possible number of outcomes (in outcome set $U$).

Note that sometimes the probability of event $E$ occurring is written as either $P(E)$, $Pr(E)$ or $pr(E)$.

If, for example, an ordinary six-sided die is to be rolled, the (equally likely) outcome set, $U$, is {1,2,3,4,5,6} and the event 'even number' has event set {2,4,6}. Therefore the theoretical probability of obtaining an even number is easily calculated as

$$p(\text{even number}) = \frac{n(\text{even number})}{n(U)} = \frac{3}{6} = 0.5.$$

## 13. Example 4 (The *theoretical probability* of given events)

*Question*

A wholesale stationer stocks heavy (2B), medium (HB), fine (2H) and extra fine (3H) pencils which come in packs of 10. Currently in stock are 2 packs of 3H, 14 packs of 2H, 35 packs of HB and 8 packs of 2B. If a pack of pencils is chosen randomly for inspection, what is the probability that they are a) medium b) heavy c) not very fine d) neither heavy nor medium.

*Answer*

Since the pencil pack is being chosen randomly, each separate pack of pencils (of any type) can be regarded as a single equally likely outcome. Thus we have an equally likely outcome set and the definition for theoretical probability can be used. The number of outcomes in total is the number of pencil packs, which is:

$2 + 14 + 35 + 8 - 59$. i.e. $n[U] = 59$.

a) The probability of choosing a medium pencil pack is given by the number of medium pencil packs divided by the total number of pencil packs.

That is, $p(\text{medium}) = \frac{n(\text{medium})}{n(U)} = \frac{35}{59} = 0.593$ (3D).

b) $p(\text{heavy}) = \frac{n(\text{heavy})}{n(U)} = \frac{8}{59} = 0.136$ (3D).

c) The number of pencil packs that are not very fine is 14+35+8 = 57.

Therefore, $p(\text{not very fine}) = \dfrac{n(\text{not very fine})}{n(U)} = \dfrac{57}{59} = 0.966$ (3D).

d) 'Neither heavy nor medium' is equivalent to 'fine or very fine', and there is 2+14 = 16 of these pencil packs.

Thus, $p$(neither heavy nor medium)

$$= \frac{n(\text{neither heavy nor medium})}{n(U)} = \frac{16}{59} = 0.271 \text{ (3D)}$$

## 14. Definition of empirical probability

Sometimes an outcome set for an experiment is not known or, if it is, the outcomes might not be equally likely. In this type of case it is not possible to use the definition of theoretical probability given, and so another method of calculating probability is required. This alternative method is based on the results of performing the experiment a number of times yielding a frequency distribution of events or outcomes, and using these results to calculate what is known as empirical probability which is now defined.

**Empirical (relative frequency) probability**

If $E$ is an event of an experiment that has been performed a relatively large number ($\Sigma f$) of times to give a frequency distribution then the empirical probability of event $E$ occurring when the experiment is performed one more time is written as $p(E)$ and given by:

$$\text{p(E)} = \frac{\text{number of times the event occurred}}{\text{number of times the experiment was performed}}$$

$$= \frac{f(E)}{\Sigma f}$$

where: $f(E)$ is the number of times that event $E$ has occurred

$\Sigma f$ is the total frequency = no. of performances of experiment.

In words, the empirical probability (sometimes known as subjective probability) of an event $E$ occurring is simply the proportion of times that event $E$ actually occurred when the experiment was performed. In a statistical context, the word 'empirical' means 'as seen in practice' or 'as experienced'. Thus an empirical approach to a statistical problem involves performing some experiment a number of times and using the results as a guide to what will be expected to happen in general.

For example, if, out of 60 orders received so far this financial year, 12 were not completely satisfied, the proportion 1260 = 0.2 is the (empirical) probability that the next order received will not be completely satisfied.

## 15. Example 5 (*Empirical probability*)

*Question*

A number of families of a particular type were measured by the number of children they contain to give the following frequency distribution.

| Number of children | 0 | 1 | 2 | 3 | 4 | 5 or more |
|---|---|---|---|---|---|---|
| Number of families | 12 | 28 | 22 | 8 | 2 | 2 |

Use this information to calculate the (relative frequency) probability that another family of this type will have (a) 2 (b) 3 or more (c) less than 2 children.

*Answer*

Here, $\Sigma f = 74$ (the number of times the experiment 'determine the number of children in a single family' has been repeated).

a) The probability that another family of this type will have just 2 children is the proportion of families that have two children.

i.e. $p(\text{2 children}) = \dfrac{f(\text{2 children})}{\Sigma f} = \dfrac{22}{74} = 0.297$ (3D).

b) From the given distribution, $f(\text{3 or more children}) = 8+2+2 = 12$.

Therefore, $p(\text{3 or more children}) = \dfrac{12}{74} = 0.162$ (3D).

c) $p(\text{less than 2 children}) = \dfrac{f(\text{less than 2 children})}{\Sigma f} = \dfrac{40}{74} = 0.541$ (3D).

## 16. Comparison of theoretical and empirical probability

a) *Theoretical probability.*

This can be regarded as the standard measure of probability. If there was a choice of either of the two methods, theoretical probability is the one that is most usefully employed. The main drawback here however, is that the outcome set must have equally likely outcomes, and this is not always the case in practice. For example, if the probability of a defective item coming off a production line was required, there is no way of calculating this precisely without looking at past production records or taking an immediate sample, both of which are using an empirical approach.

b) *Empirical probability.*

This method is usually employed when it is not possible to calculate probability in the theoretical sense. That is, when the outcome set is either difficult or impossible to define or the outcomes are not equally likely. Sometimes this method of calculating probability is regarded as an estimate of the (unknown) theoretical probability.

## 17. Further rules of probability

In order to extend the structure within which the probabilities of events occurring can be calculated, the following three rules (which are reasonable and fairly obvious) should be understood and remembered.

a) *Probability limits*. The probability of any event $E$ occurring must lie between 0 and 1 inclusive. Symbolically, this can be written as: $0 \le p(E) \le 1$

   i. If $p(E) = 0$, then $E$ is known as an *impossible event*. Examples of impossible events are a negative daily production from an assembly line or an employee of a workforce who is less than 1 year old!

   ii. If $p(E) = 1$, then $E$ is known as a *certain event*. Examples of certain events are an employee being either male or female.

b) *Total probability rule*. The sum of the probabilities of all possible outcomes of an experiment must total exactly 1. This is equivalent to saying that when an experiment is performed, one of the outcomes must occur. Symbolically, we can write: $\Sigma\, p = 1$ (for all outcomes). For example, if the lorries in a haulage fleet are of type $A$, $B$, $C$ and $D$ only, then if a lorry is picked at random, we must have that $p(A) + p(B) + p(C) + p(D) = 1$.

c) *Complementary rule*. If the complement of any event $E$ is written as $\overline{E}$ and defined as 'event $E$ does NOT occur', then the probabilities of $E$ and $\overline{E}$ must total 1. Alternatively, subtracting the probability of any event occurring from 1 will give the probability of the event NOT occurring.

   Symbolically, $p(\overline{E}) = 1 - p(E)$

As a simple example of the use of the complementary rule, suppose that it is known from past experience that only 1 out of 200 letters sent through a company's internal mailing system are 'lost'.

That is, $p(\text{letter is lost}) = 0.005$.

Then, $p(\text{letter is not lost}) = 1 - p(\text{letter is lost}) = 1 - 0.005 = 0.095$

To summarise:

**Some elementary rules of probability**

Probability limits: $0 \le p(E) \le 1$

Total probability rule: $\sum p = 1$ (for all outcomes)

Complementary rule: $p(\overline{E}) = 1 - p(E)$

## 18. The addition rule for probabilities

If $A$ and $B$ are two events, then the meaning of $p(A \text{ or } B)$ is quite precise in probability terms. It stands for the probability that either $A$ or $B$ or both occur. The addition rule states that if $A$ and $B$ are two mutually exclusive events, then the probability that $A$ or $B$ occurs (i.e. $p(A \text{ or } B)$) can be calculated as the sum of $p(A)$ and $p(B)$.

**The addition rule**

If $A$ and $B$ are any two mutually exclusive events of an experiment, then:

$$p(A \text{ or } B) = p(A) + p(B).$$

## 19. Example 6 (Probabilities of *mutually exclusive* and *complementary* events)

*Question*

The Purchase department has analysed the number of orders placed by each of the 5 departments in the company by type for this financial year as given in Table 1.

| *Order Type* | *Department* | | | | | *Total* |
|---|---|---|---|---|---|---|
| | *Sales* | *Purchase* | *Production* | *Accounts* | *Maintenance* | |
| Consumables | 10 | 12 | 4 | 8 | 4 | 38 |
| Equipment | 1 | 3 | 9 | 1 | 1 | 15 |
| Special | 0 | 0 | 4 | 1 | 2 | 7 |
| Total | 11 | 15 | 17 | 10 | 7 | 60 |

*Table 1*

An error has been found in one of these orders. What is the probability that the incorrect order:

a) was for consumables?
b) was not for consumables?
c) came from Maintenance?
d) came from Production?
e) came from Maintenance or Production?
f) came from neither Maintenance nor Production?
g) was an equipment order from Purchase?

*Answer*

Note that none of the above probabilities could be calculated without the given table. That is, only empirical probabilities are feasible in this type of situation. Also note that the above information is in the form of a two-way frequency distribution. If information by department only is required, then only the departmental totals (11, 15, 17, ... etc) need be considered, the constituent figures for order types being ignored.

a) Since 38 of the 60 orders are for consumables, then the probability that the incorrect order is for consumables is calculated as:

$p(\text{consumables}) = \frac{38}{60} = 0.633$ (3D).

b) $p(\text{not consumables}) = 1 - p(\text{consumables})$ [using the complementary rule]

$= 1 - 0.633 = 0.367$ (3D).

c) There are 7 maintenance orders out of the 60.

Thus: $p(\text{maintenance}) = \frac{7}{60} = 0.117$ (3D).

d) Similarly: $p(\text{production}) = \frac{17}{60} = 0.283$ (3D).

e) $p(\text{maintenance or production}) = p(\text{maintenance}) + p(\text{production})$ [mutually exclusive events]

$= 0.117 + 0.283$ [from (c) and (d)] $= 0.400$ (3D).

f) $p$(neither maintenance nor production)
$= 1 - p$(maintenance or production) [using the complementary rule]
$= 1 - 0.4$ [from (e)] $= 0.600$ (3D).

g) Using the information in the table, it can be seen that just 3 equipment orders have been raised by Purchase.
Therefore, $p$(equipment order from Purchase) $= \frac{3}{60} = 0.05$

## 20. The multiplication rule for probabilities

If $A$ and $B$ are two events, then the meaning of $p(A \text{ and } B)$ is quite precise in probability terms. It stands for the probability that both $A$ and $B$ occur. Further if $A$ and $B$ are two independent events, then the probability that $A$ and $B$ occur, i.e. $p(A \text{ and } B)$, can be calculated as the product, or multiple, of $p(A)$ and $p(B)$.

**The multiplication rule**

If $A$ and $B$ are any two independent events, then:

$$p(A \text{ and } B) = p(A) \times p(B).$$

As an example, suppose, in any week, the probability of an assembly line failing is 0.03 and the probability of a raw material shortage is 0.1. If also these two events are known to be independent of each other, then the probability of an assembly line failing *and* a raw material shortage is given by $0.1 \times 0.03 = 0.003$.

## 21. Compound experiments and events

Sometimes it is convenient to bring two experiments or situations together as if they were a single experiment. This is known as a *compound experiment*.

Suppose experiment $X$ has only two outcomes $a$ and $b$, i.e. has outcome set $U_X=\{a,b\}$, and experiment $Y$ has an outcome set $U_Y=\{1,2,3\}$. The outcome set for compound experiment $XY$ can be described in words as 'all the possible outcomes when experiments X and $Y$ are both performed' and written as:

$$U_{XY}=\{a1,a2,a3,b1,b2,b3\}.$$

Compound experiments can be formed from as many experiments as is necessary and/or convenient and are particularly useful structures for ensuring that all possibilities are considered when calculating probabilities in certain situations.

Example 7 demonstrates this type of situation and Example 8 shows how compound experiments can be naturally used in calculating probabilities.

## 22. Example 7 (The use of a *compound experiment*)

If a product comes off an assembly line, made up of two components $C$ and $D$, each of which could be faulty, the following compound outcome set gives all combinations for components $C$ and $D$ being OK or faulty.

$$U = \{cd, \bar{c}d, c\bar{d}, \bar{c}\bar{d}\}$$

where: $cd$ means 'both components C and $D$ are OK'
$\bar{c}d$ means 'component C is faulty; component $D$ is OK' ... etc

The event 'at least one component is faulty' is represented by $\{\bar{c}d, c\bar{d}, \bar{c}\bar{d}\}$ and the event 'at most one component is faulty' is represented by $\{cd, \bar{c}d, c\bar{d}\}$ ... and so on.

## 23. Tree diagrams

Tree diagrams can be useful in laying out alternatives when independent events are involved.

Consider the following. A lorry makes just two trips per day and on each trip is likely to carry a heavy load ($H$) with probability 0.25. Assuming the load carried in the second part of the day is independent of the previous load, what is the probability that the lorry carries just one heavy load on a particular day?

The following tree diagram lays out all the possibilities and calculates the associated probabilities.

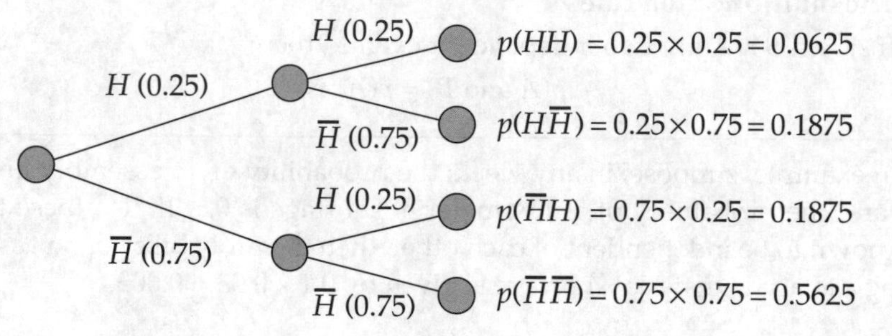


The probability required is clearly the sum of the middle two = 0.1875 + 0.1875 = 0.375

Alternatively, we could have proceeded:

$$p(\text{just 1 heavy load}) = p(H\bar{H} \text{ or } \bar{H}H)$$
$$= p(H\bar{H}) + p(\bar{H}H) \text{ etc (as above)}$$

## 24. Example 8 *(General probability)*

*Question*

A firm is independently working on two separate jobs. There is a probability of only 0.3 that either of the jobs will be finished on time.

Find the probability that:

a) both  b) neither  c) just one  d) at least one

of the jobs is finished on time.

*Answer*

Put $T$ = job on time and $L = \bar{T}$ = job NOT on time.

$p(T) = 0.3$ and $p(L) = 1 - 0.3 = 0.7$

a) $p$(both finished on time) $= p(T.T) = p(T) \times p(T) = (0.3)(0.3) = 0.09$

b) $p$(neither job finished on time) $= p(L.L) = (0.7)(0.7) = 0.49$

c) $p$(*just one* job finished on time)
$= p(L.T \text{ or } T.L)$ [both cases MUST be considered]
$= p(L.T) + p(T.L)$ [mutually exclusive events]
$= (0.7)(0.3) + (0.3)(0.7) = 0.42$

d) Thus: p(at least one job finished on time)
= 1 – p(neither job finished on time)
$= 1 - 0.49 = 0.51$

## 25. Example 9 *(General probability)*

### *Question*

Two vacuum cleaner salesmen *A* and *B* must each make two calls per day, one in the morning and one in the afternoon. *A* has probability 0.4 of selling a cleaner on any call, while *B* (a novice) has probability 0.1 of a sale. *A* works independently of *B* and, for each salesman, morning and afternoon results are independent of each other. Find the probability that, in one day:

a) *A* sells two cleaners
b) *A* sells just one cleaner
c) *B* makes at least one sale
d) Between them, *A* and *B* make exactly one sale.

### *Answer*

**Note**: Some symbols for describing events is usually necessary with problems, as seen in Example 8. Where the symbols are quite obvious, there is normally no need to explain them (for example, in an examination). If the symbols are not obvious or their use might be confusing, it is important to explain them, as can be seen in the following solutions.

a) $p$(*A* sells two cleaners in a day)
$= p(Am \text{ and } Aa)$ [$Am$ = *A* makes a sale in the *morning* etc]
$= p(Am) \times p(Aa)$ [independent events]
$= (0.4)(0.4) = 0.16$

b) $p$(*A* sells just one cleaner)
= p(A sells in the morning or the afternoon but not both)
$= p(Am.\overline{A}.a \text{ or } \overline{A}m.Aa)$
$= (0.4)(1 - 0.4) + (1 - 0.4)(0.4)$
$= (0.4)(0.6) + (0.6)(0.4) = 0.48$

c) Here,we need: $p$(*B* makes at least one sale)
= 1 – p(B makes no sales)
$= 1 - p(\overline{B}m.\overline{B}a)$
$= 1 - (1 - 0.1)(1 - 0.1)$
$= 1 - (0.9)(0.9) = 0.19$

d) p(A and B make only one sale between them)

$= p(A0.B1 \text{ or } A1.B0)$

[$A0 = A$ makes 0 sales in a day, $B1 = B$ makes 1 sale in a day etc]

$= p(A0) \times p(B1) + p(A1) \times p(B0)$

But $p(A0) = (1 - 0.4)(1 - 0.4)$ [No sale in the morning or the afternoon]

$= 0.36$

Similarly, $p(B0) = (1 - 0.1)(1 - 0.1) = 0.81$

Also, from b), $p(A1) = 0.48$

Similarly, $p(B1) = 0.1 \times (1 - 0.1) + (1 - 0.1) \times 0.1 = 0.18$

Thus, $p(A$ and $B$ make only one sale between them)

$= (0.36)(0.18) + (0.48)(0.81)$

$= 0.4536$

## 26. Summary

a) A statistical experiment is a situation, specially set up or occurring naturally, which is used to gain information.

b) An outcome set for an experiment is a specification of all the different results that are possible when an experiment is performed. Equally likely outcomes are such that no one of them has any more chance of occurring than any other.

c) An event of an experiment is any subset of the outcome set.

d) i. Mutually exclusive events are any two events of an experiments that cannot happen at the same time (i.e. their event sets do not overlap).

ii. Independent events are any two events that cannot affect each other (i.e. they are defined on two physically different experiments).

e) Theoretical probability (of some event $E$) is probability that is calculated without an experiment being performed and is defined as:

$$p(E) = \frac{\text{number of different ways that the event can occur}}{\text{number of different outcomes of the experiment}} = \frac{n(E)}{n(U)}$$

f) Empirical probability (of some event $E$) is calculated, based on the results of repeated performances of an experiment and is defined as:

$$p(E) = \frac{\text{number of times the event occurred}}{\text{number of times the experiment was performed}} = \frac{f(E)}{\Sigma f}$$

g) Further probability rules include:

i. $0 \leq p(E) \leq 1$

ii. if: $p(E) = 0$, $E$ is called an impossible event

$p(E) = 1$, $E$ is called a certain event.

iii. The sum of the probabilities of all the outcomes of an experiment must total 1. That is, $\Sigma\, p = 1$.

iv. $p(E$ does not occur$) = 1 - p(E)$.

h) Addition rule: $p(A \text{ or } B) = p(A) + p(B)$ [if $A$ and $B$ are mutually exclusive events]

Multiplication rule: $p(A \text{ and } B) = p(A) \times p(B)$ [if $A$ and $B$ are independent events]

i) A compound experiment is the joining together of two independent experiments for convenience, in order to simplify the listing of complex events.

## 27. Student self review questions

1. What is a statistical experiment? Give some examples. [3]
2. What is meant by the expression 'equally likely outcomes'? [4]
3. How is an event of an experiment defined? [5]
4. What are mutually exclusive events? [7]
5. What are independent events? [9]
6. Why cannot mutually exclusive events also be independent? [9]
7. What is the difference between theoretical and empirical (relative frequency probability? [11,12]
8. Between what limits must the probability of any event lie? [17(a)]
9. What is the 'total probability' rule? [17(b)]
10. What is the 'complementary rule' of probability? [17(c)]
11. What is the 'addition' rule of probability and for what type of events is it valid? [18]
12. What is the 'multiplication' rule of probability and for what type of events is it valid? [20]
13. What is a compound experiment and why is it used? [21]

## 28. Student exercises

1. Which of the following pairs of events are mutually exclusive?
   a) A company car is
      i. less than six months old
      ii. less than a year old.
   b) An item of stock
      i. needs re-ordering
      ii. does not need re-ordering.
   c) A company is
      i. limited
      ii. a plc.
   d) A salesman makes
      i. at least one sale
      ii. at most one sale.
2. Which of the following pairs of events are independent?
   a) An employee is
      i. over 30 years of age
      ii. has over 12 years of work experience.
   b) A company is
      i. limited
      ii. a plc.
   c) A company is in profit

   i. last year
   ii. this year.
  d) An invoice
   i. has been passed for payment
   ii. has not been received.
  e) A manufactured product is
   i. defective
   ii. for export.

3. A haulage contractor has 3 type *A*, 2 type *B* and 7 type *C* lorries available for deliveries, all of which are used equally frequently. What is the probability that a lorry delivering a load will be:
  a) of type *B*
  b) not of type *C*
  c) of type *A* or *C*?

4. *MULTI-CHOICE*. A fair 6-sided die is thrown twice. What is the probability that the sum of the numbers shown is 7?
  a) $\frac{1}{12}$  b) $\frac{1}{6}$  c) $\frac{1}{4}$  d) $\frac{1}{2}$

5. *MULTI-CHOICE*. A high-security research establishment is protected by three independent alarm systems. The protection system will operate correctly as long as at least one of the alarm systems is working when the security system is activated. The probability that any one system will fail at a trial is 1 in 100. The probability that the security system fails when activated is:
  a) 1 in 100  b) 3 in 100  c) 1 in 10,000  d) 1 in 1,000,000

6. A firm has tendered for two independent contracts. It estimates that it has probability 0.4 of obtaining contract *A* and probability 0.1 of obtaining contract *B*. Find the probability that the firm:
  a) obtains both contracts
  b) neither of the contracts
  c) obtains exactly one contract.

7. A production line is known to produce 12% of items that are imperfect, one quarter of which are rejected. If three items are selected randomly from the line, find the probability that:
  a) the first item is imperfect
  b) the first item is rejected
  c) the second item is rejected
  d) none of the items are rejected
  e) at least one item is imperfect.

8. The following data relate to the number of sales made by a company over a number of weeks:

| Weekly sales | up to 10 | 10 to 19 | 20 to 29 | 30 to 39 | 40 and over |
|---|---|---|---|---|---|
| Number of weeks | 2 | 12 | 22 | 10 | 4 |

Calculate the (empirical) probability that next week the firm will make the following number of sales:

a) at least 20
b) no more than 39
c) between 10 and 29 inclusive.

9. Three machines produce 50%, 40% and 10% respectively, of the total production. The respective percentages of defective production are 2%, 2% and 5%.
   a) If an item is selected at random from the collective output of the machines, find the probability that it is defective.
   b) If two items are selected at random from the collective output, find the probability that at least one of them is defective.
10. There are three stages of a testing process past which every product must pass. Out of 3500 products which were tested, the failures were 148 at stage 1, 130 at stage 2 and 72 at stage 3. Use this information to answer the following.
   a) What is the probability that a randomly selected product will pass the test?
   b) What is the probability that a randomly selected product will get beyond the second test stage?
   c) Approximately, how many products should be tested in order that 620 successfully pass?

# 31 Conditional probability and expectation

## 1. Introduction

The initial part of the chapter introduces expectation, which is the probability equivalent of the arithmetic mean. The rest of the chapter is concerned with conditional probability and shows how it can be applied to solving problems in probability that are more involved than those hitherto covered. This leads on to the topic of Bayes theorem and a discussion of how conditional probability and Bayes theorem can be used to solve practical problems.

## 2. Expectation

*Expectation* (or *expected value*) is the arithmetic mean of a given set of values. It is calculated using probabilities *instead of frequencies*.

Consider the data of Table 1, relating to weekly accidents at a factory.

*Table 1 Accidents at a factory*

| Number of accidents per week ($x$) | 0 | 1 | 2 | 3 | 4 | Total |
|---|---|---|---|---|---|---|
| Number of weeks ($f$) | 10 | 18 | 15 | 6 | 1 | 50 |
| ($fx$) | 0 | 18 | 30 | 18 | 4 | 70 |

$$\text{Mean number of accidents per week} = \frac{\sum fx}{\sum f} = \frac{70}{50} = 1.4.$$

The frequencies in the above table can be converted to probabilities (i.e. proportions) by dividing each by the total frequency of 50, to give Table 2.

*Table 2*

| Number of accidents per week ($x$) | 0 | 1 | 2 | 3 | 4 | Total |
|---|---|---|---|---|---|---|
| Probability ($p$) | 0.20 | 0.36 | 0.30 | 0.12 | 0.02 | 1 |
| ($px$) | 0 | 0.36 | 0.60 | 0.36 | 0.08 | 1.40 |

In Table 2, each $x$-value has been multiplied by its probability to obtain a total of 1.4, *which is the mean of the distribution* (as can be seen from the previous calculation). We say that the expected number of accidents per week (or the *expected value of* $x$) is 1.4.

## 3. Formula for expected value (expectation)

**Expected value of $x$**

If some variable $x$ has its values specified with associated probabilities $p$, then:

$$\text{expected value of } x = \sum px.$$

In other words, an expected value (or expectation) is obtained by multiplying each original value by its probability and adding the results.

Note that, as with the arithmetic mean, an expected value would not normally be the same as one of the original values.

## 4. Example 1 *(Expected value)*

*Question*

An ice-cream salesman divides his days into 'sunny', 'medium' or 'cold'. He estimates that the probability of a sunny day is 0.2 and that 30% of his days are cold. He has also calculated that his average revenue on the three types of day is £220, £130 and £40 respectively. If his average total costs per day are £80, calculate his expected profit per day.

*Answer*

We require expected profit per day. Thus we need to calculate the different values of profit that are possible (for the three types of day specified) and their respective probabilities.

First, $p$(Sunny day) = 0.2; $p$(Cold day) = 30% = 0.3.

Thus, $p$(Medium day) = 1 – 0.2 – 0.3 = 0.5

Since total costs are the same for any day (£80), the profit that the salesman makes on each of the three types of day are:

| | |
|---|---|
| Sunny day: | £220 – £80 = £140; |
| Medium day: | £130 – £80 = £50, |
| Cold day: | £40 – £80 = –£40 (i.e. a *loss* of £40). |

A *probability table* (Table 3) can now be constructed.

*Table 3 Probability table*

| | *Sunny* | *Medium* | *Cold* |
|---|---|---|---|
| Profit (£) | 140 | 50 | –40 |
| Probability | 0.2 | 0.5 | 0.3 |

From Table 3: expected profit/day = £(140)(0.2) + (50)(0.5) + (–40)(0.3) = £41

## 5. Example 2 *(Expected values)*

*Question*

On a particular day, a trader expects the sales of cauliflowers to follow the pattern:

| Sales | 0 | 100 | 200 | 300 |
|---|---|---|---|---|
| Probability | 0.1 | 0.4 | 0.3 | 0.2 |

Cauliflowers cost him 30p each and he sells them at 60p, any left over at the end of a day are sold to a local farmer at 5p each. If he buys 200 cauliflowers one morning, calculate his expected profit for that day.

*Answer*

Each cauliflower sold makes a profit of 30p (£0.30), leftovers making a loss of 25p (£0.25) each. (In the following calculations, the money units are in £'s.)

*At sales level 0*:

200 cauliflowers will be left over, giving total profit 200(–0.25) = –50.
Thus, expected profit at sales level 0 = (0.1)(–50) = –5.

*At sales level 100*:

100 cauliflowers sold give a profit of 100(0.30) = 30.
Thus, expected profit = (0.4)(30) = 12.
100 cauliflowers left over give a profit of 100(–0.25) = –25
Thus, expected profit = (0.4)(–25) = –10
Therefore, expected profit at sales level 100 = 12 – 10 = 2.

*At sales level 200*:

200 cauliflowers sold give a profit of 200(0.30) = 60.
Thus, expected profit at sales level 200 = (0.3)(60) = 18.

*At sales level 300 (which cannot be met)* only 200 can be sold:

200 cauliflowers sold give a profit of 60 (from above).
Thus, expected profit at a nominal sales level 300 = (0.2)(60) = 12.

The results are also shown in tabular form (Table 4).

*Table 4*

| | | | | | *Total* |
|---|---|---|---|---|---|
| Sales | 0 | 100 | 200 | 300 | |
| Probability | 0.1 | 0.4 | 0.3 | 0.2 | 1 |
| Profit (£) | –50 | 5 | 60 | 60 | |
| Expected profit (£) | –5 | 2 | 18 | 12 | 27 |

The expected profit for the day, if 200 cauliflowers are bought, is £27.

[Note that if a different number of cauliflowers were bought, the values in the table above would change, yielding a different expected profit. The calculation of the expected profit for purchases of 100 and 300 cauliflowers is left as an exercise. See exercise 4 at the end of the chapter.]

## 6. Conditional probability

Suppose a contract is put out to tender and four firms (*A*, *B*, *C* and *D* say) submit proposals. *A*, *B* and *C* are local firms, *D* is a national firm and each firm has an equal chance of winning the contract. So, given no further information, Pr(*A* wins contract) = $\frac{1}{3}$, since each of the four firms have an equal chance of winning.

However, suppose now that we are given the information that a local firm has obtained the contract (but no more than this). What this extra information enables us to do is to remove firm *D* from the field, since (not being a local firm) they could not have won the contract.

Therefore, based on the information given, Pr(*A* wins contract) = $\frac{1}{3}$ (since *A* is now just one out of three. This is usually written as:

Pr(*A* wins contract / local firm has won contract)

and described as 'the probability that $A$ wins the contract given that a local firm has won the contract'.

This is known as *conditional probability*.

## 7. Definition of conditional probability

**Conditional probability**

A *conditional probability*, involving two events $A$ and $B$, can be written in the form:

$$\Pr(A/B)$$

and translated as: 'the probability of event $A$ occurring conditional on $B$ having occurred'.

For an experiment having *equally likely outcomes*, the conditional probability of $A$ given $B$, $\Pr(A/B)$, can be calculated by dividing the number of outcomes in $A$ that are also in $B$ by the number of outcomes in $B$. That is:

**Definition of conditional probability (equally likely outcome set)**

$$\Pr(A/B) = \frac{n(A \text{ and } B)}{n(B)}$$

Events $A$ and $B$ must interact in some way (i.e. depend on each other) for conditional probability to be different to ordinary probability. For example, if $A$ = 'becoming an accountant' and $B$ = 'having flat feet', then we would have that $\Pr(A/B) = \Pr(A)$, since, becoming an accountant is (one would expect) independent of whether one had flat feet or not.

The following examples demonstrate the calculation of conditional probabilities.

## 8. Example 3 (*Conditional probabilities* using information from sets)

Figure 1 shows a Venn diagram with the numbers of elements in the FOUR defined areas of two intersecting sets.

The Venn diagram in Figure 1 overleaf is representing an experiment with an equally-likely outcome set $S$ having two events $A$ and $B$ defined.

From the figure, $n(\text{Outcome set}) = 12 + 4 + 10 + 24 = 50$;

$n(A) = 12 + 4 = 16$; $n(B) = 10 + 4 = 14$; $n(A \cap B) = 4$.

Now, unconditionally, $\Pr(A) = \frac{16}{50}$ (since there are 16 outcomes in event $A$) and, similarly, $\Pr(B) = \frac{14}{50}$.

But, *given that event B has occurred*, $B$ can now be considered as the new (or revised) outcome set for the experiment.

*Figure 1 Outcome set S with sets A and B enumerated*

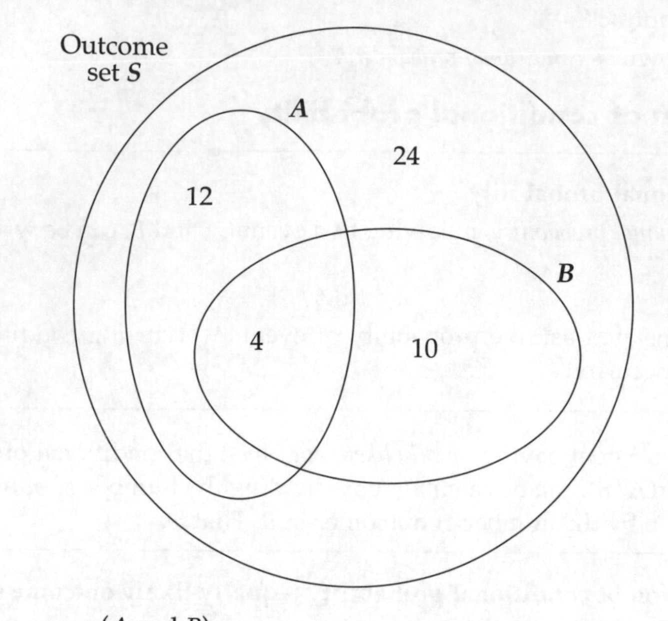


So: $\Pr(A/B) = \dfrac{n(A \text{ and } B)}{n(B)}$

$= \dfrac{4}{14}$

since there are 4 elements in that part of $A$ which is also in $B$ and 14 elements in $B$ (the revised outcome set).

In a similar fashion: $\Pr(B/A) = \dfrac{4}{16}$

## 9. Example 4 (*Conditional probabilities* in a given situation)

### *Question*

Of 8 *equal* candidates for a job, 3 are qualified accountants, 4 are graduates and 2 have neither of these qualifications. Find:

a) the probability that a graduate gets the job;
b) given that a qualified accountant has got the job, the probability that he is a graduate;
c) the probability that a qualified accountant gets the job, given that a graduate did not get the job.

### *Answer*

From the given information, a Venn diagram can be drawn and is shown in Figure 2.

a) Since there are 4 graduates out of the 8 candidates: $\Pr(\text{graduate}) = \dfrac{4}{8} = 0.5$
b) Out of the 3 qualified accountants, only 1 is a graduate.

Thus, $\Pr(\text{graduate}/\text{qualified accountant}) = \dfrac{1}{3} = 0.33$ (2D)

c) 4 of the candidates are non-graduates, and of these, 2 are qualified accountants.

Therefore, Pr(qualified accountant/non-graduate) = $\frac{2}{4}$ = 0.5.

*An alternative to a Venn diagram* for representing this situation is a table, as shown at Table 5.

Re-working c) above using this table, we require: $\Pr(C/\overline{G})$

*Figure 2*

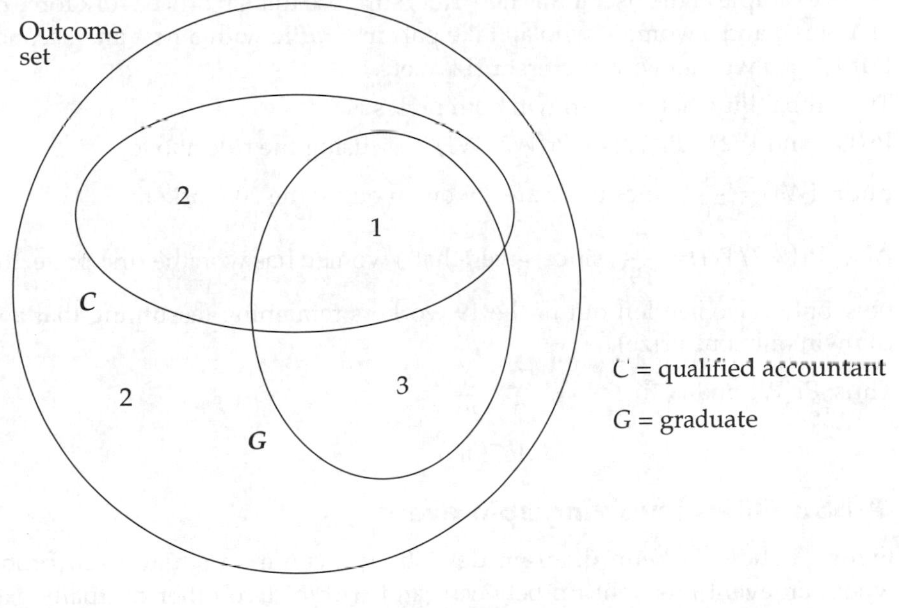


*Table 5*

| | C | $\overline{C}$ | Total |
|---|---|---|---|
| G | 1 | 3 | 4 |
| $\overline{G}$ | 2 | 2 | 4 |
| Total | 3 | 5 | 8 |

*Given that event G has not occurred*, in probability terms, is equivalent to taking account of the second row only. In other words, *all other information* in the table can be ignored.

Hence, given that we are concerned with the information in row 2 only, there are just 2 C's out of the total of 4.

Therefore, $\Pr(C/\overline{G}) = \frac{2}{4}$

= 0.5 (as before).

## 10. The general multiplication rule for probability

The multiplication rule for probability (covered in the previous chapter) can be generalised, by bringing in conditional probability, in the following way.

> **The general multiplication rule**
>
> $$\Pr(A \text{ and } B) = \Pr(A) \times \Pr(B/A)$$
>
> for any events $A$ and $B$.

**Note**: If $A$ and $B$ are *independent events*, $\Pr(B/A) = \Pr(B)$, and the above rule just reverts to the ordinary multiplication rule: $\Pr(A \text{ and } B) = \Pr(A) \times \Pr(B)$.

As an example of the use of this new rule, suppose that an office work force consists of 12 men and 8 women, who all take part in a raffle with a first and second prize. Put $W1$ = a woman wins the first prize ...etc.

The probability that women win both prizes is:

$\Pr(W1 \text{ and } W2) = \Pr(W1) \times \Pr(W2/W1)$ [using the rule above]

But $\Pr(W1) = \dfrac{8}{20}$, since there are 8 women out of the 20 workers.

Also, $\Pr(W2/W1) = \dfrac{7}{19}$, since, given that a woman has won the first prize, there are now only 7 women left out of the 19 workers remaining (assuming that a worker can win only one prize).

Thus: $\Pr(W1 \text{ and } W2) = \dfrac{7}{19} \times \dfrac{8}{20}$

$= 0.147(3D)$.

## 11. Probabilities involving split events

Figure 3 shows a Venn diagram describing a common situation in probability, where an event $E$ is split up between (and within) two other mutually exclusive events $A_1$ and $A_2$, say.

*Figure 3*

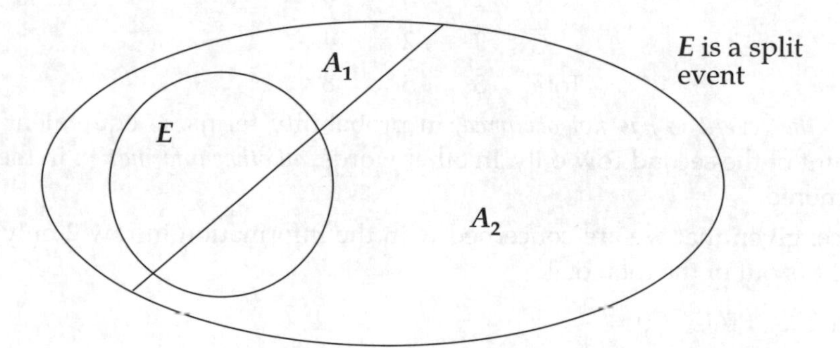


Often, A1 and A2 are complementary events. i.e. $\sum \Pr(A_i$

For example, the profitability of a company might depend on whether the economic climate is defined as good or not. In this case, we could have:

$E = P$ (profitable); $A_1 = G$ (good climate); $A_2 = \overline{G}$ (not a good climate).

Given that certain probabilities involving a split event are known, the following split event rule can be used to determine the probability of event $E$ occurring.

**A split event rule**

$$Pr(E) = Pr(A_1).Pr(E/A_1) + Pr(A_2).Pr(E/A_2)$$

where event $E$ is split between the two mutually exclusive $A$ events.

For the example above, if it is known that a good climate occurs only 30% of the time and the probability that the company makes a profit in a good climate is 0.8 and in a bad climate is 0.1, we thus have the following probabilities:

$Pr(G) = 0.3; Pr(\overline{G}) = 0.7; Pr(P/G)=0.8; Pr(P/\overline{G}) = 0.1$

Pr(company makes a profit)

$= Pr(P)$

$= Pr(G).Pr(P/G) + Pr(\overline{G}).Pr(P/\overline{G})$ [using split event rule]

$= (0.3)(0.8) + (0.7)(0.1) = 0.24 + 0.07 = 0.31$

The split event rule can be extended as follows:

**The generalised split event rule**

If $E$ is an event which is split up between and within the mutually exclusive events $A_1$, $A_2$, $A_3$, ... etc, then:

$$Pr(E) = Pr(A_1).Pr(E/A_1) + Pr(A_2).Pr(E/A_2) + Pr(A_3).Pr(E/A_3) + ...$$

$$= \sum Pr(A_i).Pr(E / A_i).$$

Thus, if event $L$ is split up between the four events $P$, $Q$, $R$ and $S$, the rule would take the form:

$Pr(L) = Pr(P).Pr(L/P) + Pr(Q).Pr(L/Q) + Pr(R).Pr(L/R) + Pr(S).Pr(L/S)$

An example follows of the use of the rule for an event split up between three other events.

## 12. Example 5 (The probability of a *split event*)

*Question*

A firm thinks that winning the contract for a new job (event $W$) has probability 0.9, 0.7 or 0.1 depending respectively on whether it finishes the current job early ($E$), on time ($T$) or late ($L$).

Management has been told that the chance of finishing the current job early or on time are, respectively, 5% and 20%. What chance does the firm have of winning the contract for the new job?

*Answer*

Note here, that event $W$ is split up into the three events $E$, $T$ and $L$. Further, they *are* mutually exclusive, since only one of 'early', 'on time' and 'late' can occur at any one time.

We are given that $Pr(E)=0.05$ and $Pr(T)=0.2$.

Thus $\Pr(L) = 1 - 0.05 - 0.2 = 0.75$.

We are also given that $\Pr(W/E) = 0.9$, $\Pr(W/T) = 0.7$ and $\Pr(W/L) = 0.1$.

Thus, using the (extended) split event rule:

$$\begin{aligned}\Pr(W) &= \Pr(E).\Pr(W/E) + \Pr(T).\Pr(W/T) + \Pr(L).\Pr(W/L)\\ &= (0.05)(0.9) + (0.2)(0.7) + (0.75)(0.1) = 0.045 + 0.140 + 0.075 = 0.26\end{aligned}$$

In other words, based on the information given, the firm has only a 26% chance of winning the contract for the new job.

## 13. Bayes Theorem

Bayes Theorem is a formula which can be thought of as 'reversing' conditional probability. That is, it finds a conditional probability $(A/B)$ given, among other things, its inverse $(B/A)$.

> **Bayes Theorem**
>
> If $A$ and $B$ are two events of an experiment, then:
>
> $$\Pr(A/B) = \frac{\Pr(A).\Pr(B/A)}{\Pr(B)}$$

**Note:** Sometimes we need to use the split event rule to obtain $\Pr(B)$ in Bayes Theorem.

As an example of using Bayes Theorem, if the probability of meeting a building contract date is 0.8,the probability of good weather is 0.5 and the probability of meeting the date *given* good weather is 0.9, we can calculate the probability that there was good weather given that the contract date was met.

Put $G$ = 'good weather' and $M$ = 'contract date is met'.

We are given that: $\Pr(M)=0.8$, $\Pr(G)=0.5$ and $\Pr(M/G)=0.9$.

We need to find $\Pr(G/M)$, which is straightforward from Bayes Theorem.

We have: $$\Pr(G/M) = \frac{\Pr(G).\Pr(M/G)}{\Pr(M)} = \frac{0.5 \times 0.9}{6.1} = 0.5625$$

## 14. Tabular solution of conditional probability problems

Many problems involving conditional probability, Bayes Theorem situations included, can be represented and then solved using a tabular form. This involves representing the situation given in terms of *expected number of occurrence*, rather than in probability form.

This is demonstrated below, using the problem given in the previous section.

We are given that: $\Pr(M)=0.8$, $\Pr(G)=0.5$ and $\Pr(M/G)=0.9$

Taking a total of 100 occurrences, the date would have been met on 80 of these [$\Pr(M)=0.8$] and good weather would happen on 50 [$\Pr(G)=0.5$]. Also, out of the 50 good weather occurrences, the date would have been met on 45 occasions [$\Pr(M/G)=0.9$ i.e. $0.9 \times 50=45$]. Table 6 shows these values incorporated, together with the remaining values which are easily obtained by subtraction.

*Table 6*

| | *Date met (M)* | *Date not met* | *Total* |
|---|---|---|---|
| Good Weather (*G*) | 45 | 5 | 50 |
| Bad Weather | 35 | 15 | 50 |
| Total | 80 | 20 | 100 |

Using the table, out of 80 times that the date was met, there were 45 times when the weather was good.

Thus, $\Pr(G/M) = \dfrac{45}{80} = 0.5625.$

This gives the same result as Bayes Theorem gave in the previous section.

## 15. Example 6 (Solving a *conditional probability* problem)

*Question*

A stock market analyst makes *wrong decisions* with probability 0.2 and he has just advised you to buy some stock. If you previously believed that the stock in question had a 70% chance of success, what must this percentage be revised to in the light of the analyst's advice?

*Answer*

Both the tabular and (Bayes Theorem) formula methods will be used to solve this problem.

a) *Tabular method*

For the table, we have the analyst's advice (either buy or not) ranged against the performance of the stock (succeed or fail).

The given (unconditional) probability of success is 70%. That is, the stock would succeed 70 times out of 100 (and fail the other 30). Out of the 70 successes, the analyst's advice would be not to buy on $0.2 \times 70 = 14$ (since he makes a wrong decision with probability 0.2). Also, out of the 30 failures, the analyst's advice would be to buy (again, making the wrong decision) on $0.2 \times 30 = 6$ of these. Using these calculated values, Table 7 can be made up as follows.

*Table 7*

| | | *Analyst's advice* | | |
|---|---|---|---|---|
| | | *Buy* | *Not buy* | *Total* |
| | Success | 56 | 14 | 70 |
| *Result* | Failure | 6 | 24 | 30 |
| | Total | 62 | 38 | 100 |

We require Pr(Stock success / Analyst says buy).

From the table above, the analyst would say buy on 62 occasions, of which 56 would result in success.

Therefore, $\Pr(\text{Stock success / Analyst says buy}) = \dfrac{56}{62} = 0.903$ (3D).

In other words, the revised probability of success is 90.3%.

b) *Formula method*

Put $S$='Success' and $F$='Failure', with regard to the progress of the stock. Put $B$='Buy' and $N$='Do not buy', with regard to the analyst's advice.
We require Pr(Stock success / Analyst says buy) = $\Pr(S/B)$.
Using the Bayes Theorem formula:

$$\Pr(S/B) = \frac{\Pr(S).\Pr(B/S)}{\Pr(B)} = \frac{\Pr(S).\Pr(B/S)}{\Pr(S).\Pr(B/S) + \Pr(F).\Pr(B/F)}$$

We are given that $\Pr(S)= 0.7$ (since there is a 70% chance that the stock will succeed).
Therefore, $\Pr(F) = 1 - 0.7 = 0.3$.
Also, Pr(analyst makes wrong decision) = $\Pr(B/F) = 0.2$,
and, Pr(analyst makes right decision) = $\Pr(B/S) = 1 - 0.2 = 0.8$

i.e. $\Pr(S/B) = \frac{(0.7)(0.8)}{(0.7)(0.8) + (0.7)(0.8)} = \frac{0.56}{0.56 + 0.56} = 0.903$ (3D), which agrees with the result in a).

## 16. Summary

a) The expectation (or expected value) is an arithmetic mean of a given set of values using probabilities instead of frequencies. The formula for calculating it is: expected value = $\sum px$
That is, each value is multiplied by its probability of occurrence and the results added.

b) The conditional probability that event $A$ occurs, *given that event B has occurred* is written as:
$p(A/B)$. For an experiment with an equally likely outcome set:

$$\Pr(A/B) = \frac{n(A \text{ and } B)}{n(B)}$$

c) The general multiplication rule for probabilities is:

$$\Pr(A \text{ and } B) = \Pr(A) \times \Pr(B/A) \quad \text{for } any \text{ events } A \text{ and } B.$$

d) If event $E$ is split up between and within the mutually exclusive events $A_1$, $A_2$, $A_3$, ..., then:

$$\Pr(E) = \Pr(A_1).\Pr(E/A_1) + \Pr(A_2).\Pr(E/A_2) + \Pr(A_3).\Pr(E/A_3) + ...$$

e) Bayes theorem calculates a conditional probability $\Pr(A/B)$, given its inverse $\Pr(B/A)$ using the formula:

$$\Pr(A/B) = \frac{\Pr(A).\Pr(B/A)}{\Pr(B)}$$

## 17. Student self review questions

1. What is 'expectation' and how is it calculated? [2,3]
2. Explain precisely what $\Pr(A/B)$ means. [7]
3. How is $\Pr(A/B)$ calculated when an experiment has an equally likely outcome set? [7]

4. Under what conditions does $\Pr(A/B) = \Pr(A)$? [7]
5. What is the general multiplication rule for probability? [10]
6. Write down the split event rule for calculating $\Pr(E)$, where $E$ is split between and within the mutually exclusive events $A$ and $B$. [11]
7. What is Bayes Theorem? [13]

## 18. Student exercises

1. For the following frequency distribution:

| Profit per quarter ($x$, £000) | 20 | 30 | 40 | 50 |
|---|---|---|---|---|
| Number of quarters ($f$) | 8 | 26 | 14 | 2 |

   a) Calculate, $\bar{x}$, the arithmetic mean (profit per quarter).
   b) Convert the frequencies to probabilities (proportions) and show that the *expected* profit per quarter is the same as the arithmetic mean, found in a).

2. Calculate the expectation of $x$ for the following distribution.

| ($x$) | 8 | 9 | 10 | 11 | 12 | 13 | 14 | 15 |
|---|---|---|---|---|---|---|---|---|
| Probability ($p$) | 0.10 | 0.15 | 0.15 | 0.25 | 0.20 | 0.10 | 0 | 0.05 |

3. The probability that it rains on any day is 0.3. A pipe-laying firm has a contract to lay 500 metres of pipe. They normally expect to lay 20 metres per day, unless it is raining, when their workrate is cut by 60%. What is the expected number of working days that the contract will take to complete?
4. Rework Example 2, drawing up a table and calculating expected profit for the day, using:
   a) 100 cauliflowers bought;
   b) 300 cauliflowers bought.
   How many cauliflowers should the trader buy, and why?
5. Given that, for the events $A$ and $B$: $n(A)$=25, $n(B)$=15 and $n(A \text{ and } B)$=10, calculate:
   a) $\Pr(A/B)$
   b) $\Pr(B/A)$.
6. A company has taken a ballot of its workforce, asking them to choose between two alternative new schemes ($A$ and $B$) for claiming travelling expenses. 60% of the workforce support scheme $A$, 50% of these owning a car. If only 30% of those who support scheme $B$ own a car, calculate the probability that if a worker is chosen at random, he will be:
   a) a car owner;
   b) a car-owning, scheme $A$ supporter;
   c) a car owner, given that he supports scheme $A$;
   d) a scheme $A$ supporter, given that he owns a car.
7. Machine $A$ produced 15 of the day's output of wodgets and machine $B$ produced the other 5. If 2 wodgets are selected at random from the day's output, find the probability that:
   a) they both came from machine $A$;
   b) they both came from the same machine;
   c) there was one from each machine;
   to three decimal places.

8. 30% of full-time and 10% of part-time workers are in the company's private medical scheme. If 25% of all workers are part-time, what is the probability that a worker chosen at random will be in the scheme?
9. A company owns a fleet of 20 cars, each having either manual or automatic transmission and either 2 or 4 doors. 13 cars are 2-door models and, of these, 12 have automatic transmission. There are only 4 cars with manual transmission. If a car is picked at random from the fleet, calculate the probability that it is:
   a) automatic;
   b) 4-door;
   c) automatic or 2-door;
   d) automatic and 2-door;
   e) automatic/4-door;
   f) 4-door/automatic.
10. Test drilling equipment has forecast (whether or not a proposed new coal seam will provide economic quantities of coal) incorrectly 20% of the time. The last 30 seams that were mined produced only 10 that were economic. The current test drilling has just given a favourable indication. What is the probability that this seam will be economic?

# Examination examples (with worked solutions)

*Question 1*

The independent probabilities that the three sections of a costing department will encounter a computer error are respectively 0.1, 0.2 and 0.3 each week. Calculate the probability that there will be:

i. at least one computer error;
ii. one and only one computer error;

encountered by the costing department next week.

*CIMA*

*Answer*

(i) Pr(at least one error) $= 1 - \text{Pr(no errors)}$
$= 1 - (1–0.1)(1–0.2)(1–0.3)$
$= 1 - (0.9)(0.8)(0.7) = 0.496$

(ii) An error in one only can be obtained in three separate ways. Namely, an error in the first section but not in the other two or an error in the second but not in the first or third or an error in the third but not in the first or second. Thus:

Pr(one and only one error)
$= (0.1)(1–0.2)(1–0.3) + (1–0.1)(0.2)(1–0.3) + (1–0.1)(1–0.2)(0.3)$
$= 0.056 + 0.126 + 0.216 = 0.398$

*Question 2*

a) State briefly whether the following pair of events are independent, and why.
   i. Winning two successive prizes in the monthly Premium Bond draws.
   ii. Earning a large salary and paying a large amount of income tax.
   iii. Being drunk while driving and having an accident.
   iv. Being an accountant and having large feet.
   v. Any two mutually exclusive events.

b) A manufacturer assembles a toy from four independently produced components, each of which has a probability of 0.01 of being defective. What is the probability of a toy being defective?

c) A factory has a machine shop in which three machines (*A*, *B* and *C*) produce 100cm aluminium tubes. An inspector is equally likely to sample tubes from *A* and *B*, and three times as likely to select tubes from *C* as he is from *B*. The defective rates from the three machines are:

*A* 10% *B* 10% *C* 20%

What is the probability that a tube selected by the inspector:

i. is from machine *A*;
ii. is defective;
iii. comes from machine *A*, given that it is defective?

*CIMA*

*Answer*

a) i. Premium bonds, once purchased, are redrawn every month. Thus, winning one month will not affect winning any other month.

ii. Income tax is calculated as a percentage of gross salary and thus is dependent on salary. Therefore not independent.

iii. It is well known that drinking impairs the faculties generally. Hence, it will affect driving ability in particular. Therefore not independent.

iv. These cannot be connected in any way. Therefore independent.

v. Two events being mutually exclusive means that they are events from the same experiment. So, if one of them occurs, the other cannot. Therefore not independent.

b) Pr(any component defective) = 0.01.

Thus Pr(any component OK) = 1– 0.01 = 0.99.

Pr(toy defective) = Pr(one or more components defective)

= 1 – Pr(all components OK)

= $1 - (0.99)^4 = 1 - 0.9606$

= 0.0394 (4D)

c) If the inspector samples 100 tubes, we would expect on average 20*A*, 20*B* and 60*C* to be selected. 10% of the *A* tubes will be defective = 10% of 20 = 2. Similarly, 10% of the 20*B* tubes = 2 will be defective and 20% of the 60*C* tubes = 12 will be defective.

The following table can be set up:

| | A | B | C | Total |
|---|---|---|---|---|
| OK | 18 | 18 | 48 | 84 |
| Defective | 2 | 2 | 12 | 16 |
| Total | 20 | 20 | 60 | 100 |

i. Pr(*A*) = 0.2 (since 20 of the 100 tubes are *A*);

ii. Pr(defective) = 0.16 (since 60 of the 100 tubes are defective);

iii. $\Pr(A/\text{defective}) = \frac{2}{16} = 0.125$ (since 2 of the 16 defective tubes are *A*).

# Examination questions

1. In 1986, a company estimated that 65% of its customers were female and, in that year, three customers entered one of their shops. Assuming that these customers were not related or acquaintances, find the probability that they were:
   i. all female,
   ii. all male, and
   iii. at least one male.

*ICSA*

2. A company has a security system comprising four electronic devices (*A*, *B*, *C* and *D*) which operate independently. Each device has a probability of failure of 0.1. The four electronic devices are arranged so that the whole system operates properly if at least one of *A* or *B* functions, and at least one of *C* or *D* functions.
   What is the probability that the whole system operates properly?

*CIMA*

3. a) A company tenders for two contracts *A* and *B*. The probability that it will obtain contract *A* is $\frac{1}{3}$ and the probability that it will obtain contract *B* is $\frac{1}{4}$.
   Find the probability that the company:
   i. will obtain both contracts; ii. will obtain only one contract.

   b) A company has a large number of typists. A survey shows that 30 can use a word processor, 25 are audio-typists and 28 are shorthand writers. Of the typists who are shorthand writers, 3 are audio-typists and can use a word processor, 5 are audio-typists and cannot use a word processor, 9 can use a word processor but are not audio-typists, 6 of the audio-typists can use a word processor but are not shorthand writers.
   i. Represent this information on a Venn diagram.
   ii. How many typists were involved in the survey?
   iii. How many typists have only one skill?

*ICSA*

4. a) State clearly by what is meant by two events being statistically independent.
   b) In a certain factory which employs 10,000 men, 1% of all employees have a minor accident in a given year. Of these, 40% had safety instructions whereas 90% of all employees had no safety instructions. What is the probability of an employee being accident-free:
   i. given that he had no safety instructions?
   ii. given that he had safety instructions?
   c) A company runs a special lottery. A box contains 100 tickets, an *unknown* number of which are winning tickets, that number having been selected from random number tables from the integers 1 to 15 inclusive. A ticket is picked at random from the box. What is the probability that it is a winning ticket?

*CIMA*

5. A company has to decide which advertising method to use in marketing a product in the next period. Advertising method *A*, which is currently used, costs £10,000/ week and is expected only to maintain sales. From past experience of using method *A*, the likelihood of weekly sales in the next period (with *A*) has been estimated as shown:

| Unit weekly sales | Probability of weekly sales if *A* is used |
|---|---|
| 0–200 | 0 |
| 200–400 | 0.3 |
| 400–600 | 0.4 |
| 600–800 | 0.2 |
| 800–1,000 | 0.1 |

Advertising method *B* would cost £37,000/ week and sales would be expected to be 50% higher than with *A*. Advertising method C would cost £60,000/week and sales would be expected to be 100% higher than with *A*.

The current contribution, taking *A* into account, is £100/ unit sold.

You are required to use a simple decision tree with expected values, to recommend the best action to take and to comment on the answer.

*CIMA*

6. For the past 200 days, the sales of bread from a bakery have been as follows:

| Daily sales (loaves) | Number of days |
|---|---|
| 0 | 10 |
| 100 | 60 |
| 200 | 60 |
| 300 | 50 |
| 400 | 20 |

a) What is the expected (mean) sales of bread?
b) The bakery's production costs are £0.20 per loaf, price is £0.40 per loaf, and any bread unsold at the end of the day is contracted to a local farmer who pays £0.10 per loaf. Draw up a suitable profit table for each sales/production combination.
c) Compute the expected profit arising from each level of production.
d) State with reasons the optimum production level for the bakery.

*CIMA*

7. An electronics company has discovered a new sensor for detecting smoke which could prove valuable for the early prevention of fire. At present it is extremely expensive and the company has to decide whether or not to continue with its development. Costs incurred up to this point may be regarded as irrelevant to this decision.

It would cost £4 million to develop the sensor fully. The scientists estimate that, at the end of this development phase, the sensor would be 'very superior', 'superior' to the existing technology with probabilities of 0.3, 0.5 and 0.2 respectively. The company could then market the new product, at a cost of £1 million, or shelve the idea.

If the new sensor is marketed there is a risk of increased competition from the other major suppliers. This is estimated to be 80% if it is 'very superior', 50% if it is 'superior', but only 10% if it is 'not superior'.

The management accountant has estimated the revenue (£ million) from the sales of the new sensor to be as follows:

| | Very superior | Superior | Not superior |
|---|---|---|---|
| Increased competition | 15 | 5 | 1 |
| Same competition | 25 | 10 | 2 |

All financial values have been discounted to the present.
The decision tree structure for this problem is given below.

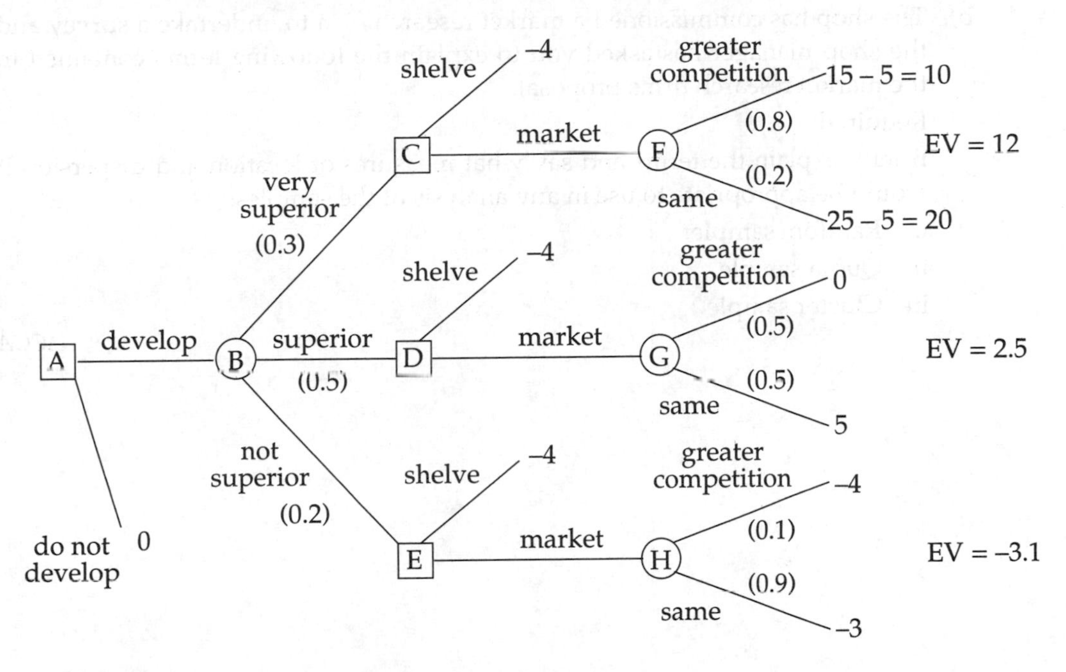


You are required, using the diagram above, to

a) complete the decision tree with probabilities, costs and revenues incorporated appropriately;
b) calculate relevant expected values and include these as appropriate on the tree;
c) recommend the action the company should take at A, C, D, E and explain the basis of your decisions.

*CIMA*

8. a) A retail shop has space to display one of two different products, coded *A* and *B*. The demand for each product can take three levels, high, medium or low. The shop manager has assessed the probabilities of each level of demand and the corresponding profit. The details are given below.

| Product *A* | Level of demand | Probability | Profit |
|---|---|---|---|
| | High | 0.2 | £2 |
| | Medium | 0.5 | £1.5 |

The profit for low demand is £0.75

| Product *B* | Level of demand | Probability | Profit |
|---|---|---|---|
| | High | 0.3 | £1.5 |
| | Medium | 0.3 | £0.5 |

The profit for low demand is £1.0.

Required:

Using the criterion of maximum expected profit what product should the shop display?

b) The shop has commissioned a market research firm to undertake a survey and the shop manager has asked you to explain the following terms contained in the market research firms proposal.

Required:

Briefly explain the terms and say what measures of location and dispersion it would be appropriate to use in any analysis of the samples.

i. Random sample.

ii. Quota sample.

iii. Cluster sample.

*ACCA*

# Part 9 Further probability

This part of the text is concerned with some further aspects of probability.

In an earlier part of the book (chapter 4), frequency distributions were introduced and described. In practice, certain types of distributions occur so frequently that they are given special names. It is with three of these that this part of the manual is concerned and they are examined in terms of probabilities (of events that might happen) rather than frequencies (of events that did happen).

Chapter 32 covers permutations and combinations and how they relate to calculating probabilities. A study of this topic is necessary for an understanding of the material in the following chapter.

Two special discrete distributions, the Binomial and Poisson, are described in chapter 33. Particular attention is paid to their relation to particular types of business situations and, within these, how probabilities are calculated.

Chapter 34 covers the single most important distribution in Statistics, known as the Normal distribution. Examples are given of applications of the distribution to business and how special tables are utilised in order to calculate probabilities. Also covered are confidence limits for the mean and a proportion and an introduction to significance tests.

# 32 Combinations and permutations

## 1. Introduction

This chapter is concerned with the ways of selecting and arranging groups of items, known as combinations and permutations. These are described and their use in solving various types of problems in probability demonstrated.

## 2. Combinations and permutations

> **Permutations and combinations**
>
> A *combination* is a selection of distinct items
>
> A *permutation* is a combination arranged in a particular way.

For example, given the items A, B, C and D to choose from:
AB is one selection (combination) of two items; AC, AD, BC, BD and CD are the other possible combinations of two items.

Also, AB is a combination of two items, for which there are *two* separate permutations AB and BA; ABC is a combination of three items for which there are *six* separate permutations, namely ABC, ACB, BAC, BCA, CAB, CBA.

**Note 1**. Re-arranging the items within a combination does not give a different combination. However, re-arranging the items within a permutation gives a different permutation. For example, ABC is the *same* (three letter) combination as ACB or BAC. However, ABC, ACB and BAC are three *different* permutations of the three letters A, B and C.

**Note 2**. Remember that it is in *permutations* that it matters which order the items are put into.

## 3. Example 1 (Listing some *permutations* and *combinations*)

a) Four firms P, Q, R and S submit tenders for two jobs 1 and 2, each of which must go to a different firm. List the possible ways that the jobs can be allocated.
Writing PQ as 'P gets job 1 and Q gets job 2', we are interested in *permutations* here, since PQ will have a different meaning to QP.
The list is: PQ QP PR RP PS SP
QR RQ QS SQ RS SR
i.e. the two jobs can be allocated in 12 different ways.

b) Out of five people in an office, A, B, C, D and E say, just three are to be selected to go to an exhibition. In how many ways can the three be chosen?
Here, the selection BCD will be the same as the selection CBD or BDC. Thus we are interested in combinations.
The list of combinations is: ABC ABD ABE ACD ACE ADE
BCD BCE BDE
CDE
i.e. there are 10 different ways that the three people can be chosen.

## 4. Factorials

In probability, it is not so much a listing of possible combinations and permutations that is important, but *how many* there are. Before looking at formulae that will give the numbers of combinations and permutations that there are in particular situations, it is necessary to understand what a factorial is.

A *factorial* of a positive whole number is written as '!' (an exclamation mark) and defined as follows:

$3! = 3 \times 2 \times 1 = 6$; $\quad 4! = 4 \times 3 \times 2 \times 1 = 24$; $\quad 6! = 6 \times 5 \times 4 \times 3 \times 2 \times 1 = 720$;

In general:

**Factorial numbers**

$$n! = n \times (n-1) \times (n-2) \times \ldots \times 2 \times 1$$

**Notes:** 1. $n$ must be a positive whole number

2. $n!$ is spoken as '$n$ factorial'

3. $0! = 1! = 1$.

## 5. The combination formula

The number of different combinations of $r$ items from $n$ distinct items is written as ${}^nC_r$ and calculated as follows.

**The combination formula**

$${}^nC_r = \frac{n!}{r!(n-r)!}$$

For example:

i. $${}^5C_3 = \frac{5!}{3!(5-3)!} = \frac{5!}{3!2!} = \frac{5\times4\times3\times2\times1}{(3\times2\times1)(2\times1)}$$

$$= \frac{5\times4}{2\times1} \text{ [3×2×1 cancels top and bottom]}$$

$= 10$ (agreeing with the result of Example 1(b)).

That is, there are 10 different combinations of 3 from 5.

ii. $${}^6C_2 = \frac{6!}{2!(6-2)!} = \frac{6!}{2!4!} = \frac{6\times5\times4\times3\times2\times1}{(2\times1)(3\times3\times2\times1)} = \frac{6\times5}{2\times1} = 15$$

That is, there are 15 different combinations of 2 from 6.

**Note** that ${}^nC_1 = n$, for $n$ as any positive whole number.

For example: ${}^5C_1 = 5$; ${}^6C_1 = 6$ and so on.

Also, ${}^nC_0 = 1$, for $n$ as any positive number. i.e. ${}^4C_0 = {}^3C_0 = 1$.

## 6. The permutation formula

The *number* of different permutations of $r$ items from $n$ distinct items is written as ${}^nP_r$ and calculated as follows.

**The permutation formula**

$$^nP_r = \frac{n!}{(n-r)!}$$

For example:

i. $^4P_2 = \frac{4!}{(4-2)!} = \frac{4!}{2!} = \frac{4\times3\times2\times1}{2\times1} = 12$ (agreeing with the result of Example 1(a)).
That is, there are 12 different permutations of 2 from 4.

ii. $^7P_3 = \frac{7!}{(7-3)!} = \frac{7!}{4!} = \frac{7\times6\times5\times4\times3\times2\times1}{4\times3\times2\times1} = 7\times6\times5 = 210.$
That is, there are 210 different permutations of 3 from 7.

## 7. Example 2 (*Permutations* and *probability*)

*Question*

a) How many ways are there of arranging 3 different jobs between 5 men, where any man can do only one job?
b) What is the probability that man A will be doing job 1?

*Answer*

a) We require a number of *permutations*, since, putting ABC as man A does job 1, man B does job 2, etc, it is clear that different arrangements of the letters mean different men to different jobs.
Thus, number of arrangements possible $= {}^5P_3 = \frac{5!}{2!} = 60.$

b) First, we need the number of permutations that satisfies AXX, where X means 'any man except A'.
The 'X's' can be filled by any 2 of the other 4 men, and thus done in

$$^4P_2 = \frac{4!}{2!} = 12 \text{ ways.}$$

Therefore, $\Pr(\text{man A does job 1}) = \frac{n(\text{man A does job 1})}{n(\text{arranging jobs between the 5 men})} = \frac{12}{60} = 0.2.$

Note that the above probability could have been calculated more easily by realising that each man has an equal chance of being allocated any job (probability 0.2 each!). However, the above calculations demonstrate the way that permutations can be used in probability problems.

## 8. Example 3 (*Combinations* and *probability*)

*Question*

A committee of 4 must be chosen from 3 women and 4 men. Calculate:

a) in how many ways the committee can be chosen;
b) in how many ways 2 men and 2 women can be chosen;
c) Pr(committee consists of 2 men and 2 women);
d) Pr(committee consists of at *least* 2 women).

*Answer*

It is only selections (*not* arrangements) that we are interested in here. That is, we are concerned with *combinations*.

a) We need to choose 4 people from 7.

This can be done in: $^7C_4 = \frac{7!}{4!3!} = 35$ ways.

b) With 4 men to choose from, 2 men can be selected in $^4C_2 = \frac{4!}{2!2!} = 6$ ways.

Similarly, 2 women can be selected in $^3C_2 = 3$ ways.

Thus they can both be selected in a total of $6 \times 3 = 18$ ways.

c) $\text{Pr(2 men and 2 women on committee)} = \frac{\text{n(2 men and 2 women)}}{\text{n(any 4 people)}} = \frac{18}{35}$

[from a) and b) above].

d) Pr(committee includes at *least* 2 women)

= Pr(2 men & 2 women *or* 1 man & 3 women)

= Pr(2 men & 2 women) + Pr(1 man & 3 women) [mutually exclusive events]

$= \frac{^4C_2\,^3C_2}{^7C_4} + \frac{^4C_1\,^3C_3}{^7C_4}$ [using techniques from (a), (b) and (c)]

$= \frac{6\times3}{35} + \frac{4\times1}{35} = \frac{22}{35}$.

## 9. Example 4 (*combinations* and *probability*)

*Question*

An equipment test is repeated on three separate occasions. The probability that the test is successful on each occasion is 0.35. Calculate the probability that, out of the three tests, there are:

a) 0 b) 1 c) 2 d) 3 successes in total.

Tabulate the number of successes against their respective probabilities, as calculated in a) to d) above.

*Answer*

Pr(success on any occasion) = 0.35.

Therefore Pr(failure) = 1 – 0.35 = 0.65.

For convenience, we will put SFF as 'success at 1st test, failure at 2nd test, failure at 3rd test' etc.

a) Pr(no successes) = Pr(3 failures) = Pr(FFF) = $(0.65)^3$ = 0.275 (3D).

b) Pr(1 success) = Pr(SFF or FSF or FFS)

= Pr(SFF) + Pr(FSF) + Pr(FFS) [mutually exclusive events]

= (0.35)(0.65)(0.65) + (0.65)(0.35)(0.65) + (0.65)(0.65)(0.35)

= $(0.35)(0.65)^2 + (0.35)(0.65)^2 + (0.35)(0.65)^2$

[Notice that the three separate terms are identical, each giving the probability of 1 success and 2 failures. There are three of them of course because a success can be at test number 1, 2 or 3. Put another way, there are just 3 combinations of 1 from 3. i.e. $^3C_1 = 3$.]

That is, Pr(1 success) = $^3C_1.(0.35)(0.65)^2 = 3(0.35)(0.65)^2 = 0.444$ (3D).

c) We require Pr(2 successes). Using a similar argument as in (b) above, the probability of 2 successes and 1 failure (in any order) is: $(0.35)^2(0.65)$.

Now, 2 successes (out of the 3 tests) can occur in $^3C_2 = 3$ different ways. i.e. at tests 1 and 2, 1 and 3 or 2 and 3.

Therefore, Pr(2 successes) $= {}^3C_2\,(0.35)^2(0.65) = 3(0.35)^2(0.65) = 0.239$ (3D).

d) Pr(3 successes) $= {}^3C_3\,(0.35)^3$ [using a similar argument to above]

$= (0.35)^3 = 0.043$ (3D).

Tabulating the above results gives:

| Number of successes | 0 | 1 | 2 | 3 |
|---|---|---|---|---|
| Probability | 0.275 | 0.444 | 0.239 | 0.043 |

[Distributions generated in this way are called *binomial probability* distributions.]

## 10. Summary

a) A combination is a selection of distinct items.

b) A permutation is a combination that is arranged in a particular way.

c) $n!$ ($n$ factorial) $= n \times (n-1) \times (n-2) \times \ldots \times 3 \times 2 \times 1$

d) The number of different combinations of $r$ from $n$ distinct items is given by:

$$^nC_r = \frac{n!}{r!(n-r)!}$$ [the *combination* formula]

e) The number of different permutations of $r$ from $n$ distinct items is given by:

$$^nP_r = \frac{n!}{(n-r)!}$$ [the *permutation* formula]

## 11. Student self review questions

1. What is a combination? [2]
2. What is a permutation? [2]
3. Is it in combinations or permutations that it matters what *order* the items are put in? [2]
4. What is a factorial? [4]
5. How is the number of different combinations of $r$ items from $n$ distinct items calculated? [5]
6. How is the number of different permutations of $r$ items from $n$ distinct items calculated? [6]

## 12. Student exercises

1. List all the possible combinations of 2 letters from the five letters A, B, C, D and E.
2. Evaluate:
   a) $5!$ b) $2! \times 3!$ c) $\frac{6!}{3!}$ d) $\frac{10!}{6!4!}$
3. Evaluate:
   a) ${}^5C_1$ b) ${}^6C_3$ c) ${}^4P_3$ d) ${}^6P_2$
4. On a particular day, a haulage contractor has 2 loads to deliver to separate customers. He has 4 trucks, of which 2 are new. If the trucks are chosen randomly for any particular delivery:
   a) In how many ways can the trucks be allocated the deliveries?
   b) In how many ways can the trucks be allocated the deliveries such that exactly one new truck is used?
   c) What is the probability that exactly 1 new truck is used?
5. A team of 5 is to be chosen from 4 men and 5 women to work on a special project.
   a) In how many ways can the team be chosen?
   b) In how many ways can the team be chosen to include just 3 women?
   c) What is the probability that the team includes just 3 women?
   d) What is the probability that the team includes at least 3 women?
   e) What is the probability that the team includes more men than women?

# 33 Binomial and poisson distributions

## 1. Introduction

The binomial and poisson are two important distributions in Statistics. This chapter describes how they are used and gives formulae for calculating probabilities associated with them. Also described is how the poisson distribution can be used as an approximation to the binomial.

## 2. A binomial distribution

Suppose a salesman makes one call per day and considers the call successful if he sells goods worth over £200. In the course of a five day working week, he can make either 0, 1, 2, 3, 4 or 5 successful calls. If he charts his progress of the number of successful calls per week over a 48-week year, the following frequency distribution might be obtained.

| Number of successful calls ($x$) | 0 | 1 | 2 | 3 | 4 | 5 | Total |
|---|---|---|---|---|---|---|---|
| Number of weeks ($f$) | 10 | 25 | 10 | 2 | 2 | 1 | 48 |

This is known as a *binomial* (frequency) *distribution*.

The particular characteristic that makes it binomial is that it describes the number of 'successes' obtained when a number of identical 'trials' of an experiment are performed.

For the above distribution, we have:

*Trial* = making a single call (on one day);

*Trial success* = a sale over £200;

*Number of trials* = 5 (i.e. 5 calls in a week).

This type of environment is sometimes called a *binomial situation*. The above frequency distribution has been obtained by repeating (physically carrying out) the defined trial 48 times and recording the number of successes per week over a period of 48 weeks.

**Binomial situation**

A *binomial situation* can be recognised by the following characteristics.

a) The existence of a trial (of an experiment) which is defined in terms of the two states 'success' and 'failure'

b) Identical trials are repeated a number of times, yielding a number of successes.

**Note 1.** Success/failure can be interpreted in any way that is convenient. For example: situation good/bad, item defective/OK, company profit/loss and so on.

**Note 2.** The variable ($x$) in a binomial frequency distribution will *always* be a whole number in the range 0 to $n$ (where $n$ is the number of repeated trials); the frequency ($f$) is the number of times that each of the $x$ successes occurs. Thus, in the above distribution, the number of successful calls in the week can range between 0 at least and 5 at most.

## 3. Example 1 (Formation of a binomial *frequency* distribution)

A quality control inspector takes periodic samples of 10 products from the output of a machine and examines them critically. He then records the number of defectives found in the sample. The following data show the number of defectives found in 30 successive samples.

11021 31002 02152
02011 02114 00413

A random sample is a classical situation upon which binomial situations can be based, since identical trials (i.e. selecting an item at random a number of times) are present. It only needs the identification of a 'success' to complete the picture. In this case, we can regard a success (from the inspector's point of view) as a defective product. Thus we have the following binomial situation:

Trial = selecting a product randomly
Trial success = a defective product
Number of trials = 10 (i.e. size of sample)

The data given are the results of repeating this experiment 30 times.

The corresponding binomial (frequency) distribution is tabulated below.

| Number of defectives in a sample of 10 | 0 | 1 | 2 | 3 | 4 | 5 | Total |
|---|---|---|---|---|---|---|---|
| Number of samples | 9 | 10 | 6 | 2 | 2 | 1 | 30 |

## 4. Binomial probability formula

The above two examples were demonstrating binomial *frequency* distributions, where the binomial situation was physically performed a number of times. Often however, we are interested only in the *probability* of a number of successes. To this end, the following formula can be used.

> **The binomial probability formula**
>
> Given a binomial situation with $p$ = probability of success at any trial and $n$ = number of trials, the probability of obtaining $x$ successes is given by:
>
> $$\Pr(x) = {}^{n}C_{x}.p^{x}(1-p)^{n-x}$$
>
> where $x$ can take any one of the values 0, 1, 2, 3, ... , $n$.

**Note**: Sometimes the symbol '$q$' is used instead of $1-p$.

If the probabilities for all values of $x$ are calculated and tabulated against their respective values of $x$, the result is known as a *binomial probability distribution.*

## 5. Use of the binomial probability formula

The above formula is used to find specific probabilities by substituting the number of successes required for $x$.

For example:

Putting $x = 0$ gives: $\Pr(0) = \Pr(\text{no successes}) = {}^{n}C_{0}.p^{0}.(1-p)^{n-0} = (1-p)^{n}$
[since ${}^{n}C_{0} = 1$ and $p^{0} = 1$]

Putting $x = 1$ gives: $\Pr(1) = \Pr(1 \text{ success}) = {}^{n}C_1.p^1.(1-p)^{n-1}$
Putting $x = 2$ gives: $\Pr(2) = \Pr(2 \text{ successes}) = {}^{n}C_2.p^2.(1-p)^{n-2}$
Putting $x = 3$ gives: $\Pr(3) = \Pr(3 \text{ successes}) = {}^{n}C_3.p^3.(1-p)^{n-3}$
... and so on.

Note that the number of successes can never exceed the number of trials. That is, if 5 trials are performed, there can be 5 successes at most.

As a particular example, if we are given a binomial situation with 3 trials ($n = 3$) and the probability of success at any trial is 0.35 ($p = 0.35$), then substituting for these values of $n$ and $p$ in the above gives:

$\Pr(0) = {}^{3}C_0.(0.35)^0.(1-0.35)^{3-0} = (0.65)^3 = 0.275$ (3D)
$\Pr(1) = {}^{3}C_1.(0.35)^1.(1-0.35)^{3-1} = 3(0.35)(0.65)^2 = 0.444$ (3D)
$\Pr(2) = {}^{3}C_2.(0.35)^2.(1-0.35)^{3-2} = 3(0.35)^2(0.65) = 0.239$ (3D)
$\Pr(3) = {}^{3}C_3.(0.35)^3.(1-0.35)^{3-3} = (0.35)^3 = 0.043$ (3D)

[Notice that these probabilities agree with the results of Example 4 in the previous chapter, which used another method of obtaining them.]

## 6. Example 2 (Calculating *binomial probabilities*)

*Question*

From past records, the probability that a machine will need correcting adjustments during a day's production run is 0.2.

If there are 6 of these machines running on a particular day, find the probability that:

a) no machines need correcting;
b) just one machine needs correcting;
c) exactly two machines need correcting;
d) more than two machines need correcting.

*Answer*

First, it is necessary to identify the binomial situation precisely.

*Trial* = identifying a particular machine;
*Trial success* = the machine needs a correcting adjustment;
*Number of trials* = 6 (the number of machines in use).

Thus, $n$=6 and we are given that:

$p = \Pr(\text{success})$
$= \Pr(\text{a machine needs correcting})$
$= 0.2.$

The binomial probability formula is $\Pr(x) = {}^{n}C_x.px.(1-p)^{n-x}$

a) Putting $x = 0$ (together with $n = 6$ and $p = 0.2$) in the above formula gives:
$\Pr(0) = \Pr(\text{no machines need correcting}) = (1-0.2)^6 = (0.8)^6 = 0.262$ (3D)

b) Putting $x = 1$ in the formula gives:
$\Pr(1) = \Pr(\text{one machine needs correcting})$
$= {}^{6}C_1.(0.2)(1\text{–}0.2)^5 = 6(0.2)(0.8)^5 = 0.393$ (3D)

c) Putting $x = 2$ in the formula gives:
Pr(2)= Pr(two machines need correcting)
$= {}^6C_2.(0.2)^2.(1–0.2)^4 = 15(0.2)^2(0.8)^4 = 0.246$ (3D)

d) Pr(more than 2 machines need correcting)
= 1 – Pr(2 or less machines need correcting)
= 1 – Pr(0 or 1 or 2 machines need correcting)
= 1 – [Pr(0) + Pr(1) + Pr(2)]
= 1 – [0.262 + 0.393 + 0.246] from a), b) and c).
= 1 – 0.901 = 0.099 (3D).

## 7. Mean and variance of the binomial distribution

The binomial probabilities calculated in section 5 (with $n$=3 and $p$=0.35) are tabulated below, where $x$ is the number of successes.

| $x$ | 0 | 1 | 2 | 3 |
|---|---|---|---|---|
| $p$ | 0.275 | 0.444 | 0.239 | 0.043 |

[Note that the fact that the above probabilities add to 1.001 (and not exactly 1) is due to rounding to 3D.]

The expectation (or mean) of this distribution is given by:
$\Sigma\, px = 0(0.275) + 1(0.444) + 2(0.239) + 3(0.043) = 1.05$ (2D).

Notice also that $np = 3(0.35) = 1.05$ (i.e. the same as above). This is no coincidence, since the mean of a binomial (probability) distribution is always given by $np$. There is a similar result for the variance, as stated below.

**Mean and variance of binomial**

Given a binomial distribution with $n$ = number of trials and $p$ = probability of success at each trial, then:

$$\text{mean} = np$$
$$\text{variance} = np(1-p).$$

For the binomial distribution given, we already have that the mean is 1.05.
Using the formula above, variance $= np(1-p) - 3(0.35)(0.65) = 0.68$(2D).

## 8. Example 3 (Calculating binomial probabilities, *mean* and *variance*)

*Question*

A manufacturer sets up the following sampling scheme for accepting or rejecting large crates of identical items of raw material received. He takes a random sample of 20 items from the crate. If he finds more than two defective items in the sample, he rejects the crate; otherwise he accepts it. It is known that approximately 5% of these type of items received are defective.

a) Calculate the proportion of crates that will be rejected.
b) Calculate the mean and variance of the number of defectives in a sample of 20.

*Answer*

This is a binomial situation, since a random sample is involved. Thus, using the information given:

*Trial* = selecting a single item from the crate;
*Trial success* = the item selected is defective;
*Number of trials* = 20 (the size of the sample).

That is, $n = 20$ and $p = \Pr(\text{an item is defective}) = 5\% = 0.05$.
Also, $1 - p = 0.95$

a) The *proportion* of crates that will be rejected is another way of saying the *probability* that a single crate will be rejected.
Thus, Pr(a single crate will be rejected)
$= \Pr(\text{more than 2 defective items in the sample})$ [given]
$= 1 - \Pr(0, 1 \text{ or } 2 \text{ defectives in the sample})$
$= 1 - [\Pr(0)+\Pr(1)+\Pr(2)]$
$= 1 - [{}^{20}C_0(0.95)^{20} + {}^{20}C_1(0.05)(0.95)^{19} + {}^{20}C_2(0.05)^2(0.95)^{18}]$
$= 1 - [1(0.95)^{20} + 20(0.05)(0.95)^{19} + 190(0.05)^2(0.95)^{18}]$
$= 1 - [0.358 + 0.377 + 0.189]$
$= 0.076.$

b) Since $n = 20$ and $p = 0.05$, we have:
mean number of defectives in a sample $= np = 20(0.05) = 1$;
variance of number of defectives in sample $= np(1-p) = 20(0.05)(0.95) = 0.95$.

## 9. A poisson distribution

If the number of telephone calls coming to a telephone switchboard each minute is counted and recorded over a number of successive minutes, the following frequency distribution might result:

| Number of calls received per minute interval | ($x$) | 0 | 1 | 2 | 3 | 4 | 5 | 6 + | Total |
|---|---|---|---|---|---|---|---|---|---|
| Number of minute intervals | ($f$) | 11 | 24 | 14 | 7 | 2 | 1 | 1 | 60 |

This is known as a *poisson distribution*. It describes the number of 'events' that occur within some given interval. The important characteristic of the poisson distribution is that the events in question (in the above case, the calls received) must occur at random. That is, they must be independent of one another. Also they must be what is described as 'rare'. That is, in any particular point in the interval, the probability of an event occurring must be *very low*.

In the example given, we normally assume that calls are 'rare' and come to a switchboard randomly in any minute interval.

For the given distribution, we have:

*Event* = call coming to a switchboard.
*Defined interval* = one minute.

This environment is sometimes called a *poisson situation* (or *poisson process*). The frequency distribution has been obtained by counting the number of events that occurred each minute over 60 successive minutes.

**Poisson situation**

A *poisson situation* or process can be recognised by the following characteristics:

a) the existence of events that:
   i. occur at random
   ii. are 'rare'
b) an interval (of time or space) is defined, within which events can occur.

**Note**: The variable ($x$), which describes the number of events occurring within the defined interval, in a poisson frequency distribution will *always* be a whole number (including 0); the frequency ($f$) is the number of times that each of the $x$ events occurs. In the above calls distribution, the number of calls received in a minute interval (variable $x$) has smallest possible value 0 but has no upper boundary (unlike a binomial distribution).

## 10. Example 4 (Some *poisson distributions*)

a) Event = an accident on a particular stretch of road
Interval = one month

| Number of accidents | 0 | 1 | 2 | 3 | 4 | 5 | 6 | 7 | 8 or more |
|---|---|---|---|---|---|---|---|---|---|
| Number of months | 0 | 3 | 8 | 14 | 12 | 9 | 4 | 2 | 1 |

b) Event = a minor flaw in manufactured cloth
Interval = one metre length

| Number of minor flaws | 0 | 1 | 2 | 3 | 4 or more |
|---|---|---|---|---|---|
| Number of metre lengths | 10 | 14 | 4 | 2 | 1 |

Notice that an interval does not have to be one involving time. In b) above, the interval is one of length.

## 11. Poisson probability formula

Given a poisson situation, it is possible to calculate the probability of any number of events occurring in the defined interval. This can only be done however, if the mean number of events per interval is known.

**The poisson probability formula**

Given a poisson situation with $m$ = mean number of events per interval, the probability of $x$ events occurring is given by:

$$\Pr(x) = e^{-m} \frac{m^x}{x!}$$

where $x$ can take any one of the values 0, 1, 2, 3, ... .

The letter '$e$' represents a special mathematical constant (having approximate value 2.718).

Appendix 3 (exponential tables) gives values of $e^{-m}$ for various values of $m$.

For example, reading from the tables, the row marked '2.6' and the column marked '0.03' yields the value 0.0721. This means that $e^{-2.63} = 0.0721$.
Similarly, $e^{-1.48} = 0.2276$ and $e^{-3.71} = 0.0245$. [These values should be checked.]

## 12. Use of the poisson probability formula

As with the binomial formula, the above poisson formula is used to find specific probabilities by substituting the number of events required for $x$.
As a demonstration of its use, suppose that a poisson-distributed event had a mean of 1.72 (within some interval).
Putting $x = 0$ in the above formula gives:

$$\text{Pr(no events occurring in the interval)} = \text{Pr}(0) = e^{-1.72}\frac{1.72^0}{0!} = 0.1791 \text{ [from tables]}$$

Putting $x = 1$ in the above formula gives:

$$\text{Pr(1 event occurring in the interval)} = \text{Pr}(1) = e^{-1.72}\frac{1.72^1}{1!} = (0.1791)(1.72)$$

$$= 0.308 \text{ (3D)}$$

$$\text{Similarly, Pr}(2) = \text{e}-1.72\frac{1.72^2}{2!} = (0.1791)(1.4792)$$

$$= 0.265 \text{ (3D)}$$

... and so on.

## 13. Notes on the poisson distribution

a) A poisson interval can be *adjusted provided the mean is adjusted accordingly*. For example, if we are given that the average number of deliveries to a large warehouse is 1.8 per day, then of course the poisson probability formula, with mean 1.8, can be used to calculate the probabilities of 0, 1, 2, ... etc deliveries on any particular day. However, if we wanted to calculate the probabilities of 0, 1, 2, ... etc deliveries in any 2-*day period*, the poisson probability formula would need to be used *with a mean of* 2(1.8)=3.6, and so on. Example 5 demonstrates this technique.

b) The variance of a poisson probability distribution is *always equal to the mean*.

c) Many electronic calculators have a special function key for either $e^x$, $e^{-x}$ or both. Students should test the use of these keys (if present) with reference to the exponential tables in Appendix 3.

## 14. Example 5 (Calculating *poisson probabilities*)

*Question*

Customers arrive randomly at a department store at an average rate of 3.4 per minute. Assuming the customer arrivals form a poisson distribution, calculate the probability that:

a) no customers arrive in any particular minute;
b) exactly one customer arrives in any particular minute;
c) two or more customers arrive in any particular minute;
d) one or more customers arrive in any 30-second period.

*Answer*

For a), b) and c), the interval for this poisson situation is one minute, with a given mean of 3.4. Notice, however, that the interval in d) is just 30 seconds, thus the mean here must be adjusted to $\frac{3.4}{2} = 1.7$.

a) Pr(no customers arrive in one minute) $= \Pr(0) = e^{-3.4}\frac{3.4^0}{0!} = e^{-3.4}$
$= 0.0334$.

b) $\Pr(1) = e^{-3.4}\frac{3.4^1}{1!} = (0.0334)(3.4)$
$= 0.1136$.

c) Pr(2 or more) = 1 – Pr(0 or 1) = 1 – [Pr(0)+Pr(1)]
= 1 – [0.0334 + 0.1136] [from (a) and (b)]
= 0.8530.

d) Here, as mentioned earlier, the interval has been changed to 30 seconds. Thus the mean must be adjusted to 1.7. In other words, we would expect 1.7 customers to arrive on average in any 30-second period.
Thus, Pr(1 or more) = 1 – Pr(0)
$= 1 - e^{-1.7}$ [from probability formula using mean = 1.7]
= 1 – 0.1827 = 0.8173.

## 15. Poisson approximation to the binomial

Given a binomial situation, if $n$ is large (greater than 30) and $p$ is small (less than 0.1), the poisson distribution can be used as an *approximation* to the binomial distribution for calculating probabilities. Given these conditions, the poisson mean would be taken as $np$.

a) The approximation is used because it is generally found easier to calculate poisson probabilities (when $n$ is large) than binomial probabilities;
b) The difference between poisson and binomial probabilities is very small under the above conditions. The larger $n$ is and the smaller $p$ is, the better is the approximation.

The following example shows the use of this approximation.

## 16. Example 6 (The use of the *poisson approximation* to the binomial)

*Question*

Items produced from a machine are known to be 1% defective. If the items are boxed into lots of 200, what is the probability of finding that a single box has 2 or more defectives?

*Answer*

Notice that this is a binomial situation, since examining the contents of a box is equivalent to taking a sample of 200.
We have then, $n = 200$ and $p$ = Pr(defective) = 1% = 0.01.

Thus, for the poisson approximation, mean = $np$ = 200(0.01) = 2.

Therefore, Pr(2 or more defectives) = 1 – Pr(0 or 1 defectives)

$$\begin{aligned} &= 1 - [\Pr(0)+\Pr(1)] \\ &= 1 - [e^{-2} + e^{-2}.(2)] \\ &= 1 - [\ 0.1353 + (0.1353)(2)\ ] \\ &= 1 - [\ 0.1353 + 0.2706\ ] \\ &= 0.5941 \end{aligned}$$

That is, about 59% of boxes would be expected to have 2 or more defectives.

Note that using the binomial probability formula, we have:

$\Pr(0) = {}^{200}C_0(0.01)^0(0.99)^{200} = 0.99^{200} = 0.1340$

$\Pr(1) = {}^{200}C_1(0.01)^1(0.99)^{199} = 200(0.01)(0.99)^{199} = 0.2707$

Thus, Pr(2 or more) = 1 – (0.1340+0.2707) = 0.5953.

This demonstrates that the above poisson approximation (0.5941) is quite good.

## 17. Summary

a) A binomial situation is a situation where a number of identical 'trials' are performed, each one of which can result in either 'success' or 'failure'.

b) A binomial frequency distribution can be described as a distribution of the number of successes obtained when a binomial situation is repeated a number of times.

c) Given a binomial situation with: $n$ = number of trials performed and $p$ = probability of success at each trial, then the binomial probability formula is:

$$\Pr(x) = {}^nC_x.p^x(1-p)^{n-x}$$

for $x = 0, 1, 2, 3, \ldots, n$.

d) For a binomial probability distribution:

Mean = $np$; Variance = $np(1-p)$ or $npq$ [where $q = 1-p$)]

e) A poisson situation (or process) is a situation where 'rare events' occur randomly in some interval.

f) A poisson frequency distribution is obtained by observing the number of random events that occur in repeated intervals.

g) Given a poisson situation with $m$ = mean number of events in the interval, then the poisson probability formula formula is:

$$\Pr(x) = e^{-m}\, e^{-m}\, \frac{m^x}{x!}$$

for $x = 0, 1, 2, 3, \ldots, n$.

h) A poisson interval can be adjusted provided the mean is adjusted accordingly.

i) For a poisson probability distribution, mean=variance.

j) In a binomial situation, the poisson distribution can be used as an approximation if:

(1) $n$ is large (greater than 30);
(2) $p$ is small (less than 0.01).

## 18. Student self review questions

1. What feature characterises a binomial distribution? [2]
2. What is a binomial situation and what is the significance of symbols $n$ and $p$? [2]
3. Write down the binomial probability formula for calculating the probability of obtaining $x$ successes. [4]
4. How can the mean and variance of a binomial probability distribution be calculated (in terms of $n$ and $p$)? [7]
5. What feature characterises a poisson distribution? [9]
6. What is a poisson situation and why is the mean important? [9]
7. Write down the poisson probability formula for calculating the probability that $x$ events will occur in some defined interval. [11]
8. What special relationship is there between the mean and variance of a poisson distribution? [13(b)]
9. What conditions are necessary in order that the poisson distribution can be used to replace the binomial distribution? [15]

## 19. Student exercises

1. As a work study exercise, a team of 6 workmen engaged on a job were observed at different times of the day and the following data obtained.

| Number of workmen idle | 0 | 1 | 2 | 3 |
|---|---|---|---|---|
| Number of occasions | 7 | 14 | 8 | 1 |

   a) On how many occasions were the workmen measured?

   b) Identify the trial, trial success and number of trials for the binomial situation on which the above distribution is based.

2. Calculate and tabulate the probabilities of 0, 1, 2, 3 and 4 successes for a binomial situation with $n = 4$ and $p = 0.25$.
3. A company Minibus has 7 passenger seats and on a routine run it is estimated that any passenger seat will be filled with probability 0.42.

   a) What is the mean and variance of the binomial distribution of the number of passengers on a routine run?

   b) Calculate the probability (to 3D) that, on a routine run:

      i. there will be no passengers;

      ii. there will be just one passenger;

      iii. there will be exactly two passengers;

      iv. there will be at least three passengers.

4. Calculate and tabulate the poisson probabilities of 0, 1, 2, 3 and 4 or more random events occurring in an interval for which the mean number of events is 2.3.
5. A firm which produces half-inch diameter rubber hose estimates that on average there are 0.4 flaws per 10-metre length. Assuming flaws occur randomly, what is the probability that:

   a) there are no flaws in a 10-metre length;

   b) there is more than 1 flaw in a 10-metre length;

   c) there are more than 2 flaws in a 20-metre length?

6. From a haulage company's records, only 4% of deliveries are made in error. If 20 trips are planned next week, calculate the probability that there will be at least two erroneous deliveries, using:
   a) the binomial distribution;
   b) the poisson distribution.
7. Six candidates apply for a job and are separately interviewed. Based on past experience, any candidate has a two-thirds probability of passing to a second interview.
   a) What is the probability that all the candidates pass to a second interview?
   b) What is the probability that exactly 5 candidates pass on to a second interview?
   c) The successful candidates are all invited to the reception area of the main offices, ready for a second interview. If the reception area has only three seats, what is the probability that there are not enough seats for these candidates?
8. A set of customer accounts is known to have errors in about 3% of them. An auditor's operational scheme for a set of accounts of this size is to sample 10 of them randomly and, if no errors are found, to pass the accounts as acceptable. If any error is found in the sample, all accounts are then inspected. The auditor *always* finds any errors in accounts *that are checked*.
   a) What is the probability that the accounts are passed?
   b) What is the probability that all the errors in the accounts will be detected?
   c) What is the minimum sample size necessary to ensure that the auditor has a better than even chance of discovering all the errors in the accounts?

# 34 Normal distribution

## 1. Introduction

This chapter describes the single most important distribution in Statistics; the Normal distribution. Some examples of the many situations in which it occurs are given initially, the rest of the chapter being concerned with methods of calculating probabilities. Its use as an approximation to a binomial distribution under particular conditions is given in the latter sections.

## 2. The Normal distribution

The *Normal distribution* is the name given to a type of distribution of *continuous* data that occurs frequently in practice. It is a distribution of 'natural phenomena', such as:

1) weights (of products produced by a machine, of people working in a factory);

2) heights (of buildings, animals or people);
3) lengths (of uniform manufactured products, of various makes of car);
4) times (taken to get to work each day, taken to complete a standard job); and so on.

The main characteristics of the distribution are:

a) It has a *symmetric* (frequency) curve about the mean of the distribution. In other words, one half of the curve is a mirror- image of the other.
b) The majority of the values tend to cluster about the mean, with the greatest frequency at the mean itself.
c) The frequencies of the values taper away (symmetrically) either side of the mean, giving the curve a characteristic 'bell-shape'.

Figure 1 overleaf shows the shapes of typical Normal curves.

## 3. Example 1 (Some *Normal distributions* that might occur in practice)

a) The weight of the contents of tins of peas measured by a quality control section.

| *Weight (gm)* | *Number of tins* |
|---|---|
| under 424.900 | 1 |
| 424.900 – 424.925 | 2 |
| 424.925 – 424.950 | 6 |
| 424.950 – 424.975 | 14 |
| 424.975 – 425.000 | 33 |
| 425.000 – 425.025 | 46 |
| 425.025 – 425.050 | 18 |
| 425.050 – 425.075 | 7 |
| 425.075 – 425.100 | 3 |
| 425.100 – and over | 1 |

Notice that the largest frequencies cluster around the central values of the distribution and taper away on either side

*Figure 1 The shapes of typical Normal curves*

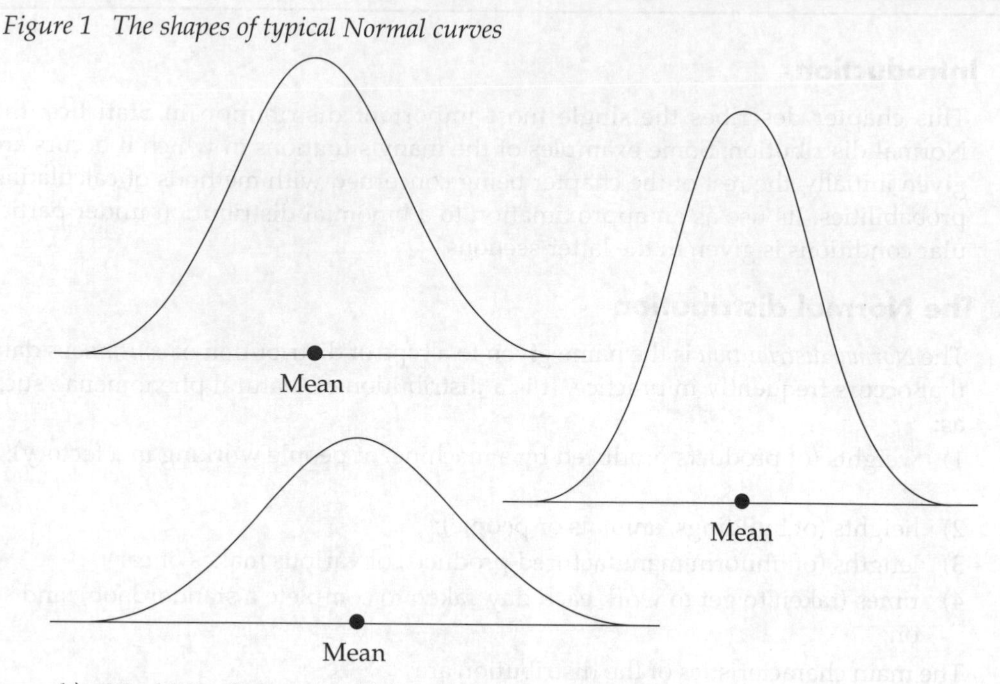


b) Weekly orders received by a company.

| Number of orders received | 10 – 14 | 15 – 19 | 20 – 24 | 25 – 29 | 30 – 34 | 35 – 39 |
|---|---|---|---|---|---|---|
| Number of weeks | 3 | 7 | 15 | 20 | 9 | 4 |

Again, notice that the majority of numbers of orders per week cluster around the central values 20 to 34.

## 4. Normal distribution probabilities

Because the Normal distribution is continuous, it is not possible to find the probability of precise values, because they are not attainable (this was mentioned earlier in chapter 3). It is only possible to find the probability for *ranges of values*. Thus, although it would not be possible to evaluate the probability that a wooden peg, manufactured by a machine, has length 2.1 cms, we could evaluate the probability that its length lay between 2.1 and 2.2 cms or between 2.05 and 2.15 cms (given relevant information).

The information necessary to evaluate probabilities for ranges of values for a Normal distribution is the values of:

a) the mean ($m$) and

b) the standard deviation, ($s$)

of the distribution in question. Both of these statistics *need to be known*.

The procedure used in calculating probabilities associated with a Normal distribution requires a knowledge of the use of:

i. Z-scores and

ii. Z (or Standard Normal) tables.

These are covered in the following two sections.

## 5. Z-scores

Suppose we know that the lengths of steel pins produced by a machine are distributed Normally with mean $m = 20$ cms and standard deviation $s = 0.1$ cms. We wish to find the probability that a randomly selected pin is less than 20.1 cms in length.

The first step in the process is to calculate the 'Z-score' for 20.1.

This is done by:

i. *subtracting the mean* and then
ii. *dividing by the standard deviation.*

In this case, the Z-score for 20.1 is: $z = \dfrac{20.1 - 20}{0.1} = \dfrac{0.1}{0.1} = 1$

> **Calculating a Z-score**
>
> The *Z-score* for any value $x$ of a normal distribution, having mean $m$ and standard deviation $s$, is given by:
>
> $$z = \frac{x - m}{s}$$
>
> This process is sometimes known as *standardising* the $x$-value.

The problem posed above: Pr(length of pin < 20.1) has now been changed to the problem of finding: $\Pr(Z < 1)$.

Z-score probabilities are found through the use of special tables. These tables are described in the following section.

## 6. Standard Normal (*Z*) tables

*Standard Normal* tables are used to calculate Z-score probabilities. They are shown at Appendix 4 and give the probability that a Z-score will have a value *between zero and the one specified.*

The diagram at the head of the standard normal tables shows how any probability read from the table can be represented as an area under the 'Standard Normal curve'. The Standard Normal distribution is in fact a Normal distribution having a mean, $m = 0$ and a standard deviation, $s = 1$.

The total area under the Standard Normal curve is always 1 (representing total probability).

Returning to our problem from the previous section, we need to find $\Pr(Z<1)$.

Reading from tables, $Z = 1.0$ gives a probability of 0.3413.

In other words, the probability that a Z-score will lie between 0 and 1 is 0.3413.

But, since total probability must be 1, then $\Pr(Z<0) = 0.5$.

Thus $\Pr(Z<1.0) = \Pr(Z<0) + \Pr(0<Z<1) = 0.5 + 0.3413 = 0.8413$.

Figure 2(a) shows the pictorial representation of $\Pr(Z<1) = 0.8413$.

*Figure 2 Diagrammatic representation of* Pr(Z<1) = 0.8413

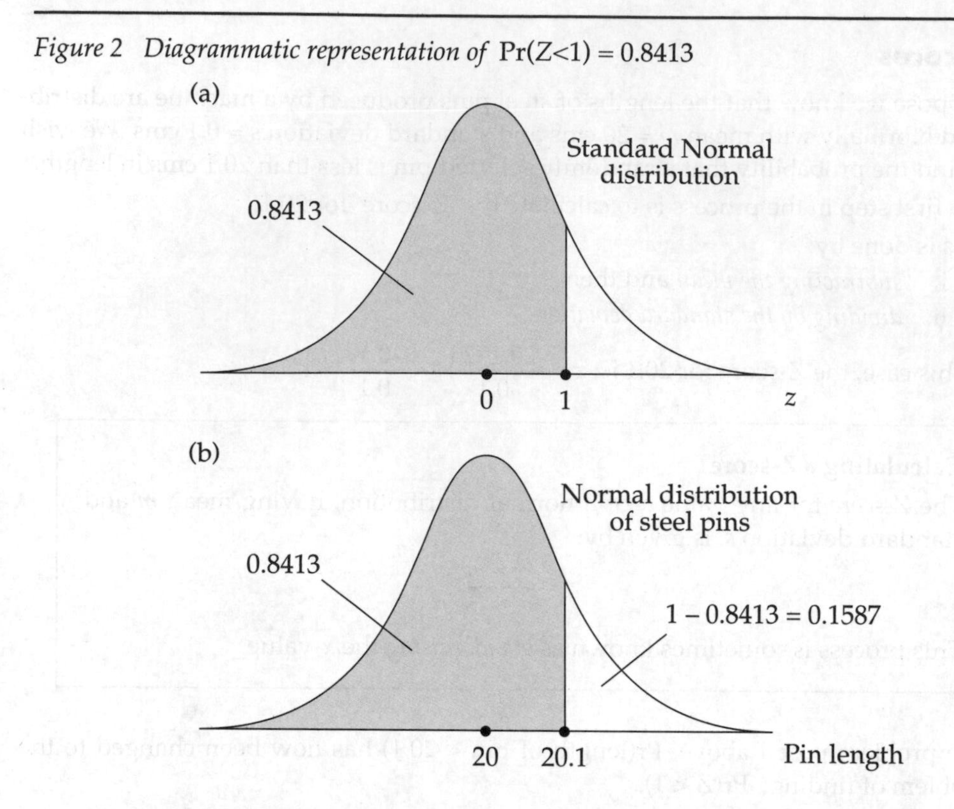


Figure 2(b) shows how the original problem can now be represented pictorially. That is, the proportion of steel pins that would be expected to be under 20.1 cms in length is 0.8413.

More specifically, Pr(pin length < 20.1) = 0.8413.

Also from Figure 2(b), the proportion of the area under the curve to the right of 20.1 = 1 – 0.8413 = 0.1587.

In other words, the proportion of steel pins that are expected to be *greater* than 20.1 is 0.1587.

Symbolically, Pr(pin length > 20.1) = 0.1587.

## 7. Example 2 (Calculating a normal probability using *Standard Normal tables*)

*Question*

Weights of bags of potatoes are Normally distributed with mean 5 lbs and standard deviation 0.2 lbs. The potatoes are delivered to a supermarket, 200 bags at a time.

a) What is the probability that a random bag will weigh more than 5.5 lbs?

b) How many bags, from a single delivery, would be expected to weigh more than 5.5 lbs?

*Answer*

a) We require Pr(bag weight > 5.5).

Standardising 5.5 gives $z = \dfrac{5.5 - 5}{236} = 2.5$

For z = 2.5, Standard Normal tables yield a probability of 0.4938.

Thus, Pr(Z<2.5) = 0.9938 = Pr(bag weight < 5.5).

Therefore, Pr(bag weight > 5.5) = 1 – 0.9938 = 0.0062.

b) From the result of (a), the proportion of all bags that would be expected to have a weight greater than 5.5 lbs is 0.0062. But each delivery consists of 200 bags. So we would 'expect' 0.0062 × 200 = 1.24 bags to have a weight greater than 5.5 lbs.

In practical terms (as a whole number), only 1 bag would be expected to weigh more than 5.5 lbs.

## 8. Extending the use of tables

Returning to the original steel pin lengths normal distribution (having mean 20 cms and standard deviation 0.1 cms), suppose we required the probability that a pin was longer than 19.9 cms.

That is, we require Pr(pin length > 19.9).

Standardising 19.9 gives $z = \dfrac{19.9 - 20}{0.1} = \dfrac{-0.1}{0.1} = -1$

In other words, in terms of Z, we require Pr(Z>–1). But this *must* be the same as Pr(Z<1). The reason for this is shown pictorially in Figure 3 and explained further now.

*Figure 3 Symmetric regions under a normal distribution*

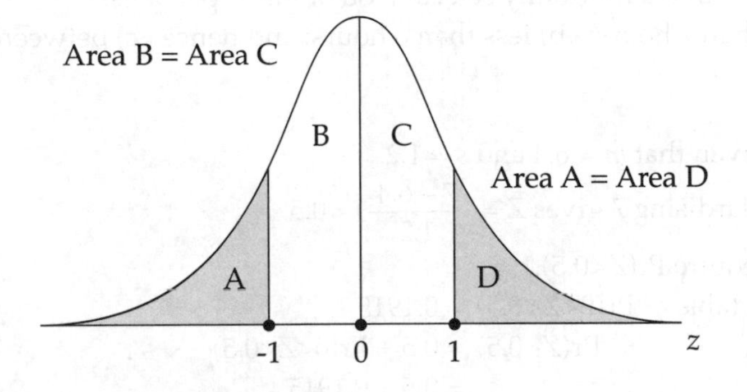


Because a Normal distribution is symmetric, Area B = Area C and Area A = Area D.

Therefore: Area A + Area B + Area C = Area B + Area C + Area D

Pr(Z < 1) = Pr(Z > –1)

Thus, Pr(pin length > 19.9) = Pr(Z>–1)
= Pr(Z<1) [using the above rule]
= 0.8413 [from tables]

Using diagrams much like the above, it is easy to manipulate Standard Normal probabilities into any form required.

That is, whatever the value of $a$: $Pr(Z<a) = 0.5 + Pr(0<Z<a)$ [the table value of $a$]

and: $Pr(Z>a) = 0.5 - Pr(0<Z<a)$ [total probability = 1]

Also: $Pr(Z<-a) = Pr(Z>a)$ [by symmetry]

Similarly $Pr(Z>-a) = Pr(Z<a)$ [by symmetry]

For example, starting off with a Z-score of 1.35 say.

Reading from tables gives $Pr(0<Z<1.35) = 0.4115$ (table probability)

Thus, $Pr(Z<1.35) = 0.5 + 0.4115 = 0.9115$

and: $Pr(Z>1.35) = 0.5 - 0.4115 = 0.0885$

Also: $Pr(Z<-1.35) = Pr(Z>1.35) = 0.0885$

Finally: $Pr(Z>-1.35) = Pr(Z<1.35) = 0.9115$

## 9. Summary of procedure for calculating Normal probabilities

The procedure for calculating any Normal probability is:

STEP 1 Standardise the Normal value in question to obtain a Z-score (see section 5).

STEP 2 Use standard Normal tables (see section 6).

STEP 3 (If necessary) manipulate the probability obtained (see section 8).

## 10. Example 3 (Calculating probabilities for a *Normal distribution*)

*Question*

The time taken to complete jobs of a particular type is known to be Normally distributed with mean 6.4 hours and standard deviation 1.2 hours. What is the probability that a randomly selected job of this type takes:

a) less than 7 hours; b) less than 6 hours; and hence c) between 6 and 7 hours?

*Answer*

We are given that $m = 6.4$ and $s = 1.2$.

a) Standardising 7 gives $Z = \dfrac{7-6.4}{1.2} = 0.5$.

We require $Pr(Z<0.5)$.

From tables, $Pr(0<Z<0.5) = 0.1915$

Thus,
$$\begin{aligned} Pr(Z<0.5) &= 0.5 + Pr(0<Z<0.5) \\ &= 0.5 + 0.1915 \\ &= 0.6915 \end{aligned}$$

That is, the probability that a randomly selected job takes less than 7 hours is 0.6915.

b) Standardising 6 gives $Z = \dfrac{6-6.4}{1.2} = -0.33$.

Here, we require $Pr(Z<-0.33) = Pr(Z>0.33) = 0.5 - Pr(0<Z<0.33)$
$= 0.5 - 0.1293 = 0.3707$

That is, the probability that a randomly selected job takes less than 6 hours is 0.3707.

c) The probability that $T$ (time to complete job) lies between 6 and 7 can be written in the form $\Pr(6<T<7)$.
But $\Pr(6<T<7) = \Pr(T<7) - \Pr(T<6)$.
Figure 4 shows this relationship pictorially.

*Figure 4 Demonstrating the relationship* $\Pr(6<T<7) = \Pr(T<7) - \Pr(T<6)$

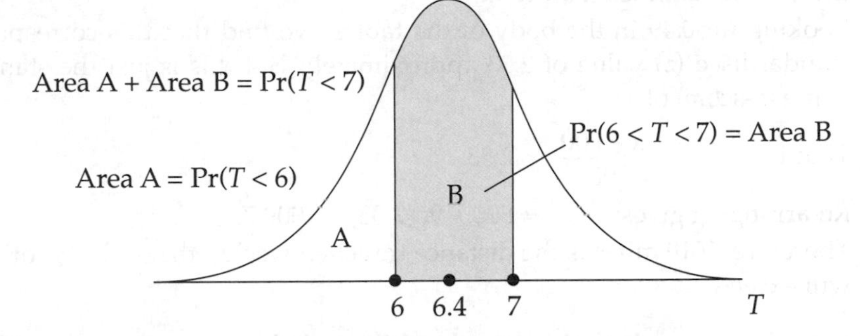


So, $\Pr(6<T<7) = 0.6915 - 0.3707$ [from a) and b) above] $= 0.3208$.
That is, the probability that a randomly selected job takes between 6 and 7 hours is 0.3208 or 32%.

## 11. Example 4 (Calculating probabilities for a given Normal distribution)

*Question*

Company records show that the weekly distance travelled by their salesmen is approximately normally distributed with mean 800 miles and standard deviation 90 miles. The sales manager considers that salesmen who travel less than 600 miles in one week are performing poorly.

a) If the company employs 200 salesmen, how many would be expected to perform poorly in a particular week?
b) The sales manager wishes to identify the number of miles travelled in one week, above which only 1% of salesmen are expected to exceed. What weekly mileage is this?

*Answer*

a) First, we need to calculate the probability that any one salesman will perform poorly in a particular week, then multiplying this probability (proportion) by 200 will give the expected number required.
Putting $D$ = distance travelled in one week by any salesman, we need to find $\Pr(D<600)$, the criterion for a poor performance.

Standardising 600 gives $z = \dfrac{600-800}{90} = \dfrac{-200}{90} = -2.22$ (2D)

Thus we require $\Pr(Z<-2.22)$.
Tables give that $\Pr(0<Z<2.22) = 0.4868$.
Hence, $\Pr(Z<-2.22) = \Pr(Z>2.2) = 0.5 - 0.4868 = 0.0132$.
Therefore we would expect $200(0.0132) = 2.64$ (i.e. 3) salesmen to perform poorly in any one week.

b) This is the type of normal problem which involves using the tables in reverse order. That is, we are given a probability, 1% here, and we need to find the Normal value to which it corresponds.
Now, if only 0.01 (1%) of salesmen are expected to exceed $x$ miles per week, it must be that 0.99 of salesmen will travel less than $x$ miles per week (where $x$ is the number of miles required).
Looking for 0.49 in the body of the tables, we find that this corresponds to a standardised ($z$) value of 2.33 approximately. But this is just the standardised value ($z$-score) of $x$.

That is, $\dfrac{x-800}{90} = 2.33.$

Re-arranging gives: $x = 800 + 90(2.33) = 1009.7.$

Therefore, 1010 miles is the distance travelled weekly that only 1% of salesmen will exceed.

## 12. Normal approximation to the binomial

Recall from the previous chapter that when a binomial situation has a large value of $n$, associated probabilities can be awkward to calculate. We also saw that if $p$ was small, the poisson distribution can be used as an approximation to the binomial.

The Normal distribution can also be used as an approximation to the binomial when $n$ is large and *when p is not too small or large.*

It is always necessary to know the mean and standard deviation for a Normal distribution. In this particular situation, these are taken as the mean and standard deviation of the binomial that is being approximated to.

> **Normal approximation to the binomial**
>
> In a binomial situation when $n$ is large and $p$ is not too small or large, the Normal distribution can be used to approximate to the binomial distribution using:
>
> $m$ = mean of Normal = $np$
>
> $s$ = standard deviation of Normal = $\sqrt{np(1-p)}$

**Note.** $n>30$ is normally considered large. The larger the value of $n$ however, the smaller or larger the value of $p$ (between 0 and 1) can be allowed to go.

For example, if we had a binomial situation with $n$=100 and $p$=0.35, the Normal approximation could be used with:

mean, $m = np = 100(0.35) = 35$

and standard deviation, $s = \sqrt{np(1-p)} = \sqrt{100(0.35)(0.65)} = 4.77$ (2D)

## 13. Example 5 (*Normal approximation* to the binomial)

*Question*

Bolts are manufactured by a machine and it is known that approximately 20% are outside certain tolerance limits. If a random sample of 200 is taken, find the probability that more than 50 bolts will be outside the limits.

*Answer*

Here we have a binomial situation with:

*Trial* = a single bolt
*Trial success* = a bolt outside tolerance limits
*Number of trials* = 200 (i.e. size of sample)

Thus, we have $n = 200$ and $p = \Pr(\text{success}) = \Pr(\text{bolt outside limits}) = 20\% = 0.2$

So that, since $n$ is large and $p$ is not too small, we can use the Normal approximation.

For the Normal, mean, $m = np = 200 \times 0.2 = 40$

$$s = \sqrt{np(1-p)} = \sqrt{200(0.2)(0.8)} = 5.657 \text{ (3D)}$$

In other words, the number of bolts outside limits ($B$, say) has an approximate Normal distribution with mean 40 and standard deviation 5.657.

We require

Pr(more than 50 bolts outside limits)

= Pr(51 or 52 or 53 or 54 or ...etc)

= Pr(between 50.5 and 51.5 or between 51.5 and 52.5 or ... etc)

[Notice that when a Normal (continuous) distribution is used to approximate to a (discrete) binomial distribution, exact values should be replaced by ranges]

= Pr(B>50.5)

Standardising 50.5 gives $z = \dfrac{50.5 - 40}{5.657} = 1.86$

Thus we require Pr(Z>1.86). Tables give Pr(0<Z<1.86) = 0.469 (3D).

Therefore, Pr(Z>1.86) = 0.5 – 0.469 = 0.031.

Thus, there is probability 0.031 (3%) that more than 50 bolts will be outside the tolerance limits.

## 14. Example 6 (*Normal approximation* to the binomial)

*Question*

From past records, about 40% of a firm's orders are for export. Their record for exports is 48% in one particular financial quarter. If they expect to satisfy about 80 orders in the next financial quarter, what is the probability that they will break their previous export record?

*Answer*

If we put an export order as a 'success', this is a binomial situation with:

*Trial* = an order
*Trial success* = an order for export
*Number of trials* = 80

Thus $n = 80$ and $p = \Pr(\text{success}) = \Pr(\text{export order}) = 40\% = 0.4$.

But, since $n$ is large and $p$ is not too small or large, we can use the Normal approximation with:

$m$ = Normal mean = $np = 80(0.4) = 32$,

and $s$ = Normal standard deviation = $\sqrt{np(1-p)} = \sqrt{80(0.4)(0.6)} = 4.38$ (2D)

In other words, the number of orders for export ($E$, say) has an approximate Normal distribution with mean 32 and standard deviation 4.38.

Now, in order that the previous record is broken, the firm needs more than 48% of the 80 orders.

That is they need $0.48 \times 80 = 38.4$ for export.

Thus we require $\Pr(E>38.4)$.

Standardising 38.4 gives $z = \dfrac{38.4-32}{5.657} = 1.46$ (2D)

Therefore we require $\Pr(Z>1.46)$. Now, tables give $\Pr(0<Z<1.46) = 0.4279$.

Therefore, $\Pr(Z>1.46) = 0.5 - 0.4279 = 0.0721$.

In other words, there is a 7% chance that the previous export record will be broken.

## 15. Confidence limits

The study of Normal distributions helps in the process of estimating certain population measures based only on the results of small samples. Two measures of interest (for the syllabuses covered in this text) are the mean of a population and a population proportion.

For example, given that a sample of 20 invoices yielded a mean value of £253, what can we say about the actual (and unknown) population invoice mean value? It is quite clear the chance that the true invoice mean was £253 is very small and thus the mean sample value on its own yields very little information of use. The ideal is to enable something useful to be obtained with a high degree of probability (or confidence). This is where confidence limits come in.

> **Confidence limits** specify a range of values within which some unknown population value (mean or proportion) lies with a stated degree of confidence. They are always based on the results of a sample.

The following statement is typical of what needs to be able to be calculated and understood:

'95% confidence limits for the lead time of a particular type of stock item order are 4.1 to 7.4 working days'

The above would have been calculated on the basis of a representative sample and is stating that there is a 95% probability that the true (unknown) mean lead time lies between 4.1 and 7.4 working days.

## 16. Confidence limits for a mean

The following gives the technique for evaluating a confidence interval for an unknown population mean.

> **Confidence interval for a mean**
>
> Given a random sample from some population, a confidence interval for the (unknown) population mean is:
>
> $$\bar{x} \pm z\frac{s}{\sqrt{n}}$$
>
> where: $\bar{x}$ = the sample mean
> $s$ = the sample standard deviation
> $n$ = the sample size
> $z$ = confidence factor (1.64 for 90%; 1.96 for 95%; 2.58 for 99%).

**Note:** $\frac{s}{\sqrt{n}}$ is known as the *standard error* of the mean.

As an example of the use of these confidence limits, suppose a sample of 100 invoices yielded a mean gross value of £45.50 and standard deviation £3.24.

Here, $n = 100$, $\bar{x} = 45.5$ and $s = 3.24$.

We would calculate a 95% confidence interval as follows:

95% interval $= \bar{x} \pm z\frac{s}{\sqrt{n}}$ [Here $z = 1.96$ for a 95% interval

$= 45.50 \pm (1.96)(0.324) = 45.50 \pm 0.635 = (44.865, 46.135)$

In words, there is a 95% probability that the mean of the complete population of invoices (from which the sample was taken) is between 44.9 and 46.1.

## 17. Confidence limits for a proportion

The following gives the technique for evaluating a confidence interval for an unknown population proportion.

> **Confidence interval for a proportion**
>
> Given a random sample from some population, a confidence interval (CI) for the (unknown) population mean is:
>
> $$p \pm z\sqrt{\frac{p(1-p)}{n}}$$
>
> where: $p$ = the sample proportion
> $n$ = the sample size
> z = confidence factor (1.64 for 90%; 1.96 for 95%; 2.58 for 99%).

**Note:** $\sqrt{\frac{p(1-p)}{n}}$ is known as the *standard error* of a proportion.

As an example of the use of these confidence limits, suppose 4 faulty components are discovered in a random sample of 20 finished components taken from a production line. What statement can we make about the defective rate of all finished components?

Notice that here $p = \frac{4}{20} = 0.2$ and $n = 20$.

Thus, 95% CI for the overall rate is given by:

$$p \pm z\sqrt{\frac{p(1-p)}{n}} = 0.2 \pm (1.96)\sqrt{\frac{(0.2)(0.8)}{20}}$$

$$= 0.2 \pm (1.96)(0.089)$$

$$= 0.2 \pm 0.175 \text{ (2D)}$$

$$= (0.025, 0.375)$$

Therefore we can state that there is a 95% chance that the defective rate of finished components lies between 0.025 and 0.375.

## 18. Example 7 (Confidence limits)

*Question*

The same fuel was tested on 21 similar cars under identical conditions. Fuel consumption was found to have a mean of 41.6 mpg with a standard deviation of 3.2 mpg. Only 14 of the cars were found to completely satisfy current exhaust emission guidelines.

a) Calculate a 95% CI for fuel consumption for this type of car.
b) Calculate a 99% CI for the percentage of similar cars that completely satisfy current exhaust emission guidelines.

*Answer*

a) Given: $\bar{x} = 41.6$; $s = 3.2$ and $n = 21$. Also $z = 1.96$ for 95% confidence. Therefore, a 95% CI for the unknown population mean is given by:

$$\bar{x} \pm z\frac{s}{\sqrt{n}} = 41.6 \pm 1.96\frac{3.2}{\sqrt{21}} = 41.6 \pm 1.37\text{(2D)} = (40.23, 42.97)$$

b) Sample $p = \frac{14}{21} = 0.667$. Also $z = 2.58$ for 99% confidence.

Therefore, a 99% CI for the unknown population proportion is given by:

$$p \pm z\sqrt{\frac{p(1-p)}{n}} = 0.667 \pm 2.58\sqrt{\frac{(0.667)(0.333)}{21}}$$

$$= 0.667 \pm (2.58)(0.103)$$

$$= 0.667 \pm 0.265$$

$$= (0.402, 0.932)$$

## 19. Test of significance for the mean

Tests of significance are directly connected to confidence limits and are based on normal distribution concepts.

Suppose we suspect that the value of type A customer monthly orders has changed from last year. Last year's type A customer average monthly order was £234.50. We now take a random sample of 20 customers and calculate a mean of £241.52 and standard deviation £13.92. Is the difference significant? That is, is the value 241.52 far enough away from 234.50 to suspect that things have changed (or not)?

The following statement gives a structure for answering the above question.

**Test for a mean**

To test whether a sample of size $n$, with mean $\bar{x}$ and standard deviation $s$, could be considered as having been drawn from a population with mean $\mu$, the test statistic:

$$z = \frac{\bar{x} - \mu}{\left(\frac{s}{\sqrt{n}}\right)}$$

must lie within the range –1.96 to +1.96.

Note: $\mu$ is the greek letter *m*, pronounced 'mew'.

In the test, we are looking for evidence of a difference between $\bar{x}$ and $\mu$.

The evidence is found if $z$ lies outside the above stated limits.

If $z$ lies within the limits we say 'there is no evidence' that the sample mean is different to the population mean.

For example, in the above situation, we have: $\bar{x} = 241.52$; $\mu = 234.50$; $s = 13.92$; $n = 20$.

$$\text{Thus, test statistic } z = \frac{\bar{x} - \mu}{\left(\frac{s}{\sqrt{n}}\right)} = \frac{241.52 - 234.5}{\left(\frac{13.92}{\sqrt{20}}\right)} = 2.26$$

Thus there is *evidence of a difference*, since $z$ lies *outside* the range –1.96 to +1.96. This can now be translated as 'there is evidence that the value of type A customer monthly orders has changed'.

## 20. Example 8

*Question*

A manager is convinced that a new type of machine does not affect production at the company's major shop floor. In order to test this, 12 samples of this week's hourly output is taken and the average production per hour is measured as 1158 with a standard deviation 71. Given that the output per hour averaged 1196 before the machine was introduced, test the manager's conviction.

*Answer*

Notice that the sample is measuring now and the population is measuring what was before. Thus evidence of a difference between sample and population would show evidence of a change.

Here: $\bar{x} = 1158$; $\mu = 1196$; $s = 71$; $n = 12$.

$$\text{Test statistic } z = \frac{\bar{x}-\mu}{\left(\frac{s}{\sqrt{n}}\right)} = \frac{1158-1196}{\left(\frac{71}{\sqrt{12}}\right)} = -1.85, \text{which lies } \textit{within } -1.96 \text{ to } +1.96.$$

Thus there is *no evidence of any difference* between sample and population. The manager's conviction is therefore supported.

## 21. Summary

a) The Normal distribution is a continuous distribution of 'natural phenomena', describing such things as weight, height, length and time.

b) Its frequency curve is symmetric, with the majority of values clustered about its centre, the mean, and the rest tailing away giving it a characteristic 'bell' shape.

c) In order to calculate probabilities for a particular Normal distribution, both the mean and the standard deviation of the distribution need to be known. Probabilities can only be evaluated for a range of values.

d) Standard Normal tables give the probabilities associated with a Normal distribution having a mean of zero and a standard deviation of 1 (known as the Standard Normal distribution).

e) Standard Normal tables only give $\Pr(Z<x)$, where $x>0$, but THREE other forms of probability, namely $\Pr(Z>x)$, $\Pr(Z<-x)$ and $\Pr(Z>-x)$ can be deduced.

f) Probabilities associated with any Normal distribution can be obtained by:
   i. 'standardising' the Normal value concerned (by subtracting the mean and dividing by the standard deviation), then
   ii. using standard Normal tables to identify the required probability, and (if necessary)
   iii. manipulating this probability in an appropriate way.

g) The Normal distribution can be used as an approximation to the binomial distribution when:
   i. $n$ is large (greater than 30);
   ii. $p$ is not too small or large (the closer to 0.5 the better).

   When used as an approximation in this way, the Normal distribution has:

$$\text{mean} = np; \qquad \text{standard deviation} = \sqrt{np(1-p)}$$

h) Confidence limits specify a range of values within which some unknown population value (mean or proportion) lies with a stated degree of confidence.

$$\text{Confidence limits for the mean: } \bar{x} \pm z\frac{s}{\sqrt{n}}$$

$$\text{Confidence limits for a proportion: } p \pm z\sqrt{\frac{p(1-p)}{n}}$$

   where: $z$ = confidence factor (1.96 for 95%; 2.58 for 99%)

i) A significance test for the mean of a sample is carried out by calculating the value of the test statistic:

$$z = \frac{\bar{x} - \mu}{\left(\frac{s}{\sqrt{n}}\right)}$$

and comparing it with range –1.96 to +1.96. There is evidence of a difference between $\bar{x}$ and $\mu$ (the population mean) only if $z$ lies outside the given range.

## 22. Student self review questions

1. Give some examples of the type of data that form a Normal distribution. [2,3]
2. Describe the main characteristics of a Normal frequency curve. [2]
3. Why cannot the Normal distribution be used to calculate the probability that a *specific value* of a variable is obtained? [4]
4. What information is required to be known about a particular Normal distribution before *any* probabilities can be calculated? [4]
5. What is a Z-score? [5]
6. What are Standard Normal tables? [6]
7. What is the normal probability 'rule of thumb' for interpreting table probabilities? [8]
8. What are the three steps in the procedure for calculating a probability associated with any Normal distribution? [9]
9. Under what conditions can the Normal distribution be used as an approximation to the binomial distribution? [12]
10. What do confidence limits measure? [15]
11. Describe how a test of significance of a sample mean is carried out. [19]

## 23. Student exercises

1. Use Standard Normal (Z) tables to calculate the following:
   a) i) Pr(Z<0.38) ii) Pr(Z<1.22) iii) Pr(Z<2.73)
   b) i) Pr(Z>1.09) ii) Pr(Z>2.90) iii) Pr(Z>0.13)
   c) i) Pr(Z<–1.10) ii) Pr(Z<–3.2) iii) Pr(Z<–1.93)
   d) i) Pr(Z>–2.67) ii) Pr(Z>–0.76)
2. Use Standard Normal (Z) tables to calculate the following:
   a) i) Pr(Z<–1.12) ii) Pr(Z>1.87) and hence iii) Pr(–1.12<Z<1.87)
   b) i) Pr(Z<0.88) ii) Pr(Z<2.16) and hence iii) Pr(0.88<Z<2.16)
   c) i) Pr(Z<–2.33) ii) Pr(Z>–0.41) and hence iii) Pr(–2.33<Z<–0.41)
3. A Normal distribution (N) has mean 140 and standard deviation 8. Find:
   a) Pr(*N*<156) b) Pr(*N*<132) c) Pr(132<*N*<156).
4. A Normal distribution has mean 56 and standard deviation 12. If an item is picked at random from the distribution, what is the probability that its value will be:
   a) less than 30;
   b) greater than 80;
   c) between 30 and 80.

5. The weights of metal components produced by a machine are distributed Normally with mean 14 lbs and standard deviation 0.12 lbs.
   a) What is the probability that a component sampled randomly from production will have a weight
      i. greater than 14.3 lbs;
      ii. less than 13.7 lbs?
   b) Components are rejected if their weights are outside the limits 13.7 to 14.3 lbs. If the machine produces 500 components per day, how many would be expected to be rejected?
6. The time taken to complete a particular type of job is distributed approximately normally with mean 1.8 hours and standard deviation 0.1 hours.
   a) If 'normal-time' work finishes at 6.00pm and a job is started at 4.00pm, what is the probability that the job will need overtime payments?
   b) What estimated completion time (to the nearest minute) should be set so that there is a 90% chance that the job is completed on time?
7. Use the Normal approximation to the binomial to calculate the probability of getting more than 10 defectives in a random sample of 100 taken from a population which is 8% defective. (Remember that the equivalent of value 11 in a binomial distribution is the *range* 10.5 to 11.5 in a Normal distribution.)
8. From experience, a travel firm knows that only 92% of people who book a Spanish holiday with them actually turn up. The firm deliberately overbooks by 8 their allocation of 172 places on a particular Spanish holiday. What is the probability that there will be dissatisfied customers?
9. The mean inside diameter of the washers produced by a machine is 1 cm with a standard deviation of 0.01 cm. The diameters are approximately normally distributed. Tolerance limits (i.e. central limits set either side of the mean, outside which washers having this diameter are rejected) are set such that only 2% of washers should be rejected. What are the tolerance limits?
10. A sample investigation of the number of successful annual sales of product A made by salesmen employed in a particular industry yielded mean 15.3 and standard deviation 6.1.

    Find a) 95% b) 99% confidence limits for the mean.
11. A random sample of 25 components from a production line yielded 4 outside quality limits. Give 95% confidence limits for the proportion of all output that lies outside the quality limits.
12. Seven accounting workers in a particular industry were interviewed and some data is given below:

| Employee | 1 | 2 | 3 | 4 | 5 | 6 | 7 |
|---|---|---|---|---|---|---|---|
| Salary | 21,650 | 18,400 | 22,500 | 22,150 | 22,150 | 17,680 | 21,400 |
| Qualified | Yes | Yes | No | Yes | Yes | Yes | No |

    a) Calculate the mean and standard deviation
    b) Give a 95% confidence limit for the mean salary
    c) Give a 95% confidence interval for the proportion of all employees of this type who are qualified.

13. Could a sample of size 25 with mean 14.3 and standard deviation 2.6 have been drawn from a population with mean 15?
14. In the past average weekly earnings in a factory have averaged £302. A random sample of 9 employees yielded a mean of £287 with standard deviation £16. Is there any evidence of a difference in pay?
15. The life of electric light bulbs from a particular manufacturer have an average life of 800 hours. A sample of 25 bulbs was taken and found to have mean life 850 hours with standard deviation 80 hours.
    a) Is there evidence that the sample bulbs have performed differently to the population norm?
    b) What is the largest whole-number value that the sample mean could have taken that would have reversed the above decision?

## Examination example (with worked solution)

*Question*

The specification for the length of an engine is a minimum of 99mm and a maximum of 104.4mm. A batch of parts is produced that is normally distributed with a mean of 102mm and a standard deviation of 2mm. Parts cost £10 to make. Those that are too short have to be scrapped; those too long are shortened, at a further cost of £8. You are required

a) to find the percentage of parts which are (i) undersize, (ii) oversize;
b) to find the expected cost of producing 1,000 usable parts;
c) to calculate and to explain the implications of changing the production method so that the mean is halfway between the upper and lower specification limits. (*The standard deviation remains the same.*)

*CIMA*

*Answer*

a) i. The probability of an undersize part is $\Pr(x \le 99)$ where $x$ is its length (mm)

To transform this we have, $z = \dfrac{x - \mu}{\sigma} = \dfrac{99 - 102}{2} = -1.5$

Hence probability required $= \Pr(z \le -1.5) = 0.5 - 0.4332 = 0.0668$, from normal tables.

That is, 6.68% of the parts will be undersize.

ii. In the same way, standardising $x = 104.4$: gives $z = \dfrac{104.4 - 102}{2} = 1.2$

Hence the probability of a part being oversize is:

$\Pr(z \ge 1.2) = 0.5 - 0.3849 = 0.1151$

Thus 11.51% of the parts will be oversize.

b) If 1,000 parts are produced, 67 would be expected to be undersize, 115 oversize. Thus 933 parts are usable (the 67 undersize ones being scrapped), at a cost of: 885 parts at £10 and 115 parts at £18 (those originally oversize, now shortened) = £10,920.

Hence 1,000 usable parts would have an expected production cost of:

$$10920 \times \frac{1000}{933} = £11,704.18$$

(Alternatively, produce $\dfrac{1,000}{0.9332}$ parts which yield 1000 usable items.

The final cost is the same.)

c) If the mean is changed to 101.7mm, then the symmetry of the normal distribution will mean that the same proportion of parts will be too long as too short. To compute this proportion:

$$z = \frac{104.4 - 101.7}{2} = 1.35$$

and $\quad P(z \ge 1.35) = 0.5 - 0.4115 = 0.0885$

Hence, of 1,000 parts produced, 88.5 (on average) will be too long and 88.5 will be too short.

Thus 911.5 parts will be usable, at a production cost of (911.5 × 10) (88.5 × 18) = £10,708.

The cost of producing 1,000 usable parts would therefore rise to $\frac{1000}{911.5} \times 10{,}708 = £11{,}747.67$.

(Alternatively, produce $\frac{1000}{0.9115} = 1097$ parts which yield 1000 usable items. The final cost is the same.)

# Examination questions

1. The number of invoices being processed through a sales ledger, grouped by value, are shown below. The standard deviation is £1,600.
   You are required
   a) to draw a histogram to represent these data;
   b) assuming these data to represent a simple random sample, to find 90% confidence limits for the mean value of all invoices being processed through this sales ledger;
   c) if twenty of these invoices contain errors, to find a 95% confidence interval for the overall error rate on invoices processed through this sales ledger.

| *Value of invoice* £ | *Value of invoice* £ | *Number of invoices* |
|---|---|---|
| At least | Less than | |
| 0 | 500 | 20 |
| 500 | 1,000 | 40 |
| 1,000 | 2,000 | 80 |
| 2,000 | 4,000 | 150 |
| 4,000 | 5,000 | 60 |
| 5,000 | 6,000 | 30 |
| 6,000 | 7,000 | 20 |
| 7,000 and over | | 0 |
| Total number | | 400 |

*CIMA*

2. a) A company produces batteries whose lifetimes are normally distributed with a mean of 100 hours. It is known that 90% of the batteries last at least 40 hours.
      i. Estimate the standard deviation lifetime.
      ii. What percentage of batteries will not last 70 hours?
   b) A company mass produces electronic calculators. From past experience it knows that 90% of the calculators will be in working order and 10% will be faulty if the production process is working satisfactorily. An inspector randomly selects 5 calculators from the production line every hour and carries out a rigorous check.
      i. What is the probability that a random sample of 5 will contain at least 3 defective calculators?
      ii. A sample of 5 calculators is found to contain 3 defectives; do you consider the production process to be working satisfactorily?

*CIMA*

3. a) Your company requires a special type of inelastic rope which is available from only two suppliers. Supplier A's ropes have a mean breaking strength of 1,000 kgs with a standard deviation of 100 kgs. Supplier B's ropes have a mean breaking strength of 900 kgs with a standard deviation of 50 kgs. The distribution of the breaking strengths of each type of rope is Normal. Your company requires that the breaking strength of a rope be not less than 750 kgs.
      All other things being equal, which rope should you buy, and why?

b) One per cent of calculators produced by a company is known to be defective. If a random sample of 50 calculators is selected for inspection, calculate the probability of getting no defectives by using:
   i. the Binomial distribution;
   ii. the Poisson distribution.

*CIMA*

4. a) In 1985 a company has found that, in a certain category, the values of claims (£) are Normally distributed with a mean of 200 and a variance of 2,500. What percentage of these claims are:
   i. under £100;
   ii. under £150;
   iii. over £350?

   b) Ruritania airlines deliberately overbooks its Jumbo flights because it knows from experience that on the average only 90% of passengers who book for a given flight actually turn up for that flight. The airline takes 290 bookings for the 275 available seats. Assuming the number of arrivals is Normally distributed and passengers arrive independently, find the approximate probability that for a given flight there will be too many booked passengers turning up for the seats available.

*CII*

5. a) Bus-Hire Limited has two coaches which it hires out for local use by the day. The number of demands for a coach on each day is distributed as a poisson distribution, with a mean of 2 demands.
   i. On what proportion of days is neither coach used?
   ii. On what proportion of days is at least one demand refused?
   iii. If each coach is used an equal amount, on what proportion of days is 'one' particular coach not in use?

   b) Steel rods are manufactured to a specification of 20 cm length and are acceptable only if they are within the limits of 19.9 cm. and 20.1 cm. If the lengths are Normally distributed, with mean 20.02 cm and standard deviation 0.05 cm, find the percentage of rods which will be rejected as
   i. undersize and ii. oversize.

*CIMA*

6. a) Discuss the importance of the Normal distribution.

   b) Daily electricity power consumption in a large office block is Normally distributed with a mean of 10,000 kilowatts and a standard deviation of 2,000 kilowatts. What is the probability that the consumption of electricity on a given day is:
   i. Greater than 13,000 kilowatts;
   ii. Less than 8000 kilowatts; iii. Between 7,500 and 14,000 kilowatts?

*ICSA*

7. Applicants for a certain job are given an aptitude test. Past experience shows that the scores from the test are normally distributed with a mean of 60 points and a standard deviation of 12 points. What percentage of candidates would be expected to pass the test if a minimum score of 75 is required?

*CIMA 30%*

8. (a) Samples of 10 boxes of cornflakes are randomly selected from a production line and weighed. It is found that the mean number of underweight boxes in such a sample of 10 is 1.6. Assuming that the majority of individual boxes are of acceptable weight, what then is the probability that a box is underweight ?

   (b) In a sample of 6 such boxes, calculate the probability that :–

   (i) no boxes

   (ii) 2 boxes

   (iii) at least 3 boxes

   are underweight.

   (c) If a sample of 100 such boxes is taken, use an appropriate method of approximation to calculate the probability that, in the sample,

   (i) less than 10

   (ii) more than 28

   (iii) between 16 and 24

   are underweight.

*ICSA*

# Part 10 Specialised business applications

This part of the manual covers some specialised business applications that do not form a natural part of any previous part.

Chapter 35 introduces linear inequalities and describes how they are manipulated in order to maximise (or minimise) certain types of business function.

Matrices are introduced in chapter 36, dealing with their form, manipulation and use in the representation of certain business situations and in solving 2 by 2 simultaneous linear equations.

The structure, management and control of stock is known as Inventory Control. Chapter 37 introduces this topic and looks at two important models, aspects of inventory costs and graphs.

Finally, chapter 38 briefly covers network planning and analysis.

# 35 Linear inequalities

## 1. Introduction

The chapter begins by defining inequalities and gives some examples of their graphs. This is followed by a description of a major application, linear programming, and the graphical techniques available for solving related problems.

## 2. The need for inequalities

Often, in practical situations, it is not possible to match available resources with a planned schedule of activities. For example, even though a factory might have 25 men available for 8 hours on a particular day with a total of 200 hours work to be done, one man cannot be in two places at once and machine cleaners are not necessarily machine operators.

In other words, in real life, production processes *cannot always be described* in terms of equations such as:

'200 men available, each for one hour = 200 hours of productive work'

or 'total materials available = total materials used'.

There is *often a need* to describe situations in terms of:

'there is a *maximum* of 500 manhours available'

or 'at least 45 units of product X are to be produced'.

This is an area where *inequalities* rather than equations are used and are described in the following section. The main object of this chapter is to solve what is known as a linear programming problem, which requires the ability to understand and graph inequalities. The following three sections are included to this end.

## 3. Linear inequalities

a) Whereas equations are two expressions joined by an '=' sign, inequalities are two expressions joined by one of the 'inequality' signs. These are:

$<$ (less than)

$>$ (greater than)

$\leq$ (less than or equal to)

$\geq$ (greater than or equal to).

b) Linear equations (i.e. $y = x + 2$, $3x - 2y = 12$, $y = 6$) *define straight lines*. Linear inequalities always define *regions* of the $x$–$y$ plane. (Note that the '$x$–$y$ plane' can be thought of physically as the piece of graph paper on which an $x$–axis and $y$–axis are drawn.)

For example, see Figure 1(a). The line $x = 2$ splits the $x$–$y$ plane up into two distinct regions; one to the left of the line, the other to the right of the line. All the points in the region to the right of the line have their $x$–values greater than 2. Thus, this region is described by the inequality $x>2$. Similarly, the inequality $x<2$ describes the region to the left of the line. Also, consider Figure 1(b). The line $y = x$ is shown on the graph. Again, this line splits the $x$–$y$ plane up into two distinct regions. One region (shown on the graph) is $y<x$, the other is $y>x$. [Test this yourself by selecting *any* point in the marked region and examining the $x$ and $y$ values. The $y$-value *will always be less* than the $x$ value.]

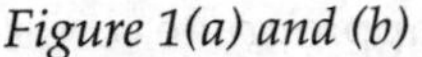
*Figure 1(a) and (b)*

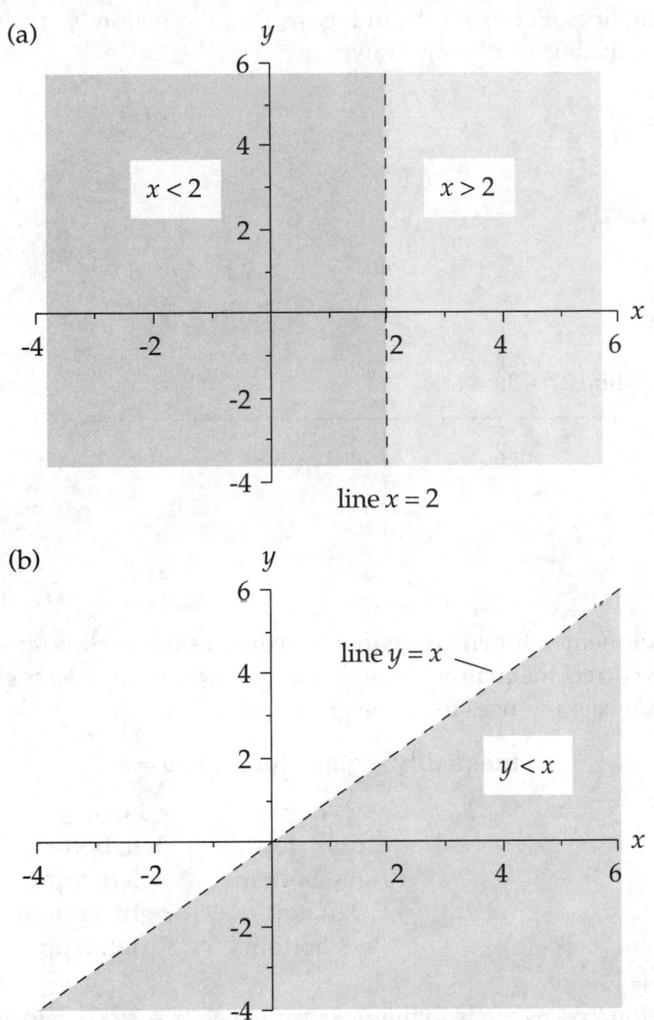


Figure 1(c) shows the line $2x + 3y = 6$ and the marked region to the top and right of the line shows the area of the plane for which $2x + 3y$ is greater than 6, i.e. $2x + 3y > 6$.

The obvious question is 'how can we be sure that this region is $2x + 3y > 6$ (and not, say, $2x + 3y < 6$)?' A foolproof (but relatively laborious) method is to use the *substitution technique as follows*:

1) Select *any* point in the region. [We will take the point (2,2) i.e. $x = 2$ and $y = 2$.]
2) Ensure that the inequality is *satisfied*. In other words, we need to show that the value of $2x + 3y$ is greater than 6 for the point selected. [Substituting $x = 2$ and $y = 2$ into $2x + 3y$ gives $2(2) + 3(2) = 10$, which is greater than 6. Thus the region in question is correctly described by the inequality $2x + 3y > 6$.]

Notice that (linear) inequalities are always bounded by their appropriate corresponding lines. For example, in Figure 1(b), the region '$y < x$' cannot be defined without the line $y=x$ being drawn.

*Figure 1(c)*

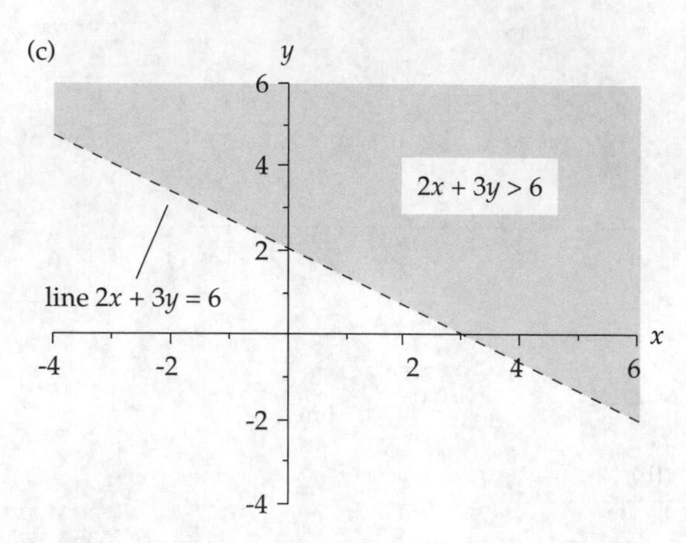


c) Another method for ensuring that the correct side of a line ($ax + by = c$) is chosen for the correct inequality (i.e. either $ax + by < c$ or $ax + by > c$) is known as the *sign technique* and uses the table given below.

**Inequality regions for $ax + by = c$**

| $a$ | $b$ | $>$ or $\geq$ | $<$ or $\leq$ |
|---|---|---|---|
| + | + | right, top | left, bottom |
| + | – | right, bottom | left, top |
| – | + | left, top | right, bottom |
| – | – | left, bottom | right, top |

For example:

the region $2x + 3y > 6$ is to the *right and top* of $2x + 3y = 6$ (since both $a$ and $b$ are positive)

and $3x - 30y < 30$ is to the *left and top* of $3x - 30y = 30$ (since $a$ is positive and $b$ is negative).

## 4. Feasible regions

a) Several business production constraints can be graphically demonstrated using their separate respective inequalities to produce a *feasible region* of operation.

> A *feasible region* is defined as that region of the $x$-$y$ plane which satisfies *all* of the inequalities under consideration.

An example of how a feasible region is constructed is given below, using two business constraints and their corresponding inequalities.

b) Suppose that a small factory produces two products X and Y, only one of which can be produced at a time within a 50-hour week. Given that product X takes 2 hours to make, while product Y takes only one hour, the time restriction (of 50 hours) means that the following inequality must be satisfied:

$$2x + y \le 50 \text{ [TIME constraint]}$$

(where $x$ is the number of X products produced in one week and $y$ the number of Y produced per week).

Given further that product Y uses 3 times the amount of material that X uses and there is only enough material supply per week to produce the equivalent of 90 X products, the following must also be true:

$$x + 3y \le 90 \text{ [MATERIAL constraint]}$$

Using the rule given in section 3 c) for identifying the region $ax + by \le c$:

i. $2x + y \le 50$ must lie to the left and bottom of $2x + y = 50$, since $a = 2$ and $b = 1$ (i.e. both positive).

ii. Similarly, $x + 3y \le 90$ must also lie to the left and bottom of $x + 3y = 90$, since $a = 1$ and $b = 3$ (both positive).

These two constraints taken together will define a feasible region of the weekly numbers of X and Y products possible that satisfy both given conditions. This is shown in Figure 2, with the feasible region shaded.

*Figure 2*

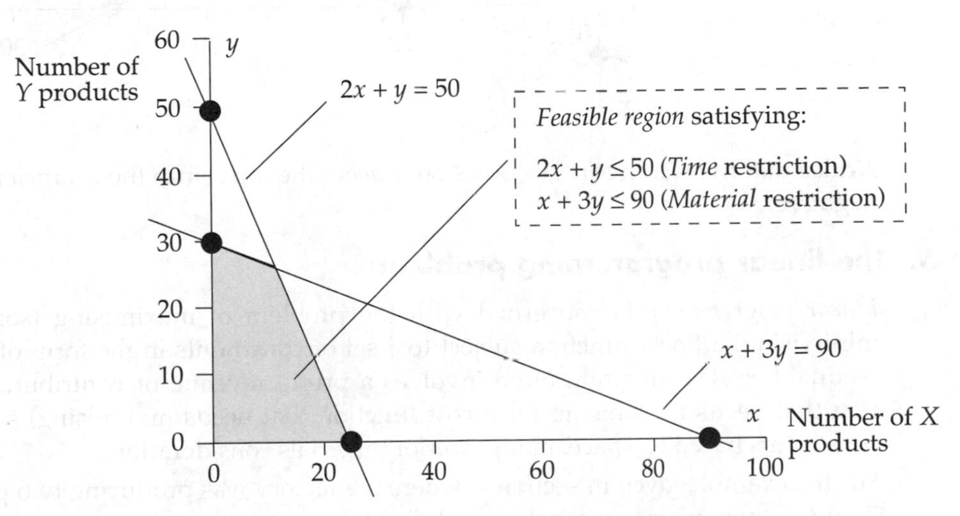


**Note**: in situations similar to the above, where each variable ($x$ and $y$) is representing a number of products, only positive $x$ and $y$ values are relevant. In other words, it is only necessary to consider the quadrant for both $x$ and $y$ positive as drawn above.

## 5. Example 1 (Plotting a *feasible region* defined by a set of *inequalities*)

To display the region defined by the following inequalities:

$$\begin{aligned} 2x + 10y &\le 50 \\ x &\ge 10 \\ 3x - 30y &\le 30 \end{aligned}$$

First, plot the graphs of the three corresponding equations.

Now, $2x + 10y = 50$ crosses respective axes at $x = 25$ and $y = 5$, $x = 10$ is a vertical line crossing the $x$-axis at $x = 10$ and $3x - 30y = 30$ crosses respective axes at $x = 10$ and $y = -1$.

The lines are plotted and the defined region is highlighted in Figure 3.

*Figure 3*

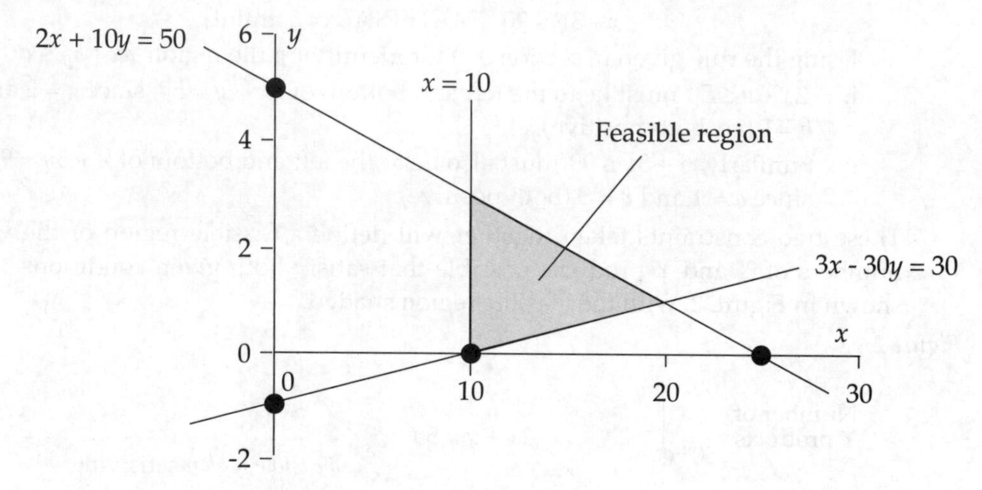


Notice that the region for $3x - 30y \le 30$ is *above* the line, since the coefficient of $y$ is negative.

## 6. The linear programming problem

*Linear programming* is concerned with the problem of maximising (sometimes minimising) a linear function subject to a set of constraints in the form of a set of inequalities. The situation often involves a profit, revenue or contribution function that needs maximising (or a cost function that needs minimising) subject to constraints based on space, manpower or materials considerations.

For the example given in section 4, where the factory was producing two products X and Y, there were two constraints defined:

a) a time constraint: $2x + y \le 50$;

b) a material constraint: $x + 3y \le 90$.

If the contribution from each unit of product X was £4 and the contribution from each unit of product Y was £3, the total contribution from the two products could be put in the form $4x + 3y$, a linear function.

The *solution of the problem* involves finding the values of $x$ and $y$ (the numbers of products X and Y to be produced) that will maximise the linear contribution function. (This particular problem is worked through and solved in Example 2.)

Although there are various methods of approach for solving this type of problem, the only one relevant to this manual is the graphical method. This is demonstrated in Example 2 and a summary of the technique is given in section 8.

## 7. Example 2 (Solution to a *linear programming* problem)

In the example given in section 4, there were two products X and Y that were subject to:

$$2x + y \leq 50 \quad \text{(time constraint)}$$

and $$x + 3y \leq 90 \quad \text{(material constraint)},$$

where $x$ and $y$ were the amounts of products X and Y respectively that could be produced.

If product X sells for £4 and Y for £3, find the number of each of products X and Y to be produced (i.e. $x$ and $y$) so that the total revenue is maximised.

Figure 4 (which reproduces Figure 2) shows the lines corresponding to the time and material constraints given above and the corresponding feasible region (which is shaded).

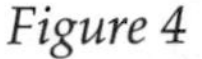

*Figure 4*

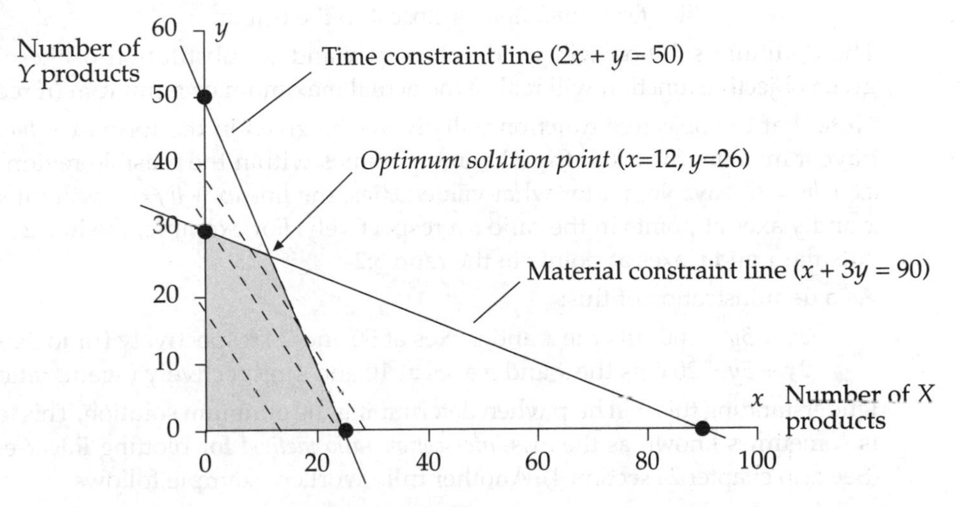


We need to maximise the total revenue function, which is given by $4x + 3y$ (using the information above). That is, we need to find the values of $x$ and $y$, called the *optimum solution*, which:

i. lie in the feasible region, and
ii. maximise the revenue function $4x + 3y$.

Now, consider the line $4x + 3y = \text{R}$. No matter what value R is, $4x + 3y = \text{R}$ will intersect the $x$ and $y$ axes in the ratio 3:4 (i.e. 3 units on the $x$-axis to 4 units on the $y$-axis). Some of these parallel lines have been drawn (using dashes) in Figure 4. That dashed line which lies furthest away from the origin ($x = 0$, $y = 0$) *but which still lies within the feasible region* is the line that needs to be identified, since the corner of the feasible region (or vertex) that it intersects is the solution required.

From Figure 4, the optimum point (solution) which maximises the revenue function is seen to be $x = 12$ and $y = 26$. In other words, 12 $X$'s and 26 $Y$'s need to be produced in order to maximise revenue, subject to the given time and material constraints.

## 8. Graphical solution to the linear programming problem

This section summarises the procedure for solving the linear programming problem.

Given a linear function to maximise (or minimise), called the *objective function*, and a set of constraints in the form of inequalities:

STEP 1 Plot the lines corresponding to each of the given inequalities.

STEP 2 Define the *feasible region* as that region satisfying all the given inequalities.

STEP 3 Determine the *optimum solution* as that point which:

a) lies in the feasible region (always at a vertex), and

b) intersects a line parallel to the given objective function causing it to be maximised (or minimised).

In practical terms this will mean finding that line parallel to the objective function which:

i. *for a maximum*, is furthest away from the origin

ii. *for a minimum*, is closest to the origin.

The optimum solution will yield values of $x$ and $y$. Substitution of these into the given objective function will realise the actual maximum or minimum (if required).

**Note** that the objective function will always be given in the form $ax + by$ and will have a meaningful value for all $x$ and $y$ values within the feasible region. That is, $ax + by = C$, say. No matter what value C has, the line $ax + by = C$ will intersect the $x$ and $y$ axes at points in the ratio $b{:}a$ respectively. For example, the line $2x + 5y = C$ cuts the $x$ and $y$ axes at points in the ratio 5:2.

As a demonstration of this:

$2x + 5y = 100$ cuts the $x$ and $y$ axes at 50 and 20 respectively (ratio 5:2)

$2x + 5y = 20$ cuts the $x$ and $y$ axes at 10 and 4 respectively (again, ratio 5:2).

Understanding this will help when determining the optimum solution. This technique is sometimes known as the *axis-intersection ratio method* for plotting linear equations (See also chapter 24 section 4). Another fully worked example follows.

## 9. **Example 3** (Solution to a *linear programming* problem)

*Question*

A production line can be set up to produce either product $X$ or product $Y$. The following table gives the breakdown for each product.

| *Product* | *Labour (minutes)* | *Materials (lbs)* | *Testing minutes* |
|---|---|---|---|
| X | 30 | 2 | 3 |
| Y | 15 | 4 | 4 |

In any one week only 30 hours of labour and 280 lbs of material is available, and owing to cost and availability, the testing equipment must be used for at least 4 hours. Also, because of existing orders, at least 20 $X$ products must be produced. The contribution from each unit of $X$ produced is £12 and each unit of $Y$, £9.

Find:

a) the weekly production that will maximise contribution and calculate this resulting contribution;

b) the weekly production that minimises contribution and calculate this resulting contribution;

c) the percentage utilisation of the available labour for both minimum and maximum contribution.

*Answer*

Putting $x$ and $y$ as the number of products of X and $Y$ to be produced in one week, and converting all times to minutes, the four constraints given can be expressed as:

| | | |
|---|---|---|
| Labour: | $30x + 15y \leq 1800$ | (Line meets axes at $x$=60 and $y$=120) |
| Materials: | $2x + 4y \leq 280$ | (Line meets axes at $x$=140 and $y$=70) |
| Testing: | $3x + 4y \geq 240$ | (Line meets axes at $x$=80 and $y$=60) |
| Sales: | $x \geq 20$ | |

The contribution function is given by $12x + 9y$ and this needs to be maximised and minimised. Figure 5 shows the four constraint lines plotted, the feasible region based on the four constraints and some contribution lines (dotted).

*Figure 5*

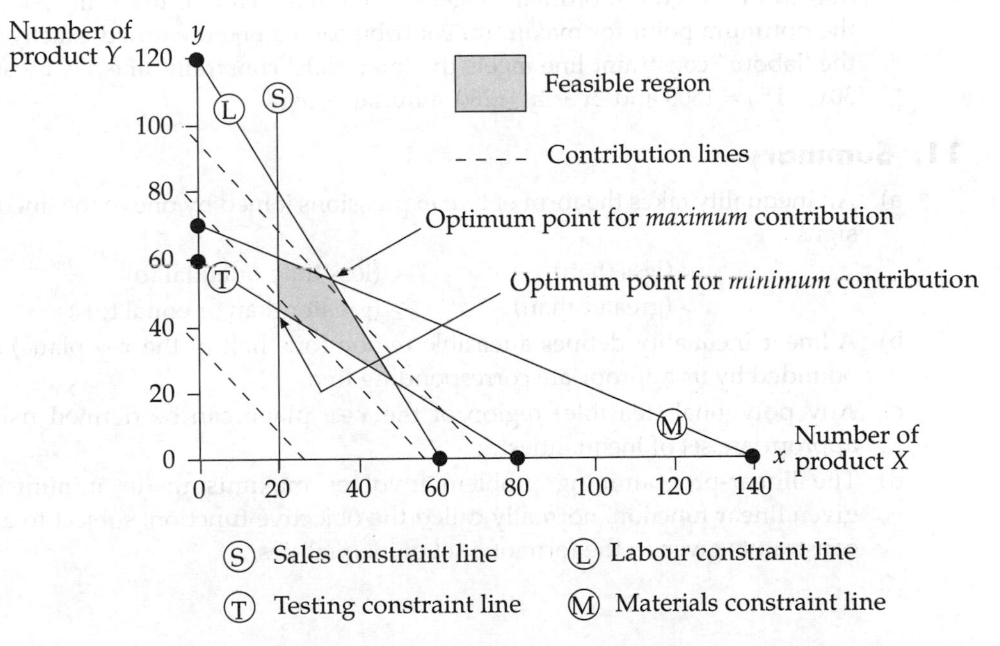

a) The optimum point for maximum contribution is marked on the diagram in Figure 5 at $x = 33$ and $y = 53$. Substituting these two values into the contribution function $(12x + 9y)$ will give a maximum contribution of $12(33) + 9(53) = £873$/wk.

b) The optimum point for minimum contribution is also marked on the diagram at $x = 20$ and $y = 45$. Thus the minimum contribution is $12(20) + 9(45) = £645$/wk.

c) At minimum contribution, $x = 20$ and $y = 45$, and thus the labour used (from the labour function $30x + 15y$) is $30(20) + 15(45) = 1275$ minutes. This represents $\frac{1275}{1800} \times 100 = 70.8\%$ of available labour.

At maximum contribution, $x = 33$ and $y = 53$, and labour used is $30(33) + 15(53) = 1785$ minutes. This represents $\frac{1785}{1800} \times 100 = 99.2\%$ of available labour.

## 10. Practical notes on linear programming

a) In problems of this type (where a graphical solution is obtained):

   i. there will be at most two variables ($x$ and $y$ in the examples) involved, and

   ii. both of the variables will be non-negative, so that any diagrams drawn will only use the positive $x$ and $y$ quadrant.

b) The objective function (to be maximised and/or minimised) should be identified using the axes-intersection ratio method described in section 8. 'Trial and error' using a ruler is the best method of identifying the optimum point(s). It is usual to show two or three placings of this line (dotted in the previous examples).

c) If necessary, rather than reading the $x$ and $y$ values that form the optimum solution from the graph, ordinary algebraic methods can be used. In Example 3, the optimum point for maximum contribution can be identified exactly where the 'labour' constraint line meets the 'materials' constraint line. i.e. by solving $30x + 15y = 1800$ and $2x + 4y = 280$ simultaneously.

## 11. Summary

a) An inequality takes the form of two expressions joined by one of the 'inequality signs':

| | |
|---|---|
| $<$ (less than) | $\leq$ (less than or equal to) |
| $>$ (greater than) | $\geq$ (greater than or equal to) |

b) A linear inequality defines a feasible region (one half of the $x$–$y$ plane) and is bounded by its appropriate corresponding line.

c) Any polygonal (feasible) region of the $x-y$ plane can be defined using an appropriate set of inequalities.

d) The linear-programming problem involves maximising (or minimising) a given linear function, normally called the objective function, subject to a set of constraints given in the form of a set of inequalities.

e) The graphical solution to a linear programming problem involves:
   i. graphing the constraint inequalities to form a feasible region,
   ii. choosing that line which is parallel to the objective function and within the feasible region (at a vertex) which:
      is furthest away from the origin (for a maximum)
      is closest to the origin (for a minimum).

f) The basis for the graphical solution to a linear programming problem described here is subject to the two conditions:
   i. only two variables are involved, and
   ii. each variable can take only non-negative values.

## 12. Points to note

a) Only linear inequalities have been discussed here. Obviously other general inequalities can be used in appropriate situations. However, the linear programming technique described above is only valid for constraints based on linear inequalities and a linear objective function.

b) When there are many constraints and/or there are more than two variables involved in a linear programming problem, other methods must be used. The most common of these is known as the *simplex method* (but outside the scope of this manual).

c) Two methods are available for ensuring that the region defined by an inequality is the correct side of the corresponding defining line:
   i. the substitution technique,
   ii. the sign technique.

d) When presenting the graphical solution to a linear programming problem, there is usually a lot of information that could be shown. The danger is that the diagram becomes too cluttered and confusing and this should be avoided, particularly in an examination. The best approach is:
   i. always shade the feasible region,
   ii. label the constraint lines with letters or numbers only (a key can be provided underneath the diagram if necessary),
   iii. mark the optimum point(s) clearly and
   iv. only dash (or dot) in that objective function line that is either maximum or minimum.

## 13. Student self review questions

1. How does a linear inequality differ from a linear equation? [3]
2. Give some examples of constraints which can be described by linear inequalities in a business context. [4]
3. Describe what a feasible region is, when related to a given set of inequalities. [4]
4. What is the 'linear programming problem'? [6]
5. Given a feasible region together with a defined objective function, describe briefly how an optimum solution is found. [7]

6. What restrictions are placed on the variables used in a linear programming problem? [10]

## 14. Student exercises

1. If $x$ and $y$ are both $\geq 0$, describe in words the position of the defined (feasible) region for the following inequalities in relation to their corresponding lines.
   a) $4x - 2y \leq 100$
   b) $x + 3y \geq 20$
   c) $4x + 3y \leq 120$; $y \geq 20$
   d) $10x + 10y \leq 300$; $2x + 6y \geq 60$; $x \geq 10$

   Show the regions corresponding to a), b), c) and d) graphically.

2. The following four inequalities are supposed to define a four-sided polygonal region (where the highest coordinate of any vertex is 80), but one of the inequalities has had its sign reversed in error.

   $$x+y \geq 80; \quad x \leq 40; \quad 12x + 6y \geq 480; \; y \geq 20$$

   Correct this sign and give the coordinates of each of the four vertices of the defined region.

3. Find the values of $x$ and $y$ that maximise the function $2x + 4y$ subject to the constraints:

   $$0 \leq x \leq 400; \quad 0 \leq y \leq 300; \quad x+y \leq 600.$$

   Calculate also the maximum value of the function.

4. Subject to the conditions:

   $$2x + 3y \leq 120 \text{ and } 2x + y \leq 60 \text{ with } x,y \text{ non-negative}$$

   find the values of $x$ and $y$ that maximise the function $2x + 2y$ and calculate its maximum value.

5. A company makes two kinds of lampshades. Shade $A$ is high quality and shade $B$ is of lower quality. The respective contributions are £0.40 and £0.50 per shade. Each shade of type $A$ requires twice as much time as a shade of type $B$ and, if all shades were of type $B$, the company could make 1000 per day. The supply of material is sufficient for only 800 shades per day ($A$ and $B$ combined). Shade $A$ requires a fancy support and only 400 per day are available. There are only 700 supports a day available for shade $B$.
   a) Set up the four constraints (two on supports, one on material and one on time), using $x$ and $y$ respectively as the number of shades of type $A$ and $B$ made, and identify the contribution function.
   b) Represent the constraints graphically, identifying the feasible region and hence find the optimum solution for $x$ and $y$ that maximises contribution. What is the maximum daily contribution?

6. A factory produces two products $X$ and $Y$ each of which must pass through two production processes $A$ and $B$. Product $X$ requires 10 hours in $A$ and 8 in $B$, while product $Y$ requires 10 hours in $A$ and 3 in $B$. The maximum capacity of process $A$ for a certain period is 12000 hours and, due to cost and manpower agreements, process $B$ must be used for at least 4800 hours over the same period. Due to fixed commitments, at least 600 $Y$ products must be produced, while a maximum of 500 $X$ products will be produced over the given period.

   a) Set up the constraints for processes $A$ and B and the constraints for the production levels of $X$ and $Y$ (where $x$ and $y$ respectively are the numbers of $X$ and $Y$ produced).

   b) If $X$ has a contribution of £100 and $Y$ a contribution of £140, find the points that maximise and minimise total contribution and give the value of these contributions.

   c) Answer part (b) if the contributions of $X$ and $Y$ are exchanged.

# 36 Matrices

## 1. Introduction

This chapter introduces matrices as structures within which numeric data can be stored and manipulated.

There is an explanation of how data is referenced within them and of their size and description. The rules for adding and multiplying matrices are listed and demonstrated and some special matrices are then described which enable simultaneous equations to be represented and solved.

Finally transition matrices are defined, demonstrated and used in practical situations.

## 2. A matrix structure

*A matrix* can be thought of as a rectangular table of values conveying numerical information. A business might have the number of employees in each of its departments classified as to their gender.

For example:

| | *Male* | *Female* |
|---|---|---|
| Sales | 14 | 8 |
| Purchasing | 6 | 6 |
| Stock control | 8 | 2 |
| Production | 21 | 18 |

The actual matrix is a *framework which holds the data values*; the labels or categories Sales, Purchasing, ...etc and Male, Female are there simply to identify precisely what the data are.

Thus, if we call the whole matrix set X, for convenience, we have:

$$\text{matrix } X = \begin{bmatrix} 14 & 8 \\ 6 & 6 \\ 8 & 2 \\ 21 & 18 \end{bmatrix}$$

where it is understood that the first row is Sales, the second refers to Purchasing, and so on. Similarly, the first column is numbers of male employees, the second female employees. Thus it is most important to keep each element in its correct position in the matrix.

## 3. Identifying data within a matrix

Matrices can consist of as many rows and columns as the situation requires, but no matter how small or large a matrix is, it is important to be able to reference a particular data item (element or cell) within the matrix in a logical way.

This is done by identifying which row and column it is in. $x_{i,j}$ means 'the data held at the intersection of *row i and column j* of matrix X'.

For example, using the matrix data of the previous section:

$x_{1,1} = 14$ (The value held in the cell at the intersection of row 1 and column 1 is 14)

$x_{1,2} = 8$ (The value held in the cell at the intersection of row 1 and column 2 is 8)

$x_{2,1} = 6$ (The value held in the cell at the intersection of row 2 and column 1 is 6)
... etc.

Given that the nature of the given data is precisely known, $x_{1,2}=8$ can be interpreted in words as 'there are 8 people who work in Sales (Dept 1) who are female (Gender 2)'

The following section summarises the work done so far, defines a square matrix and introduces the idea of the size of a matrix.

## 4. Definition and description of a matrix

A **matrix** is a simple mathematical structure that holds numerical information in rectangular (tabular) form.

1. A matrix can consist of any number of complete rows and columns.
2. The value at the intersection of a row and column is referred to as a *cell, element* or *data item*.
3. The values of a matrix are normally enclosed within brackets (to identify the values as a complete set.
4. If a matrix is labelled as *A*, then *i, j* is the value held in the cell at the intersection of row *i* and column *j*.
5. A matrix which has *a* rows and *b* columns is called an '*a* by *b*' matrix. In this case, '*a* by *b*', which can also be written as $a \times b$, is said to be the *size* of the matrix.
6. A *square* matrix is one that has the same number of rows as columns.

**Note** that a matrix can be enclosed by any type of bracket although normally they are either round ( ... ) or square [ ... ]

## 5. Example 1 (Examples of *matrices* of various *sizes*)

a) $\begin{pmatrix} 2 & 4 & 1 \\ 3 & 2 & 6 \end{pmatrix}$ is a $2 \times 3$ matrix, where 2 is the number of rows, 3 is the number of columns.

b) $\begin{bmatrix} 3 & 6 \\ 1 & -2 \\ 1 & 4 \\ 2 & 2 \end{bmatrix}$ is a $4 \times 2$ matrix

c) $\begin{bmatrix} 2 \\ 1 \\ 3 \end{bmatrix}$ is a 3 by 1 matrix (sometimes called a *column* matrix)

d) $(6 \;\; 2 \;\; 4 \;\; 1)$ is a 1 by 4 matrix (sometimes called a *row* matrix).

## 6. Addition and subtraction of matrices

If necessary, two matrices can be added or subtracted. For example, suppose a company has just two warehouses A and B, each carrying stocks of a particular item as follows:

$$A = \begin{array}{c} \\ \text{Grey} \\ \text{White} \end{array}\begin{array}{c} \text{Large Small} \\ \begin{bmatrix} 12 & 4 \\ 8 & 2 \end{bmatrix} \end{array} \quad \text{and } B = \begin{array}{c} \\ \text{Grey} \\ \text{White} \end{array}\begin{array}{c} \text{Large Small} \\ \begin{bmatrix} 3 & 2 \\ 1 & 5 \end{bmatrix} \end{array}$$

These two matrices can be added as follows:

$$A + B = \begin{bmatrix} 12 & 4 \\ 8 & 2 \end{bmatrix} + \begin{bmatrix} 3 & 2 \\ 1 & 5 \end{bmatrix} = \begin{bmatrix} 12+3 & 4+2 \\ 8+12 & 2+5 \end{bmatrix}$$

$$= \begin{array}{c} \\ \text{Grey} \\ \text{White} \end{array}\begin{array}{c} \text{Large Small} \\ \begin{bmatrix} 15 & 6 \\ 9 & 7 \end{bmatrix} \end{array}$$

which represents the stocks of the item for the company as a whole.

Notice that, in the above matrix addition, corresponding elements have been added together. In other words, the element in row 1, column 1 (12) of the first matrix has been added to the element in row 1, column 1 (3) of the second matrix to form the element at row 1, column 1 (12+3 = 15) of the resultant matrix, and so on.

Subtraction is performed in a similar way. For example:

$$\begin{bmatrix} 15 & 6 \\ 9 & 7 \end{bmatrix} - \begin{bmatrix} 3 & 2 \\ 1 & 5 \end{bmatrix} = \begin{bmatrix} 15-3 & 6-2 \\ 9-1 & 7-5 \end{bmatrix}$$

$$= \begin{bmatrix} 12 & 4 \\ 8 & 2 \end{bmatrix}$$

These procedures are summarised in the following section, together with some rules and notation.

## 7. Definition of matrix addition and subtraction

> **Addition and subtraction of matrices**
>
> Two matrices can be added (or one matrix subtracted from another) only if they have identical sizes.
>
> Addition is performed by adding together corresponding elements. Similarly for subtraction.

**Note**: Although two matrices may be of the right size to be added mathematically, this does not guarantee that their sum is meaningful. For example, a 3 × 2 matrix describing the numbers of employees in each of three firms by their sex added to a 3 × 2 matrix describing the number of salesmen's cars for the three firms by type of car would be totally meaningless in any practical sense.

**8. Example 2** (*addition* and *subtraction* of matrices)

A lorry/town matrix has the following structure representing the *number of trips made* by a particular lorry to a particular town in a certain week.

$$\text{Lorry number} \begin{matrix} \\ 1 \\ 2 \\ 3 \end{matrix} \overset{\begin{matrix}\text{Town} \\ \text{A} \quad \text{B}\end{matrix}}{\begin{bmatrix} X & X \\ X & X \\ X & X \end{bmatrix}}$$

If $W_i$ is the lorry/town matrix for week $i$, and

$$W_1 = \begin{bmatrix} 4 & 0 \\ 0 & 0 \\ 2 & 3 \end{bmatrix}; \quad W_2 = \begin{bmatrix} 3 & 1 \\ 2 & 1 \\ 0 & 2 \end{bmatrix}; \quad W_3 = \begin{bmatrix} 2 & 0 \\ 2 & 4 \\ 2 & 0 \end{bmatrix}$$

find the value of, and describe in words the meaning of:

a) $W_{1+2}$ (where $W_{1+2} = W_1 + W_2$) b) $W_{1+2+3}$ c) $W_{1+2+3} - W_2$

*Solution*

a) $W_{1+2} = W_1 + W_2 = \begin{bmatrix} 4 & 0 \\ 0 & 0 \\ 2 & 3 \end{bmatrix} + \begin{bmatrix} 3 & 1 \\ 2 & 1 \\ 0 & 2 \end{bmatrix}$

$$= \begin{bmatrix} 7 & 1 \\ 2 & 1 \\ 2 & 5 \end{bmatrix}$$

which is the number of trips made in weeks 1 and 2 combined.

b) $W_{1+2+3} = W_{1+2} + W_3 = \begin{bmatrix} 7 & 1 \\ 2 & 1 \\ 2 & 5 \end{bmatrix} + \begin{bmatrix} 2 & 0 \\ 2 & 4 \\ 2 & 0 \end{bmatrix}$

$$= \begin{bmatrix} 9 & 1 \\ 4 & 5 \\ 4 & 5 \end{bmatrix}$$

which is the number of trips made in the first 3 weeks combined.

c) $W_{1+2+3} - W_2 = \begin{bmatrix} 9 & 1 \\ 4 & 5 \\ 4 & 5 \end{bmatrix} - \begin{bmatrix} 3 & 1 \\ 2 & 1 \\ 0 & 2 \end{bmatrix}$

$$= \begin{bmatrix} 6 & 0 \\ 2 & 4 \\ 4 & 3 \end{bmatrix}$$

which is the number of trips made in weeks 1 and 3 combined.

Note that the result of $W_{1+2+3} - W_2$ would have given exactly the same result as $W_1 + W_3$.

## 9. Multiplying a matrix by a number

Sometimes it is necessary to multiply a complete set (or matrix) of values by some number. For example, suppose a salesman's daily expenses are standardised and split up into Travelling, Food and Other. These could be represented as a single row matrix with three elements. Multiplying each element of the matrix by 5 will give a new matrix, representing standard weekly expenses.

Thus, a matrix can be multiplied by any number with the effect that each element of the matrix is multiplied by that number.

A simple example is given now, where it should be noted that often a 'star' (*) is used to signify multiplication when matrices are involved.

If $A=\begin{pmatrix} 4 & 3 \\ 7 & 11 \end{pmatrix}$, then: $4^*A = 4^*\begin{pmatrix} 4 & 3 \\ 7 & 11 \end{pmatrix} = \begin{pmatrix} 4\times4 & 4\times3 \\ 4\times7 & 4\times11 \end{pmatrix} = \begin{pmatrix} 16 & 12 \\ 28 & 44 \end{pmatrix}$

Notice that exactly the same result would have been obtained by adding matrix A up four times. That is, $A + A + A + A = 4A$.

However, matrices can also be multiplied by numbers other than whole numbers, and an example follows, giving a practical business interpretation.

## 10. Example 3 (Multiplying a matrix by a number)

The following matrix, X, gives the price of comparable articles by size and type (in £).

$$X = \begin{matrix} \\ A \\ B \\ C \end{matrix} \begin{matrix} \text{Small} & \text{Large} \\ \begin{bmatrix} 0.40 & 0.58 \\ 0.32 & 0.56 \\ 0.42 & 0.60 \end{bmatrix} \end{matrix}$$

*If all prices were to increase by 50%*, the new price matrix, Y, could be represented as:

$$Y = (1.5)^* \begin{bmatrix} 0.40 & 0.58 \\ 0.32 & 0.56 \\ 0.42 & 0.60 \end{bmatrix} = \begin{bmatrix} 1.5\times0.40 & 1.5\times0.58 \\ 1.5\times0.32 & 1.5\times0.56 \\ 1.5\times0.42 & 1.5\times0.60 \end{bmatrix}$$

i.e. $$Y = \begin{bmatrix} 0.60 & 0.87 \\ 0.48 & 0.84 \\ 0.63 & 0.90 \end{bmatrix}$$

## 11. Multiplying matrices together

Both the need for and the technique used in multiplying two matrices together is demonstrated in the following text.

Suppose matrix X is used to hold the stock balances of items A, B and C and matrix Y to hold the unit cost (£) of each item as follows:

$$X = \begin{matrix} A & B & C \\ [12 & 23 & 7] \end{matrix} \qquad Y = \begin{matrix} A \\ B \\ C \end{matrix}\begin{bmatrix} 2.50 \\ 0.50 \\ 1.20 \end{bmatrix}$$

a natural question to ask is, 'is there a way to multiply the two matrices together to obtain the total costs associated with all the items held'?

The answer is 'yes'!

Since there are 12 $A$ items @ £2.50 each, the total cost associated with the $A$ items is 12 × £2.50 = £30.00. Similarly, the total cost associated with the other two items can be calculated.

That is, total cost (in £) associated with all items = 12(2.50) + 23(0.50) + 7(1.20) = 49.90.

In other words:

$$X^*Y = \begin{bmatrix} 12 & 23 & 7 \end{bmatrix} * \begin{bmatrix} 2.50 \\ 0.50 \\ 1.20 \end{bmatrix} = 12(2.50) + 23(0.50) + 7(1.20) = 49.90.$$

In the above matrix multiplication, a row (matrix) was 'multiplying' a column (matrix) only. For matrices that have a number of rows and columns however, the same principle as above can be used in multiplication. That is, each *row* of the first matrix 'multiplies' each *column* of the second matrix.

The steps involved in general matrix multiplication are carefully laid out in the following sections.

## 12. Factors involved in matrix multiplication

Suppose that matrix $A$ is to be multiplied by matrix $B$ to obtain matrix $C$. Then we write $C = A^*B$. The condition necessary for the multiplication of $A$ and $B$ to be allowed and the procedure involved in the calculation of $C$ are now listed.

a) **Condition.**

Each row of $A$ must have the same number of elements as each column of B. This is equivalent to $A$ having the same number of columns as $B$ has rows.

For example, suppose $A$ is a 4 × 2 matrix and $B$ is a 2 × 3 matrix. Consider:

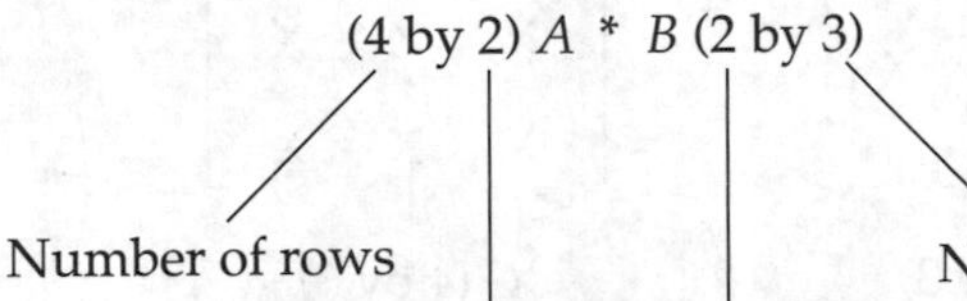


This means that $A^*B$ is *suitable* for multiplication. Notice however that $B^*A$ is *not suitable* for multiplication, since:

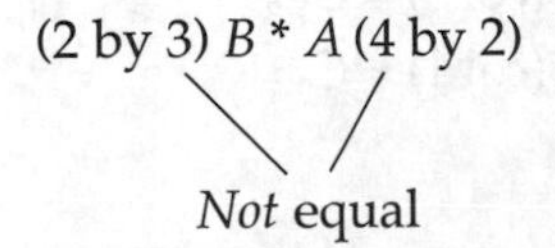

b) **Procedure.** Each row of $A$ must be 'multiplied' by each column of $B$ to give a *single element* of C.

For example, if $\begin{bmatrix} 6 & 5 \end{bmatrix}$ is a row of A and $\begin{bmatrix} 8 \\ 1 \end{bmatrix}$ is a column of B, the 'multiplication' is carried out as:

multiply $6 \times 8 = 48$

$$\begin{bmatrix} 6 & 5 \end{bmatrix} * \begin{bmatrix} 8 \\ 1 \end{bmatrix}$$ then add $48 + 5 = 53$

multiply $5 \times 1 = 5$

This procedure is shown in context next, together with some detail concerning the resultant matrix C = $A$*$B$.

## 13. Pictorial demonstration of matrix multiplication

This section demonstrates the procedure involved in multiplying row 1 of matrix $A$ (a 4 by 2 matrix) by column 1 of $B$ (a 2 by 3 matrix) to give the value of the cell at the intersection of row 1 and column 1 of matrix C = $A$*$B$.

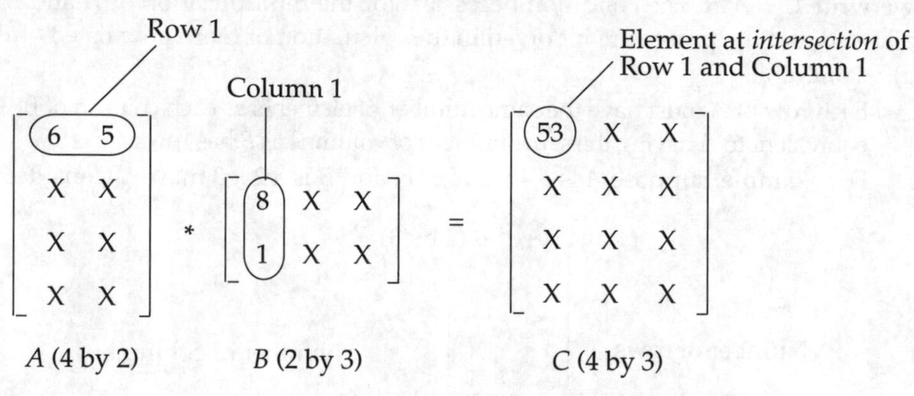


The multiplication shown above is $(6 \times 8) + (5 \times 1) = 48 + 5 = 53$.

i.e. $A(\text{row } 1) * B(\text{col } 1) = C_{1,1}$

Similarly $A(\text{row } 1) * B(\text{col } 2) = C_{1,2}$

$A(\text{row } 1) * B(\text{col } 3) = C_{1,3}$

$A(\text{row } 2) * B(\text{col } 1) = C_{2,1}$

...

...

and up to $A(\text{row } 4) * B(\text{col } 3) = C_{4,3}$

Notice that the resultant matrix C has the *same number of rows* as A and the *same number of columns* as B.

## 14. The matrix multiplication size diagram

In order to check:

a) whether two matrices are compatible for multiplication in terms of their size, and

b) (if they *are* compatible), the size of the resultant matrix,

it is very useful to bring the following diagram to mind, which demonstrates the matrix multiplication given in section 12.

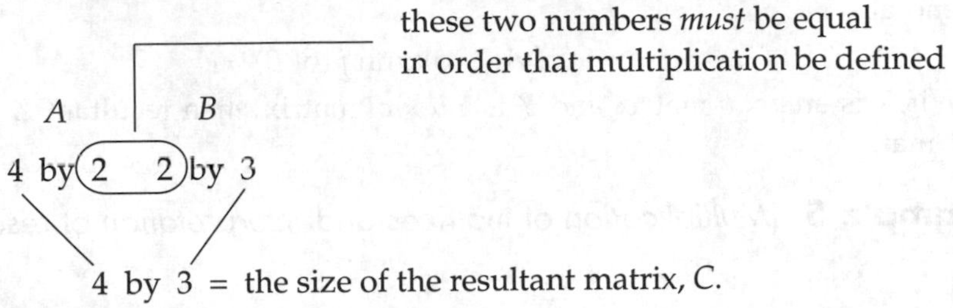

```
                                 these two numbers must be equal
                  ┌───────────── in order that multiplication be defined
        A         B
    4 by (2      2) by 3
        \          /
     4 by 3 = the size of the resultant matrix, C.
```

## 15. Example 4 (matrix multiplication)

*Question*

Multiply $X = \begin{Bmatrix} 2 & 4 & 2 \\ 1 & 2 & 3 \end{Bmatrix}$ by $Y = \begin{Bmatrix} 4 \\ 2 \\ 6 \end{Bmatrix}$ to give matrix $Z$.

*Answer*

Notice immediately that $X$ is $2 \times 3$ and $Y$ is $3 \times 1$. Thus $X^*Y$ is defined (since $X$ has the same number of columns as $Y$ has rows, namely 3) and the resultant matrix, $Z$, will be $2 \times 1$.

$$Z = \begin{Bmatrix} 2 & 4 & 2 \\ 1 & 2 & 3 \end{Bmatrix} * \begin{Bmatrix} 4 \\ 2 \\ 6 \end{Bmatrix} = \begin{Bmatrix} (2\times4)+(4\times2)+(2\times6) \\ (1\times4)+(2\times2)+(3\times6) \end{Bmatrix} = \begin{Bmatrix} 28 \\ 26 \end{Bmatrix}$$

The definition of and procedure for matrix multiplication is now summarised and followed by another example.

## 16. Condition and procedure for matrix multiplication

> **Matrix multiplication**
> Two matrices can be multiplied from left to right on the condition that the number of columns in the left-hand matrix equals the number of rows in the right-hand matrix.

Matrix multiplication is performed by each row of the left-hand matrix in turn multiplying each column of the right-hand matrix.

The procedure for multiplying row $i$ by column $j$ is:

STEP 1 Row $i$ is superimposed on column $j$.

STEP 2 Each respective pair of superimposed cells are multiplied together.

STEP 3 The multiples are added to form a single value.

STEP 4 This value is entered into the cell of the resultant matrix at the intersection of the $i$-th row and $j$-th column.

That is, if matrix $A$ is to multiply matrix $B$ to give the resultant matrix $C$, then, in general:

$$\text{row } i \text{ (of } A) * \text{column } j \text{ (of } B) = c_{i,j}$$

Also, if $X$ is an $a \times b$ matrix and $Y$ is a $b \times c$ matrix, then resultant $Z = X^*Y$ is an $a \times c$ matrix.

## 17. Example 5 (*Multiplication* of matrices and *interpretation* of result)

*Question*

A firm keeps details of component parts used in the make-up of each product (matrix X) and products made on each day of the week (matrix Y) as follows:

$$\begin{array}{c} \\ \text{Mon} \\ \text{Tue} \\ \text{Wed} \\ \text{Thu} \\ \text{Fri} \end{array} \overset{\textit{Products}}{\begin{array}{cc} 1 & 2 \end{array}} \begin{bmatrix} 0 & 1 \\ 2 & 2 \\ 3 & 2 \\ 1 & 1 \\ 1 & 0 \end{bmatrix} = Y \qquad \text{Products} \begin{array}{c} 1 \\ 2 \end{array} \overset{\textit{Parts}}{\overset{A \ B \ C}{\begin{bmatrix} 3 & 2 & 1 \\ 1 & 4 & 2 \end{bmatrix}}} = X$$

Using matrix multiplication, find a matrix that describes the number of component parts used on each day of the week.

*Answer*

There are two ways to multiply the two matrices $X$ and $Y$. Either (i) $X^*Y$ or (ii) $Y^*X$

$X^*Y$ is $(2 \times 3) * (5 \times 2)$, which means that multiplication is NOT possible since 3 does not match 5.

However, $Y^*X$ is: (5 by 2) * (2 by 3) which ARE compatible

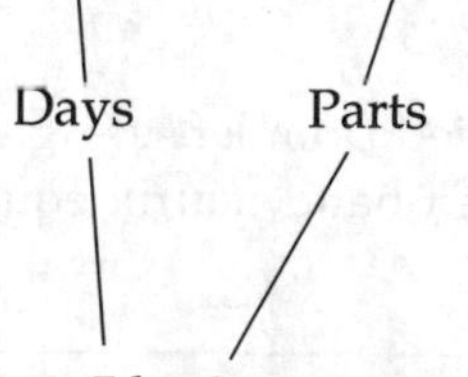


Thus, $Y^*X$ will be a 5 by 3 matrix

$$Y*X = \begin{bmatrix} 0 & 1 \\ 2 & 2 \\ 3 & 2 \\ 1 & 1 \\ 1 & 0 \end{bmatrix} * \begin{bmatrix} 3 & 2 & 1 \\ 1 & 4 & 2 \end{bmatrix} = \begin{bmatrix} (0\times3)+(1\times1) & (0\times2)+(1\times4) & (0\times1)+(1\times2) \\ (2\times3)+(2\times1) & (2\times2)+(2\times4) & (2\times1)+(2\times2) \\ (3\times3)+(2\times1) & (3\times2)+(2\times4) & (3\times1)+(2\times2) \\ (1\times3)+(1\times1) & (1\times2)+(1\times4) & (1\times1)+(1\times2) \\ (1\times3)+(0\times1) & (1\times2)+(0\times4) & (1\times1)+(0\times2) \end{bmatrix}$$

*Parts*

| | A | B | C |
|---|---|---|---|
| Mon | 1 | 4 | 2 |
| Tue | 8 | 12 | 6 |
| Wed | 11 | 14 | 7 |
| Thu | 4 | 6 | 3 |
| Fri | 3 | 2 | 1 |

## 18. Additional features of matrices

A particular use of matrices is in the solving of simultaneous equations. However, in order to cover this application, it is necessary to describe three more important features of matrices.

a) The *unit matrix* is a *square* matrix whose elements have value 0 except for those on the main diagonal (top left to bottom right) which have value 1. It is usually denoted by I.

$$I_2 = \begin{bmatrix} 1 & 0 \\ 0 & 1 \end{bmatrix} \qquad I_3 = \begin{bmatrix} 1 & 0 & 0 \\ 0 & 1 & 0 \\ 0 & 0 & 1 \end{bmatrix}$$

The 2 *by* 2 *unit matrix* The 3 *by* 3 *unit matrix*

b) The *determinant* of a (2 by 2) matrix $A$ is denoted by $|A|$ or Det $A$ and is a *number*, defined as follows.

If $A = \begin{bmatrix} a & b \\ c & d \end{bmatrix}$ then $|A| = ad - bc$ (a number).

For example, if $A = \begin{bmatrix} 3 & 4 \\ 1 & 2 \end{bmatrix}$

then $|A| = (3)(2) - (1)(4) = 6 - 4 = 2$

and if $C = \begin{pmatrix} 2 & 6 \\ 1 & 4 \end{pmatrix}$

then $|C| = (2)(4) - (-1)(6) = 8 + 6 = 14$.

c) The *inverse* of a *square* matrix $A$ is denoted by $A^{-1}$ and satisfies the relationship: $A * A^{-1} = A^{-1} * A = I_2$ (the 2 by 2 unit matrix).

In other words, if $A = \begin{bmatrix} 3 & 4 \\ 1 & 2 \end{bmatrix}$, then $A^{-1}$ is the matrix below having elements as '?' and which satisfies the relationship:

$$\begin{bmatrix} a & b \\ c & d \end{bmatrix} * \begin{bmatrix} ? & ? \\ ? & ? \end{bmatrix} = \begin{bmatrix} ? & ? \\ ? & ? \end{bmatrix} * \begin{bmatrix} a & b \\ c & d \end{bmatrix} = \begin{bmatrix} 1 & 0 \\ 0 & 1 \end{bmatrix}$$

The inverse of a 2 by 2 matrix can be specifically calculated as follows.

If $A = \begin{bmatrix} a & b \\ c & d \end{bmatrix}$, then $A^{-1} = \dfrac{1}{|A|}\begin{bmatrix} d & -b \\ -c & a \end{bmatrix}$

## 19. Example 6 (The *inverse* of a 2 × 2 matrix)

*Question*

Find the inverse of matrix $A = \begin{Bmatrix} 4 & 5 \\ 2 & 3 \end{Bmatrix}$ and check the result, using the relationship

$A * A^{-1} = \begin{Bmatrix} 1 & 0 \\ 0 & 1 \end{Bmatrix}$

*Answer*

Firstly, $\text{Det}\begin{Bmatrix} 4 & 5 \\ 2 & 3 \end{Bmatrix} = (4 \times 3) - (5 \times 2) = 12 - 10 = 2.$

i.e. $\text{Det } A = |A| = 2.$

Thus: $A^{-1} = \begin{Bmatrix} 4 & 5 \\ 2 & 3 \end{Bmatrix}^{-1} = \dfrac{1}{|A|}\begin{Bmatrix} 3 & -5 \\ -2 & 4 \end{Bmatrix} = \dfrac{1}{2}\begin{Bmatrix} 3 & -5 \\ -2 & 4 \end{Bmatrix} = \begin{Bmatrix} \frac{3}{2} & \frac{-5}{2} \\ \frac{-2}{2} & \frac{4}{2} \end{Bmatrix} = \begin{Bmatrix} \frac{3}{2} & \frac{-5}{2} \\ -1 & 2 \end{Bmatrix}$

*For the check*: $A * A^{-1} = \begin{Bmatrix} 4 & 5 \\ 2 & 3 \end{Bmatrix} * \begin{Bmatrix} \frac{3}{2} & \frac{-5}{2} \\ -1 & 2 \end{Bmatrix} = \begin{Bmatrix} 4(\frac{3}{2}) + 5(-1) & 4(\frac{-5}{2}) + 5(2) \\ 2(\frac{3}{2}) + 3(-1) & 2(\frac{-5}{2}) + 3(2) \end{Bmatrix}$

$= \begin{Bmatrix} 6-5 & -10+10 \\ 3-3 & -5+6 \end{Bmatrix} = \begin{Bmatrix} 1 & 0 \\ 0 & 1 \end{Bmatrix}$

## 20. Solving simultaneous equations

As already mentioned, the information given in the last few sections is necessary in order to solve simultaneous equations. Although matrices can be used to solve any size of systems of simultaneous equations (i.e. 2 × 2, 3 × 3, 4 × 4 etc), for the purposes of this manual only 2 × 2 equations are considered. The technique is described in the following procedure.

## 21. Solving 2 by 2 simultaneous equations using matrices

The procedure for solving the 2 by 2 simultaneous equations

$$\begin{cases} ax + by = c \\ dx + ey = f \end{cases}$$

is now given.

STEP 1 Write the equation in the matrix form: $\begin{pmatrix} a & b \\ d & e \end{pmatrix} * \begin{pmatrix} x \\ y \end{pmatrix} = \begin{pmatrix} c \\ f \end{pmatrix}$

(**Note**: The above matrix equation is exactly equivalent to the original pair of simultaneous equations. A useful and practical exercise is to prove it!)

STEP 2 Solve for $x$ and $y$ using: $\begin{pmatrix} x \\ y \end{pmatrix} = \begin{pmatrix} a & b \\ d & e \end{pmatrix}^{-1} * \begin{pmatrix} c \\ f \end{pmatrix}$

The right–hand side of the above equation will reduce to a 2 by 1 matrix, containing the values of $x$ and $y$.

## 22. Example 7 (Solving 2x2 *simultaneous equations* using *matrices*)

*Question*

Solve the simultaneous equations: $5x + 9y = -30$
$6x - 2y = 28$

using matrices.

*Answer*

STEP 1 Putting the equations into matrix form gives:

$$\begin{bmatrix} 5 & 9 \\ 6 & -2 \end{bmatrix} * \begin{bmatrix} x \\ y \end{bmatrix} = \begin{bmatrix} -30 \\ 28 \end{bmatrix}$$

STEP 2 $$\begin{bmatrix} x \\ y \end{bmatrix} = \begin{bmatrix} 5 & 9 \\ 6 & -2 \end{bmatrix}^{-1} * \begin{bmatrix} -30 \\ 28 \end{bmatrix} = \frac{1}{5(-2)-9(6)} \begin{bmatrix} -2 & -9 \\ -6 & 5 \end{bmatrix} * \begin{bmatrix} -30 \\ 28 \end{bmatrix} = \frac{1}{-64} \begin{bmatrix} -192 \\ 320 \end{bmatrix}$$

i.e. $\begin{bmatrix} x \\ y \end{bmatrix} = \begin{bmatrix} 3 \\ -5 \end{bmatrix}$

Thus the required solutions are: $x = 3$ and $y = -5$.

## 23. Transition matrices

Suppose that a construction company divides its working days up into only 'wet' and 'not wet' and, for a period of 25 working days, recorded the weather conditions to obtain the following (W=wet; N=not wet):

NWNNN WWNNW NWWNN NWNNN NWNNN.

If the company is particularly interested in the way that one day's weather *might affect* the weather on the next day, they can organise the above data into tabular form by describing the number of times, for instance, a wet day followed a wet day (pattern WW, which occurred twice) or the number of times a dry day followed a wet day (WN, occurring 6 times).

This information follows and is known as a *transition matrix*.

$$\begin{array}{ll} & \textit{Weather conditions} \\ & \textit{on the following day} \\ & \quad\; \text{N} \quad \text{W} \\ \begin{array}{r}\textit{Weather conditions} \\ \textit{on a particular day}\end{array} & \begin{array}{c}\text{N} \\ \text{W}\end{array}\begin{pmatrix} 10 & 6 \\ 6 & 2 \end{pmatrix} \end{array}$$

In words, on 10 occasions the weather kept 'not wet' from one day to the next, while on 6 occasions it changed from 'not wet' to 'wet'. It changed from 'wet' to 'not wet' on 6 occasions also and on only 2 occasions did it stay 'wet' from one day to the next.

The two alternatives, *N* and *W*, are technically known as *states*.

The above matrix can be changed into a matrix of *proportions* (or *probabilities*) by dividing each element by its row total, as follows:

$$\begin{array}{ll} & \textit{Weather conditions} \\ & \textit{on the following day} \\ & \quad\; \text{N} \quad \text{W} \\ \begin{array}{r}\textit{Weather conditions} \\ \textit{on a particular day}\end{array} & \begin{array}{c}\text{N} \\ \text{W}\end{array}\begin{pmatrix} \frac{10}{16} & \frac{6}{16} \\ \frac{6}{8} & \frac{2}{8} \end{pmatrix} \end{array}$$

The cell in the top left is translated as: the proportion of days that changed from 'not wet' on one day to 'not wet' on the next $= \Pr(\text{N} \rightarrow \text{N}) = \dfrac{10}{16}$

Similarly: $\Pr(\text{N} \rightarrow \text{W}) = \dfrac{6}{16}$; $\Pr(\text{W} \rightarrow \text{N}) = \dfrac{6}{8}$; $\Pr(\text{W} \rightarrow \text{W}) = \dfrac{2}{8}$

A proportional transition matrix gives the proportion of times that states (in the above case either wet or not wet) change in a particular defined period of time.

## 24. Definition of a transition matrix

> A **transition matrix** is a square matrix which gives the number or proportion of times that some process changed from one state to another during some period.
>
> The states and time period involved must be specified.

**Notes.** 1. A proportional transition matrix has the sum of the proportions in each row adding to 1.

2. If $T$ is some transition matrix, then $T^*T = T^2$ is another transition matrix with proportions relevant to changes in states over 2 time periods. Similarly, $T^*T^*T = T^3$ is a transition matrix with proportions relevant to changes in states over 3 time periods, and so on.

## 25 Example 8 (Formation of a *transition matrix*)

*Question*

A firm has two lorries and keeps records of the number that are serviceable for a whole day over 42 successive days for a particular period as follows (reading from left to right):

222 121 011 212 221 201 122

122 211 012 122 201 002 122

a) Obtain the transition matrix, $T$, for the proportions of 0, 1 or 2 lorries being serviceable on the following day, *given* the number (0, 1 or 2) serviceable on one particular day.

b) Find $T^2$ also, and use this to find, *if there is one serviceable lorry today*, the proportion of times that there will be at least one serviceable lorry in two days time.

*Answer*

a) Reading the data from left to right gives: 2 → 2; 2 → 2; 2 → 1; 1 → 2; 2 → 1; ...etc leading to the following numbers of changes of numbers of serviceable lorries from one day to the next:

| | | Number of lorries serviceable the following day | | | |
|---|---|---|---|---|---|
| | | 0 | 1 | 2 | *Total* |
| *Number of lorries serviceable on a particular day* | 0 | 1 | 4 | 1 | 6 |
| | 1 | 3 | 3 | 9 | 15 |
| | 2 | 2 | 8 | 10 | 20 |

Writing each value as a proportion of the row total gives the following transition matrix:

$$T = \begin{matrix} 0 \\ 1 \\ 2 \end{matrix} \overset{\begin{matrix} 0 & 1 & 2 \end{matrix}}{\begin{bmatrix} \frac{1}{6} & \frac{4}{6} & \frac{1}{6} \\ \frac{3}{15} & \frac{3}{15} & \frac{9}{15} \\ \frac{2}{20} & \frac{8}{20} & \frac{10}{20} \end{bmatrix}} = \begin{bmatrix} 0.167 & 0.667 & 0.167 \\ 0.200 & 0.200 & 0.600 \\ 0.100 & 0.400 & 0.500 \end{bmatrix}$$

b)

$$T^2 = T^*T = \begin{bmatrix} 0.167 & 0.667 & 0.167 \\ 0.200 & 0.200 & 0.600 \\ 0.100 & 0.400 & 0.500 \end{bmatrix} * \begin{bmatrix} 0.167 & 0.667 & 0.167 \\ 0.200 & 0.200 & 0.600 \\ 0.100 & 0.400 & 0.500 \end{bmatrix}$$

i.e.

$$T^2 = \begin{matrix} 0 \\ 1 \\ 2 \end{matrix} \overset{\begin{matrix} 0 & 1 & 2 \end{matrix}}{\begin{bmatrix} 0.178 & 0.311 & 0.511 \\ 0.133 & 0.413 & 0.454 \\ 0.147 & 0.346 & 0.507 \end{bmatrix}}$$

(Note that the sum of the proportions in each row adds to 1, as required for a transition matrix.)

The proportion of times that there will be at least one serviceable lorry in two days time (if there is one serviceable lorry today), is equivalent to:

Pr(1 → 1 or 1 → 2) after two transitions = 0.413 + 0.454 (from row 2) = 0.867

## 26. Summary

a) A matrix is a table of values conveying numerical information. The data items held within it are sometimes referred to as elements or cells.

b) Any cell of matrix $X$ can be referenced uniquely using the notation $x_{i,j}$, which is translated as 'the cell lying at the intersection of row $i$ and column $j$'.

c) The size of a matrix is determined by its number of rows and columns.

   i. A matrix having $a$ rows and $b$ columns is said to have size $a$ by $b$.

   ii. A *square* matrix is one having the same number of rows as columns.

   iii. A matrix having just one column is sometimes called a *column matrix*. A *row matrix* contains only one row.

d) Matrix addition and subtraction.

   i. Two matrices can be added or subtracted only if they are of identical size.

   ii. Two matrices are added by adding corresponding elements.

   iii. One matrix can be subtracted from another by subtracting corresponding elements.

e) If $A$ is a matrix and $b$ is any number, $b*A$ means 'multiply each element of matrix $A$ by the number $b$'.

f) Matrix multiplication.

   i. Two matrices can be multiplied (from left to right) only if the number of columns of the left matrix = the number of rows of the right matrix.

   ii. If two matrices ($A$ and $B$) are multiplied together, the resultant matrix has the same number of rows as $A$ and the same number of columns as $B$.

   iii. The elements of the resultant matrix are obtained by 'multiplying' each row of the first matrix by each column of the second matrix.

   iv. If $C = A*B$, then $c_{i,j}$ = row $i$ (of $A$) × column $j$ (of $B$).

g) A *unit matrix* is a square matrix whose elements have value 0 except for those on the main diagonal (top left to bottom right) which have value 1. It is usually denoted by $I$.

h) The *determinant* of matrix $A = \begin{bmatrix} a & b \\ c & d \end{bmatrix}$ is written as Det $A$ or $|A|$ and is a number having value $ad-bc$

i) The inverse of a matrix $A$ is written as $A^{-1}$ and satisfies the relationship:

$$A * A^{-1} = A^{-1} * A = I_2 \text{ (the 2 by 2 unit matrix).}$$

j) Solving 2 × 2 simultaneous equations.

The simultaneous equations: $\begin{cases} ax + by = c \\ dx + ey = f \end{cases}$ can be solved using matrices by:

   i. writing the equations in the matrix form: $\begin{pmatrix} a & b \\ d & e \end{pmatrix} * \begin{pmatrix} x \\ y \end{pmatrix} = \begin{pmatrix} c \\ f \end{pmatrix}$

ii. and solving using: $\begin{pmatrix} x \\ y \end{pmatrix} = \begin{pmatrix} a & b \\ d & e \end{pmatrix}^{-1} * \begin{pmatrix} c \\ f \end{pmatrix}$

k) A *transition matrix* is a square matrix which gives the number or proportion of times that some process will change from one state to another in a defined period of time. The sum of the proportions in each row must add to 1.

## 27. Points to note

a) Sometimes row and column matrices are known as *vectors*.
b) Division of one matrix by another is *not possible*.
c) When multiplying two numbers together, it does not matter which order the numbers are multiplied, i.e. 4 × 5 = 5 × 4 = 20. *This is not the case with matrices.* That is, if $A$ and $B$ are any two matrices, $A*B$ does not normally equal $B*A$, even if both $A*B$ and $B*A$ are defined for multiplication.
d) The solution of 3 × 3, 4 × 4, etc simultaneous equations can be performed using matrices, but involves a level of complexity not thought suitable for the syllabuses that this manual covers.

## 28. Student self review questions

1. What is a matrix? [2]
2. How is the cell at the intersection of row 3 and column 1 of matrix $D$ referenced uniquely? [3]
3. What precisely is meant by the size of a matrix? [4,5]
4. What condition must be met before two matrices can be added together and how is addition performed? [6,7]
5. What condition must be satisfied before two matrices can be multiplied together? [11]
6. If an $a$ by $b$ matrix is multiplied by a $c$ by $d$ matrix, what is the size of the resultant matrix? [13,14,16]
7. Explain how to calculate the determinant of a 2 by 2 matrix. [18(b)]
8. How is the inverse of a 2 by 2 matrix determined and what relationship must it have with the 2 by 2 unit matrix? [18(a)(c)]
9. How are the simultaneous equations $ax + by = c; dx + ey = f$ written in their equivalent matrix form? [21]
10. What form do the elements of a proportional transition matrix take and how is the matrix used and manipulated? [23,24]

## 29. Student exercises

1. If $M = \begin{pmatrix} 2 & 3 \\ -1 & 4 \\ 8 & 1 \end{pmatrix}$ and $N = \begin{pmatrix} 2 & 8 & 2 \\ 1 & 2 & -2 \\ 3 & 4 & -6 \end{pmatrix}$, identify the following elements:

a) $m_{3,2}$ b) $m_{1,2}$ c) $n_{2,2}$ d) $n_{2,3}$

2. Perform the following matrix arithmetic:

a) $\begin{bmatrix} 2 & 1 \\ 4 & 1 \end{bmatrix} + \begin{bmatrix} 8 & 2 \\ 1 & -3 \end{bmatrix}$  b) $\begin{bmatrix} 1 & 4 \\ 3 & 2 \\ 4 & 6 \end{bmatrix} - \begin{bmatrix} 0 & 2 \\ 1 & 2 \\ 5 & 1 \end{bmatrix}$

c) If $A = \begin{pmatrix} 1 & 1 & 2 \\ 2 & 4 & 1 \end{pmatrix}$ and $B = \begin{pmatrix} 1 & 2 & -1 \\ 4 & 2 & -1 \end{pmatrix}$, identify the matrix $C = 3A - 2B$.

3. If $A = \begin{bmatrix} 1 & 1 & 2 \\ 2 & 4 & 1 \end{bmatrix}$, $B = \begin{bmatrix} 2 \\ 4 \\ 1 \end{bmatrix}$ and $C = \begin{bmatrix} 3 & 1 & 2 \end{bmatrix}$

evaluate any of the following matrices that are possible:

$A^*B, A^*C, B^*C, B^*A, C^*A, C^*B$ and $A^*B^*C$.

4. For the matrices *defined in exercise* 3, *A* is describing the numbers of each of three machines owned by each of two firms; *B* is describing the standard running costs per hour of each of the three machines; C is describing the number of hours of use needed on each machine to satisfy a particular job. For each of the four matrix multiplications that were possible (in the answer to the previous exercise), state in words the meaning of the matrix obtained or state whether it has no practical meaning.

5. Three jobs, A, B and C, require a succession of different processes (J, K and L) in a particular sequence in order to complete. Job A requires LKLK; job B requires LJL; job C requires JKLJKLJ.

a) Write down the 3 × 3 matrix X, which describes the number of times each job is in each process.

Each separate process normally requires setting-up (S), machining (M) and finishing (F) stages. However, there is no S time needed for L, and K requires no F time. Processes J and K both need 10 minutes for S, and J requires one hour for M and 10 minutes for F. L takes 20 minutes for F and M takes up half an hour for K and 10 minutes for L.

b) Write down a 3 × 3 matrix *Y*, which describes the time (in mins) for each process to complete each stage.

c) Evaluate the matrix $X^*Y$ (if necessary re-arrange the rows and columns of one of them to ensure they are compatible for multiplication).

d) Which job takes the longest time to complete and how long does it take?

The cost (in £/minute) for S is 0.10, for M, 0.40 and for F, 0.05.

e) If *Z* is the cost matrix for each stage, evaluate either $X^*Y^*Z$ or $Z^*X^*Y$ (depending on which is compatible for multiplication), and state the meaning of the result.

6. If $A = \begin{pmatrix} 2 & 1 \\ 1 & 2 \end{pmatrix}$ and $B = \begin{pmatrix} 3 & 2 \\ 4 & -1 \end{pmatrix}$ find: a) $|A|$ b) $|B|$ c) $|A+B|$ d) $|A^*B|$

7. Using matrices *A* and *B* from Question 6, find:

a) $A^{-1}$ b) $B^{-1}$ c) $(A+B)^{-1}$ d) $(A^*B)^{-1}$

8. Solve the following simultaneous equations using a matrix method. Check your results.

   a) $4x+2y=8;\ 3x+y=5$ b) $3x-y=0;\ 4x+2y=20$
   c) $p+q=1;\ q-2p=7$ d) $3x+2y-5.4;\ 2x\ \ y=-1.3$

9. Matrix $M$ below describes the number of items A, B and C stored at warehouses X and Y and matrix $N$ describes the cost (in pence per day) for storing (S) and maintaining (M) the items.

$$M = \begin{array}{c} \\ X \\ Y \end{array} \begin{array}{c} \begin{array}{ccc} A & B & C \end{array} \\ \begin{bmatrix} 14 & 12 & 41 \\ 52 & 0 & 60 \end{bmatrix} \end{array} \qquad N = \begin{array}{c} \\ A \\ B \\ C \end{array} \begin{array}{c} \begin{array}{cc} S & M \end{array} \\ \begin{bmatrix} 2 & 0.5 \\ 2 & 0.5 \\ 1.5 & 1.5 \end{bmatrix} \end{array}$$

   a) Evaluate the matrix $3(M^*N)$ and interpret its meaning.

   Now assume:

   i. that matrix $M$ shows the stock position at the beginning of day 1
   ii. the following stock movements occur:
   Withdrawals on day 2: 2B from warehouse X; 23A from warehouse Y.
   Deliveries on day 3: 8B and 12C to warehouse X; 12B to warehouse Y.
   iii. full costs for a day are incurred if stock is held for any part of a day

   b) Evaluate the 2 × 3 matrix $T$ as the stock (cost-effective) movements on day 3, using *all* the information given in (ii).
   c) Write down (but do not evaluate) a matrix expression to describe the storage and maintenance costs of items A, B and C at warehouses X and Y for the period day 1 to day 5 inclusive.

10. Product $x$ has fixed costs of £60 and variable costs of £6 per product; product y has fixed costs of £35 and variable costs of £8 per product.

   If $C = \begin{bmatrix} 60 & 6 \\ 35 & 8 \end{bmatrix}$ and $Q = \begin{bmatrix} 1 \\ q \end{bmatrix}$ where C is the *cost coefficient* matrix:

   a) evaluate, and explain the significance of, the matrix product $C^*Q$.

   Given now that products $x$ and $y$ sell at £9 and £10 respectively:

   b) write down the *revenue coefficient* matrix, $R$, and
   c) evaluate the matrix $R^*Q - C^*Q$ and explain its significance.

   Putting $P = R^*Q - C^*Q$, $A = [1\ \ 0]$ and $B = [0\ \ 1]$,

   d) solve the matrix equation $A^*P = B^*P$ and interpret the value of $q$ obtained.

# 37 Inventory control

## 1. Introduction

This chapter takes an elementary look at inventory control. Initially, costs associated with inventories and terminology is considered, followed by inventory graphs. Re-order level and periodic review inventory control systems are described, together with two common inventory models – the 'basic' and 'gradual replenishment'. Inventory cost graphs are covered for the basic model.

## 2. The need for an inventory

A firm's inventory can be described as the totality of stocks of various kinds. These include basic raw materials, partly-finished goods and materials, sub-assemblies, office and workshop supplies and finished goods.

The reasons why an inventory must be carried by a firm are various, including:

a) anticipating normal demand;
b) taking advantages of bulk-purchase discounts;
c) meeting emergency shortages (due to industrial strikes for example);
d) as a natural part of the production process (cooling of metals, maturing of spirits etc);
e) absorbing wastages and unpredictable fluctuations;
f) obsolete items and over-ordering (due possibly to badly organised stock control).

## 3. Costs associated with inventories

Besides the obvious *buying costs* (i.e. the total price of the materials), the holding of stock necessarily involves what are known as inventory costs which can be broadly split into three categories.

a) *Ordering* (*or Replenishment*) *costs*. These involve transport, clerical and administrative costs associated with the physical movement of bought-in external goods. However, where the goods are manufactured internally to the organisation, there are alternative initial costs to be borne with each production run known as set-up costs.
b) *Holding* (*or Carrying*) *costs*. These include the following:
    i. Stores costs (staffing, equipment maintenance, handling);
    ii. Storage overheads (heat, light, rent etc);
    iii. Cost of capital tied up in inventory;
    iv. Insurance, security and pilferage;
    v. Deterioration or breakages.
c) *Stockout costs*. These are the costs (sometimes not easily quantifiable) associated with running out of stock. They include penalty payments, loss of goodwill (possibly affecting number and size of future orders), idle manpower and machines, etc. Of course one of the main reasons why inventories are held is to avoid just such costs.

## 4. Inventory control purpose

**Objective of inventory control**

The objective of inventory control is to maintain a system which will minimise total costs and establish:

a) the optimum amount of stock to be ordered

b) the period between orders.

## 5. Some terminology

Before proceeding further, there are certain terms used in inventory control which need to be defined briefly. Some of these will be described further in later sections.

a) *Lead time*. The time between ordering goods and their replenishment (i.e. physically ready for use). Note that orders may be internal (requiring a production run) or external.

b) *Economic ordering quantity* (*EOQ*). This is a specially derived value which gives the external order quantity that *minimises total inventory costs.*

c) *Economic batch quantity* (*EBQ*). This is an adapted EOQ which gives the size of the internal production run that *minimises total inventory costs.*

d) *Safety stock.* A term used to describe the stock held to cover possible deviations in demand or supply during the lead time. Sometimes known as *buffer* or *minimum* stock.

e) *Maximum stock.* A level used as an indicator above which stocks are too high.

f) *Reorder level.* A level of stock which, when reached, signals a replenishment order.

g) *Reorder quantity*. The level of a replenishment order. This will often be the EOQ or EBQ.

## 6. The inventory graph

Of primary importance in Inventory Control is knowledge, and through this, control, of the amount of stock held at any time. The purpose of an *inventory graph* is to give this knowledge in diagrammatic form by plotting the relationship between the quantity of stock held ($q$) and time ($t$).

Figure 1 shows a general inventory graph with various significant features labelled, with an initial inventory of 100 items which was replenished by a further 100 items continuously over a time period. For the next time period there was no activity, but at time point 2, 100 items were demanded, followed, over the next two time periods, by a continuous demand which used up the last 100 items. As soon as this stockout position was reached, a further 150 items were delivered.

*Figure 1 Inventory graph*

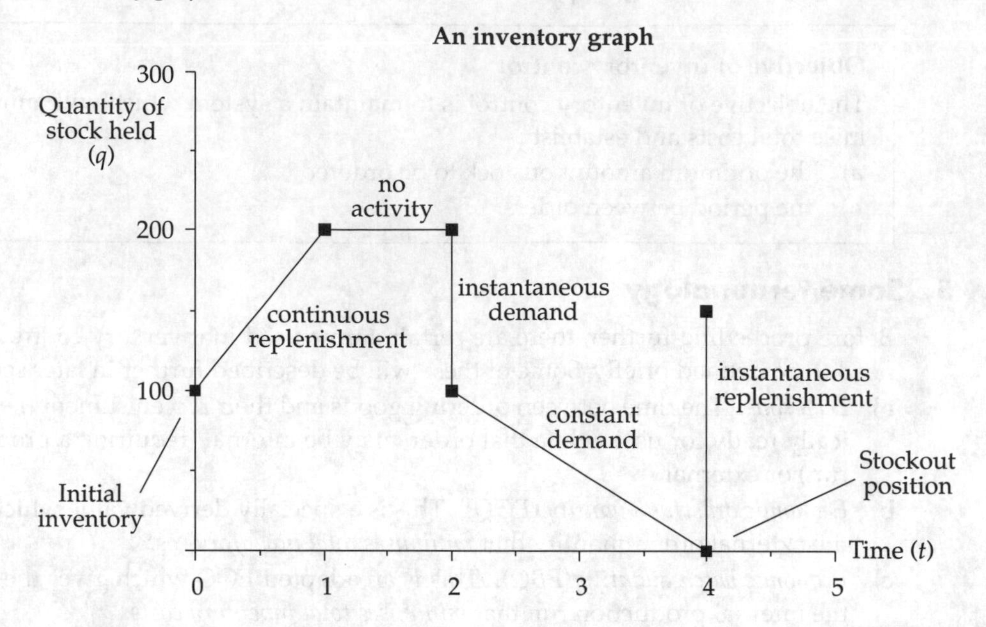


Notice the difference between continuous replenishment (or constant demand) and instantaneous replenishment (or demand) of stock items.

## 7. Example 1 (Drawing an inventory *graph*)

*Question*

A particular item of stock has an initial inventory of 600. A particular production line requires the items to be drawn (continuously) from stores at a steady rate of 200 per day. As soon as a stockout is reached, a batch of 600 items is moved in overnight from another source to replenish the inventory. Sketch the inventory graph for a period of 9 days.

*Answer*

Notice that stockout occurs after 3 days and thus we have a cyclic situation over the whole 9-day period. Figure 2 shows the graph for the information given.

## 8. Some further terminology

a) *Inventory cycle.* This is the part of an inventory graph which regularly repeats itself in a cyclic process. For example, for the inventory in Figure 2, the inventory cycle is that part of the graph between 0 and 3 days. The cycle (where it exists) will always include:

i. an ordering (i.e. replenishment) component and

ii. a demand component.

*Figure 2 Inventory graph*

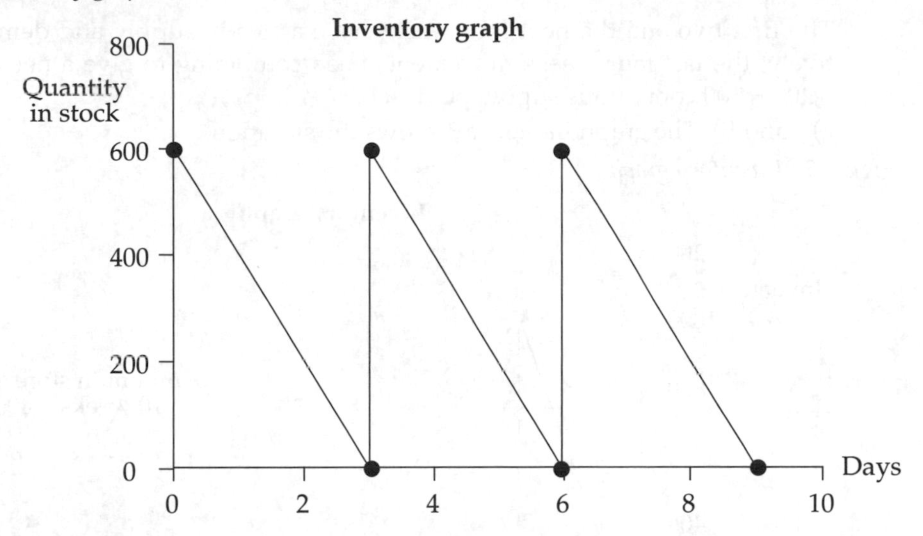


b) *Length of (inventory) cycle.* This is simply the length of time over which an inventory cycle extends. The cycle length of the process shown in Figure 2 is 3 days.

c) *Average inventory level.* The level is calculated using the following formula:

$$\text{Average inventory level} = \frac{\text{Total area under graph}}{\text{Total time}}$$

The best way to calculate the area under the graph is to split it up into right-angled triangles (and/or rectangles) and use the rule:

area of triangle = 0.5 × base × height; area of rectangle = base × height.

For example, in Figure 2, the graph is already composed of right-angled triangles. Thus for each triangle: area = 0.5 × 3 × 600 = 900. Total area is 3 × 900 = 2700.

Therefore, average inventory level $\frac{2700}{9} = 300$

## 9. Example 2 (Drawing an *inventory graph*)

### *Question*

An inventory situation over a period of 10 weeks was as follows.

*Beginning of period*: No initial inventory.

*First two weeks*: Goods supplied at continuous rate of 500 per week. No withdrawals.

*Next four weeks*: Goods withdrawn at constant rate of 250 per week. No input.

*Next four weeks*: Goods required at the constant rate of 200 per week and also goods supplied to store continuously at 300 per week.

a) Sketch the inventory graph for the 10 week period;

b) determine the amount in store at the end of the period;

c) calculate the average inventory level for the 10 week period.

*Answer*

The first two and the next four weeks have a steady supply and demand respectively; the last four weeks movements need combining to give a net value of 300 – 200 = 100 continuous supply per week.

a) and b) The graph in Figure 3 shows the situation.

*Figure 3 Inventory graph*

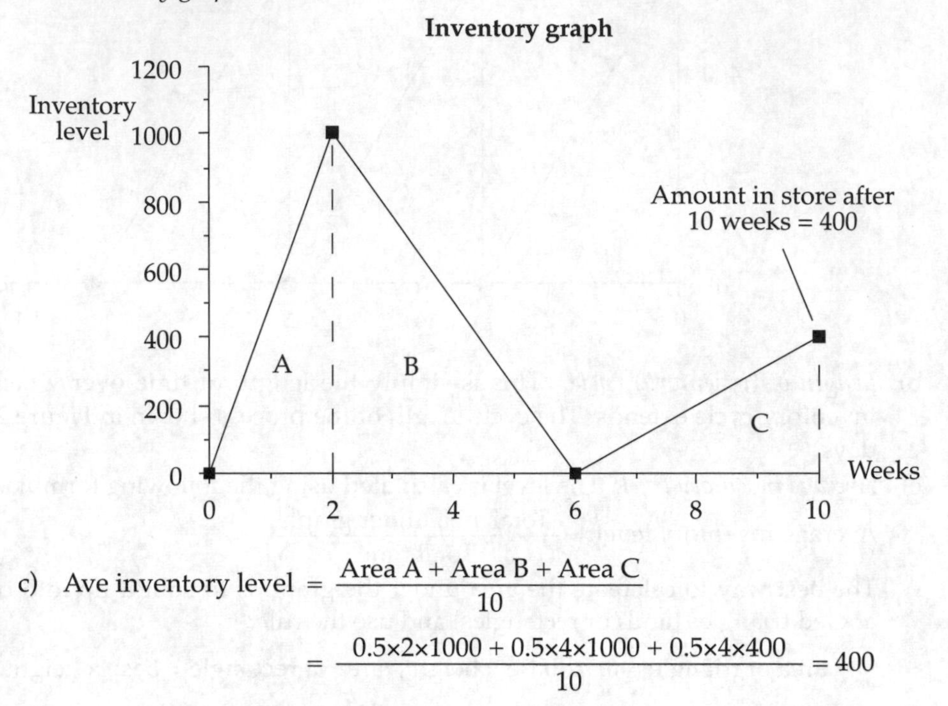


c) $$\text{Ave inventory level} = \frac{\text{Area A} + \text{Area B} + \text{Area C}}{10}$$

$$= \frac{0.5 \times 2 \times 1000 + 0.5 \times 4 \times 1000 + 0.5 \times 4 \times 400}{10} = 400$$

## 10. Inventory control systems

There are two standard systems for controlling inventory.

a) *Re-order level system.* For each stock item, this system sets a fixed quantity of stock (normally the EOQ) which is ordered every time the level of stock meets (or falls below) the calculated re-order level.

b) *Periodic review system.* For each stock item, this system sets a review period, at the end of which the stock level of the item is brought up to a predetermined value.

The following two sections describe these systems in more detail.

## 11. Re-order level system

This control system is the one that is most commonly used, and the two basic inventory models discussed in the remainder of the chapter are of this type. This particular system will generally result in lower stocks, items will be ordered in more economic quantities (via a calculated EOQ) and it is more responsive to fluctuations in demand compared with the periodic review system.

The system sets the value of three important levels of stock as warning or action triggers for management.

a) *Re-order level.* This is an *action* level of stock which causes a replenishment order (normally the EOQ) to be placed. Given a particular period of time (i.e. day, week or year), it is calculated as:

$$L_{RO} = \text{maximum usage (per period)} \times \text{maximum lead time (in periods)}$$

b) *Minimum level.* This is a *warning* level set such that only in extreme cases (i.e. above average demand or late replenishment) should it be breached. It is calculated as:

$$L_{Min} = \text{Re-order level} - (\text{normal usage} \times \text{average lead time})$$

c) *Maximum level.* This is another *warning* level set such that only in extreme cases (i.e. low levels of demand) should it be breached. It is calculated as:

$$L_{Max} = \text{Re-order level} + \text{EOQ} - (\text{minimum usage} \times \text{minimum lead time})$$

For example, suppose for a particular inventory, we had:

i. the weekly minimum, normal and maximum usage as 600, 1000 and 1400 respectively;
ii. the lead time varying between 4 to 8 weeks (average = 6 weeks); and
iii. the normal ordering quantity (EOQ) as 20,000.

Then: Re-order level = 1400 × 8 = 11200;

Minimum stock level = 11200 – 1000 × 6 = 5200;

Maximum stock level = 11200 + 20000 – 600 × 4 = 28800.

## 12. Periodic review system

Although not considered from here on in this chapter, this system has its merits. For example:

i. stock positions are reviewed periodically and thus there is little chance of them becoming obsolete;
ii. economies can be made when many items are ordered at the same time or in the same sequence.

The inventory graph for this control system is worth considering through an example. Suppose a company sets a predetermined stock level of 5000 for a particular product. Figure 4 overleaf shows three cycles of activity.

The following points should be noted from Figure 4.

i. Stock is reviewed and orders placed at days 3, 8 and 13. That is, regularly, every 5 days.
ii. It can be seen that the lead time is 3 days. i.e. order 1 is placed at day 3 and replenished at day 6; order 2 is placed at day 8 and replenished at day 11 and so on.
iii. The amount of each order is always the difference between the current stock level and the predetermined level.

Thus, order 1 (at day 3, when stock level is at 2800) is placed for 5000 – 2800 = 2200 products. These are replenished at day 6, causing the stock level to rise from 1900 to 1900 + 2200 = 4100 products.

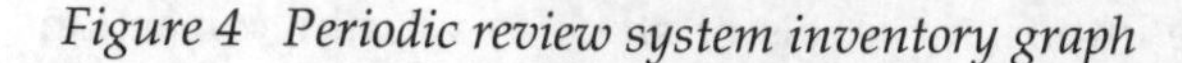
*Figure 4 Periodic review system inventory graph*

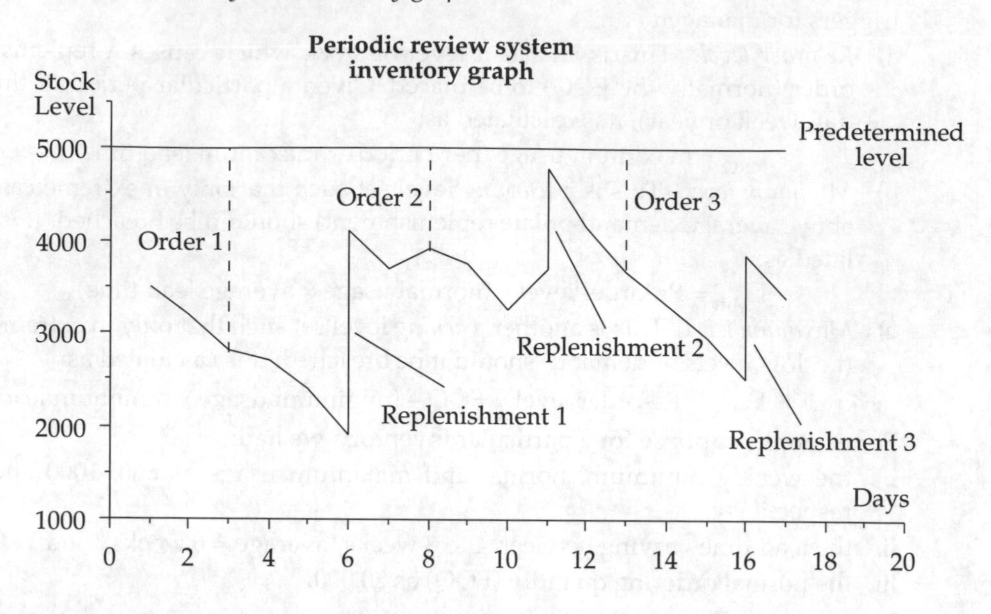


## 13. Inventory models

For the purposes of this manual, it is necessary to be aware of two inventory situations or models which are both (i) cyclic and (ii) described over a period of a year. They are:

a) The *basic* model
b) The *adapted basic* model (with gradual replenishment)

These are described over the next few sections together with their standard inventory graphs.

## 14. The basic model

This model assumes the following characteristics:

i. The demand rate ($D$ = number of items / year) is constant and continuous over a given period and all demand is satisfied.
ii. The ordering cost ($Co$ = £/cycle) is constant and *independent of the quantity ordered*.
iii. Only one type of stock item is considered and its cost ($P$ = £/item) is constant.
iv. The holding cost ($C_h$ = £/item) is the cost of carrying one article in stock for one year.
v. The quantity ordered per cycle ($q$) is supplied to store *instantaneously* whenever the inventory level falls to zero.

The standard inventory graph for the basic model is shown at Figure 5.

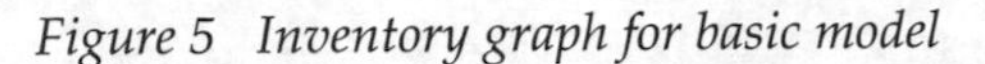

*Figure 5 Inventory graph for basic model*

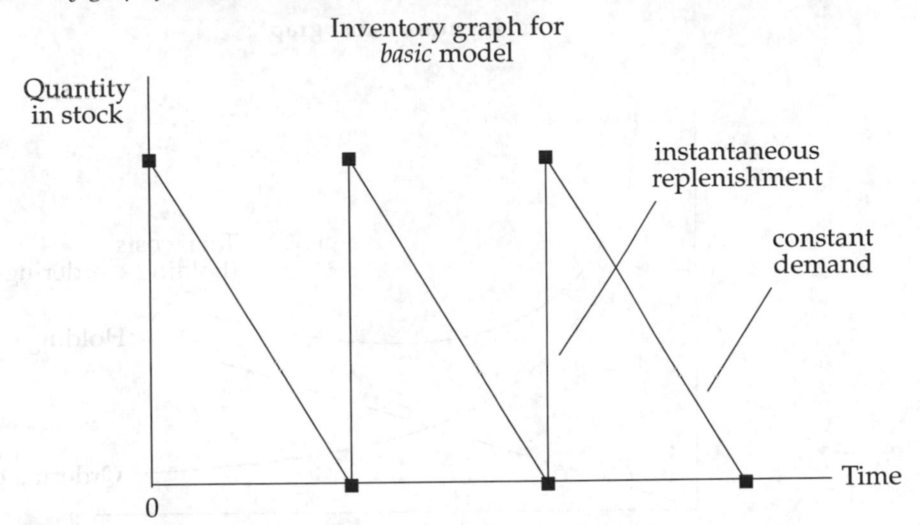


## 15. Cost equation for basic model

Here, we derive a cost equation to describe annual costs for the basic model.

Putting the general order quantity as variable $q$, and using some results from the previous section, we have:

Total annual inventory cost = C, say

but:

$C$ = total ordering cost + total holding cost

= number of orders/year × order cost

+ average inventory level × holding cost/item

i.e. $C = \frac{D}{q}.C_o + \frac{q}{2}.C_h$ [$C_o$= annual ordering cost; $C_h$ = annual holding cost]

As $q$ gets larger, so: annual ordering cost becomes *smaller*

annual holding cost becomes *larger*.

These two functions (ordering and holding costs) are sketched on the graph in Figure 6 overleaf, together with total yearly cost.

It can be shown (see Derivation 1, in section 23 later) that the total annual inventory cost is minimised when the order quantity, $q$, takes the following value. It is known as the *Economic Ordering Quantity* (*EOQ*).

$$\text{EOQ} = \sqrt{\frac{2DC_o}{C_h}}$$

Notice also from the graph that the minimum total cost occurs *where ordering cost = holding cost*. This is always true and is sometimes more useful to work with than the total cost graph since the latter is usually 'flat-bottomed', making estimation difficult.

*Figure 6 Inventory cost graphs*

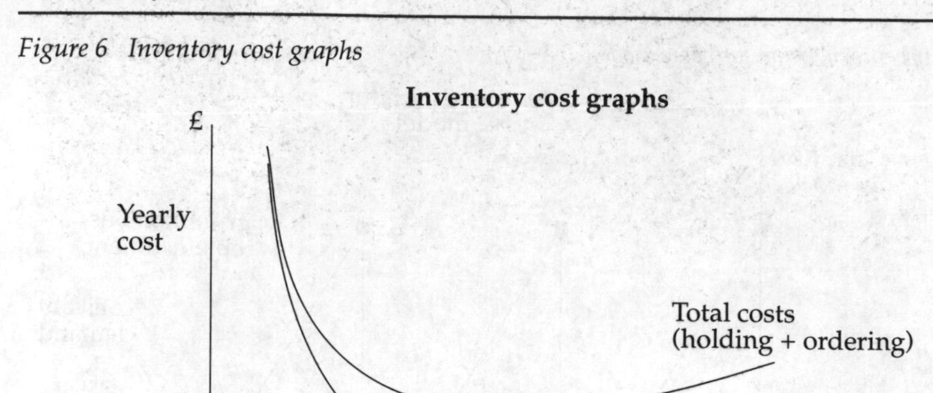


## 16. Definitions and formulae for basic model

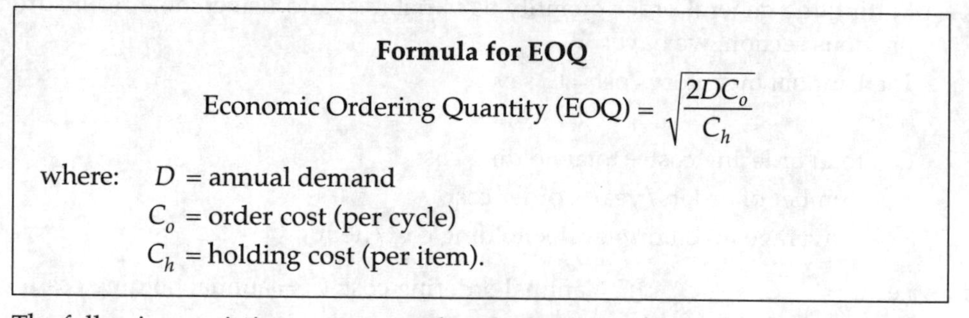

**Formula for EOQ**

$$\text{Economic Ordering Quantity (EOQ)} = \sqrt{\frac{2DC_o}{C_h}}$$

where: $D$ = annual demand

$C_o$ = order cost (per cycle)

$C_h$ = holding cost (per item).

The following statistics are commonly calculated for this model:

a) *Number of orders per year* $= \dfrac{\text{Yearly demand}}{\text{EOQ}}$

b) *Length of cycle (days)* $= \dfrac{\text{Number of days per year}}{\text{Number of orders per year}}$

c) *Average inventory level* $= \dfrac{\text{EOQ}}{2}$

## 17. Example 3 (*basic inventory model* problem using graph and formulae)

*Question*

A commodity has a steady rate of demand of 2000 per year. Placing an order costs £10 and it costs 10p to hold a unit for a year.

a) On a graph plot order and holding costs for a year, together with total costs.

b) Use the graph to estimate the order quantity (EOQ) that minimises total costs. Confirm this value using the EOQ formula.

c) Find the number of orders placed per year and the length of the inventory cycle.

*Answer*

Here, $D = 2000$, $C_o = 10$ and $C_h = 0.1$ (all units in £).

a) Table 1 shows the layout of calculations for the required costs and Figure 7 represents their graphs.

*Table 1 Layout of calculations*

| $q$ | *Annual ordering cost* $\frac{D}{q}.C_o = \frac{2000}{q}$ (a) | *Annual holding cost* $\frac{q}{2}.C_h = 0.05q$ (b) | *Total cost* (a) + (b) |
|---|---|---|---|
| 100 | 200.0 | 5.0 | 205.0 |
| 200 | 100.0 | 10.0 | 110.0 |
| 300 | 66.7 | 15.0 | 81.7 |
| 400 | 50.0 | 20.0 | 70.0 |
| 500 | 40.0 | 25.0 | 65.0 |
| 600 | 33.3 | 30.0 | 63.3 |
| 700 | 28.6 | 35.0 | 63.6 |
| 800 | 25.0 | 40.0 | 65.0 |
| 900 | 22.2 | 45.0 | 67.2 |
| 1000 | 20.0 | 50.0 | 70.0 |

*Figure 7 Inventory cost graph for a component*

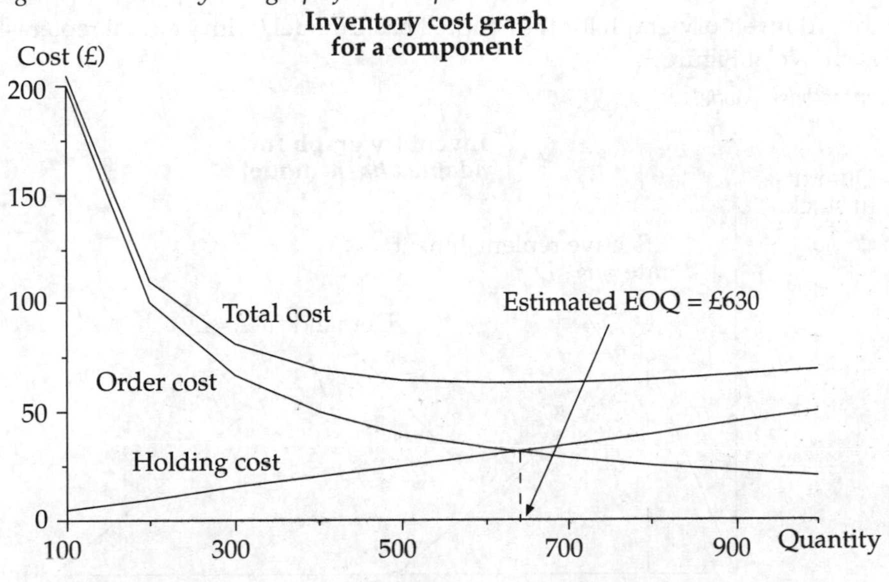

b) Using the EOQ formula gives: $EOQ = \sqrt{\frac{2DC_o}{C_h}} = \sqrt{\frac{2(2000)(10)}{0.1}} = 632.5$ (1D)

(which agrees well with the estimate from the graph).

c) $\text{Number of orders / year} = \frac{\text{Yearly demand}}{\text{EOQ}} = \frac{2000}{632.5} = 3.2$

$\text{Cycle length} = \frac{\text{Number of days per year}}{\text{Number of orders per year}} = \frac{365}{3.2} = 114 \text{ days}$

## 18. Adapted model (with gradual replenishment)

In the basic model, it was assumed that a complete replenishment order could be put into store instantaneously. This of course fits the situation where orders are received from an external source (i.e. an outside supplier) and the whole batch is delivered at one time. However, if stock is received from a production line, particularly one that is internal to the organisation, it is quite likely that finished articles are received *continuously* over a period, that is, stock is subject to *gradual replenishment*.

This adapted basic model (sometimes known as the *production run* model) assumes exactly the same characteristics as the basic model, particularly with reference to constant demand rate $D$ and the two cost factors $C_o$ and $C_h$, except for the fact that:

i. A *production run* is started every time the level of inventory falls to zero and stops when $q$ items have been produced (supplied). The run lasts for time $t$ and is known as the *run time*.

ii. The quantity ordered per cycle is now known as the *run size*, and items are supplied at rate $R$ per annum. Thus, the *effective* replenishment rate is $R–D$ items.

Essentially then, this model concerns order quantities that are raised *internally* (via a production run) as opposed to the assumed *external* orders for the previous model.

The standard inventory graph for the adapted basic model (with gradual replenishment) is shown at Figure 8.

*Figure 8 Adapted basic model*

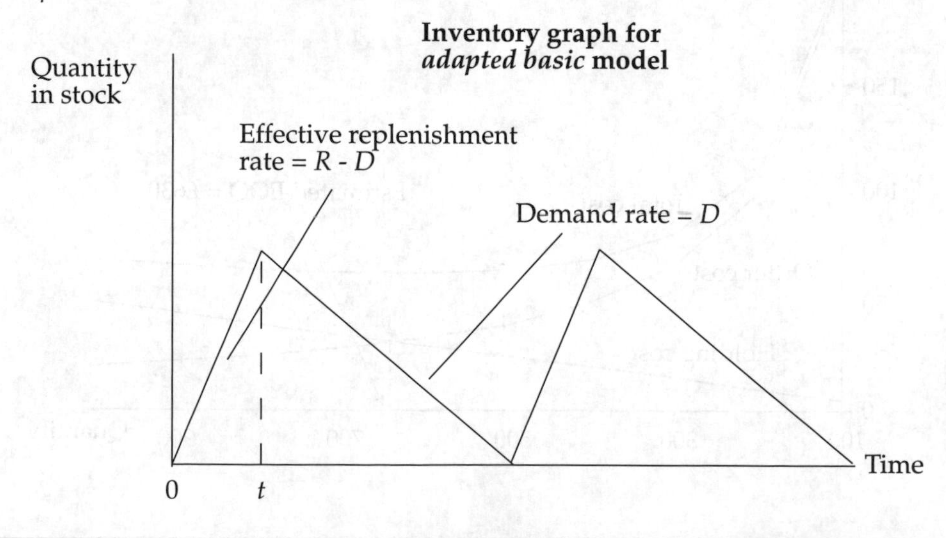

## 19. Definitions and formulae for adapted basic model

The formula for the value of the run size that minimises inventory costs for this model is now given.

**Formula for EBQ**

$$\text{Economic Batch Quantity (EBQ)} = \sqrt{\frac{2DC_o}{C_h\left(1-\frac{D}{R}\right)}}$$

where: $D$ = annual demand
$R$ = annual production rate
$C_o$ = order cost (per cycle)
$C_h$ = holding cost (per item).

Notice that the optimum run size is known as the *Economic Batch Quantity* (*EBQ*) and $C_o$ is known as the *setup cost*.

Notice also the factor of $\left(1-\frac{D}{R}\right)$ in the denominator is different to that of the EOQ. There are a few notes on this in Derivation 2, later in section 24.

The following optimal statistics, are used with the adapted model.

a) *Number of runs per year* $= \dfrac{\text{Yearly demand}}{\text{EBQ}}$

b) *Length of cycle* (*days*) $= \dfrac{\text{Number of days per year}}{\text{Number of runs per year}}$

c) *Run time* (*days*) $= \dfrac{\text{EBQ} \times \text{Number of days per year}}{\text{Annual production rate}}$

d) *Peak inventory level* = Effective replenishment rate × Run time

e) *Average inventory level* = $\frac{1}{2}$ × Peak inventory level.

## 20. Example 4 (*adapted model* problem)

*Question*

A manufacturing process requires a continuous supply of 3000 items per year from store, which is replenished by production runs, each of which operate at the constant rate of 5000 items per year. Each production run has a set-up cost of £18 and the holding cost per item per annum is 5p.
Calculate the EBQ and use it to find the number of runs per year, length of cycle, run time, peak inventory level and average inventory level. Draw an inventory graph showing two complete inventory cycles and clearly label all points.

*Answer*

$D$ = demand rate = 3000; $R$ = production rate = 5000; $C_o$ = setup cost = 18;

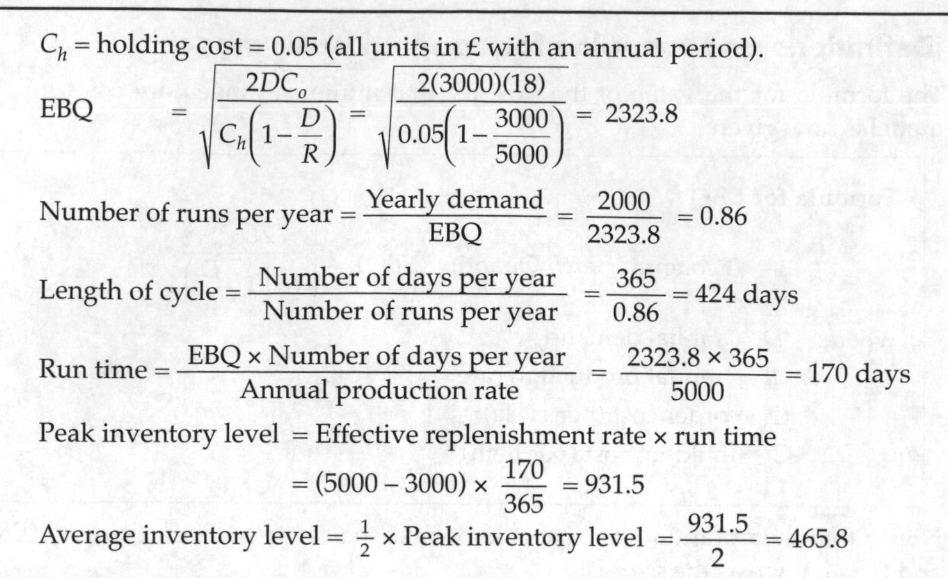

$C_h$ = holding cost = 0.05 (all units in £ with an annual period).

$$\text{EBQ} = \sqrt{\frac{2DC_o}{C_h\left(1-\frac{D}{R}\right)}} = \sqrt{\frac{2(3000)(18)}{0.05\left(1-\frac{3000}{5000}\right)}} = 2323.8$$

$$\text{Number of runs per year} = \frac{\text{Yearly demand}}{\text{EBQ}} = \frac{2000}{2323.8} = 0.86$$

$$\text{Length of cycle} = \frac{\text{Number of days per year}}{\text{Number of runs per year}} = \frac{365}{0.86} = 424 \text{ days}$$

$$\text{Run time} = \frac{\text{EBQ} \times \text{Number of days per year}}{\text{Annual production rate}} = \frac{2323.8 \times 365}{5000} = 170 \text{ days}$$

$$\text{Peak inventory level} = \text{Effective replenishment rate} \times \text{run time}$$

$$= (5000 - 3000) \times \frac{170}{365} = 931.5$$

$$\text{Average inventory level} = \tfrac{1}{2} \times \text{Peak inventory level} = \frac{931.5}{2} = 465.8$$

*Figure 9 Inventory graph*

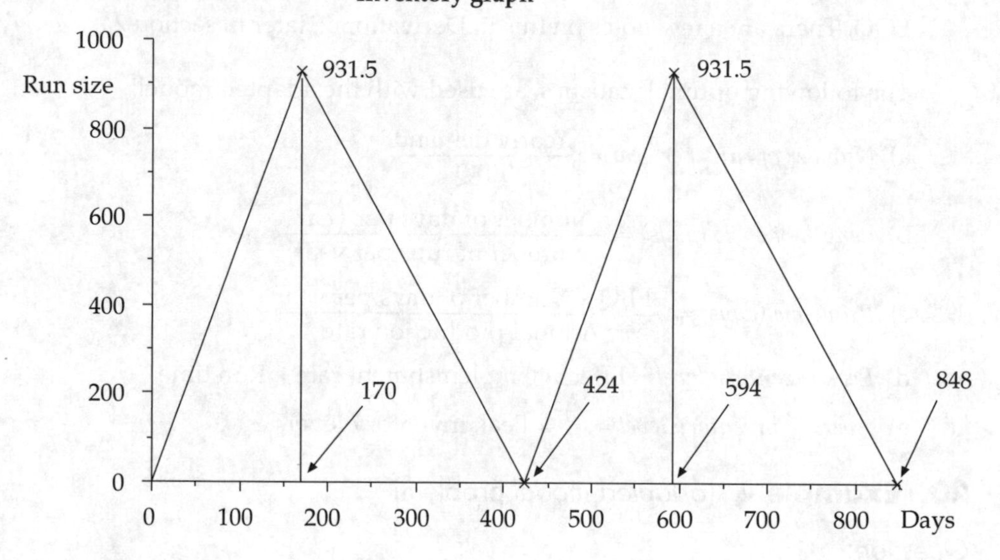


## 21. Summary

a) A firms inventory can be described as the totality of stocks of various kinds and the reasons why it is carried include: anticipating demand, bulk purchasing, absorbing wastages and over-ordering.

b) Inventory costs can be split into:

   i. Ordering (or Replenishment) costs. These can either be administrative costs (for external orders) or setup costs (for internal production runs).

   ii. Holding (or Carrying) costs. These include things such as: stores costs and overheads, deterioration, insurance and security.

iii. Stockout costs. These are associated with loss of goodwill or future orders due to running out of stock.

c) The objective of inventory control is to maintain a system which will minimise total costs and establish the optimum amount of stock to be ordered and the period between orders.

d) Some terms used in inventory control are:

i. Lead time – the time between ordering goods and their replenishment.

ii. Economic Ordering Quantity (EOQ) – the (external) order quantity that minimises total inventory costs.

iii. Economic Batch Quantity (EBQ) – the size of the (internal) production run that minimises total inventory costs.

iv. Safety stock – minimum stock level for control purposes.

v. Maximum stock – another control stock level.

vi. Re-order level – actual stock level that, when reached, causes an order to be placed.

vii. Re-order quantity – amount of stock that is ordered via the re-order level. It is normally the EOQ (or EBQ).

e) An inventory graph plots the relationship between the amount of stock held and time.

f) Some other terms used in inventory control are:

i. Inventory cycle – is the part of an inventory graph that regularly repeats itself.

ii. Average inventory level is calculated as:

$$\text{Average inventory level} = \frac{\text{Total area under graph}}{\text{Total time}}$$

g) A re-order level inventory control system sets a fixed quantity of stock (normally EOQ) to be ordered every time the stock level meets or falls below the calculated re-order level. Advantages (over periodic review system) are lower stocks, more economic quantities ordered and that it is more responsive to fluctuations in demand.

h) A periodic review inventory control system sets a review period, at the end of which the stock level of the item is brought up to a predetermined value. Advantages (over re-order level system) include stock positions reviewed periodically and economies may be made when items are ordered at the same time or in the same sequence.

i) The basic inventory model assumes that: demand is constant and satisfied, ordering cost is independent of quantity ordered, orders are supplied to store instantaneously and only one type of stock item is considered.

j) For the basic model: Annual (inventory) costs = total ordering cost + total holding cost.

In formula terms: $C = \frac{D}{q}.C_o + \frac{q}{2}.C_h$ and $EOQ = \sqrt{\frac{2DC_o}{C_h}}$

k) The adapted basic inventory model (with gradual replenishment) is similar to the basic model except that stock is replenished gradually in a continuous fashion. Stock input is considered normally from an internal production run.

l) For the adapted model: Economic Batch Quantity (EBQ) = $\sqrt{\dfrac{2DC_o}{C_h\left(1-\dfrac{D}{R}\right)}}$

## 22. Points to note

a) The control systems and models covered in this chapter are very basic. In reality, inventories might be mixtures of these or other more complex systems. However, the concepts covered are important ones and should be understood by students.

b) The two models covered assumed that demand and lead times were not variable. In cases of uncertainty, where demand can fluctuate or lead times vary, a *safety stock value* can be added to a re-order level.

## 23. Derivation 1 – EOQ formula

In section 15, the cost equation for the basic model was given as:

$$C = \frac{D}{q}.C_o + \frac{q}{2}.C_h$$

which can be rearranged to give: $C = D.C_o.q^{-1} + \dfrac{C_h}{2}.q$

But to find the value of $q$ that minimises $C$, we need to solve $\dfrac{dC}{dq} = 0$

Now, $\dfrac{dC}{dq} = -D.C_o.q^{-2} + \dfrac{C_h}{2}$

$$= \frac{-D.C_o}{q^2} + \frac{C_h}{2}$$

Thus $\dfrac{dC}{dq} = 0$ gives $\dfrac{D.C_o}{q^2} = \dfrac{C_h}{2}$

i.e. $q^2 = \dfrac{2.D.C_o}{C_h}$

Therefore, $q = \sqrt{\dfrac{2DC_o}{C_h}}$

which is the Economic Ordering Quantity (EOQ).

## 24. Derivation 2 – EBQ formula

As already mentioned, the cost equation for the adapted model is similar to that for the basic model and is given by:

$$C = \frac{D}{q}.C_o + \frac{q}{2}.C_h\left(1 - \frac{D}{R}\right)$$

Notice that the only difference is that $C_h$ is adapted to $C_h\left(1-\frac{D}{R}\right)$, which is less than $C_h$, since $\left(1-\frac{D}{R}\right)< 1$. This takes into account the fact that the number of items held is less for this model.

Thus, the EOQ formula needs $C_h$ changed to read $C_h\left(1-\frac{D}{R}\right)$ in order to obtain the EBQ formula.

That is: $\text{EBQ} = \sqrt{\dfrac{2DC_o}{C_h\left(1-\frac{D}{R}\right)}}$

## 25. Student self review questions

1. What is an inventory? Give some reasons why a firm should carry one. [2]
2. Name the main categories of cost in holding an inventory. [3]
3. What is the purpose of inventory control? [4]
4. What is lead time? [5(a)]
5. Explain the difference between EOQ and EBQ. [5(b,c)]
6. What is a re-order level? [5(f)]
7. What is an inventory cycle? [8(a)]
8. Given an inventory graph, how is the average inventory level calculated? [8(c)]
9. Explain the difference between the *re-order level* and *periodic review* inventory control systems. [10]
10. What advantages does the re-order level system have over the periodic review system? [11]
11. How is a re-order level calculated when using the re-order level system? [11]
12. On an inventory cost graph, what is the significance of the point where holding costs and ordering costs are equal? [15]
13. What is the 'run time' when using the adapted basic inventory model? [18]

## 26. Student exercises

1. The initial inventory of a certain component is 12000. A particular manufacturing process requires the components to be drawn from stores at a steady rate of 3000 per week. As soon as a stockout is reached, a batch of 12000 items is moved overnight from another factory to replenish the inventory. Sketch the inventory graph for a period of 12 weeks.
2. At the beginning of a period of 9 days, a certain inventory stood at 500 items. Over the next two days, manufacturing demands required withdrawals of stock items at the constant rate of 200 per day. Over the next four days, demand remained at a constant rate of 200 per day but continuous replenishment began at 350 items per day. At the end of day 6, 400 items were withdrawn from stock and during day 7 there were no stock movements. For the final two days of the period, demands required withdrawals at a constant rate of 50 per day.
   a) Draw an inventory graph for the 9-day period.
   b) Find the level of the inventory at the end of the period.
   c) Calculate the average inventory level over the whole period.

3. A manufacturer has to supply 16,500 articles annually at a steady rate. The ordering cost is £8 per order, the price per unit is £2.60 and the holding costs are 25% (of price). Assume the basic inventory model is appropriate.
   a) Plot the annual order costs and the annual holding costs on the same graph and *use the two curves* to estimate the EOQ.
   b) Verify the above EOQ value by using the formula.
   c) Calculate the inventory cycle length and the number of runs per year.
4. A builder uses bricks at a more or less constant rate of 600,000 per year. The carrying cost of a brick is 0.1p and the ordering cost is £6 per order. Determine the EOQ, optimum cycle length and the optimum number of orders per year.
5. A wholesaler has a steady demand for an article of 300 per quarter. He buys from the factory at a cost of £3 per item and the cost of ordering is £8 per order. The stock holding costs amount to 25% per annum of stock value. How often should an order be made out?
6. A company uses steel pins at a constant rate of 5000 standard boxes per year. The pins cost £2 per box, each order for a consignment of boxes costs £20 and carrying costs are estimated at 10% of original cost price.
   a) How frequently should orders for boxes of pins be placed and what quantity should be ordered each time?
   b) If the actual costs are £50 to place an order and 15% for carrying costs, the optimal policy would clearly change. How much is the company losing every year because of imperfect cost information?
7. An inventory system follows the adapted basic model form (with gradual replenishment). The demand rate is constant at 1600 articles per year, the price per article is £1.80, the holding cost is 10% of price per year and the setup cost is £12 per run. If the production rate is 5000 articles per year, calculate:
   a) the optimum run size (EBQ), the run time, the cycle length and the average inventory level.
   b) Draw an inventory graph showing the run size and average inventory level clearly.
8. An inventory system is supplied by an internal production line which can produce 25,000 articles per year. However, demand per annum is estimated at only one-fifth this production rate. The setup cost for the line is £200 per run and the holding cost is 25p per article per year. What is the Economic Batch Quantity for this system and how often will the production line be activated for a continuous run?
9. A contractor has to supply 1000 bolts per day to a manufacturer. He finds that on a production run he can produce 2500 per day. The cost of holding a bolt for one year is 3.6p and the setup cost of a production run is £35. How frequently should the runs be made?

# 38 Network planning and analysis

## 1. Introduction

Network analysis covers the way we plan large projects such as major construction work, research and development projects, the computerisation of systems etc.

It is usual to distinguish between network planning and analysis.

## 2. Network planning

Network planning is the task of breaking down a project into its constituent activities and deciding on their time relationships. This in itself is a valuable exercise since it can help management to establish objectives and pinpoint courses of action to follow. The resultant chart in some respects forms a policy statement.

The planning stages consist of listing the activities and the time necessary to complete them, organising their order of completion and drawing the associated network.

The immediately following sections describe some of the concepts involved.

## 3. Activities and events

An activity may be a task which requires effort (for example, drawing up a set of accounts) or it may involve simply waiting for something to happen (for example, waiting for a delivery).

The choice of the set of activities to use for a particular project depends on the level of the analysis that is required.

An event signifies a point in time (milestone) in the progress of the project in question. All activities must begin and end with an event.

Graphically, we represent activities by arrowed lines (normally from left to right) and events by small circles.

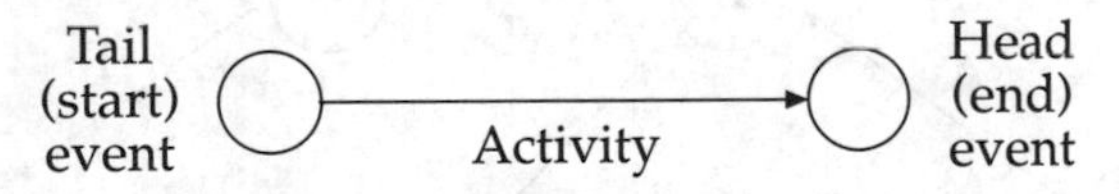


In general we will label events with capital letters and events with numbers. Sometimes however it is convenient to label activities using start and end events. For example, if an activity lies between events 2 and 4 then it might be described as activity 2–4.

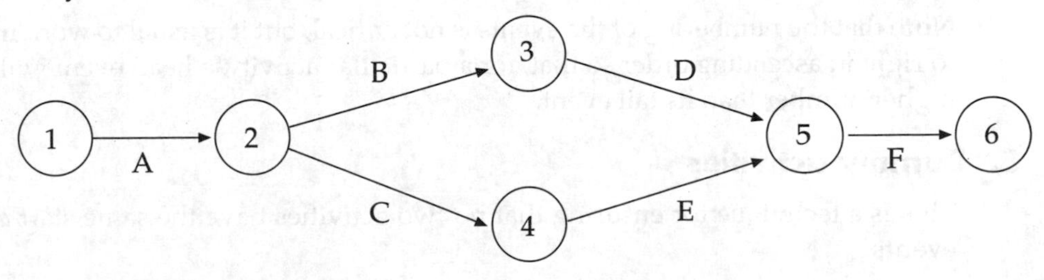

The previous diagram illustrates particular relationships of significance;

a) activities B and C cannot begin until A has finished;

b) activity F cannot begin until both D and E are completed;

c) there is just one start and end event (this must always be the case).

Before drawing a network, it is advisable to construct an activity list showing, for each activity, the preceding activity. The following *precedence table* describes the simple network above.

| *Activity* | *Preceding activity* |
|---|---|
| A | – |
| B,C | A |
| D | B |
| E | C |
| F | D,E |

## 4. Example 1 (Drawing a network plan from an activity list)

*Question*

Draw the network for the following activity list:

| *Activity* | *Preceding activity* |
|---|---|
| A,B | – |
| C,D | A |
| E | B |
| F | C |
| G | D,E |

*Answer*

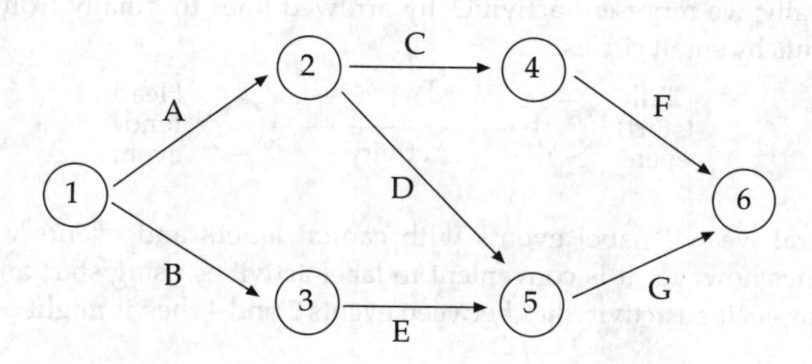


**Note** that the numbering of the events is not critical, but it is usual to work from left to right in ascending order so that, for a particular activity, a head event will have a higher number than its tail event.

## 5. Dummy activities

This is a technique for ensuring that no two activities have the same start *and* end events.

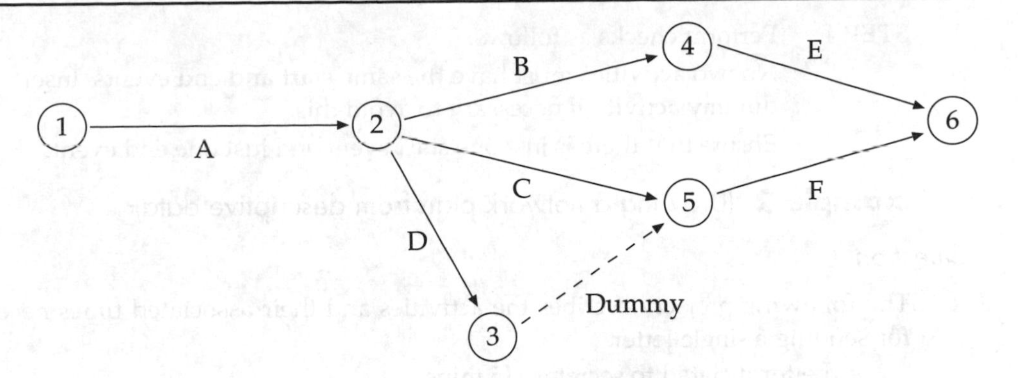


Notice in the above diagram that activities C and D would have the same start and end events except for the extra event 3 and its associated dummy activity (3 → 5). Another reason for using a dummy activity is to preserve the correct sequence of activities. Consider the following precedence table and its network

| *Activity* | *Preceding activity* |
|---|---|
| ... | ... |
| E,F | D |
| H | F |
| G | E,F |
| I | G,H |
| ... | ... |

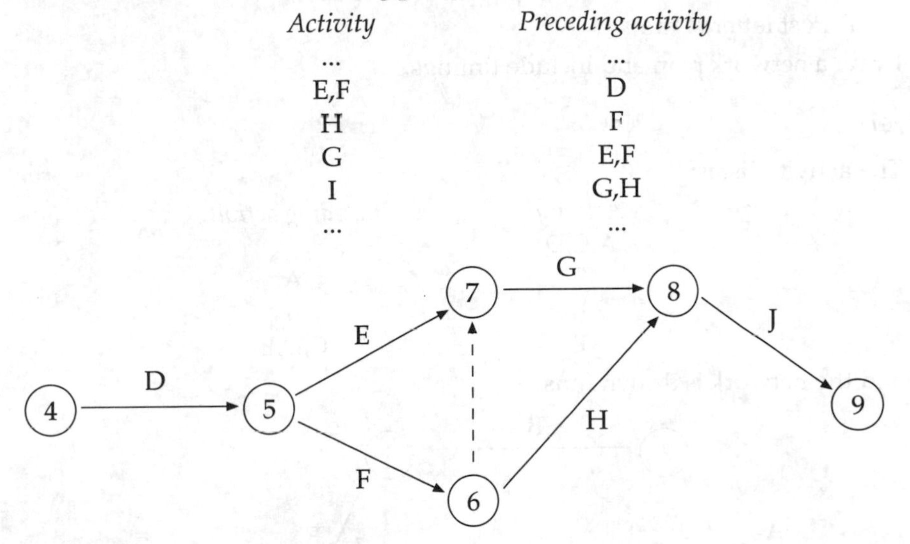


Notice that F must come before both G and H and thus needs a dummy activity. The necessity for this type of dummy activity is easily seen from a precedence table – there will be more than one instance of an activity listed on the right hand side. There will always be *one less* dummy activity than there are instances of the same activity.

Dummy activities always have zero time associated with them.

## 6. Summary of network planning procedures

STEP 1 Write down activities and their times.

STEP 2 Decide on activity order and precedents.

STEP 3 Sketch out network. It is not unusual to redo a network at least once before it is correct.

STEP 4 Perform checks as follows:

No two activities must have the same start and end events. Insert dummy activity if necessary to avoid this.

Ensure that there is just one start event and just one end event.

## 7. Example 2 (Drawing a network plan from descriptive data)

*Question*

The following project describes the activities and their associated times necessary for sending a single letter.

A. Letter dictated to secretary (5 mins)
B. Letter typed (10 mins)
C. Envelope addressed by clerk (3 mins)
D. Envelope stamped by clerk (1 min)
E. Clerk puts letter in envelope and seals (1 min)
F. Post letter (4 mins)

Draw a network plan and include timings.

*Answer*

The activity list is:

| *Activity* | *Preceding activity* |
|---|---|
| A,C,D | – |
| B | A |
| E | B |
| F | C,D,E |

and the network is drawn thus:

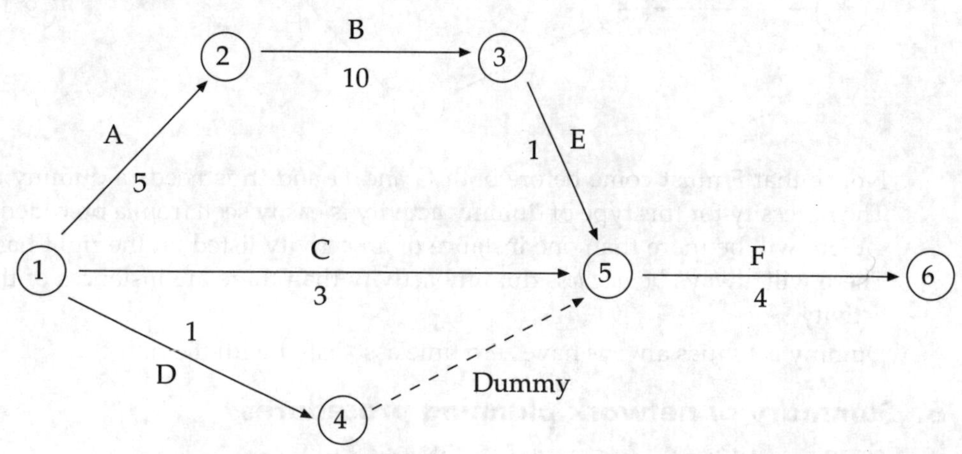

## 8. Critical path 1

This section begins the basic analysis stage.

If we are given the times of all activities for a particular project, we can evaluate the time taken for each of the possible paths through the network. The longest of these is known as the *critical path*. The total time on this path is the shortest time within which the project can be completed.

Consider the simple network of Example 2 above. There are three possible paths and for each of these, the time lengths are shown below:

| *Path* | *Time length* | *Total* | |
|---|---|---|---|
| ABEF | 5+10+1+4 | 20 | |
| CF | 3+4 | 7 | |
| D()F | 1+4 | 5 | () signifies dummy activity |

Here the largest value is 20. Thus ABEF is the critical path.

To summarise:

> **Critical path**
>
> If all the paths through a network are identified and timed, the path with the longest time is known as the *critical path*.

**Notes.**

a) The critical path need not be unique.

b) In order to shorten the length of a project, the duration time of an activity on the critical path must be shortened. In other words, reducing the duration time of an activity which is not on the critical path will not affect the total project time.

## 9. Earliest and latest event times

Although a critical path can be identified using the method above, it is a relatively tedious manual process. A much more systematic method, suitable for computerisation, involves calculating the earliest and latest times for events.

The notation for marking events is shown below.

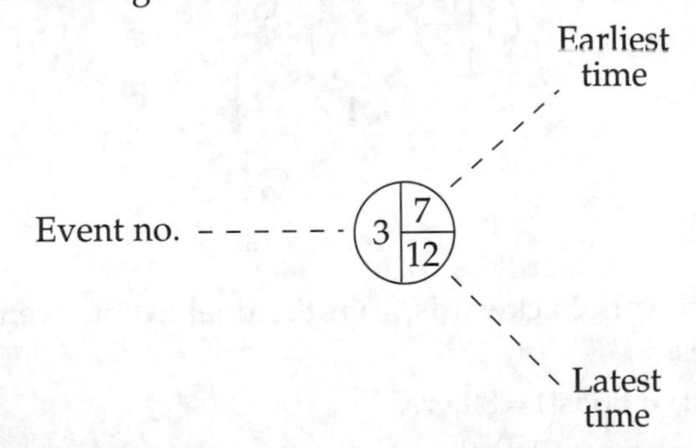


A simple example will demonstrate the technique that should be employed. Consider the network in Figure 1 below.

*Figure 1*

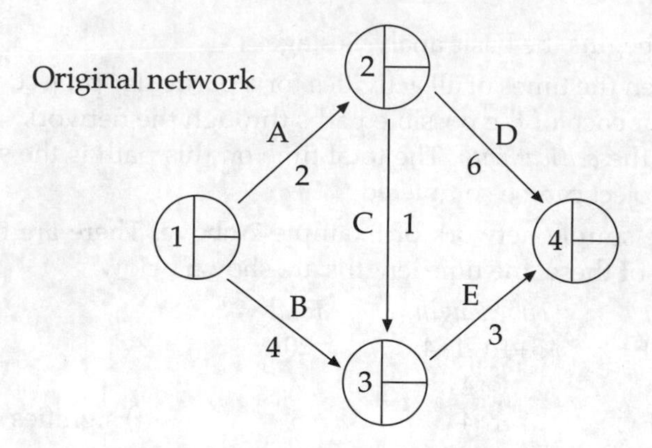


We need to enter earliest times. The procedure is to work from left to right. Enter '0' in event 1 early.

The basic technique now is to go to each event in turn adding the activity time to the previous event early time. Thus for example, event 2 early = E(2) = E(1) + A = 0 + 2 = 2.

Note however that if there is a choice, that is if there is *more than one arrow head* leading to the event, it is necessary to *take the largest* of the possible early times. This is the case now with E(3). Here, there are 2 heads, C and B.

C gives 2 + 1 = 3 whereas B gives 0 + 4 = 4. Thus choose the larger, i.e. E(3) = 4.

Similarly, E(4) is the largest of D (2 + 6 = 8) and E (4 + 3 = 7).

i.e. E(4) = 2 + 6 = 8 = project duration time. Figure 2 shows the values entered.

*Figure 2*

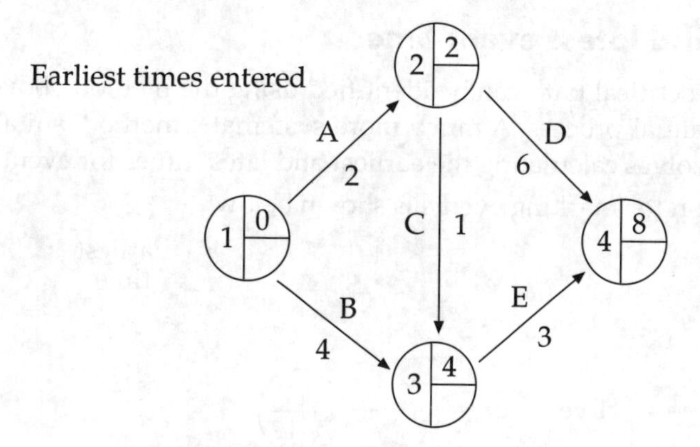


Now we need to work backwards from the final event, event 4, and fill in all the latest event times.

Begin with E(last) = L(last) = 8 here.

Now the rule is reversed. To calculate the event late time = L(event) for each event we need to subtract the activity time from L(previous). Thus, for event 3, L(3) = L(4) – E = 8 – 3 = 5.

However, for event 2 there are two possibilities (i.e. there are 2 tails).

The rule here is the reverse of the early event choice rule. Here, if there is more than one tail leading from the event, it is necessary to take the smallest.

For event 2, there is the choice of 8 – 6 = 2 or 5 – 1 = 4. Clearly, using the above rule, L(2) = 2.

Check this rule for the last event, event 1 and check with Figure 3.

*Figure 3*

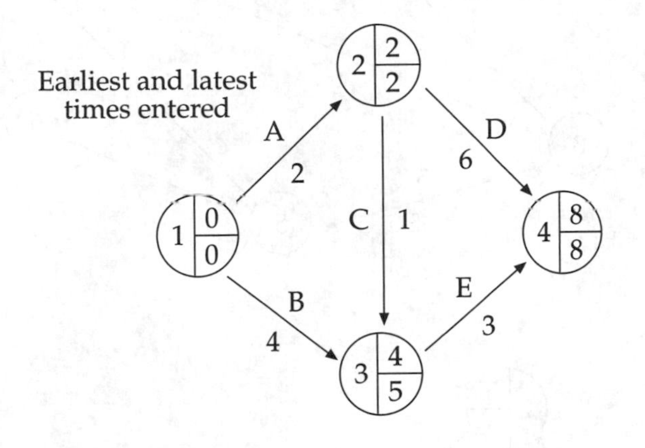


To summarise:

**Earliest and latest times.**

*Earliest (E) calculation*: E(event) = E(previous event) + activity time

Note if there are 'head' alternatives, choose the largest.

*Latest (L) calculation*: L(event) = L(next event) – activity time

Note if there are 'tail' alternatives, choose the smallest.

## 10. Critical path 2

A critical path always has E = L for all events.

However, if a particular path has E = L for all events, it is not a critical path unless each activity duration is the difference between its head and tail event times.

## 11. Example 3

*Question*

The following network gives the event numbers and activity times of a project. Redraw the network, showing earliest and latest event times for all events and thus highlight the critical path.

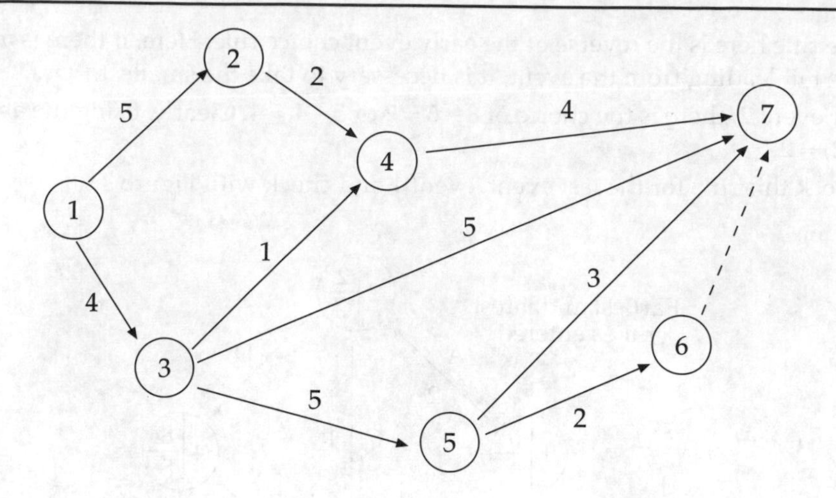


*Answer*

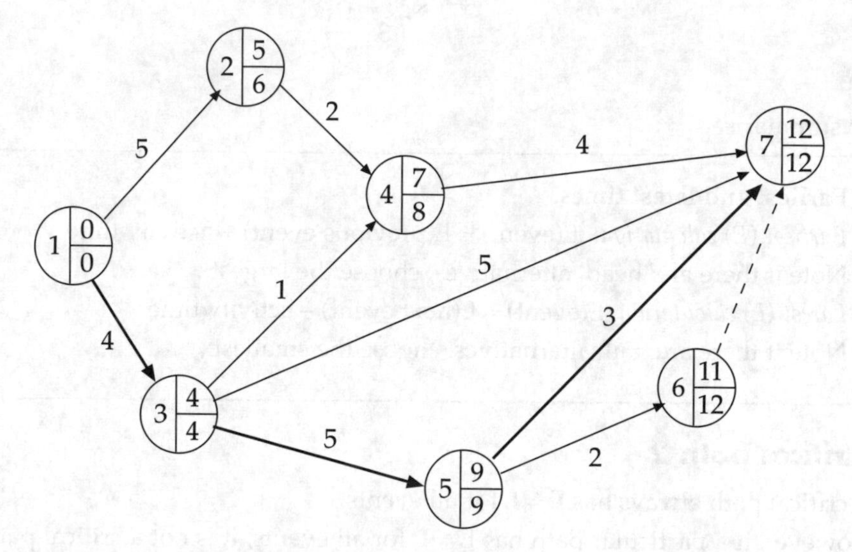


The critical path, 1–3–5–7 is highlighted in bold.

## 12. Float

If an activity is not on a critical path, it must be possible to increase its duration time without increasing the total project time. Of course, it can only be permitted a certain amount of extra duration time before a new critical path is created and thus an increase in the total project time.

This extra time is known as float. There are three different types of float; total, free and independent.

The following extract from a network will help to explain how they are defined.

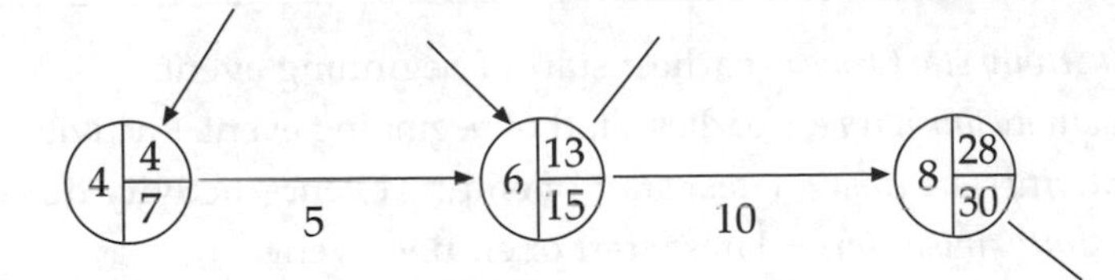


Total float is the amount of spare time available given the most favourable conditions. It is defined as follows.

| | |
|---|---|
| *Total float* | = maximum available time – activity duration time<br>= latest time of end event – earliest time of start event – activity duration time |

For activity 4 – 6 above: Total float = 15 – 4 – 5 = 6.

For activity 6 – 8 above: Total float = 30 – 13 – 10 = 7.

Free float is the amount of the total float which can be used up without affecting the float of succeeding activities. It is defined as follows.

| | |
|---|---|
| *Free float* | = duration between the earliest times of both events – activity duration time |

For activity 4 – 6 above: Free float = 13 – 4 – 5 = 4.

For activity 6 – 8 above: Free float = 28 – 13 – 10 = 5.

Independent float is the amount of time by which the activity duration can be expanded without affecting the floats of succeeding or preceding activities.

| | |
|---|---|
| *Independent float* | = minimum available time – activity duration time<br>= earliest time of end event – latest time of start event – activity duration time |

For activity 4 – 6 above: Independent float = 13 – 7 – 5 = 1.

For activity 6 – 8 above: Independent float = 28 – 15 – 10 = 3.

Note that the critical path activities have no float associated with them.

## 13. Tabular presentation of a network

It is usual to present a network in tabular form, showing activity name, duration, earliest and latest start and finish times and the three floats described above.

The following example shows such a table.

It is however worth noting the following relationships.

> *Earliest activity start time* = earliest start of beginning event.
> *Earliest activity finish time* = earliest start of beginning event + activity duration.
> *Latest activity start time* = latest start of ending event – activity duration.
> *Latest activity finish time* = latest start of ending event.

## 14. Example 4

*Question*

Give a tabular presentation of the following network.

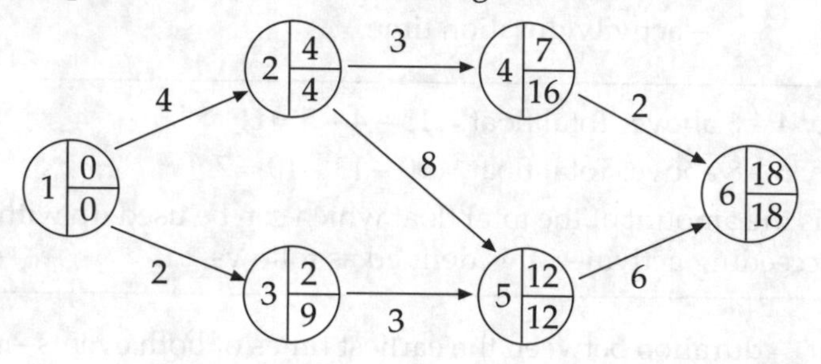


*Answer*

| *Activity* | *Duration* | *Start* | | *Finish* | | *Floats* | | | |
|---|---|---|---|---|---|---|---|---|---|
| | | *E* | *L* | *E* | *L* | *Tot* | *Free* | *Ind* | |
| 1 – 2 | 4 | 0 | 0 | 4 | 4 | 0 | 0 | 0 | cp |
| 1 – 3 | 2 | 0 | 7 | 2 | 9 | 7 | 0 | 0 | |
| 2 – 4 | 3 | 4 | 13 | 7 | 16 | 9 | 0 | 0 | |
| 2 – 5 | 8 | 4 | 4 | 12 | 12 | 0 | 0 | 0 | cp |
| 3 – 5 | 3 | 2 | 9 | 5 | 12 | 7 | 7 | 0 | |
| 4 – 6 | 2 | 7 | 16 | 9 | 18 | 9 | 9 | 0 | |
| 5 – 6 | 6 | 12 | 12 | 18 | 18 | 0 | 0 | 0 | cp |

## 15. Student self review questions

1. In a network, how are activities and events defined? [3]
2. What is a precedence table? [3]
3. Why are dummy activities necessary? [5]
4. What is a critical path through a network? [8]
5. Define total, free and independent floats. [12]

## 16. Student exercises

1. Set up a precedence table and draw the network for the following project.
   B – Attend meeting; C – Write up minutes; A – Arrange when minutes can be typed; D – Have minutes typed out; E – Distribute minutes.
2. The activities A to K in the network specified below have duration times as follows:

| A | B | C | D | E | F | G | H | I | J | K |
|---|---|---|---|---|---|---|---|---|---|---|
| 8 | 4 | 3 | 6 | 2 | 8 | 9 | 7 | 5 | 2 | 3 |

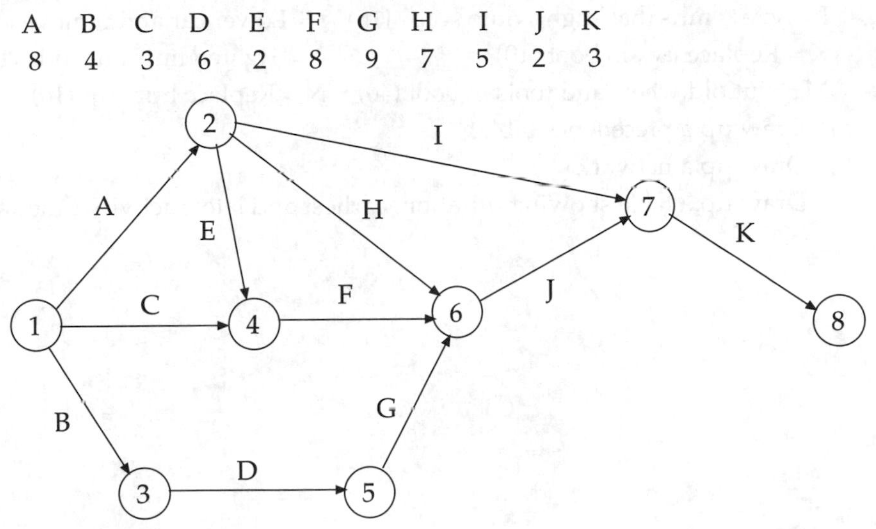


   a) Write out the lists of all possible paths through the network and calculate their time lengths.
   b) Determine the total project time and thus the critical path.
3. For the following precedence table:

| *Activity* | *Preceding activity* |
|---|---|
| A,B | – |
| C | A |
| D,E,F | B |
| G | C,D |
| H,I | F |

   and duration times A = 5, B = 4, C = 2, D = 1, E = 5, F = 5, G = 4, H = 3, I = 2:
   a) draw up a network,
   b) identify the paths through the network and their durations and thus:
   c) identify the critical path and its duration.
4. For the network of exercise 2, determine the earliest and latest event times for all events and check the critical path.
5. For the network of exercise 2, draw up a table showing activity name (in the form 1–2), duration, earliest and latest times and floats.

6. The following activities describe the project 'changing a wheel on a car'. The bracketed figures are timings in seconds.

A = Take jack/tools from boot (40)
B = Remove hub cap (30)
C = Loosen wheel nuts (50)
D = Place jack under car (25)
E = Lift car using jack (20)
F = Get spare wheel from boot (25)
G = Remove wheel nuts and wheel (20)
H = Place spare wheel onto studs (10)
I = Screw nuts (half-tight) onto studs (15)
J = Lower car and remove jack (25)
K = Replace jack in boot (10)
L = Tighten nuts on studs (12)
M = Put old wheel and tools in boot (40)
N = Replace hub cap (10)

a) Draw up a precedence table.
b) Draw up a network.
c) Draw up a table showing duration, earliest and latest activity times and floats.

# **Examination example** (with worked solution)

*Question*

In a machine shop a company manufactures two types of electronic component, X and Y, on which it aims to maximise the contribution to profit. The company wishes to know the ideal combination of X and Y to make. All the electronic components are produced in three main stages:

Assembly, Inspection & Testing, and Packing.

In Assembly each X takes 1 hour and each Y takes 2 hours.

Inspection & Testing takes 7.5 minutes for each X and 30 minutes for each Y, on the average, which includes the time required for any faults to be rectified.

In total there are 600 hours available for assembly and 100 hours for inspection and testing each week. At all stages both components can be processed at the same time.

At the final stage the components require careful packing prior to delivery. Each X takes 3 minutes and each Y takes 20 minutes on average to mount, box and pack properly. There is a total of 60 packing hours available each week.

The contribution on X is £10 per unit and on Y is £15 per unit. For engineering reasons not more than 500 of X can be made each week. All production can be sold.

a) State the objective function in mathematical terms.
b) State the constraints as equations/inequalities.
c) Graph these constraints on a suitable diagram, shading the feasible region.
d) Advise the company on the optimal product mix and contribution.

*CIMA*

*Answer*

a) If the company manufactures $x$ of component X, $y$ of component Y per week, the objective is to maximise profit contribution, £Z, where
$Z = 10x + 15y$.

b) The constraints on production are

| | | |
|---|---|---|
| Assembly time: | $x + 2y \leq 600$ | (1) |
| Inspection, testing time: | $\frac{x}{8} + \frac{y}{2} \leq 100$ | |
| or | $x + 4y \leq 800$ | (2) |
| Packing time: | $\frac{x}{20} + \frac{y}{3} \leq 60$ | |
| or | $3x + 20y \leq 3{,}600$ | (3) |
| Engineering: | $x \leq 500$ | (4) |
| Common sense: | $x, y \geq 0$ | |
| | $x, y$ integers | |

c) See figure.

d) Considering only the vertices of the feasible region:

| *Vertex* | *(x, y)* | $Z = 10x + 15y$ |
|---|---|---|
| A | (0,180) | 2,700 |
| B | (200,150) | 4,250 |
| C | (400,100) | 5,500 |
| D | (500,50) | 5,750* |
| E | (500,0) | 5,000 |

* Hence profit contribution can be maximised at £5,750 per week by manufacturing 500 of type X and 50 of type *Y* per week.

**Note**: The co-ordinates of vertices B, C and D can be found either by reading off an accurate graph or by solving simultaneous equations.

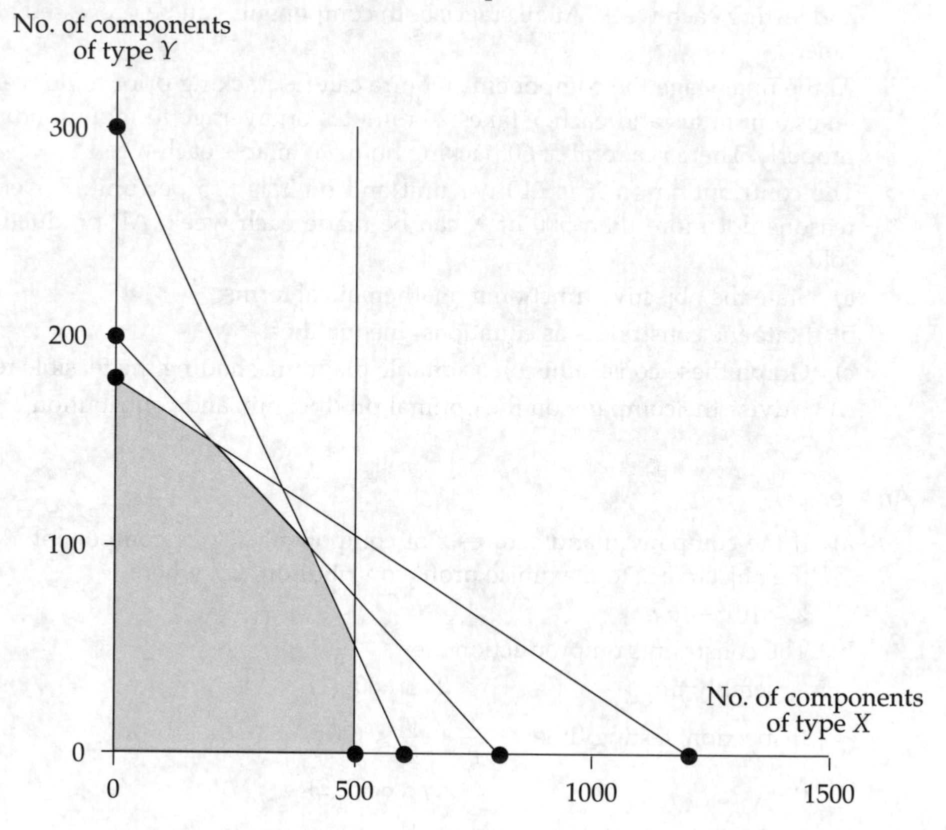

# Examination questions

1. A company has an advertising budget for Brand X of £100,000. It must decide how much to spend on television advertising and how much on newspaper advertisements. From past experience each advertisement is expected to achieve extra sales of Brand X as follows:

   | | |
   |---|---|
   | Newspaper advertisement | 400 units |
   | Television spot | 1,000 units |

   The gross profit on sales is £10 a unit. For contractual reasons, the company can spend up to £70,000 on either form of advertising. For marketing balance, at least half as many newspaper advertisements as television spots are required. Each television spot and newspaper advertisement costs £5,000 and £2,000 respectively. The objective is to maximise expected contribution (gross profit less advertising costs).

   You are required

   a) to find the expected contributions for (i) a television spot, (ii) a newspaper advertisement;
   b) to state the objective function and constraints in mathematical terms;
   c) to find the company's optimum contribution by graphical means, and to comment on your answer.

   *CIMA*

2. a) To select trainee accountants a firm uses a two stage selection process. If successful at the first stage the applicant passes to the second stage which is an interview with the head of the accounting department. It has been suggested that an intelligence test be used to supplement the first stage interview. In order to look at the effectiveness of the test it will initially be used to supplement the first stage interview.
   The probability of failing the test is 0.8.
   The probability of someone being appointed having passed the test is 0.3
   The probability of someone being appointed having failed the test is 0.1.
   Required:
      i. What is the probability of a candidate being appointed?
      ii. If a candidate was not appointed, what is the probability that they passed the test?
      iii. If, on average, 15% of applicants are called for first interview, what is the overall probability of an applicant being appointed?

   b) A company has a 25% share of the market for its products. For each product there are three processes in its production. These three processes are shared by the three main products. The table below gives the amount of each process required to produce a unit of each product.

| | *Process* | | |
|---|---|---|---|
| | P1 | P2 | P3 |
| Product requirements (hours per unit) | | | |
| Product $x$ (1) | 1 | 1 | 2 |
| Product $x$ (2) | 1 | 2 | 2 |
| Product $x$ (3) | 2 | 4 | 3 |

The company requires to fully utilise each process.

Profit for $x$ (1) is £4 per unit, for $x$ (2) £4 per unit and for $x$ (3) is £2 per unit.

Required:

i. Formulate the above situation as a matrix problem.

ii. If the company must produce 2,000 units of $x$ (1), 1,000 units of $x$ (2) and 4,500 units of $x$ (3) to fully utilise each process, how much monthly profit will it make?

iii. The company can add to the availability for each process and increase its output of $x$ (1) to 3,000 units, of $x$ (2) to 1,500 units and of $x$ (3) to 5,000 units. If the company could maintain its market share what would the total annual value of the market be in these circumstances?

*ACCA*

3. A company needs to purchase a number of small printing presses, of which there are two types, $X$ and $Y$. Type $X$ costs £4000, requires two operators and occupies 20 square metres of floor space. Type $Y$ costs £12000, also requires two operators but occupies 30 square metres. The company has budgeted for a maximum expenditure on these presses of £120,000. The print shop has 480 square metres of available floor space, and work must be provided for at least 24 operators. It is proposed to buy a combination of presses $X$ and $Y$ that will maximise production, given that type $X$ can print 150 sheets per minute and type $Y$, 300 per minute.

You are required to:

a) write down all the equations/inequalities which represent the cost and space conditions. The labour conditions are given by $2X + 2Y \geq 24$;

b) draw a graph to represent this problem, shading any unwanted regions;

c) use the graph to find the number of presses $X$ and $Y$ the company should buy to achieve its objective of maximum production;

d) state the figure of maximum production and the total cost of the presses in this case.

*CIMA*

4. a) i. Calculate the rates of interest that give a break-even position if the equation for net present value (£$P$) is given by $P = 192 - 28r + r^2$, where $r$ is the discount factor.

ii. Explain the significance of the values of $r$ that you obtain in part (i) with respect to profit and loss in the context of net present value.

b) Socsport plc is a company in the wholesale trade selling sportswear and stocks two brands, $A$ and $B$, of football kit, each consisting of a shirt, a pair of shorts and a pair of socks. The costs for brand $A$ are £5.75 for a shirt, £3.99 for a pair of shorts and £1.85 for a pair of socks and those for brand $B$ are £6.25 for a shirt,

£4.48 for a pair of shorts and £1.97 for a pair of socks. Three customers *X*, *Y* and *Z* demand the following combinations of Brands: X, 36 kits of brand *A* and 48 kits of brand *B*; *Y*, 24 kits of brand *A* and 72 kits of brand *B*; *Z*, 60 kits of brand *A*.

Required:

i. Express the costs of brands *A* and *B* in matrix form, then the demands of the customers, X, *Y* and Z in matrix form.
ii. By forming the product of the two matrices that you obtain in the previous part, deduce the detailed costs to each of the customers.

c) As a result of recent price increases, *Z* has ceased to be a customer of Socsport, but the demands of *X* and *Y* remain the same. Socsport no longer stocks socks of either brand.

Required:

Re-express the demands of *X* and *Y* in matrix form and find its inverse matrix.

*ACCA*

5. The following table gives data for a simple project:

| *Activity* | *Preceding activity* | *Duration (days)* |
|---|---|---|
| A | – | 3 |
| B | – | 3 |
| C | – | 7 |
| D | A | 1 |
| E | D,J | 2 |
| F | B | 2 |
| G | C | 1 |
| H | E,F,G | 1 |
| J | B | 1 |

You are required

a) to draw a network diagram for the project;
b) to draw up a table, with a list of activities, durations, earliest start and finish times, latest start and finish times, and total floats;
c) to state and to explain the critical path.

*CIMA*

6. Your company has to buy immediately two types of table for its canteens. A maximum sum of £24,000 is available for this purpose. A Type X table costs £40 and seats four people. A Type *Y* table costs £30 and seats two people. Seating for at least 1,800 people is required. There must be at least as many Type *Y* tables as Type *X* because the tables are to be used for a variety of functions in the canteens. For reasons of maintenance, storage etc, the company wishes to buy the smallest total number of tables to meet its requirements.

You are required to

a) state the company's objective function;
b) state all the constraints (equations/inequalities);
c) draw a graph of these constraints, shading any unwanted regions;
d) recommend the number of each type of table the company should buy, justifying your answer.

*CIMA*

7. There are three types of breakfast meal available in supermarkets known as brand BM1, brand BM2 and brand BM3. In order to assess the market, a survey was carried out by one of the manufacturers. After the first month the survey revealed that 20% of the customers purchasing brand BM1 switched to BM2 and 10% of the customers purchasing brand BM1 switched to BM3. Similarly after the first month of the customers purchasing brand BM2, 25% switched to BM1 and 10% switched to BM3 and of the customers purchasing brand BM3 5% switched to BM1 and 15% switched to BM2.

   Required:

   i. Display in a matrix S, the patterns of retentions and transfers of customers from the first to the second month, expressing percentages in decimal form.
   ii. Multiply matrix S by itself (that is form $S^2$).
   iii. Interpret the results you obtain in part (ii) with regard to customer brand loyalty.

*ACCA*

8. Your departmental manager has asked for help in the design of a stock control system for a client. The following data is available for the system. The rate at which the client uses the stock is 1,000 units per year. His order cost is £10 per order. The cost of storing one unit of the stock for a year is £2.

   Required:

   a) If the cost, £C, of an ordering policy is given by

   $$C = D.\frac{c_1}{q} + q.\frac{c_2}{2}$$

   where $D$ is the demand per period
   $q$ is the quantity ordered
   $c_1$ is the cost (£) of placing an order
   $c_2$ is the cost (£) of holding a unit of stock for a period.

   Derive the formula for the order quantity which minimises the cost of the ordering policy.

   b) For batch delivery system determine the order quantity which minimises cost.
   c) Draw a schedule of order size against ordering cost, stock holding cost and total cost for order sizes of 0,20,40,...160.
   d) Use the schedule of c) to construct a graph of ordering cost, holding cost and total cost against order size.
   e) What assumptions are made in this model of stock control?

*ACCA*

9. A chocolate manufacturer produces two kinds of chocolate bar, *X* and *Y*, which are made in three stages: blending, baking and packaging. The time, in minutes, required for each box of chocolate bars is as follows:

| | Blending | Baking | Packaging |
|---|---|---|---|
| *X* | 3 | 5 | 1 |
| *Y* | 1 | 4 | 3 |

The blending and packaging equipment is available for 15 machine hours and the baking equipment is available for 30 machine hours. The contribution on each box of X is £1 and on each box of *Y*, £2. The machine time may be used for either *X* or *Y* at all times it is available. All production may be sold.

You are required to

a) state the equations/inequalities which describe the production conditions;

b) draw a graph of these equations/inequalities and hence find how many boxes of each chocolate bar the manufacturer should produce to maximise contribution;

c) state this maximum contribution and comment on your answer.

*CIMA*

# Answers to student exercises

## Chapter 2 – Sampling and data collection

1. a) is the correct answer.

   c) is a sophisticated and expensive form of random sampling and b) and d) are quasi random sampling techniques.

2. (d) is correct. Picking a random starting point and then choosing every nth item to coincide with the proportion desired is the precise way systematic samples are structured.

3. Primary: current average pay and estimated new average pay; workers attitudes to new scheme; pay structures in similar companies; official trade union views. Secondary: hours worked and average pay in this industry and over all industries.

4. General: available time and manpower; complexity of enquiry; a census might be feasible if information can be obtained by observation only. Probably a sample would suffice unless the survey was official. Is there a sampling frame available?

5. (d) is correct, since it is the only one that gives each employee of the whole company an equal chance of being chosen.

6. a) Probably cluster
   b) Stratified would be essential, due to the nature of the enquiry (personnel records would be a good sampling frame)
   c) Systematic
   d) Multi-Stage/Cluster
   e) Simple random (from school records sampling frame).

7. a) 150 (b) 95

   c)

| | Own | Public | Other | TOTAL |
|---|---|---|---|---|
| Manual | 60 | 75 | 15 | 150 |
| Non-manual | 25 | 20 | 5 | 50 |
| TOTAL | 85 | 95 | 20 | 200 |

   d) Better to spread sample over more than one location and occasion, since: factory gate B might be used by only certain types of workers; shift workers who finish before 5 o'clock will not be considered. Also, workers might resent the inconvenience and the statistical worker would find extracting the information difficult with so many people moving quickly. It will also be very difficult to count moving cars and interview people at the same time. A short simple census would be much better (random sampling would not be appropriate here).

8. 1) A pointless question 2) Satisfactory 3) Which dog food?
   4) Exactly what does 'amount' mean and over what period of time?
   5) Difficult memory question (how accurate does the respondent have to be?) Better is one of, say, four boxes to be ticked.
   6) Better to give a set of house categories and ask respondents to choose one.

10. People who are against the scheme are much more likely to write in. Thus no conclusions can be drawn. An unbiased sampling technique (based in the area immediately surrounding the proposed scheme) would be appropriate.

## Chapter 3 – Data and their accuracy

1. c) is the correct answer. Counting the number of times an event occurs will always give a discrete (precise) value. Times, weights and ages can never be calculated precisely, only approximated. a), b) and d) are examples of continuous data.
2. a) is the correct answer. The largest value that 8.2 can take is 8.25 and the largest value that 16 can take is 16.5, giving 24.75 as the largest value that their sum can take.
3. a) Univariate, variable='time to complete job', continuous, numeric.
   b) Bivariate, variable 1='job title', non-numeric, discrete;
   variable 2='age', numeric, continuous.
   c) Bivariate, variable 1='location', non-numeric, discrete;
   variable 2='no. of employees', numeric, discrete.
   d) Univariate, variable='dept. name', non-numeric, discrete.
   e) Multivariate (5-variable):
   variable 1='average wage', numeric, discrete;
   variable 2='manual/non-manual', non-numeric, discrete;
   variable 3='sex', non-numeric, discrete;
   variable 4='industry', non numeric, discrete;
   variable 5='year', numeric, discrete.
4. (a) £148 360 (b) 23 000 tons (c) 3.2 mm (d) £16 (e) 30 months (f) £18 600
5. b) is the correct answer. The form of sampling used has no particular bearing on the way data is measured and thus its accuracy. Therefore c) and d) are misleading statements.
6. (b) is correct. Kappa range is (95000, 105000) and Lambda (180000, 220000). Thus, range of joint stock is (275000, 325000) which has a possible highest error of 25,000 of the estimated 300,000 valuation = 25000/300000 = 0.083 = 8.3%
7. (a) [192,206] (b) [56,96] (c) [159.71,172.15] (2D)
   (d) [457.54,502.38] (2D) (e) [0,0.75] (2D)
8. (a) £600
   (b) [0,£1320]. Buy = [180,220]. Per item worst = Buy at £16, Sell at £16, Profit=0; Per item best = Buy at £14, Sell at £20, Profit=£6. Thus profit range = £[180×0, 220×6].
9. (a) [1776,2028] (b) (i) [£3463.20,£4157.40] (ii) [£5239.20,£6185.40]
   (c) [£1081.80,£2722.20] (d) £3800; min profit = 28.5%, max profit = 71.6%
10. [104 269,105 816] in £000.
11. (a) [53,107] (b) 80

## Chapter 4 – Frequency distributions and charts

1.

| Number of vans unavailable | 0 | 1 | 2 | 3 | 4 | 5 | 6 | 7 | 8 | Total |
|---|---|---|---|---|---|---|---|---|---|---|
| Number of days | 12 | 21 | 11 | 9 | 2 | 3 | 1 | 0 | 1 | 60 |

Comments: The most common number of vans unavailable was 1 and on only 12 occasions were all vans available. On most days there were not more than 3 vans unavailable and at no time was there more than 8 vans unavailable.

2. (a) Since the data is discrete (counted), the limits of separate classes should not meet.
   (b) There are too few classes given.
   (c) The distribution of frequencies is not good.
   (d) The last two classes are out of order.
   (e) A better structure would be to have classes such as: 500–509, 510–519, 520–524, 525–529, 530–534, 535–539, 540–559, 560–579.

3. (a)

| Class | Limits lower | Limits upper | Boundaries lower | Boundaries upper | Width | Mid-point |
|---|---|---|---|---|---|---|
| 0 to 4 | 0 | 4 | 0 | 4.5 | 5 | 2 |
| 5 to 9 | 5 | 9 | 4.5 | 9.5 | 5 | 7 |
| 10 to 19 | 10 | 19 | 9.5 | 19.5 | 10 | 14.5 |
| 20 to 29 | 20 | 29 | 19.5 | 29.5 | 10 | 24.5 |
| 30 to 49 | 30 | 49 | 29.5 | 49.5 | 20 | 39.5 |

   (b) 12(2)+28(7)+9(14.5)+7(24.5)+2(39.5) = 601.

4.

| Diameter of bolt | Number of bolts |
|---|---|
| 1.90 – 1.94 | 3 |
| 1.95 – 1.99 | 9 |
| 2.00 – 2.04 | 17 |
| 2.05 – 2.09 | 10 |
| 2.10 – 2.14 | 9 |
| 2.15 – 2.19 | 5 |
| 2.20 – 2.24 | 5 |
| 2.50 and over | 2 |

6. (b) 4143 (c) Total number of car-hours per day = 400 ×12 = 4800.
   Thus utilization = $\frac{3485}{236}$ = 86%

8. (a) 19 (b) 14 (c) 238 hours (approx)

9.

| 0-100 | 100-200 | 200-300 | 300-400 | 400-500 | 500-600 | 600-700 | 700-800 | 800-900 | 900-1000 |
|---|---|---|---|---|---|---|---|---|---|
| 1 | 1 | 8 | 8 | 12 | 25 | 19 | 14 | 9 | 3 |

10.

| Repayments (under, £000) | 0.4 | 0.8 | 1.2 | 1.6 | 2.0 | 2.4 | 2.8 | 3.2 |
|---|---|---|---|---|---|---|---|---|
| % of home owners | 2 | 17 | 42 | 69 | 87 | 95 | 98 | 100 |

   (i) 30 (ii) 38

11. (a)

| | | | | | | | |
|---|---|---|---|---|---|---|---|
| Cumulative % repayments | 0.3 | 6.9 | 25.3 | 53.1 | 76.9 | 89.9 | 95.6 | 100 |
| Cumulative % number of home owners | 2 | 17 | 42 | 69 | 87 | 95 | 98 | 100 |

   (c) 37

12.

| | | | | | | |
|---|---|---|---|---|---|---|
| Cumulative % net output | 11 | 37 | 49 | 65 | 78 | 100 |
| Cumulative % number of firms | 40 | 72 | 87 | 94 | 98 | 100 |

   The largest 20% of firms account for approximately 57% of all net output.

13.

| | | | | | | | | | |
|---|---|---|---|---|---|---|---|---|---|
| Cumulative % income before tax | 1 | 6 | 15 | 35 | 60 | 72 | 82 | 89 | 100 |
| Cumulative % number of incomes | 4 | 16 | 30 | 56 | 80 | 89 | 95 | 98 | 100 |

## Chapter 5 – General charts and graphs

3. Line diagram or simple bar chart. The latter should have each bar 'broken' in some way to emphasise the break of scale, making the former preferable. The shifts appear to be in cycles of three.
4. Multiple line diagram and multiple bar chart. The former is preferable, particularly since the three variable values are conveniently separate.
5. Neither; they both show different aspects of the data. Both have their place, depending on whether absolute values and totals or relative comparisons are more important.
7. Radius 2 = 1.17 × radius 1; radius 3 = 1.25 × radius 1. Angles (in degrees):
   Circle 1 = 50, 81, 166, 63;
   Circle 2 = 49, 62, 177, 71; Circle 3 = 47, 53, 171, 90.

## Chapter 6 – Arithmetic mean

1. (a) 80.4 (1D) (b) 0.507 (3D)   2. (a) 19.91 (2D) (b) 4.1 (1D)
3. (a) is correct. 'Expected' = mean = [0(6)+1(3)+4(2)+4(3)+2(4)+1(5)]/20 = 1.8.
4. 25.2 (1D)   5. 37.04 yrs (2D)
6. 425.0012 gms (4D). The consumer is getting reasonable value since the mean is (just) over the advertised contents weight of 425 gms.
7. (a) 23.625 (b) 40   8. (a) 16.7 (b)(i.) 24.3 (ii.) 20.5 (c) 62
9. d) is incorrect. The mean CAN always be calculated, no matter how large the set of values is.

## Chapter 7 – Median

1. (b) is correct. The items in size order are: 35 35 36 36 36 37 38 40 40 42 43 and the middle item is the 6th, which is 37.
2. (a) 3.75 (b) 78.5
3. (a) 22 (b) Extreme values are present at the lower end of the distribution. (c) The bags would probably be packed to some nominal weight, which would mean that the actual weights would be fairly symmetrically distributed. Thus the mean would be an ideal average.
4. 126.75. A class width of 9 is not a very desirable one from the point of view of an observer attempting to comprehend the general nature of the data. A much better class structure would be 90–99, 100–109, etc.
5. 1.79 hours (2D).   6. 580 hours.

## Chapter 8 – Mode and other measures

1. (a) 11 (b) 2 and 3 (bi-modal) (c) 15
2. 21.8 (1D). Since the distribution is fairly well skewed, the mode is giving a value that is too high for practical purposes. The median is probably more suitable here.
3. Mode=1. This type of data is ideally suited to a modal average, since in normal circumstances the most typical number of children can be put to more practical use than a non-typical mathematical value given, for instance, by the mean.
4. (c) is incorrect. Refer to Figure 3 for the relative positions of the mean, median and mode in a right-skewed distribution.
5. (a) 2972 (b) 2769 (c) 2744 (d) 456 (e) 7. Distribution is moderately skewed.
6. 3;2.88;2.77.
7. 75.4 customers per hour.
8. 3.2%
9. 8.21%
10. (a) 58.3 (b) 56.3
11. (a) £20/sq yd (b) £32/sq yd.

## Chapter 9 – Measures of dispersion and skewness

1. 25; 8.375
2. Before: mean=£115.50; md=£22.90. After: mean=£98.50; md=£30.20.
3. (a) 4.1 (1D) (b) 0.81 (2D) (c) 4 (d) 0.77 (2D)
4. 5.0 (1D)

## Chapter 10 – Standard deviation

1. 1.63 (2D)
2. (c) is correct. All values have been converted to percentages by multiplying by 4. The standard deviation is measured in the same units as the petty cash amounts, thus the new mean and standard deviation also should be multiplied by 4.
3. (a) 6.6; 6; 8; 1.9 (1D) (b) 0.3 (1D) (c) Range/6 = 1.3 (1D). Sd is 1.9 (larger than 1.3) since the distribution is right skewed.
4. (c) is correct. CV = (mean/sd)×100% = 25%.
5. (a) 36.32 (2D); 9.95 (2D). (b) 16.42 and 56.22.
6. (a) 0.16712; 0.00221 (b) cv(1)=1.32%; cv(2)=1.82% (c) 0.16571 (d) New mean increased to 0.16800. Sd unaltered at 0.00300. New cv(2)=1.79%. Second sample still more variable.

## Chapter 11 – Quantiles and the quartile deviation

1. (a) is correct. All the others are measured in the same units as the data.
2. Median=30; Q1=28; Q3=32; qd=2
3. 1; 1
4. (a) 35.07; 9.74 (b) 0.07
5. (b) is correct. The diagram shows a 'less than' ogive and C corresponds to a point 600/800 = 75% along the distribution of values.
6. (a) 10.90; 11.98 (b) Only 30% of bulbs last longer than 11.98 days.
7. Classes are 0.5–1.5, 1.5–2.5, ... etc. Median=4.15 wks.

## Chapter 12 – Linear functions and graphs

1. (b) is correct.
2. (a) 4;13 (b) 3;–12 (c) 2;1.5 (d) 0.25;0.5
3. (Graphs NOT given in these answers)
4. (a) 1 (b) –6 (c) 0.1
5. (a) $y = 1 + x$ (b) $y = 6 - x$ (c) $y = 0.5x - 1$
6. $y = 2x - 4$
7. (a) is correct. The relationship can be re-arranged into the form $y = 4 - 2x$ which shows the gradient is –2.
8. (c) is correct. Only equation (c) satisfies the two points x=12, y=2 and x=8, y=0

## Chapter 13 – Regression techniques

1. Mean value = (7,5); Estimate = 6.7 (approx).
2. $y = 0.51 + 0.64x$; Estimate = 6.9 (approx).
3. $y = -0.6 + 0.9x$; 48.9
4. $y = 1.94x + 10.83$
5. (d) is correct. Taking the regression line in the form $y = mx + c$, c = 6, since the lines meets the $y$-axis at 6. Also the orientation of the line shows that the gradient (m) is negative with ratio $-6/6 = -1$.
6. $x$=capacity, $y$=price. $y = 1236.9 + 3.0192x$.
7. (a) There is a clear case of independent versus dependent variable here, since overtime worked will presumably depend on how much work there is to do (i.e. how many orders have been received). Thus $x$=orders and $y$=overtime. Line is $y = -2.26 + 0.40x$ and $y(100)$=38 (i.e. 38 overtime hours are needed to support 100 orders).
   (b) We require the number of orders that corresponds to 35 hours total overtime, which is obtained by substituting $y$=35 into the above line. The value of $x$ obtained (number of orders) is the least that would be acceptable as a criterion for taking on a new employee.

## Chapter 14 – Correlation techniques

1. 0.98(2D)
2. (a) is correct. Value +1 is perfect positive correlation and –1 is perfect negative correlation. No correlation has value zero.
3. (a) is correct. Remember that positive correlation shows a pattern stretching from bottom left to top right. Also, the stronger the correlation is, the more the points resemble a straight line.
4. 0.60
5. 0.97. This shows that 97% of the variation in profit (before taxation) is due to variations in turnover. Clearly a result to be expected, since no other significant factors should be expected.
6. $r$=0.93. A high positive correlation coefficient seems appropriate. A cost on output regression line could then be extrapolated with some degree of confidence.
7. (a) 0.49 (b) 0.33
8. $r'$=0.51. Only a moderate degree of positive correlation, and with the coefficient of determination ($r'^2$) = 0.26, this means that only 26% of the movement of the share price can be attributed to the FT Index. Thus the FT Index can be considered as only an approximate indicator for the particular share price.

9. $r'$=0.67. A moderate level of agreement. They both clearly agree on the best and the worst suppliers and in between there is no strong measure of disagreement. No particular conclusions about quality of suppliers can be deduced since the two managers would be assumed to weigh various attributes of suppliers differently, according to their own needs and functions.
10. (a) is incorrect. Data must always be numeric for the product moment correlation coefficient to be calculated.

## Chapter 15 – Time series model

1. Extreme weather (storms, heat waves); stock shortages; special events (fairs, shows); new competition in the area; bank holidays; national strikes; special sales offers; 'flu epidemics.
2. Company sales are highly seasonal, with the peak during the summer. Autumn sales are beginning to catch the value of sales in the winter quarter. The general trend in sales is upwards.

## Chapter 16 – Time series trend

1. 15.4, 15.2, 15.0, 14.8, 14.6, 14.4, 14.2, 14.0, 13.8, 13.6
2. (a) is correct. A 4-point centered average will always leave two time points at either end of the data without a plotted point.
3. (a) 9.7,14.0,11.7,11.3,8.3,10.3,15.0,12.7,12.7,9.3,11.3,16.7,14.3 (1D)
   (b) 10.8,11.0,11.2,11.2,11.6,11.8,12.0,12.2,12.4,13.0,13.2
   The moving average of period 5 is the correct one to use for the trend, since the data cycle over groups of 5 exactly.
5. The moving average trend, beginning at Year 4 is (to nearest unit):
   323, 328, 332, 339, 346, 357, 366, 375, 384, 390, 392.
6. 3.3, 3.4, 3.5, 3.6, 3.7, 3.9, 4.0, 4.1, 4.2, 4.3, 4.4, 4.5

## Chapter 17 – Seasonal variation and forecasting

1. (a) Seasonal factors: 13.8, 23.8, –41.3, 3.8
   Seasonally adjusted figures: 126, 126, 136, 131, 136, 146, 146, 161, 166, 166, 176, 171
   (b) Seasonal factors: 1.09, 1.17, 0.71, 1.02
   Seasonally adjusted figures: 127, 128, 137, 126, 137, 147, 145, 168, 166, 165, 171, 182
2. (a) Season 1 = –2.2; season 2 = 0.2; season 3 = –1.5; season 4 = 10.6; season 5 = –7.1
   (b) 10.2,10.8,11.5,10.4,11.1,11.2,11.8,11.5,12.4,12.1,12.2,12.8,12.5,15.4,13.1
   (c) 11.7, 14.4, 12.9, 25.2, 7.8
3. (a) 0.78, 0.95, 0.98, 1.29 (b) 318, 302, 319 (to nearest unit)
   (c) 289, 301, 403 (to nearest unit)
4 (a) –0.8, 0.5, –0.6, 0.9 (b) 3.8, 5.2, 4.2, 5.8
5. (c) is correct. The model is: y = t × S or Sales = Trend × Seasonal.
   ie 1600 = Trend × 0.8
   Thus, trend value = 1600/0.8 = 2,000.

6. (a) is correct. The seasonal (proportion) values must add to 4, which gives Q4 seasonal value as 0.5. Alternatively, the percentage seasonal variation should add to zero, giving Q4 seasonal variation as –50%.
7. (a) is correct. Trend for Q1 is known as 2,000 (from Q5). Similarly, Q2 trend = 4400/2 = 2,200 and Q3 trend = 1680/0.7 = 2,400. Clearly, Q4 trend = 2,600. Thus Q4 sales = trend × seasonal factor = 2600 × 0.5 = 1,300.
8. Seasonally adjusted values:
   Additive: 6.4, 9.1, 8.6, 8.4, 8.9, 8.1, 8.2, 8.8, 8.2, 9.2, 8.9, 9.8 (%)
   Multiplicative: 7.1, 9.2, 8.6, 8.2, 8.8, 8.1, 8.2, 8.8, 8.3, 9.3, 9.0, 10.2 (%)
   Forecast for 1983, quarter 1 = 13% (additive and multiplicative)
   The multiplicative model would be better since figures are percentages.

## Chapter 18 – Index relatives

1. (c) is correct, since $120^*(1.05)^3$ = £162.07.
2. Price relative = 90; quantity relative = 130; expenditure relative = 117.
3.

| | Mar | Apr | May | Jun | Jul | Aug | Sep | Oct |
|---|---|---|---|---|---|---|---|---|
| (a) | 100 | 88.7 | 90.1 | 73.2 | 76.1 | 102.8 | 111.3 | 96.5 |
| (b) | 110.9 | 98.4 | 100 | 81.3 | 84.4 | 114.1 | 123.4 | 107.0 |
| (c) | 97.3 | 86.3 | 87.7 | 71.2 | 74.0 | 100 | 108.2 | 93.8 |

4.

| | Mar | Apr | May | Jun | Jul | Aug | Sep | Oct | Nov |
|---|---|---|---|---|---|---|---|---|---|
| Fixed base | 100 | 103.8 | 108.2 | 109.4 | 108.4 | 106.1 | 103.5 | 104.2 | 101.5 |
| Chain base | 103.8 | 104.2 | 101.1 | 99.2 | 97.8 | 97.5 | 100.8 | 97.4 | – |

5. (a)

| | 19X1 | 19X2 | 19X3 | 19X4 | 19X5 | 19X6 | 19X7 | 19X8 | 19X9 |
|---|---|---|---|---|---|---|---|---|---|
| Chain base index | – | 105.2 | 98.0 | 112.0 | 114.3 | 94.5 | 119.8 | 102.8 | 102.0 |
| Amount harvested | – | – | – | 587 | – | 634 | 760 | 781 | 797 |

6.

| | 19X0 | 19X1 | 19X2 | 19X3 | 19X4 | 19X5 | 19X6 | 19X7 |
|---|---|---|---|---|---|---|---|---|
| Index for firm (19X2=100) | 101 | 96 | 100 | 107 | 98 | 98 | 103 | 107 |
| National index (19X2=100) | 90 | 89 | 100 | 104 | 97 | 96 | 100 | 104 |

7.

| | | | | | | | | | | |
|---|---|---|---|---|---|---|---|---|---|---|
| Whole economy | 100 | 102 | 103 | 103 | 104 | 106 | 105 | 107 | 106 | 107 |
| Coal and coke | 100 | 157 | 176 | 178 | 189 | 191 | 193 | 196 | 196 | 204 |

8.

| Time point | 1 | 2 | 3 | 4 | 5 | 6 | 7 |
|---|---|---|---|---|---|---|---|
| Real index | 100 | 99.7 | 102.1 | 110.7 | 111.6 | 111.9 | 112.8 |

9.

| | 19X5 | 19X6 | 19X7 | 19X8 | 19X9 | 19Y0 | 19Y1 |
|---|---|---|---|---|---|---|---|
| Fixed base | 100 | 99.5 | 89.2 | 98.2 | 99.0 | 98.3 | 105.5 |
| Chain base | – | 99.5 | 89.6 | 110.1 | 100.8 | 99.3 | 107.3 |

10. (a) is correct. Over the period, inflation has increased by 210/180=16.7% and money wages by 115/100=15%. Thus (i) is correct. Also, 5% compounded for 3 years yields $1.05^3$ = 1.157 or 15.7%. Thus (ii) is incorrect.
11. (b) is correct. By calculating the RVI (see section 13) we obtain $\frac{115}{100} \times \frac{180}{210} \times 100$ which represents a 1.43% decrease.

## Chapter 19 – Composite index numbers

1. (a) is correct. 7(130)+3X=10(127) is rearranged to give 3X=360. Thus X=120.
2. (a) 105.4 (b) 104.9
3. $I_{Feb}$=110.2; $I_{Mar}$=101.2; Volumes increased overall by 10% from Jan to Feb, then fell by almost the same amount from Feb to Mar.
4. 101.7
5. $L_2$=110.6; $L_3$=176.6.
6. (a)

| | 19X0 | 19X1 | 19X2 |
|---|---|---|---|
| Wheat | 0.99 | 0.99 | 0.98 |
| Barley | 0.19 | 0.19 | 0.24 |
| Oats | 2.70 | 3.15 | 3.48 |

(b) $L_{19X1}$=107.0; $L_{19X2}$=112.4

(c) $P_{19X1}$=107.3; $P_{19X2}$=112.3

7. $P_{cost}$=107.7; $P_{quantity}$=109.3

## Chapter 21 – Interest and depreciation

1. 44; 345
2. 47.4
3. –560
4. 2187; 3280
5. 18.60
6. 70
7. £1650
8. (a) £924.20 (b) £4578.47
9. £873.22
10. 14
11. 24.57%
12. £9929.97
13. (d) is correct. APR is calculated at 1% compounded for 12 periods (of 1 month). This is $(1.01)^{12}$ = 1.1268 yielding 12.7%.
14. (a) 18.81% (b) 19.25% (c) 19.56%
15. (b) is correct. Using the depreciation formula in section 22 gives $5 = 46(1-i)^3$. This can be rearranged to give i = 0.4772, giving the average rate as 47.7%.
16. £35895
17. (a) is correct. £20,000 × $(0.78)^3$ = £9,491 and £20,000 × $(0.82)^3$ = £11,027.
18. (a) £55188.71 (b) 17.57%
19. £4774.42

## Chapter 22 – Present value and investment appraisal

1. (a) £986.27 (b) £1134.85 (c) £3583.79
2. PV of £10000 in 2 years is £9053.99. Therefore pay £9000 now.
3. £15686.62
4. PV of debt = £2524 + o/heads of £250 gives total cost of £2774. So the minimum selling price = £277.40 per set.
5. The real comparison is between £800 now or £1000 in 1 year. This represents a discount rate of 25%, which is higher than any standard investment rate. Hence, £1800 cash is well spent here!
6. NPV = £5366.80.
7. At rate 18%, NPV is £14069. Purchase is recommended.
8. At 18%, NPV = £14069; at 25%, NPV = –£3359. IRR estimate = 23.7%.
9. (a) NPV1 = £11314.70; NPV2 = £12816.00. Choose project 2 on NPV.
   (b) NPV1 = –£232.20; NPV2 = –£1202.50. IRR1 = 19.9%; IRR2 = 19.6%. Choose project 1 on IRR.
   (c) The two projects are different in scale and also have different flow patterns. (Note however that with a discount rate of 20%, the NPV criterion would choose project 1 since it has the least deficit.)
10. (a) –£3490 (b) The value of the project will not be enough to repay the loan.

## Chapter 23 – Annuities

1. (a) £2933.30 (b) £3167.96
2. £317.62
3. (d) is correct. Calculation = £17,000/(0.06) = £283,333.
4. (a) £9124.69 (b) £14419.43 (c) £21739.13
5. (b) is correct. Use the formula in section 11 with P=£200,000, n=15 and i=0.06
6. (a) £13224.31

   (b)

| Year | 1 | 2 | 3 | 4 | 5 |
|---|---|---|---|---|---|
| Interest paid (£) | 7062.00 | 6045.22 | 4860.67 | 3480.67 | 1872.97 |

7. (a) £2196.69 (b) £183.06 per month can be invested as an ordinary annuity at 0.75%/mth to yield £2289.64 at year end. Therefore excess = £92.95 = 0.5% of original principal.
8. Payment is £3686.45
9. Depreciation charge is £2490.56
10. £7767.23; £14784.46

## Chapter 24 – Functions and graphs

1. (a) is correct. $\frac{20}{0.2}\times\frac{1}{r}=\frac{20}{0.2}\times\frac{1}{100}=\frac{20}{20}=1$
2. $x$=0.82; $y$=9.53
3. (a) $x$=100, $y$=200 (b) $x$=3, $y$=4 (c) $x$=–2, $y$=4 (d) $x$=12, $y$=–12
4. Plotted points:

| $x$ | –4 | –3 | –2 | –1 | 0 | 1 | 2 |
|---|---|---|---|---|---|---|---|
| $y$ | 9 | 0 | –5 | –6 | –3 | 4 | 15 |

   (i) $-3 < x < 0.5$ (ii) $x < -3$ and $x > 0.5$

5. Plotted points:

| $x$ | –1 | 0 | 1 | 2 | 3 | 4 | 5 |
|---|---|---|---|---|---|---|---|
| $14-3x$ | 17 | 14 | 11 | 8 | 5 | 2 | –1 |
| $2x^2-12x+14$ | 28 | 14 | 4 | –2 | –4 | –2 | 4 |

   $x$=0, $x$=4.5

6. Plotted points:

| $x$ | –2 | –1 | 0 | 1 | 2 | 3 | 4 |
|---|---|---|---|---|---|---|---|
| $2x^2-4x-5$ | 11 | 1 | –5 | –7 | –5 | 1 | 11 |
| $9+5x-x^2$ | –5 | 3 | 9 | 13 | 15 | 15 | 13 |

   $x$=–1, $x$=4

7. (a) $x^3-4x^2+x+6$

   (b) Plotted points:

| $x$ | –1 | 0 | 1 | 2 | 3 | 4 |
|---|---|---|---|---|---|---|
| $x^3-4x^2+x+6$ | 0 | 6 | 4 | 0 | 0 | 10 |

   Curve meets $x$-axis at $x$=–1, $x$=2 and $x$=3.

   (c) $x$=0 and $x$=1.5

8. Plotted points:

| $x$ | 1 | 1.25 | 1.5 | 1.75 | 2 | 2.25 | 2.5 |
|---|---|---|---|---|---|---|---|
| $y$ | –0.5 | 0.19 | 0.5 | 0.44 | 0 | –0.81 | –2 |

   (i) 1200 and 2000 (2S) (ii) max profit = £520, production level = 1600 (2S)

9. (a) Plotted points:

| $x$ | 1 | 2 | 3 | 4 | 5 | 6 |
|---|---|---|---|---|---|---|
| $y$ | 315 | 220 | 191 | 180 | 175 | 173 |

   (b) This gives total running costs per week for $x$ lorries (c) £45500

## Chapter 25 – Linear equations

1. (a) is correct. To solve, equate to give $3x-2 = x+2$ yielding $x=2$, $y=4$.
2. (a) $x=6$ (b) $x=13$ (c) $x=-1$ (d) $x=5$ (e) $x=5$ (f) $x=0.5$ (g) $x=-2$ (h) $x=-2.5$ (i) $x=-15$ (j) $x=2$ (k) $x=5$
3. $x=3$
4. $y=1.5$
5. (a) $x=3, y=1$ (b) $x=3, y=5$ (c) $x=1, y=2$ (d) $x=4, y=1$ (e) $x=7, y=3$
6. (a) $x=5, y=3$ (b) $x=3, y=1.5$ (c) $x=y=4$ (d) $x=-0.1, y=0.4$
7. 6
8. (a) $10(0.9)x+3y=37.5$; $7x+4y=40$ (b) $x=2, y=6.5$
9. 45*A* and 15*B* tables
10. (a) $x=2, y=-1, z=5$ (b) $x=2, y=3, z=-1$ (c) $x=-1, y=10, z=5$
11. 25*X*, 5*Y* and 20*Z* products.

## Chapter 26 – Quadratic and cubic equations

1. (a) –3,4 (b) –1/3,5/2 (c) –2,2 (d) 0,0.4 (e) –1.5,8 (f) 3,4 (g) 2,3.5 (h) –2,2.5
2. (d) is correct. For the formula: a=1, b=–2 and c=–24. Inside the root, the calculation needs to be made carefully as: $\sqrt{[(-2)^2 - 4(1)(-24)]} = \sqrt{[4 + 96]} = \sqrt{100} = 10$.
3. (a) 1.5,2.5 (b) 2.38,4.62 (c) 3,4 (d) No solutions (e) –0.72,1.39 (f) 0.37,3.63 (g) 3.05,7.06 (h) –1.32, 0.57
4. (a) Plotted points:

| $x$ | 0 | 1 | 2 | 3 | 4 | 5 | 6 | 7 | 8 |
|---|---|---|---|---|---|---|---|---|---|
| $y$ | 4.5 | 1 | –0.5 | 0 | 2.5 | 7 | 13.5 | 22 | 32.5 |

   (b) (i) 1.5,3 (ii) 0.4,4.1 (iii) 1.3,4.2 (iv) 0.4,7.2
5. (a) –4.65,0.65 (b) –2.58,0.58
6. 0,1,2
7. Plotted points:

| $x$ | –6 | –5 | –4 | –3 | –2 | –1 | 0 | 1 | 2 | 3 |
|---|---|---|---|---|---|---|---|---|---|---|
| $y=2x^3+8x^2-18x-25$ | –61 | 15 | 47 | 47 | 27 | –1 | –25 | –33 | –13 | 47 |
| $y=2x+20$ | | 10 | | | | 18 | | | 24 | |

8. (a) Plotted points:

| $x$ | 2 | 2.5 | 3 | 3.5 | 4 | 4.5 | 5 |
|---|---|---|---|---|---|---|---|
| $x^3-10.55x^2+36.4x-40.8$ | –2.20 | –0.11 | 0.45 | 0.24 | 0 | 0.49 | 2.45 |

   solutions = 2.55,4,4

   (b) solutions are –0.5,0.5,3

   (c) Plotted points:

| $x$ | –2 | –1 | 0 | 1 | 2 | 3 | 4 | 5 |
|---|---|---|---|---|---|---|---|---|
| $-20+8x+12x^2-3x^3$ | 36 | –13 | –20 | –3 | 20 | 31 | 12 | –55 |

   solutions = –1.4,1.1,4.3
9. –0.97

## Chapter 27 – Differentiation and integration

1. (a) $8x$ (b) $12x^3+4x$ (c) $60x-10$ (d) $8 - \frac{2}{x^2}$ (e) $2x+3$ (f) $4.5x^2$ (g) $2 - \frac{4}{x^2}$
2. (a) $\frac{dy}{dx} = 10x-2$, $\frac{d^2y}{dx^2} = 10$ (b) $\frac{dy}{dx} = -10+12x-6x^2$, $\frac{d^2y}{dx^2} = 12-12x$

   (c) $\frac{dy}{dx} = 9x^2+4$, $\frac{d^2y}{dx^2} = 18x$
3. Turning point is a minimum at $x=2, y=-8$
4. $x=0.33$, $y=3.81$ (max) and $x=5$, $y=-47$ (min)

5. (a) $3x^3+C$ (b) $0.5x^4+C$ (c) $2x^4-x^3+4x^2-10x+C$ (d) $\frac{-1}{x^2}$  6. $y=2x^2-3x+6$

7. (a) $P=11x-x^2-24$ (b) 8000 (i.e. $x=8$) (c) 5500 (i.e. $x=5.5$) (d) £625 (i.e. $P=6.25$)

## Chapter 28 – Cost, revenue and profit functions

1. (b) is correct. C = 4 + 0.8(2) = 4 + 1.6 = 5.6 (000) = 5,600.
2. (d) is correct.
3. (a) $P=18x-20-4x^2$ (b) 225 units produced give a total profit of £250 (c) A loss of £6000
4. Cost-minimising yearly level of production is 4309 items. Total cost is £16661.27
5. $p_r$(£)=13.75–0.175$q$, where $q$ is the quantity demanded  6. £8100
7. (a) £2.50 (b) (i) $\frac{2000}{x}+2$ (ii) $0.5-\frac{2000}{x}$ (c) 4000 units (d) 5000 units (e) 2000 units
8. (a) $8.9-0.0005x$ (b) $7.1x-0.0005x^2-15000$ (c) £10,205 at a sales level of 7100 (d) £5.35

## Chapter 29 – Set theory and enumeration

1. (a) False. (b) False, C is a subset of A. (c) True. (d) False, smallest value must be $n$[A]–7. (e) False. (f) True (g) True.
2. (a) {$a,b,c,d,e,f$} (b) {$a,b,c,d,e,g,h$} (c) {$c,d,e$} (d) {$c,d,e$} (e) {$a,c,d,e,f,g,h$} (f) {$a,c,d,e$} (g) {$a,b,f,g,h,i$} (h) {$b,i$} (i) {$a,b,f,g,h,i$}
3. (a) U = {$r,s,t,u,v,w,x,y$} = set of all products stocked.
   (b) {$r,t,v,w,x$} = set of products bought by either C or D. (c) E∩B = E = {$r,v,w,x$} = set of products bought by both B and E. Also, all products bought by E are also bought by B (E is a subset of B).
   (d) {$s,u,v,w,y$} = products that are never bought by C.
   (e) A∪C = {$r,s,t,v,x$} and B′ = {$s,u,y$}. Thus (A∪C)∩B′ = {s} = the products bought by either A or C that are never bought by B.
   (f) {$r$} = the products bought by all regular customers.
   (g) {$u,y$} = products never bought by any of the regular customers.
4. $ab$=12, $a$=16.  5. 5  6. (a) 5 (b) 158 (c) 72 (d) 42 (e) 10
7. (a) 19 (b) 15 (c) 3 (d) 2

## Chapter 30 – Introduction to probability

1. b;c  2. c generally; e  3. (a) $\frac{1}{6}$ (b) $\frac{5}{12}$ (c) $\frac{5}{6}$
4. (b) is correct. There are 36 combinations altogether and of these just 6 (6 and 1, 1 and 6, 5 and 2, 2 and 5, 3 and 4, 4 and 3) add to 7. Thus probability $=\frac{6}{36}=\frac{1}{6}$.
5. (d) is correct. All three systems must fail with probability $\left(\frac{1}{100}\right)^3$.
6. (a) 0.04 (b) 0.54 (c) 0.42
7. (a) 0.12 (b) 0.03 (c) 0.03 (d) 0.913 (3D) (e) 0.319 (3D)
8. (a) 0.72 (b) 0.92 (c) 0.68  9. (a) 0.023 (b) 0.045 (3D)
10. (a) 0.9 (b) 0.921 (3D) (c) 689 (to nearest unit)

## Chapter 31 – Conditional probability and expectation

1. (a) £32,000 (b) 0.16, 0.52, 0.28, 0.04 2. 10.85
3. 36.25 days (using EXPECTATION); 30.5 days (using other techniques). The latter is strictly correct.
4. (a) expected profit = £24.50 (b) expected profit = £18.50. 200 cauliflowers should be bought, since this leads to the largest profit (£27) in a day.
5. (a) $\frac{2}{3}$ (b) $\frac{2}{5}$ 6. (a) 0.42 (b) 0.30 (c) 0.50 (d) 0.71 (2D)
7. (a) 0.553 (b) 0.605 (c) 0.395
8. 0.25 9. (a) 0.80 (b) 0.35 (c) 0.85 (d) 0.60 (e) 0.57 (2D) (f) 0.25 10. $\frac{2}{3}$.

## Chapter 32 – Permutations and combinations

1. AB, AC, AD, AE, BC, BD, BE, CD, CE, DE. (10 altogether.)
2. (a) 120 (b) 12 (c) 120 (d) 210 3. (a) 5 (b) 20 (c) 24 (d) 30 4. (a) 6 (b) 4 (c) $\frac{2}{3}$
5. (a) 126 (b) 60 (c) $\frac{60}{126} = 0.476$ (3D) (d) $\frac{81}{126} = 0.643$ (3D) (e) $\frac{45}{126} = 0.357$ (3D)

## Chapter 33 – Binomial and poisson distributions

1. (a) 30 (b) Trial = selecting a workman; trial success = finding a workman idle; number of trials = 6 (the number of workmen involved).
2.

| $x$ | 0 | 1 | 2 | 3 | 4 |
|---|---|---|---|---|---|
| p | 0.316 | 0.422 | 0.211 | 0.047 | 0.004 |

3. (a) 2.94; 1.71 (2D) (b) (i) 0.022 (ii) 0.112 (iii) 0.243 (iv) 0.623 (3D)
4.

| $x$ | 0 | 1 | 2 | 3 | 4 |
|---|---|---|---|---|---|
| p | 0.1003 | 0.2306 | 0.2652 | 0.2033 | 0.2006 |

5. (a) 0.670 (b) 0.062 (c) 0.047 (3D) 6. (a) 0.190 (b) 0.191 (3D)
7. (a) 0.088 (3D) (b) 0.263 (3D) (c) 0.68 8. (a) 0.737 (b) 0.263 (3D) (c) 23.

## Chapter 34 – Normal distribution

1. (a) (i) 0.6480 (ii) 0.8888 (iii) 0.9968 (b) (i) 0.1379 (ii) 0.0019 (iii) 0.4483
   (c) (i) 0.1357 (ii) 0.0007 (iii) 0.0268 (d) (i) 0.9962 (ii) 0.7764
2. (a) (i) 0.1314 (ii) 0.0307 (iii) 0.8379 (b) (i) 0.8106 (ii) 0.9846 (iii) 0.1740
   (c) (i) 0.0099 (ii) 0.6591 (iii) 0.3310
3. (a) 0.9772 (b) 0.1587 (c) 0.8185 4. (a) 0.0150 (b) 0.0228 (c) 0.9622
5. (a) (i) 0.0062 (ii) 0.0062 (b) 6
6. (a) 0.0228 (b) 5.56pm 7. 0.1788
8. 0.03
9. Lower limit = 0.9767 cms; upper limit = 1.0233 cms.
10. a) $15.3 \pm (1.96)(0.682) = (13.96, 16.64)$ b) $15.3 \pm (2.58)(0.682) = (13.54, 17.06)$
11. (0.016, 0.304)
12. a) 20.85; 1.816 b) $20.85 \pm 1.34 = (19.51, 22.19)$ c) $0.714 \pm 0.335 = (0.379, 1.0)$

13. $z = -1.35$; no evidence of difference. Therefore yes, sample could have been drawn from given population.
14. $z = -2.81$; evidence of difference. Pay appears to have changed.
15. a) $z = 3.125$; evidence of difference.
    b) $\bar{x} = 832$ gives $z = 2.0$; $\bar{x} = 831$ gives $z = 1.938$. Thus 831 is the largest whole-number value.

## Chapter 35 – Linear inequalities

1. (a) To the left, top of line $4x-2y=100$ (b) To the right, top of line $x+3y=20$
   (c) To the left, bottom of line $4x+3y=120$ and to the top of line $y-20$
   (d) To the left, bottom of line $10x+10y=300$ and to the right, top of line $2x+6y=60$ and to the right of line $x=10$.
2. The first inequality should read $x+y<=80$. The four vertices are (40,20), (40,40), (0,80) and (30,20).
3. $x=y=300$; maximum value = 1800. 4. $x=15$, $y=30$; maximum value = 90
5. (a) $x<=400$, $y<=700$, $x+y<=800$, $2x+y<=1000$. Contribution function is $0.4x+0.5y$
   (b) $x=100$, $y=700$. Maximum contribution is £390/day.
6. (a) Process A: $10x+10y<=12000$. Process B: $8x+3y>=4800$. $Y$: $y>=600$ $X$: $x<=500$.
   (b) Coordinates for maximum (240,960). Maximum = £158,400. Coordinates for minimum (375,600). Minimum = £121,500. (c) Coordinates for maximum = (500,700). Maximum = £140,000. Coordinates for minimum (375,600). Minimum = £112,500.

## Chapter 36 – Matrices

1. (a) 1 (b) 3 (c) 2 (d) –2
2. (a) $\begin{bmatrix} 10 & 3 \\ 5 & -2 \end{bmatrix}$ (b) $\begin{bmatrix} 1 & 2 \\ 2 & 0 \\ -1 & 5 \end{bmatrix}$ (c) $\begin{bmatrix} 4 & 8 & 5 \\ -11 & 5 & 5 \end{bmatrix}$
3. A*B = $\begin{bmatrix} 8 \\ 21 \end{bmatrix}$. A*C is NOT defined. B*C = $\begin{bmatrix} 6 & 2 & 4 \\ 12 & 4 & 8 \\ 3 & 1 & 2 \end{bmatrix}$

   B*A is NOT defined. C*A is NOT defined. C*B = 12 (a 1×1 matrix is a number).

   A*B*C = $\begin{bmatrix} 24 & 8 & 16 \\ 63 & 21 & 42 \end{bmatrix}$
4. A*B = total cost per hour for each firm if all machines were utilised. B*C has no practical meaning. C*B = total cost involved for the specified job. A*B*C has no practical meaning.
5. (a) $X = \begin{array}{c} \\ A \\ B \\ C \end{array}\begin{array}{c} \begin{array}{ccc} J & K & L \end{array} \\ \begin{bmatrix} 0 & 2 & 2 \\ 1 & 0 & 2 \\ 3 & 2 & 2 \end{bmatrix} \end{array}$ (b) $Y = \begin{array}{c} \\ J \\ K \\ L \end{array}\begin{array}{c} \begin{array}{ccc} S & M & F \end{array} \\ \begin{bmatrix} 10 & 60 & 10 \\ 10 & 30 & 0 \\ 0 & 10 & 20 \end{bmatrix} \end{array}$

(c) $X*Y = \begin{array}{c} \\ A \\ B \\ C \end{array} \begin{array}{c} \begin{array}{ccc} S & M & F \end{array} \\ \begin{bmatrix} 20 & 80 & 40 \\ 10 & 80 & 50 \\ 50 & 260 & 70 \end{bmatrix} \end{array}$ (d) Job C takes 380 minutes.

(e) Here, $Z = \begin{bmatrix} 0.10 \\ 0.40 \\ 0.05 \end{bmatrix}$ and $X*Y*Z$ (compatible for multiplication)

$= \begin{bmatrix} 36.0 \\ 35.5 \\ 112.5 \end{bmatrix}$ giving the costs of jobs A, B and C respectively.

6. (a) 3 (b) –11 (c) –10 (d) –33

7. (a) $\begin{bmatrix} \frac{2}{3} & -\frac{1}{3} \\ -\frac{1}{3} & \frac{2}{3} \end{bmatrix}$ (b) $\begin{bmatrix} \frac{1}{11} & \frac{2}{11} \\ \frac{4}{11} & -\frac{3}{11} \end{bmatrix}$ (c) $\begin{bmatrix} -\frac{1}{10} & \frac{3}{10} \\ \frac{5}{10} & -\frac{5}{10} \end{bmatrix}$

(d) $\begin{bmatrix} 0 & \frac{1}{11} \\ \frac{1}{3} & -\frac{10}{33} \end{bmatrix}$

8. (a) $x$=1, $y$=2 (b) $x$=2, $y$=6 (c) $p$=–2, $q$=3 (d) $x$=0.4, $y$=2.1

9. (a) 3(M*N), in £, is $\begin{array}{c} \\ X \\ Y \end{array} \begin{array}{c} \begin{array}{cc} S & M \end{array} \\ \begin{bmatrix} 3.41 & 2.24 \\ 5.84 & 3.48 \end{bmatrix} \end{array}$

and gives the storage and maintenance cost at warehouses $X$ and $Y$ for each item held over a period of three days (assuming there are no stock movements).

(b) $T = \begin{bmatrix} 0 & 6 & 12 \\ -23 & 12 & 0 \end{bmatrix}$ (c) 2(M*N) + 3[(M+T)*N]

10. (a) $C*Q = \begin{bmatrix} 60 & 6q \\ 35 & 8q \end{bmatrix}$ which describes the total cost functions for products $x$ and $y$

(b) $R = \begin{bmatrix} 0 & 9 \\ 0 & 10 \end{bmatrix}$ (c) $R*Q - C*Q = \begin{bmatrix} 3q-60 \\ 2q-35 \end{bmatrix}$ which describes the profit functions for products $x$ and $y$. (d) q=35, giving the level of production that yields the same total profit (of £15) for each product.

## Chapter 37 – Inventory control

1.

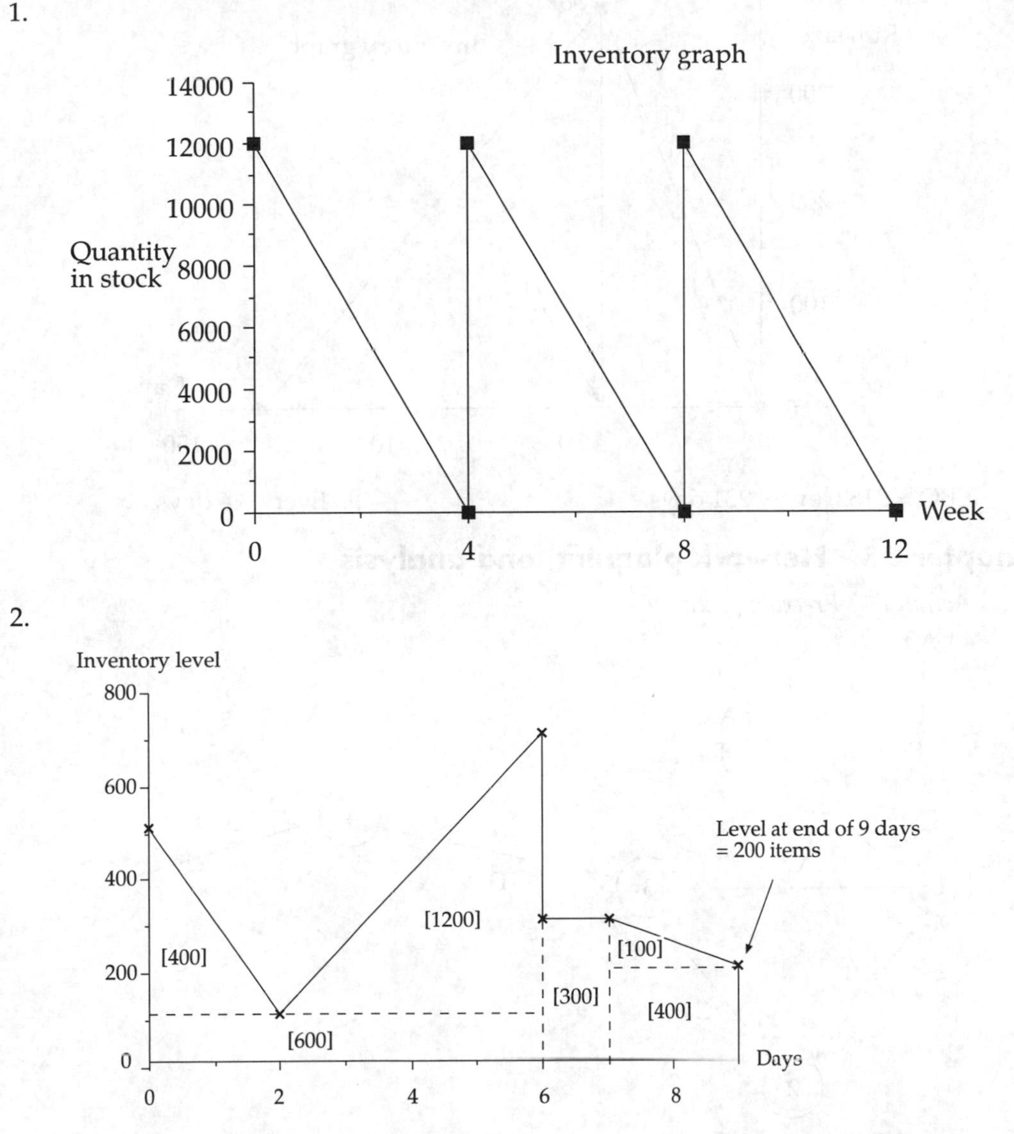


Constituent areas under the curve are shown in square brackets and they total 3000. Thus, the average inventory level over the whole period = 3000/9 = 333.

3. (a) and (b) EOQ = 637; (c) 14 days, 25.9 runs. 4. EOQ = 84852, no. of orders = 7; length of cycle = 51.6 days.

5. Order every 49 (48.6) days. 6. (a) Every 73 days; EOQ = 1000 boxes (b) £187.30 (94% increase in costs).

7. (a) 560 (b) 41 days (c) 128 days (d) 191

(b)

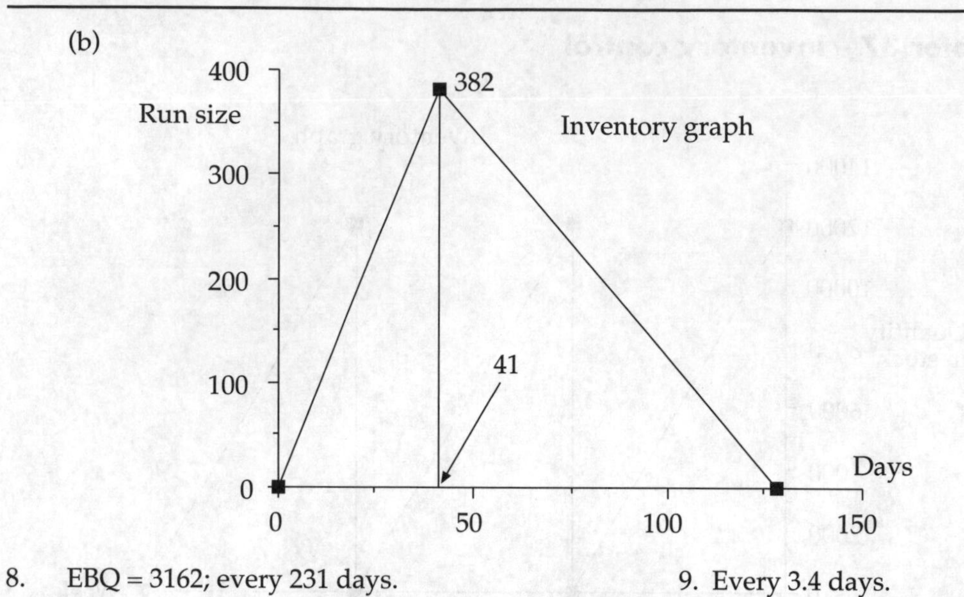


8. EBQ = 3162; every 231 days.

9. Every 3.4 days.

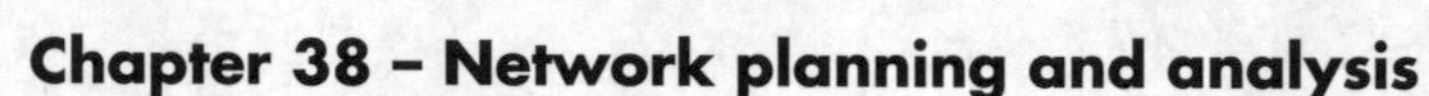

## Chapter 38 – Network planning and analysis

1.

| *Activity* | *Preceding activity* |
|---|---|
| A,B | – |
| C | B |
| D | A,C |
| E | D |

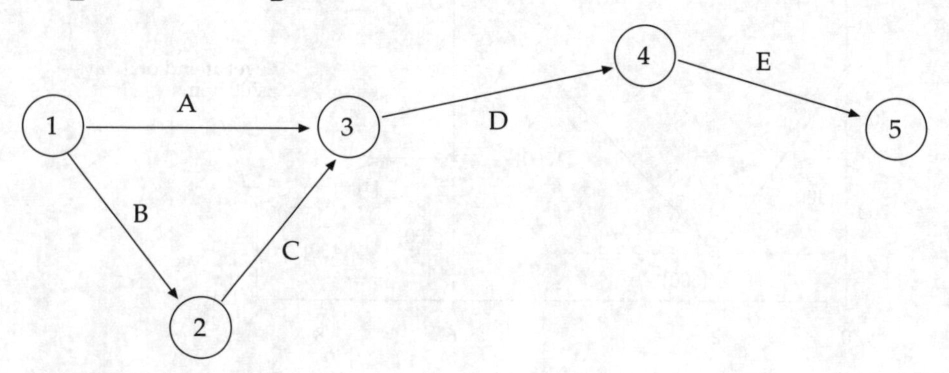


2. a) AIK = 8+5+3 = 16; AHJK = 8+7+2+3 = 20; AEFJK = 8+2+8+2+3 = 23; CFJK = 3+8+2+3 = 16; BDGJK = 4+6+9+2+3 = 24
   b) Total project time = 24. Critical path is BDGJK.
3. a) See diagram on next page
   b) ACG = 11; BDG = 9; BE = 9; BFH = 12; BFI[] = 11
   c) Critical path is BFH with duration time 12.

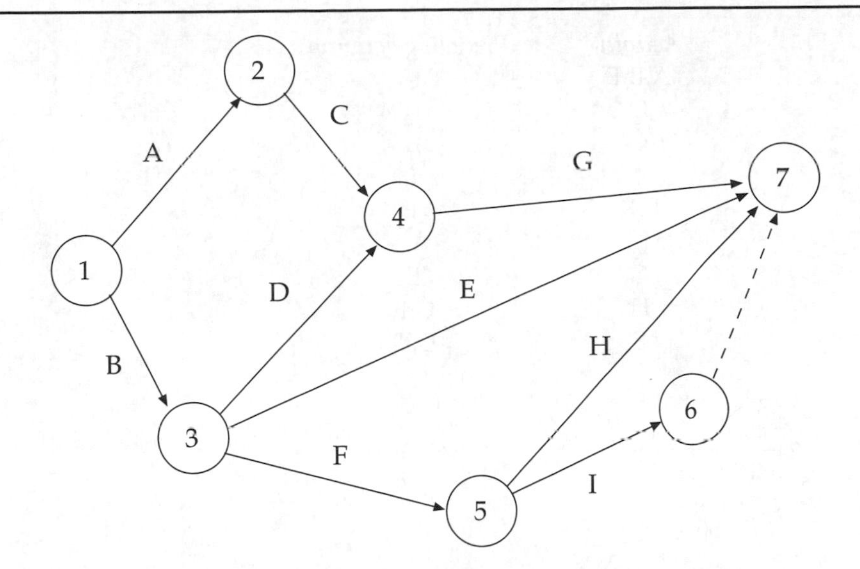


4.

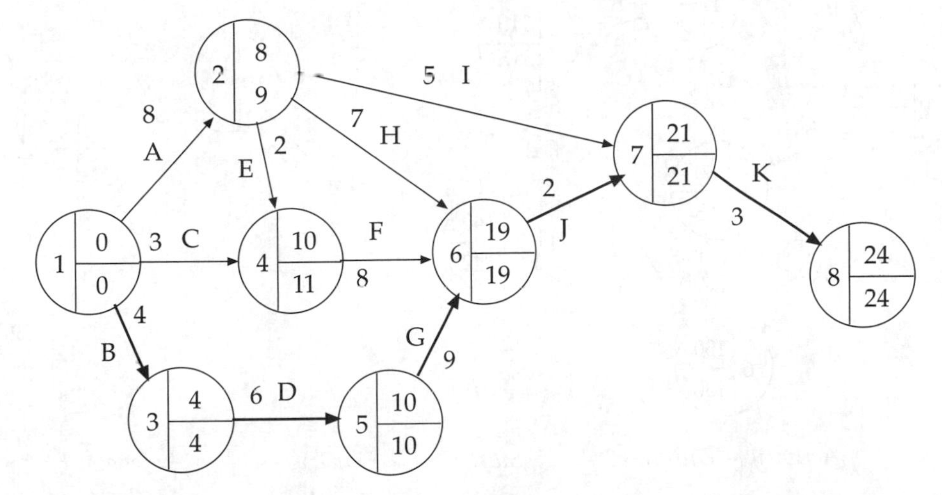


The critical path (BDGJK) has been *enboldened* and agrees with the result of exercise 2.

5.

| *Activity* | *Duration* | *Start* | | *Finish* | | *Floats* | | |
|---|---|---|---|---|---|---|---|---|
| | | E | L | E | L | Tot | Free | Ind |
| 1–2 | 8 | 0 | 1 | 8 | 9 | 1 | 0 | 0 |
| 1–3 | 4 | 0 | 0 | 4 | 4 | 0 | 0 | 0 |
| 1–4 | 3 | 0 | 8 | 3 | 11 | 8 | 7 | 7 |
| 2–4 | 2 | 8 | 9 | 10 | 11 | 1 | 0 | 0 |
| 2–6 | 7 | 8 | 12 | 15 | 19 | 4 | 4 | 3 |
| 2–7 | 5 | 8 | 16 | 13 | 21 | 8 | 8 | 7 |
| 3–5 | 6 | 4 | 4 | 10 | 10 | 0 | 0 | 0 |
| 4–6 | 8 | 10 | 11 | 18 | 19 | 1 | 1 | 0 |
| 5–6 | 9 | 10 | 10 | 19 | 19 | 0 | 0 | 0 |
| 6–7 | 2 | 19 | 19 | 21 | 21 | 0 | 0 | 0 |
| 7–8 | 3 | 21 | 21 | 24 | 24 | 0 | 0 | 0 |

6. a)

| *Activity* | *Preceding activity* |
|---|---|
| A,B,F | – |
| D | A |
| C | B |
| J | I |
| L,K | J |
| M,N | L |
| G | E |
| H | G,F |
| E | D,C |

b)

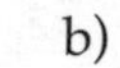

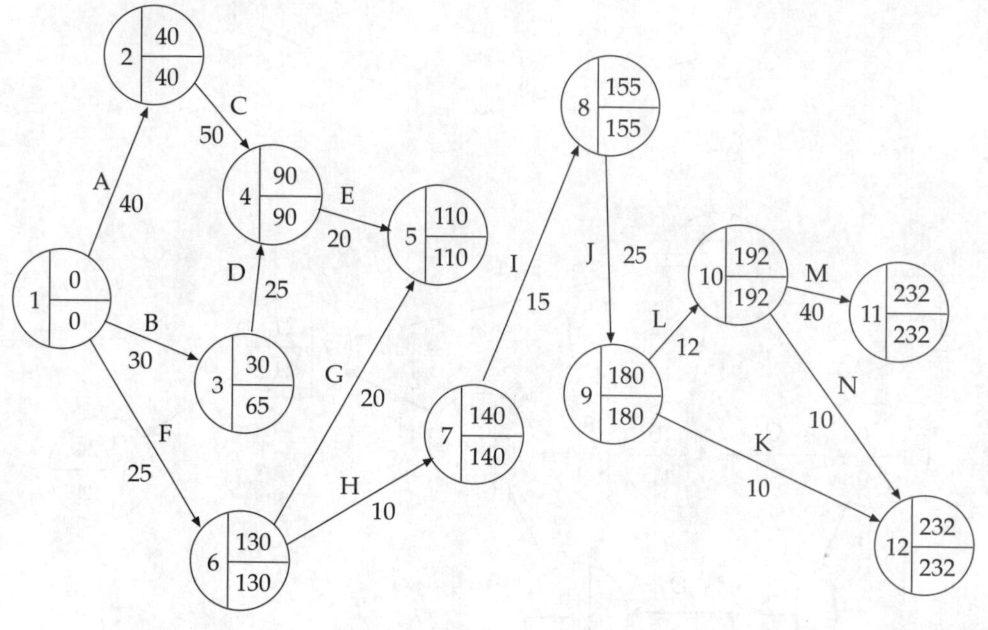


c)

| *Activity* | *Duration* | *Start* | | *Finish* | | *Floats* | | |
|---|---|---|---|---|---|---|---|---|
| | | E | L | E | L | Tot | Free | Ind |
| A | 40 | 0 | 0 | 40 | 40 | 0 | 0 | 0 |
| B | 30 | 0 | 35 | 30 | 65 | 35 | 0 | 0 |
| C | 50 | 40 | 40 | 90 | 90 | 0 | 0 | 0 |
| D | 25 | 30 | 65 | 55 | 90 | 35 | 35 | 0 |
| E | 20 | 90 | 90 | 110 | 110 | 0 | 0 | 0 |
| F | 25 | 0 | 105 | 25 | 130 | 105 | 105 | 105 |
| G | 20 | 110 | 110 | 130 | 130 | 0 | 0 | 0 |
| H | 10 | 130 | 130 | 140 | 140 | 0 | 0 | 0 |
| I | 15 | 140 | 140 | 155 | 155 | 0 | 0 | 0 |
| J | 25 | 155 | 155 | 180 | 180 | 0 | 0 | 0 |
| K | 10 | 180 | 222 | 190 | 232 | 42 | 42 | 42 |
| L | 12 | 180 | 180 | 192 | 192 | 0 | 0 | 0 |
| M | 40 | 192 | 192 | 232 | 232 | 0 | 0 | 0 |
| N | 10 | 192 | 222 | 202 | 232 | 30 | 30 | 30 |

# Answers to examination questions

## Part 1

### Question 1

*Simple random sampling*. A method of sampling whereby each member of the population has an equal chance of being chosen. Normally, random sampling numbers are used to select individual items from some defined sampling frame.

*Stratification*. This is a process which splits a population up into as many groups and sub-groups (strata) as are of significance to the investigation. It can be used as a basis for quota sampling, but more often is associated with stratified (random) sampling. Stratified sampling involves splitting the total sample up into the same proportions and groups as that for the population stratification and then separately taking a simple random sample from each group. For example, employees of a company could be split into male/female, full-time/part-time and occupation category.

*Quota sampling*. A method of non-random sampling which is popular in market research. It uses street interviewers, armed with quotas of people to interview in a range of groups, to collect information from passers-by. For example, obtaining peoples' attitudes regarding the worth of secondary double glazing.

*Sample frame*. This is a listing of the members of some target population which needs to be used in order to select a random sample. An example of a sampling frame would be a stock list, if a random sample was required from current warehouse stock.

*Cluster sampling*. This is another non-random method of sampling, used where no sampling frame is in evidence. It consists of selecting (randomly) one or more areas, within which all relevant items or subjects are investigated. For example, a cluster sample could be taken in a large town to interview tobacconists.

*Systematic sampling*. A quasi-random method of sampling which involves examining or interviewing every *n*-th member of a population. Very useful method where no sampling frame exists, but population members are physically in evidence and ordered. For example, items coming off a production line. It is virtually as good as random sampling except where the items or members repeat themselves at regular intervals, which could lead to serious bias.

### Question 2

(a) A postal questionnaire is a much cheaper and more convenient method of collecting data than the personal interview and often very large samples can be taken. However, much more care must be taken in the design of the questions, since there will be no help to hand if questions seem ambiguous or personal to the respondent. Also the response rate is very low, sometimes less than 20%, but this can sometimes be made larger by free gifts or financial incentives.

The personal interview has the particular advantage that difficult or ambiguous questions can be explained as well as the fact that an interviewer can make allowances or small adjustments according to the situation. Also, the questionnaire will be filled in as required. Disadvantages of this method include the cost, the fact that large samples cannot generally be undertaken and the training of interviewers.

(b) Simple random sampling has the particular advantage that the method of selection (normally through the use of random sampling numbers) is free from bias. That is, each member of the population has an equal chance of being chosen as part of the sample. However, it cannot be guaranteed that the sample itself is truly representative of the population. For example, if a human population being sampled comprised 48% males, it is unlikely that the sample would reflect this percentage exactly.

Quota sampling is not a random sampling method and thus is generally at a disadvantage with regard to obtaining information that can claim to be representative. However, if the population has been stratified reasonably well, the street interviewer is experienced and conscientious and the questioning sites have been well thought out, it could be argued that, in certain localised situations, a quota sample could be very representative. For example, to gauge peoples opinions of a new shopping centre or to find out the views of theatregoers about a particular theatre.

*Question 3*

(a) (i) See pie chart.

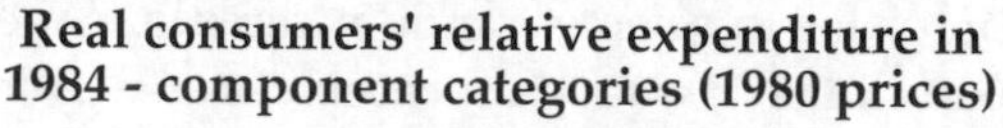


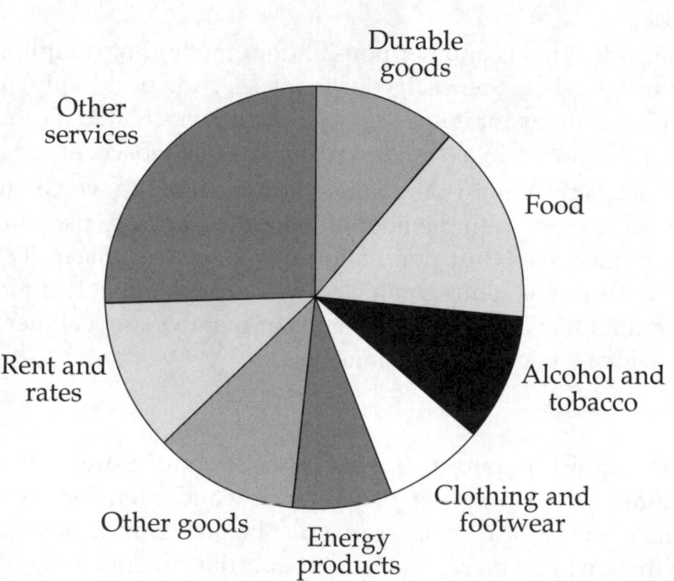


(ii) Other goods: books, toys, toiletries, transport. Other services: insurance, recreation, entertainment, (private) dental/health care.

(b) See line diagram.

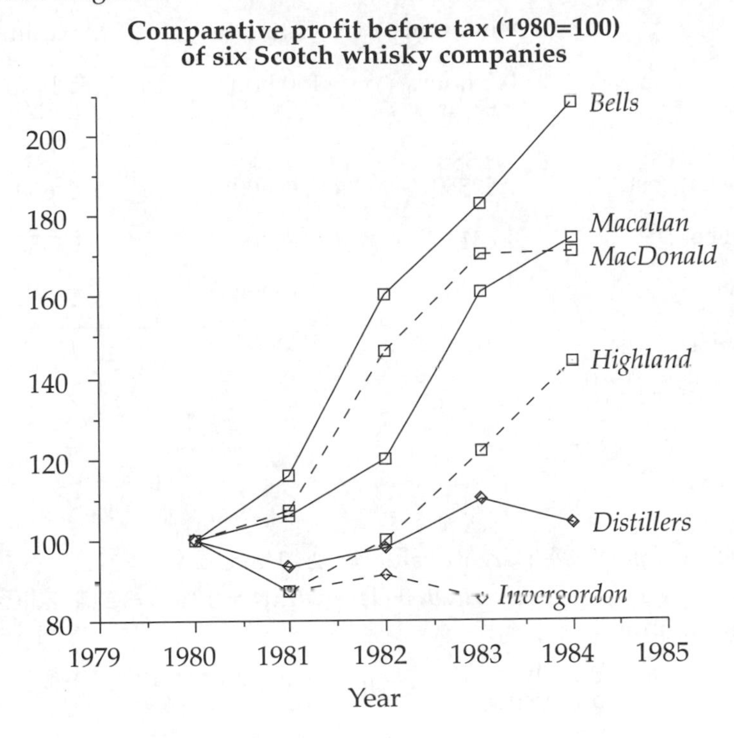


## Question 4

(a) i. An absolute error is the difference between an estimated value and its true value. In most cases, only a maximum absolute error will be able to be calculated. For example, if a company's yearly profit was quoted as £252,000 (to the nearest £1000), the maximum absolute error would be £500.

ii. A relative error is an absolute error expressed as a percentage of the given estimated value. Thus in the example above, the maximum relative error in the company's yearly profit is:

$$\frac{500}{252{,}000} \times 100\% = 0.2\%$$

iii. A compensating error is an error that is made when 'fair'rounding has been carried out. For example, the numbers of people employed in each of a number of factories might well be rounded fairly, to the nearest 1000, say. When numbers, subject to compensating errors, are added, the total relative error should be approximately zero.

iv. Biased errors are made if rounding is always carried out in one direction. For example, when people's ages are quoted, they are normally rounded *down* to the lowest year. The error in the sum of numbers that are subject to biased errors is relatively high.

(b)

| | Minimum | Estimate | Maximum |
|---|---|---|---|
| Time | 145 hours | 150 hours | 155 hours |
| Wage rate | £4/hr | £4/hr | £4.40/hr |
| Labour cost | £580 | £600 | £682 |
| Material cost | £2550 | £2600 | £2650 |
| Total cost | £3130 | £3200 | £3332 |
| Quote | £4000 | £4000 | £4000 |
| PROFIT | £668 | £800 | £870 |

## Question 5

(a) Smallest value = 347; largest value = 469. Thus, range = 122.
Since five classes are required, a class width of 122÷5 = 24.4, adjusted up to 25, seems appropriate.

| Weekly production | Tally | Number of weeks |
|---|---|---|
| 345 to 369 | IIII IIII IIII I | 16 |
| 370 to 394 | IIII III | 8 |
| 395 to 419 | IIII | 4 |
| 420 to 444 | I | 1 |
| 445 to 469 | IIII IIII I | 11 |
| | Total | 40 |

(b) To construct the ogive, *cumulative frequency* needs to be plotted against *class upper bounds*.

| Weekly production (upper bound) | Cumulative number of weeks |
|---|---|
| 369.5 | 16 |
| 394.5 | 24 |
| 419.5 | 28 |
| 444.5 | 29 |
| 469.5 | 40 |

The ogive is shown in the figure following.

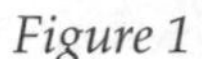

*Figure 1*

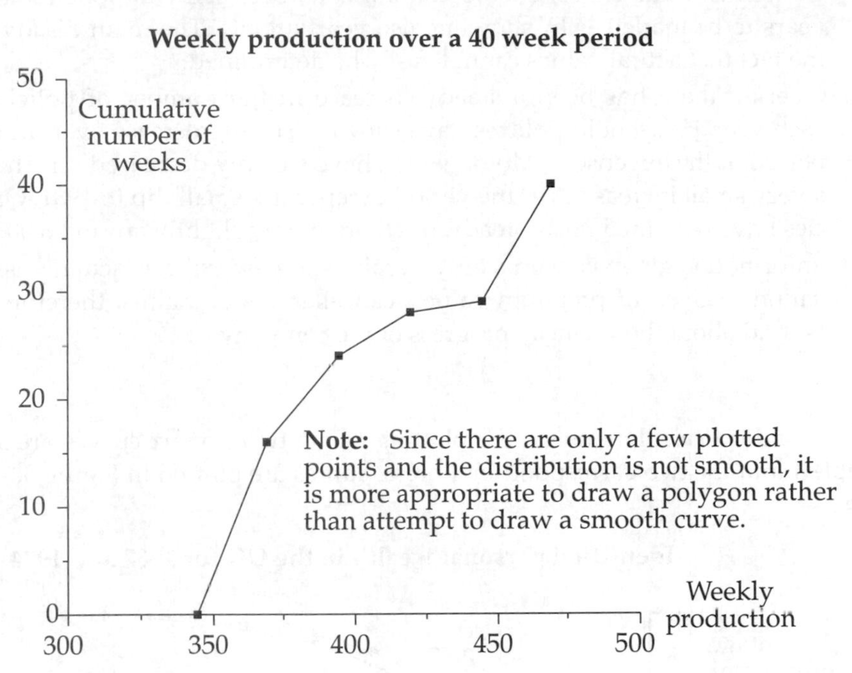


## *Question 6*

(a) Although generally a component time series is best represented by a (cumulative) line diagram, in this case, since there are so few time points, a component bar chart has more impact. The chart is drawn in Figure 2.

*Figure 2*

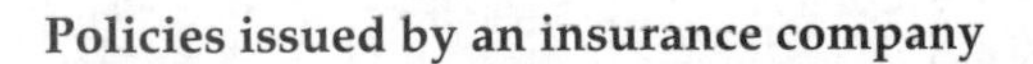


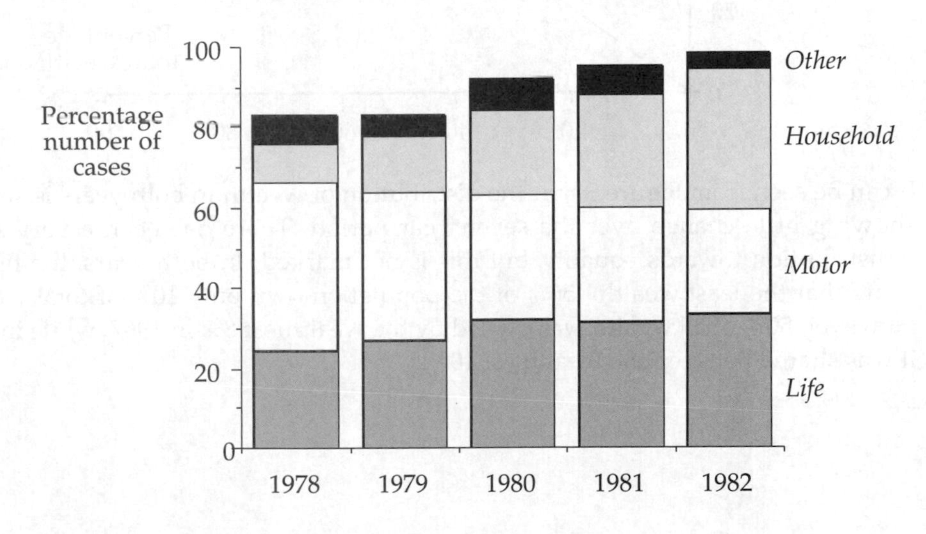

(b) Component bar charts enable comparisons between components across the years to be made easily, showing also yearly totals. The main disadvantage is the fact that actual values cannot easily be determined.

(c) Overall, there has been a steady increase in the number of policies issued each year. Household policies have shown a steady increase over the five year period at the expense of Motor, which have steadily decreased. Life has shown a very small increase over the period except for a small dip in 1981. Other policies have remained fairly steady, fluctuating only slightly around 6,000.

The information given concerns only numbers of new policies actually issued. No indication is given of premium values, cancellations or claims, therefore nothing can be said about the financial progress of the company.

## Question 7

The standard calculations for the plotting of the two Lorenz curves are shown in Table 1 and the two corresponding Lorenz curves are plotted in Figure 3.

*Figure 3*

**Identified personal wealth in the UK for 1967 and 1974**

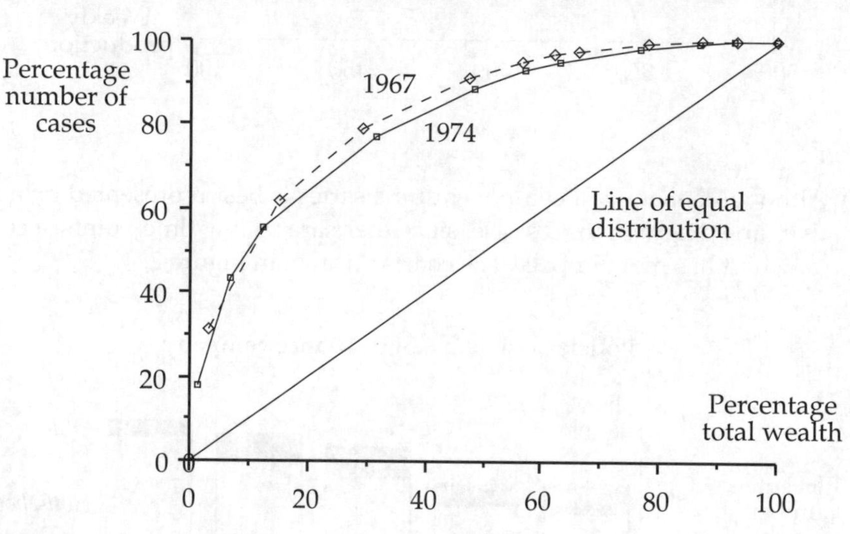


It can be seen from Figure 3 that the distribution of wealth in both years is similar, showing little change over the seven year period. There has been a very small redistribution towards equality, but this is not marked. In both years, the figures show that the least wealthy 50% of the population own only 10% of total wealth. However, 50% of all wealth was owned by the wealthiest 8% in 1967, while in 1974 it was shared between the wealthiest 10%.

*Table 1*

| Range of wealth (£000) | Number of cases | | Total wealth | | Number of cases | | Total wealth | |
|---|---|---|---|---|---|---|---|---|
| | % | cum % | % | cum % | % | cum % | % | cum % |
| 0 to 1 | 31.2 | 31.2 | 3.4 | 3.4 | 18.1 | 18.1 | 1.3 | 1.3 |
| 1 to 3 | 30.5 | 61.7 | 11.7 | 15.1 | 25.4 | 43.5 | 5.5 | 6.8 |
| 3 to 5 | 17.1 | 78.8 | 13.9 | 29.0 | 11.8 | 55.3 | 5.5 | 12.3 |
| 5 to 10 | 12.6 | 91.4 | 18.3 | 47.3 | 21.9 | 77.2 | 19.3 | 31.6 |
| 10 to 15 | 3.6 | 95.0 | 9.1 | 56.4 | 11.5 | 88.7 | 16.9 | 48.5 |
| 15 to 20 | 1.6 | 96.6 | 5.7 | 62.1 | 4.0 | 92.7 | 8.5 | 57.0 |
| 20 to 25 | 0.9 | 97.5 | 4.1 | 66.2 | 2.2 | 94.9 | 6.1 | 63.1 |
| 25 to 50 | 1.6 | 99.1 | 11.8 | 78.0 | 3.4 | 98.3 | 13.8 | 76.9 |
| 50 to 100 | 0.6 | 99.7 | 9.0 | 87.0 | 1.2 | 99.5 | 9.8 | 86.7 |
| 100 to 200 | 0.2 | 99.9 | 6.1 | 93.1 | 0.4 | 99.9 | 5.8 | 92.5 |
| over 200 | 0.1 | 100 | 6.9 | 100 | 0.1 | 100 | 7.5 | 100 |

## *Question 8*

(i) The component bar chart for the given data is drawn in Figure 4.

*Figure 4*

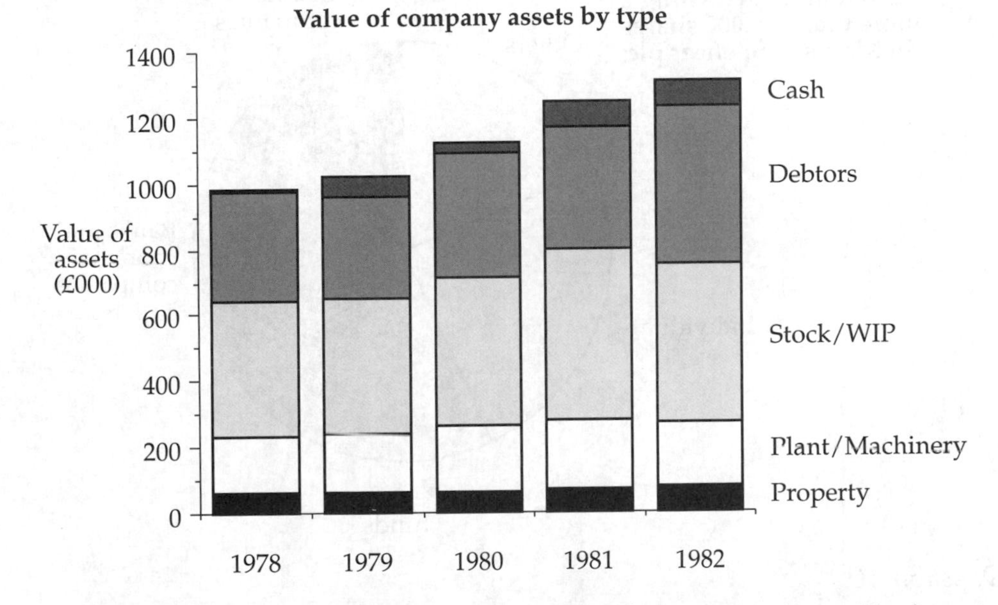


(ii) Overall, the total value of the given assets has increased steadily from just under £1m in 1978 to £1.3m in 1982. The most significant increase has been the debtors component, which has caught up with the stock and work-in-progress component, even though the latter has also increased. The property component shows very small increases, while plant and machinery shows small increases in the first four years and a decrease in the fifth year. Although the cash component has fluctuated over the five year period, it has shown an increase and is now comparable with property.

## Question 9

(a) (i) *Pictogram*. A representation that is easy to understand for a non-sophisticated audience. However, it cannot represent data accurately or be used for any further statistical work.

(ii) *Simple bar chart*. One of the most common forms of representing data which can be used for time series or qualitative frequency distributions. It is easy to understand and can represent data accurately. However, data values are not easily determined.

(iii) *Pie chart*. A type of chart which can have a lot of impact. Used mainly where the classes need to be compared in relative terms. However, they involve fairly technical calculations.

(iv) *Simple line diagram*. The simplest and most popular form of representing time series. They are easy to understand and represent data accurately. However, data values are not easily determined.

(b) A pie chart is one of the charts that could be drawn for the given data and is shown at Figure 5. Note however that a simple bar chart could equally well represent the data.

*Figure 5*

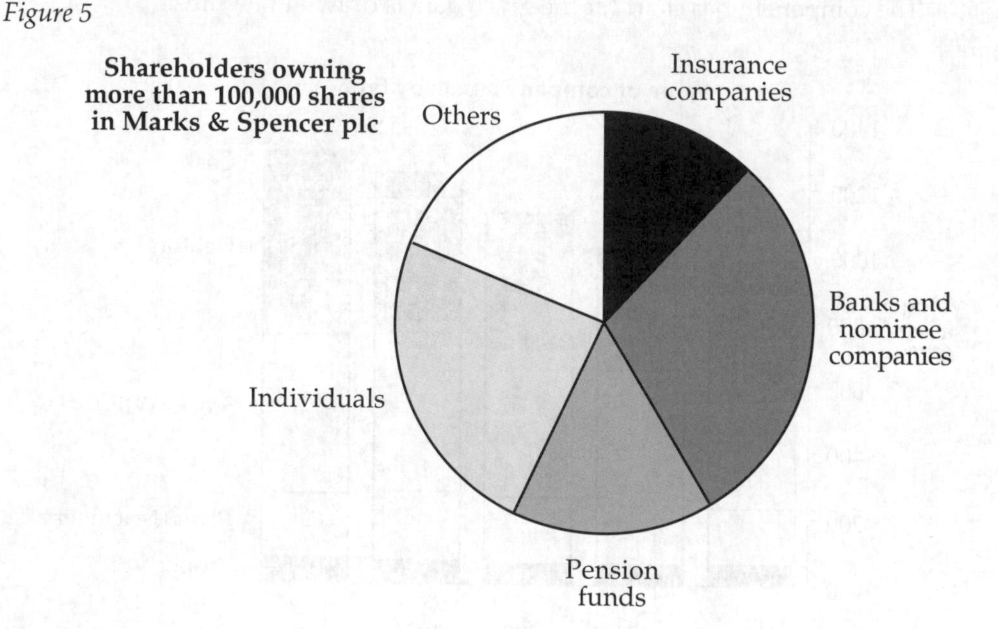


## Question 10

The company can make and sell 10,000 ± 2,000 units in the year

The selling price will lie in the range £50 ± £5 per unit

Thus the maximum revenue is 12,000 × £55 = £660,000

The minimum revenue is 8,000 × £45 = £360,000

The estimated revenue is 10,000 × £50 = £500,000

The maximum error from the estimated revenue is £660,000 – £500,000 = £160,000

$$\text{and relative error} = \frac{150{,}000}{500{,}000} \times 100\% = 32\%$$

The ranges of the various costs are:

| | min | | max | |
|---|---|---|---|---|
| materials | £147,000 | to | £153,000 | |
| wages | £95,000 | to | £105,000 | |
| marketing | £45,000 | to | £55,000 | |
| miscellaneous | £45,000 | to | £55,000 | |
| Total: | £332,000 | to | £368,000 | est = £350,000 |

Maximum error from the estimated costs = £18,000

and relative error = $\frac{80,000}{350,000} \times 100\% = 5.1\%$

Maximum contribution = £660,000 – £332,000 = £328,000

Minimum contribution = £360,000 – £368,000 = –£8,000

the estimated contribution = £150,000

The maximum error from the estimated contribution is £328,000 – £150,000 = £178,000

Which gives the relative error $\frac{178,000}{150,000} \times 100\ \% = 118.7\%$

The maximum contribution of £328,000 arises when 12,000 units are made and sold

Therefore contribution/unit = $\frac{£328,000}{12,000}$ = £27.33/unit

The minimum contribution of – £8,000 arises when 8,000 units are made and sold

Therefore contribution/unit = $\frac{-£8,000}{8,000}$ = – £1

The estimated contribution /unit = £15

The maximum error from the estimated contribution/unit is £15 – ( –£1) = £16

Therefore, relative error = relative error as $\frac{16}{15} \times 100\% = 106.7\%$.

## Part 2

### *Question 1*

| True limits | | | | | | | |
|---|---|---|---|---|---|---|---|
| Lower | Upper | Mid-point | $f$ | fx | $x - \bar{x}$ | $(x - \bar{x})^2$ | $f(x - \bar{x})^2$ |
| 0 | 5 | 2.5 | 39 | 97.5 | –34.16 | 1,166.91 | 45,509.31 |
| 5 | 15 | 10 | 91 | 910 | –26.66 | 710.75 | 64,678.75 |
| 15 | 30 | 22.5 | 122 | 2,745 | –14.16 | 200.51 | 24,462.22 |
| 30 | 45 | 37.5 | 99 | 3,712.5 | 0.84 | 71 | 70.29 |
| 45 | 65 | 55 | 130 | 7,150 | 18.33 | 335.99 | 43,678.70 |
| 65 | 75 | 70 | 50 | 3,500 | 33.34 | 1,111.56 | 55,578.00 |
| 75 | 95 | 85 | 28 | 2,380 | 48.34 | 2,336.76 | 65,429.28 |
| *Total* | | | *559* | *20,495* | | | *299,406.55* |

The upper class limit of the final class is such that the class width is double that of the preceding class.

Mean, $\bar{x} = 20,495/559 = 36.66$

$$\text{Standard deviation} = \sqrt{\frac{299,406.55}{559}} = 23.14$$

For the histogram plot, since the class widths are uneven, we would need to scale the frequencies to plot heights for all those bars that are not standard width. This is due to the fact that the area of histogram bars (NOT height – except for classes that are all the same width) should represent frequency. Since most of the classes are of different widths, the calculations will be impractical and *not advisable in an examination*.

[AUTHOR NOTE: Clearly the examiner has made an error. This can be verified by reference to the suggested solution published by the examining board (ACCA) which shows the heights of bars representing frequencies. The correct histogram is too tedious to calculate and draw and thus is not represented here!]

| Lower $x$ | Upper $x$ | $f$ | % | Cumulative % |
|---|---|---|---|---|
| 0 | 5 | 39 | 6.98 | 6.98 |
| 5 | 15 | 91 | 16.28 | 23.26 |
| 15 | 30 | 122 | 21.82 | 45.08 |
| 30 | 45 | 99 | 17.71 | 62.79 |
| 45 | 65 | 130 | 23.26 | 86.05 |
| 65 | 75 | 50 | 8.94 | 94.99 |
| 75 | 95 | 28 | 5.01 | 100 |

Median = 34. See following graph.

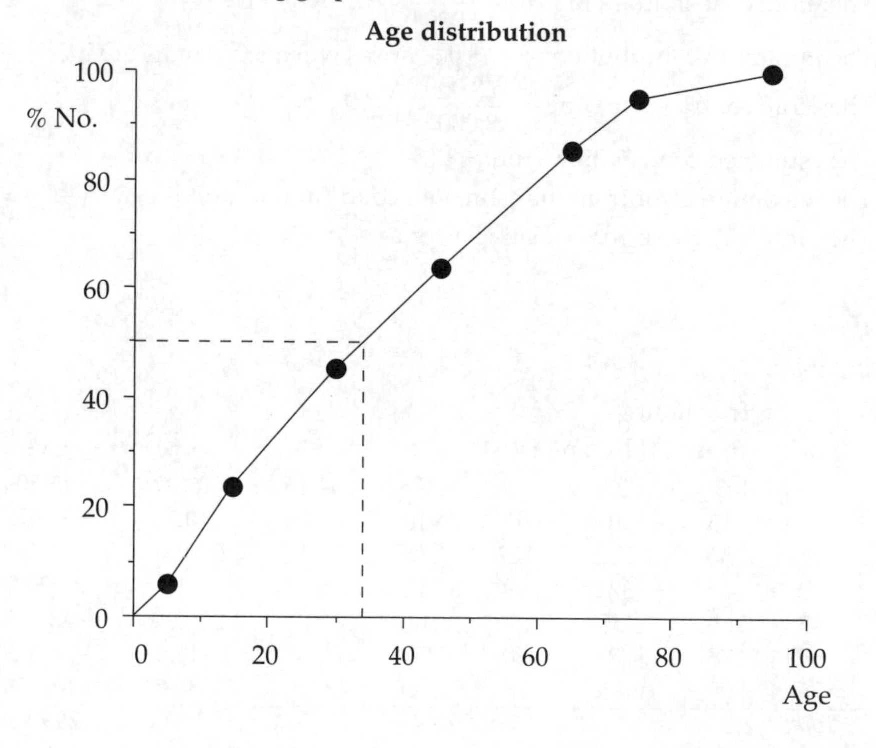

## Question 2

(a) (i) Only 100 – 36.4 = 63.6% received training. Hence, required percentage

$$= \frac{36.2}{63.6} \times 100 = 56.9\%$$

(ii) Similarly, $\frac{7.2}{100 - 48.2} = 13.9\%$ received training.

(b) *Non-apprentices receiving training*:

| | Male | | Female | |
|---|---|---|---|---|
| | % | cum % | % | cum % |
| 1 to 2 | 9.9 | 9.9 | 9.8 | 9.8 |
| 3 to 8 | 28.1 | 38.1 | 43.3 | 53.1 |
| 9 to 26 | 30.0 | 68.1 | 29.5 | 82.6 |
| 27 to 52 | 12.3 | 80.4 | 7.9 | 90.6 |
| 53 to 104 | 11.1 | 91.5 | 6.7 | 97.3 |
| 105 or more | 8.5 | 100.0 | 2.7 | 100.0 |

**Note**: Each male percent in the above is calculated using the given table % as a percentage of 100 57.7.

For example, $9.9 = \frac{4.2}{100 - 57.7} \times 100$. Similarly for females.

**Length of training of non-apprentices**

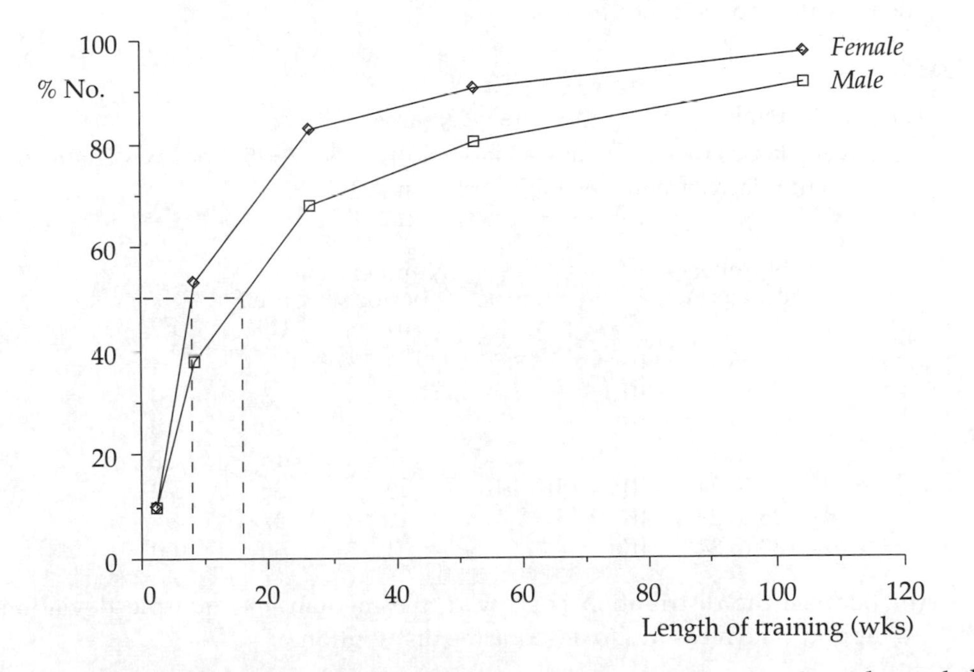


From the graph, the median for male non-apprentices is 16 weeks and the median for female non-apprentices is 8 wks.

(c) All apprentices receive at least 1 year's training. Non-apprentices receive 10 to 11 weeks training on average. Males receive more training than females in general. Although a greater proportion of all females receive no training at all, more female non-apprentices receive *some* training than their male counterparts. A much greater proportion of males are apprenticed (37%) than females (8%).

## Question 3

| Group | Mid-point $x$ | $f$ | $fx$ | $fx^2$ |
|---|---|---|---|---|
| 30 but less than 35 | 32.5 | 17 | 552.5 | 17956 |
| 35 but less than 40 | 37.5 | 24 | 900.0 | 33750 |
| 40 but less than 45 | 42.5 | 19 | 807.5 | 34319 |
| 45 but less than 50 | 47.5 | 28 | 1330.0 | 63175 |
| 50 but less than 55 | 52.5 | 19 | 997.5 | 52369 |
| 55 but less than 60 | 57.5 | 13 | 747.5 | 42981 |
| | | 120 | 5335.0 | 244550 |

$$\overline{x} = \frac{\sum fx}{\sum f} = \frac{5335}{120} = 44.5 \text{ milliseconds}$$

$$s = \sqrt{\frac{\sum fx^2}{\sum f} - \left(\frac{\sum fx}{\sum f}\right)^2} = \sqrt{\frac{244550}{559120} - \left(\frac{5335}{120}\right)^2} = 7.8 \text{ milliseconds}$$

Since the data are grouped, and thus the original access times are not known, both the measures above are estimates.

## Question 4

(a) Smallest value = 3; largest value = 33; range = 30.
Seven classes will each have a class width of 30÷7 = 5 (approx). The formation of a cumulative frequency table is shown at Table 2.

*Table 1*

| Number of rejects | | Number of periods (f) | Cum f (F) | F% |
|---|---|---|---|---|
| 0 to 4 | II | 2 | 2 | 4 |
| 5 to 9 | III | 3 | 5 | 10 |
| 10 to 14 | IIII | 4 | 9 | 18 |
| 15 to 19 | ~~IIII~~ II | 7 | 16 | 32 |
| 20 to 24 | ~~IIII~~ ~~IIII~~ ~~IIII~~ ~~IIII~~ | 20 | 36 | 72 |
| 25 to 29 | ~~IIII~~ ~~IIII~~ I | 11 | 47 | 94 |
| 30 to 34 | III | 3 | 50 | 100 |

(b) Because the distribution is skewed, the median and quartile deviation are appropriate measures to describe the distribution.

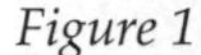

*Figure 1*

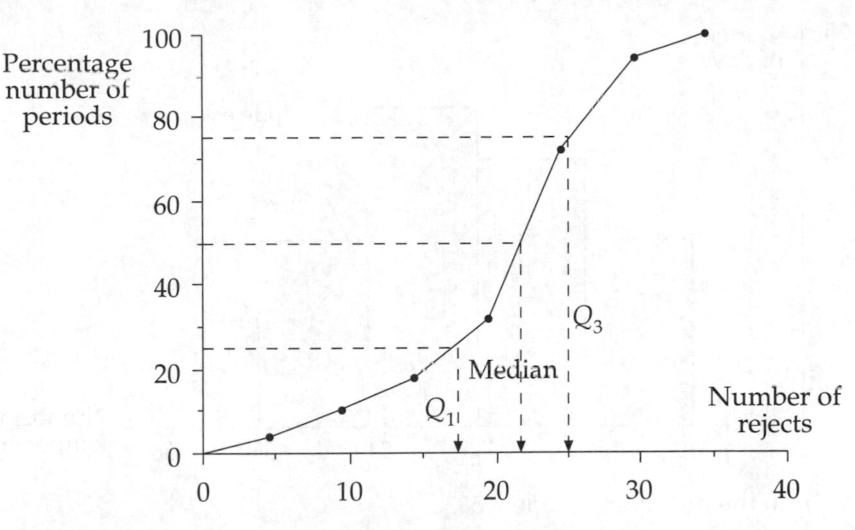


From the graph at Figure 2: $Q_1 = 17.5$; median = 21.5; $Q_3 = 25$.

Therefore, quartile deviation $= \dfrac{Q_3 - Q_1}{2} = \dfrac{150 + 120}{2} = 3.8$

(c) The median of 21.5 describes the average number of rejects in each five minute period = 260/hr (approx). The quartile deviation measures the variability in the number of rejects from one five minute period to the next. In particular, we expect 50% of rejects to lie within 21.5 ± 3.8 in one five minute period.

## Question 5

(a) Smallest value = 510; largest value = 555; range = 45. For 5 classes, each class should have width of 45÷5 = 9 (but 10 is better!)

| Number of components | | f |
|---|---|---|
| 510-519 | ~~IIII~~ II | 7 |
| 520-529 | ~~IIII~~ ~~IIII~~ | 10 |
| 530-539 | ~~IIII~~ ~~IIII~~ II | 12 |
| 540-549 | ~~IIII~~ II | 7 |
| 550-559 | IIII | 4 |

(b) (c) Figure 3 shows the histogram and, from it, the calculation of the mode.

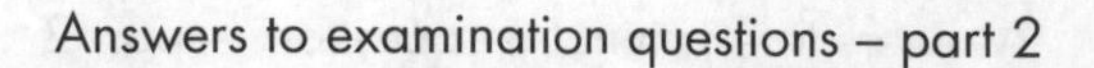


*Figure 2*

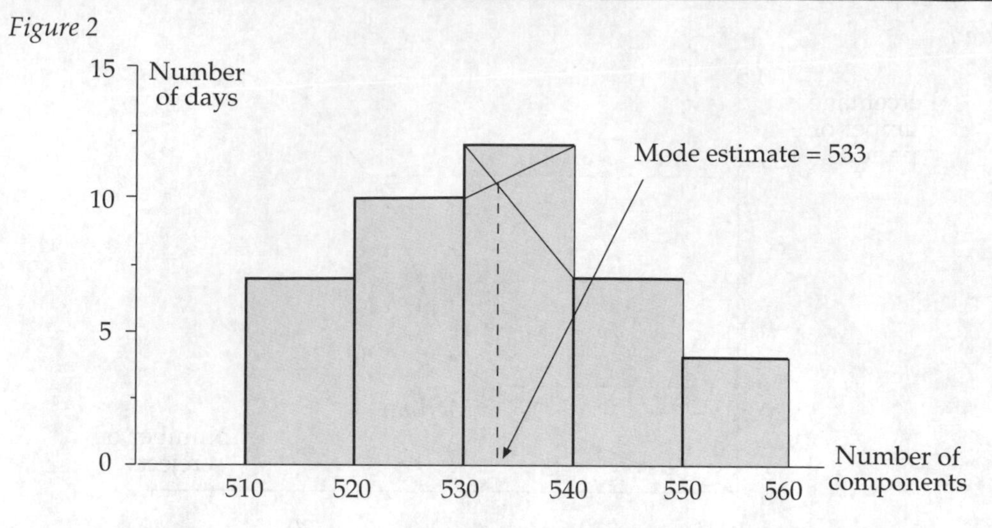


From the histogram, mode = 533.

(d) (e)

| $x$ | $f$ | $fx$ | $fx^2$ |
|---|---|---|---|
| 514.5 | 7 | 3601.5 | 1852971.7 |
| 524.5 | 10 | 5245 | 2751002.5 |
| 534.5 | 12 | 6414 | 3428283 |
| 544.5 | 7 | 3811.5 | 2075361.7 |
| 554.5 | 4 | 2218 | 1229881 |
| | 40 | 21290 | 11337499 |

$$\overline{x} = \frac{\sum fx}{\sum f} = \frac{21290}{40} = 532.25$$

$$s = \sqrt{\frac{\sum fx^2}{\sum f} - \left(\frac{\sum fx}{\sum f}\right)^2} = \sqrt{\frac{11337499}{40} - \left(\frac{21290}{40}\right)^2} = 12.14 \text{ (2D)}$$

(f) Mode = 533; mean = 532.25; sd = 12.14. Mode>mean implies *slight* left skew, which can just be made out from the frequency distribution.

## *Question 6*

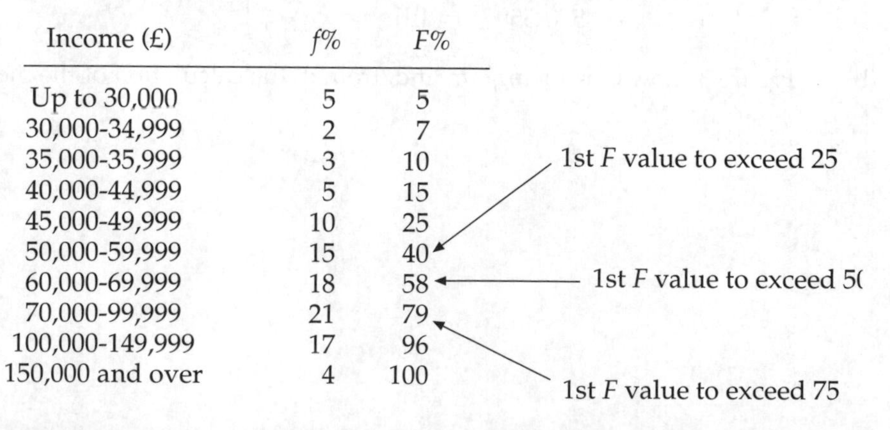

| Income (£) | $f$% | $F$% |
|---|---|---|
| Up to 30,000 | 5 | 5 |
| 30,000-34,999 | 2 | 7 |
| 35,000-35,999 | 3 | 10 |
| 40,000-44,999 | 5 | 15 |
| 45,000-49,999 | 10 | 25 |
| 50,000-59,999 | 15 | 40 |
| 60,000-69,999 | 18 | 58 |
| 70,000-99,999 | 21 | 79 |
| 100,000-149,999 | 17 | 96 |
| 150,000 and over | 4 | 100 |

Since the data is skewed and the first and last classes are open-ended, the median and quartile deviation are the most suitable measures of location and dispersion. The interpolation formula is used below to calculate the measures.

$$Q_1 = 49999.5 + \frac{25-25}{15} \times 10000 = 49{,}999.5\ (50{,}000)$$

$$\text{Median} = 59999.5 + \frac{50-40}{18} \times 10000 = 65{,}555.1\ (65{,}555)$$

$$Q_3 = 69999.5 + \frac{75-58}{21} \times 30000 = 94{,}285.1\ (94{,}285)$$

$$\text{Quartile deviation} = \frac{94{,}285 - 50{,}000}{2} = 22{,}142.$$

*Question 7*

(a) Average rates of increase are usually found using the geometric mean.
For example:

| Year | 1 | 2 | 3 | 4 |
|---|---|---|---|---|
| Rate of increase | 2.3% | 3.8% | 1.9% | 4.2% |

The appropriate multipliers are 1.023, 1.038, 1.019 and 1.042

Thus average multiplier = Geometric mean

$= \sqrt{1.023 \times 1.038 \times 1.019 \times 1.042}$

Therefore, average rate of increase = 3.05%

(b) In a skewed distribution, particularly where only a few values are contained at just one end, the median is the appropriate average to use since it largely ignores extremes and it would be giving the information that 50% of all values are less, and 50% more, than the median value.

(c) Since the speeds need to be averaged *over the same distance*, the harmonic mean is the appropriate average.

$$\text{hm} = \frac{2}{\frac{1}{30} + \frac{1}{60}} = 40 \text{ mph}$$

(However, if the speeds needed to be averaged *over the same time*, the arithmetic mean would be used, giving am = $\frac{30 + 60}{2}$ = 45 mph.)

(d) An average would not be appropriate at all here, since clearly some of the ships would not be able to pass under a bridge built to this height. The height necessary needs to be (at least) the largest value in the distribution.

(e) A weighted mean would be appropriate here. If there were $n_1$ skilled and $n_2$ unskilled workers, the income for each of the workers could be calculated as:

$$\frac{4500 \times n_1 + 3500 \times n_2}{n_1 + n_2}$$

(f) A simple mean is all that is required.

i.e. mean amount = $\frac{\text{Total profits to be allocated}}{\text{Number of employees}}$

*Question 8*

The table showing calculations is given below.

| | Mid-point $(x)$ | $(f)$ | $(fx)$ | $(fx^2)$ |
|---|---|---|---|---|
| 40 to 60 | 50 | 5 | 250 | 12500 |
| 60 to 80 | 70 | 7 | 490 | 34300 |
| 80 to 100 | 90 | 7 | 630 | 56700 |
| 100 to 120 | 110 | 18 | 1980 | 217800 |
| 120 to 140 | 130 | 23 | 2990 | 388700 |
| 140 to 160 | 150 | 14 | 2100 | 315000 |
| 160 to 180 | 170 | 10 | 1700 | 289000 |
| 180 to 220 | 200 | 16 | 3200 | 640000 |
| | | 100 | 13340 | 1954000 |

(i) $\bar{x} = \frac{\sum fx}{\sum f} = \frac{13340}{100} = 133.4$

$$s = \sqrt{\frac{\sum fx^2}{\sum f} - \left(\frac{\sum fx}{\sum f}\right)^2} = \sqrt{\frac{1954000}{100} - \left(\frac{13340}{100}\right)^2} = 41.77 \text{ (2D)}$$

(ii) $\text{cv}(1) = \frac{41.77}{133.4} \times 100 = 31.3\%$; $\text{cv}(2) = \frac{29.33}{88.0} \times 100 = 33.3\%$.

(iii) Distribution 2 is relatively more variable

## Part 3

*Question 1*

(a)

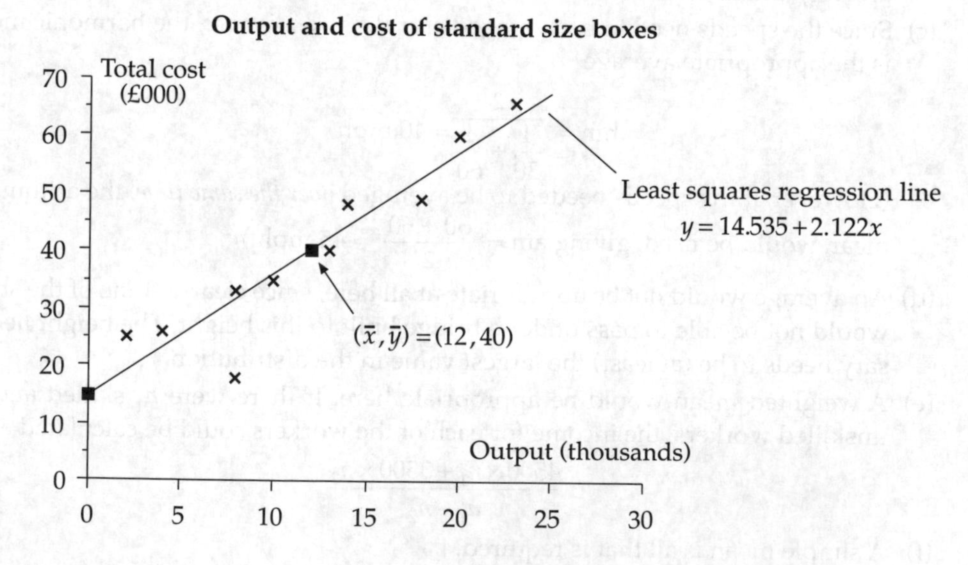

(b) Week 8's figures of 8000 output at a total cost of £18000 are distinctly out of line with the rest of the data. This is clearly due to special circumstances, perhaps a cheap off-loading of old stock.

(c) Any regression line fitted to a set of bivariate data must pass through the mean point $(\bar{x}, \bar{y})$.

In this case, $\bar{x} = \frac{120}{10} = 12$ and $\bar{y} = \frac{400}{10} = 40$.

(d) Let the regression line be in the form: $y = a + bx$. Using the least squares technique, we have:

$$b = \frac{n\sum xy - \sum x \sum y}{n\sum x^2 - \left(\sum x\right)^2} = \frac{10(5704) - 120(400)}{10(1866) - (120)^2} = \frac{9040}{4260} = 2.122 \text{ (3D)}$$

$$a = \frac{\sum y}{n} - b\frac{\sum x}{n} = 40 - (2.122)12 = 14.535 \text{ (3D)}$$

i.e. least squares line of $y$ (total cost) on $x$ (output) is $y = 14.535 + 2.122x$

For the graph plot, $y$-intercept is 14.535 and the line must pass through the point (12,40) from part (c) above.

(e) The fixed costs of the factory is just the value of the $y$-intercept point of the regression line = 14.536 or £14536.

(f) If 25000 standard boxes are produced, then the regression line can be used to estimate the total costs as follows:

Estimated total costs = 14.535 + (2.122)(25) in £000 = £67585.

*Question 2*

(a) See the diagram below. Since both sets of data are close to their respective regression lines, correlation is quite good (and positive). The average turnover for multiples is higher than that for co-operatives, as evidenced by the higher figures, and, since the gradient of the multiple line is larger, multiples have also the higher marginal turnover.

(b) Putting $X$=500 into both regression lines gives:

multiples: $Y = -508.5 + (4.04)(500) = 1511.5$ i.e. a turnover of £1501m.

co-operatives: $Y = 22.73 + (0.67)(500) = 357.73$ i.e. a turnover of £350m.

Since correlation is high and both estimates have been interpolated, a good degree of accuracy might be expected.

(c) As mentioned in (a), the marginal turnover for multiples is higher than for co-operatives. Specifically, for multiples between 253 and 952 stores, each extra store generates a turnover of £4.04m.; for co-operatives between 210 and 575 stores, each extra store generates a turnover of £0.67m.

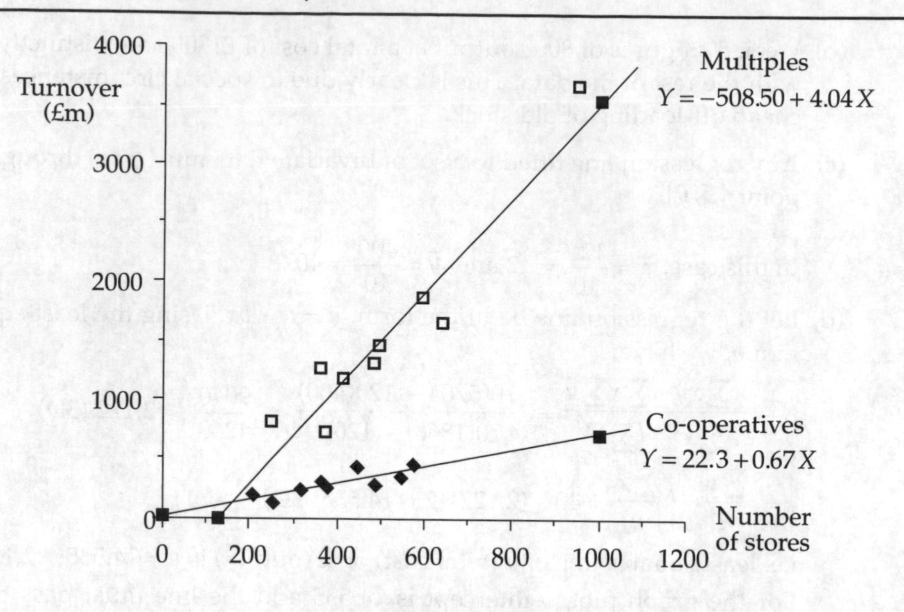


## *Question 3*

(a) This statement is correct. Correlation does not attempt to measure the cause and effect that may exist between two variables, only the *strength of the mathematical relationship*. However, if a causal relationship exists between two variables, there should be a fairly high degree of correlation present.

Example 1: $x$ = Milk consumption; $y$ = Number of violent crimes. Clearly there will be high correlation due to higher population, but obviously no causation!

Example 2: $x$ = Distance travelled by salesman; $y$ = Number of sales made. Here a causal relationship is very probable with a resultant high correlation coefficient.

(b) Table for calculations:

| Colour TV licences (millions) $x$ | Cinema admissions (millions) $y$ | $xy$ | $x^2$ | $y^2$ |
|---|---|---|---|---|
| 5.0 | 134 | 670.0 | 25.0 | 17956 |
| 6.8 | 138 | 938.4 | 46.24 | 19044 |
| 8.3 | 116 | 962.8 | 68.89 | 13456 |
| 9.6 | 104 | 998.4 | 92.16 | 10816 |
| 10.7 | 103 | 1102.1 | 114.49 | 10609 |
| 12.0 | 126 | 1512.0 | 144.0 | 15876 |
| 12.7 | 112 | 1422.4 | 161.29 | 12544 |
| 12.9 | 96 | 1238.4 | 166.41 | 9216 |
| 78.0 | 929 | 8844.5 | 818.48 | 109517 |

$$r = \frac{n\sum xy - \sum x \sum y}{\sqrt{\left(n\sum x^2 - \left(\sum x\right)^2\right)\left(n\sum y^2 - \left(\sum y\right)^2\right)}}$$

$$= \frac{(8)(8844.5) - (78)(929)}{\sqrt{\left((8)(818.48) - 78^2\right)}\sqrt{\left((8)(109517) - 929^2\right)}}$$

$$= \frac{-1706}{(21.537)(114.433)} = -0.692$$

A moderately high degree of negative correlation, showing that as the number of colour licences increases so the number of cinema admissions decreases.

A causal relationship seems reasonable here, and with $r^2 = 0.48$ (2D), this demonstrates that approximately 50% of the variation in cinema attendances is explained by variations in the number of colour licences.

## *Question 4*

(a) See the figure.

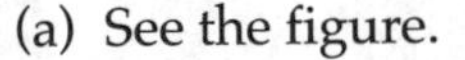

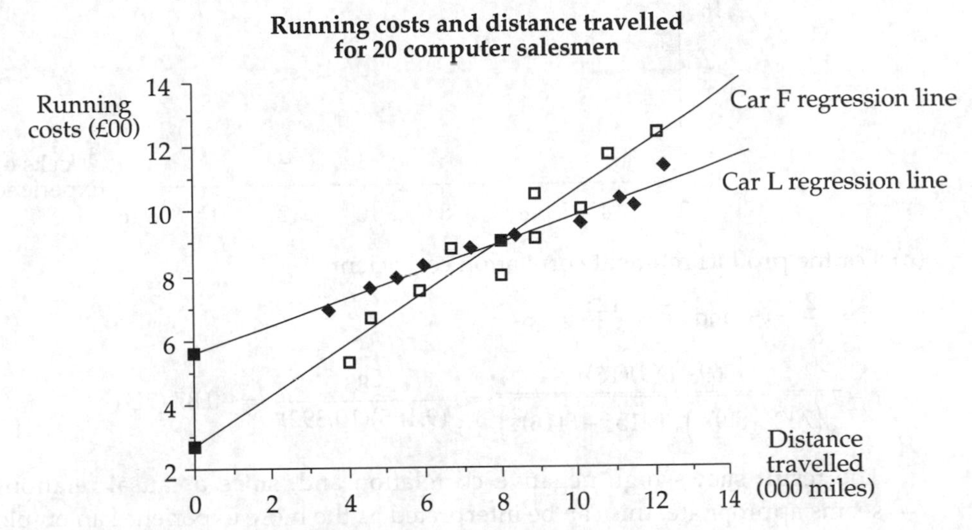


For the regression line plots in the figure,

Car F: intercept on $y$-axis is 2.65 and line must pass through (8,9).

Car L: intercept on $y$-axis is 5.585 and line must pass through (8,9).

(b) Car F: 2.65 is the initial (or fixed) running costs (£00) and 0.794 is the extra cost (£00) for each further one thousand miles travelled.

Car L: 5.585 is the initial cost and 0.427 is the extra cost for each further one thousand miles travelled.

(c) It is necessary to minimise the *average* cost per car for the two different types, taking into account the new average distance travelled = 1.5 $x$ 8 = 12 (000 miles).

For type F: Average cost = 2.65 + (0.794)12 = 12.18 = £1218.

For type L: Average cost = 5.585 + (0.427)12 = 10.71 = £1071.

Therefore car L is cheaper on average.

(d) Using car L, the average cost for one car (with 10% extra costs) is given by (1.1)(5.585 + (0.427)12) = 11.7799 = £1177.99. Thus, expected total running costs for 5 cars is 5 × £1177.99 = £5889.95.

## Question 5

(a)

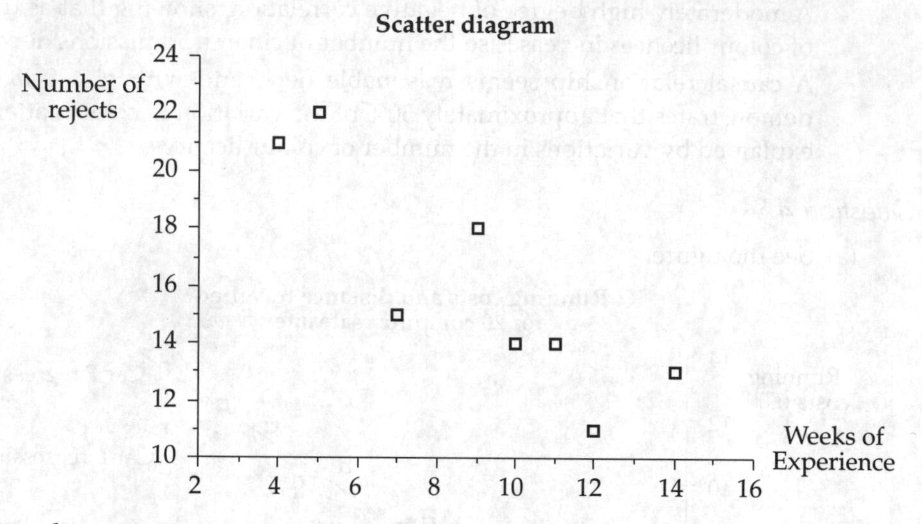


(b) For the product moment correlation coefficient:

$$\bar{x} = \frac{72}{8} = 9 \text{ and } \bar{y} = \frac{128}{8} = 16$$

$$r = \frac{1069 - 8(9)(16)}{\sqrt{(732 - 8(9)^2)}\sqrt{(2156 - 8(16)^2)}} = \frac{-83}{(9.165)(10.392)} = -0.87 \text{ (2D)}$$

The result shows high negative correlation and, since a causal relationship seems appropriate, this can be interpreted as the more experience an employee has in wiring components, the fewer the number of rejects to be expected.

c) Assuming a least squares line of the form $y = a + bx$, $a$ and $b$ are calculated as follows:

$$b = \frac{1069 - 8(9)(16)}{4\times3\times2\times1} = \frac{-83}{84} = -0.988 \text{ (3D) and } a = 16 - (-0.988)\times9 = 24.892 \text{ (3D)}$$

The least squares regression line of $y$ (rejects) on $x$ (experience) is thus:

$$y = 24.892 - 0.988x.$$

After one week of experience ($x$=1), the expected number of rejects is given by:

$$y = 24.892 - 0.988(1) = 23.9\text{(1D) or 24 (to nearest whole number).}$$

## Question 6

(i) Table for calculations:

| Value | rank | Value | rank | $d^2$ | Value | rank | $d^2$ |
|---|---|---|---|---|---|---|---|
| 15 | 3 | 13 | 2 | 1 | 16 | 2 | 1 |
| 19 | 5 | 25 | 5 | 0 | 19 | 3.5 | 2.25 |
| 30 | 7 | 23 | 4 | 9 | 26 | 6 | 1 |
| 12 | 2 | 26 | 6 | 16 | 14 | 1 | 1 |
| 58 | 8 | 48 | 8 | 0 | 65 | 8 | 0 |
| 10 | 1 | 15 | 3 | 4 | 19 | 3.5 | 6.25 |
| 23 | 6 | 28 | 7 | 1 | 27 | 7 | 1 |
| 17 | 4 | 10 | 1 | 9 | 22 | 5 | 1 |
| | | | | 40 | | | 12.5 |

(1) Coefficient for actual and forecast 1:

$$r' = 1 - \frac{6(40)}{8(63)} = 0.52$$

(2) Coefficient for actual and forecast 2:

$$r' - 1 - \frac{6(12.5)}{8(63)} = 0.85$$

(ii) Clearly, forecasting method 2 is superior.

## Question 7

(a) Briefly, regression describes the mathematical (linear) relationship between two variables while correlation describes the strength of this linear relationship.

(b) (i) See the figure on the following page.
The figure clearly shows that as the number of colour licences increases, so the number of cinema attendances decreases.

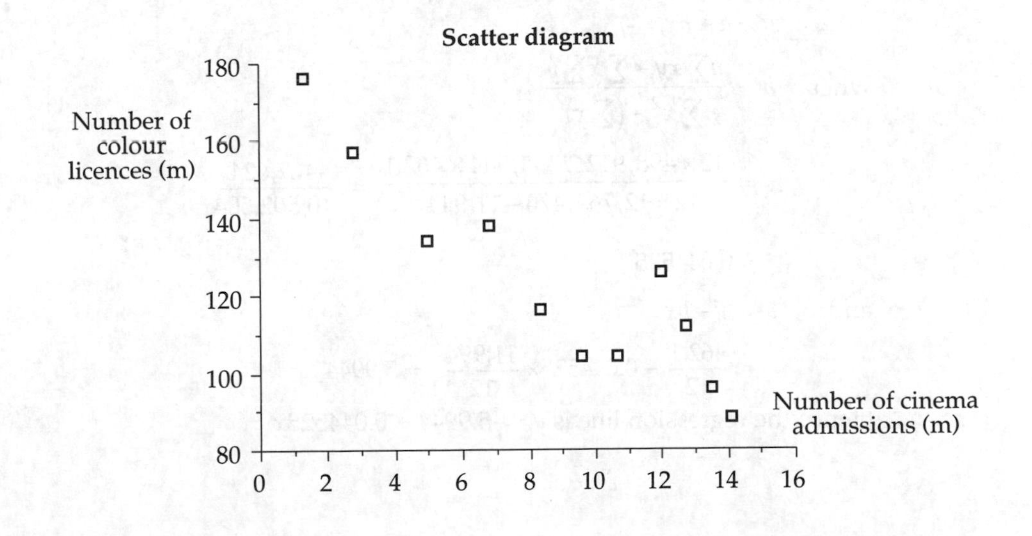

(ii)

| Number of TV licences (m) | rank | Number of cinema admissions (m) | rank | $d^2$ |
|---|---|---|---|---|
| 1.3 | 1 | 176 | 11 | 100 |
| 2.8 | 2 | 157 | 10 | 64 |
| 5.0 | 3 | 134 | 8 | 25 |
| 6.8 | 4 | 138 | 9 | 25 |
| 8.3 | 5 | 116 | 6 | 1 |
| 9.6 | 6 | 104 | 3.5 | 6.25 |
| 10.7 | 7 | 104 | 3.5 | 12.25 |
| 12.0 | 8 | 126 | 7 | 1 |
| 12.7 | 9 | 112 | 5 | 16 |
| 13.5 | 10 | 96 | 2 | 64 |
| 14.1 | 11 | 88 | 1 | 100 |
| | | | | 414.5 |

Rank correlation coefficient: $r' = 1 - \dfrac{6(414.5)}{11(120)}$

$= -0.88$

The above coefficient is showing strong inverse (or negative) correlation and, since there is every reason to believe that there is a causal relationship here, the hypothesis seems reasonable.

## *Question 8*

(a) As the regression equation of profit on sales is required, put profit = $y$ and sales = $x$

The equation is:

$y = a + bx,$

where: $b = \dfrac{n\sum xy - \sum x \sum y}{n\sum x^2 - \left(\sum x\right)^2}$

$= \dfrac{12 \times 498{,}912.2 - 11{,}944 \times 462.1}{12 \times 12{,}763{,}470 - 11{,}944^2} = \dfrac{467{,}624}{10{,}502{,}504}$

$= 0.044525$

and: $a = \overline{y} - b\overline{x}$

$= \dfrac{462.1}{12} - 0.044525 \times \dfrac{11{,}994}{12} = 5.9944$

Hence the regression line is $y = -5.9944 + 0.044525x$

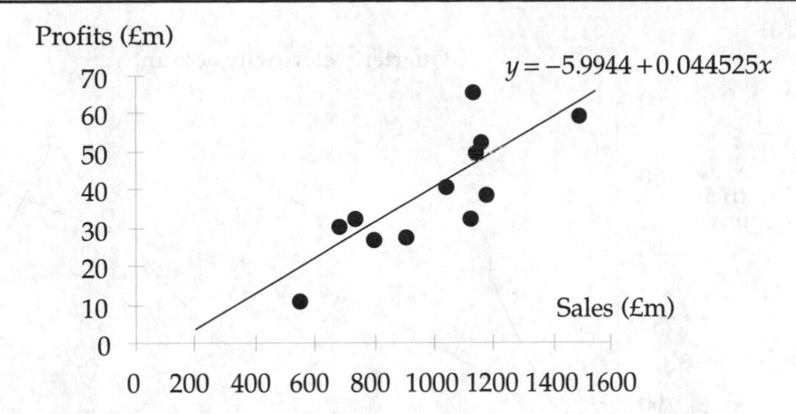


If the sales are £1,000 million, then $x$ = 1,000 and the regression equation gives:

$$y = -5.9944 + 0.044525 \times 1000 = 38.5306$$

That is, profits = £38.5 million

(b) The regression line can be interpreted as indicating that, within the range of the data, each £1 million of sales generates £44,525 profit.

Assuming there are no changes in background circumstances, the forecast can be considered fairly reliable. First of all, the graph indicates a good correlation between profits and sales and so any forecasts produced by the regression line are likely to be reliable. (In fact the correlation coefficient, $r$, is approximately 0.8). Further, the forecast is an interpolation (the $x$-value line within the range of the given data), which is a further indication of reliability.

## Part 4

### *Question 1*

(a)(c)

| | Account (£) ($y$) | Moving total | Moving average | Centred moving average ($t$) | Seasonal variation | Deseasonalised data |
|---|---|---|---|---|---|---|
| 1982 Q2 | 662 | | | | -39 | 701 |
| Q3 | 712 | | | | 45 | 667 |
| | | 2850 | 712.50 | | | |
| Q4 | 790 | | | 719.5 | 51 | 739 |
| | | 2906 | 726.50 | | | |
| 1983 Q1 | 686 | | | 740.1 | -56 | 742 |
| | | 3015 | 753.75 | | | |
| Q2 | 718 | | | 760.8 | -39 | 757 |
| | | 3071 | 767.75 | | | |
| Q3 | 821 | | | 774.9 | 45 | 776 |
| | | 3128 | 782.00 | | | |
| Q4 | 846 | | | 790.0 | 51 | 795 |
| | | 3192 | 798.00 | | | |
| 1984 Q1 | 743 | | | 798.8 | -56 | 799 |
| | | 3198 | 799.50 | | | |
| Q2 | 782 | | | 803.3 | -39 | 821 |
| | | 3228 | 807.00 | | | |
| Q3 | 827 | | | 814.8 | 45 | 782 |
| | | 3290 | 822.50 | | | |
| Q4 | 876 | | | 830.0 | 51 | 825 |
| | | 3350 | 837.50 | | | |
| 1985 Q1 | 805 | | | 843.6 | -56 | 861 |
| | | 3399 | 849.75 | | | |
| Q2 | 842 | | | | -39 | 881 |
| Q3 | 876 | | | | 45 | 831 |

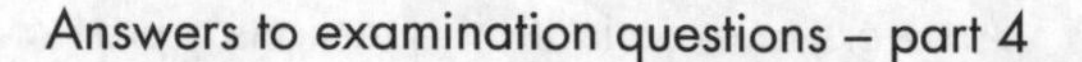


(b)

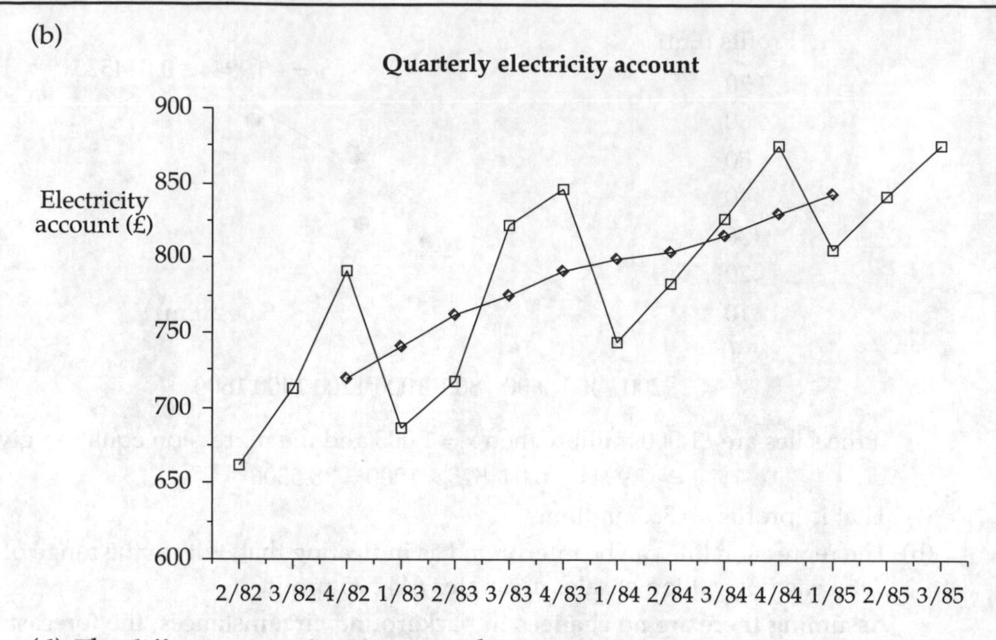


(d) The difference in values between the first and last trend values is:

843.6–719.5 = 124.1

Thus, the average increase between trend values is

$$\frac{124.1}{9} = 14 \text{ approximately.}$$

*Estimated trend values*

1985 Q4: 843.6+3(14) = 886

1986 Q1: 886+14 = 900.

Therefore, adding the respective seasonal variation value, we have:

*Forecast values*

1985 Q4: 886 + 51 = £937

1985 Q4: 900 – 56 = £844.

## *Question 2*

(a) and (b) Main table of calculations:

| | | Moving total | Moving average | Centred moving average ($t$) | Deviation ($y$-$t$) | Seasonal variation ($s$) | Deseasonalised data ($y$-$s$) |
|---|---|---|---|---|---|---|---|
| 73 1 | 100 | | | | | -10.9 | 110.9 |
| 2 | 125 | | | | | 11.0 | 114.0 |
| | | 454 | 113.50 | | | | |
| 3 | 127 | | | 114.000 | 13.000 | 12.6 | 114.4 |
| | | 458 | 114.50 | | | | |
| 4 | 102 | | | 114.875 | -12.875 | -12.7 | 114.7 |
| | | 461 | 115.25 | | | | |
| 74 1 | 104 | | | 115.625 | -11.625 | -10.9 | 114.9 |
| | | 464 | 116.00 | | | | |
| 2 | 128 | | | 116.625 | 11.375 | 11.0 | 117.0 |
| | | 469 | 117.25 | | | | |
| 3 | 130 | | | 118.000 | 12.000 | 12.6 | 117.4 |
| | | 475 | 118.75 | | | | |
| 4 | 107 | | | 119.125 | -12.125 | -12.7 | 119.7 |
| | | 478 | 119.50 | | | | |
| 75 1 | 110 | | | 119.875 | -9.875 | -10.9 | 120.9 |
| | | 481 | 120.25 | | | | |
| 2 | 131 | | | 120.250 | 10.750 | 11.0 | 120.0 |
| | | 481 | 120.25 | | | | |
| 3 | 133 | | | 120.125 | 12.875 | 12.6 | 120.4 |
| | | 480 | 120.00 | | | | |
| 4 | 107 | | | 120.125 | -13.125 | -12.7 | 119.7 |
| | | 481 | 120.25 | | | | |
| 76 1 | 109 | | | | | -10.9 | 119.9 |
| 2 | 132 | | | | | 11.0 | 121.0 |

*Seasonal variation calculations*

| | Q1 | Q2 | Q3 | Q4 | |
|---|---|---|---|---|---|
| 1973 | | | 13.000 | –12.875 | |
| 1974 | –11.625 | 11.375 | 12.000 | –12.125 | |
| 1975 | –9.875 | 10.750 | 12.875 | –13.125 | |
| Totals | –21.500 | 22.125 | 37.875 | –38.125 | |
| Averages (1D) | –10.8 | 11.1 | 12.6 | –12.7 | (Total=+0.2) |
| Adjustments | –0.1 | –0.1 | – | – | |
| Seasonal variation | –10.9 | 11.0 | 12.6 | –12.7 | |

(c)

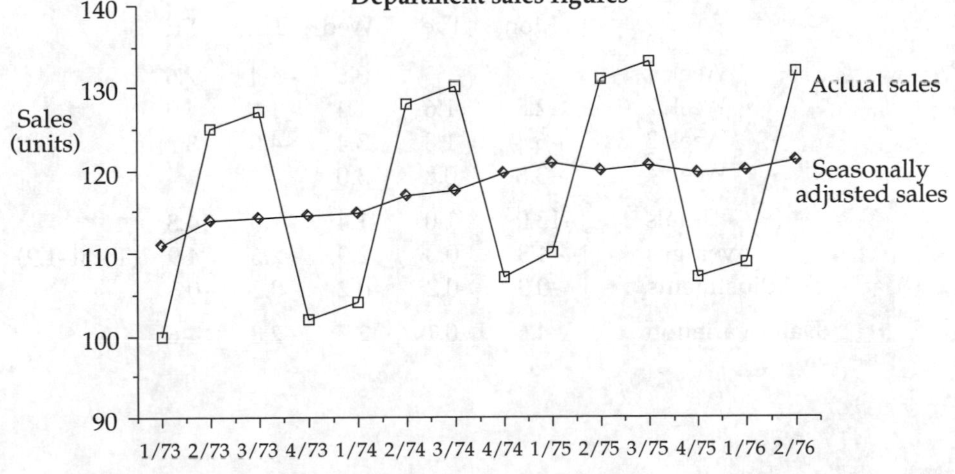

(d) Both trend and seasonally adjusted values show a steady increase up to the beginning of 1975, when they levelled out. Seasonal patterns are well marked and continue throughout the whole period. Adjustments to seasonal averages were very small, leading to the conclusion that there was very little residual variation other than random factors.

## Question 3

(a) Table of main calculations.

| | | Output ($y$) | 5-day moving total | Trend ($t$) | Variation ($y$-$t$) |
|---|---|---|---|---|---|
| Week 1 | Mon | 187 | | | |
| | Tue | 203 | | | |
| | Wed | 208 | 1022 | 204.4 | 3.6 |
| | Thu | 207 | 1042 | 208.4 | -1.4 |
| | Fri | 217 | 1047 | 209.4 | 7.6 |
| Week 2 | Mon | 207 | 1049 | 209.8 | -2.8 |
| | Tue | 208 | 1048 | 209.6 | -1.6 |
| | Wed | 210 | 1043 | 208.6 | 1.4 |
| | Thu | 206 | 1038 | 207.6 | -1.6 |
| | Fri | 212 | 1040 | 208.0 | 4.0 |
| Week 3 | Mon | 202 | 1042 | 208.4 | -6.4 |
| | Tue | 210 | 1041 | 208.2 | 1.8 |
| | Wed | 212 | 1043 | 208.6 | 3.4 |
| | Thu | 205 | 1049 | 209.8 | -4.8 |
| | Fri | 214 | 1054 | 210.8 | 3.2 |
| Week 4 | Mon | 208 | 1059 | 211.8 | -3.8 |
| | Tue | 215 | 1071 | 214.2 | 0.8 |
| | Wed | 217 | 1070 | 214.0 | 3.0 |
| | Thu | 217 | | | |
| | Fri | 213 | | | |

(b) See graph on opposite page

(c)

| | Mon | Tue | Wed | Thu | Fri | |
|---|---|---|---|---|---|---|
| Week 1 | | | 3.6 | –1.4 | 7.6 | |
| Week 2 | –2.8 | –1.6 | 1.4 | –1.6 | 4.0 | |
| Week 3 | –6.4 | 1.8 | 3.4 | –4.8 | 3.2 | |
| Week 4 | –3.8 | 0.8 | 3.0 | | | |
| Totals | –13.0 | 1.0 | 11.4 | –7.8 | 14.8 | |
| Averages | –4.3 | 0.3 | 2.9 | –2.6 | 4.9 | (Total–1.2) |
| Adjustments | –0.3 | –0.2 | –0.2 | –0.2 | –0.3 | |
| Daily variation | –4.6 | 0.1 | 2.7 | –2.8 | 4.6 | |

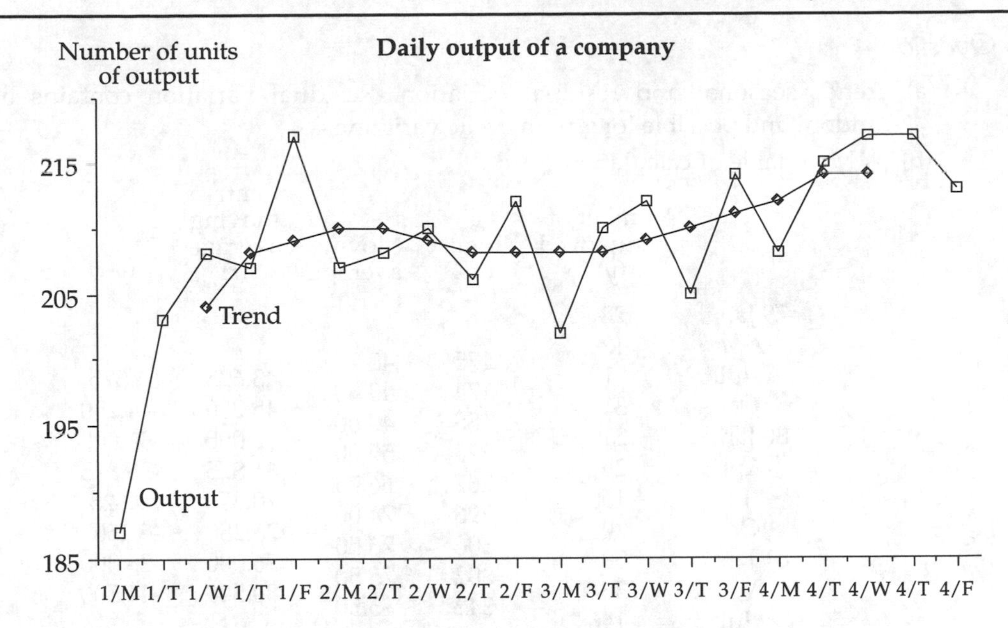


(d) Using the calculated (moving average) trend values, the average daily increase in trend can be calculated as: $\dfrac{214.0 - 204.4}{15} = 0.64$.

Trend value for Week 5 (Monday) = 214.0 + 3(0.64) = 215.9 (1D).

Trend value for Week 5 (Tuesday) = 215.9 + 0.64 = 216.6 (1D).

Forecast output for Week 5 (Monday) = 215.9 – 4.6 = 211 (to nearest unit).

Forecast output for Week 5 (Tuesday) = 216.6 + 0.1 = 217 (to nearest unit).

(e) No forecast can ever be confidently made, since it is based only on past evidence and there can be no guarantee that the trend projection is accurate or that the daily variation figures used will be valid for future time points. Only general experience and a particular knowledge of the given time series environment would help further in determining the accuracy of the given forecasts.

*Question 4*

(a) Trend, seasonal and residual variation. Residual variation contains both random and possible long-term cyclic variations.

(b) (i) Main table of calculations.

| | Number of unemployed ($y$) | Totals of 4 | Moving average | Centred moving average ($t$) | ($y$-$t$) |
|---|---|---|---|---|---|
| 79 Jan | 22 | | | | |
| Apr | 12 | 175 | 43.75 | | |
| Jul | 110 | 174 | 43.50 | 43.625 | 66.375 |
| Oct | 31 | 188 | 47.00 | 45.250 | -14.250 |
| 80 Jan | 21 | 228 | 57.00 | 52.000 | -31.000 |
| Apr | 26 | 267 | 66.75 | 61.875 | -35.875 |
| Jul | 150 | 296 | 74.00 | 70.350 | 79.625 |
| Oct | 70 | 306 | 76.50 | 75.250 | -5.250 |
| 81 Jan | 50 | 302 | 75.50 | 76.000 | -26.000 |
| Apr | 36 | 342 | 85.50 | 80.500 | -44.500 |
| Jul | 146 | | | | |
| Oct | 110 | | | | |

Calculations for seasonal variation:

| | Jan | Apr | Jul | Oct |
|---|---|---|---|---|
| 1979 | | | 66.375 | –14.250 |
| 1980 | –31.000 | –35.875 | 79.625 | –5.250 |
| 1981 | –26.000 | –44.500 | | |
| Totals | –57.000 | –80.375 | 146.000 | –19.500 |
| Averages | –28.2 | –40.2 | 73.0 | –9.8 (Tot=–5.5) |
| Adjustments | +1.4 | +1.4 | +1.4 | =1.3 |
| Seasonal variation | –27.1 | 74.4 | 74.4 | –8.5 |

(ii) Seasonally adjusted values: 1981, Jan = $y - s = 50 - (-27.1) = 77.1$

1981, Apr = $36 - (-38.3) = 74.3$

## Part 5

*Question 1*

(i) An index number enables the value of some economic commodity to be compared over some defined time period. It is expressed in percentage terms, using a base of 100.

(ii) (iii) are shown in the following table:

| Year | Average salary (£) | Year on year increase (%) | Retail Price Index (1975=100) | Year on year increase (%) | Revalued salary (1985 base) |
|---|---|---|---|---|---|
| 1977 | 9500 | | 135.1 | | 19464.10 |
| 1978 | 10850 | 14.2 | 146.2 | 8.2 | 20542.27 |
| 1979 | 13140 | 21.1 | 165.8 | 13.4 | 21936.98 |
| 1980 | 14300 | 8.8 | 195.6 | 18.0 | 20236.40 |
| 1981 | 14930 | 4.4 | 218.8 | 11.9 | 18887.68 |
| 1982 | 15580 | 4.4 | 237.7 | 8.6 | 18142.80 |
| 1983 | 16200 | 4.0 | 248.6 | 4.6 | 18037.65 |
| 1984 | 16800 | 3.7 | 261.0 | 5.0 | 17817.01 |
| 1985 | 17500 | 4.2 | 276.8 | 6.1 | 17500.00 |

**Note**: $£17{,}817.01 = \dfrac{16800 \times 276.8}{261.0}$ ; $£18{,}037.65 = \dfrac{16200 \times 276.8}{248.6}$ ; etc

(iv) Except for the first two years, the increase in prices has outstripped the increase in salary. The revalued salary shows that (in real terms) the systems analysts are being paid less now than in any of the previous nine years.

*Question 2*

(a) The Retail Prices Index can be used by retailers to compare their own average price increases with those that consumers are subject to. Wage-earners often use the RPI (although the Tax and Price Index is more relevant) to compare their wage increases with the increases in prices. Trades Union use the value of the RPI to negotiate price increases with employers.

The Producer Price Indices can be used by retailers to compare the prices they are paying for their goods. It can also be used by consumers as a long term warning (nine months or so) of trends that will inevitably be felt in the RPI.

The Index of Output of the Production Industries is used as a general guide to measure the changes in the level of production in the UK.

(b) Putting July 1979 as year 0 etc, we have:

| | $W_0 \times E_0$ | $W_1 \times E_0$ | $W_0 \times E_2$ | $W_2 \times E_2$ |
|---|---|---|---|---|
| $Q$ | 300 | 350 | 240 | 320 |
| $R$ | 120 | 130 | 180 | 210 |
| $S$ | 140 | 170 | 70 | 90 |
| $T$ | 90 | 110 | 180 | 240 |
| Total | 650 | 760 | 670 | 860 |

(i) Laspeyre index for year 1 $= L_1 = \dfrac{\sum W_1 E_0}{\sum W_0 E_0} = \dfrac{760}{650} \times 100 = 116.9$

(ii) Paasche index for year 2 $= P_2 = \dfrac{\sum W_2 E_2}{\sum W_0 E_2} = \times\ 100 = 128.4$

(iii) The indices given can be base-changed to 1979. Thus:

$$L_{1/0} = \frac{187.4}{156.3} \times 100 = 119.9; \quad L_{2/0} = \frac{203.4}{156.3} \times 100 = 130.1$$

These indices, when compared with the company's indices, show that the wage rates of the company are lagging slightly behind the Chemical and Allied Industry's rates by about two points.

## Question 3

(a) In 1974 (on average, per week) 4 hours overtime was worked, which is equivalent to 4 ×1.5=6 normal hours. Thus, dividing the average weekly earnings by 46 (the equivalent normal hours worked per week) will give the normal rate per hour as 40.19÷46=£0.87. Multiplying this by 40 will thus yield the normal weekly rate of 40× 0.87=£34.95. This must be done for each year.

i.e. average normal weekly hours = 40 + (ave hours worked – 40) × 1.5

$$\text{and normal weekly rate} = \text{ave earnings} \times \frac{40}{\text{average normal weekly hours}}$$

| Year | Average weekly earnings | Average hours worked | Equivalent normal weekly hours | Normal weekly rate |
|---|---|---|---|---|
| 1974 | 40.19 | 44 | 46 | 34.95 |
| 1975 | 52.65 | 45 | 47.5 | 44.34 |
| 1976 | 62.03 | 45 | 47.5 | 52.24 |
| 1977 | 70.20 | 46 | 49 | 57.31 |
| 1978 | 76.83 | 46 | 49 | 62.72 |
| 1979 | 91.90 | 46 | 49 | 75.02 |
| 1980 | 107.51 | 45 | 47.5 | 90.53 |
| 1981 | 121.95 | 43 | 44.5 | 109.62 |

(b)

| RPI | 80.5 | 100.0 | 116.5 | 135.0 | 146.2 | 165.8 | 195.6 | 218.9 |
|---|---|---|---|---|---|---|---|---|
| Log | 1.91 | 2.00 | 2.07 | 2.13 | 2.16 | 2.22 | 2.29 | 2.34 |

**Semi-logarithmic graph of the retail price index**

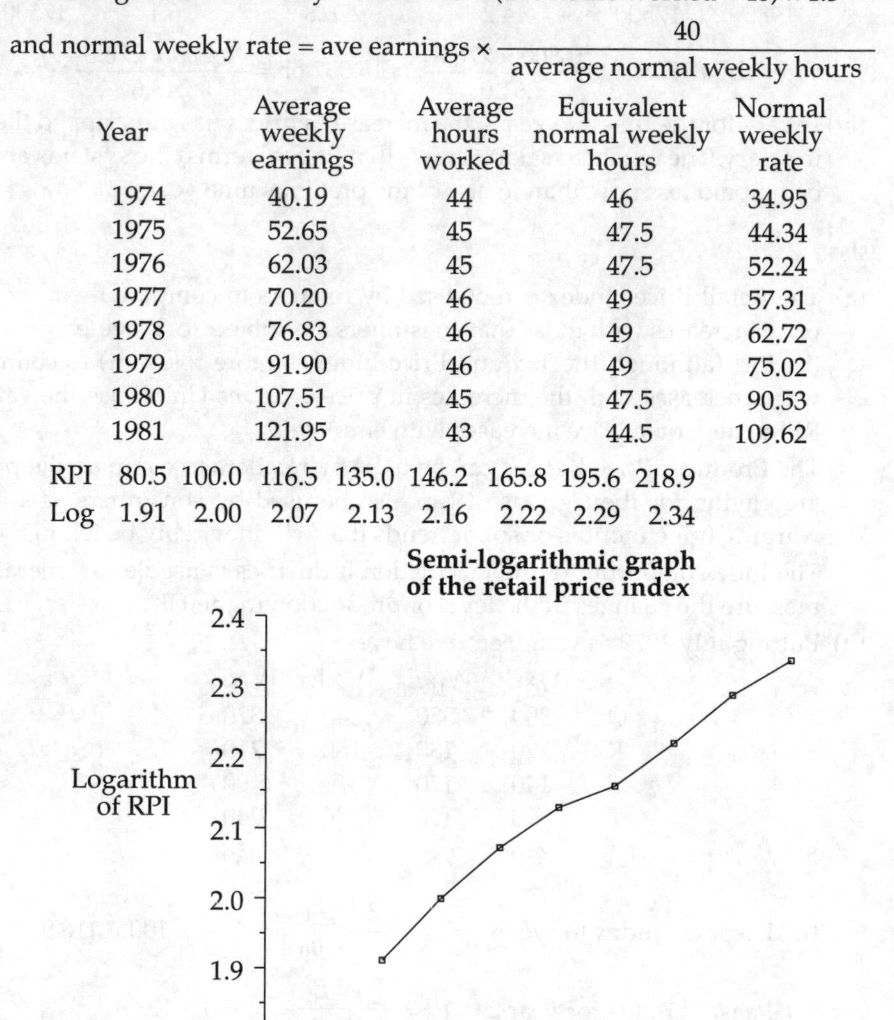


The above logarithms are plotted against the relevant year to form a semi-logarithmic graph which is shown in the figure above.

(c) Since the semi-graph in Figure 1 is an approximate straight line, this demonstrates that the rate of increase of the RPI is constant.

(d) A deflated normal weekly rate can be obtained by dividing each normal weekly rate by the value of the RPI for that year and multiplying back by 100 to bring the value back to the correct form.

e.g. deflated normal weekly rate for 1974 = $34.95 \times \frac{100}{80.5}$ = £43.42

| Year | 1974 | 1975 | 1976 | 1977 | 1978 | 1979 | 1980 | 1981 |
|---|---|---|---|---|---|---|---|---|
| Deflated normal weekly rate (£) | 43.42 | 44.34 | 44.84 | 42.45 | 42.90 | 45.25 | 46.28 | 50.08 |

(e) If an index of real wages is calculated, it will enable a comparison between the increase in prices and real wages to be made.

## Question 4

| Item | Weight (w) | Index (I) | (1) (wI) | (2) (wI) | (3) (wI) |
|---|---|---|---|---|---|
| Mining and quarrying | 41 | 361 | 14801 | | |
| Manufacturing | | | | | |
| Food, drink and tobacco | 77 | 106 | 8162 | 8162 | 8162 |
| Chemicals | 66 | 109 | 7194 | 7194 | 7194 |
| Metal | 47 | 72 | 3384 | 3384 | 3384 |
| Engineering | 298 | 86 | 25628 | 25628 | 25628 |
| Textiles | 67 | 70 | 4690 | 4690 | 4690 |
| Other manufacturing | 142 | 91 | 12922 | 12922 | 12922 |
| Construction | 182 | 84 | 15288 | 15288 | |
| Gas, electricity and water | 80 | 115 | 9200 | 9200 | |
| | 1000 | | 101269 | 86468 | 61980 |

(i) (1) All industries index is given by: $\frac{101269}{1000} = 101.3$

(2) All industries except mining and quarrying index is: $\frac{86468}{1000 - 41} = 90.2$

(3) Manufacturing industries index is: $\frac{61980}{1000 - 41 - 182 - 80} = 88.9$

(ii) The high mining and quarrying index of 361 was severely offset by its small weight in the relatively low value of 101.3 for the overall index in (1). However, the index of only 90.2 in (2) shows the significance of mining and quarrying (particularly North Sea oil) to industrial production in the UK. The low manufacturing index of 88.9 in (3) is due to the fact that the three largest weights are assigned to relatively low indices.

## Question 5

FOOD: The movement in these weights can be accounted for by the increased affluence of our society, which results in a much greater pool of disposable income left after the basic necessities (of which food is one of the most important) have been acquired. Also, since it is reasonable to assume that we are not buying less food in

1981 than we were in 1961, it means that (in relative terms) food is now cheaper. Since food has such a high weighting, the value of its index will bear the most significant effect on the RPI itself.

HOUSING: The increase in expenditure is probably due to two factors. First, a significant part of the extra disposable income is being spent on housing. Second, housing is more expensive in real terms. Changes in such things as mortgage rates and rents will now have a more significant effect on the RPI than was previously the case.

CLOTHING: The decrease in expenditure is probably due to cheap imports, since again we can only suppose that we are buying at least as much clothing in 1981 as we were in 1961.

TRANSPORT: The dramatic increase in transport costs are probably due to the increased mobility we now have as a society. We travel much further both to and from our place of occupation and also for leisure and recreation purposes. Changes in petrol prices and car tax will now have much more effect on the RPI than they did previously.

## Part 6

### *Question 1*

(a) Set P = £40,000, for convenience.

Time 0: amount owed = $P$.

*After 1 quarter*, 4% is added to amount owing, giving $P(1 + 0.04) = PR$

X is paid, leaving amount owed = $PR - X$.

*After 2 quarters* 4% is added to amount owing, and X is paid.

Amount now owed $= (PR - X)R - X$

$= PR^2 - XR - X$ or £$(40{,}000\ R^2 - XR - X)$

(b) After 3 quarters, amount owed $= (PR^2 - XR - X) - R - X$

$= PR^3 - XR^2 - XR - X$

And so on, until, after 80 quarters, the amount owed

$= PR^{80} - XR^{79} - XR^{78} - \dots - X$

$= PR^{80} - X(R^{79} + R^{78} + \dots + R + 1)$

$= PR^{80} - X\left(\dfrac{R^{80}-1}{R-1}\right)$ from the geometric progression formula.

Now, as the mortgage is to be paid off in this period, this amount owed must be zero, and so:

$$PR^{80} = X\left(\frac{R^{80}-1}{R-1}\right)$$

and: $$P.R^{80}.\left(\frac{R-1}{R^{80}-1}\right) = X$$

$0.0418P = X$ (from tables) [*]

Since $P$ = £40,000, the quarterly repayment is £1,672.56.

(c) Using [*], if $P$ is doubled from £40,000 to £80,000, the factor 0.0418 would be unaltered, and so the repayment figure would double to 2X.

*Question 2*

(a) We are given that: $P$=12000; $i$=0.15; $n$=5. Putting the amortization payment as A, we must have that:

$$12000 = \frac{A}{1.15} + \frac{A}{1.15^2} + \dots + \frac{A}{1.15^5}$$
$$= A(0.86957 + 0.75614 + 0.65752 + 0.57175 + 0.49718)$$
$$= A(3.35216)$$

Therefore, $A = \frac{12000}{3.35216} = 3579.79$. That is, amortization payment = £3579.79

The amortization schedule is tabulated as follows:

| Year | Amount outstanding (beginning) | Interest | Payment |
|---|---|---|---|
| 1983 | 12000.00 | 1800.00 | 3579.79 |
| 1984 | 10220.21 | 1533.03 | 3579.79 |
| 1985 | 8173.45 | 1226.02 | 3579.79 |
| 1986 | 5819.68 | 872.95 | 3579.79 |
| 1987 | 3112.84 | 466.93 | 3579.79 |
| 1988 | (0.02) | | |

(b) Here, there are two interest rates. The investment rate, $j$=0.1 and the borrowing rate, $i$=0.15. Also, $P$=12000 and $n$=5.

The calculations for the sinking fund payment (ordinary annuity) is given in the following.

The debt will amount to 12000(1.15)5 = £24,136.29 after 5 years. Thus, the sinking fund must mature to this amount. If A is the annual deposit into the fund, then we must have that:

$$24136.29 = A + A(1.1) + A(1.1)\ 2 + A(1.1)\ 3 + A(1.1)\ 4$$
$$= A(1 + 1.1 + 1.21 + 1.331 + 1.4641) = A(6.1051)$$

Therefore, $A = \frac{24136.29}{6.1051} = £3953.46$

The Sinking Fund schedule is tabulated as follows:

| Year | Debt outstanding | Interest on debt | Deposit | Amount in fund | Interest on fund |
|---|---|---|---|---|---|
| 1983 | 12000.00 | 1800.00 | 0 | 0 | 0 |
| 1984 | 13800.00 | 2070.00 | 3953.46 | 3953.46 | 395.35 |
| 1985 | 15870.00 | 2380.50 | 3953.46 | 8302.27 | 830.23 |
| 1986 | 18250.50 | 2737.58 | 3953.46 | 13085.96 | 1308.60 |
| 1987 | 20988.08 | 3148.21 | 3953.46 | 18348.02 | 1834.80 |
| 1988 | 24136.29 | | 3953.46 | 24136.28 | |

(c) Discounted cash flow table for calculation of NPV:

| Year | Net cash flow | Discount Factor (10%) | Discount Factor (15%) | Present value (10%) | Present value (15%) |
|---|---|---|---|---|---|
| 1983 | (12000) | 1.0000 | 1.0000 | (12000.00) | (12000.00) |
| 1984 | 6600 | 0.9091 | 0.8696 | 6000.06 | 5739.36 |
| 1985 | 6000 | 0.8264 | 0.7561 | 4958.40 | 4536.60 |
| 1986 | 4500 | 0.7513 | 0.6575 | 3380.85 | 2958.75 |
| 1987 | (1000) | 0.6830 | 0.5718 | (683.00) | (571.80) |
| 1988 | (2600) | 0.6209 | 0.4972 | (1614.34) | (1292.72) |
| | | | NVP | 41.97 | (629.81) |

(d) Using the formula method to determine the IRR, we have:

$I_1$=10; $N_1$=41.97; $I_2$=15; $N_2$=–629.81

and $\text{IRR} = \dfrac{N_1 I_2 - N_2 I_1}{N_1 - N_2} = \dfrac{(41.97)(15) - (-629.81)(10)}{41.97 - (-629.81)} = \dfrac{629.55 + 6298.1}{671.78}$

giving IRR = 10.3%.

The IRR gives the rate which makes NPV=0.

*Question 3*

(a) This can be calculated using a schedule as follows:

| Year | Amount in fund | Interest | Total in fund | |
|---|---|---|---|---|
| 1 | 10000 | 1000 | 11000 | |
| 2 | 21000 | 2100 | 23100 | |
| 3 | 33100 | 3310 | 36410 | |
| 4 | 46410 | 4641 | 51051 | = value |

(b) To calculate present value:

| Year | Net savings | 11% discount factor | Present value |
|---|---|---|---|
| 1 | 2000 | 0.9009 | 1801.81 |
| 2 | 2000 | 0.8116 | 1623.20 |
| 3 | 2000 | 0.7312 | 1462.40 |
| 4 | 2000 | 0.6587 | 1317.40 |
| | | | 6204.80 |

(c) This can again be calculated using a schedule:

| Year | Total amount invested | Interest (10%) | Total | |
|---|---|---|---|---|
| 1 | 1000.00 | 100.00 | 1100.00 | |
| 2 | 1600.00 | 160.00 | 1760.00 | |
| 3 | 2260.00 | 226.00 | 2486.00 | |
| 4 | 2986.00 | 298.60 | 3284.60 | |
| 5 | 3784.60 | 378.46 | 4163.06 | = acc sum |

(d) This requires the value of an amortization annuity. Given that: $P = 20000$; $n = 20$; $i = 0.14$. If the yearly repayment is A, then:

$$20000 = \frac{A}{1.14} + \frac{A}{1.14^2} + \ldots + \frac{A}{1.14^{20}}$$
$$= A.\left(\frac{1}{1.14} + \frac{1}{1.14^2} + \ldots + \frac{1}{1.14^{20}}\right)$$

But the terms in the bracket form a gp with $a = \frac{1}{1.14}$ and $r = \frac{1}{1.14}$.

The sum to 20 terms of this gp is given by:

$$\frac{\frac{1}{1.14}\left[1-\left(\frac{1}{1.14}\right)^{20}\right]}{1-\frac{1}{1.14}} = \frac{0.8772 \times 0.9272}{0.1228} = 6.6233$$

Therefore, $20000 = A(6.6233)$, giving $A = \frac{20000}{6.6233} = £3019.65$

*Question 4*

(a) The amount of the mortgage must sum to the present value of all the payments made. If $A$ is the annual payment, then:

$$10000 = \frac{A}{1.12} + \frac{A}{1.12^2} + \ldots + \frac{A}{1.12^5}$$

Thus: $10{,}000 = A(0.8929 + 0.7972 + 0.7118 + 0.6355 + 0.5674)$
$= 3.6048A$

Therefore, $A = \frac{10000}{3.6048} = £2774.10$

(b)

| Year | Outstanding debt | Interest paid | Payment | Principal repaid |
|---|---|---|---|---|
| 1 | £10,000.00 | £1,200.00 | £2,774.10 | £1,574.10 |
| 2 | £8,425.90 | £1,011.11 | £2,774.10 | £1,762.99 |
| 3 | £6,662.91 | £799.55 | £2,774.10 | £1,974.55 |
| 4 | £4,688.37 | £562.60 | £2,774.10 | £2,211.49 |
| 5 | £2,476.87 | £297.22 | £2,774.10 | £2,476.87 |

(c) After 5 years, the debt will amount to: $£10{,}000(1.12)^5 = £17{,}623.42$, which is therefore the amount that the fund must mature to. Putting A as the annual premium into the fund, gives:

$17{,}623.42 = A(1.15) + A(1.15)^2 + \ldots + A(1.15)^5$
$= A(1.15 + 1.3225 + 1.5209 + 1.7490 + 2.0114)$
$= 7.7538A$

Thus, $A = £2272.89$

(d)

| Year | Debt Outstanding | Interest on Debt | Deposit | Amount in Fund | Interest on Fund |
|---|---|---|---|---|---|
| 1 | £10,000.00 | £1,200.00 | £2,272.89 | £2,272.89 | £340.93 |
| 2 | £11,200.00 | £1,344.00 | £2,272.89 | £4,886.72 | £733.01 |
| 3 | £12,544.00 | £1,505.28 | £2,272.89 | £7,892.62 | £1,183.89 |
| 4 | £14,049.28 | £1,685.91 | £2,272.89 | £11,349.41 | £1,702.41 |
| 5 | £15,735.19 | £1,888.22 | £2,272.89 | £15,324.71 | £2,298.71 |
| 6 | £17,623.42 | | | £17,623.42 | |

## *Question 5*

(a) This question refers to repayments including both capital and interest; that is, amortization. With $P = 100{,}000$ $n = 4$ and $i = 0.12$, the amortization payment, $A$ say, must satisfy:

$$100{,}000 = \frac{A}{1.12} + \frac{A}{1.12^2} + \frac{A}{1.12^3} + \frac{A}{1.12^4}$$
$$= A(0.89286 + 0.79719 + 0.71178 + 0.63552)$$
$$= A(3.03735).$$

Therefore, $A = \dfrac{100000}{3.03735} = £32{,}923.44.$

(b) The reducing balance depreciation formula can be used, namely: $D = B.(1-i)^n$ where $D = 1000$, $B = 50000$ and $n = 5$. Here, $i$ is to be determined.

Re-arranging gives $\dfrac{D}{B} = (1-i)^n$ or $1-i = \left(\dfrac{D}{B}\right)^{\frac{1}{n}}$

So that: $1-i = \left(\dfrac{1000}{50000}\right)^{\frac{1}{5}} = (0.02)^{0.2} = 0.457$

Therefore $i = 1 - 0.457 = 0.543 = 54.3\%$

(c) Working in time periods of one quarter (year), we have $A = P.(1+i)^n$ where: $P = 1000$, $A = 3000$

and $i = \dfrac{0.12}{4} = 0.03$ (per quarter).

Re-arranging gives: $(1+i)^n = \dfrac{A}{P}$

and substituting: $(1.03)^n = \dfrac{3000}{1000} = 3$

Using logarithms: $n.\log(1.03) = \log(3)$

Therefore $n = \dfrac{\log(3)}{\log(1.03)} = \dfrac{0.4771}{0.01284} = 37$ (approximately).

Thus, number of years $= \dfrac{37}{4} = 9.25.$

*Question 6*

(a) Cash flow table:

| End year | Machinery | Maintenance | Revenue | Net |
|---|---|---|---|---|
| 0 | (75000) | | | (75000.00) |
| 1 | | (1000) | 20000.00 | 19000.00 |
| 2 | | (1100) | 21500.00 | 20400.00 |
| 3 | | (1210) | 23112.50 | 21902.50 |
| 4 | | (1331) | 24845.94 | 23514.94 |
| 5 | | | 26709.38 | 26709.38 |
| 6 | 1250 | | | 1250.00 |

(b) and (c) Discounted cash flow table:

| Year | Net flow | Discount factor (10%) | Discount factor (15%) | Present value (10%) | Present value (15%) |
|---|---|---|---|---|---|
| 0 | (75000.00) | 1.0000 | 1.0000 | (75000.00) | (75000.00) |
| 1 | 19000.00 | 0.9091 | 0.8696 | 17272.90 | 16522.40 |
| 2 | 20400.00 | 0.8264 | 0.7561 | 16858.55 | 15424.44 |
| 3 | 21902.50 | 0.7513 | 0.6575 | 16455.34 | 14400.89 |
| 4 | 23514.94 | 0.6830 | 0.5718 | 16060.70 | 13445.84 |
| 5 | 26709.38 | 0.6209 | 0.4972 | 16583.85 | 13279.90 |
| 6 | 1250.00 | 0.5645 | 0.4323 | 705.62 | 540.37 |
| | | | | 8936.99 | (1386.15) |

(d)

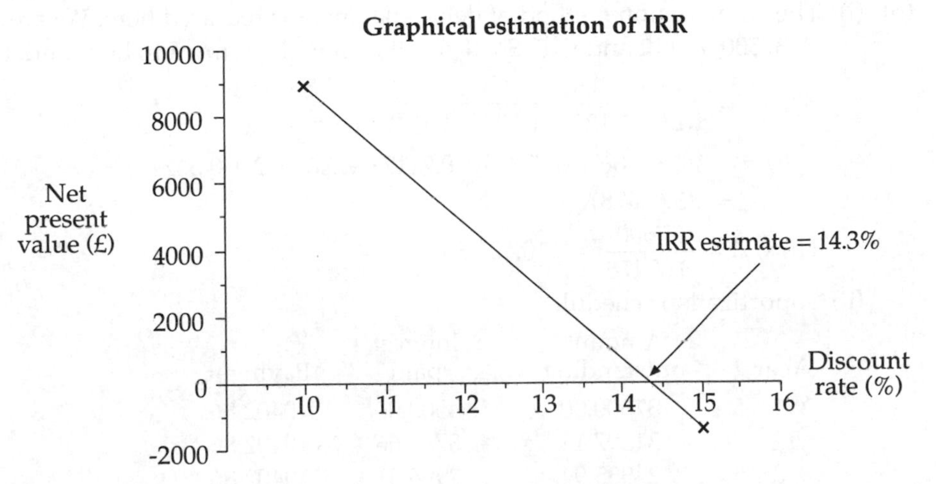


(e) The value of 14.3% for the IRR can be interpreted as the rate of return that the project earns.

## Question 7

The calculations for the net present value for each one of the three decisions are tabulated below.

| Discount factor at 12% | Decision (i) Net cost | Present value | Decision (ii) Net cost | Present value | Decision (iii) Net cost | Present value |
|---|---|---|---|---|---|---|
| 1 | 95000 | 95000 | 52000 | 52000 | 32000 | 32000 |
| 0.8929 | 20000 | 17857 | 27000 | 24107 | 32000 | 28571 |
| 0.7972 | 20000 | 15944 | 27000 | 21524 | 32000 | 25510 |
| 0.7118 | 20000 | 14236 | 27000 | 19218 | 32000 | 22777 |
| 0.6355 | 20000 | 12710 | 27000 | 17159 | 32000 | 20337 |
| 0.5674 | –10000 | –5674 | –10000 | –5674 | 0 | 0 |
| | | 150073 | | 128334 | | 129195 |

Decision (ii) is the least costly and thus, on pure financial grounds, should be chosen. However, this does not take into account the usefulness of the computer over the next five years. For example, could the old machine run newly developed software?

## Question 8

(a) (i) After four years (i.e. 8 x 6 months):
$A = P.(1+i)^n = 2750(1.035)^8 = £3621.22$.

(ii) $(1.035)^2 - 1 = 1.071 - 1 = 0.071$ or 7.1%.

(b) (i) The amortization method of debt repayment is required here. We are given: $P$=37500, $i$=0.12 and $n$=5. So, if $A$ is the annual payment to be found, then:

$$37500 = \frac{A}{1.12} + \frac{A}{1.12^2} + \frac{A}{1.12^3} + \frac{A}{1.12^4} + \frac{A}{1.12^5}$$

$$= A(0.89286 + 0.79719 + 0.71178 + 0.63552 + 0.56743)$$

$$= A(3.60478)$$

Thus, $A = \frac{37500}{3.60478} = £10,402.86$.

(ii) Amortization schedule:

| Year | Amount outstanding | Interest paid | Payment |
|---|---|---|---|
| 1 | 37500.00 | 4500.00 | 10402.86 |
| 2 | 31597.14 | 3791.66 | 10402.86 |
| 3 | 24985.94 | 2998.31 | 10402.86 |
| 4 | 17581.39 | 2109.77 | 10402.86 |
| 5 | 9288.29 | 1114.59 | 10402.86 |
| Balance | 0.03 | | |

(c) (i) Borrowing rate, $i = 12\% = 0.12$; Investment rate, $j = 8\% = 0.08$. Principal amount borrowed, $P = £37500$, $n = 5$ (years).

Notice that the payments into the fund are in advance. i.e. the payments form a due annuity.

The debt will amount to £37500(1.12)5 = £66,087.81 after 5 years. Thus the fund must mature to this amount. Putting $A$ as the annual deposit into the fund, we must have that:

$$
\begin{aligned}
66087.81 &= A(1.08) + A(1.08)^2 + A(1.08)^3 + A(1.08)^4 + A(1.08)^5 \\
&= A(1.08 + 1.1664 + 1.25971 + 1.36049 + 1.46933) \\
&= A(6.33593).
\end{aligned}
$$

Therefore, $A = \dfrac{66087.81}{6.33593} = £10{,}430.64.$

(ii) Schedule:

| Year | Debt outstanding | Interest paid | Deposit | Amount in fund | Interest earned |
|---|---|---|---|---|---|
| 1 | 37500.00 | 4500.00 | 10430.64 | 10430.64 | 834.45 |
| 2 | 42000.00 | 5040.00 | 10430.64 | 21695.73 | 1735.66 |
| 3 | 47040.00 | 5644.80 | 10430.64 | 33862.03 | 2708.96 |
| 4 | 52684.80 | 6322.18 | 10430.64 | 47001.63 | 3760.13 |
| 5 | 59006.98 | 7080.84 | 10430.64 | 61192.40 | 4895.39 |
| 6 | 66087.82 | | | 66087.79 | |

## Question 9

(a)

| Time (year) | Receipt (£) | Present value at start of annuity (£) |
|---|---|---|
| 0.5 | 1,500 | $\dfrac{1500}{1045}$ |
| 1.0 | 1,500 | $\dfrac{1500}{1045^2}$ |
| 1.5 | 1,500 | $\dfrac{1500}{1045^3}$ |
| 10.0 | 1,500 | $\dfrac{1500}{1045^{20}}$ |

The net present value of the annuity is thus a geometric progression with $n = 20$ terms, first term 'a' $= \dfrac{1500}{1.045}$ and ratio 'r' $= \dfrac{1}{1.045}$

The sum is therefore $= \dfrac{\dfrac{1500}{1.045}\left(1 - \dfrac{1}{1.045^{20}}\right)}{1 - \dfrac{1}{1.045}}$

It is therefore worth paying up to £19,512, for the annuity.

(b) If we denote the quarterly amount by £$X$

after 1 quarter, value of amount paid $= X$

after 2 quarters, value of amounts paid $= X(1 + 0.025) + X$

after 3 quarters, value of amounts paid $= X(1 + 0.025)^2 + X$

after 100 quarters:

$$X(1.025)^{99} + X(1.025)^{98} + ... + X = \frac{X(1.025^{100} - 1)}{1.025 - 1} = 432.5488X$$

This must balance the value the original loan has reached after 100 quarters at 2.5% per quarter.

By the compound interest formula, this is 50,000 $(1 + 0.025)^{100}$ = £590,686

Hence: 432.5488X = 590,686 giving X = 1365.5939

Thus the quarterly repayments are £1,365.59

(c) As in (b), after 59 months, the value of the scheme will have reached $300\ (1.01)^{59} + 300\ (1.01)^{58} + ... + 300$

(Note that £300 is paid immediately, at time 0, in this case)

$$\text{This value} = \frac{300(1.01^{60} - 1)}{1.01 - 1} = 24{,}500.91$$

Now, adding on the interest for the final month, the final value is 24,500.91 × 1.01 = £24,746 (to nearest £)

$$\text{The real value of the investment is } \frac{24{,}746}{1.05^5} = £19{,}389 \text{ (to nearest £)}$$

(d) In (a), the administrative and other charges usually involved with annuities may vary in the future.

In (b), it is very common for mortgage interest rates to vary, thereby varying the quarterly instalments.

In (c), the scheme may pay more than the minimum 1% or the assumption of 5% inflation may prove accurate.

## *Question 10*

(a) The NPVs at 10% cost of capital for the two machines are shown

| **Machine A** | | Net | **10%** | Present |
|---|---|---|---|---|
| | | Flow | Discount | value |
| Year | Note | (£000) | factor | (£000) |
| 0 | | -100 | 1.0000 | -100.00 |
| 1 | a | -60 | 0.9091 | -54.55 |
| 2 | | 40 | 0.8264 | 33.06 |
| 3 | | 40 | 0.7513 | 30.05 |
| 4 | | 40 | 0.6830 | 27.32 |
| 5 | | 40 | 0.6209 | 24.84 |
| 6 | b | 60 | 0.5645 | 33.87 |
| | | Net present value | | -5.41 |

| **Machine B** | | Net | **10%** | Present |
|---|---|---|---|---|
| | | Flow | Discount | value |
| Year | Note | (£000) | factor | (£000) |
| 0 | | -120 | 1.0000 | -120.00 |
| 1 | a | -70 | 0.9091 | -63.64 |
| 2 | | 50 | 0.8264 | 41.32 |
| 3 | | 50 | 0.7513 | 37.57 |
| 4 | | 50 | 0.6830 | 34.15 |
| 5 | | 50 | 0.6209 | 31.05 |
| 6 | b | 74 | 0.5645 | 41.77 |
| | | Net present value | | 2.22 |

Note a: Includes balance of cost of machine. Note b: Includes scrap value.

(b) From the figures shown, it is clear that machine B should be chosen since it has the higher NPV. It should be noted that machine A does not even return as much as 10% on the overall investment.

The assumptions made in the above recommendation are that the company can can afford the outlays inherent in the above structure. In the comparison above, no account has been taken of relative risks and it is clear that machine B, being the more expensive, is the riskier investment.

| Machine B | Year | Note | Net Flow (£000) | 11% Discount factor | Present value (£000) |
|---|---|---|---|---|---|
| | 0 | | -120 | 1.0000 | -120.00 |
| | 1 | a | -70 | 0.9009 | -63.06 |
| | 2 | | 50 | 0.8116 | 40.58 |
| | 3 | | 50 | 0.7312 | 36.56 |
| | 4 | | 50 | 0.6587 | 32.94 |
| | 5 | | 50 | 0.5935 | 29.67 |
| | 6 | b | 74 | 0.5346 | 39.56 |
| | | | | Net present value | -3.75 |

Note a: Includes balance of cost of machine. Note b: Includes scrap value.

(c) The table above shows the net flows at an 11% cost of capital.

The diagram below shows the NPVs obtained from 10 and 11% costs of capital plotted and the estimate of the IRR is seen to be approximately 10.4%.

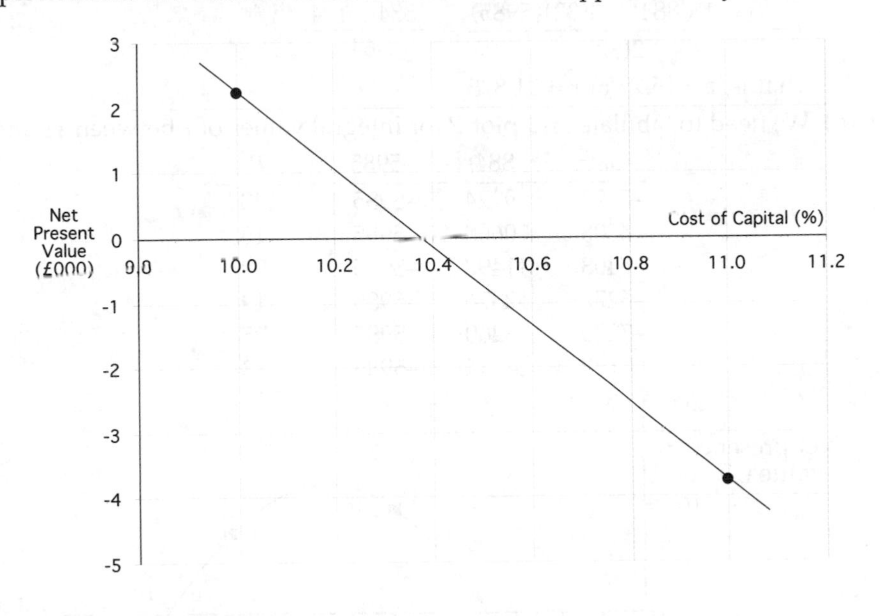


## Part 7

*Question 1*

a), b) Materials cost £0.50 per poster, that is $0.50N$ (£); Labour costs £15 per hour, and $N$ posters take $\frac{N}{300} + 2$ hours to produce. Thus the labour costs are:

$$15.\left(\frac{N}{300} + 2\right) = \frac{N}{20} + 30 \text{ (£)}$$

Administration costs are £10 per hundred posters plus £50; that is:

$$10.\frac{N}{100} + 50 = \frac{N}{10} + 50 \text{ (£)}$$

Adding (i), (ii) and (iii), total costs are given by:

$$C = 0.50N + \frac{N}{20} + 30 + \frac{N}{10} + 50 = 0.65N + 80 \text{ (£)}$$

Hence producing 1,000 posters will cost: 0.65 × 1000 + 80 = £730

The formula for $C$ indicates a fixed cost of £80 and a variable cost of £0.65 per poster produced.

(c) i. If the cost is £500, then $500 = 0.65N + 80$. Hence $= \dfrac{420}{0.65} = 646$ posters

ii. If the cost is $N$, then $N = 0.65\text{N} = 80$.

Hence $N = \dfrac{80}{0.35} = 229$ posters

## Question 2

(a) We need to solve the equation $P = -32r^2 + 884r - 5985 = 0$

i.e. to solve $32r^2 - 884r + 5985 = 0$

Using the formula, with $a$=32, $b$=–884 and $c$=5985, gives:

$$r = \frac{884 \pm \sqrt{884^2 - 4(32)(5985)}}{2(32)} = \frac{884 \pm 124}{64}$$

That is, $r = 15.75$ or $r = 11.875$

(b) We need to tabulate and plot $P$ for integral values of $r$ between 11 and 16.

| $-32r^2$ | $884r$ | –5985 | $P$ |
|---|---|---|---|
| –3872 | 9724 | –5985 | –133 |
| –4608 | 10608 | –5985 | 15 |
| –5408 | 11492 | –5985 | 99 |
| –6272 | 12376 | –5985 | 119 |
| –7200 | 13260 | –5985 | 75 |
| –8192 | 14144 | –5985 | –33 |

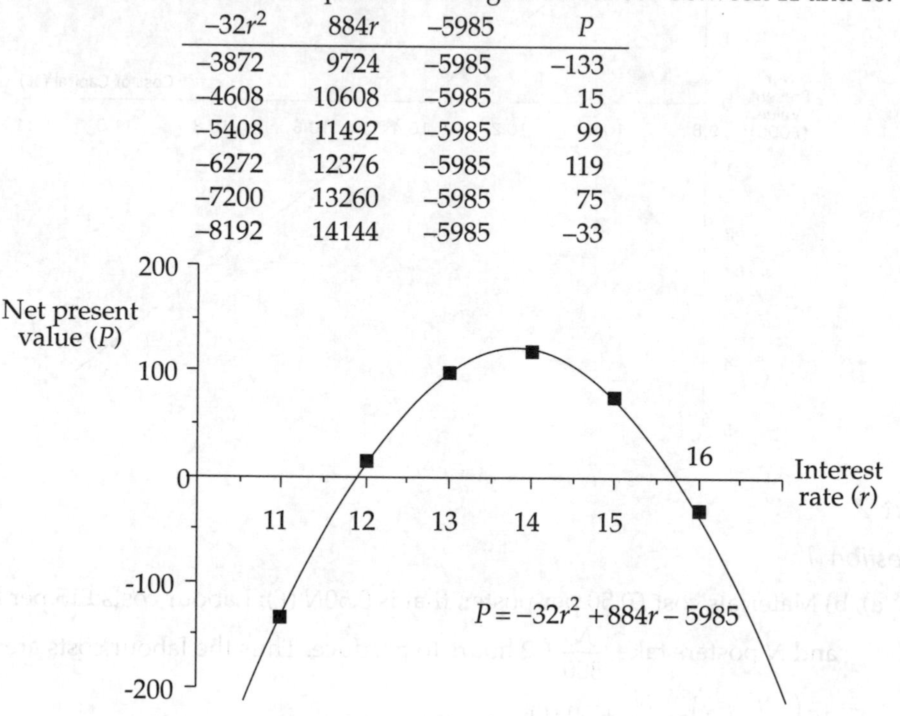


(c) With interest rates between 11.875% and 15.75%, the project yields a positive NPV and thus is worthwhile.

(d) Now, $P = -32r^2 + 884r - 5985$.

Thus: $\dfrac{dP}{dr} = -64r + 884$ and $\dfrac{dP}{dr} = 0$ when $64r = 884$. i.e. $r = 13.81$

Also, $\dfrac{d^2P}{dr^2} = -64$, and since this is negative, $r = 13.81$ signifies a max value of $P$.

Thus, the maximum value of $r$ is $-32(13.81)^2 + 884(13.81) - 5985 = £120.125$.

## Question 3

(a) Let $n$ be the old number of passengers and $f$ be the old fare.
Therefore, old revenue = $nf$.
A 30% increase in passengers gives new number of passengers as $1.3 \times n$; a 10% decrease in fare gives the new fare as $0.9 \times f$.
Thus, new revenue = new number of passengers × new fare
$= (1.3)(0.9)nf$
$= (1.17)nf$ (i.e. 0.17 increase in old revenue).
Thus percentage increase in revenue is 17%

(b) New fare = $f\left(1-\dfrac{x}{100}\right)$ and new number of passengers = $n\left(1+\dfrac{2x}{100}\right)$

Therefore, new revenue $= nf\left(1-\dfrac{x}{100}\right)\left(1+\dfrac{2x}{100}\right)$

Thus, the multiplier of $nf$ $= \left(1-\dfrac{x}{100}\right)\left(1+\dfrac{2x}{100}\right)$

$$= 1-\frac{x}{100}+\frac{2x}{100}-\frac{2x^2}{10000}$$

$$= 1+\frac{x}{100}-\frac{2x^2}{10000}$$

$$= 1 + 0.01x - 0.0002x^2.$$

(c) We need to find the value of $x$ that maximises the multiplier:
$M = 1 + 0.01x - 0.0002x^2$.

But $\dfrac{dM}{dx} = 0.01 - 0.0004x$ and when $\dfrac{dM}{dx} = 0$, $0.0004x = 0.01$.

Thus, the value of $x$ that maximises $M$ is $x = \dfrac{0.01}{0.0004} = 25$.

When $x$=25, M $= 1 + (0.01)(25) - (0.0002)(25)^2$
$= 1 + 0.25 - 0.125$
$= 1.125$

Therefore, percentage increase in revenue is 12.5%.

## Question 4

(a) The selling price is £15/unit. Therefore the revenue, $R = 15x$.
The costs, $C = 800 + 5x + 0.009x^2$.
Thus, profit, $P = R - C = 15x - (800 + 5x + 0.009x^2)$
i.e. $P = 10x - 800 - 0.009x^2$.

(i) We require the range of $x$ such that $P \geq 200$.
i.e. such that $10x - 800 - 0.009x^2 \geq 200$ or $10x - 1000 - 0.009x^2 \geq 0$.
Solving $10x - 1000 - 0.009x^2 = 0$ will give the 'critical' points for $x$.

Here, to use the formula, $a$=–0.009, $b$=10 and $c$=–1000, and:

$$x = \frac{-10 \pm \sqrt{10^2 - 4(-0.009)(-1000)}}{2(-0.009)} = \frac{-10 \pm \sqrt{64}}{-0.018}$$

Therefore $x$=1000 or $x$=111.1 (1D).

Now, since the graph of $y = 10x - 1000 - 0.009x^2$ is a reverse 'U' (mountain) curve, values of $x$ between 111.1 and 1000 will give $y \geq 0$ as required.

So that, a weekly profit of at least £200 will be provided if the weekly production is between 111.1 and 1000 units.

(ii) The calculations for the graph of $P = 10x - 800 - 0.009x^2$ are tabulated:

| $x$ | 50 | 100 | 200 | 500 | 800 | 1000 | 1200 |
|---|---|---|---|---|---|---|---|
| $10x$ | 500 | 1000 | 2000 | 5000 | 8000 | 10000 | 12000 |
| –800 | –800 | –800 | –800 | –800 | –800 | –800 | –800 |
| $-0.009x^2$ | –22.5 | –90 | –360 | –2250 | –5760 | –9000 | –12960 |
| $y$ | –322.5 | 110 | 840 | 1950 | 1440 | 200 | –1760 |

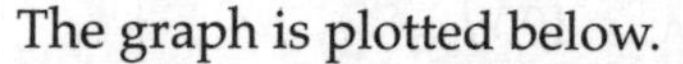

The graph is plotted below.

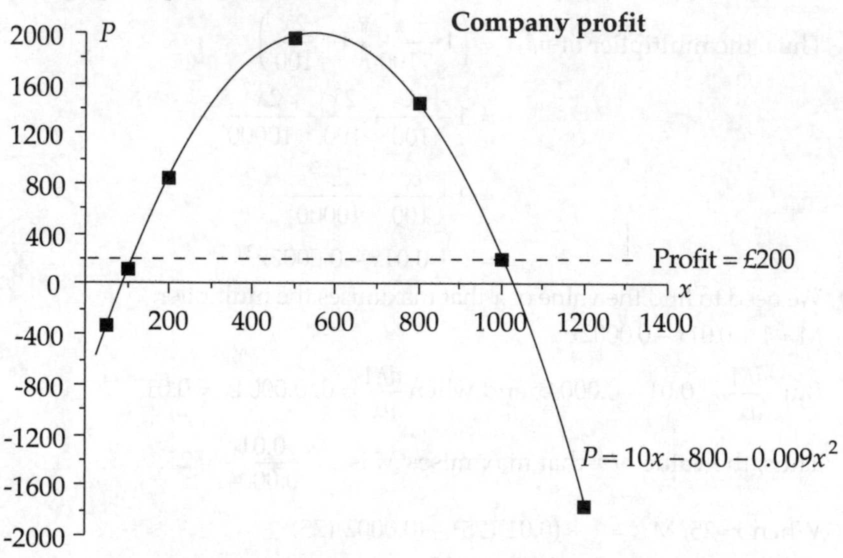


(b) Using the formula for reducing balance depreciation: $D = B(1\text{–}i)$

we have: $23500 = 32000(1\text{–}i)^6$ giving $1\text{–}i = \left(\frac{23500}{32000}\right)^{\frac{1}{6}}$

Hence, $1\text{–}i = 0.9498$. Therefore, $i$=0.0502. The rate of depreciation is thus 5.02%.

*Question 5*

(a) Total cost, $C = x^2 + 16x + 39$.

Average cost per unit $= \frac{C}{x} = x + 16 + \frac{39}{x}$

The calculations for the plot of the average cost per unit are tabulated below.

| $x$ | 0 | 1 | 2 | 3 | 4 | 5 | 6 | 7 | 8 |
|---|---|---|---|---|---|---|---|---|---|
| $x + 16$ | 16 | 17 | 18 | 19 | 20 | 21 | 22 | 23 | 24 |
| $\frac{39}{x}$ | $\infty$ | 39 | 19.5 | 13 | 9.75 | 7.8 | 6.5 | 5.6 | 4.9 |
| $\frac{C}{x}$ | $\infty$ | 56 | 37.5 | 32 | 29.75 | 28.8 | 28.5 | 28.6 | 28.9 |

and the graph is shown in the figure.

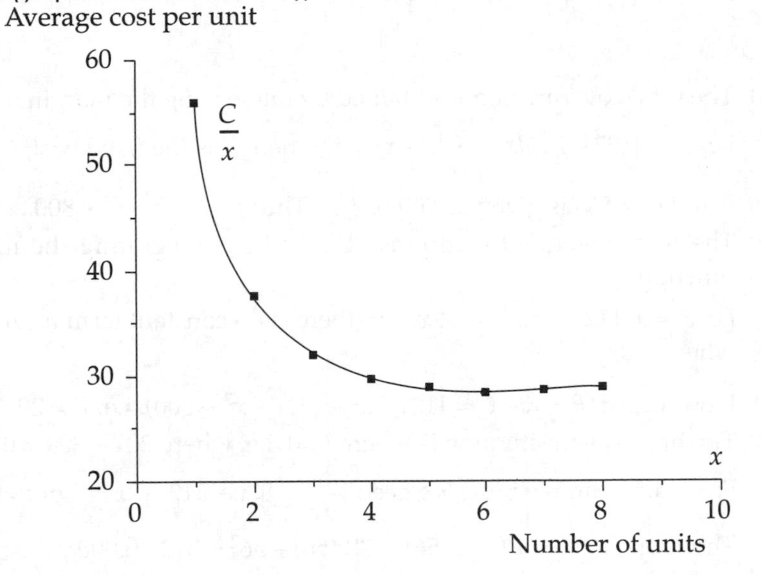


(b) (i) Demand function: $p = x^2 - 24x + 117.$

Total revenue: $R = p.x = x^3 - 24x^2 + 117x.$

(ii) R is maximised where $\frac{dR}{dr} = 0$ and $\frac{dR}{dr} = 3x^2 - 48x + 117.$

Therefore $\frac{dR}{dr} = 0$ where $3x^2 - 48x + 117 = 0.$

Solving the previous quadratic equation using the formula with $a$=3, $b$=–48 and $c$=117 gives:

$$x = \frac{48 \pm \sqrt{48^2 - 4(3)(117)}}{2(3)} = \frac{48 \pm \sqrt{900}}{6} = \frac{48 \pm 30}{6} \text{ i.e. } x=3 \text{ or } x=13.$$

Now, $\frac{d^2R}{dx^2} = 6x - 48$, and when $x$=3, $\frac{d^2R}{dx^2} = 6(3) - 48 = -30.$

Thus $R$ is a maximum when $x$=3.

Price at $x$=3 is $p(3) = 3^2 - 24(3) + 117 = £54/\text{unit}.$

(iii) $\frac{dp}{dx} = 2x - 24.$

Therefore, elasticity of demand

$$= \left(\frac{x^2 - 24x + 117}{x}\right)\left(\frac{1}{2x - 24}\right) = \frac{x^2 - 24x + 117}{2x(x - 12)}$$

and at $x$=3 (the maximum revenue point): elasticity

$$= \frac{3^2 - 24(3) + 117}{2(3)(3 - 12)} = \frac{54}{-54} = -1$$

*Question 6*

(a) The total cost function is obtained by integrating the marginal cost function.

i.e. $C = \int (92 - 2x)dx = 92x - x^2 + K$(where $K$ is the fixed cost).

Fixed cost ($K$) is given as 800 (£000). Thus $C = 92x - x^2 + 800$.

(b) The total revenue function is obtained by integrating the marginal revenue function.

i.e. $R = \int (112 - 2x)dx = 112x - x^2$ (there is no constant term for revenue since $R$=0 when $x$=0)

(c) Now, profit, $P = R - C = 112x - x^2 - (92x - x^2 + 800)$. i.e. $P = 20x - 800$.

The break-even situation is where $P$=0. i.e. where $20x - 800 = 0$ or $x$=40.

(d) For maximum revenue, we need $\frac{dR}{dx} = 0$. i.e. $112 - 2x = 0$ or $x$=56.

Therefore, $R_{max} = R(\text{at } x=56) = 112(56) - 56^2 = 3136$ (£000).

For maximum costs, we need $\frac{dC}{dx} = 0$. i.e. $92 - 2x = 0$ or $x$=46.

Therefore, $C_{max} = C(\text{at } x=46) = 92(46) - 46^2 + 800 = 2916$ (£000).

(e) The calculations for the plots of the two graphs are tabulated below.

| $x$ | 0 | 10 | 20 | 30 | 40 | 50 | 60 |
|---|---|---|---|---|---|---|---|
| $-x^2$ | 0 | –100 | –400 | –900 | –1600 | –2500 | –3600 |
| $92x$ | 0 | 920 | 1840 | 2760 | 3680 | 4600 | 5520 |
| 800 | 800 | 800 | 800 | 800 | 800 | 800 | 800 |
| C | 800 | 1620 | 2240 | 2660 | 2880 | 2900 | 2720 |
| $-x^2$ | 0 | –100 | –400 | –900 | –1600 | –2500 | –3600 |
| $112x$ | 0 | 1120 | 2240 | 3360 | 4480 | 5600 | 6720 |
| $R$ | 0 | 1020 | 1840 | 2460 | 2880 | 3100 | 3120 |

The two graphs are plotted in the figure on the following page.

From the graphs, it is clear that, since costs are falling while revenue is still increasing between the break-even production point and the maximum production point, the profit is increasing. Hence, maximum profit is obtained at maximum production.

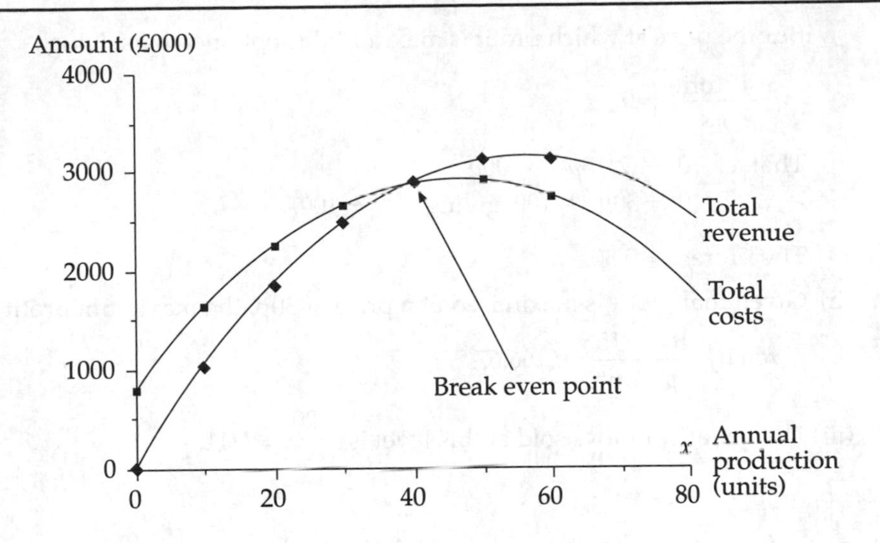


*Question 7*

(a) Put $x$ as the old price of the ticket. Then the number of tickets which could be purchased previously is $\frac{2850}{x}$ and the number which could be purchased after the

£6 price increase is $\frac{2850}{x+6}$

The price increase results in a reduction of 36 in the number of tickets which can be purchased. Therefore:

$$\frac{2850}{x} = \frac{2850}{x+6} + 36$$

$$2850(x+6) = 2850x + 36x(x+6)$$

$$2850x + 17100 = 2850x + 36x^2 + 216x$$

$$0 = 36x^2 + 216x - 17100$$

$$0 = x^2 + 6x - 475$$

Thus $x = -25$ (not feasible) or $x = 19$

The percentage increase in price is therefore $\frac{6}{19} \times 100 = 31.58\%$

(b) Quantity sold is $\frac{100}{p^2}$, price ($P$) is £p and cost ($C$) is 15p per toy.

Now, profit = quantity sold × (price – cost)

$$= \frac{100}{p^2} \times (P - C)$$

$$= 100p^{-1} - 15p^{-2}$$

(i) $\frac{d\text{Profit}}{dx} = 100p^{-2} + 2 \times 15p^{-3}$

But the price at which profit is maximised is obtained by solving

$$\frac{d\text{Profit}}{dx} = 0$$

That is, $0 = (2)15p^{-3} - 100p^{-2}$

$0 = 30p^{-1} - 100$ giving $\frac{30}{p} = 100.$

Therefore, $p = 0.30$

(ii) Given that profit is maximised at a price of 30p, the maximum profit is given by $\frac{100}{p} - \frac{15}{p^2} = £166.67$

(iii) The quantity of toys sold at this level is $\frac{100}{0.3^2} = 1111.$

## *Question 8*

(a) $TC = \int(x^2 - 28x + 211)dx = \frac{x^3}{3} - 14x^2 + 211x + c$

When $x = 0$, $TC = c$. and from the question when $x = 0$, $TC = 10$, so $c = 10$.

Therefore: $TC = \frac{x^3}{3} - 14x^2 + 211x + 10$

(b) $TR = (200 - 8x)x = 200x - 8x^2$

(c) Profit, $P = TR - TC = 200x - 8x^2 - \left(\frac{x^3}{3} - 14x^2 + 211x + 10\right) = \frac{x^3}{3} + 6x^2 - 11x - 10$

$$\frac{dP}{dx} = -x^2 + 12x - 11 = 0 \text{ for critical values.}$$

$-x^2 + 12x - 11 = 0 = x^2 - 12x + 11$

$(x - 11)(x - 1) = 0$

$x = 1$ or $11$ (can also use the quadratic formula)

$\frac{d^2R}{dx^2} = -2x + 12$ and when $x$=1, $\frac{d^2R}{dx^2} = 10$ (minimum),

when $x$=11, $\frac{d^2R}{dx^2} = -10$ (maximum)

Therefore profit is maximised when output is 11.

(d) $MR = \frac{dTR}{dx} = 200 - 16x$. When $x$=0, $MR$=200. When $MR$=0, $x$=12.5.

When $x$=0, $MC$=211.

(e)

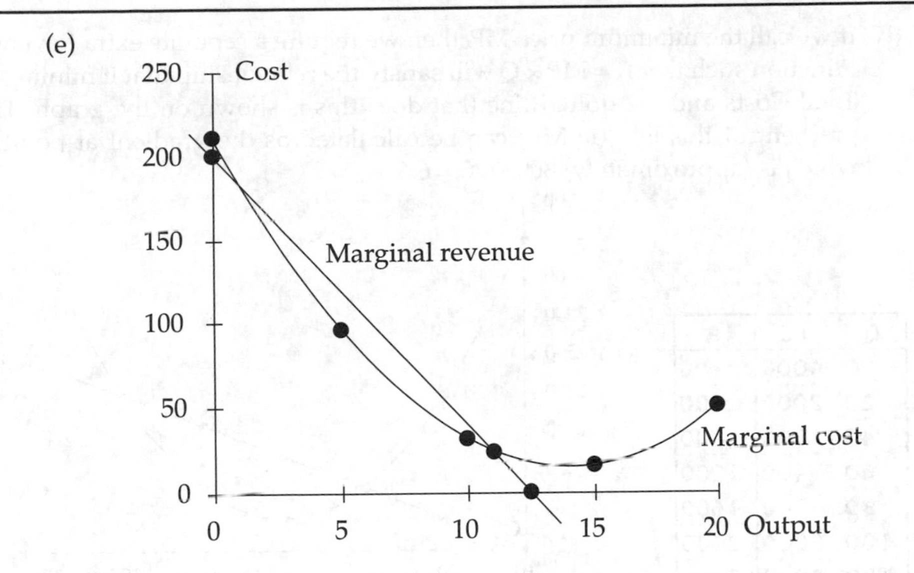


*Question 9*

(a) (i)

Given $C = aQ^2 - bQ + c$ and substituting for the information given, we have:

$$2900 = 100a - 10b \quad \text{....... 1}$$

$$800 = 1600a - 40b + c \quad \text{....... 2}$$

$$2000 = 10000a - 100b + c \quad \text{....... 3}$$

1 – 2 : $2100 = -1500a + 30b$ ....... 4

3 – 2 : $1200 = 8400a + 60b$ ....... 5

4 × 2 : $4200 = -3000a + 60b$ ....... 6

5 + 6 : $540000 = 5400a$. Therefore: $a = 1$.

Substitute for a in 4 : $2100 = -1500 + 30b$ Therefore: $b = 120$.

Substitute for a, b in 1 : $2900 = 100 - 1200 + c$ Therefore $c = 4000$

Check for a,b and c in 2 : $800 = 1600 - 4800 + 4000$. OK!

Thus: $C = Q^2 - 120Q + 4000$

(ii) The following table shows the values calculated for Total Costs ($TC$) for $Q$ in the range [0,120] and the subsequent graph of $TC$ plotted. It is easily seen that the quantity that minimises Total Costs (£400) is $Q$=60.

(iii) The Revenue Function ($R$) is shown tabulated and then plotted on the graph. The profit range required is such that $R > TC$ and the graph shows this is true for $Q$ between 40 and 100.

(iv) If we call the minimum price MP, then we require a separate extra Revenue Function such that R = MP × Q will satisfy the relationship that it minimises Total Costs and the dotted line that does this is shown on the graph. The gradient of this line (ie MP) can be calculated as the gradient at point A which is (approximately) 800/120 = 6.6.

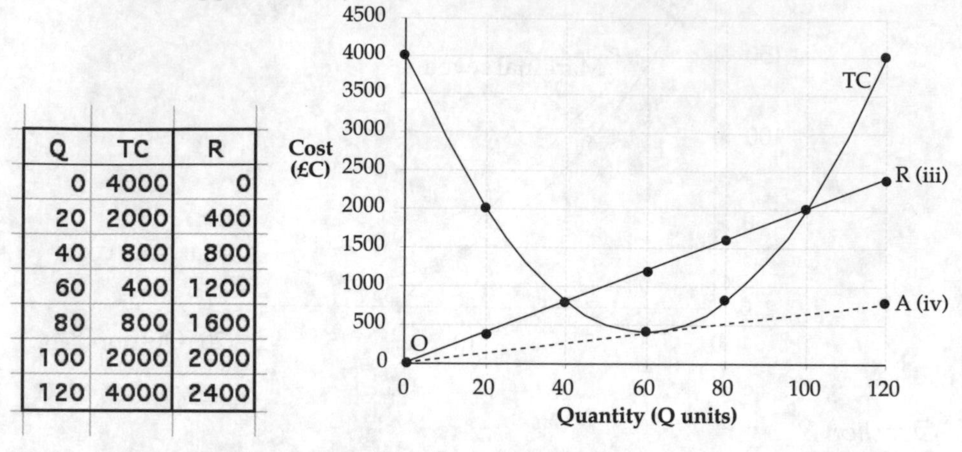

| Q | TC | R |
|---|---|---|
| 0 | 4000 | 0 |
| 20 | 2000 | 400 |
| 40 | 800 | 800 |
| 60 | 400 | 1200 |
| 80 | 800 | 1600 |
| 100 | 2000 | 2000 |
| 120 | 4000 | 2400 |



(b) Future Total Cost (FTC) = $Q^2 - 120Q + 8000$

The Net Present Value of this = NPV(FTC) = $\dfrac{Q^2 - 120Q + 8000}{(1.1)^{10}}$

= 0.38554 × NPV(FTC) from discount tables.

(i) These values are shown tabulated below and then plotted as shown. The production quantities for which real costs are the same are seen to be in the range [14, 106]

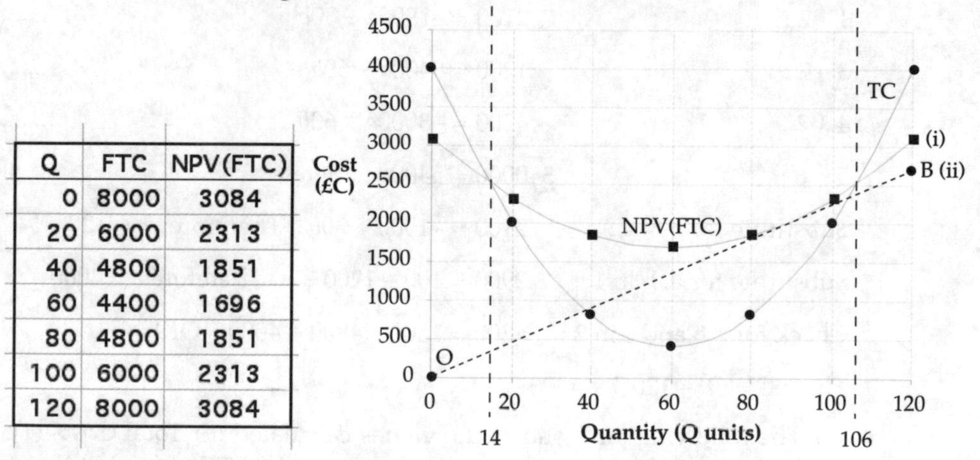

| Q | FTC | NPV(FTC) |
|---|---|---|
| 0 | 8000 | 3084 |
| 20 | 6000 | 2313 |
| 40 | 4800 | 1851 |
| 60 | 4400 | 1696 |
| 80 | 4800 | 1851 |
| 100 | 6000 | 2313 |
| 120 | 8000 | 3084 |



(ii) The minimum price that the product must be sold at in 10 years can be calculated by finding the gradient at point B = $\dfrac{2700}{120}$ = 22.5 (approx) and then inflating this to 22.5 × $(1.1)^{10}$ = 58.36.

## Part 8

### *Question 1*

Here, Pr(Female) = Pr(F) = 0.65 and Pr(Male) = Pr(M) = 0.35.

(i) Pr(all female) = Pr(FFF) = $(0.65)^3$ = 0.275 (3D)

(ii) Pr(all male) = Pr(MMM) = $(0.35)^3$ = 0.043 (3D)

(iii) Pr(at least one male) = 1 – Pr(none male)
= 1 – Pr(all female)
= 1 – 0.275 [from (i) above]
= 0.725 (3D)

### *Question 2*

Pr(system operates properly)
= Pr(at least one of A or B functions and at least one of C or D functions)
= Pr(at least one of A or B functions) × Pr(at least one of C or D functions)
= [1 – Pr(both A and B fail)] × [1 – Pr(both B and C fail)]
= [1 – (0.1)(0.1)] × [1 – (0.1)(0.1)] = [1 – 0.01] × [1 – 0.01]
= [0.99] × [0.99] = 0.9801

### *Question 3*

(a) Pr(obtaining both contracts) = Pr(obtaining A and obtaining B)
= Pr(obtaining A) × Pr(obtaining B)

$$= \frac{1}{3} \times \frac{1}{4}$$

$$= \frac{1}{12} = 0.083 \text{ (3D)}$$

(b) (i) The Venn diagram is shown in the figure on the following page.

(ii) Number of typists who can use word processors
= $n(W) = 30 = w + 6 + 3 + 9$.
Thus, $w$=12. Similarly, $n(A) = 25 = a + 5 + 3 + 6$.
So that, $a$=11. And $n(S) = 28 = s + 9 + 3 + 5$. Therefore, $s$=11.
Therefore, the total number of typists is the sum of the number of elements in the seven distinct areas in Figure 1 = 9+3+5+6+12+11+11 = 57.

(iii) The number of typists with just one skill is $a$+$w$+$s$ = 34.

### *Question 4*

(a) Two events are statistically independent if the occurrence (or not) of one in no way affects the occurrence (or not) of the other.

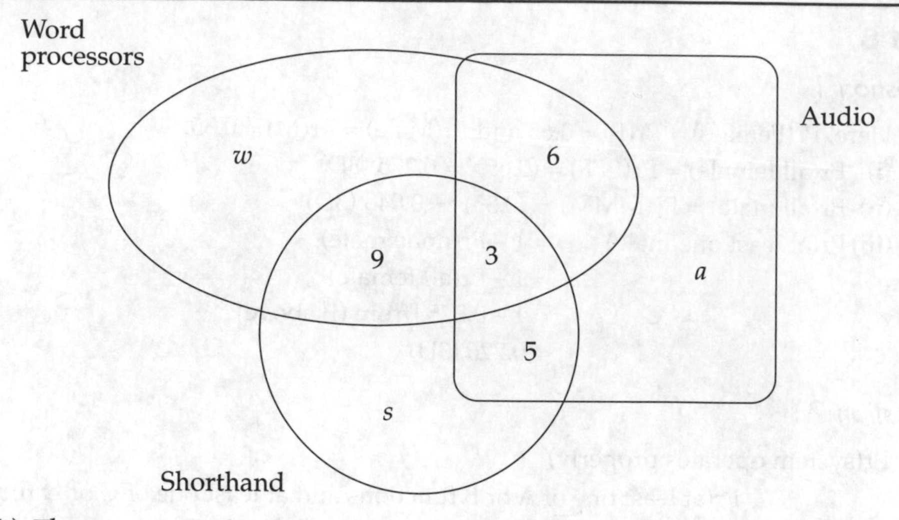


(b) There are two factors involved here, minor accidents (yes/no) and safety instructions (yes/no). Using a numeric approach for the solution of the problem means evaluating the number of men that fall into each one of the four categories defined (see table).

Total of men who had minor accidents = 1% of 10,000 = 100. Thus, 40% of 100 = 40 had safety instructions. Also, 90% of the 10,000 = 9000 had safety instructions. Using these figures, the table below can be filled as follows:

| | | *Minor accidents* | | |
|---|---|---|---|---|
| | | Yes | No | Total |
| Safety Instructions | Yes | 40 | 960 | 1000 |
| | No | 60 | 8940 | 9000 |
| | Total | 100 | 9900 | 10000 |

(i) Pr(No minor accidents / no safety instructions) = 0.993 (3D)

(ii) Pr(No minor accidents / safety instructions) = 0.96

(c) If there are $x$ winning tickets out of 100, then:

Pr(win / $x$ tickets) = $\frac{x}{100}$ with $x$ possible between 1 and 15.

But $Pr(x=1) = Pr(x=2) = ... = Pr(x=15) = \frac{1}{15}$

Therefore, $Pr(win) = \Sigma[\, Pr(x).Pr(win/x)\,]$

$$= Pr(x=1).Pr(win/x=1) + Pr(x=2).Pr(win/x=2) + ... ... + Pr(x=15).Pr(win/x=15)$$

$$= \frac{1}{15}\times\frac{1}{100} + \frac{1}{15}\times\frac{2}{100} + ... + \frac{1}{15}\times\frac{15}{100}$$

$$= \frac{1}{1500}(1 + 2 + ... + 15) = \frac{120}{1500} = 0.08$$

### *Question 5*

If advertising method A is used in the next period, the expected weekly sales can be calculated:

| week sales, $x$ units (mid-points) | probability, $p$ | $x.p$ |
|---|---|---|
| 100 | 0 | 0 |
| 300 | 0.3 | 90 |
| 500 | 0.4 | 200 |
| 700 | 0.2 | 140 |
| 900 | 0.1 | 90 |
| | | 520 |

Hence the expected weekly sales are 520 units.
If advertising B is used, this would increase expected weekly sales by 50%, to 780 units.
If advertising C is used,this would increase expected weekly sales by 100%, to 1,040 units.
The decision tree showing the options open to the company in the next period is:

advertising A — expected contribution = 520 × £100 = £52,000/wk

advertising B — expected contribution = 780 × £100 – £27,000 = £51,000/wk

advertising C — expected contribution = 1,040 × £100 – £50,000 = £54,000/wk

(**Note**: The contribution of £100/unit is based on using A. The figures of £27,000 and £50,000 in the above are therefore the additional advertising expenditure, above that of A.)

Thus based on expected values, the best option is to use advertising C.

The expected profit from A is a more reliable estimate as it is based on past experiance, but even this depends on the assumption that what has happened in the past will happen again during the next period. The expected profits for B and C are more speculative as they are based on estimates (possibly subjective) of the affects of different, new types of advertising on sales.

All other things being equal, the expected profits for B and C are equally unreliable, and so C would seem to be a better option than B. It is far more debatable whether C is indeed a 'better' option than A: all the expected profits are so close that only a small variation in any of the estimates could alter their relative values. For example, if C's sales are only 2% down on those expected, the company will be worse off than sticking with A. This, combined with the relatively higher reliability of the expected value for A, would lead many organisations to choose A. Only the more risk-seeking would choose C.

## *Question 6*

(a) To find the expected (mean) sales.

| Number of loaves sold ($x$) | 0 | 100 | 200 | 300 | 400 | Total |
|---|---|---|---|---|---|---|
| Number of days ($f$) | 10 | 60 | 60 | 50 | 20 | 200 |
| Probability ($p$) | 0.05 | 0.30 | 0.30 | 0.25 | 0.10 | |

Expected number of loaves sold $= \sum$ (number of loaves × probability)

$\sum px = 0(0.05)+100(0.30)+200(0.30)+300(0.25)+400(0.10) = 205.$

(b)

| PROFIT | | Production | | | |
|---|---|---|---|---|---|
| | | 100 | 200 | 300 | 400 |
| Sales | 0 | –10 | –20 | –30 | –40 |
| | 100 | 20 | 10 | 0 | –10 |
| | 200 | 20 | 40 | 30 | 20 |
| | 300 | 20 | 40 | 60 | 50 |
| | 400 | 20 | 40 | 60 | 80 |

(c) Expected profit = actual profit × probability. The following table shows expected profit. i.e. each profit value from the table in (b) multiplied by the probability of the respective sales level given.

*EXPECTED PROFIT TABLE*

| Sales Value | Pr | *Production* 100 | 200 | 300 | 400 |
|---|---|---|---|---|---|
| 0 | 0.05 | –0.5 | –1 | –1.5 | –2 |
| 100 | 0.30 | 6 | 3 | 0 | –3 |
| 200 | 0.30 | 6 | 12 | 9 | 6 |
| 300 | 0.25 | 5 | 10 | 15 | 12.5 |
| 400 | 0.10 | 2 | 4 | 6 | 8 |
| Expected profit | | 18.5 | 28 | 28.5 | 21.5 |

(d) The optimum level of production is 300 loaves per day, since this has the highest expected profit of £28.50 per day.

## Question 7

EV = expected value

At C: EV (shelve) = –4

EV (market) = (0.8 × 10) + (0.2 × 20) = 12

∴ market

At D: EV (shelve) = –4

EV (market) = (0.5 × 0) + (0.5 × 5) = 2.5

∴ market

At E: EV (shelve) = –4

EV (market) = (–4 × 0.1) + (–3 × 0.9) = –3.1

∴ market

The 'rolled-back' tree is shown on the opposite page.

At A: EV (do not develop) = 0

EV (develop) = (0.3 × 12) + (0.5 × 2.5) + (1.2 × – 3.1) = 4.23

∴ The company should develop.

All of the above decisions have been based on the criterion of maximising expected values (of profits). At F and G, the decisions are clear-cut, since the choices in both cases are between one option which at least breaks even and one which loses £4m. The decision is also clear-cut at E, because one option guarantees losing £4m, while the other can lose at most £4m, but might incur a reduced loss of £3m.

At A, the decision is more debatable, as the 'develop' option could lead to a loss (at E), while 'not develop' guarantees no loss. The chances of the loss are, however, small (20%) and the potential gains are large, and so many companies would be willing to take the risk of developing.

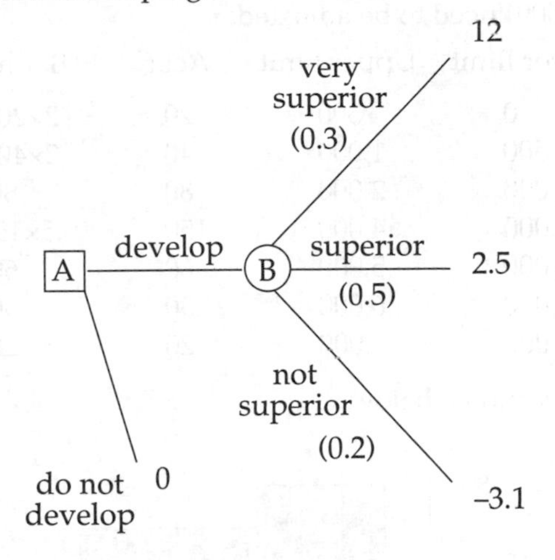


## *Question 8*

(a) For product A

Probability of low demand = 1 – 0.7 = 0.3

Expected profit £ =0.2 × 2 + 0.5 × 1.5 + 0.3 × 0.75 = 0.4 + 0.75 + 0.225 = 1.375

For product B

Probability of medium demand = 1 – 0.4 = 0.6

Expected profit £ =0.3 × 1.5 + 0.6 × 1 + 0.1 × 1.5 = 0.45+ 0.6 + 0.05 = 1.1

The shop should display product A.

(b) i. A random sample is one where each member of the population has an equal chance of being selected for the sample. The appropriate measures for such a sample are the mean and standard deviation.

ii. In a quota sample, all the members of the population do not have an equal chance of being selected for the sample. An interviewer is required to select a number of interviewees who may be required to have certain characteristics. As the selection is not random the appropriate statistical measures are thc median and semi-interquartile range.

iii. A cluster is one where the population is divided into sub-groups which may represent geographical areas. A sample is then chosen at random within each cluster. As the sampling is random the mean and standard deviation may be used. Alternatively a cluster may be chosen at random and all the members of the cluster interviewed, again the mean and standard deviation are suitable measures of location and dispersion.

## Part 9

### *Question 1*

(a) The bar heights of the classes that have a width different to the standard (taken here as 1,000) need to be adjusted.

| Lower limit | Upper limit | Act f | Bar height |
|---|---|---|---|
| 0 | 500 | 20 | 2×20=40 |
| 500 | 1,000 | 40 | 2×40=80 |
| 1,000 | 2,000 | 80 | 80 |
| 2,000 | 4,000 | 150 | 0.5×150=75 |
| 4,000 | 5,000 | 60 | 60 |
| 5,000 | 6,000 | 30 | 30 |
| 6,000 | 7,000 | 20 | 20 |

The chart is shown below.

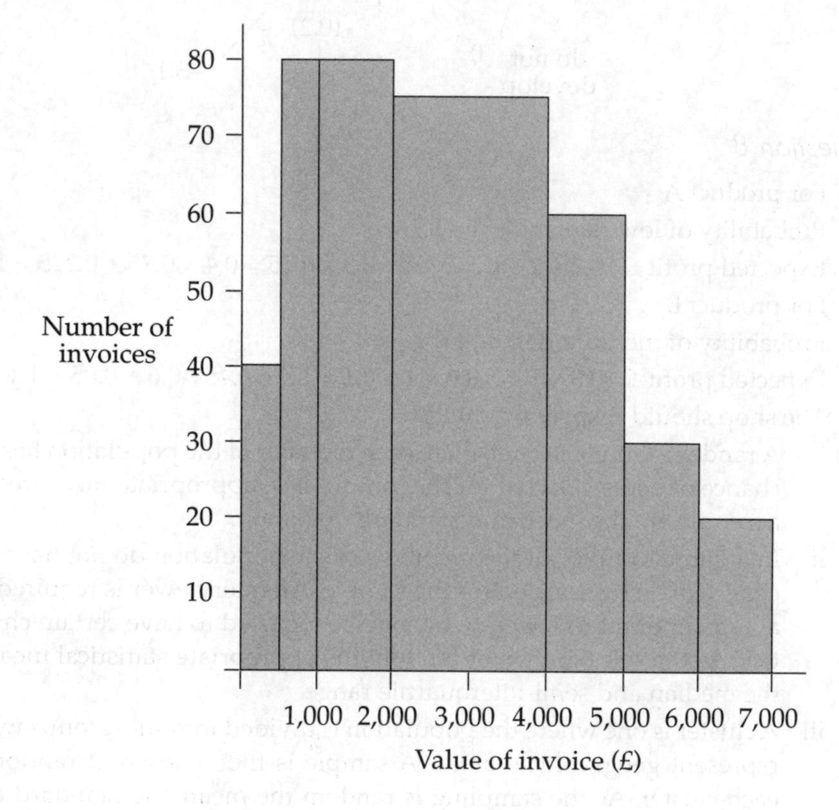

(b) Using the mid-points of each class as a representative figure

| $x$ £ | $f$ | $fx$ |
|---|---|---|
| 250 | 20 | 5,000 |
| 750 | 40 | 30,000 |
| 1,500 | 80 | 120,000 |
| 3,000 | 150 | 450,000 |
| 4,500 | 60 | 270,000 |
| 5,500 | 30 | 165,000 |
| 6,500 | 20 | 130,000 |
| | 400 | 1,170,000 |

The mean value is $\bar{x} = \frac{\sum fx}{\sum f} = \frac{1{,}170{,}000}{400} = £2{,}925.$

Note that $s$ = standard deviation = £1,600 (given)

Thus a 90% CI is: $\bar{x} = 1.64\frac{s}{\sqrt{n}} = 1.64\frac{1600}{\sqrt{400}} = 2925 \pm 131.2 = (27938.8, 3056.2)$

(c) If 20 invoices (out of 400) contain errors then $p = \frac{1}{20} = 0.05$

Thus a 95% confidence interval is:

$$p \pm 1.96\sqrt{\frac{p(1-p)}{n}} = 0.05 \pm 1.96\sqrt{\frac{(0.05)(0.95)}{400}}$$

$$= 0.05 \pm 0.02$$

$$= (0.03, 0.07)$$

## *Question 2*

(a) The lifetimes ($L$, say) are distributed Normally with mean, $m$=100 and standard deviation, $s$ unknown.

(i) If 90% of the batteries last at least 40 hours, we have: $\Pr(L>40) = 0.9$.

Standardising gives $\Pr\left(Z > \frac{40-100}{s}\right) = 0.9$. i.e. $\Pr\left(Z > \frac{-60}{s}\right) = 0.9$

Now, from Standard Normal tables, a table probability of 0.9 yields $Z = 1.28$.

Hence, $\frac{-60}{s} = 1.28$, giving $s = 46.88$ (2D).

(ii) We require Pr($L$<70).

Standardising gives: $\Pr\left(Z < \frac{70-100}{46.88}\right) = \Pr(Z < -0.64)$

$= 1 - 0.7389$ [from tables]

$= 0.2611$

Thus, 26% of batteries will not last 70 hours.

(b) This is a binomial situation with:

$n$ = size of sample = 5 and $p$ = Pr(defective) = 0.1.

(i) We require Pr(at least 3 defectives) = Pr(3 or 4 or 5 defectives)

= Pr(3) + Pr(4) + Pr(5)

= 5C3(0.1)3(0.9)2 + 5C4(0.1)4(0.9)1 + 5C5(0.1)5(0.9)2

= 10(0.1)3(0.9)2 + 5(0.1)4(0.9)1 + (0.1)5

= 0.0081 + 0.00045 + 0.00001

= 0.00856

= 0.01 (2D).

(ii) The result of (i) shows that only one time in a hundred should there be 3 or more defectives out of a total of 5 calculators examined. The fact that this has happened after one sample (i.e. 1 out of 1) must throw suspicion on the original assumption that only 10% of calculators are defective. The conclusion would be that the process is not working satisfactorily.

## *Question 3*

(a) A has a Normal distribution with $m$=1000 kgs and $s$ =100 kgs; B has a Normal distribution with $m$ =900 kgs and $s$=50 kgs; the breaking strength must be at least 750 kgs (given).

Thus we need to find those ropes which have the greatest probability of a breaking strength greater than 750 kgs.

Pr(A>750) = $\Pr\left(Z > \frac{750-1000}{100}\right)$ [standardising] = Pr(Z>–2.5) = 0.9938 [tables]

Pr(B>750) = $\Pr\left(Z > \frac{750-900}{50}\right)$ [standardising] = Pr(Z>–3) = 0.9987 [tables]

Thus, supplier B's ropes should be bought.

(b) This is a binomial situation with $n$=50 and $p$=Pr(defective)=0.01.

(i) Using the binomial distribution:

Pr(0 defectives) = $^{50}C_0(0.01)^0(0.99)^{50} = (0.99)^{50} = 0.605$ (3D).

(ii) Using the poisson distribution, the mean is calculated as the mean of the binomial.

i.e. $m = np = 50(0.01) = 0.5$. Therefore, Pr(0 defectives) = $e^{-0.5} = 0.607$ (3D).

*Question 4*

(a) We are given a Normal distribution of claims, with $m = 200$ and $s = \sqrt{2500} = 50$

(i) $\Pr(\text{claim}<100) = \Pr\left(Z > \frac{100-200}{50}\right)$ [standardising] $= \Pr(Z<-2) = 1 - 0.9772$

[from tables] = 0.0228.

Thus, only about 2% of claims will be under £100.

(ii) $\Pr(\text{claim}<150) = \Pr\left(Z > \frac{150-200}{50}\right)$ [standardising] $= \Pr(Z<-1) = 1 - 0.8413$

[from tables] = 0.1587.

Thus, about 16% of claims will be under £150.

(iii) $\Pr(\text{claim}>350) = \Pr\left(Z > \frac{350-200}{50}\right)$ [standardising] $= \Pr(Z>-3) = 1 - 0.9987$

[from tables] = 0.0013.

That is, only about 0.1% of claims will be over £350.

(b) This is a binomial situation with $p$ = Pr(passenger turns up) = 0.9 and $n$ = 290 (number of bookings taken). Using the Normal approximation, we have:

$m = \text{mean} = n.p = 290(0.9) = 261$

$s = \text{standard deviation} = \sqrt{[n.p.(1-p)]} = \sqrt{[290(0.9)(0.1)]} = 5.11$ (2D)

We need the probability that the number of passengers who turn up (P, say) will exceed 275 (i.e. 276 or 277 or ... etc), where it should be carefully noted that 276 for a binomial is the equivalent of the range 275.5 to 276.5 for a Normal distribution.

Thus, we require: $\Pr(P>275.5) = \Pr\left(Z > \frac{275.5-261}{5.11}\right)$ [standardising]

$= \Pr(Z>2.84)$

$= 1 - 0.9977$ [from tables] $= 0.0023$.

*Question 5*

(a) We are given a poisson situation with mean, $m$=2 (demands for a coach each day). NOTE: Proportions and probabilities are identical concepts.

(i) Pr(neither coach used) = Pr(0 demands) = $e^{-2} = 0.1353$.

(ii) At least 1 demand refused means that at least 3 demands have been made (since there are 2 coaches available)

Thus we require Pr(at least 3 demands)

= 1 – Pr(0 or 1 or 2 demands)

= 1 – [Pr(0) + Pr(1) + Pr(2)]

$= \left[0.1353 + 2.e^{-1} + \frac{2^2}{2}.e^{-2}\right]$

= 1 – [0.1353 + 0.2706 + 0.2706]

= 0.324 (3D).

(iii) Consider coach A. If no coaches are demanded, then coach A will not be used. This will happen proportion Pr(0) = 0.1353 of the time. If 1 coach is demanded, there is only a 50% chance that it will be A that is not used. This will happen with probability: 0.5× Pr(1) = 0.5× 0.2706 = 0.1353. Thus the total proportion of times that coach A will not be used is: 0.1353 + 0.1353 = 0.271 (3D).

(b) The lengths (*L*, say) are given as normal with mean, *m*=20.02 and standard deviation, *s*=0.05 cm.

(i) $\Pr(\text{L will be undersize}) = \Pr(L<19.9) = \Pr\left(Z<\dfrac{19.1-20.02}{0.05}\right) = \Pr(Z<-2.4)$

$= 1 - 0.9918 \text{ [from tables]} = 0.0082.$

i.e. 0.8% of rods will be undersize.

(ii) $\Pr(L \text{ will be oversize}) \quad = \Pr(L>20.1) = \Pr\left(Z<\dfrac{20.1-20.02}{0.05}\right) = \Pr(Z>1.6)$

$= 1 - 0.9452 \text{ [from tables]} = 0.0548.$

Thus, approximately 5.5% of rods will be oversize.

## *Question 6*

(a) The Normal distribution is known as the distribution of 'natural phenomena' and is the most commonly occurring continuous distribution. Heights, weights, lengths and times commonly form Normal distributions, which are characterized by their 'bell-shaped' curves.

(b) Consumption (C, say) is given as a Normal distribution with $m$ = 10,000 and $s$ = 2000.

(i) $\Pr(C>13000) = \Pr\left(Z<\dfrac{13000-10000}{2000}\right)$ [standardising]

$= \Pr(Z>1.5) = 1 - 0.9332 \text{ [from tables]}$

$= 0.0668.$

(ii) $\Pr(C<8000) = \Pr\left(Z<\dfrac{8000-10000}{2000}\right)$ [standardising]

$= \Pr(Z<-1) = 1 - 0.8413 \text{ [from tables]}$

$= 0.1587.$

(iii) To find $\Pr(7500<C<14000) = 1 - \Pr(C<7500) - \Pr(C>14000)$.

But $\Pr(C<7500) = \Pr\left(Z<\dfrac{7500-10000}{2000}\right)$ [standardising]

$= \Pr(Z<-1.25) = 1 - 0.8944 \text{ [from tables]}$

$= 0.1056$

Also $\Pr(C>14000) = \Pr\left(Z<\dfrac{14000-10000}{2000}\right)$ [standardising]

$= \Pr(Z>2) = 1 - 0.9772 \text{ [from tables]}$

$= 0.0228.$

Therefore, $\Pr(7500<C<14000) = 1 - 0.1056 - 0.0228 = 0.8716.$

*Question 7*

Putting S as the scores Normal variable, with mean 60 and standard deviation 12, we require:

$\Pr(S > 75) = \Pr\left(Z < \frac{75-60}{60}\right) = \Pr(Z > 1.25) = 1 - \Pr(Z < 1.25) = 1 - 0.8944 = 0.1056$

*Question 8*

(a) Mean in sample of 10 = 1.6 Therefore p(box underweight) = 1.6/10 = 0.16

(b) We have n=6 and p(underweight)=0.16 and a binomial situation.

(i) $p(0) = {}^6C_0.(0.16)^0(0.84)^6 = 0.3513$

(ii) $p(2) = {}^6C_2.(0.16)^2(0.84)^4 = 15(0.056)(0.4979) = 0.1911$

Also $p(1) = {}^6C_1.(0.16)^1(0.84)^5 = 0.4015$

(iii) p(at least 3) = 1 – p(0, 1 or 2)
= 1 – (0.3513 + 0.1911 + 0.4015)
= 0.0561

(c) Since n=100, we can approximate to a Normal distribution with mean = np = 100(0.16) = 16 and sd = $\sqrt{100(0.16)(0.84)} = 3.67$

(i) p(<10) = p(<9.5) with continuity correction

Standardising 9.5 gives $z = \frac{9.5-16}{3.67} = -1.77$

Thus: p(<10) = p(Z<–1.77)
= 1 – p(Z<1.77)
= 1 – 0.9938
= 0.01

(ii) p(>28) = p(>28.5) with continuity correction

Standardising 28.5 gives $z = \frac{28.5-16}{3.67} = 3.41$

Thus: p(>28) = p(Z>3.41)
= 1 – p(Z<3.41)
= 1 – 1
= 0

(iii) p(between 16 and 24) = p(<24) – p(<16)
= p(<23.5) – p(15.5) with continuity correction

Standardising 23.5 gives $z = \frac{23.5-16}{3.67} = 2.04$

Standardising 15.5 gives $z = \frac{15.5-16}{3.67} = -0.14$

p(between 16 and 24) = p(Z<2.04) – p(Z<–0.14)
= p(Z<2.04) – [1 – p(Z<0.14)]
= p(Z<2.04) + p(Z<0.14) – 1
= 0.98 + 0.54 – 1
= 0.52

## Part 10

### *Question 1*

a) i. A TV spot will generate 1000 extra sales at a gross profit of £10/unit, and will cost £5,000.

The contribution is thus: $1,000 \times 10 - 5,000 = £5,000$

ii. In the same way, the contribution of a newspaper advertisement is $400 \times 10 - 2,000 = £2,000$

b) Suppose the company buys $x$ TV spots and $y$ newspaper advertisements.
The objective is to maximise contribution, $z = 5,000x + 2,000y$ (£).
The constraints are: advertising budget:

$$5,000x + 2,000y \leq 100,000 \text{ or } 5x + 2y \leq 100 \quad (1)$$

maximum to be spent on each mode:

$$5,000x \leq 70,000 \quad \text{or } x \leq 14 \quad (2)$$

$$2,000y \leq 70,000 \quad \text{or } y \leq 35 \quad (3)$$

for marketing balance: $y \geq \frac{1}{2}x$ or $2y \geq x$ (4)

c)

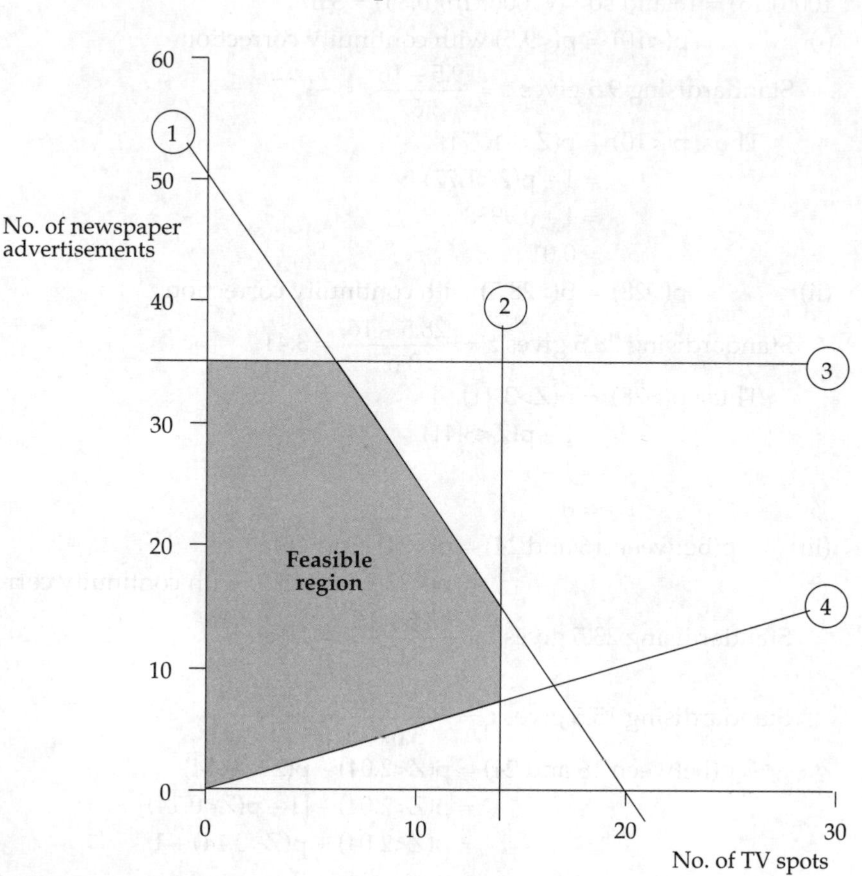

Vertex B will clearly represent an advertising mix which will generate a higher contribution than that of A. Similarly, mix C will generate a higher contribution than mix D. Only these two vertices of the feasible area are therefore considered.

| Vertex | $(x,y)$ | $z = 5{,}000x + 2{,}000y$ (£) |
|---|---|---|
| B | (6,35) | 100,000 |
| C | (14,15) | 100,000 |

Thus the maximum contribution which can be generated by these promotions is £100,000. This can be achieved by any mix on the line (1) of the graph, between B and C, provided $x$ and $y$ are whole numbers.

These are:

| $x$ (number of TV spots) | $y$ (number of newspaper advertisements) |
|---|---|
| 6 | 35 |
| 8 | 30 |
| 10 | 25 |
| 12 | 20 |
| 14 | 15 |

It can be seen that there are *five* different ways of achieving the maximum contribution. This gives the company added flexibility, in that it can choose one from the five which gives extra benefits. For example, if the company felt that television advertising gave a high profile to all its products (not just Brand X), then it could choose the last of the combinations listed, and still maximise X's contribution.

## Question 2

The purpose of this question is to test the candidates knowledge of probability and of matrix notation.

a) i. Let F represent the event of failing the test and A the event of being appointed.

Then P(F) = 0.8 hence P (not F) = 1 – P(F) = 1 – 0.8 = 0.2

From the question P(A/not F) = 0.3 and P(A / F) = 0.1

The probability of being appointed, P(A), is

P(A) = P(A/F)P(F) + P(A/not F)P(not F) = 0.1 × 0.8 + 0.3 × 0.2 = 0.14

ii. P(not A) = 1 – P(A) = 1 – 0.14 = 0.86

P(not A/F) = 1 – P(A/not F) = 1 – 0.3 = 0.7

P(not A intersection not F) = P(not A/not F) × P(not F) = 0.7 × 0.2 = 0.14

P(not F/not A) = P(not A intersection not F)/P(not A)

$$= \frac{0.14}{0.86} = \frac{7}{43} = 0.163$$

iii. Probability of being appointed for all applicants

= probability of being selected for first interview × probability of being selected = 0.15 × 0.14 = 0.021

b) i. $x(1) + x(2) + 2x(3) = 12{,}000$

$x(1) + 2x(2) + 4x(3) = 22{,}000$

$2x(1) + 2x(2) + 3x(3) = 19{,}500$

The above equations in matrix format are: $\begin{bmatrix} 1 & 1 & 2 \\ 1 & 2 & 4 \\ 2 & 2 & 3 \end{bmatrix} * \begin{bmatrix} x(1) \\ x(2) \\ x(3) \end{bmatrix} = \begin{bmatrix} 12{,}000 \\ 22{,}000 \\ 19{,}500 \end{bmatrix}$

ii. Profit (£) = 4 × 2,000 + 4 × 1,000 + 2 × 4,500 = 21,000

iii. Profit (£) = 4 × 3,000 + 4 × 1,500 + 2 × 5,000 = 28,000

Annual market value (£) = 28,000 × 4 × 12 = 112,000 × 12 = 1,344,000

*Question 3*

(a) The conditions under which production is possible are:

Labour: $2X + 2Y \geq 24$ or $X + Y \geq 12$

Cost: $4000X + 12000Y \leq 120000$ or $X + 3Y \leq 30$

Space: $20X + 30Y \leq 480$ or $2X + 3Y \leq 48$

The (production) function to be maximised is $150X + 300Y$.

(b) See the figure below.

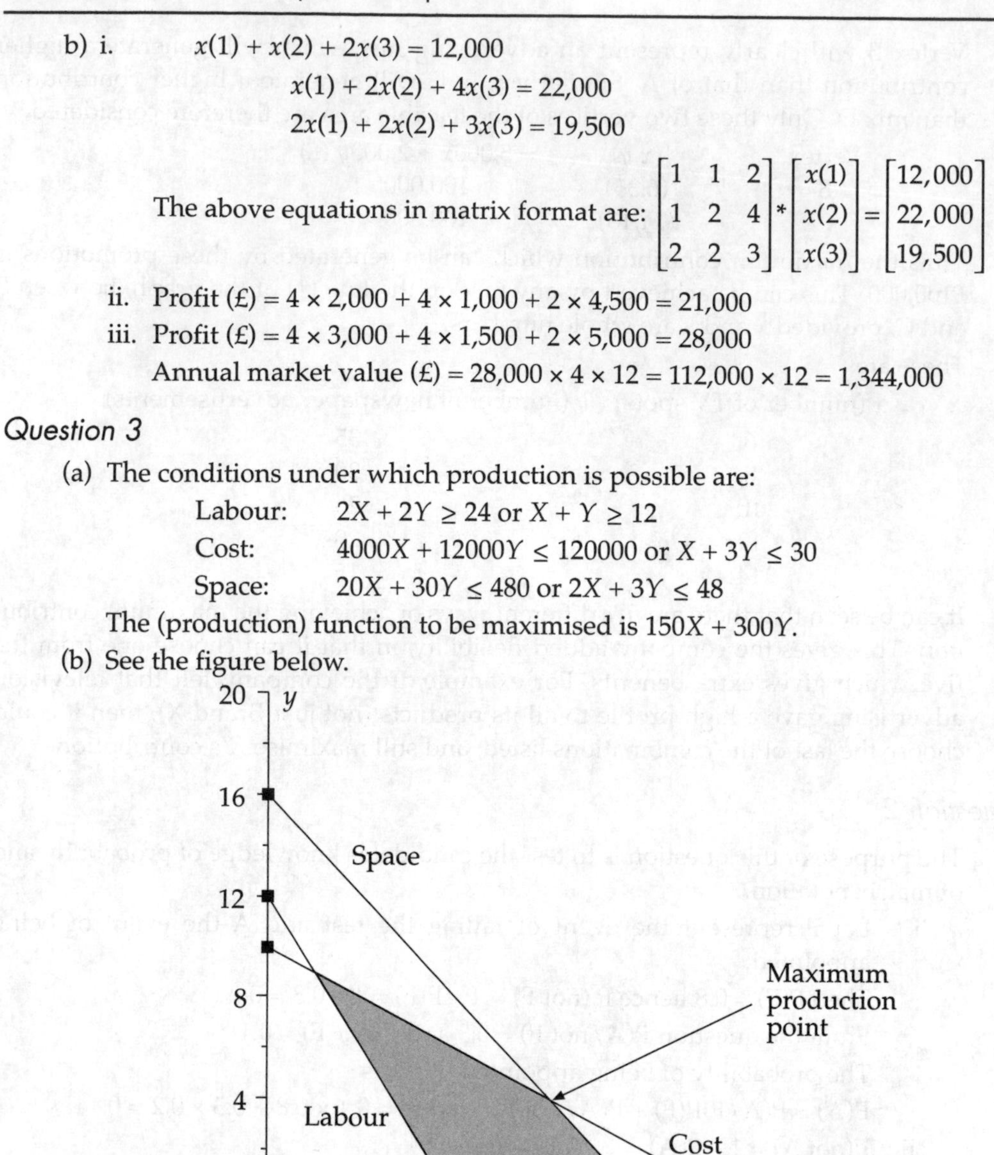


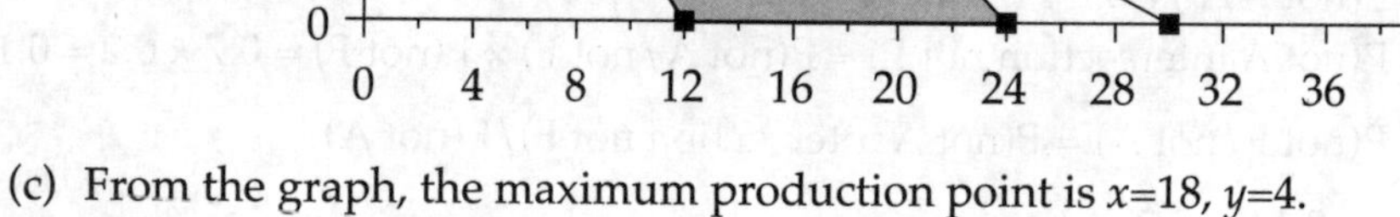


(c) From the graph, the maximum production point is $x$=18, $y$=4.
That is, 18 $X$ presses and 4 $Y$ presses should be bought.

(d) Maximum production is 150(18) + 300(4) = 3900 sheets/minute.
Total cost of presses is 18(4000) + 4(12000) = £120,000.

*Question 4*

(a) (i) If $P = 192 - 28r + r^2$, then the break even point is where P = 0.
i.e. where $r^2 - 28r + 192 = 0$.
Solution by factorization: (r–16)(r–12)=0. Therefore r=12 or r=16.
Solution by formula: (with a=1, b=–28 and c=192).

$$r = \frac{28 \pm \sqrt{28^2 - 4(1)(192)}}{2(1)} = \frac{28 \pm \sqrt{16}}{2} = \frac{28+4}{2} \text{ or } \frac{28-4}{2} = 16 \text{ or } 12$$

(ii) Consider a graph of $P = 192 - 28r + r^2$. It must cross the $r$-axis at the two points $r$=12 and $r$=16. But $P$ is a 'U-shaped' parabola, which means that $P$ must be negative between the values $r$=12 and $r$=16 and positive for all other values of $r$. Thus, the particular project in question makes a loss if the discount rate is between 12% and 16% and a profit for all other values.

(b) (i) If C is the cost matrix and D is the demand matrix, then:

$$C = \begin{array}{l} \text{Shirts} \\ \text{Shorts} \\ \text{Socks} \end{array} \overset{\begin{array}{cc} A & B \end{array}}{\begin{bmatrix} 5.75 & 6.25 \\ 3.99 & 4.48 \\ 1.85 & 1.97 \end{bmatrix}} \text{ and } D = \begin{array}{l} A \\ B \end{array} \overset{\begin{array}{ccc} X & Y & Z \end{array}}{\begin{bmatrix} 36 & 24 & 60 \\ 48 & 72 & 0 \end{bmatrix}}$$

(ii)
$$C^*D = \begin{bmatrix} 5.75 & 6.25 \\ 3.99 & 4.48 \\ 1.85 & 1.97 \end{bmatrix} * \begin{bmatrix} 36 & 24 & 60 \\ 48 & 72 & 0 \end{bmatrix}$$

$$= \begin{array}{l} \text{Shirts} \\ \text{Shorts} \\ \text{Socks} \end{array} \overset{\begin{array}{ccc} X & Y & Z \end{array}}{\begin{bmatrix} 507 & 588 & 345 \\ 358.68 & 418.32 & 239.40 \\ 161.16 & 186.24 & 111.00 \end{bmatrix}}$$

(c) The new demand matrix is:

$$D = \begin{array}{l} A \\ B \end{array} \overset{\begin{array}{cc} X & Y \end{array}}{\begin{bmatrix} 36 & 24 \\ 48 & 72 \end{bmatrix}}$$

and its inverse can be calculated as follows:

$$D^{-1} = \frac{1}{(36)(72) - (48)(24)} \begin{bmatrix} 72 & -24 \\ -48 & 36 \end{bmatrix} = \begin{bmatrix} 72/1440 & -24/1440 \\ -48/1440 & 36/1440 \end{bmatrix}$$

$$= \begin{bmatrix} 0.05 & -0.017 \\ -0.033 & 0.025 \end{bmatrix}$$

## Question 5

a)

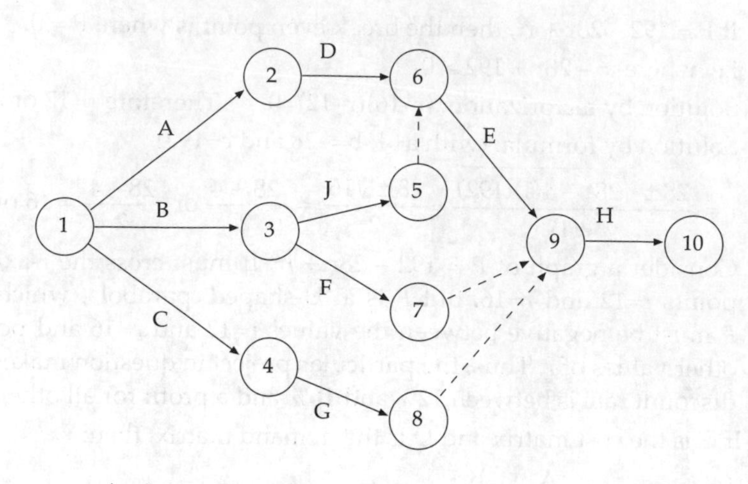


b)

| | Activity | Duration | Earliest start | Earliest finish | Latest start | Latest finish | Total Float |
|---|---|---|---|---|---|---|---|
| A | 1–2 | 3 | 0 | 3 | 2 | 5 | 2 |
| B | 1–3 | 3 | 0 | 3 | 2 | 5 | 2 |
| C | 1–4 | 7 | 0 | 7 | 0 | 7 | 0 |
| D | 2–6 | 1 | 3 | 4 | 5 | 6 | 2 |
| J | 3–5 | 1 | 3 | 4 | 5 | 6 | 2 |
| F | 3–7 | 2 | 3 | 5 | 6 | 8 | 3 |
| G | 4–8 | 1 | 7 | 8 | 7 | 8 | 0 |
| | 5–6 | 0 | 4 | 4 | 6 | 6 | 2 |
| E | 6–9 | 2 | 4 | 6 | 6 | 8 | 2 |
| | 7–9 | 0 | 5 | 5 | 8 | 8 | 3 |
| | 8–9 | 0 | 8 | 8 | 8 | 8 | 0 |
| H | 9–10 | 1 | 8 | 9 | 8 | 9 | 0 |

c) The critical path is C–G–H (Alternatively 1–4–8–9–10). This path determines the minimum project completion time (9 days). All activities on the critical path must be completed without delay if the project is to be completed in 9 days.

## Question 6

a) If the company buys $x$ tables of type $X$ and $y$ tables of type $Y$, then the objective function is $x + y$, which needs to be minimised.

b) The constraints are given as follows:

| | | |
|---|---|---|
| Money: | $40x + 30y$ | $\leq 24000$ |
| Seating capacity: | $4x + 2y$ | $\geq 1800$ |
| Mixture of $x$ and $y$: | $x - y$ | $\leq 0$ |

c) The graphs of the constraints are shown in the figure, with the feasible region marked.

d) The optimum solution can be found by evaluating $x + y$ for each of the three vertices of the feasible region.

For the vertex at $x$=300, $y$=300, $x + y = 600$

For the vertex at $x$=150, $y$=600, $x + y = 750$

For the vertex at $x$=342.9, $y$=342.9, $x + y = 685.8$.

Thus the minimum total number of tables needed is 600, 300 of type $X$ and 300 of type $Y$.

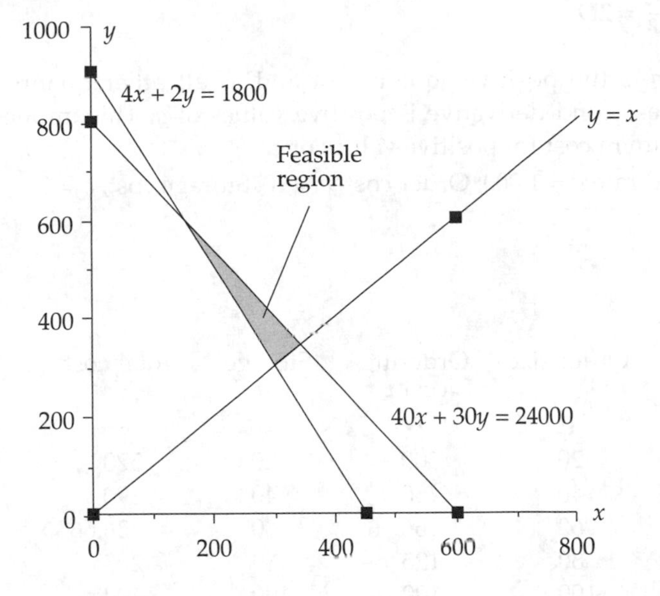


## *Question 7*

(i) The matrix showing the pattern of retention and transfer from the first to the second month is:

$$S = \begin{matrix} & \text{BM1} & \text{BM2} & \text{BM3} \\ \text{BM1} \\ \text{BM2} \\ \text{BM3} \end{matrix} \begin{bmatrix} 0.70 & 0.20 & 0.10 \\ 0.25 & 0.65 & 0.10 \\ 0.05 & 0.15 & 0.80 \end{bmatrix}$$

(ii) The product of matrix S with itself is:

$$\begin{bmatrix} 0.70 & 0.20 & 0.10 \\ 0.25 & 0.65 & 0.10 \\ 0.05 & 0.15 & 0.80 \end{bmatrix} * \begin{bmatrix} 0.70 & 0.20 & 0.10 \\ 0.25 & 0.65 & 0.10 \\ 0.05 & 0.15 & 0.80 \end{bmatrix} = \begin{bmatrix} 0.5450 & 0.2850 & 0.1700 \\ 0.3425 & 0.4875 & 0.1700 \\ 0.1125 & 0.2275 & 0.6600 \end{bmatrix}$$

(iii) The resulting matrix can be interpreted as follows.

Of the original customers who buy BM1, 54.5% will remain loyal to the brand in month 3, 28.5% will have switched to BM2 and 17% will have switched to BM3.

Of the original customers who buy BM2, 48.75% will remain loyal to the brand in month 3, 34.25% will have switched to BM1 and 17% will have switched to BM3.

Of the original customers who buy BM3, 66% will remain loyal to the brand in month 3, 11.25% will have switched to BM1 and 22.75% will have switched to BM2.

## Question 8

(a) $C = D\frac{c_1}{q} + q\frac{c_2}{2}$; $\frac{dC}{dq} = -D\frac{c_1}{q^2} + \frac{c_2}{2} = 0$ for critical values gives $q = \sqrt{2c_1\frac{D}{c_2}}$

and $\frac{d^2C}{dq^2} = 2D\frac{c_1}{q^3}$

When $q$ is the positive square root and as all other quantities are positive then the second derivative is positive values of $q$. This implies that there is a minimum in cost for positive values of $q$.

(b) Demand rate $d$ = 1,000; Order cost, $c_1$= 10; Storage cost, $c_2$= 2.

$\text{EOQ} = \sqrt{2c_1\frac{d}{c_2}}$

(c)

| Order size | Ordering cost £ | Storage cost £ | Total cost £ |
|---|---|---|---|
| 0 | 0 | ∞ | ∞ |
| 20 | 500 | 20 | 520 |
| 40 | 250 | 40 | 290 |
| 60 | 166.66 | 60 | 226.66 |
| 80 | 125 | 80 | 205 |
| 100 | 100 | 100 | 200 |
| 120 | 83.33 | 120 | 203.33 |
| 140 | 71.42 | 140 | 211.42 |
| 160 | 62.5 | 160 | 22.5 |

(d) Order cost, storage cost and total cost against order size. See Figure Q8(d).

(e) The model assumes that demand is regular, that stock delivery is instantaneous and that there is no chance of stockout.

## Question 9

(a) Let $X$ be the number of boxes of $X$ produced and $y$ be the number of boxes of $Y$ produced.

The given restrictions (constraints) are tabulated below as linear inequalities.

| | |
|---|---|
| Blending: | $3x + y < 900$ |
| Baking: | $5x + 4y < 1800$ |
| Packaging: | $x + 3y < 900$ |

We need to maximise the objective function, $x + 2y$.

(b) The constraints are graphed in Figure Q9(b).

The optimum solution can be read from the graph as $x$=164 and $y$=245. That is, 164 $X$ chocolate bars and 245 $Y$ chocolate bars should be produced in order to maximise contribution.

(c) The maximum contribution is obtained by substituting the optimal solutions into the objective function, $x + 2y$, to give 164 + 2(245) = 654. That is, the maximum contribution is £654.

*Figure Q8(d)*

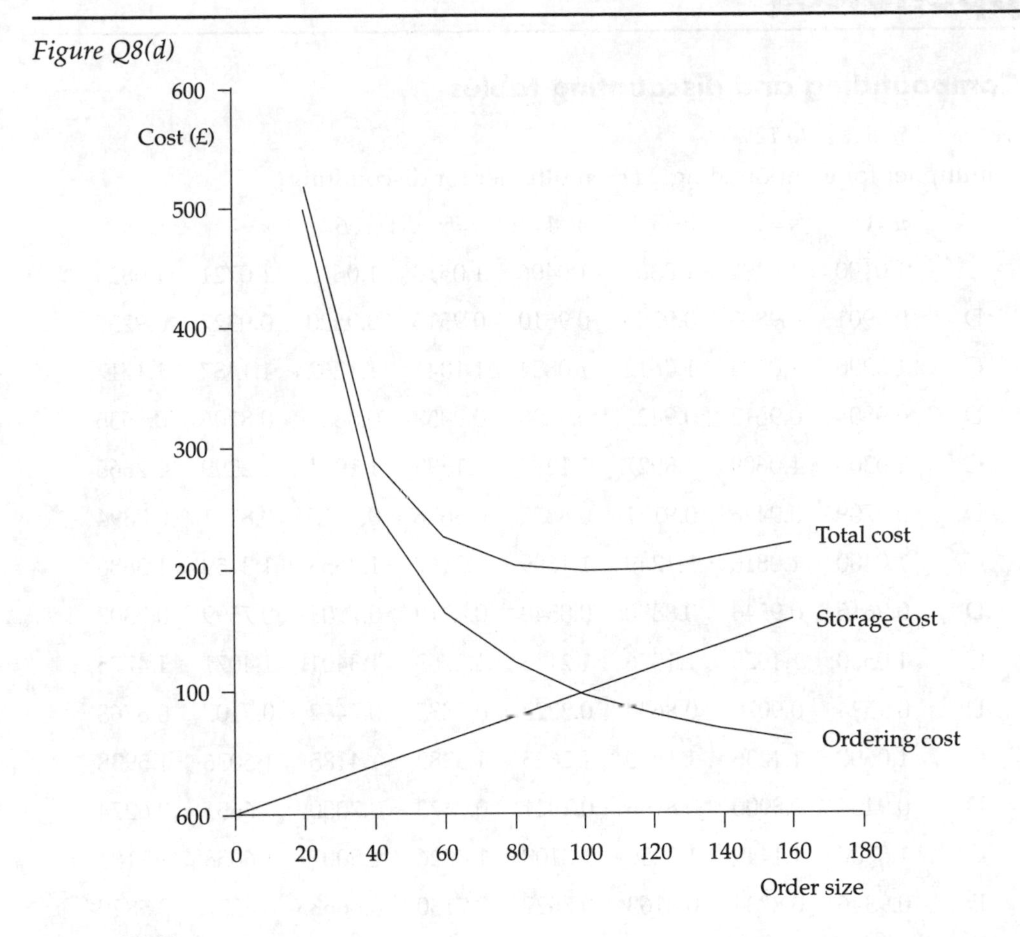


*Figure Q9(b)*

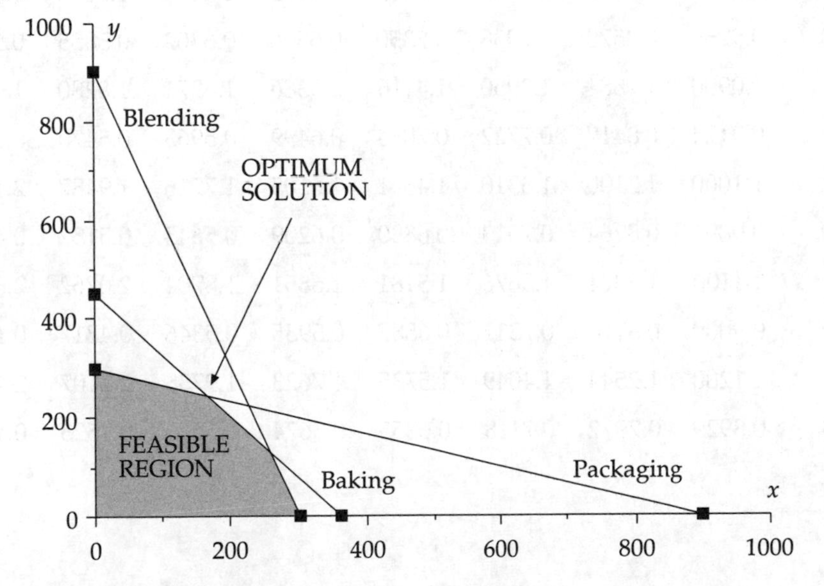

# Appendices

## 1 Compounding and discounting tables

*Range: n = 1 to 8, 1% to 12%*

[*C* = multiplier for compounding; *D* = multiplier for discounting]

| | | *n*=1 | *n*=2 | *n*=3 | *n*=4 | *n*=5 | *n*=6 | *n*=7 | *n*=8 |
|---|---|---|---|---|---|---|---|---|---|
| 1% | C | 1.0100 | 1.0201 | 1.0303 | 1.0406 | 1.0510 | 1.0615 | 1.0721 | 1.0829 |
| | D | 0.9901 | 0.9803 | 0.9706 | 0.9610 | 0.9515 | 0.9420 | 0.9327 | 0.9235 |
| 2% | C | 1.0200 | 1.0404 | 1.0612 | 1.0824 | 1.1041 | 1.1262 | 1.1487 | 1.1717 |
| | D | 0.9804 | 0.9612 | 0.9423 | 0.9238 | 0.9057 | 0.8880 | 0.8706 | 0.8535 |
| 3% | C | 1.0300 | 1.0609 | 1.0927 | 1.1255 | 1.1593 | 1.1941 | 1.2299 | 1.2668 |
| | D | 0.9709 | 0.9426 | 0.9151 | 0.8885 | 0.8626 | 0.8375 | 0.8131 | 0.7894 |
| 4% | C | 1.0400 | 1.0816 | 1.1249 | 1.1699 | 1.2167 | 1.2653 | 1.3159 | 1.3686 |
| | D | 0.9615 | 0.9246 | 0.8890 | 0.8548 | 0.8219 | 0.7903 | 0.7599 | 0.7307 |
| 5% | C | 1.0500 | 1.1025 | 1.1576 | 1.2155 | 1.2763 | 1.3401 | 1.4071 | 1.4775 |
| | D | 0.9524 | 0.9070 | 0.8638 | 0.8227 | 0.7835 | 0.7462 | 0.7107 | 0.6768 |
| 6% | C | 1.0600 | 1.1236 | 1.1910 | 1.2625 | 1.3382 | 1.4185 | 1.5036 | 1.5938 |
| | D | 0.9434 | 0.8900 | 0.8396 | 0.7921 | 0.7473 | 0.7050 | 0.6651 | 0.6274 |
| 7% | C | 1.0700 | 1.1449 | 1.2250 | 1.3108 | 1.4026 | 1.5007 | 1.6058 | 1.7182 |
| | D | 0.9346 | 0.8734 | 0.8163 | 0.7629 | 0.7130 | 0.6663 | 0.6227 | 0.5820 |
| 8% | C | 1.0800 | 1.1664 | 1.2597 | 1.3605 | 1.4693 | 1.5689 | 1.7138 | 1.8509 |
| | D | 0.9259 | 0.8573 | 0.7938 | 0.7350 | 0.6806 | 0.6302 | 0.5835 | 0.5403 |
| 9% | C | 1.0900 | 1.1881 | 1.2950 | 1.4116 | 1.5386 | 1.6771 | 1.8280 | 1.9926 |
| | D | 0.9174 | 0.8417 | 0.7722 | 0.7084 | 0.6499 | 0.5963 | 0.5470 | 0.5019 |
| 10% | C | 1.1000 | 1.2100 | 1.3310 | 1.4641 | 1.6105 | 1.7716 | 1.9487 | 2.1436 |
| | D | 0.9091 | 0.8264 | 0.7513 | 0.6830 | 0.6209 | 0.5645 | 0.5132 | 0.4665 |
| 11% | C | 1.1100 | 1.2321 | 1.3676 | 1.5181 | 1.6851 | 1.8704 | 2.0762 | 2.3045 |
| | D | 0.9009 | 0.8116 | 0.7312 | 0.6587 | 0.5935 | 0.5346 | 0.4817 | 0.4339 |
| 12% | C | 1.1200 | 1.2544 | 1.4049 | 1.5735 | 1.7623 | 1.9738 | 2.2107 | 2.4760 |
| | D | 0.8929 | 0.7972 | 0.7118 | 0.6355 | 0.5674 | 0.5066 | 0.4523 | 0.4039 |

*Range: n = 1 to 8, 13% to 25%*
[C = multiplier for compounding; D = multiplier for discounting]

| | | n=1 | n=2 | n=3 | n=4 | n=5 | n=6 | n=7 | n=8 |
|---|---|---|---|---|---|---|---|---|---|
| 13% | C | 1.1300 | 1.2769 | 1.4429 | 1.6305 | 1.8424 | 2.0820 | 2.3526 | 2.6584 |
| | D | 0.8850 | 0.7831 | 0.6931 | 0.6133 | 0.5428 | 0.4803 | 0.4251 | 0.3762 |
| 14% | C | 1.1400 | 1.2996 | 1.4815 | 1.6890 | 1.9254 | 2.1950 | 2.5023 | 2.8526 |
| | D | 0.8772 | 0.7695 | 0.6750 | 0.5921 | 0.5194 | 0.4556 | 0.3996 | 0.3506 |
| 15% | C | 1.1500 | 1.3225 | 1.5209 | 1.7490 | 2.0114 | 2.3131 | 2.6600 | 3.0590 |
| | D | 0.8696 | 0.7561 | 0.6575 | 0.5718 | 0.4972 | 0.4323 | 0.3759 | 0.3269 |
| 16% | C | 1.1600 | 1.3456 | 1.5609 | 1.8106 | 2.1003 | 2.4364 | 2.8262 | 3.2784 |
| | D | 0.8621 | 0.7432 | 0.6407 | 0.5523 | 0.4761 | 0.4104 | 0.3538 | 0.3050 |
| 17% | C | 1.1700 | 1.3689 | 1.6016 | 1.8739 | 2.1924 | 2.5652 | 3.0012 | 3.5115 |
| | D | 0.8547 | 0.7305 | 0.6244 | 0.5337 | 0.4561 | 0.3898 | 0.3332 | 0.2848 |
| 18% | C | 1.1800 | 1.3924 | 1.6430 | 1.9388 | 2.2878 | 2.6996 | 3.1855 | 3.7589 |
| | D | 0.8475 | 0.7182 | 0.6086 | 0.5158 | 0.4371 | 0.3704 | 0.3139 | 0.2660 |
| 19% | C | 1.1900 | 1.4161 | 1.6852 | 2.0053 | 2.3864 | 2.8398 | 3.3793 | 4.0214 |
| | D | 0.8403 | 0.7062 | 0.5934 | 0.4987 | 0.4190 | 0.3521 | 0.2959 | 0.2487 |
| 20% | C | 1.2000 | 1.4400 | 1.7280 | 2.0736 | 2.4883 | 2.9860 | 3.5832 | 4.2998 |
| | D | 0.8333 | 0.6944 | 0.5787 | 0.4823 | 0.4019 | 0.3349 | 0.2791 | 0.2326 |
| 21% | C | 1.2100 | 1.4641 | 1.7716 | 2.1436 | 2.5937 | 3.1384 | 3.7975 | 4.5950 |
| | D | 0.8264 | 0.6830 | 0.5645 | 0.4665 | 0.3855 | 0.3186 | 0.2633 | 0.2176 |
| 22% | C | 1.2200 | 1.4884 | 1.8158 | 2.2153 | 2.7027 | 3.2973 | 4.0227 | 4.9077 |
| | D | 0.8197 | 0.6719 | 0.5507 | 0.4514 | 0.3700 | 0.3033 | 0.2486 | 0.2038 |
| 23% | C | 1.2300 | 1.5129 | 1.8609 | 2.2889 | 2.8153 | 3.4628 | 4.2593 | 5.2389 |
| | D | 0.8130 | 0.6610 | 0.5374 | 0.4369 | 0.3552 | 0.2888 | 0.2348 | 0.1909 |
| 24% | C | 1.2400 | 1.5376 | 1.9066 | 2.3642 | 2.9316 | 3.6352 | 4.5077 | 5.5895 |
| | D | 0.8065 | 0.6504 | 0.5245 | 0.4230 | 0.3411 | 0.2751 | 0.2218 | 0.1789 |
| 25% | C | 1.2500 | 1.5625 | 1.9531 | 2.4414 | 3.0518 | 3.8147 | 4.7684 | 5.9605 |
| | D | 0.8000 | 0.6400 | 0.5120 | 0.4096 | 0.3277 | 0.2621 | 0.2097 | 0.1678 |

*Range: n = 9 to 16, 1% to 12%*

[C = multiplier for compounding; *D* = multiplier for discounting]

| | | *n*=9 | *n*=10 | *n*=11 | *n*=12 | *n*=13 | *n*=14 | *n*=15 | *n*=16 |
|---|---|---|---|---|---|---|---|---|---|
| 1% | C | 1.0937 | 1.1046 | 1.1157 | 1.1268 | 1.1381 | 1.1495 | 1.1610 | 1.1726 |
| | D | 0.9143 | 0.9053 | 0.8963 | 0.8874 | 0.8787 | 0.8700 | 0.8613 | 0.8528 |
| 2% | C | 1.1951 | 1.2190 | 1.2434 | 1.2682 | 1.2936 | 1.3195 | 1.3459 | 1.3728 |
| | D | 0.8368 | 0.8203 | 0.8043 | 0.7885 | 0.7730 | 0.7579 | 0.7430 | 0.7284 |
| 3% | C | 1.3048 | 1.3439 | 1.3842 | 1.4258 | 1.4685 | 1.5126 | 1.5580 | 1.6047 |
| | D | 0.7664 | 0.7441 | 0.7224 | 0.7014 | 0.6810 | 0.6611 | 0.6419 | 0.6232 |
| 4% | C | 1.4233 | 1.4802 | 1.5395 | 1.6010 | 1.6651 | 1.7317 | 1.8009 | 1.8730 |
| | D | 0.7026 | 0.6756 | 0.6496 | 0.6246 | 0.6006 | 0.5775 | 0.5553 | 0.5339 |
| 5% | C | 1.5513 | 1.6289 | 1.7103 | 1.7959 | 1.8856 | 1.9799 | 2.0789 | 2.1829 |
| | D | 0.6446 | 0.6139 | 0.5847 | 0.5568 | 0.5303 | 0.5051 | 0.4810 | 0.4581 |
| 6% | C | 1.6895 | 1.7908 | 1.8983 | 2.0122 | 2.1329 | 2.2609 | 2.3966 | 2.5404 |
| | D | 0.5919 | 0.5584 | 0.5268 | 0.4970 | 0.4688 | 0.4423 | 0.4173 | 0.3936 |
| 7% | C | 1.8385 | 1.9672 | 2.1049 | 2.2522 | 2.4098 | 2.5785 | 2.7590 | 2.9522 |
| | D | 0.5439 | 0.5083 | 0.4751 | 0.4440 | 0.4150 | 0.3878 | 0.3624 | 0.3387 |
| 8% | C | 1.9990 | 2.1589 | 2.3316 | 2.5182 | 2.7196 | 2.9372 | 3.1722 | 3.4259 |
| | D | 0.5002 | 0.4632 | 0.4289 | 0.3971 | 0.3677 | 0.3405 | 0.3152 | 0.2919 |
| 9% | C | 2.1719 | 2.3674 | 2.5804 | 2.8127 | 3.0658 | 3.3417 | 3.6425 | 3.9703 |
| | D | 0.4604 | 0.4224 | 0.3875 | 0.3555 | 0.3262 | 0.2992 | 0.2745 | 0.2519 |
| 10% | C | 2.3579 | 2.5937 | 2.8531 | 3.1384 | 3.4523 | 3.7975 | 4.1772 | 4.5950 |
| | D | 0.4241 | 0.3855 | 0.3505 | 0.3186 | 0.2897 | 0.2633 | 0.2394 | 0.2176 |
| 11% | C | 2.5580 | 2.8394 | 3.1518 | 3.4985 | 3.8833 | 4.3104 | 4.7846 | 5.3109 |
| | D | 0.3909 | 0.3522 | 0.3173 | 0.2858 | 0.2575 | 0.2320 | 0.2090 | 0.1883 |
| 12% | C | 2.7731 | 3.1058 | 3.4785 | 3.8960 | 4.3635 | 4.8871 | 5.4736 | 6.1304 |
| | D | 0.3606 | 0.3220 | 0.2875 | 0.2567 | 0.2292 | 0.2046 | 0.1827 | 0.1631 |

*Range: n = 9 to 16, 13% to 25%*

[C = multiplier for compounding; *D* = multiplier for discounting]

| | | *n*=9 | *n*=10 | *n*=11 | *n*=12 | *n*=13 | *n*=14 | *n*=15 | *n*=16 |
|---|---|---|---|---|---|---|---|---|---|
| 13% | C | 3.0040 | 3.3946 | 3.8359 | 4.3345 | 4.8980 | 5.5348 | 6.2543 | 7.0673 |
| | D | 0.3329 | 0.2946 | 0.2607 | 0.2307 | 0.2042 | 0.1807 | 0.1599 | 0.1415 |
| 14% | C | 3.2519 | 3.7072 | 4.2262 | 4.8179 | 5.4924 | 6.2613 | 7.1379 | 8.1372 |
| | D | 0.3075 | 0.2697 | 0.2366 | 0.2076 | 0.1821 | 0.1597 | 0.1401 | 0.1229 |
| 15% | C | 3.5179 | 4.0456 | 4.6524 | 5.3503 | 6.1528 | 7.0757 | 8.1371 | 9.3576 |
| | D | 0.2843 | 0.2472 | 0.2149 | 0.1869 | 0.1625 | 0.1413 | 0.1229 | 0.1069 |
| 16% | C | 3.8030 | 4.4114 | 5.1173 | 5.9360 | 6.8858 | 7.9875 | 9.2655 | 10.7480 |
| | D | 0.2630 | 0.2267 | 0.1954 | 0.1685 | 0.1452 | 0.1252 | 0.1079 | 0.0930 |
| 17% | C | 4.1084 | 4.8068 | 5.6240 | 6.5801 | 7.6987 | 9.0075 | 10.5387 | 12.3303 |
| | D | 0.2434 | 0.2080 | 0.1778 | 0.1520 | 0.1299 | 0.1110 | 0.0949 | 0.0811 |
| 18% | C | 4.4355 | 5.2338 | 6.1759 | 7.2876 | 8.5994 | 10.1472 | 11.9737 | 14.1290 |
| | D | 0.2255 | 0.1911 | 0.1619 | 0.1372 | 0.1163 | 0.0985 | 0.0835 | 0.0708 |
| 19% | C | 4.7854 | 5.6947 | 6.7767 | 8.0642 | 9.5964 | 11.4198 | 13.5895 | 16.1715 |
| | D | 0.2090 | 0.1756 | 0.1476 | 0.1240 | 0.1042 | 0.0876 | 0.0736 | 0.0618 |
| 20% | C | 5.1598 | 6.1917 | 7.4301 | 8.9161 | 10.6993 | 12.8392 | 15.4070 | 18.484 |
| | D | 0.1938 | 0.1615 | 0.1346 | 0.1122 | 0.0935 | 0.0779 | 0.0649 | 0.0541 |
| 21% | C | 5.5599 | 6.7275 | 8.1403 | 9.8497 | 11.9182 | 14.4210 | 17.4494 | 21.1138 |
| | D | 0.1799 | 0.1486 | 0.1228 | 0.1015 | 0.0839 | 0.0693 | 0.0573 | 0.0471 |
| 22% | C | 5.9874 | 7.3046 | 8.9117 | 10.8722 | 13.2641 | 16.1822 | 19.7423 | 24.0856 |
| | D | 0.1670 | 0.1369 | 0.1122 | 0.0920 | 0.0754 | 0.0618 | 0.0507 | 0.0415 |
| 23% | C | 6.4439 | 7.9259 | 9.7489 | 11.9912 | 14.7491 | 18.1414 | 22.3140 | 27.4462 |
| | D | 0.1552 | 0.1262 | 0.1026 | 0.0834 | 0.0678 | 0.0551 | 0.0448 | 0.0364 |
| 24% | C | 6.9310 | 8.5944 | 10.6571 | 13.2148 | 16.3863 | 20.3191 | 25.1956 | 31.2426 |
| | D | 0.1443 | 0.1164 | 0.0938 | 0.0757 | 0.0610 | 0.0492 | 0.0397 | 0.0320 |
| 25% | C | 7.4506 | 9.3132 | 11.6415 | 14.5519 | 18.1899 | 22.7374 | 28.4217 | 35.5271 |
| | D | 0.1342 | 0.1074 | 0.0859 | 0.0687 | 0.0550 | 0.0440 | 0.0352 | 0.0281 |

## 2 Random sampling numbers

| | | | | | | | | | |
|---|---|---|---|---|---|---|---|---|---|
| 33865 | 04131 | 78302 | 22688 | 79034 | 01358 | 61724 | 98286 | 97086 | 21376 |
| 09356 | 09387 | 52825 | 93134 | 21731 | 93956 | 85324 | 68767 | 49490 | 11449 |
| 98243 | 37636 | 64825 | 43091 | 24906 | 13545 | 90172 | 31265 | 81457 | 93108 |
| 99052 | 61857 | 33938 | 86339 | 63531 | 77146 | 33252 | 81388 | 28302 | 18960 |
| 00713 | 24413 | 36920 | 03841 | 48047 | 04207 | 50930 | 84723 | 07400 | 81109 |
| 34819 | 80011 | 17751 | 03275 | 92511 | 70071 | 08183 | 72805 | 94618 | 46084 |
| 20611 | 34975 | 96712 | 32402 | 90182 | 94070 | 94711 | 94233 | 06619 | 34162 |
| 64972 | 86061 | 04685 | 53042 | 82685 | 45992 | 19829 | 45265 | 85589 | 83440 |
| 15857 | 73681 | 24790 | 20515 | 01232 | 25302 | 30785 | 95288 | 79341 | 54313 |
| 80276 | 67053 | 99022 | 36888 | 58643 | 96111 | 77292 | 03441 | 52856 | 95035 |
| 30548 | 51156 | 63914 | 64139 | 14596 | 35541 | 70324 | 20789 | 29139 | 66973 |
| 53530 | 79354 | 75099 | 89593 | 36449 | 66618 | 32346 | 37526 | 20084 | 52492 |
| 77012 | 18480 | 61852 | 82765 | 29602 | 10032 | 78925 | 71953 | 21661 | 95254 |
| 04304 | 40763 | 24847 | 07724 | 99223 | 77838 | 09547 | 47714 | 13302 | 17121 |
| 76953 | 39588 | 90708 | 67618 | 45671 | 19671 | 92674 | 22841 | 84231 | 59446 |
| 34479 | 85938 | 26363 | 12025 | 70315 | 58971 | 28991 | 35990 | 23542 | 74794 |
| 28421 | 16347 | 66638 | 25578 | 70404 | 67367 | 14730 | 37662 | 64669 | 16752 |
| 58160 | 17725 | 97075 | 99789 | 24304 | 63100 | 22123 | 83692 | 92997 | 58699 |
| 96701 | 73743 | 82979 | 69917 | 34993 | 36495 | 47023 | 48869 | 50611 | 61534 |
| 55600 | 61672 | 99136 | 73925 | 30250 | 12533 | 46280 | 03865 | 88049 | 13080 |
| 55850 | 38966 | 46303 | 37073 | 42347 | 36157 | 44357 | 52065 | 66913 | 06284 |
| 47089 | 83871 | 51231 | 32522 | 41543 | 22675 | 89316 | 38451 | 78694 | 01767 |
| 26035 | 86173 | 11115 | 22083 | 12083 | 43374 | 66542 | 23518 | 05372 | 33892 |
| 74920 | 35946 | 21149 | 70861 | 13235 | 02729 | 57485 | 23895 | 80607 | 11299 |
| 44498 | 00498 | 31354 | 39787 | 65919 | 61889 | 17690 | 10176 | 94138 | 95650 |
| 80045 | 71846 | 17840 | 23670 | 77769 | 84062 | 52850 | 20241 | 06073 | 20083 |
| 15828 | 95852 | 12124 | 95053 | 09924 | 91562 | 09419 | 27747 | 84732 | 81927 |
| 04100 | 75759 | 37926 | 70040 | 80884 | 48939 | 65228 | 60075 | 45056 | 56399 |
| 69257 | 48373 | 58911 | 78549 | 63693 | 43727 | 81058 | 53301 | 85945 | 54890 |
| 33915 | 26034 | 08166 | 59242 | 03881 | 88690 | 92298 | 48628 | 02698 | 94249 |
| 83497 | 62761 | 68609 | 85811 | 40695 | 08342 | 67386 | 63470 | 85643 | 68568 |
| 46466 | 15977 | 69989 | 90106 | 01432 | 59700 | 13163 | 56521 | 96687 | 41390 |
| 03573 | 87778 | 27696 | 35147 | 54639 | 20489 | 03688 | 72254 | 28402 | 98954 |
| 02046 | 44774 | 31500 | 30232 | 27434 | 14925 | 65901 | 34521 | 94104 | 54935 |
| 68736 | 12912 | 02579 | 34719 | 09568 | 21571 | 91111 | 81307 | 97866 | 76483 |

## 3 Exponential tables. Values of $e^{-m}$

*Range: m = 0 to 2.4*

| $m$ | 0.00 | 0.01 | 0.02 | 0.03 | 0.04 | 0.05 | 0.06 | 0.07 | 0.08 | 0.09 |
|---|---|---|---|---|---|---|---|---|---|---|
| 0.0 | 1.0000 | 0.9900 | 0.9802 | 0.9704 | 0.9608 | 0.9512 | 0.9418 | 0.9324 | 0.9231 | 0.9139 |
| 0.1 | 0.9048 | 0.8958 | 0.8869 | 0.8781 | 0.8694 | 0.8607 | 0.8521 | 0.8437 | 0.8353 | 0.8270 |
| 0.2 | 0.8187 | 0.8106 | 0.8025 | 0.7945 | 0.7866 | 0.7788 | 0.7711 | 0.7634 | 0.7558 | 0.7483 |
| 0.3 | 0.7408 | 0.7334 | 0.7261 | 0.7189 | 0.7118 | 0.7047 | 0.6977 | 0.6907 | 0.6839 | 0.6771 |
| 0.4 | 0.6703 | 0.6637 | 0.6570 | 0.6505 | 0.6440 | 0.6376 | 0.6313 | 0.6250 | 0.6188 | 0.6126 |
| 0.5 | 0.6065 | 0.6005 | 0.5945 | 0.5886 | 0.5827 | 0.5769 | 0.5712 | 0.5655 | 0.5599 | 0.5543 |
| 0.6 | 0.5488 | 0.5434 | 0.5379 | 0.5326 | 0.5273 | 0.5220 | 0.5169 | 0.5117 | 0.5066 | 0.5016 |
| 0.7 | 0.4966 | 0.4916 | 0.4868 | 0.4819 | 0.4771 | 0.4724 | 0.4677 | 0.4630 | 0.4584 | 0.4538 |
| 0.8 | 0.4493 | 0.4449 | 0.4404 | 0.4360 | 0.4317 | 0.4274 | 0.4232 | 0.4190 | 0.4148 | 0.4107 |
| 0.9 | 0.4066 | 0.4025 | 0.3985 | 0.3946 | 0.3906 | 0.3867 | 0.3829 | 0.3791 | 0.3753 | 0.3716 |
| 1.0 | 0.3679 | 0.3642 | 0.3606 | 0.3570 | 0.3535 | 0.3499 | 0.3465 | 0.3430 | 0.3396 | 0.3362 |
| 1.1 | 0.3329 | 0.3296 | 0.3263 | 0.3230 | 0.3198 | 0.3166 | 0.3135 | 0.3104 | 0.3073 | 0.3042 |
| 1.2 | 0.3012 | 0.2982 | 0.2952 | 0.2923 | 0.2894 | 0.2865 | 0.2837 | 0.2808 | 0.2780 | 0.2753 |
| 1.3 | 0.2725 | 0.2698 | 0.2671 | 0.2645 | 0.2618 | 0.2592 | 0.2567 | 0.2541 | 0.2516 | 0.2491 |
| 1.4 | 0.2466 | 0.2441 | 0.2417 | 0.2393 | 0.2369 | 0.2346 | 0.2322 | 0.2299 | 0.2276 | 0.2254 |
| 1.5 | 0.2231 | 0.2209 | 0.2187 | 0.2165 | 0.2144 | 0.2122 | 0.2101 | 0.2080 | 0.2060 | 0.2039 |
| 1.6 | 0.2019 | 0.1999 | 0.1979 | 0.1959 | 0.1940 | 0.1920 | 0.1901 | 0.1882 | 0.1864 | 0.1845 |
| 1.7 | 0.1827 | 0.1809 | 0.1791 | 0.1773 | 0.1755 | 0.1738 | 0.1720 | 0.1703 | 0.1686 | 0.1670 |
| 1.8 | 0.1653 | 0.1637 | 0.1620 | 0.1604 | 0.1588 | 0.1572 | 0.1557 | 0.1541 | 0.1526 | 0.1511 |
| 1.9 | 0.1496 | 0.1481 | 0.1466 | 0.1451 | 0.1437 | 0.1423 | 0.1409 | 0.1395 | 0.1381 | 0.1367 |
| 2.0 | 0.1353 | 0.1340 | 0.1327 | 0.1313 | 0.1300 | 0.1287 | 0.1275 | 0.1262 | 0.1249 | 0.1237 |
| 2.1 | 0.1225 | 0.1212 | 0.1200 | 0.1188 | 0.1177 | 0.1165 | 0.1153 | 0.1142 | 0.1130 | 0.1119 |
| 2.2 | 0.1108 | 0.1097 | 0.1086 | 0.1075 | 0.1065 | 0.1054 | 0.1044 | 0.1033 | 0.1023 | 0.1013 |
| 2.3 | 0.1003 | 0.0993 | 0.0983 | 0.0973 | 0.0963 | 0.0954 | 0.0944 | 0.0935 | 0.0926 | 0.0916 |
| 2.4 | 0.0907 | 0.0898 | 0.0889 | 0.0880 | 0.0872 | 0.0863 | 0.0854 | 0.0846 | 0.0837 | 0.0829 |

*Range: m = 2.5 to 5.0*

| $m$ | 0.00 | 0.01 | 0.02 | 0.03 | 0.04 | 0.05 | 0.06 | 0.07 | 0.08 | 0.09 |
|---|---|---|---|---|---|---|---|---|---|---|
| 2.5 | 0.0821 | 0.0813 | 0.0805 | 0.0797 | 0.0789 | 0.0781 | 0.0773 | 0.0765 | 0.0758 | 0.0750 |
| 2.6 | 0.0743 | 0.0735 | 0.0728 | 0.0721 | 0.0714 | 0.0707 | 0.0699 | 0.0693 | 0.0686 | 0.0679 |
| 2.7 | 0.0672 | 0.0665 | 0.0659 | 0.0652 | 0.0646 | 0.0639 | 0.0633 | 0.0627 | 0.0620 | 0.0614 |
| 2.8 | 0.0608 | 0.0602 | 0.0596 | 0.0590 | 0.0584 | 0.0578 | 0.0573 | 0.0567 | 0.0561 | 0.0556 |
| 2.9 | 0.0550 | 0.0545 | 0.0539 | 0.0534 | 0.0529 | 0.0523 | 0.0518 | 0.0513 | 0.0508 | 0.0503 |
| 3.0 | 0.0498 | 0.0493 | 0.0488 | 0.0483 | 0.0478 | 0.0474 | 0.0469 | 0.0464 | 0.0460 | 0.0455 |
| 3.1 | 0.0450 | 0.0446 | 0.0442 | 0.0437 | 0.0433 | 0.0429 | 0.0424 | 0.0420 | 0.0416 | 0.0412 |
| 3.2 | 0.0408 | 0.0404 | 0.0400 | 0.0396 | 0.0392 | 0.0388 | 0.0384 | 0.0380 | 0.0376 | 0.0373 |
| 3.3 | 0.0369 | 0.0365 | 0.0362 | 0.0358 | 0.0354 | 0.0351 | 0.0347 | 0.0344 | 0.0340 | 0.0337 |
| 3.4 | 0.0334 | 0.0330 | 0.0327 | 0.0324 | 0.0321 | 0.0317 | 0.0314 | 0.0311 | 0.0308 | 0.0305 |
| 3.5 | 0.0302 | 0.0299 | 0.0296 | 0.0293 | 0.0290 | 0.0287 | 0.0284 | 0.0282 | 0.0279 | 0.0276 |
| 3.6 | 0.0273 | 0.0271 | 0.0268 | 0.0265 | 0.0263 | 0.0260 | 0.0257 | 0.0255 | 0.0252 | 0.0250 |
| 3.7 | 0.0247 | 0.0245 | 0.0242 | 0.0240 | 0.0238 | 0.0235 | 0.0233 | 0.0231 | 0.0228 | 0.0226 |
| 3.8 | 0.0224 | 0.0221 | 0.0219 | 0.0217 | 0.0215 | 0.0213 | 0.0211 | 0.0209 | 0.0207 | 0.0204 |
| 3.9 | 0.0202 | 0.0200 | 0.0198 | 0.0196 | 0.0194 | 0.0193 | 0.0191 | 0.0189 | 0.0187 | 0.0185 |
| 4.0 | 0.0183 | 0.0181 | 0.0180 | 0.0178 | 0.0176 | 0.0174 | 0.0172 | 0.0171 | 0.0169 | 0.0167 |
| 4.1 | 0.0166 | 0.0164 | 0.0162 | 0.0161 | 0.0159 | 0.0158 | 0.0156 | 0.0155 | 0.0153 | 0.0151 |
| 4.2 | 0.0150 | 0.0148 | 0.0147 | 0.0146 | 0.0144 | 0.0143 | 0.0141 | 0.0140 | 0.0138 | 0.0137 |
| 4.3 | 0.0136 | 0.0134 | 0.0133 | 0.0132 | 0.0130 | 0.0129 | 0.0128 | 0.0127 | 0.0125 | 0.0124 |
| 4.4 | 0.0123 | 0.0122 | 0.0120 | 0.0119 | 0.0118 | 0.0117 | 0.0116 | 0.0114 | 0.0113 | 0.0112 |
| 4.5 | 0.0111 | 0.0110 | 0.0109 | 0.0108 | 0.0107 | 0.0106 | 0.0105 | 0.0104 | 0.0103 | 0.0102 |
| 4.6 | 0.0101 | 0.0100 | 0.0099 | 0.0098 | 0.0097 | 0.0096 | 0.0095 | 0.0094 | 0.0093 | 0.0092 |
| 4.7 | 0.0091 | 0.0090 | 0.0089 | 0.0088 | 0.0087 | 0.0087 | 0.0086 | 0.0085 | 0.0084 | 0.0083 |
| 4.8 | 0.0082 | 0.0081 | 0.0081 | 0.0080 | 0.0079 | 0.0078 | 0.0078 | 0.0077 | 0.0076 | 0.0075 |
| 4.9 | 0.0074 | 0.0074 | 0.0073 | 0.0072 | 0.0072 | 0.0071 | 0.0070 | 0.0069 | 0.0069 | 0.0068 |
| 5.0 | 0.0067 | 0.0067 | 0.0066 | 0.0065 | 0.0065 | 0.0064 | 0.0063 | 0.0063 | 0.0062 | 0.0061 |

## 4 Standard Normal distribution tables

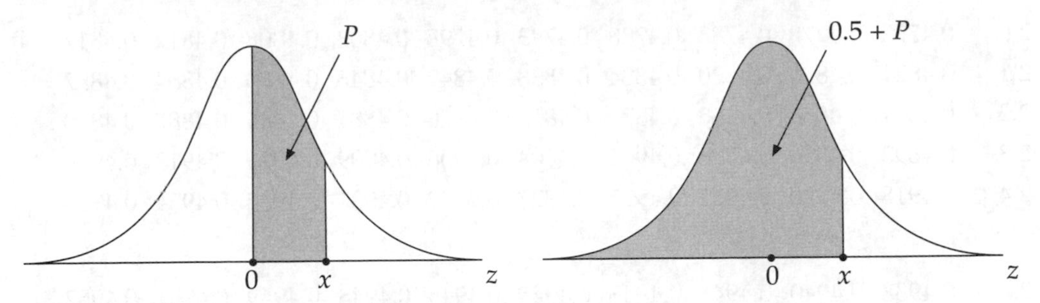


The table following gives the probability ($P$) that a Standard Normal variable lies between 0 and $x$.

This is equivalent to the shaded area in the left-hand figure.

To obtain the probability shown in the shaded area in the right-hand figure, 0.5 needs to be added to $P$ as shown.

| $x$ | 0.00 | 0.01 | 0.02 | 0.03 | 0.04 | 0.05 | 0.06 | 0.07 | 0.08 | 0.09 |
|---|---|---|---|---|---|---|---|---|---|---|
| 0.0 | 0.0000 | 0.0040 | 0.0080 | 0.0120 | 0.0160 | 0.0199 | 0.0239 | 0.0279 | 0.0319 | 0.0359 |
| 0.1 | 0.0398 | 0.0438 | 0.0478 | 0.0517 | 0.0557 | 0.0596 | 0.0636 | 0.0675 | 0.0714 | 0.0754 |
| 0.2 | 0.0793 | 0.0832 | 0.0871 | 0.0910 | 0.0948 | 0.0987 | 0.1026 | 0.1064 | 0.1103 | 0.1141 |
| 0.3 | 0.1179 | 0.1217 | 0.1255 | 0.1293 | 0.1331 | 0.1368 | 0.1406 | 0.1443 | 0.1480 | 0.1517 |
| 0.4 | 0.1554 | 0.1591 | 0.1628 | 0.1664 | 0.1700 | 0.1736 | 0.1772 | 0.1808 | 0.1844 | 0.1879 |
| 0.5 | 0.1915 | 0.1950 | 0.1985 | 0.2019 | 0.2054 | 0.2088 | 0.2123 | 0.2157 | 0.2190 | 0.2224 |
| 0.6 | 0.2258 | 0.2291 | 0.2324 | 0.2357 | 0.2389 | 0.2422 | 0.2454 | 0.2486 | 0.2518 | 0.2549 |
| 0.7 | 0.2580 | 0.2612 | 0.2642 | 0.2673 | 0.2704 | 0.2734 | 0.2764 | 0.2794 | 0.2823 | 0.2852 |
| 0.8 | 0.2881 | 0.2910 | 0.2939 | 0.2967 | 0.2996 | 0.3023 | 0.3051 | 0.3078 | 0.3106 | 0.3133 |
| 0.9 | 0.3159 | 0.3186 | 0.3212 | 0.3238 | 0.3264 | 0.3289 | 0.3315 | 0.3340 | 0.3365 | 0.3389 |
| 1.0 | 0.3413 | 0.3438 | 0.3461 | 0.3485 | 0.3508 | 0.3531 | 0.3554 | 0.3577 | 0.3599 | 0.3621 |
| 1.1 | 0.3643 | 0.3665 | 0.3686 | 0.3708 | 0.3729 | 0.3749 | 0.3770 | 0.3790 | 0.3810 | 0.3830 |
| 1.2 | 0.3849 | 0.3869 | 0.3888 | 0.3907 | 0.3925 | 0.3944 | 0.3962 | 0.3980 | 0.3997 | 0.4015 |
| 1.3 | 0.4032 | 0.4049 | 0.4066 | 0.4082 | 0.4099 | 0.4115 | 0.4131 | 0.4147 | 0.4162 | 0.4177 |
| 1.4 | 0.4192 | 0.4207 | 0.4222 | 0.4236 | 0.4251 | 0.4265 | 0.4279 | 0.4292 | 0.4306 | 0.4319 |
| 1.5 | 0.4332 | 0.4345 | 0.4357 | 0.4370 | 0.4382 | 0.4394 | 0.4406 | 0.4418 | 0.4429 | 0.4441 |
| 1.6 | 0.4452 | 0.4463 | 0.4474 | 0.4484 | 0.4495 | 0.4505 | 0.4515 | 0.4525 | 0.4535 | 0.4545 |
| 1.7 | 0.4554 | 0.4564 | 0.4573 | 0.4582 | 0.4591 | 0.4599 | 0.4608 | 0.4616 | 0.4625 | 0.4633 |
| 1.8 | 0.4641 | 0.4649 | 0.4656 | 0.4664 | 0.4671 | 0.4678 | 0.4686 | 0.4693 | 0.4699 | 0.4706 |
| 1.9 | 0.4713 | 0.4719 | 0.4726 | 0.4732 | 0.4738 | 0.4744 | 0.4750 | 0.4756 | 0.4761 | 0.4767 |

| $x$ | 0.00 | 0.01 | 0.02 | 0.03 | 0.04 | 0.05 | 0.06 | 0.07 | 0.08 | 0.09 |
|---|---|---|---|---|---|---|---|---|---|---|
| 2.0 | 0.4772 | 0.4778 | 0.4783 | 0.4788 | 0.4793 | 0.4798 | 0.4803 | 0.4808 | 0.4812 | 0.4817 |
| 2.1 | 0.4821 | 0.4826 | 0.4830 | 0.4834 | 0.4838 | 0.4842 | 0.4846 | 0.4850 | 0.4854 | 0.4857 |
| 2.2 | 0.4861 | 0.4864 | 0.4868 | 0.4871 | 0.4875 | 0.4878 | 0.4881 | 0.4884 | 0.4887 | 0.4890 |
| 2.3 | 0.4893 | 0.4896 | 0.4898 | 0.4901 | 0.4904 | 0.4906 | 0.4909 | 0.4911 | 0.4913 | 0.4916 |
| 2.4 | 0.4918 | 0.4920 | 0.4922 | 0.4925 | 0.4927 | 0.4929 | 0.4931 | 0.4932 | 0.4934 | 0.4936 |
| 2.5 | 0.4938 | 0.4940 | 0.4941 | 0.4943 | 0.4945 | 0.4946 | 0.4948 | 0.4949 | 0.4951 | 0.4952 |
| 2.6 | 0.4953 | 0.4955 | 0.4956 | 0.4957 | 0.4959 | 0.4960 | 0.4961 | 0.4962 | 0.4963 | 0.4964 |
| 2.7 | 0.4965 | 0.4966 | 0.4967 | 0.4968 | 0.4969 | 0.4970 | 0.4971 | 0.4972 | 0.4973 | 0.4974 |
| 2.8 | 0.4974 | 0.4975 | 0.4976 | 0.4977 | 0.4977 | 0.4978 | 0.4979 | 0.4979 | 0.4980 | 0.4981 |
| 2.9 | 0.4981 | 0.4982 | 0.4982 | 0.4983 | 0.4984 | 0.4984 | 0.4985 | 0.4985 | 0.4986 | 0.4986 |

| 3.0 | 3.1 | 3.2 | 3.3 | 3.4 | 3.5 | 3.6 | 3.7 | 3.8 | 3.9 |
|---|---|---|---|---|---|---|---|---|---|
| 0.4987 | 0.4990 | 0.4993 | 0.4995 | 0.4997 | 0.4998 | 0.4998 | 0.4999 | 0.4999 | 0.5000 |

# Index

Note that the references below refer to chapters and sections. Thus 20:11 refers to chapter 20, section 11.